Pathogenic *Treponema*
Molecular and Cellular Biology

Edited by

Justin D. Radolf

University of Connecticut Health Center,
Farmington, CT 06030-3715, USA

Sheila A. Lukehart

University of Washington School of Medicine,
Seattle, WA 98195, USA

Caister Academic Press
32 Hewitts Lane
Wymondham
Norfolk NR18 0JA
England

www.caister.com

British Library Cataloguing-in-Publication Data
A catalogue record for this book is available from the British Library
ISBN: 1-904455-10-7

Cover image kindly provided by Dr David L. Cox depicts *T. pallidum* attached to a monolayer of cotton-tail rabbit epithelial (Sf1Ep) cells, 750X. Several flat epithelial cells can be seen in this micrograph. Frequently small clusters of organisms are observed attached to cells, as well as individual spirochetes.

Printed and bound in Great Britain

This book is dedicated, with gratitude, to James N. Miller, Ph.D., for his many contributions to the field of treponemal research, for nurturing our early careers, and for his continuing support and wisdom.

J.D.R. and S.A.L.

Contents

Contributors

Jorge L Benach
Center for Infectious Diseases
Department of Molecular Genetics and Microbiology
Stony Brook University
248 Centers for Molecular Medicine
Stony Brook
NY 11794-5120, USA

Julie A. Boylan
Laboratory of Zoonotic Pathogens
NIAID, NIH
Rocky Mountain Laboratories
903 S. 4th St.
Hamilton
MT 59840, USA

Paul J. Brett
Laboratory of Zoonotic Pathogens
NIAID, NIH
Rocky Mountain Laboratories
903 S. 4th St.
Hamilton
MT 59840, USA

John A. Breznak
Department of Microbiology & Molecular Genetics
6190 Biomedical and Physical Sciences Bldg.
Michigan State University
East Lansing
MI 48824-4320, USA

Caroline E. Cameron
Department of Medicine
Box 357185
University of Washington
1959 NE Pacific St.
Room I-104E
Seattle
WA 98195, USA

Arturo Centurion-Lara
Department of Medicine
Box 359779
University of Washington
Harborview Medical Center
325 9th Avenue
Seattle
WA 98104, USA

Nyles W. Charon
Department of Microbiology, Immunology, and Cell Biology
Health Sciences Center
Box 9177,West Virginia University
Morgantown
WV 26506-9177, USA

David L. Cox
Treponema Immunobiology and Syphilis Serology Activity
NCHSTP / DSTDP / LRRB
Centers for Disease Control and Prevention
Mailstop D-13
1600 Clifton Road NE
Atlanta
GA 30333, USA

Floyd E. Dewhirst
Department of Molecular Genetics
The Forsyth Institute
140 Fenway
Boston
MA 02115-3799, USA

Richard P. Ellen
CIHR Group in Matrix Dynamics
Dental Research Institute
University of Toronto
Faculty of Dentistry
124 Edward Street
Toronto
Canada M5G 1G6

Emily Erbelding
Division of Infectious Diseases
Johns Hopkins University School of Medicine
830 E. Monument St
Rm 445
Baltimore
MD 21205, USA

Jeffrey Ebersole
Center for Oral Health Research
College of Dentistry
D544 Dental Science Buidling 0297
University of Kentucky
Lexington
KT 40536, USA

Frank C. Gherardini
Laboratory of Zoonotic Pathogens
NIAID, NIH
Rocky Mountain Laboratories
903 S. 4th St.
Hamilton
MT 59840, USA

Ulf B. Göbel
Institut für Mikrobiologie und Hygiene
Charité - Universitätsmedizin Berlin
Dorotheenstrasse 96
D-10117 Berlin
Germany

Stuart F. Goldstein
Department of Genetics
Cell Biology and Development
6-160 Jackson Hall
University of Minnesota
Minneapolis
MN 55455, USA

Karsten R. O. Hazlett
Center for Immunology
and Microbial Disease
Albany Medical Center
Albany
NY 12208, USA

Stanley C. Holt
Department of Periodontology
The Forsyth Institute
140 The Fenway
Boston
MA 02115-3799, USA

Jacques Izard
The Forsyth Institute
140 The Fenway
Boston
MA 02115-3799, USA

Howard K. Kuramitsu
Departments of Oral Biology and
Microbiology
State University of New York
3435 Main Street
Buffalo
NY 14214, USA

Chunhao Li
Department of Microbiology
Immunology and Cell Biology
Health Sciences Center
Box 9177,West Virginia University
Morgantown
WV 26506-9177, USA

Ronald J. Limberger
Wadsworth Center
David Axelrod Institute
120 New Scotland Ave. PO Box 22002
Albany
NY 12201-2002, USA

Sheila A. Lukehart
Department of Medicine
Box 359779
University of Washington School of Medicine
Harborview Medical Center
325 Ninth Avenue
Seattle
WA 98104, USA

Renate Lux
University of California
Los Angeles
School of Dentistry
10833 LeConte Avenue
Los Angeles
CA90095-1668, USA

Laura J. McGough
Division of Infectious Diseases
Johns Hopkins Bayview Medical Center
4940 Eastern Avenue
B-3-North
Baltimore
MD 21224-2780, USA

Daniel M. Musher
Division of Infectious Diseases
Michael E. Debakey Veterans Affairs Medical
Center
Baylor College of Medicine
Houston
TX 77030, USA

Steven J. Norris
Departments of
Pathology & Laboratory Medicine
Microbiology and Molecular Genetics
University of Texas Medical School at
Houston
6431 Fannin St.
Houston
TX 77030, USA

Bruce J. Paster
Department of Molecular Genetics
The Forsyth Institute
140 Fenway
Boston
MA 02115-3799, USA

Justin D. Radolf
Departments of Medicine and
Genetics and Developmental Biology
University of Connecticut Health Center
Farmington
CT 06030-3715, USA

Ronald F. Schell
Wisconsin State Laboratory of Hygiene
University of Wisconsin
465 Henry Mall
Madison
WI 53706, USA

Nicolas W.J. Schröder
Pediatrics Infectious Diseases
Cedars Sinai Medical Center
110 George Burns Road
Los Angeles
CA 90048, USA

Ralf R. Schumann
Institut für Mikrobiologie und Hygiene
Charité - Universitätsmedizin Berlin
Dorotheenstrasse 96
D-10117 Berlin
Germany

Wenyuan Shi
School of Dentistry
Molecular Biology Institute
University of California
Los Angeles
CA 90095-1668, USA

Lola V. Stamm
Department of Epidemiology
School of Public Health
McGavran Hall
The University of North Carolina
S. Columbia Street
Chapel Hill
NC 27599-7435, USA

Darren J. Trott
School of Veterinary Science
The University of Queensland
St. Lucia
Queensland
Australia

George M. Weinstock
Human Genome Sequencing Center
Baylor College of Medicine
One Baylor Plaza
Houston
TX 77030, USA

Preface

The Pathogenesis and Immunology of Treponemal Infection addressed a pressing need when it was published in 1983. More than twenty-five years had elapsed since Turner and Hollander had published their classic book *Biology of the Treponematoses*, and much had changed in the world of microbiology, including the status of the treponemal field. As recounted by R.F. Schell and D.M. Musher in the preface to their landmark book, syphilis had spawned a "long and venerable line of distinguished investigators" but, at the dawn of the molecular era, had fallen out of the mainstream of microbiological research. Ironically, the technical factors cited as major contributors to this state of affairs, the inability to cultivate *Treponema pallidum* on artificial medium and the paucity of animal models, are equally as pertinent today though, fortunately, they are less formidable barriers to progress. By pulling together the existing corpus of knowledge concerning *T. pallidum*, syphilis pathogenesis, and syphilis immunology, the 1983 work helped to establish the modern science of treponematology; its publication was very timely for those of us who at that time were just beginning careers in treponemal research. Indeed, the current volume can be viewed as the successor to the Musher/Schell compilation. Quite fittingly, though coincidentally, this book marks the centenary of the discovery of "Spirochaeta pallida" by Schaudinn and Hoffman, and documents the many successes, as well as failures, that have transpired during the first hundred years of contemporary syphilis research.

Today, the need for a re-examination of the field is at least as urgent as when Drs. Schell and Musher compiled their volume. At that time, "treponeme" and "treponemal infection" were virtually synonymous with *Treponema pallidum* and syphilis. With approximately 12 million new cases per year, syphilis is the most prevalent spirochetal infection on a global scale; its ability to facilitate the transmission of HIV further heightens its importance as a major public health threat. The chapters in this book devoted to *T. pallidum*, syphilis pathogenesis, syphilis immunology, and the history of syphilis underscore the continuing relevance of *T. pallidum* as a bacterial pathogen and our ongoing fascination with the sociologic and scientific aspects of a disease that has afflicted humans and influenced human behavior for centuries. Nevertheless, *T. pallidum* can no longer lay claim to being the center of the spirochetal universe; even within the more circumscribed world of treponemes, *T. pallidum* must now share the limelight. The current volume documents the dramatic expansion in our understanding of treponemal diversity, cellular and molecular biology, and diseases in the 22 years since the publication of the Schell and Musher book. Coupled with an explosion in our understanding of treponemal phylogeny, ultrastructure, and physiology is the availability of genomic sequences that are enabling researchers to address the fundamental question of what makes a treponeme a treponeme. Rather than diminishing the importance of *T. pallidum*, the many "non-syphilis" chapters in this book illustrate how comparative genomics now enables us to view the syphilis spirochete in a much broader physiologic, genetic and evolutionary context. Given the technical difficulties in working with *T. pallidum*, these chapters also suggest exciting possibilities for using cultivatable treponemes as surrogates for *T. pallidum*.

Drs. Musher and Schell lamented the fact that, in the years following World War II, treponemal research had become limited to the laboratories of a few devotees and that treponemes were largely being ignored. The current book highlights the remarkable resurgence of interest in treponemes and treponemal diseases during the past two decades. At the same time, it is fair to say that the growth in this field has not kept pace with most other areas of microbiology and bacterial pathogenesis. Undoubtedly, the reputations of treponemes, even those that can be cultivated, as recalcitrant organisms, have frightened potential investigators. A major objective of the current volume is to prove that the powerful molecular and genetic tools developed since the publication of the Schell and Musher book are indeed capable of extracting from these wily organisms their elusive secrets. At the same, it also candidly acknowledges that there is much to be done and learned. It is hoped that the current volume will stimulate a new generation of aspiring microbiologists to become captivated by these fascinating organisms, as the current editors have been for decades.

Justin D. Radolf
Sheila A. Lukehart
May 2006

Introduction

*Jorge L. Benach**

Introduction

There is no other disease that has captured our imagination and creativity more than syphilis. Beginning with the historical lore of its origins, to its numerous and suggestive common names, and the notable historical personages who were thought to be afflicted with the "Great Pox," syphilis has always been close to the human psyche. Venereal syphilis was first reported as a new disease in Italy at the end of the 15th century, during the time of Spanish exploration and colonization of the American continent. Thus, many experts believe that this disease, and its causative organism *Treponema pallidum*, sailed from the Americas to Europe. There is anatomical and molecular fossil evidence supporting the existence of treponemal infections among native Americans in the pre-Columbian period, and perhaps more compelling, such evidence is scant among Europeans for the same period of time. Increasingly more specific and sensitive molecular probes may resolve this historical puzzle conclusively. Nonetheless, syphilis reached epidemic proportions in, and spread throughout, Europe during the early part of the 16th century. This epidemic, coincident with increasing contact between Europe and the Americas, follows the classic pattern of rapid spread of a newly introduced disease. In the centuries following its recognition in Europe, syphilis has had an extraordinary history. It has spread to all parts of the globe, and has undergone major and dramatic changes in incidence. Reliable historical accounts disclose a very high prevalence in the 19th century in both Europe and America, with 10-20% of the population thought to be infected. By the time of the First World War, serological tests provided accurate estimates of the prevalence of syphilis, and the newly discovered arsenic compound, salvarsan, was available for treatment. The ability to identify cases and treat them, together with institution of contact tracing and the public education campaign led by Surgeon General Thomas Parran, led to decreases in the incidence of this disease over several decades. The discovery that penicillin was highly effective against *Treponema pallidum*, and the widespread availability of this drug following World War II, led to sharp decreases

*For correspondence email jbenach@notes.cc.sunysb.edu

in infectious cases during the early 1950's. Since that time, syphilis incidence has been characterized by regular fluctuations, not unlike its relapsing-remitting clinical course. In 2001, syphilis incidence reached its lowest level since World War II, and eradication of syphilis from the United States became a stated goal of the Centers for Disease Control. Although significant success was achieved by targeted interventions among African Americans largely in the South, the advent of highly active anti-retroviral therapy has resulted in new high risk sexual behaviors among men who have sex with men, with a consequent resurgence of syphilis in this population.

Of all the treponematoses, syphilis historically has garnered the most attention because of its protean manifestations as well as the remarkable and still unsolved medical, social, and historical questions that it poses. The natural history of syphilis is played out in decades, and the chronicity of the illness continues to baffle researchers. Numerous theories have been proposed to explain the mechanisms which enable *T. pallidum* to persist within human and experimental hosts, and discussing these is a favorite pastime of syphilis researchers. A number of chapters in this book are devoted to *T. pallidum,* to its pathogenesis, immunology, metabolism, and history. This book discusses the state-of-the-art in the understanding of these issues, and evaluates older hypotheses based upon new data.

T. pallidum has unique features that it is believed help to explain the chronic nature of syphilitic infection. This organism is an obligate human parasite that has no other hosts. This means, fundamentally, that this spirochete should be extremely well adapted to its only host. But, is it? We know, for example, that this organism does not have classical virulence factors, and none were noted in the genome. Several chapters in this book discuss the most current models to explain how this organism causes disease. One reason for the unique pathogenesis of syphilis is that *T. pallidum* has an outer membrane that is mostly lipid with a paucity of integral proteins. Based on this premise, it could be argued that many of the secrets of syphilis lie on the outer membrane of the spirochete. This topic of outer membrane architecture and topology of *T. pallidum* is covered extensively here, with a complementary analysis of the molecular architecture of other cultivable and non-cultivable members of *Treponema*. There is much biochemical and ultrastructural evidence to bolster the argument that the outer membrane of this spirochete is relatively inert antigenically. This unique outer membrane may also contain cardiolipin, the phospholipid which is the basis of the VDRL test, the first in the two-step diagnostic algorithm used for serodiagnosis of syphilis. It is interesting to note that cardiolipin reactivity in anti-phospholipid syndrome was found to be actually directed to a contaminating peptide fraction, but in syphilis, the cardiolipin reactivity is directed against this negatively charged phospholipid.

Antigenic variation of outer membrane proteins is one strategy for persistence of bacterial pathogens in the host. By presenting new, changing repertoires of epitopes to the immune response, the organisms can persist for a long time. Systems for antigenic variation exist in the related Borrelia spirochetes, and also in Neisseria, Mycoplasma, and even eukaryotic Trypanosomes. The work recently undertaken in antigenic variation in *T. pallidum* is centered around a paralogous family of *tpr* genes, and in one gene product in particular, TprK, that contains several variable regions. This family of genes may hold many answers to explain the staying power of *T. pallidum* in its long association with the human host.

The nonvenereal treponemal infections (yaws, bejel, pinta) were so prevalent in developing countries that, in the 1950's and 1960's, the World Health Organization undertook a global screening and mass treatment campaign in which over 60 million people were given penicillin in an effort to eradicate these infections. Today, in the face of the horror of AIDS, very few health officers are even cognizant of the nonvenereal treponematoses. Even the most basic understanding of the relationship among the agents of treponemal infection still eludes investigators. Bacterial taxonomy has not helped our understanding of these diseases. Medical students and microbiologists have been puzzled that the best-recognized and distinct treponematoses (syphilis, yaws, and bejel) are caused by members of the same species, *Treponema pallidum*. Differentiation into subspecies is based solely on the clinical description and geography of the patient from whom the strains are isolated. The causative agent of pinta, *T. carateum*, remains taxonomically a separate species by default - simply because strains were not available for the DNA reassociation kinetics studies in 1980 that showed a very close genetic relationship between strains from syphilis and yaws (bejel agents were also never tested, but were grouped within *T. pallidum*). Venereal syphilis and the nonvenereal endemic treponematoses comprise a group of syndromes that are clinically distinct, but that also have remarkable similarities. Because of the close genetic relationship among the pathogenic treponemes and the clinical similarities of their respective diseases, it is likely that many details of the pathogenesis of syphilis and the nonvenereal treponematoses are similar. As in syphilis, the clinical manifestations of endemic treponematoses have early and late stages, where early infectious lesions are followed by periods of latency (asymptomatic), alternating with symptomatic episodes. All treponemal diseases have a relapsing-remitting clinical course that begins with cutaneous or mucosal manifestations that have very similar histopathology. Venereal syphilis can affect virtually any organ system, including invasion of the central nervous system and transplacental infection of the fetus. In contrast to syphilis, congenital infection does not occur in the endemic treponematoses, and CNS invasion is rare. The reasons for the different manifestations of these related infections are not known. It has been recognized for some time that protective immunity can be induced by infection and is subspecies-specific. The lack of cross-protection provides evidence for key antigenic differences among the subspecies of *T. pallidum*. It is only now, with the availability of the genomic approaches, that specific antigens are being identified among the agents of the venereal and non-venereal treponematoses. Comparative genomics studies are under way to understand the molecular bases for the clinical features caused by each subspecies. Future research on these bacteria would benefit markedly from genomic sequences of representative *T. pallidum* subspecies.

Two formidable impediments to the study of syphilis pathogenesis and immunology remain: *T. pallidum* cannot be cultured continuously *in vitro* and therefore no means for genetic manipulation exists. Thus, the elegant gene knock-out or gene replacement studies that are conducted with many other bacteria are not possible with *T. pallidum*. Additionally, the dependence on the rabbit as the experimental animal model for syphilis has hampered this field in that inbred rabbits and specific rabbit immunological or biochemical reagents are not available. A concerted effort to sequence the laboratory rabbit genome and produce specific rabbit reagents would significantly accelerate research in this field.

Treponemal research has shifted significantly during the 20 years since the last comprehensive book on the topic was edited by Schell and Musher. This excellent book summarized the knowledge on syphilis and the nonvenereal treponemes, but did not include the oral treponeme *T. denticola* and periodontal disease. Since publication of the Schell and Musher book, the breadth of the genus Treponema has been significantly expanded to include the recognition of treponemes in many new ecological niches. Two species, *T. pallidum* and *T. paraluiscuniculi,* were known to be venereal pathogens of humans and rabbits, respectively. The treponemal community in the diseased gingiva has been shown to extend beyond *T. denticola* to include many other species of treponemes. Other species of Treponema have colonized the alimentary canal of cattle *(T. bryantii* and *T. saccharophilum),* swine intestine *(T. succinifaciens),* and the termite hindgut *(T. primita* and *T. azotonutricum).* Papillomatous digital dermatitis (PDD), caused by a mixed population of predominantly anaerobic treponemes, is an infectious, superficial dermatitis that primarily affects the feet of dairy cattle. Furthermore, three species of free-living Spirochaeta have now been shown to be in the Treponema cluster rather than within the cluster of other free-living Spirochaeta. Early free-living treponemes could be the progenitors of today's host-dependent organisms. That treponemes occupy such diverse niches ranging from free-living to obligate pathogens is an indication of the versatile nature of these bacteria. Genome reduction in some species and lateral gene transfer have resulted in the diversity that exists today within this genus.

The completed *T. pallidum* and *T. denticola* genomic sequences have transformed research on the treponemes, causing an explosion of new information that is reflected in the current volume. It would be fair to expect that an understanding of the genetic capacity of these two pathogens will help us understand, for example, why one of these bacterium, *T. denticola*, can be propagated *in vitro* while the other, *T. pallidum*, cannot. The availability of the genome of *T. denticola* will surely help in further understanding the pathogenesis of periodontitis and of the role of this organism in the polymicrobial "chain of command" that gives rise to this complex gingival disorder. Furthermore, the availability of both genomes has already proved to be synergistic. The paralogous *tpr* gene family in *T. pallidum* that encodes several putative outer membrane proteins is related to the porin-like major outer sheath protein (Msp) of *T. denticola*. This and other examples of gene relationships advancing the understanding of function of those shared gene products are considered in depth in the chapter on comparative genomics of treponemes in this book.

Consistent with the new perception of treponemal diversity, this book, appropriately, has many "non-syphilis" chapters, stressing that the various treponemes, and other related Spirochetes, have many morphological, genetic and physiological features in common. The Spirochete family is characterized by a unique cell architecture with an outer membrane, a cell membrane and a number of flagellar filaments located in the periplasm. Within the treponeme group there are cells that have a regular helical shape, cells that have irregularly shaped helices, and cells with a flat-wave morphology of the cell cylinder. It is possible that, given the difficulties in culturing *T. pallidum,* these other treponemes can be used to learn more about all of the species of Treponema. These chapters also suggest exciting possibilities for using cultivatable treponemes as surrogates for functional studies of *T. pallidum* proteins.

T. pallidum has a circular chromosome of approximately 1000 kb pairs, with a total of 1041 predicted open reading frames. Of these, biological and functional roles were annotated for only 577 genes. The remaining open reading frames represent hypothetical genes with homology to genes in other bacteria and a relatively large percentage (28%) of unique genes. This genome is among one of the smallest prokaryotic genomes, a mystery to the research community in terms of explaining how an organism with so few genetic resources can cause one of the most complex and protean bacterial infections. In the absence of other virulence factors, toxins, or secreted effector molecules, motility is a critical feature for spirochete invasiveness. Motility-associated genes are shared in both *T. pallidum* and *Borrelia burgdorferi*, consistent with the importance of motility and chemotaxis in these species. The genes encoding for proteins involved in flagellar structure and function are very similar in both organisms. Both genomes also show limited biosynthetic capabilities that require multiple nutrients from the host. Neither *B. burgdorferi* nor *T. pallidum* is capable of synthesizing enzyme co-factors, fatty acids, and nucleotides. However, in both organisms with limited biosynthetic capabilities, there is a rather large contingent of transport proteins with broad binding specificity for amino acids, carbohydrates, lipids, and inorganic molecules. An area where the physiology of *T. pallidum* and *B. burgdorferi* diverge is in their ability to handle oxidative stress. While Borrelia has genes encoding for superoxide dismutase and catalase that protect against oxygen toxicity, *T. pallidum* has only an NADH oxidase for oxygen utilization. This is intriguing as both organisms are microaerophilic with a physiologic requirement for oxygen, but *T. pallidum* is challenged in its ability to protect itself against oxidative stress.

Comparative genomics using *T. pallidum* and *B. burgdorferi* will continue to prove quite useful to learn about fundamental physiology of both organisms. It is also anticipated that lessons in pathogenesis will be learned from future comparative efforts. Both organisms have small genomes, and comparative data can act synergistically to increase our understanding of how small genome spirochetes can cause chronic and protean infections. The clinical similarities (as well as some of the differences) between syphilis and Lyme disease and relapsing fever have been obvious for years. The same can be said about the similarities that exist in the host responses to both organisms. It is important to remember that, in the early 1980's, the nascent field of Lyme borreliosis looked toward the more mature field of syphilis for guidance, and for learning to work with fastidious spirochetes. The more we learn about pathogenesis of one will obviously be of benefit to the other. Thus, these two fields are intimately linked at fundamental levels.

While there are many similarities in the genomes of Borrelia and *T. pallidum*, and also in the clinical presentations of the diseases that they cause, the recently published genome of *T. denticola* also has a significant impact in comparative studies on the evolution of the genus and in the obvious similarities and differences between both treponemes. Because of the polymicrobial milieu in which it exists, *T. denticola* alone is not thought to be the sole cause of periodontitis, but rather it is part of the diverse biofilm that resides in the periodontal pockets of individuals with this disease. *T. denticola* can adhere to different types of cells and extracellular membranes and their components, notably, collagens, fibronectin, laminin and others. Because of its participation in a polymicrobial environment, binding to other oral bacteria is also required to maintain the biofilm framework, and thus, this organism can also adhere to other bacterial components of the biofilm. *T. denticola* has a well documented arsenal of proteolytic enzymes that are likely associated with degradation of surrounding gingival tissues. The pathogenic role of *T. denticola* and other treponemes in periodontal disease is well represented in this book.

The genome of *T. denticola* has some important differences from that of *T. pallidum*. The most obvious difference is that it is almost three times larger, but curiously, has roughly a similar percent of unique genes (about one quarter of the genes). The genome of *T. denticola* clearly reflects its adaptations for thriving and surviving in the biofilm environment of the periodontal pocket, and thus it is highly specialized to survive within this highly competitive niche. About one-fourth of *T. denticola* genes have their best matches in the *T. pallidum* genome, which underscores the relatedness of these two members of the same genus. On the other hand, genome size disparity between the two species requires some consideration. It has been suggested that the *T. pallidum* genome was derived from the genome of a *T. denticola*-like spirochete or another group of ancestral host-associated treponemes by the deletion and divergence of genes through reductive evolution. Another important evolutionary consideration is that *T. denticola* may have acquired genes by lateral transfer. Exchange of genes in a prolific polymicrobial community would be the norm, whereas the host specificity of *T. pallidum* would reduce the chances for gene transfer. The evolutionary aspect of these treponemal organisms is a topic of much interest.

This book highlights recent major advances in our understanding of treponeme biology. Treponemal research in the third millennium will continue to change dramatically. The already published genomes will provide the means to identify *in silico* what may not be possible on the bench. At the bench, research with these organisms will accelerate by adapting contemporary biochemical and immunological principles to studies on the biology of treponemes. Specifically, advances in innate immunity could help to explain the earliest steps in pathogenesis as the organism survives to cause disease. Likewise, advances in the field of adaptive immunity and vaccinology could spell success in the creation of a vaccine to prevent syphilis and associated diseases caused by subspecies of *T. pallidum*.

T. denticola can be easily cultivated *in vitro*. Researchers have recently developed means for genetic manipulation of this species, making it an excellent model for spirochete research. It would be of enormous value to the field of spirochetal biology if the techniques used for the genetic manipulation of *T. denticola* could be adapted to *T. pallidum*, and even to the Borrelia. This may or may not happen. Nothing is easy in the world of spirochetes, and within this group, the Treponema are the most challenging to study in the laboratory. This very balanced, comprehensive, and optimistic book edited by Radolf and Lukehart, with significant contributions from the major researchers in this field, gives the impression that working with these organisms has been easy. Of course, we all know better, as researchers in this field need to be more creative, more patient, more intelligent, and more resourceful than researchers in other areas of microbiology. That the investigators represented here have accomplished so much is a clear indication that they, in fact, possess all of these qualities.

1

The Phylogenetic Diversity of the Genus *Treponema*

Bruce J. Paster and Floyd E. Dewhirst*

Abstract

Species of *Treponema* are generally considered as anaerobic, host-associated spirochetes that represent one of nine spirochetal genera of the phylum *Spirochaetes* as based on 16S rRNA sequence analysis. Traditionally, treponemal species have been characterized using conventional phenotypic and genotypic traits. More recently, comparative whole genomic analysis has been useful to provide insight on the evolutionary aspects of treponemes. Clonal analysis of 16S rRNA genes amplified from clinical and environmental samples has revealed hundreds of not-yet-cultivated treponemal species in the human oral cavity, the termite gut, the bovine rumen, other host-associated sources, and environmental sources, such as waste treatment plants. It is likely that there remains a considerable number of unrecognized species of *Treponema*.

1. The phylogenetic foundation of spirochetes

Based on analysis of 16S rRNA sequences, the spirochetes represent one of about 40 major bacterial phyla (Hugenholtz et al., 1998). 16S rRNA is a highly conserved molecule that is present in all prokaryotic organisms, exhibits functional constancy, and changes slowly in sequence, properties that make it extremely useful for measuring phylogenetic relationships (Woese 1987). Spirochetes are well known for their unique ultrastructure among the bacteria. They are helical or flat-wave cells with periplasmic flagella that insert subterminally at both cell ends and wrap around the protoplasmic cylinder, typically overlapping in the central region of the cell. The flagella are internal organelles of motility (Bromley and Charon, 1979; Paster and Canale-Parola 1980) as they are enclosed within the periplasmic space by the outer membrane (Holt, 1978). Other unifying phenotypic traits of the spirochetes include antibiotic resistance to rifampin (Leschine and Canale-Parola, 1986) and, with the exception of *Leptospira* spp., the use of ornithine as the diamino acid in

*For correspondence email bpaster@forsyth.org

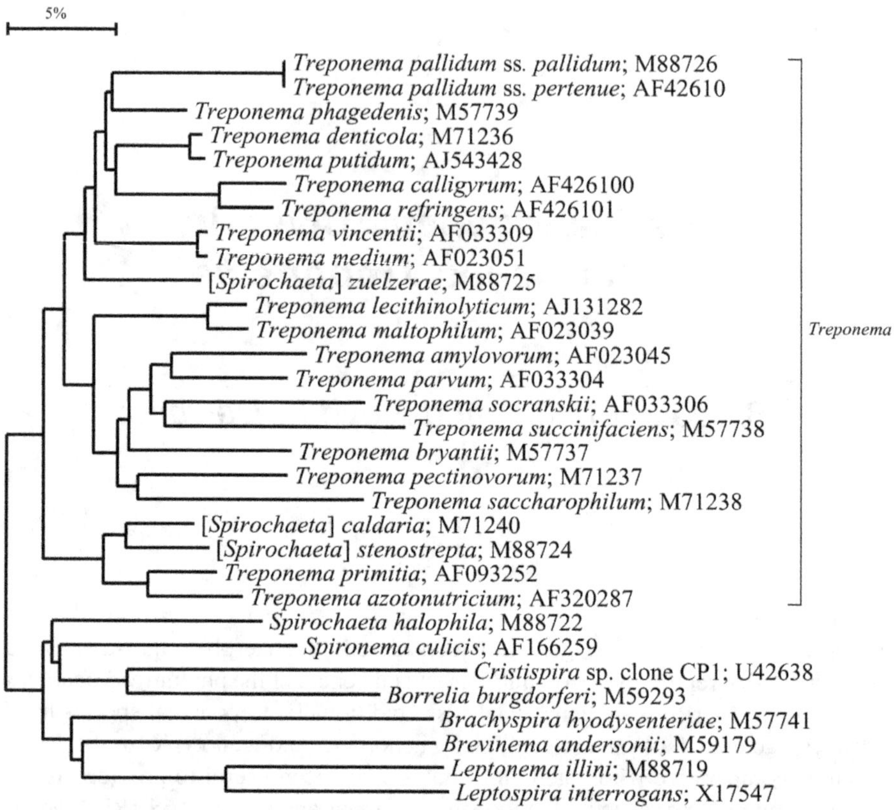

Figure 1. Phylogenetic relationships of species of *Treponema* and representatives of other spirochetal genera based upon 16S rRNA gene sequence comparisons. The sequences of the species shown may be obtained through GenBank or by request from the authors. Similarity matrices were corrected for multiple base changes by the method of Jukes and Cantor (1969). The neighbor-joining method (Saitou and Nei, 1987) was used for phylogenetic tree construction. TREECON, a software package for the Microsoft Windows environment, was used for the construction and drawing of evolutionary trees (Van de Peer et al., 1994). The scale bar represents a 5% difference in nucleotide sequence determined by taking the sum of all of the horizontal lines connecting two species. Vertical distance has no meaning.

peptidoglycan (Olsen et al., 2000). It is their unique ultrastructure, other unifying phenotypic traits, and 16S rRNA sequence analyses that define the spirochetes as a monophyletic group (Paster et al., 1991; Paster and Dewhirst, 2000).

Based on these data, it is likely that the spirochetes descended from a common spirochetal ancestor (Paster and Dewhirst, 2000). More recently, comparisons of selected DNA sequences have been shown to be useful for determination of relatedness among genomes (Zeigler, 2003). Although the genome sequences of only a limited number of spirochetes have been sequenced to date, namely, *Treponema pallidum, Treponema denticola, Borrelia burgdorferi, Leptospira interrogans*, and, to a limited extent, *Treponema lecithinoliticum* (Dewhirst, unpublished data), several genes have been shown to be spirochetal-associated (Seshardri, et al. 2004).

The spirochetes are divided into three families. The first family, *Spirochaetaceae*, contains species of the genera *Borrelia*, host-associated spirochetes transmitted to humans and animals via an insect vector (e.g., Lyme disease) (Wang et al., 1999); *Brevinema*, which include infectious spirochetes of the white-footed mouse (Defosse et al., 1995); *Cristispira*, large spirochetes from the digestive tract of bivalves (Paster et al., 1996b); the free-living *Spirochaeta* from aquatic environments (Leschine et al., 2001); "*Spironema*" from the mosquito (Paster and Dewhirst, 2000), and mammalian- and insect-associated *Treponema* (Norris et al., 2002). The family *Brachyspiraceae* (Paster and Dewhirst, 2000) contains the genus *Brachyspira*, previously known as *Serpulina*. *B. hyodysenteriae* is the etiologic agent of swine dysentery. The family *Leptospiraceae* (Johnson and Faine, 1984) contains species of *Leptonema* and *Leptospira*, the latter of which causes leptospirosis. The phylogenetic positions of these nine genera are shown in Figure 1.

The purposes of this chapter are 1) to review the phylogenetic relationships among members of the genus *Treponema*, including those that cannot yet be cultivated *in vitro*, and 2) to illustrate the vast diversity of these spirochetes. *Treponema* are typically anaerobic, host-associated spirochetes, many of which are pathogenic (Norris et al., 2002). One of the most renowned species is *T. pallidum*, the etiologic agent of syphilis. *Treponema paraluiscuniculi* is the etiologic agent of venereal rabbit spirochetosis (Bayon, 1913) and is closely related to *T. pallidum* subspecies (Baker-Zander and Lukehart, 1984; Centurion-Lara, et al., 1997). Other putative pathogens include some species of oral treponemes, such as *T. denticola* and *T. lecithinolyticum*, which have been associated with periodontal disease. Other species of *Treponema* have been isolated from the bovine rumen (*T. bryantii*, strain CA, and *T. saccharophilum*), swine intestine (*T. succinifaciens*), and the termite hindgut (*T. primitia* and *T. azotonutricum*) (Graber et al., 2004).

However, three named species of the free-living *Spirochaeta*, namely *S. stenostrepta*, *S. zuelzerae*, and *S. caldaria*, fall within the *Treponema* cluster rather than within the cluster of free-living *Spirochaeta* (Figure 1, species are noted in brackets). *S. zuelzerae*, originally characterized as a species of *Treponema* (Veldkamp, 1960), possesses the "one shared protein" antigen with *T. pallidum* (Deacon and Hunter, 1962). It has been suggested that these "free-living" spirochetes may represent precursors to host-associated treponemes or that they originated from the host but were disseminated via fecal contamination (Paster et al., 1991).

2. Phylogeny of not-yet-cultivated treponemes

Sequence analysis of 16S rRNA genes of DNA isolated from host-associated sources has identified novel species, termed phylotypes, which have not yet been cultivated *in vitro*. By using this molecular approach, the human oral cavity (Choi et al., 1994; Dewhirst et al., 2000), the hindgut of the termite (Lilburn et al., 1999; Ohkuma et al., 1999), the rumen of cows (Tajima et al., 1999; Koike et al., 2003), and the intestines of horse (Simpson et al., 2004), yak and pig (Leser et al., 2002) have shown extensive spirochetal diversity (Figure 2). For example in the human oral cavity, 76 oral species of *Treponema* have been recovered; 10 are known species, an additional 9 are cultivable, unnamed strains, and 57 (75% of the total!) are uncultivated phylotypes (Figure 3) (Paster et al., 2001). As previously

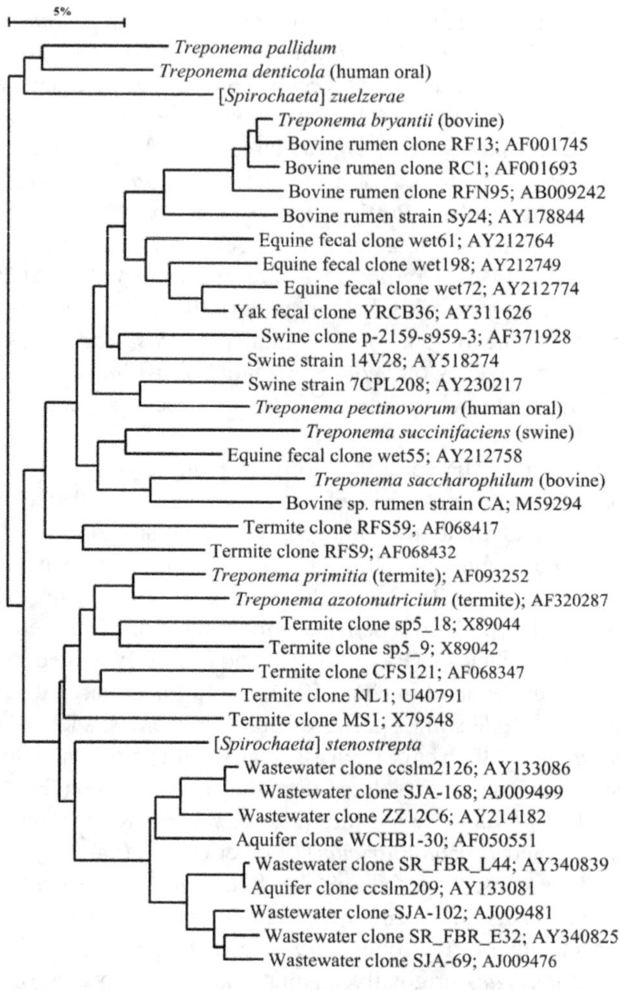

Figure 2. Phylogenetic relationships of species and uncultivated phylotypes of *Treponema* from host-associated and environmental sources.

defined and shown in Figure 3, the oral treponemes fall into 10 phylogenetic groups (Dewhirst et al., 2000; Paster et al., 2001). Using 16S rRNA-based DNA probes, treponemes in the oral cavity of dogs and cats were shown to be similar to those in the human oral cavity, including *Treponema denticola*, *T. socranskii* ssp., *T. vincentii*, *T. maltophilum*, *T. medium* in dogs and *T. maltophilum* and *T. socranskii* in cats (Valdez et al., 2000).

In addition, uncultivated treponemes, as based on 16S rRNA gene sequence analysis, have been detected in fluidized bed reactors in wastewater treatment plants (von Wintzingerode et al., 1999) and contaminated aquifers, which are sources of ground water for wells and springs (Dojka et al., 1998). The phylogenetic positions of these phylotypes in relation to other phylotypes and select known *Treponema* spp. are shown in Figure 2.

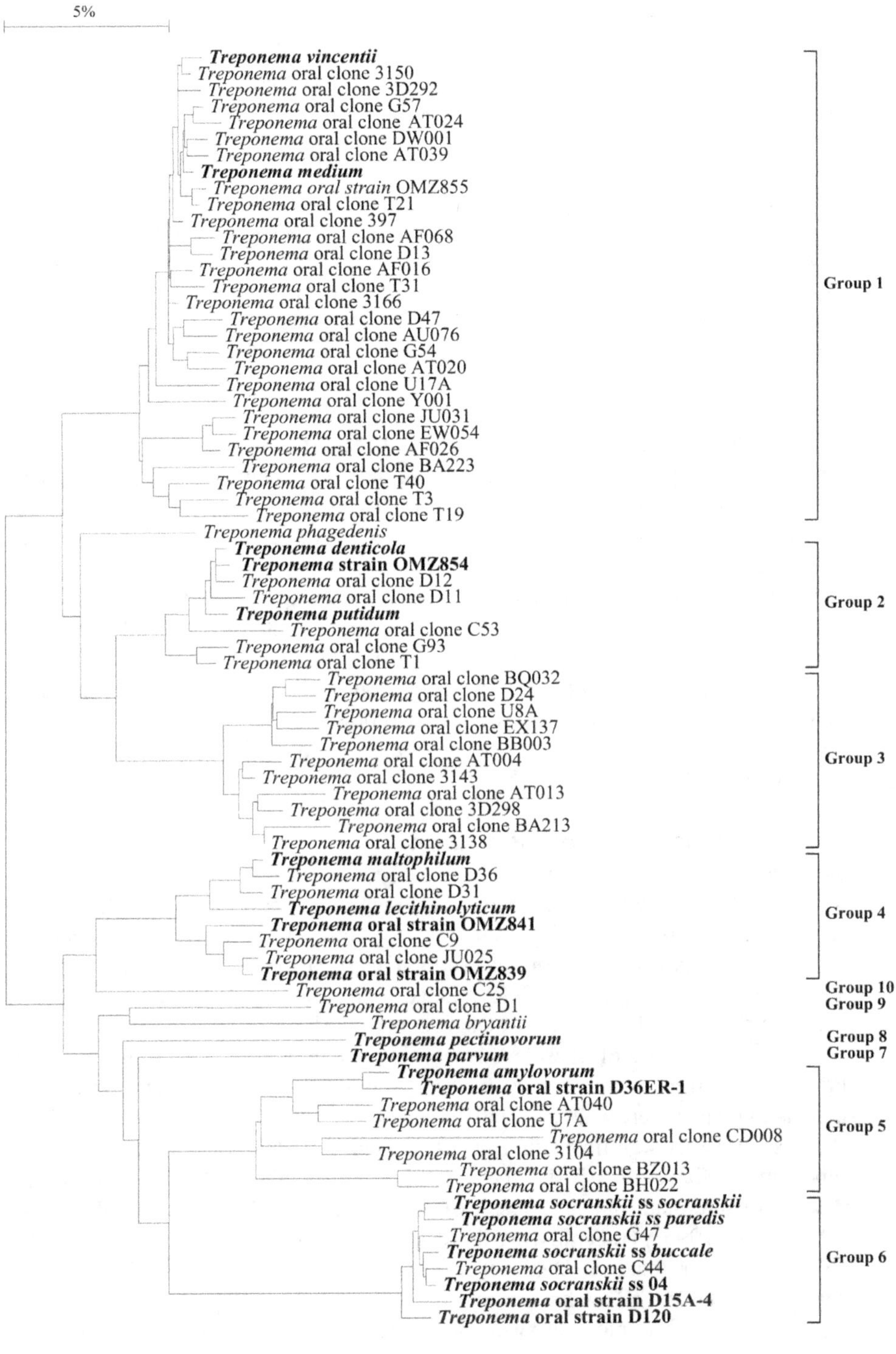

Figure 3. Phylogenetic relationships of species and uncultivated phylotypes of *Treponema* from the human oral cavity. Cultivable species are noted in bold.

Microscopic examination of termite hindguts also reveals an impressive diversity of spirochetes. Attempts to cultivate termite spirochetes have only recently been successful; these have resulted in the characterization of *T. primitia* and *T. azotonutricum* (Figure 1) (Graber et al., 2004). However, by using the molecular techniques described above, 21 novel phylotypes of *Treponema* were detected from only one termite species (Lilburn et al., 1999). A few representative termite clones and the named species are shown in Figure 2, but a detailed description of spirochetal diversity of the termite gut is described in the chapter by Breznak. Similar levels of phylogenetic diversity were observed in other termite families (Berchtold et al., 1994, Paster et al., 1996a; Lilburn et al., 1999; Ohkuma et al., 1999), however little or no overlap between termite families was observed. Considering that there are over 2,000 species of termites, the breadth of spirochetal diversity is astounding.

3. Papillomatous digital dermatitis treponemes

Papillomatous digital dermatitis (PPD) is an infection occurring in the fleshy part of the feet of cattle and sheep (reviewed in Norris et al., 2002) and is described in detail in the chapter by Stamm and Trott. Cultivated treponemal species and uncultivated phylotypes have been detected in these polymicrobial lesions and fall into at least 4 phylogenetic clusters (Figure 4). One group from PPD and contagious ovine digital dermatitis (CODD) lesions is closely related to the oral species *T. medium* and *T. vincentii*. The second group from PPD and CODD lesions clusters with the oral species *T. denticola*. The third group from PPD lesions is essentially identical to *T. phagedenis*. The last group is represented by a single strain, named *Treponema brennaborense*, which was isolated from an infected bovine lesion (Schrank et al., 1999).

4. Other taxonomic considerations

Treponema and other spirochetal genera have been characterized using conventional phenotypic and genotypic characteristics including analysis of cellular lipids, carbohydrates, enzymes, cell proteins, cytoplasmic fibrils, metabolites, genome size, base composition, restriction endonuclease analysis, multilocus enzyme electrophoresis (MEE), pulsed-field gel electrophoresis, DNA-DNA hybridization, and DNA fingerprinting (details in Olsen et al., 2000).

In general, most of these analyses have been useful for identifying and differentiating treponemal species. For example, DNA-DNA hybridization studies demonstrated that there was essentially no homology among *T. pallidum* subspecies whereas there were significant differences between *T. pallidum* subspecies and other treponemes (Miao *et al.* 1978). Other investigators used DNA-DNA homology and MEE to differentiate oral treponemes (Olsen et al., 1995; Smibert et al., 1984; Dahle et al., 1995). Generally, MEE results were close to those observed with DNA-DNA hybridization (Dahle et al., 1995). Sequence analysis of conserved genes other than 16S rRNA has also been useful to determine genetic relatedness of treponemes. For example, there were no significant differences in DNA sequences of the genes for the coding region of the 15-kDa antigen of *T. pallidum* subspecies, *T. paraluiscuniculi*, and Fribourg-Blanc strain, a poorly characterized simian isolate (Centurion-Lara et al., 1997).

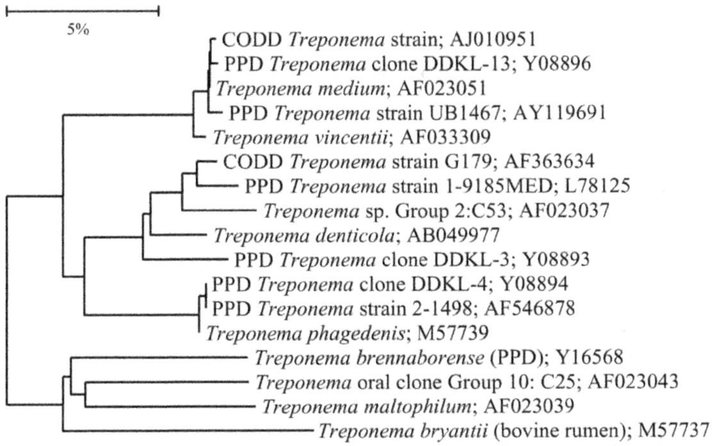

Figure 4. Phylogenetic relationships of species and uncultivated phylotypes of *Treponema* from digital dermatitis infections of sheep and cattle. (PPD) Papalomatous digital dermatitis of cattle; (CODD) contagious ovine digital dermatitis.

It should be noted that the treponemes and other spirochetal genera (e.g., *Spirochaeta*) represent deeply branching 16S rRNA-based phylogenetic groups (Figure 1). For example within the treponemes, the phylogenetic distance between some species is greater than the distance between the families *Enterobacteriaceae* and *Pasteurellaceae*. Despite this depth of branching, it had been suggested that the phylogenetic clustering of spirochetal groups, rather than a specific percent similarity, be used as a basis for defining spirochetal genera (Paster et al., 1991). This is consistent with the recommendations of two *ad hoc* committees addressing considerations in bacterial taxonomy in that phylogenetically-based taxonomic schemes should reflect phenotypic and genomic consistency (Wayne et al., 1987; Stackebrandt et al., 2002). It has been suggested that newer, innovative methods be utilized, such as sequencing of housekeeping genes, DNA profiling, and DNA arrays (Stackebrandt et al., 2002).

5. Summary

The phylogenetic diversity of treponemes is extensive with an uncultivated species segment of the population far more diverse than previously believed. More than 3/4 of the treponemal population still remains uncultivated. Consequently, culture independent techniques, such as the 16S rRNA approach, should be utilized to fully assess the full breadth of diversity of the treponemes. By using DNA hybridization techniques that target 16S rRNA genes, such as checkerboard hybridization (Paster et al., 1998) or microarrays (Boches et al., 2004), we can determine the potential role and significance that specific treponemal species play in the ecosystem. Use of newer, innovative genomic and other techniques will help to better define species of the genus *Treponema*.

Acknowledgments

This work was supported in part by Public Health Service grants DE-10374 and DE-11443 from the National Institute of Craniofacial and Dental Research.

References

Baker-Zander, S.A., and Lukehart, S.A. (1984). Antigenic cross-reactivity between *Treponema pallidum* and other pathogenic members of the family *Spirochaetaceae*. Infect. Immun. *46,* 116-121.

Bayon, H. (1913). A new species of *Treponema* found in the genital sores of rabbits. Br. Med. J. *2,* 1159.

Berchtold, M., Ludwig, W., and König, H. (1994). 16S rDNA sequence and phylogenetic position of an uncultivated spirochete from the hindgut of the termite *Mastotermes darwiniensis* Froggatt. FEMS Microbiol. Lett. *123,* 269-274.

Boches, S.K., Lee, A.M., Paster, B.J., and Dewhirst, F.E. (2004). Development of a human oral microbe identification microarray. J. Dent. Res. *83,* 2263.

Bromley, D.B., and Charon, N.W. (1979). Axial filament involvement in the motility of *Leptospira interrogans*. J Bacteriol. *137,* 1406-1412.

Centurion-Lara, A., Arroll, T., Castillo, R., Shaffer, J.M., Castro, C., Van Voorhis, W.C., and Lukehart S.A. (1997). Conservation of the 15-kilodalton lipoprotein among *Treponema pallidum* subspecies and strains and other pathogenic treponemes: genetic and antigenic analyses. Infect. Immun. *65,* 1440-1444.

Choi, B.K., Paster, B.J., Dewhirst, F.E., and Göbel, U.B. (1994). Diversity of cultivable and uncultivable oral spirochetes from a patient with severe destructive periodontitis. Infect. Immun. *62,* 1889-1895.

Dahle, U.R., Olsen, I., Tronstad, L. and Caugant, D.A. (1995). Population genetic analysis of oral treponemes by multilocus enzyme electrophoresis. Oral Microbiol. Immunol. *10,* 265-270.

Deacon, W.E., and Hunter, E.F. (1962). Treponemal antigens as related to identfication and syphilis serology. Proc. Soc. Exp. Biol. Med. *110,* 352-356.

Defosse, D.L., Johnson, R.C., Paster, B.J., Dewhirst, F.E., and Fraser, G.J. (1995). *Brevinema andersonii* gen. nov., sp. nov. an infectious spirochete isolated from the short-tailed shrew *Blarina brevicauda.* and the white-footed mouse *Peromyscus leucopus*. Int. J. Syst. Bacteriol. *45,* 78-84.

Dewhirst, F.E., Tamer, M.A., Ericson, R.E., Lau, C.N., Levanos, V.A., Boches, S.K., Galvin, J.L., and Paster, B.J. (2000). The diversity of periodontal spirochetes by 16S rRNA analysis. Oral Microbiol. Immunol. *15,* 196-202.

Dojka, M.A., Hugenholtz, P., Haack, S.K., and Pace, N.R. (1998). Microbial diversity in a hydrocarbon- and chlorinated-solvent-contaminated aquifer undergoing intrinsic bioremediation. Appl. Environ. Microbiol. *64,* 3869-3877.

Holt, S.C. (1978). Anatomy and chemistry of spirochetes. Microbiol. Rev. *42,* 114-160.

Hugenholtz, P., Goebel, B.M., and Pace, N.R. (1998). Impact of culture-independent studies on the emerging phylogenetic view of bacterial diversity. J. Bacteriol. *180,* 4765-4774.

Fraser, C.M., Norris, S.J., Weinstock, G.M., White, O., Sutton, G.G., Dodson, R., Gwinn, M., Hickey, E.K., Clayton, R., Ketchum, K.A., Sodergren, E., Hardham, J.M., McLeod, M.P., Salzberg, S., Peterson, J., Khalak, H., Richardson, D., Howell, J.K., Chidambaram, M., Utterback, T., McDonald, L., Artiach, P., Bowman, C., Cotton, M.D., and Venter, J.C. (1998). Complete genome sequence of *Treponema pallidum*, the syphilis spirochete. Science *281,* 375-388.

Graber, J.R., Leadbetter, J.R., and Breznak, J.A. (2004). Description of *Treponema azotonutricium* sp. nov. and *Treponema primitia* sp. nov., the first spirochetes isolated from termite guts. Appl. Environ. Microbiol. *70,* 1315-1320.

Johnson, R.C., and Faine, S. Genus I. *Leptospira* Noguchi *1917,* 755^{AL} (1984). In Bergey's Manual of Systematic Bacteriology, vol. *1,* N.R. Krieg and J.G. Holt, eds. (The Williams and Wilkins Co., Baltimore), pp. 49-57.

Koike, S., Yoshitani, S., Kobayashi, Y. and Tanaka, K. (2003). Phylogenetic analysis of fiber-associated rumen bacterial community and PCR detection of uncultured bacteria. FEMS Microbiol. Lett. *229,* 23-30

Jukes, T.H., and Cantor, C.R. (1969). Evolution of protein molecules. In Mammalian Protein Metabolism, vol. *3,* H.N Munro, ed. (Academic Press, Inc., New York), pp. 21-132.

Leschine, S.B, and Canale-Parola, E. 1986. Rifampin-resistant RNA polymerase in spirochetes. FEMS Microbiol. Lett. 35: 199-204.

Leschine, S.B., Paster B.J., and Canale-Parola, E. (2001). Free-Living saccharolytic spirochetes: The genus *Spirochaeta.* In The Prokaryotes: An Evolving Electronic Resource for the Microbiological Community, M. Dworkin, S. Falkow, E. Rosenberg, K-H. Schleifer, E. Stackebrandt, eds., 3rd edition, release 3.*5,* (Springer-Verlag, New York), www.prokaryotes.com.

Leser, T.D., Amenuvor, J.Z., Jensen, T.K., Lindecrona, R.H., Boye, M., and Moller, K. (2002). Culture-independent analysis of gut bacteria: the pig gastrointestinal tract microbiota revisited. Appl. Environ. Microbiol. *68,* 673-690

Lilburn, T.G., Schmidt, T.M., and Breznak, J.A. (1999). Phylogenetic diversity of termite gut spirocheaetes. Environ. Microbiol. *1,* 331-345.

Miao, R., and Fieldsteel, A.H. (1978). Genetics of *Treponema*: relationship between *Treponema pallidum* and five cultivable treponemes. J. Bacteriol. *133,* 101-107.

Norris, S.J., Paster, B.J., Moter, A., and Göbel, U.B. (2002). The Genus *Treponema.* In The Prokaryotes: An Evolving Electronic Resource for the Microbiological Community, M. Dworkin, S. Falkow, E. Rosenberg, K-H. Schleifer, E. Stackebrandt, eds., 3rd edition, release 3.*12,* (Springer-Verlag, New York), www.prokaryotes.com.

Ohkuma, M., Iida, T., and Kudo, T. (1999). Phylogenetic relationships of symbiotic spirochetes in the gut of diverse termites. FEMS Microbiol. Lett. *181,* 123-129.

Olsen I, Paster BJ, Dewhirst FE. Taxonomy of spirochetes. *Anaerobe* 2000;**6**:39-57.

Olsen, I., Fiehn, N.-E. and Johnson, J.L. (1995). Identity of 1:2:1 and 2:4:2 subgingival spirochetes by DNA hybridization. Oral Microbiol Immunol *10,* 378-381.

Paster, B.J., Bartoszyk, I.M., and Dewhirst, F.E. (1998). Identification of oral streptococci using PCR-based, reverse-capture, checkerboard hybridization. Methods Cell. Sci. *20,* 223-231.

Paster, B.J., Boches, S.K., Galvin, J.L., Ericson, R.E., Lau, C.N., Levanos, V.A., Sahasrabudhe, A., and Dewhirst, F.E. (2001). Bacterial diversity in human subgingival plaque. J. Bacteriol. *183,* 3770-3783.

Paster, B.J., and Canale-Parola, E. (1980). Involvement of periplasmic fibrils in motility of spirochetes. J. Bacteriol. *141,* 359-364.

Paster, B.J., and Dewhirst, F.E. (2000). Phylogenetic foundation of spirochetes. J. Mol. Microbiol. Biotechnol. *2,* 341-344.

Paster B.J., Dewhirst F.E., Cooke S.M., Fussing V., Poulsen L.K., and Breznak J.A. 1996a. Phylogeny of not-yet-cultured spirochetes from termite guts. Appl. Environ. Microbiol. *62,* 347-352.

Paster, B.J., Dewhirst, F.E., Weisburg, W.G., Tordoff, L.A., Fraser, G.J., Hespell, R.B., Stanton, T.B., Zablen, L., Mandelco, L., and Woese, C.R. (1991). Phylogenetic analysis of the spirochetes. J. Bacteriol. *173,* 6101-6109.

Paster, B.J., Pelletier, D.A., Weisburg, W.G., Dewhirst, F.E., Fussing, V., Poulsen, L.K., Dannenberg, S., and Schroeder, I. 1996b. Phylogenetic position of the spirochetal genus *Cristispira.* Appl. Environ. Microbiol. *62,* 942-946.

Saitou, N., and Nei, M. (1987). The neighbor-joining method: A new method for reconstructing phylogenetic trees. Mol. Biol. Evol. *4,* 406-425.

Schrank, K., Choi, B.K. Grund, S., Moter, A., Heuner, K., Nattermann, H., and Göbel, U.B. (1999). *Treponema brennaborense* sp. nov., a novel spirochaete isolated from a dairy cow suffering from digital dermatitis. Int. J. Syst. Bacteriol. *49,* 43-50.

Seshadri, R., Myers, G.S., Tettelin, H., Eisen, J.A., Heidelberg, J.F., Dodson, R.J., Davidsen, T.M., DeBoy, R.T., Fouts, D.E., Haft, D.H., Selengut, J, Ren, Q., et al. (2004) Comparison of the genome of the oral pathogen *Treponema denticola* with other spirochetegenomes. Proc Natl Acad Sci U S A. *10,* :5646-5651

Simpson, J.M., Santo Domingo, J.W. and Reasoner, D.J. (2004). Assessment of equine fecal contamination: the search for alternative bacterial source-tracking targets FEMS Microbiol. Ecol. *47,* 65-75.

Smibert, R.M, Johnson, J.L., and Ranney, R.R. (1984). *Treponema socranskii* sp. nov., *Treponema socranskii* subsp. *socranskii* subsp. nov., *Treponema socranskii* subsp. *buccale* subsp. nov., and *Treponema socranskii* subsp. *paredis* subsp. nov. isolated from the human periodontia. Int. J. Syst. Bacteriol. *34,* 457-462.

Stackebrandt, E., Frederiksen, W., Garrity, G.M., Grimont, P.A., Kampfer, P., Maiden, M.C., Nesme, X., Rossello-Mora, R., Swings, J., Truper, H.G., Vauterin, L., Ward, A.C., and Whitman, W.B. (2002). Report of the ad hoc committee for the re-evaluation of the species definition in bacteriology. Int. J. Syst. Evol. Microbiol. *52,* 1043-1047.

Tajima, K., Aminov, R.I., Nagamine, T., Ogata, K., Nakamura, M., Matsui, H., and Benno, Y. (1999). Rumen bacterial diversity as determined by sequence analysis of 16S rDNA libraries. FEMS Microbiol. Ecol. *29,* 159-169.

Van de Peer, Y., and De Wachter, R. (1994). TREECON for Windows: a software package for the construction and drawing of evolutionary trees for the Microsoft Windows environment. Comput. Applic. Biosci. *10,* 569-570.

Valdez, M., Haines, R, Riviere, K.H., Riviere, G.R., and Thomas, D.D. (2000) Isolation of oral spirochetes from dogs and cats and provisional identification using polymerase chain reaction (PCR) analysis specific for human plaque *Treponema* spp. J. Vet. Dent. *17*:23-26.

Veldkamp, H. (1960). Isolation and characteristics of *Treponema zuelzerae* nov. spec., and anaerobic, free-living spirochete. Antonie Van Leeuwenhoek *26,* 103-125.

Wang, G., van Dam, A.P., Schwartz, I., and Dankert, J. (1999). Molecular typing of *Borrelia burgdorferi* sensu lato: taxonomic, epidemiological, and clinical implications. Clin. Microbiol. Rev. *12,* 633-653.

Wayne, L.G., Brenner, D.J., Colwell, R.R., Grimont, P.A.D., Kandler, O., Krichevsky, M.I., Moore, L.H., Moore, W.E.C., Murray, R.G.E., Stackebrandt, E., Starr, M.P., and Truper, H.G. (1987). Report of the Ad Hoc Committee on Reconciliation of Approaches to Bacterial Systematics. Int. J. Syst. Bacteriol. *37,* 463-464.

Woese, CR. (1987). Bacterial evolution. Microbiol. Rev. *51,* 221-271.

Zeigler, DR. (2003). Gene sequences useful for predicting relatedness of whole genomes in bacteria. Int. J. Syst. Bacteriol. *53,* 1893-1900.

2

Comparative Genomics of Spirochetes

Steven J. Norris and George M. Weinstock*

Abstract

The recent sequencing of the genomes of several spirochetes permits a thorough analysis of the similarities and differences within this bacterial phylum. *Treponema pallidum* subsp. *pallidum* has one of the smallest bacterial genomes at 1.14 million base pairs (Mb) and has limited metabolic capabilities, reflecting its adaptation through genome reduction to the rich environment of mammalian tissue. Comparison with the genome of *Treponema denticola* reveals that, despite its larger genome size (2.84 Mb), it also has limited biosynthetic activities and is highly dependent on the gingival crevice microenvironment to provide required nutrients. The two *Borrelia* species sequenced to date also have limited 'core' metabolic functions encoded on a 0.91 Mb linear chromosome, but have acquired multiple plasmids that are important for their survival in the tick midgut and mammalian or avian species and in transmission between these two environments. The larger genomes of the two *Leptospira interrogans* strains sequenced (4.6 Mb) are highly similar and reflect extensive biosynthetic and ATP production mechanisms consistent with its ability to live free in the environment as well as infect and cause disease in mammals. Information to date indicates that the genus *Treponema* evolved from a common ancestor early in bacterial evolution (perhaps hundreds of millions of years ago), whereas the *T. pallidum* subspecies are closely related and may have diverged within tens of thousands of years.

1. Introduction

The genus *Treponema* is a branch of *Spirochetales*, which is now considered a separate phylum within the kingdom *Procaryotae* (Paster and Dewhirst, 2000). By all indications to date, spirochetes derive from a single ancient ancestor and thus represent one of the clearest examples of divergent evolution among bacteria. Studies of 16S ribosomal RNA gene sequences by Paster and colleagues (Paster et al., 2001; Paster and Dewhirst, 2000) have demonstrated that the genera *Treponema*,

*For correspondence email Steven.J.Norris@uth.tmc.edu

Table 1. Characteristics of spirochete genomes[a]

Property	***T. pallidum***	***T. denticola***	***B. burgdorferi***	***L. interrogans* Icterohemorrhagiae[b]**	***L. interrogans* Copenhageni[b]**
Chromosome conformation	single circular	single circular	single linear	two circular	two circular
Chromosome size(s) (base pairs)	1,138,012	2,843,201	910,725	4,227,185 and 350,181	4,332,241 and 358,943
Extrachromosomal elements	none	(1 circular)[c]	12 linear 10 circular[c]	none	none
Extrachromosomal content (bp)		(2,647)[d]	609,132		
Total genome size (bp)	1,138,012	2,843,201	1,519,856	4,577,366	4,627,184
G+C content (%)	52.8	37.9	28.6	36.0	34.9
Prophages or phage remnants	no	yes	yes	yes	yes
Insertion sequences	no	yes	no	yes	yes
Restriction/modification systems	no	yes	yes	yes	yes
Total protein-coding genes[e]	1,039	2,786	1,740	4,727	3,728
Assigned function	541	1,223	1,045	2,060	1,972

Conserved hypotheticals and hypotheticals	498	1,563	695	2,667	1,756
Average CDS size (bp)[f]	1017	939	992	778	947
Total coding sequence (%)	93.0	92.1	93.5	78.4	74.7
rRNA genes	2 23S, 2 16S, 2 5S	2 23S, 2 16S, 2 5S	2 23S, 2 16S, 1 5S	1 23S, 2 16S, 1 5S	2 23S, 2 16S, 1 5S
tRNA genes	45	44	34	37	37

[a] Adapted from Seshadri et al., 2004. Primary references: *T. pallidum* subsp. *pallidum* Nichols (Fraser et al., 1998): *T. denticola* 35405 (Seshadri et al., 2004); *B. burgdorferi* B31 (Casjens et al., 2000; Fraser et al., 1997; Miller et al., 2000); *L. interrogans* serotype Icterohemorrhagiae serovar Lai 56601 (Ren et al., 2003); *L. interrogans* serovar Copenhageni Fiocruz L1-130 (Nascimento et al., 2004).

[b] The genome information for *L. interrogans* represents combined data from both chromosomes (4,332,241 and 358,943 bp for serovar Lai, 4,277,185 and 350,181 bp for serovar Copenhageni). Fewer predicted genes were identified in serovar Copenhageni due to the omission of predicted genes <150 bp with no known homologs.

[c] The total number of *B. burgdorferi* B31 plasmids is derived from (Casjens et al., 2000), with the addition of the previously undetected plasmid cp9-2 (Miller et al., 2000).

[d] The 2,647 bp circular plasmid pTD1 is present in *T. denticola* 35220 (MacDougall et al., 1992), but is absent in *T. denticola* 35405 (Seshadri et al., 2004).

[e] Key for gene designations: Assigned function – significant amino acid sequence similarity with genes in other organisms (orthologs) of known function; Conserved hypothetical – similarity to orthologs of unknown function; Hypothetical – no significant similarity with other protein database entries. The number of predicted genes in the *T. pallidum* and *B. burgdorferi* chromosomes are derived from the original annotation. These numbers, particularly hypothetical and conserved hypothetical proteins, may be significantly different with updated blast searches and annotation.

[f] CDS = predicted coding sequence

Borrelia, *Leptospira*, *Brachyspira*, and *Spirochaeta* each represent distinct groups within this phylum, although some organisms previously identified as *Spirochaeta* species (*S. stenostrepa* and *S. zuelerzae*) appear to fall within the genus *Treponema*. The recent availability of the genome sequences from *Treponema pallidum* subsp. *pallidum* Nichols (Fraser et al., 1998), *Treponema denticola* 35405 (Seshadri et al., 2004), *Borrelia burgdorferi* B31 (Fraser et al., 1997), *Borrelia garinii* PBi (Glöckner et al., 2004), *Leptospira interrogans* serogroup Icterohaemorrhagiae serovar Lai (Ren et al., 2003), and *Leptospira interrogans* serogroup Borgpetersoni serovar Copenhageni (Nascimento et al., 2004) provide fertile ground for the comparative genomic analysis of spirochetes, as described in this chapter and several previous reviews (Casjens, 2000; Casjens et al., 2000; Norris et al., 2001; Norris and Weinstock, 2000; Radolf et al., 1999; Subramanian et al., 2000; Weinstock et al., 1998; 2000a; 2000b).

The characteristics of the spirochetal genomes sequenced to date are provided in Table 1. *T. pallidum* subsp. *pallidum* has the smallest overall genome size and simplest conformation of this group, with a single circular chromosome of 1.14 million basepairs (Mb). Previous studies had indicated that *T. pallidum* also contained a small, circular plasmid (Norgard and Miller, 1981), but these results were not substantiated by subsequent electrophoretic and genome sequence analyses. *T. denticola* also has a single circular chromosome, but of a much larger size (2.84 Mb); some strains also contain a 2.6 kb plasmid described as pTD1 (MacDougall et al., 1992). By comparison, the *B. burgdorferi* B31 genome is much more complex, containing a linear chromosome and 22 circular and linear plasmids (Fraser et al., 1997; Miller et al., 2000; Stevenson et al., 2000); however, the overall genome size is quite small, totaling approximately 1.5 Mb. The two strains of *L. interrogans* sequenced to date have closely related genomes with two circular chromosomes each and a total genome size of 4.6 Mb.

2. Early studies

Maio and Fieldsteel (Maio and Fieldsteel, 1978; Maio and Fieldsteel, 1980; Maio et al., 1978) performed the earliest analyses of the genetic relatedness and complexity of members of the genus *Treponema*. DNA-DNA saturation hybridization, hybridization kinetics, and temperature renaturation studies were utilized to compare *T. pallidum* subsp. *pallidum* Nichols to other treponemes. The Nichols strain was found to be highly related to the *T. pertenue* strain based on >95% cross-hybridization. As a result the yaws and endemic syphilis isolates were reclassified as *T. pallidum* subsp. *pertenue* and *T. pallidum* subsp. *endemicum*, respectively (Smibert, 1984). In contrast, <5% hybridization was observed between the Nichols strain, *T. phagedenis*, and *T. refringens*. Also, estimates of the percent G + C based on melting temperatures differed significantly among these species, and together these results provided the first evidence for significant divergence between the *T. pallidum* subspecies and other *Treponema* species. In addition, DNA from *Treponema hyodysenteriae*, as it was then called, did not hybridize with DNA from any of the *Treponema* species (Maio et al., 1978); this agent of swine dysentery was later reclassified as *Brachyspira hyodysenteriae*.

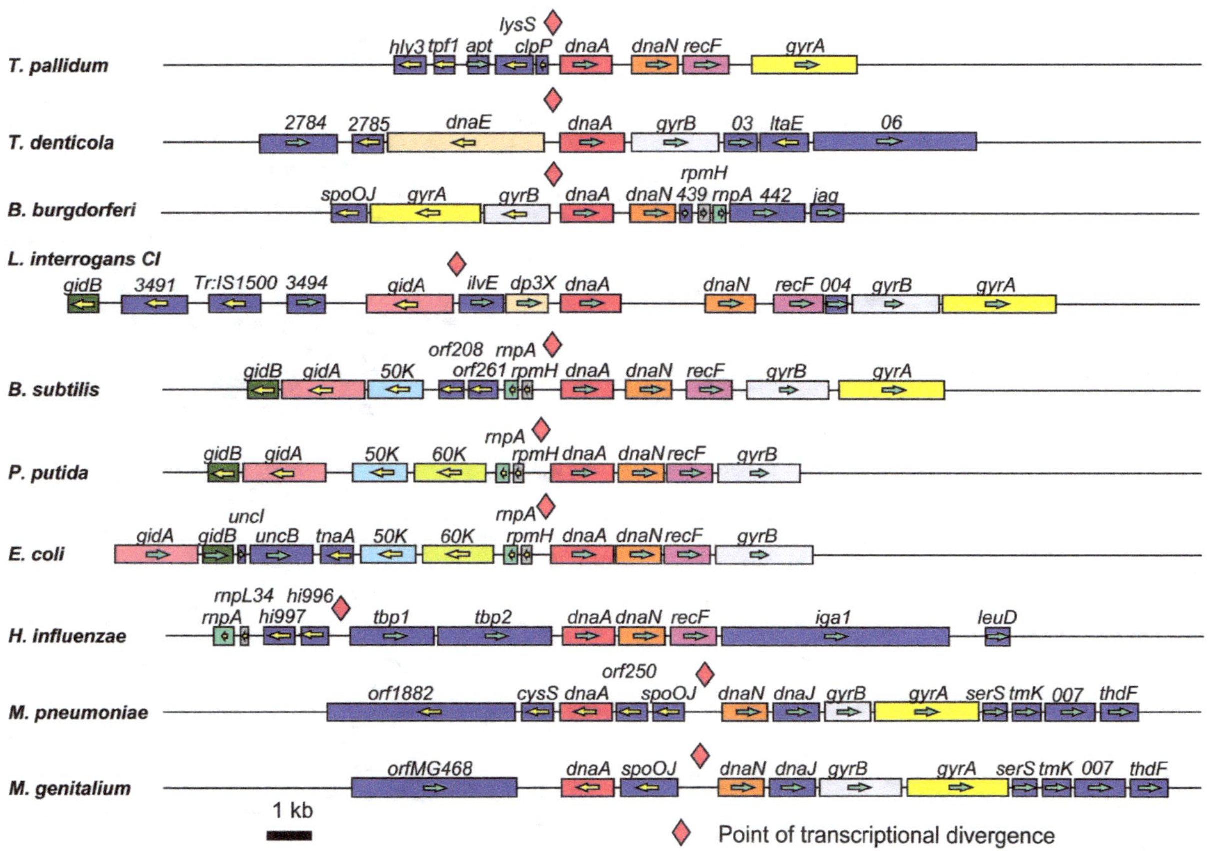

Figure 1. Comparison of putative origin of replication (*ori*) regions of spirochetes and other bacteria. Homologues of genes commonly found at bacterial origins are indicated in different colors; genes in dark blue are not commonly associated with origins. Red and yellow arrows indicate the direction of transcription, and the point of transcription divergence (usually associated with the origin) is shown by a red diamond. Organisms included in this figure are *Treponema pallidum*, *Treponema denticola*, *Borrelia burgdorferi*, *Leptospira interrogans*, *Bacillus subtilis*, *Pseudomonas putida*, *Escherichia coli*, *Haemophilus influenzae*, *Mycoplasma pneumoniae*, and *Mycoplasma genitalium*.

A

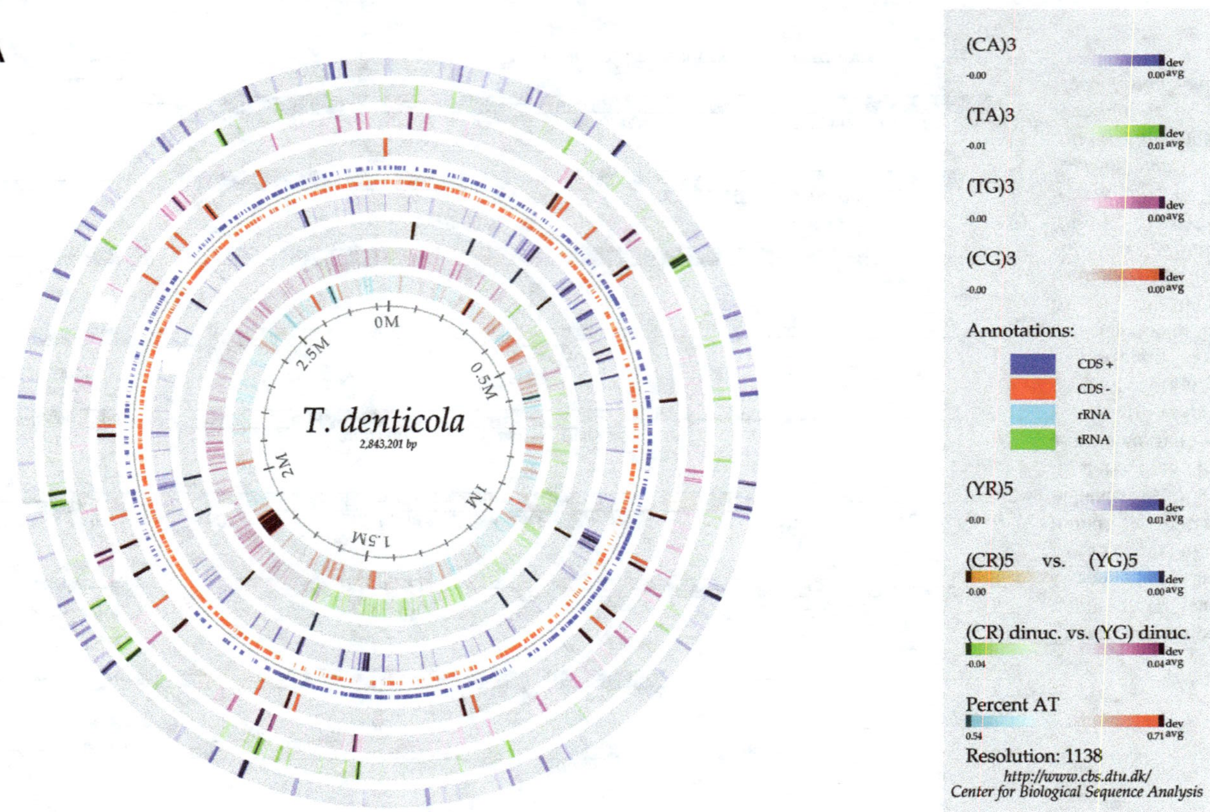

T. denticola
2,843,201 bp
0M
0.5M
1M
1.5M
2M
2.5M
(CA)3
-0.00
0.00 avg
dev
(TA)3
-0.01
0.01 avg
dev
(TG)3
-0.00
0.00 avg
dev
(CG)3
-0.00
0.00 avg
dev
Annotations:
CDS +
CDS -
rRNA
tRNA
(YR)5
-0.01
0.01 avg
dev
(CR)5 vs. (YG)5
-0.00
0.00 avg
dev
(CR) dinuc. vs. (YG) dinuc.
-0.04
0.04 avg
dev
Percent AT
0.54
0.71 avg
dev
Resolution: 1138
http://www.cbs.dtu.dk/
Center for Biological Sequence Analysis

Z-DNA ATLAS

B

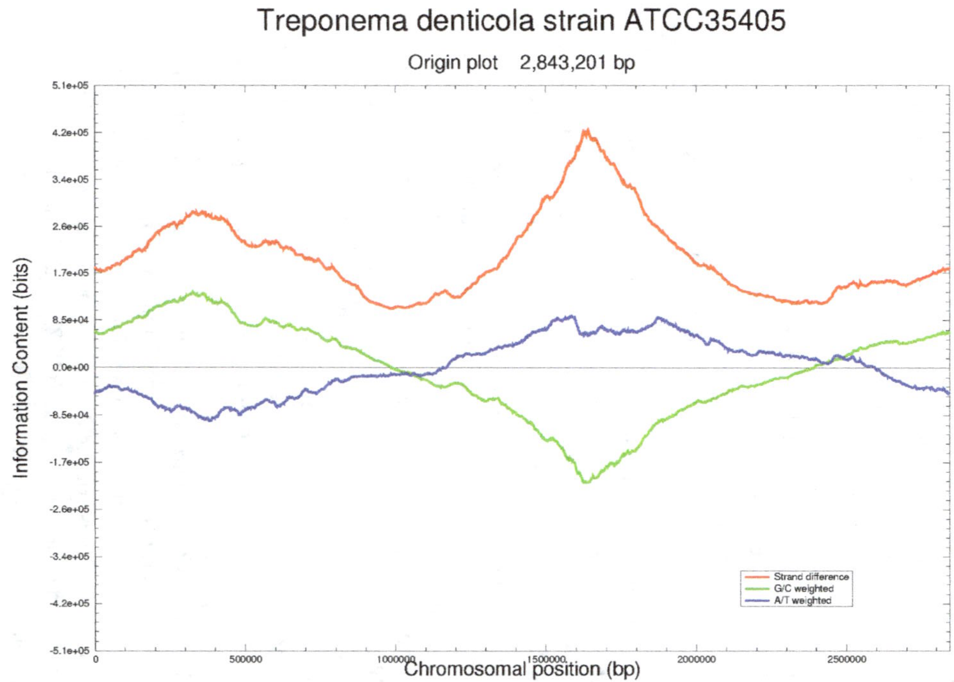

Figure 2. Z-DNA data (**A**) and GC skew data (**b**) indicating the predicted location of the *Treponema denticola* 35405 origin of replication and termini at coordinates ~1,641,000 and ~325,000, respectively. **a**) Atlas of Z-DNA related sequences. Each concentric circle represents a different property, as indicated in the key (outermost circle corresponds to the top key item). Small boxes correspond to coding sequences (CDSs) on the plus strand (blue) and the negative strand (red). The skew of cytosine-purine (CR) vs. pyrimidine-guanine dinucleotide pairs (second ring from the middle) has a high concordance with the GC skew data. (**B**) Plot of GC strand skew (green line), AT strand skew (blue line), and combined results (red line). The sharp peak at 1,641,000 bp indicates a transition corresponding to the likely origin of replication, whereas the more gradual peak at 325,000 bp is the probable location of replication termination. Figures graciously provided by David W. Ussery, Center of Biological Sequence Analysis, The Technical University of Denmark, Lyngby, Denmark. For details regarding these analyses, see Pedersen et al. (2000) or www.cbs.dtu.dk.

3. *T. pallidum*

The size of the single circular chromosome of *T. pallidum* subsp. *pallidum* Nichols was first estimated as ~900 kbp based on the summation of *Not*I, *Sfi*I, and *Srf*I fragment sizes based on pulsed field gel electrophoresis (PFGE) (Walker et al., 1991). Later, derivation of a physical map of the genome resulted in a somewhat higher estimate of 1.03 to 1.08 Mb (Walker et al., 1995). These values were lower than the actual size determined by genome sequencing, most likely because of comigration of fragments or the limited accuracy of PFGE size estimates.

The complete genome sequence of the Nichols strain was determined by standard shotgun sequencing and closure techniques and was reported in 1998 (Fraser et al., 1998). The *T. pallidum* genome is by far the simplest of all the spirochetal genomes sequenced to date by several criteria. First, it is among the smallest bacterial genomes yet characterized, with a total size of 1.14 Mb. (In bacteria, only *Mycoplasma* species have been shown to have smaller genomes, *e.g.* 0.58 Mb for the *M. genitalium* (Fraser et al., 1995); the archaeal hyperthermophile *Nanoarchaeum equitans*, an obligate symbiont of chrenarchaeal *Ignicoccus*, has the smallest microbial genome (0.49 Mb) sequenced thus far (Waters et al., 2003)). Second, the *T. pallidum* genome consists of a single circular chromosome, whereas many of the spirochetal genomes are more complex, having multiple plasmids in the case of *Borrelia* species and two chromosomes in *L. interrogans*. Third, the genomic sequence lacks recognizable restriction-modification genes, insertion sequences indicative of transposons, or any evidence of bacteriophage embedded in the genome (prophages) or phage sequence fragments. In contrast, these characteristics are present in *T. denticola* and the other spirochetes examined. The lack of these elements and the small, simple genome of *T. pallidum* suggests that its evolution as an obligate pathogen and isolation in the aseptic environment of human/animal tissue resulted in genome reduction, absence of 'infectious' elements from other bacteria, and loss of protective mechanisms against foreign DNA.

Evidence to date indicates that all *T. pallidum* subspecies are monophyletic and essentially identical in sequence, except for some gene rearrangements and point mutations. Only a single nucleotide difference was noted in the 963 bp 4D antigen gene (TP1038) between *T. pallidum* subsp. *pallidum* and *T. pallidum* subsp. *pertenue* strains, but this difference was not consistent within each subspecies (Noordhoek et al., 1989). Later studies demonstrated that single nucleotide polymorphisms (SNPs) in the genes for the 15 kDa lipoprotein (*tpp15*, TP0171), glycerophosphodiester phosphodiesterase (*gpd*, TP0257), and membrane protein Tp92 (TP0326) could be used to distinguish the *pallidum*, *pertenue*, and *endemicum* subspecies (Cameron et al., 1999; 2000; Centurion-Lara et al., 1998). A somewhat greater degree of polymorphism was observed in the rabbit venereal spirochetosis agent, *T. paraluiscuniculi*, which is also closely related to *T. pallidum*; for example, 6 nucleotide substitutions (5 of which are silent) were noted between the *gpd* genes of *T. pallidum* subsp. *pallidum* and *T. paraluiscuniculi*, while only one nucleotide difference existed between the subsp. *pallidum* and the subsp. *pertenue* or *endemicum* isolates.

The regions containing the *tpr* genes A-L (see Chapter 11, Cameron, and Chapter 12, Centurion-Lara) appear to exhibit the greatest polymorphism among *T. pallidum* isolates (Centurion-Lara et al., 2000a; 2000b; Giacani et al., 2005; Sun et al., 2004). Recent studies have shown that *tprK* undergoes sequence variation through gene conversion from a donor locus close to *tprD* (Centurion-Lara et al., 2004; LaFond et al., 2003). This variation occurred following infection of rabbits with a clonal isolate of *T. pallidum* subsp. *pallidum* Chicago, demonstrating that several new alleles arose during the course of infection. It is speculated that this variation may provide a mechanism of immune evasion in chronic syphilitic infections (Centurion-Lara et al., 2004).

In 1998, Pillay et al. developed a molecular subtyping system which distinguishes different strains of *T. pallidum* subsp. *pallidum* (Pillay et al., 1998). This system uses PCR amplification of *T. pallidum* DNA in human genital ulcer or blood specimens and two measures of heterogeneity: length polymorphism in the acid repeat protein (*arp*), which contains variable numbers of a 60 bp repeat; and RFLP analysis of *tprE*, *tprG*, and *tprJ*. *arp* corresponds to TP0433 and TP0434 in the *T. pallidum* genome sequence (Fraser et al., 1998), two neighboring hypothetical ORFs that may have been separated by a sequencing error. By analyzing 229 genital ulcer or blood specimens from the United States, Madagascar, and South Africa, 43 different subtypes of *T. pallidum* subsp. *pallidum* were identified. A lower degree of heterogeneity in *arp* and the three *tpr* genes was observed in the few subspecies *endemicum* and *pertenue* isolates examined. The molecular subtyping approach has been utilized to study an increase in syphilis incidence in Maricopa County, Arizona (Sutton et al., 2001). Genital ulcer samples yielded typable specimens by PCR in 41 of 43 (93%) of cases, whereas blood samples provided sufficient DNA in only 4 of 15 (27%) of specimens. Overall, 10 different subtypes were identified in the 45 typable specimens, with subtype 14f being predominant (53% of cases). Only two of the cases could be linked epidemiologically in this population. However, the molecular subtyping system is likely to be of value in studying the epidemiology of syphilis outbreaks.

4. *T. denticola*

T. denticola is an oral spirochete commonly found in the gingival crevice of humans and other animals, and is associated with the occurrence of periodontal disease. A physical map of the *T. denticola* strain 33520 genome was generated by PFGE and hybridization with probes by MacDougall and Saint Girons (1995). They found that the genome was a single circular chromosome ~3.0 Mb in size and, like those of *T. pallidum* and *T. phagedenis,* contained two rRNA loci. The genome sequence of *T. denticola* 35405 was determined in 2004 (Seshadri et al., 2004) and confirmed this previous analysis. The 2.84 Mb *T. denticola* 35405 genome differs in several ways from the *T. pallidum* subsp. *pallidum* genome. First, it is approximately 2.5 times larger, and yet exhibits little increase in biosynthetic capabilities (see below). Second, the G+C content is markedly different (37.9% compared to 52.8%), consistent with the genetic divergence between these *Treponema* species. Third, there is little synteny (correlation of gene order) between the two genomes, and conservation of gene order is limited to large operons such as those that encode ribosomal and flagellar proteins. Fourth, the *T. denticola* genome contains an apparently degenerate prophage (at bp 1,164,840 to 1,202,400 in the genome sequence), a 65-kb region with characteristics of a phage-mediated integration event, and several phage remnants, as well as at least

8 putative mobile elements (Seshadri et al., 2004). Lastly, the *T. denticola* genome encodes at least three type-I and –II restriction endonucleases, whereas none have been detected in the *T. pallidum* coding sequence. These last two differences reflect the competitive bacterial environment that *T. denticola* inhabits in dental plaque, in which ~500 bacterial species cohabitate at an extremely high density (Paster et al., 2001). Despite these differences, 728 of the 2,786 predicted genes in *T. denticola* have the highest similarity to *T. pallidum* ORFs, and this group corresponds with 68% of the *T. pallidum* genes. In addition, rDNA sequences and whole genome comparisons (M. McLeod and G. M. Weinstock, unpublished data) indicate that the genus *Treponema* is indeed a monophyletic group, *i.e.* branched out from a single ancient "*Treponema*-specific"ancestor. Taken together, these differences and similarities indicate that the divergence of a common ancestor into the *T. pallidum* subgroup (containing the *T. pallidum* subspecies, *Treponema carateum*, and *T. paraluiscuniculi*) and the oral treponemes and other spirochetes occurred millions of years ago, long before the evolution of humans. This aspect will be discussed in greater detail at the end of the chapter.

5. Other *Treponema*

Based on 16S rDNA sequence amplification and sequencing, dental plaque from the gingival crevice of humans contains at least 57 distinct 'clusters' of *Treponema* with >99% sequence similarity, subdivided into 10 groups with >90% identity (Dewhirst et al., 2000; Paster et al., 2001). Correia et al. (2004) determined the genome sizes of seven strains of oral treponemes by restriction with rare-cutting enzymes and PFGE (Table 2). The sizes of the genomes of other *Treponema* species examined were estimated to be somewhat smaller than that of *T. denticola*, with a range of 2.2 Mb to 2.4 Mb. Termite hindguts also contain *Treponema* species, and two of these were recently characterized as *T. azotonutricum* and *T. primitia* (Graber et al., 2004; see also Chapter 18, Breznak). The genome sizes of both of these species (3.4 to 3.9 Mb) are larger that those described thus far for other treponemes, and the G+C content of both species is within the range of 50 to 51 percent. In the same study, the genome of *Spirochaeta aurantia*, a mud-dwelling, free-living spirochete, was 3.7 Mb with a G+C content of 65.6%.

6. *Borrelia*

Borrelia are spirochetes that are maintained by transmission between arthropod and mammalian or avian hosts, and include the agents that cause Lyme borreliosis and relapsing fever. The genomes of *Borrelia* species are arguably the most complex bacterial genomes characterized to date, consisting of a single linear chromosome and multiple linear plasmids (lps) and circular plasmids (cps). The *B. burgdorferi* B31 genome consists of a 910 kb linear chromosome, 12 linear plasmids (56 to 5 kb), and 10 circular plasmids (32 kb to 9 kb) (Casjens et al., 2000; Fraser et al., 1997; Miller et al., 2000). The plasmids contain genes encoding multiple paralogous families of proteins, and most of these lack homology with predicted proteins outside the genus (Casjens et al., 2000). Both the linear and circular plasmids tend to be conserved in *Borrelia* species, and certain plasmids of *B. burgdorferi* B31 (*e.g.* lp25 and lp28-1) have been shown to be required for infection in mammals (Grimm

Table 2. Estimated sizes of spirochete genomes not yet sequenced, as determined by treatment with rare-cutting endonucleases and pulsed field gel electrophoresis

Organism (Strain)	Estimated Size (Mb)	Reference
Treponema denticola (33520)	3.0	MacDougall and Saint Girons, 1995; Correia et al., 2004
Treponema vincentii (N-9)	2.5	Correia et al., 2004
Treponema lecithinolyticum (OMZ 684)	2.3	Correia et al., 2004
Treponema maltophilum (OMZ 679)	2.3	Correia et al., 2004
Treponema amylovorum (ATCC 700288)	2.5	Correia et al., 2004
Treponema socranskii ss (04 D11B2)	2.5	Correia et al., 2004
Treponema parvum (D96NR3)	2.4	Correia et al., 2004
Treponema pectinovorum (D36DR2)	2.3	Correia et al., 2004
Treponema primitia (ZAS-1) (ZAS-2)	3.5 3.8	Graber et al., 2004
Treponema azotonutricum (ZAS-9)	3.9	Graber et al., 2004
Spirochaeta aurantia (J1)	3.7	Graber et al., 2004

et al., 2004a; Labandeira-Rey et al., 2001; 2003; Purser et al., 2003; Purser and Norris, 2000); in addition, *B. burgdorferi* can not be 'cured' of some plasmids such as cp26 and lp54, despite the fact that other plasmids are easily lost during *in vitro* culture (Barbour, 1988; Byram et al., 2004; Grimm et al., 2003). The recent advent of transformation of low-passage, infectious strains of *B. burgdorferi* has shown that the plasmid-encoded genes *pncA* (on lp25) and *ospC* (on cp26) are required for mammalian infection (Grimm et al., 2004b; Purser et al., 2003), whereas the *ospA/B* locus (on lp54) is necessary for survival and transmission in ticks (Yang et al., 2004). Based on these results, it appears that some, and perhaps all, plasmids may be required for the natural life cycle of *Borrelia* organisms; thus they have also been called 'minichromosomes' (Bergstrom et al., 1992). The linear replicons of *Borrelia* organisms have covalently closed ends, resulting from the presence of a phosphate bond between the 5′ and 3′ ends (Barbour and Garon, 1987; Casjens et al., 1997). Replication of these linear plasmids involves formation of a double circular intermediate, followed by cleavage and ligation at the telomere sequence by a site-specific resolvase, ResT, encoded on cp26 (Bankhead and Chaconas, 2004; Chaconas et al., 2001; Kobryn and Chaconas, 2002).

B. burgdorferi has been shown to produce bacteriophages upon exposure to DNA damaging agents (Eggers et al., 2000). Furthermore, one of these virions (phiBB-1) packages the cp32 plasmids, leading to the suggestion that the cp32s actually represent prophages (Eggers et al., 2000; 2001). However, it has not as yet been determined whether cp32s are competent phage genomes, defective phage genomes, or simply DNA packaged by a generalized transduction mechanism.

7. *Leptospira*

The genome sequences of two *L. interrogans* strains were determined recently (Nascimento et al., 2004; Ren et al., 2003), and confirmed the previous finding that the genome consists of two circular replicons of ~4.3 Mb and ~0.35 Mb (Zuerner et al., 1993). The large size and encoded metabolic capacities of the *Leptospira* genome relative to other pathogenic spirochetes are consistent with the ability of these organisms to both thrive in mammalian tissue and survive in environmental conditions such as contaminated water (Bharti et al., 2003). There are multiple prophages and insertion sequences in the *L. interrogans* genome, indicative of common exposure to bacteriophages and foreign DNA.

8. Genome comparisons

A number of properties distinguish the spirochetal genomes, including 1) size and conformation; 2) G+C content; 3) presence or absence of prophage, insertion sequences and restriction/modification systems; and 4) predicted origins of replication; and 5) encoded metabolic capabilities. These will be discussed separately below.

In general, the size of a bacterial genome is directly related to its coding capacity (the number of genes) and hence the complexity of its encoded activities. *T. pallidum* has the smallest spirochetal genome characterized to date, corresponding to the small number of predicted genes (1,039). This feature is indicative of reductive evolution, in which *T. pallidum* has adapted to the relatively rich and homeostatic environment of mammalian tissue. *B. burgdorferi* has a slightly higher coding capacity (1,740 genes), but most of this increase is in hypothetical plasmid-encoded genes rather than 'core' metabolic functions encoded almost entirely in the chromosome. *T. denticola* and other *Treponema* have intermediate genome sizes of 2.3 to 3.9 Mb (Tables 1 and 2). *L. interrogans* has the largest spirochetal genome examined to date, in keeping with its ability to survive in the environment and apparent acquisition of sequences from other bacteria. Note that the difference in number of predicted coding sequences (CDSs) in *L. interrogans* Icterohemorrhagiae and *L. interrogans* Copenhageni (Table 1) is due to different criteria for identification of genes, not any sequence differences in the genomes. Nascimento et al. (2004) used more stringent criteria that excluded CDSs of less than 150 bp that lacked database matches. This difference in analyses also is reflected in the average gene length (Table 1). Ussery and Hallin (2004) have commented that most bacterial genomes are over-annotated by at least 20%; they use the *L. interrogans* Icterohemorrhagiae annotation (Ren et al., 2003) as a specific example, with as much as 60% of the CDSs in the genome being falsely identified as predicted genes. The bottom line is that CDSs identified in genome annotations should not be considered 'true' genes until their expression is confirmed experimentally.

Sequence conservation, gene synteny and the presence or absence of insertions and deletions (indels) have been used as a measure of the relatedness of bacterial strains, with a close correlation of gene sequences and order corresponding to close relatedness (Ussery et al., 2004). The *T. pallidum* subspecies exhibit a high degree of synteny, based on whole genome PCR amplification and cross-hybridization studies (D. Ŝmajs, E. Sodergren, S. J. Norris, and G. M. Weinstock, unpublished studies). The chromosomes of *B. burgdorferi* B31 and *B. garinii* PBi (Fraser et al., 1997; Glöckner et al., 2004) and (separately) the two *L. interrogans* strains sequenced thus far (Nascimento et al., 2004; Ren et al., 2003) also exhibit a high degree of relatedness based on their overall DNA sequence homology and gene order, although a large relative inversion exists between the two *L. interrogans* strains. In contrast, the genomes of *T. pallidum* and *T. denticola* have essentially no preservation of gene order beyond that present in large conserved operons (Fraser et al., 1998; Seshadri et al., 2004); indeed, in terms of gene order *T. pallidum* is no more similar to *T. denticola* than to the other spirochete genome sequences examined thus far.

G+C content has also been used as a measure of bacterial relatedness. *T. pallidum* and *T. denticola*, and indeed spirochetes as a whole, are highly divergent in this regard (Table 1). The *T. pallidum* genome has a G+C content of 52.8%, whereas the *T. denticola* genome has a much lower value of 37.9%. *Borrelia* species have the lowest G+C content of the spirochetes examined to date, with a value of 28.6%. It is thought that the G+C ratio and hence codon usage is driven by spontaneous deamination of cytosine to uracil and the resulting transition of GC pairs to AT pairs during DNA replication. Uracil-DNA glycosylase (UDG) recognizes deoxyuracil in DNA and excises the uracil base, resulting in small patch excision repair and restoration of the GC base pair prior to fixation of the AT mutation during replication (Pearl, 2000). Low efficacy of UDG may therefore be related to the propensity of an organism to develop a lower G+C content; conversely, organisms that have a high G+C content or have a low tolerance for mutation may have a highly active deoxyuracil repair mechanism (Fogg et al., 2002; Venkatesh et al., 2003). UDG homologues are present in the *T. pallidum* (TP0229) and *T. denticola* (TDE2491) genomes, but are annotated as DNA polymerases. Both *L. interrogans* genomes have genes annotated as UDG (LA3668 and LIC10548); however, none of the predicted proteins of *B. burgdorferi* or *B. garinii* (which have extremely low G+C contents) have sequence similarity to UDG. At this time, the extent to which this repair mechanism is involved in the divergence of G+C content among the spirochetes has not been examined.

An unusual feature of the *T. pallidum* genome is the apparent lack of insertion sequences or prophage. The genomes of all of the other spirochetes, and indeed nearly all bacteria thus far examined, typically contain insertion sequence elements, prophage, or both. It is reasonable to assume that the lack of these elements in *T. pallidum* is due to its primary residence in human tissue, where it would not be exposed to transposons, insertion elements, or bacteriophage produced by other bacteria. Any residual elements from an ancient ancestor have apparently been 'cleaned up' through the genome reduction process.

9. Origin of replication

The origin of DNA replication (ori) of bacteria is typically characterized by 1) a point at which transcription diverges in both directions; 2) a changeover in the GC skew (G-C/G+C on one strand), with G being predominant on the leading strand for replication; 3) proximity to *dnaA* (the gene for the DNA replication initiator protein DnaA) and other genes involved in DNA replication and recombination;

Table 3. Comparison of predicted metabolic activities of spirochetes, based on genome sequence and experimental data

Metabolic Activity	***T. pallidum***	***T. denticola***	***B. burgdorferi***	***L. interrogans***
Oxygen requirement	Microaerophilic	Anaerobic	Facultative	Aerobic
Glycolysis	+	+	+	+ (lacks hexokinase)
Tricarboxylic acid cycle	-	-	-	+
Electron transport chain	-	-	-	+
Pentose monophosphate pathway	+	Lacks oxidative branch	+	Lacks oxidative branch
Amino acid biosynthesis	-	+/-	-	+
Purine and pyrimidine biosynthesis	-	-	-	+
Fatty acid biosynthesis	-	+	-	+
Lipopolysaccharide biosynthesis	-	+/-	-	+
Amino acid transport systems	+	++	+	+
Phosphotransferase system	No transporters (regulatory only?)	No transporters (regulatory only?)	Complete system	Complete system
ATPases	2 V-type	1 V-type (defective?)	1 V-type	1 F-type
DNA recombination mechanisms	*recFGJNR*	*recFGJNR*	*recBCD*	*recBCD* and *recFGJNR*
Restriction-modification systems	-	+	+	+
Chemotaxis systems	+	+	+	+

and 4) the presence of DnaA box sequences (TGTGGATAA in *E. coli*). The *T. pallidum* ori generally follows this pattern and resembles the gene sequence of *E. coli*, *Bacillus subtilis*, and *Pseudomonas putida* on one side of the predicted ori (Figure 1). However, DnaA box sequences are not localized close to the predicted ori in any of the spirochetal genomes sequenced thus far. The *T. denticola* 35405 ori does not appear be located in the *dnaA* region. The points of transcriptional and GC skew divergence are localized at ~325,000 bp and ~1,641,000 bp in the genome coordinates, with the latter representing the strongest point of divergence (Figure 2). The skewing of C-purine (CR) vs. pyrimidine-G (YG) dinucleotides (which are indicative of Z-DNA structure) and oligonucleotide sequences has been used to delineate the origins of replication of other bacteria (Worning et al., 2005; Zhang and Zhang, 2002) and provides a clear demarcation at these coordinates in the *T. denticola* genome (Figure 2B). The *gyrA* gene, which is often located near the ori of other bacteria, is at one of these points (gene coordinates = complement of 324310 – 326757). These findings suggest that the the *T. denticola* 35405 ori has relocated during evolution, perhaps as the result of a DNA rearrangement. Experimental verification will be required to verify the origin of replication, as was accomplished with the *B. burgdorferi* B31 chromosome (Picardeau et al., 1999).

10. Predicted metabolic activities

The most valuable information provided by genome sequences is a list of the predicted genes present in the organism. These data permit the formulation of a complete 'parts list' for the organism and assessment of metabolic activities, virulence factors, and other properties. The predicted metabolic activities of *T. pallidum* and *T. denticola* are considered in detail in Chapter 4 (Cox and Radolf). In this chapter, consideration is restricted to a global comparison of the predicted metabolic activities of the spirochete genomes sequenced to date: *T. pallidum*, *T. denticola*, *B. burgdorferi*, and *L. interrogans* (Table 3). The genes with predicted functions encoded by the *B. burgdorferi* and *B. garinii* genomes are extremely similar, as are the genes from the two *L. interrogans* strains; therefore representative results for each of these genera are shown in the table. As expected, *T. pallidum* has the most limited metabolic functions; nearly all biosynthetic pathways are absent, including those for amino acid, purine, pyrimidine, and fatty acid synthesis. An unexpected finding was that *T. denticola*, despite its much larger genome, is also lacking these pathways, although a somewhat higher amino acid interconversion capability is predicted. In *B. burgdorferi*, most 'housekeeping' activities are encoded on the small, 0.9 Mb chromosome, and it also is lacking most biosynthetic activities. Thus these organisms have evolved to acquire required compounds from the surrounding environment, *i.e.* human tissue for *T. pallidum*, mammalian tissue or the tick midgut in *Borrelia* species, and the gingival crevice biofilm in *T. denticola*. In contrast, *L. interrogans* has retained most of these activities, and also appears to have acquired the ability to synthesize LPS from other organisms (Bulach et al., 2000a; 2000b). Moreover, variation in the gene content in the *rfb* locus may explain the occurrence of over 200 O-antigen types in *Leptospira* species (Bulach et al., 2000b; Nascimento et al., 2004).

11. Applications of genomic data

The genome sequence of *T. pallidum* has already been used to assess the spirochete's metabolic capabilities (Norris et al., 2001), adhesins and other potential virulence determinants (Cameron, 2003; Cameron et al., 2004; Radolf et al., 1999; Weinstock et al., 2000a), and antigenic variation systems (Centurion-Lara et al., 2004). The many additional applications include a global analysis of the antigenicity of *T. pallidum* proteins (McKevitt et al., 2003) and determination of transcript levels by microarray and quantitative RT-PCR methods (Smajs et al., 2005).

12. Conclusion

By all indications, the spirochetes represent a monophyletic group that originated from a single ancient ancestor. The *T. pallidum* subspecies appear to represent an extremely closely related cluster, exhibiting only limited SNPs and rearrangements relative to each other. This limited variation is similar to the divergence of *Yersinia pestis* strains, which is estimated to have occurred over the past 20,000 years (Achtman et al., 2004; Zhou et al., 2004). Thus it is reasonable to assume that the divergence of *T. pallidum* subspecies has occurred over a similar time period. The *Borrelia* species and the *Leptospira* strains examined thus far each also have limited sequence differences, indicating that they also diverged within the range of tens of thousands of years. In contrast, *T. pallidum* and *T. denticola* exhibit vastly different genome sizes, G+C content, gene arrangements, and gene sequences. This degree of divergence is greater than that observed between *E. coli* and *Salmonella enterica* Typhimurium, which separated ~140 million years ago (Ochman and Wilson, 1987). Therefore, the initial branchpoint of *Treponema* species likely occurred hundreds of millions of years ago. The divergence of spirochetal genera of course occurred much earlier, and indeed may date back a billion years or more. The most likely scenario is that the ancestral spirochete was a free-living organism with a large genome and extensive metabolic capabilities. In the process of adapting to the relatively rich environment provided by mammalian or arthropod hosts, members of the genus *Treponema* and *Borrelia* have undergone reductive evolution and lost most biosynthetic activities. On the contrary, *Leptospira* species (and presumably free-living, non-pathogenic spirochetes) have retained a large genome, a more extensive energy metabolism including the tricarboxylic acid cycle and an electron transport chain, and the ability to synthesize most required compounds. The availability of additional spirochetal genome sequences, particularly from diverse organisms (*e.g.* free-living spirochetes such as *Spirochaeta aurantia*) would be useful in identifying the common characteristics of all spirochetes and thus catch a glimpse of their common ancestor.

Acknowledgment

Supported by grant AI-49557 from the National Institutes of Health.

References

Achtman, M., Morelli, G., Zhu, P., Wirth, T., Diehl, I., Kusecek, B., Vogler, A. J., Wagner, D. M., Allender, C. J., Easterday, W. R., et al. (2004). Microevolution and history of the plague bacillus, *Yersinia pestis*. Proc. Natl. Acad. Sci. USA. *101*, 17837-17842.

Bankhead, T., and Chaconas, G. (2004). Mixing active-site components: a recipe for the unique enzymatic activity of a telomere resolvase. Proc. Natl. Acad. Sci. USA. *101*, 13768-13773.

Barbour, A. G. (1988). Plasmid analysis of *Borrelia burgdorferi*, the Lyme disease agent. J. Clin. Microbiol. *26*, 475-478.

Barbour, A. G., and Garon, C. F. (1987). Linear plasmids of the bacterium *Borrelia burgdorferi* have covalently closed ends. Science *237*, 409-411.

Bergstrom, S., Garon, C. F., Barbour, A. G., and MacDougall, J. (1992). Extrachromosomal elements of spirochetes. Res. Microbiol. *143*, 623-628.

Bharti, A. R., Nally, J. E., Ricaldi, J. N., Matthias, M. A., Diaz, M. M., Lovett, M. A., Levett, P. N., Gilman, R. H., Willig, M. R., Gotuzzo, E., and Vinetz, J. M. (2003). Leptospirosis: a zoonotic disease of global importance. Lancet Infect. Dis. *3*, 757-771.

Bulach, D. M., Kalambaheti, T., de la Pena-Moctezuma, A., and Adler, B. (2000a). Functional analysis of genes in the *rfb* locus of *Leptospira borgpetersenii* serovar Hardjo subtype Hardjobovis. Infect. Immun. *68*, 3793-3798.

Bulach, D. M., Kalambaheti, T., de la Pena-Moctezuma, A., and Adler, B. (2000b). Lipopolysaccharide biosynthesis in *Leptospira*. J. Mol. Microbiol. Biotechnol. *2*, 375-380.

Byram, R., Stewart, P. E., and Rosa, P. (2004). The essential nature of the ubiquitous 26-kilobase circular replicon of *Borrelia burgdorferi*. J. Bacteriol. *186*, 3561-3569.

Cameron, C. E. (2003). Identification of a *Treponema pallidum* laminin-binding protein. Infect. Immun. *71*, 2525-2533.

Cameron, C. E., Brown, E. L., Kuroiwa, J. M., Schnapp, L. M., and Brouwer, N. L. (2004). *Treponema pallidum* fibronectin-binding proteins. J. Bacteriol. *186*, 7019-7022.

Cameron, C. E., Castro, C., Lukehart, S. A., and Van Voorhis, W. C. (1999). Sequence conservation of glycerophosphodiester phosphodiesterase among *Treponema pallidum* strains. Infect. Immun. *67*, 3168-3170.

Cameron, C. E., Lukehart, S. A., Castro, C., Molini, B., Godornes, C., and Van Voorhis, W. C. (2000). Opsonic potential, protective capacity, and sequence conservation of the *Treponema pallidum* subspecies *pallidum* Tp92. J. Infect. Dis. *181*, 1401-1413.

Casjens, S. (2000). Borrelia genomes in the year 2000. J. Mol. Microbiol. Biotechnol. *2*, 401-410.

Casjens, S., Murphy, M., DeLange, M., Sampson, L., van Vugt, R., and Huang, W. M. (1997). Telomeres of the linear chromosomes of Lyme disease spirochaetes: nucleotide sequence and possible exchange with linear plasmid telomeres. Mol. Microbiol. *26*, 581-596.

Casjens, S., Palmer, N., van Vugt, R., Huang, W. M., Stevenson, B., Rosa, P., Lathigra, R., Sutton, G., Peterson, J., Dodson, R. J., et al. (2000). A bacterial genome in flux: the twelve linear and nine circular extrachromosomal DNAs in an infectious isolate of the Lyme disease spirochete *Borrelia burgdorferi*. Mol. Microbiol. *35*, 490-516.

Centurion-Lara, A., Castro, C., Castillo, R., Shaffer, J. M., Van Voorhis, W. C., and Lukehart, S. A. (1998). The flanking region sequences of the 15-kDa lipoprotein gene differentiate pathogenic treponemes. J. Infect. Dis. *177*, 1036-1040.

Centurion-Lara, A., Godornes, C., Castro, C., Van Voorhis, W. C., and Lukehart, S. A. (2000a). The *tprK* gene is heterogeneous among *Treponema pallidum* strains and has multiple alleles. Infect. Immun. *68*, 824-831.

Centurion-Lara, A., LaFond, R. E., Hevner, K., Godornes, C., Molini, B. J., Van Voorhis, W. C., and Lukehart, S. A. (2004). Gene conversion: a mechanism for generation of heterogeneity in the *tprK* gene of *Treponema pallidum* during infection. Mol. Microbiol. *52*, 1579-1596.

Centurion-Lara, A., Sun, E. S., Barrett, L. K., Castro, C., Lukehart, S. A., and Van Voorhis, W. C. (2000b). Multiple alleles of *Treponema pallidum* repeat gene D in *Treponema pallidum* isolates. J. Bacteriol. *182*, 2332-2335.

Chaconas, G., Stewart, P. E., Tilly, K., Bono, J. L., and Rosa, P. (2001). Telomere resolution in the Lyme disease spirochete. EMBO J. *20*, 3229-3237.

Correia, F. F., Plummer, A. R., Paster, B. J., and Dewhirst, F. E. (2004). Genome size of human oral *Treponema* species by pulsed-field gel electrophoresis. Oral Microbiol. Immunol. *19*, 129-131.

Dewhirst, F. E., Tamer, M. A., Ericson, R. E., Lau, C. N., Levanos, V. A., Boches, S. K., Galvin, J. L., and Paster, B. J. (2000). The diversity of periodontal spirochetes by 16S rRNA analysis. Oral Microbiol. Immunol. *15*, 196-202.

Eggers, C. H., Casjens, S., Hayes, S. F., Garon, C. F., Damman, C. J., Oliver, D. B., and Samuels, D. S. (2000). Bacteriophages of spirochetes. J. Mol. Microbiol. Biotechnol. *2*, 365-373.

Eggers, C. H., Kimmel, B. J., Bono, J. L., Elias, A. F., Rosa, P., and Samuels, D. S. (2001). Transduction by phiBB-1, a bacteriophage of *Borrelia burgdorferi*. J. Bacteriol. *183*, 4771-4778.

Fogg, M. J., Pearl, L. H., and Connolly, B. A. (2002). Structural basis for uracil recognition by archaeal family B DNA polymerases. Nat. Struct Biol. *9*, 922-927.

Fraser, C. M., Casjens, S., Huang, W. M., Sutton, G. G., Clayton, R., Lathigra, R., White, O., Ketchum, K. A., Dodson, R., Hickey, E. K., et al. (1997). Genomic sequence of a Lyme disease spirochaete, Borrelia burgdorferi. Nature *390*, 580-586.

Fraser, C. M., Gocayne, J. D., White, O., Adams, M. D., Clayton, R. A., Fleischmann, R. D., Bult, C. J., Kerlavage, A. R., Sutton, G., Kelley, J. M., and et al. (1995). The minimal gene complement of *Mycoplasma genitalium*. Science *270*, 397-403.

Fraser, C. M., Norris, S. J., Weinstock, G. M., White, O., Sutton, G. G., Dodson, R., Gwinn, M., Hickey, E. K., Clayton, R., Ketchum, K. A., et al. (1998). Complete genome sequence of *Treponema pallidum*, the syphilis spirochete. Science *281*, 375-388.

Giacani, L., Sambri, V., Marangoni, A., Cavrini, F., Storni, E., Donati, M., Corona, S., Lanzarini, P., and Cevenini, R. (2005). Immunological evaluation and cellular location analysis of the TprI antigen of *Treponema pallidum* subsp. *pallidum*. Infect. Immun. *73*, 3817-3822.

Glöckner, G., Lehmann, R., Romualdi, A., Pradella, S., Schulte-Spechtel, U., Schilhabel, M., Wilske, B., Sühnel, J., and Platzer, M. (2004). Comparative analysis of the *Borrelia garinii* genome. Nucleic Acids Res. *32*, 6038-6046.

Graber, J. R., Leadbetter, J. R., and Breznak, J. A. (2004). Description of *Treponema azotonutricium* sp. nov. and *Treponema primitia* sp. nov., the first spirochetes isolated from termite guts. Appl Environ. Microbiol. *70*, 1315-1320.

Grimm, D., Eggers, C. H., Caimano, M. J., Tilly, K., Stewart, P. E., Elias, A. F., Radolf, J. D., and Rosa, P. A. (2004a). Experimental assessment of the roles of linear plasmids lp25 and lp28-1 of *Borrelia burgdorferi* throughout the infectious cycle. Infect. Immun. *72*, 5938-5946.

Grimm, D., Elias, A. F., Tilly, K., and Rosa, P. A. (2003). Plasmid stability during *in vitro* propagation of *Borrelia burgdorferi* assessed at a clonal level. Infect. Immun. *71*, 3138-3145.

Grimm, D., Tilly, K., Byram, R., Stewart, P. E., Krum, J. G., Bueschel, D. M., Schwan, T. G., Policastro, P. F., Elias, A. F., and Rosa, P. A. (2004b). Outer-surface protein C of the Lyme disease spirochete: a protein induced in ticks for infection of mammals. Proc. Natl. Acad. Sci. USA. *101*, 3142-3147.

Kobryn, K., and Chaconas, G. (2002). ResT, a telomere resolvase encoded by the Lyme disease spirochete. Mol. Cell *9*, 195-201.

Labandeira-Rey, M., Baker, E., and Skare, J. (2001). Decreased infectivity in *Borrelia burgdorferi* strain B31 is associated with loss of linear plasmid 25 or 28-1. Infect. Immun. *69*, 446-455.

Labandeira-Rey, M., Seshu, J., and Skare, J. T. (2003). The absence of linear plasmid 25 or 28-1 of *Borrelia burgdorferi* dramatically alters the kinetics of experimental infection via distinct mechanisms. Infect. Immun. *71*, 4608-4613.

LaFond, R. E., Centurion-Lara, A., Godornes, C., Rompalo, A. M., Van Voorhis, W. C., and Lukehart, S. A. (2003). Sequence diversity of *Treponema pallidum* subsp. pallidum *tprK* in human syphilis lesions and rabbit-propagated isolates. J. Bacteriol. *185*, 6262-6268.

MacDougall, J., Margarita, D., and Saint Girons, I. (1992). Homology of a plasmid from the spirochete *Treponema denticola* with the single-stranded DNA plasmids. J. Bacteriol. *174*, 2724-2728.

MacDougall, J., and Saint Girons, I. (1995). Physical map of the *Treponema denticola* circular chromosome. J. Bacteriol. *177*, 1805-1811.

Maio, R. M., and Fieldsteel, A. H. (1978). Genetics of *Treponema*: relationship between *Treponema pallidum* and five cultivable treponemes. J. Bacteriol. *133*, 101-107.

Maio, R. M., and Fieldsteel, A. H. (1980). Genetic relationship between *Treponema palidum* and *Treponema pertenue*, two noncultivable human pathogens. J. Bacteriol. *141*, 427-429.

Maio, R. M., Fieldsteel, A. H., and Harris, D. L. (1978). Genetics of *Treponema*: characterization of *Treponema hyodysenteriae* and its relationship to *Treponema pallidum*. Infect. Immun. *22*, 736-739.

McKevitt, M., Patel, K., Smajs, D., Marsh, M., McLoughlin, M., Norris, S. J., Weinstock, G. M., and Palzkill, T. (2003). Systematic cloning of *Treponema pallidum* open reading frames for protein expression and antigen discovery. Genome Res. *13*, 1665-1674.

Miller, J. C., Bono, J. L., Babb, K., El-Hage, N., Casjens, S., and Stevenson, B. (2000). A second allele of *eppA* in *Borrelia burgdorferi* strain B31 is located on the previously undetected circular plasmid cp9-2. J. Bacteriol. *182*, 6254-6258.

Nascimento, A. L., Ko, A. I., Martins, E. A., Monteiro-Vitorello, C. B., Ho, P. L., Haake, D. A., Verjovski-Almeida, S., Hartskeerl, R. A., Marques, M. V., Oliveira, M. C., et al. (2004). Comparative genomics of two *Leptospira interrogans* serovars reveals novel insights into physiology and pathogenesis. J. Bacteriol. *186*, 2164-2172.

Noordhoek, G. T., Hermans, P. W. M., Paul, A. N., Schouls, L. M., van der Sluis, J. J., and van Embden, J. D. A. (1989). *Treponema pallidum* subspecies *pallidum* (Nichols) and *Treponema pallidum* subspecies *pertenue* (CDC 2575) differ in at least one nucleotide: comparison of two homologous antigens. Microb Pathog *6*, 29-42.

Norgard, M. V., and Miller, J. N. (1981). Plasmid DNA in *Treponema pallidum* (Nichols): potential for antibiotic resistance by syphilis bacteria. Science *213*, 553-555.

Norris, S. J., Cox, D. L., and Weinstock, G. M. (2001). Biology of *Treponema pallidum*: correlation of functional activities with genome sequence data. J. Mol. Microbiol. Biotechnol. *3*, 37-62.

Norris, S. J., and Weinstock, G. M. (2000). The genome sequence of *Treponema pallidum*, the syphilis spirochete: will clinicians benefit? Curr. Opin. Infect. Dis. *13*, 29-36.

Ochman, H., and Wilson, A. C. (1987). Evolution in bacteria: evidence for a universal substitution rate in cellular genomes. J. Mol. Evol. *26*, 74-86.

Paster, B., Boches, S., Galvin, J., Ericson, R., Lau, C., Levanos, V., Sahasrabudhe, A., and Dewhirst, F. (2001). Bacterial diversity in subgingival plaque. J. Bacteriol. *183*, 3770-3783.

Paster, B., and Dewhirst, F. (2000). Phylogenetic foundation of spirochetes. J. Mol. Microbiol. Biotechnol. *2*, 341-344.

Pearl, L. H. (2000). Structure and function in the uracil-DNA glycosylase superfamily. Mutat Res. *460*, 165-181.

Pedersen, A. G., Jensen, L. J., Brunak, S., Staerfeldt, H. H., and Ussery, D. W. (2000). A DNA structural atlas for *Escherichia coli*. J. Mol. Biol. *299*, 907-930.

Picardeau, M., Lobry, J. R., and Hinnebusch, B. J. (1999). Physical mapping of an origin of bidirectional replication at the centre of the *Borrelia burgdorferi* linear chromosome. Mol. Microbiol. *32*, 437-445.

Pillay, A., Liu, H., Chen, C. Y., Holloway, B., Sturm, W., Steiner, B., and Morse, S. A. (1998). Molecular subtyping of *Treponema pallidum* subspecies *pallidum*. Sex Transm. Dis. *25*, 408-414.

Purser, J. E., Lawrenz, M. B., Caimano, M. J., Howell, J. K., Radolf, J. D., and Norris, S. J. (2003). A plasmid-encoded nicotinamidase (PncA) is essential for infectivity of *Borrelia burgdorferi* in a mammalian host. Mol. Microbiol. *48*, 753-764.

Purser, J. E., and Norris, S. J. (2000). Correlation between plasmid content and infectivity in *Borrelia burgdorferi*. Proc. Natl. Acad. Sci. USA. *97*, 13865-13870.

Radolf, J. D., Steiner, B., and Shevchenko, D. (1999). *Treponema pallidum*: doing a remarkable job with what it's got. Trends Microbiol. *7*, 7-9.

Ren, S. X., Fu, G., Jiang, X. G., Zeng, R., Miao, Y. G., Xu, H., Zhang, Y. X., Xiong, H., Lu, G., Lu, L. F., et al. (2003). Unique physiological and pathogenic features of *Leptospira interrogans* revealed by whole-genome sequencing. Nature *422*, 888-893.

Seshadri, R., Myers, G. S., Tettelin, H., Eisen, J. A., Heidelberg, J. F., Dodson, R. J., Davidsen, T. M., DeBoy, R. T., Fouts, D. E., Haft, D. H., et al. (2004). Comparison of the genome of the oral pathogen Treponema denticola with other spirochete genomes. Proc. Natl. Acad. Sci. U S A. *101*, 5646-5651.

Smajs, D., McKevitt, M., Howell, J. K., Norris, S. J., Cai, W.-W., Palzkill, T., and Weinstock, G. M. (2005). Transcriptome of *Treponema pallidum*: gene expression profile during experimental rabbit infection. J. Bacteriol. *187*, 1866-1874.

Smibert, R. M. (1984). Genus III: *Treponema* Schaudinn 1905, 1728[AL]. In Bergey's Manual of Systematic Bacteriology, N. R. Kreig, and J. G. Holt, eds. (Baltimore, The Williams and Wilkins Co.), pp. 49-57.

Stevenson, B., Porcella, S. F., Oie, K. L., Fitzpatrick, C. A., Raffel, S. J., Lubke, L., Schrumpf, M. E., and Schwan, T. G. (2000). The relapsing fever spirochete *Borrelia hermsii* contains multiple, antigen-encoding circular plasmids that are homologous to the cp32 plasmids of lyme disease spirochetes. Infect. Immun. *68*, 3900-3908.

Subramanian, G., Koonin, E. V., and Aravind, L. (2000). Comparative genome analysis of the pathogenic spirochetes *Borrelia burgdorferi* and *Treponema pallidum*. Infect. Immun. *68*, 1633-1648.

Sun, E. S., Molini, B. J., Barrett, L. K., Centurion-Lara, A., Lukehart, S. A., and Van Voorhis, W. C. (2004). Subfamily I *Treponema pallidum* repeat protein family: sequence variation and immunity. Microbes Infect. *6*, 725-737.

Sutton, M. Y., Liu, H., Steiner, B., Pillay, A., Mickey, T., Finelli, L., Morse, S., Markowitz, L. E., and St. Louis, M. E. (2001). Molecular subtyping of *Treponema pallidum* in an Arizona County with increasing syphilis morbidity: use of specimens from ulcers and blood. J. Infect. Dis. *183*, 1601-1606.

Ussery, D. W., and Hallin, P. F. (2004). Genome Update: annotation quality in sequenced microbial genomes. Microbiology *150*, 2015-2017.

Ussery, D. W., Jensen, M. S., Poulsen, T. R., and Hallin, P. F. (2004). Genome update: alignment of bacterial chromosomes. Microbiology *150*, 2491-2493.

Venkatesh, J., Kumar, P., Krishna, P. S., Manjunath, R., and Varshney, U. (2003). Importance of uracil DNA glycosylase in *Pseudomonas aeruginosa* and *Mycobacterium smegmatis*, G+C-rich bacteria, in mutation prevention, tolerance to acidified nitrite, and endurance in mouse macrophages. J. Biol. Chem. *278*, 24350-24358.

Walker, E. M., Arnett, J. K., Heath, J. D., and Norris, S. J. (1991). *Treponema pallidum* has a single, circular chromosome with a size of ~900 kilobase pairs. Infect. Immun. *59*, 2476-2479.

Walker, E. M., Howell, J. K., You, Y., Hoffmaster, A. R., Heath, J. D., Weinstock, G. M., and Norris, S. J. (1995). Physical map of the genome of *Treponema pallidum* subsp. *pallidum* (Nichols). J. Bacteriol. *177*, 1797-1804.

Waters, E., Hohn, M. J., Ahel, I., Graham, D. E., Adams, M. D., Barnstead, M., Beeson, K. Y., Bibbs, L., Bolanos, R., Keller, M., et al. (2003). The genome of *Nanoarchaeum equitans*: insights into early archaeal evolution and derived parasitism. Proc. Natl. Acad. Sci. USA. *100*, 12984-12988.

Weinstock, G. M., Hardham, J. M., McLeod, M. P., Sodergren, E. J., and Norris, S. J. (1998). The genome of *Treponema pallidum*: New light on the agent of syphilis. FEMS Microbiol. Rev. *22*, 323-332.

Weinstock, G. M., Norris, S. J., Sodergren, E. J., and Smajs, D. (2000a). Identification of virulence genes *in silico*: infectious disease genomics. In Virulence mechanisms of bacterial pathogens, K. A. Brogden, J. Roth, T. Stanton, C. Bolin, C. Minion, and M. Wannemuehler, eds. (Washington, D.C., ASM Press), pp. 251-261.

Weinstock, G. M., Smajs, D., Hardham, J., and Norris, S. J. (2000b). From microbial genome sequence to applications. Res. Microbiol. *151*, 151-158.

Worning, P., Jensen, L. J., Hallin, P. F., Staerfeldt, H.-H., and Ussery, D. W. (2005). Origin of replication in circular prokaryotic chromosomes.Environ. Microbio. *8*, 353-361.

Yang, X. F., Pal, U., Alani, S. M., Fikrig, E., and Norgard, M. V. (2004). Essential role for OspA/B in the life cycle of the Lyme disease spirochete. J. Exp. Med. *199*, 641-648.

Zhang, R., and Zhang, C. T. (2002). Single replication origin of the archaeon *Methanosarcina mazei* revealed by the Z curve method. Biochem. Biophys Res. Commun. *297*, 396-400.

Zhou, D., Han, Y., Song, Y., Tong, Z., Wang, J., Guo, Z., Pei, D., Pang, X., Zhai, J., Li, M., et al. (2004). DNA microarray analysis of genome dynamics in *Yersinia pestis*: insights into bacterial genome microevolution and niche adaptation. J. Bacteriol. *186*, 5138-5146.

Zuerner, R. L., Herrmann, J. L., and Saint Girons, I. (1993). Comparison of genetic maps for two *Leptospira interrogans* serovars provides evidence for two chromosomes and intraspecies heterogeneity. J. Bacteriol. *175*, 5445-5451.

3

Structural and Genomic Features of Treponemal Architecture

*Jacques Izard and Ronald J. Limberger**

Abstract

Treponema denticola and *Treponema phagedenis* are important models for deciphering the unique architectural and genetic features of treponemes. Darkfield and electron microscopy have delineated the general structure of treponemes as deduced from these two dimensional techniques. Advances in genetic tools and three dimensional visualization techniques are now linking cell architecture, cell ultrastructures, and genetic data. Ultrastructures are multi-component assemblies within the cell that drive specific functions critical to the cell such as cell division, cell shape and motility. A key treponemal ultrastructure is the cytoplasmic filament ribbon which is involved in the cell division process. Recently, application of new methods to the structural analysis of *T. phagedenis* has revealed novel features of the cytoplasmic filament ribbon. The observed cytoplasmic filament ribbon actually consists of independent filaments connected to each other via bridging components and anchored to the inner membrane. The nature and components of this macromolecular complex are discussed as well as the novelty of the filamentous ribbon structure. The availability of whole genome sequences from two treponemal species also allows the comparative study with other bacterial cytoskeleton-associated structures towards achieving a global picture of spirochetal cell biology.

1. Introduction

Organisms in the genus *Treponema* are prominent members of the Spirochaetaceae, a family that is characterized by a unique cell architecture (Holt et al., 1994; Holt, 1978; Hovind-Hougen, 1972; 1974; 1976; Hovind-Hougen and Birch-Andersen, 1971; Hovind-Hougen et al., 1976). All treponemes have an outer membrane, an inner membrane and a characteristic number of flagellar filaments located in the periplasmic space (Figure 1). This basic arrangement is a defining feature of the

*For correspondence email Ron.Limberger@wadsworth.org

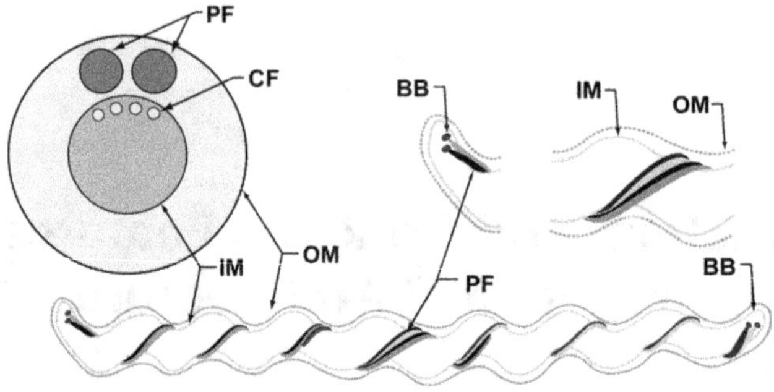

Figure 1. Diagrammatic representation of a *T. denticola* cell, showing the periplasmic location of the flagellar filaments. Also shown is an enlarged region (top right) as well as a cross-sectional view (top left). Although the size of *T. denticola* varies, a typical cell is 0.2 μm in diameter by 6 μm in length. PF= periplasmic flagella, CF=cytoplasmic filaments, BB= basal body, IM= inner membrane, OM= outer membrane. This figure is adapted from Limberger (2004) courtesy of S. Karger AG, Basel.

spirochetes (Smibert, 1984). The overall shape of treponemes appears as coiled, wavelike, helical, or serpentine although the exact structure is difficult to ascertain when viewed under a darkfield microscope. Closer examination of the cell structure reveals the variety of treponemal shapes. Within the treponeme group can be found cells that have a regular helical shape, cells that have irregularly shaped helices, and cells with a flat-wave morphology of the cell cylinder (Charon et al., 1991; 1992b; Charon and Goldstein, 2002; Cox, 1972; Ruby et al., 1997).

Although the coiled or wave-like shape and the periplasmic location of the flagellar filaments define the *Treponema* as spirochetes, the genetic and structural factors that contribute to these distinctive features have not been elucidated. However, recent advances in manipulating spirochete genes and in three-dimensional imaging capabilities have shed new light on treponemal structure. Via the former, creation of stable directed mutants in *Treponema denticola* has become an established technique. More recently, new shuttle plasmids have been developed, thus enabling complementation studies of these mutants. Genetic regulation, as well as structural and functional domain analysis, is now possible using these plasmids (Chi et al., 2002; Slivienski-Gebhardt et al., 2004). The new shuttle plasmids also enable the heterologous expression of treponemal genes, which in turn permits cross-species functionality to be assessed (Chi et al., 1999; Slivienski-Gebhardt et al., 2004). Application of electron tomography has allowed us to study cellular structures in three dimensions, thus providing new insights into treponemal molecular architecture (Izard et al., 2004).

The structure and shape of the treponemal cell are driven by their ultrastructural, multicomponent assemblies. These drive, maintain, or assist in specific functions in the cell and may extend through large regions of the cell. Such structures in *Treponema* include the flagella, the cytoplasmic filaments, the cell division septation ring, the filaments associated with rod cell-shape maintenance, and the peptidoglycan layer. This chapter will describe the contributions of these structures to the distinctive architecture of the *Treponema* and how this architecture determines their cell biology and pathogenicity.

2. Treponemal architecture

We will review the overall structure of this group of bacteria and how the individual components influence the shape of the *Treponema*. Because treponemal cell structure has been extensively detailed in other publications, we will provide only a brief summarization of the major components of the cell (Holt et al., 1994; Holt, 1978; Hovind-Hougen, 1972; 1974; 1976; Hovind-Hougen and Birch-Andersen, 1971; Hovind-Hougen et al., 1976) before focusing on the influence of these structures on cell function.

A typical treponemal cell ranges between 5 and 20 μm in length and between 0.1 and 0.5 μm in diameter, depending on the species. It possesses an outer membrane, a periplasmic space in which the flagellar filaments reside, a peptidoglycan layer, and an inner membrane. The above information was obtained by electron microscopic analysis of fixed and stained cells. New approaches using flash-frozen hydrated unstained cells are resolving those structures in a "caught in the action" state (J. Izard and C.E. Hsieh, unpublished data). This approach combined with tomographic methods can reveal the three dimensional architecture of the treponemal cell without the artifacts due to staining and processing for traditional electron microscopy (Baumeister, 2002; Grimm et al., 1998; McEwen and Frank, 2001).

All natural treponemal cells are flagellated. The number of flagella varies by species but typically ranges from 1 to 8. The number of flagella at each cell end in conjunction with cell length, diameter and wavelength, have been proposed as classification and identification criteria (Hovind-Hougen, 1976). The flagella are inserted subterminally at each end of the cell. A given cell usually contains similar numbers of flagellar filaments at each cell end. The flagellar filaments wind around the cell body toward the center of the cell and may or may not overlap with one another in the center. The overall diameter of a filament is usually within the range of 16-25 nm. The typical treponemal filament consists of multiple FlaB polypeptide species which constitute the core filament, and one or two FlaA polypeptides, which make up the filament sheath. The flagellar basal body shows a general morphological similarity to those of other bacteria, although the *Treponema* possess one fewer ring structure due to the periplasmic location of the flagellar filaments. Accordingly, there is no ring creating an anchor in the outer membrane. More detailed analyses of flagellar structure can be found elsewhere in this book as well as in several reviews (Canale-Parola, 1984a; Charon and Goldstein, 2002; Li et al., 2000; Limberger, 2004; Penn et al., 1985; Penn and Luke, 1992; see Chapter 6, Charon et al.,). The polypeptides that are important for the motility of enteric bacteria have identifiable homologues in the treponemes. However, there are additional spirochete-specific polypeptides important for these organisms' unique motility (Limberger et al., 1996).

When observed by darkfield light microscopy or electron microscopy, the treponemal cell appears spiral in shape. Some treponemes, like *T. phagedenis*, possess regular helical waves of consistent amplitude and pitch except at the cell ends which are bent or distorted (Charon et al., 1991). Others, like *T. denticola,* are more irregular in shape, yet still possess primarily a helical form (Ruby et al., 1997). The structure of *T. pallidum* is less clear. Earlier studies of slowly rotating cells by darkfield microscopy have suggested that *T. pallidum* is primarily planar in nature

(Cox, 1972). Subsequently, scanning electron microscopy analysis has suggested that *T. pallidum* is predominantly a left-handed helix (Stepan and Johnson, 1981). In clinical samples, as well as under certain conditions in cell culture, treponemes can assume a spherical shape by association of each cell end to ultimately form a spherical body. However, the formation and significance of these spherical bodies are unknown and remain part of the historical mystique of the biology of treponemes (Cox, 1972; Hampp, 1951). Because *T. pallidum* cannot be manipulated genetically in order to generate specific mutants, it is unknown whether a non-flagellated strain of *T. pallidum* would retain its flat-waved or helical shape. Further work using high resolution microscopic techniques is needed to elucidate the precise structure of *T. pallidum*. As pointed out in the Chapter 6 (Charon et al.) many spirochetes are able to assume different morphological shapes because of the flexible cell cylinder and the influence of the periplasmic flagellum.

The distortions observed at the cell ends of treponemes are due to the exertion of tortional stress by the periplasmic flagella. Non-flagellated mutants of *Treponema* species retain their helical shape and are much more regular at the cell ends (Charon et al., 1991; Ruby et al., 1997). One striking difference between *Borrelia burgdorferi* and the *Treponema* spp. is that, in the absence of flagella, *B. burgdorferi* becomes a straight rod-shaped bacterium (Motaleb et al., 2000; Sadziene et al., 1991). The presence of the flagellar filament in *B. burgdorferi* is directly related to the maintenance of the helical shape, which in this organism is independent of the membrane potential (Motaleb et al., 2000). In the *Treponema*, in the absence of flagellar filaments, there is maintenance of the overall cell morphology. However, there is a loss of the "bent-end shape" in flagellar filament deficient mutants of *T. phagedenis* (Izard et al., 1999; Limberger and Charon, 1986), and the formation of a more regular helical cell shape in *T. denticola* (Limberger et al., 1999; Ruby et al., 1997); these features are the consequence of the release of tension, modifying the well-defined helical shape of the bacterium.

Treponemal cells possess internal cytoplasmic filaments that consist of one major polypeptide, CfpA. These filaments are smaller in cross-section (5.0 x 6.0 nm) than the flagellar filaments and are found throughout the length of the cell. Their exact function is unknown, but recent evidence suggests that they are involved in the cell-division processes (Izard et al., 2001). The number of cytoplasmic filaments per cell varies by species but it usually is in the range of 4-12 (Izard et al., 1999). In cross sections of the treponemal cell, they are very difficult to discern clearly. They are best visualized by solubilization of the outer membrane followed by negative staining and examination of the entire cell length (Figure 2). Mutants that lack cytoplasmic filaments retain the normal cell shape. The cytoplasmic filaments will be described in detail later in this chapter.

3. Periplasmic flagella, outer membrane, and cell wall

The best-studied structures of the *Treponema* have been the periplasmic flagella (Figure 1). These immunogenic structures are easy to visualize by electron microscopy as well as by high resolution darkfield microscopy. They are uniquely confined to the periplasm, although they can protrude from the cell in some species, under certain circumstances (Charon et al., 1992a). Despite this protrusion, the flagellar filaments remain covered by what appears to be the outer membrane. The physiological significance of this protrusion is unclear. The details of the flagellar structure are described by our colleagues in this book (See Chapter 6, Charon et al.). The dynamics of the flagellar activities within the periplasm are unknown, but this surely must be a region of high activity given the vigorous rotation of the flagellar filaments within this narrow region.

The outer membrane of the *Treponema* has also been extensively studied using a variety of techniques. The protein components of the membrane are being identified, and their relation to host immune system are being deciphered (Wang et al., 2001). The lipid, glycolipid, and lipoprotein content of treponemal membranes are also under investigation (Haake, 2000; Kesavalu et al., 2002; Martinez-Morales et al., 2003). The details of these outer membrane analyses are presented elsewhere in this book (See Chapter 11, Cameron and Chapter 14, Holt and Ebersole).

The peptidoglycan layer of treponemal cells has not been sufficiently studied (Caimano et al., 1999; Morioka et al., 1979; Umemoto et al., 1981; Yanagihara et al., 1984). It is clear that treponemes do possess a cell wall that is at least partly responsible for maintenance of the cell shape. Analysis of the treponemal cell wall by electron microscopy has revealed an electron-dense layer of about 5 nm in thickness. The major constituents are glucosamine, muramic acid, D-glutamic acid, L- and D-alanine, ornithine, and glycine. In addition, whole-genome analysis has revealed that most of the genes necessary for the protein and sugar moieties of the peptidoglycan formation are present in *Treponema* (Fraser et al., 1998; Seshadri et al., 2004).

4. A Bacterial cytoskeleton

Traditionally, the term cytoskeleton has been limited to eukaryotic cells. However, given the large number of genomes now sequenced, the analysis of sequence similarity, and the in-depth study of molecular systems in certain organisms, it is now clear that structural and functional homologues of tubulin (FtsZ) and actin (MreB) are present virtually throughout the eubacteria as well as in some *Archaea*. Recent data show that these filaments are dynamic structures in both prokaryotes and eukaryotes.

The cytoskeleton in eukaryotes is generally accepted to be a protein framework within the cytoplasm of the cell. This framework is composed of protein complexes that each have a filament as a central component: actin microfilaments, microtubules composed of tubulin filaments, and intermediate filaments. These filaments are involved in a variety of functions, from providing support for cell extensions such as the small intestinal villi or the axons of neurons, to participating in cell motility and a large array of internal cell functions such as cell division.

The dynamic aspect of the bacterial cytoskeleton has not yet been demonstrated experimentally in spirochetes. However, the evidence for the presence of the septation ring (including FtsZ), and the shape driving filaments (including MreB) has been indirectly documented for years through electron microscopy data and more recently from genomic sequence data. We will describe, in the following paragraphs some of the features of the spirochetal cytoskeleton.

5. The cytoplasmic filament ribbon of *Treponema*

The cytoplasmic filament of *Treponema* is a ribbon-like structure (Figure 2) composed of two to six filaments (Holt, 1978; Hovind-Hougen, 1976; Izard et al., 1999). Other cytoplasmic filaments have been described in other spirochete genera. Electron microscopic evidence for this structure has been published for the following genera: *Leptonema*, *Spirochaeta*, *Pillotina*, *Hollandina*, and *Diplocalyx* (Bermudes et al., 1988; 1987; Hollande et al., 1967; Holt, 1978; Hovind-Hougen, 1979).

Electron microscopy has provided a wealth of information about the cytoplasmic filaments. Treponemal cytoplasmic filaments are located beneath the flagellar filament bundle (Hovind-Hougen, 1972; 1974; 1976; Ovcinnikov and Delektorskij, 1970), in close apposition to the cytoplasmic membrane (Eipert and Black, 1979; Hovind-Hougen, 1974; Hovind-Hougen and Birch-Andersen, 1971). The filament bundle is helical inside the cell, and the periodicity of the helix is equivalent to the cell's helical periodicity (Eipert and Black, 1979; Zemper and Black, 1978). The bundle spans the length of the cell, is not transitory, and is detectable at all steps of the cell cycle (Izard et al., 1999). During cell division, the filaments are severed. Roughly half of the length remains in each daughter cell (Eipert and Black, 1979; Izard et al., 1999).

Masuda and Kawata were the first to purify the cytoplasmic filaments of various strains of treponemes. The major constituent of *T. denticola, T. phagedenis, and T. pallidum* cytoplasmic filaments is an 82-kDa protein (Izard et al., 1999; Masuda and Kawata, 1989; You et al., 1996). Using protein sequencing, PCR, and library hybridization, You and co-workers were able to identify the open reading frame in the *T. pallidum* subsp. *pallidum* genome (Fraser et al., 1998; You et al., 1996). The gene product was named cytoplasmic filament protein A (CfpA). The nucleotide and amino-acid sequences are well conserved among *Treponema* species (Izard et al., 1999). The alignment of *cfpA* nucleotide sequences from *T. denticola*, *T. vincentii*, *T. phagedenis* and *T. pallidum* subspp. *pallidum* and *pertenue* reveals several gaps, from 3 to 15 nucleotides long, at the 3′ end of the sequences but no frameshifts, and no mutational hot spots. The amino-acid sequence identity of

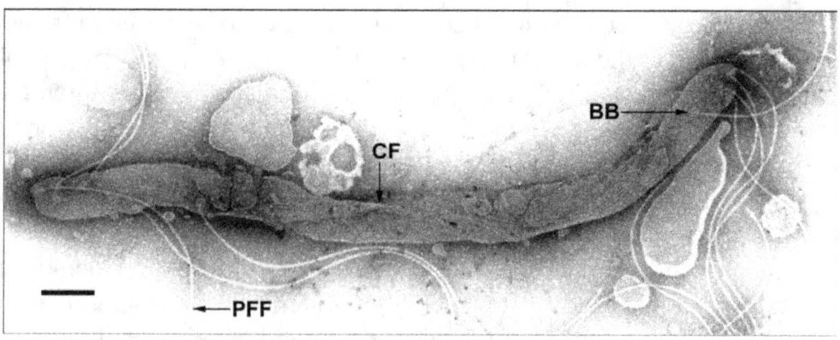

Figure 2. The cytoplasmic filament (CF) ribbon spans the entire cytoplasmic compartment. After removal of the outer membrane, the flagellar filaments are liberated and three helical turns of the cytoplasmic filament ribbon can be seen. *T. phagedenis* cells were stripped of their outer membrane, negatively stained with sodium phosphotungstate, and visualized in the electron microscope. (Izard et al., 1999). BB, flagellar basal body. Scale bar, 250 nm. Photo is reprinted courtesy of ASM Press.

CfpA among the above treponemes ranges from 83 to 100% (Izard et al., 1999). The high levels of sequence similarity are independent of the habitat or mode of transmission of the organism, or the cultivability of the organism in the laboratory. Analysis of the nucleotide and amino-acid sequences of the various cytoplasmic filament core constituents using bioinformatics has failed to identify the function of these structures.

Expression of the *cfpA* gene is regulated by a σ^{70} recognition sequence (Izard et al., 1999; You et al., 1996). The *T. phagedenis cfpA* σ^{70} promoter differs by only one nucleotide from the *Escherichia coli* consensus σ^{70} sequence (Hawley and McClure, 1983). A similar recognition sequence is used by sigma A (σ^{43}) in *Bacillus subtilis* (Graves and Rabinowitz, 1986; Moran et al., 1982). Little is known about the temporal regulation of genes in *Treponema* (Indest et al., 2000). The expression of *cfpA* is expected to be constitutive. During experimental rabbit infection by *T. pallidum* sp. *pallidum, cfpA* is one of the ten most highly transcribed genes (Smajs et al., 2005). This analysis of the *T. pallidum* transcriptome underscores the importance of the cytoplasmic filament ribbon during the infection process.

What is the function of the filamentous cytoplasmic filament bundle? The shape and location of the bundle suggest it is involved in cell structure and shape maintenance, transport, cell motility, and/or cell division. Analysis of the predicted amino-acid sequence of CfpA does not reveal any similarity with proteins of known function, or with any putative protein domain(s). To achieve a better understanding of the relation between cell function and the filamentous structure, we generated a knockout mutant of *cfpA* in *T. denticola* (Izard et al., 2001). The insertional inactivation of *cfpA* was not lethal for the cell. The *cfpA* mutant predominantly forms long chains of cells (Izard et al., 2001). Each such chain is composed of cytoplasmic cylinders of normal length under a single shared outer membrane. Each cytoplasmic cylinder is independent, as determined by electron and fluorescence microscopy. No alteration of the cell shape or structure was observed in the mutant strain (Izard et al., 2001). Consequently, the hypothesis that the cytoplasmic filament ribbon is involved in the maintenance of cell structure and shape could be eliminated. The most probable candidate for such a function is the filament structure composed of MreB, as described later in this chapter.

By dark-field microscopy, it is possible to observe the asynchronous gyration of cell ends from two individual cytoplasmic cylinders facing one another. During the rare event when a single *cfpA* mutant cell of wild-type size was observed, no obvious differences from wild-type movement could be observed, in either liquid or dense medium at 36°C (1% methylcellulose, 15 centipoises). We concluded that motility was not altered at the single cell level. The macroscopic (on agarose plates) and microscopic (in methylcellulose solutions) alteration of motility compared to the wild type could be explained by the asynchronous crankshaft-like movement of the cell. Such a phenomenon can be observed in wild-type cells in early log phase, when they form a long chain. At that time, the rate of cytoplasmic cylinder formation with complete cylinder septation is greater than the rate of the cytoplasmic cylinder detachment (cytokinesis). In the *cfpA* mutant, however, over 90% of the cell population is composed of cells having more than two cytoplasmic cylinders at all stages of cell growth. Therefore, it appears that the cytoplasmic filaments have no direct involvement in treponemal motility, despite their proximity to the flagellar filaments.

The involvement of the cytoplasmic filaments in cell division was revealed when the chromosomal DNA of the CfpA-deficient mutant was visualized using fluorescent dyes (Izard et al., 2001). In wild-type cells, the chromosomal DNA is uniformly distributed throughout the cytoplasmic cylinder (Izard et al., 2001). However, in the CfpA-deficient mutant, the DNA is condensed in distinct areas, and there is an increase in the number of anucleate cells (Izard et al., 2001). As a result of this work, we have hypothesized that the cytoplasmic filaments are involved in the cell-division process, chromosome segregation, or DNA structure localization and maintenance. At present, it is unclear how the filaments would interact with the chromosomal DNA. The evidence for active chromosome segregation after duplication seems unambiguous, even though the model of duplication and separation of the origin of replication region (*oriC*) and the termination region (*ter*) differs for *E. coli* and *B. subtilis* (Glaser et al., 1997; Jensen and Shapiro, 1999; Niki and Hiraga, 1998; Sharpe and Errington, 1999; Webb et al., 1998). The spatial and temporal mechanisms that control and direct *oriC* are unknown. However, it is likely that a hypothetical spindle or a mitotic-like apparatus and "motor" proteins are involved in this process (Sharpe and Errington, 1999). The involvement of the cytoplasmic filaments in cell-division processes and their relationship to the chromosomal DNA are under investigation.

The three-dimensional structure of the cytoplasmic ribbon has been revealed using electron tomography (Izard et al., 2004). As in medical imaging applications, a three-dimensional volume is reconstructed from a series of two-dimensional images of a given structure and recorded at various tilt angles (Crowther et al., 1970; McEwen and Marko, 1999). The advantage of this technique is that it enables one to obtain three-dimensional reconstructions of complex and irregularly shaped structures at moderately high resolution. We used negatively stained *T. phagedenis* cells that had been stripped of their outer membrane to visualize the cytoplasmic filaments. A series of tilt images were recorded in the electron microscope with a 1.5 degree angular interval, over a +/- 60 degree angular range. After alignment of the successive images, a three-dimensional volume was calculated using Fourier transforms (SPIDER program) by weighted back projection methods (Frank et al., 1996). The result of such calculations can be seen as an added feature on the Molecular Microbiology website (Izard et al., 2004). After tracing the different volumes, it was possible to analyze each structure independently (Figure 3).

The novelty of the organization of the cytoplasmic filaments resides in their unique molecular arrangement. Upon analysis of the tomographic reconstructions, we learned that the filaments are not in contact with one another, *i.e.* there is no interfacial connection. Each filament is composed of one protein, CfpA (Izard et al., 1999; Masuda and Kawata, 1989; You et al., 1996). We were able to provide more accurate measurements of the filaments: the cross-section is 5.0 x 6.0 nm ± 0.5 nm (horizontal x vertical, or width x depth) (Izard et al., 2004). Their length is the same as the cell length (Izard et al., 1999). The center-to-center spacing is 10.4 nm ± 1.8 nm. The large standard deviation indicates that there is some variability in the distance between two filaments, even though long stretches of the ribbon have been observed with a uniform spacing.

Another novel feature is that cross-bridging protein(s) are located on the cytoplasmic side of the filament ribbon. They are probably involved in filament

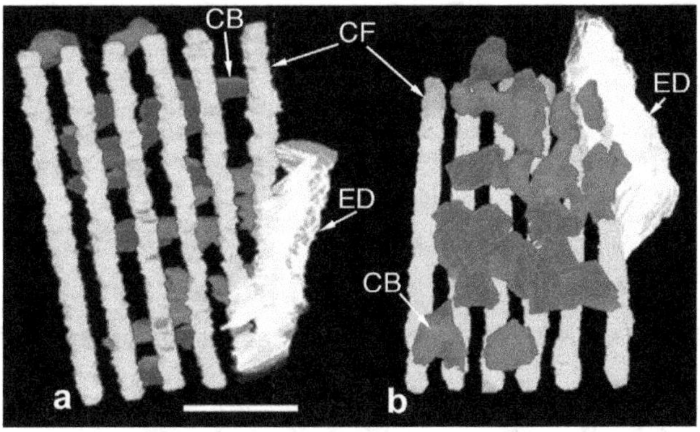

Figure 3. Reconstruction of the treponemal cytoplasmic filaments and filament bridging structures. For simplicity, six filaments were initially traced from axial views over a limited distance. A. Viewed from the periplasmic side. B. Viewed from the opposite side (cytoplasmic side). Note that the connections form an irregular lattice on one side of the filament band. CF, cytoplasmic filaments; ED, unidentified electron-dense material; CB, cross-bridging protein(s). Scale bar, 25 nm. Photo is modified from Izard et al. (2004), courtesy of Blackwell Publishing.

spacing, maintaining the structural integrity of the filament ribbon, and they likely interact with other cytoplasmic components associated with the functional ribbon. The variation in the center-to-center spacing of the filaments could arise from the loss of bridging components during sample preparation; such a loss would impart a greater flexibility to the filament organization (Figure 3) (Izard et al., 2004).

Finally, anchoring protein(s) are located on the periplasmic side of the cytoplasmic filaments. They are associated with the cytoplasmic membrane, creating a stable link between the membrane and the ribbon of filaments. The anchoring protein may be a lipoprotein or an integral membrane protein. The membrane-connector width is approximately the same as the width of the filaments (Figure 3) (Izard et al., 2004).

On the basis of these data, we could estimate the extent of the coverage of the cytoplasmic surface of the inner membrane by the cytoplasmic filament ribbon. For *T. phagedenis*, based on the center-to-center spacing of the filaments, the horizontal cross-section of the filaments, and an estimated cytoplasmic cylinder diameter of 200 nm, an estimated 3-18% (average, 7-9%) of the surface is covered (Izard et al., 2004). The membrane-associated region of the cytoplasmic filament structure occupies a significant space proximal to the inner cell membrane. Although, the abolition of the cytoplasmic filament structure by gene knockout is not lethal, cell growth, motility and cell division processes are profoundly altered, thus highlighting the physiological importance of such a structure (Izard et al., 2001).

Study of the cytoplasmic filaments and their protein partners will provide critical data on cell division processes and maintenance of chromosome structure.

6. Cell septation and filamentous structures in *Treponema*

On the basis of our work (Izard et al., 1999) (Izard J., Samsonoff W. A., Slivienski-Gebhardt L. L., and Limberger R. J., unpublished data) and data published by other groups on *Treponema* (Hovind-Hougen et al., 1976; Izard et al., 1999; Listgarten

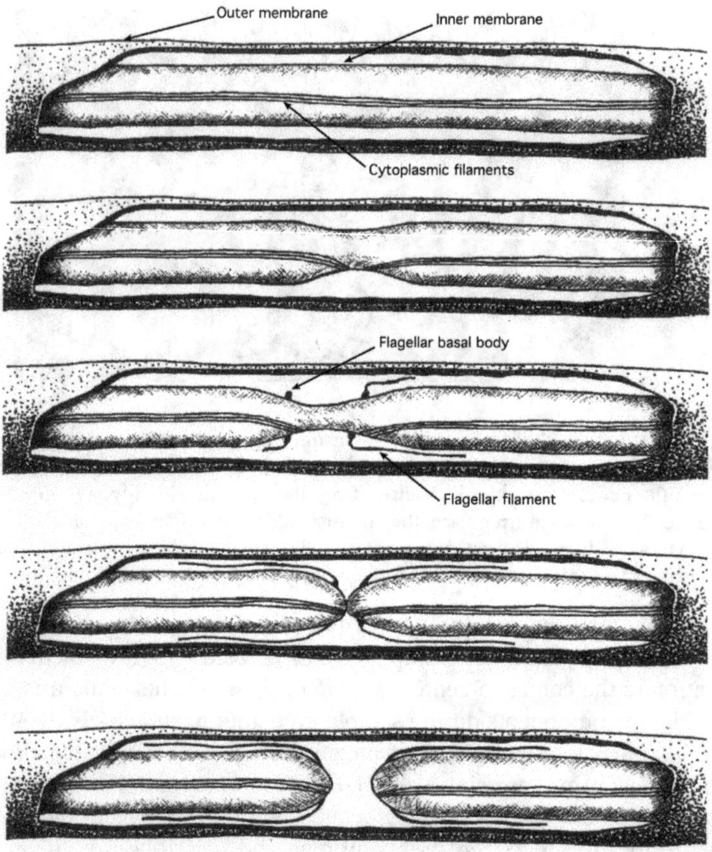

Figure 4. Temporal model of treponemal cytokinesis showing the severing of the cytoplasmic filament bundle and the formation of the flagellar apparatus. This cut-away view of the model is based on published observations (Izard et al., 1999) of treponemal cells undergoing cell division, using *T. phagedenis* cells stripped of their outer membrane, negatively stained with sodium phosphotungstate, and visualized in the electron microscope. The actual attachment site of the cytoplasmic filaments to the cell body is unclear.

and Socransky, 1964; Ryter and Pillot, 1963), and on other spirochetes (Holt, 1978; Hughes et al., 1975; Nauman et al., 1969; Ritchie and Ellinghausen, 1965), we can establish a probable sequence of events triggered by the start of the cell division cycle, and ending with the separation of two individual cells. The model is shown in Figure 4.

During the cell cycle, the cytoplasmic cylinder extends to about twice the length of a single cell, and a septum of cell division becomes positioned at the mid-cell region. During or before the positioning and formation of the FtsZ ring, the cytoplasmic filaments are severed, and their ends are anchored into the membrane. When the septum of cell division is viewed under the microscope, two independent bundles of filaments can be observed on each side. The number of filaments is the same on either side of the septum. At this stage, a nascent flagellar apparatus can be observed, with a basal body, a hook, and a short flagellar filament. While the septum is constricting, the flagellar filaments are elongating. When the septum of cell division is closed and the two cytoplasmic cylinders are separated under a common outer membrane, the flagellar filaments may continue to elongate. The final step is the separation of the two cells.

We hypothesize that the filaments within the ribbons are cut prior to the formation of the FtsZ ring. It is expected that a dedicated machinery exists to sever the cytoplasmic filament bundle and to anchor them on both sides of the septum of cell division. In the case of Mbl filaments, in *B. subtilis*, experiments have shown that the septum of division is required for the severing of the filaments to occur. The severing occurs at a very late stage of septum constriction (Carballido-Lopez and Errington, 2003a). How the treponemal cytoplasmic filaments are anchored to the inner cell membrane, and how they are severed, are remaining questions to be addressed experimentally.

7. The septation ring and FtsZ, a tubulin homologue

Treponema cells divide symmetrically during normal growth, constricting in the middle. The cytokinesis is a constriction process, as in *E. coli*. It involves the outer membrane, the peptidoglycan layer, and the inner membrane. Rupture of the outer membrane is the final event occurring in cytokinesis. In *Treponema*, during the early and log phases of bacterial growth, it is common to observe cells with two independent cytoplasmic cylinders that are rotating, showing the autonomy of the motility apparatus of each cytoplasmic cylinder, even though they are located under the same outer membrane. Before that step, the periplasmic peptidoglycan layer will have been broken down and reconstituted, providing each daughter cell with its own exoskeleton. The first step of cytokinesis is the formation of a constriction ring.

The closest bacterial model that may help to understand *Treponema* cytokinesis is *E. coli*. Both stain as Gram-negative (Canale-Parola, 1984b). They have similar membrane-periplasm-membrane-cytoplasm organizations, the major difference being the periplasmic location of the flagellar filaments in spirochetes (Holt, 1978). They are relatively distant from one another phylogenetically (Paster et al., 1991). Numerous proteins are associated with the cell division processes, a subset is associated with cytokinesis, and a group is involved in forming a filamentous structure at the septum of cell division. The key component of such a filamentous structure is FtsZ (Bi and Lutkenhaus, 1991). The FtsZ protein and associated components form the divisome (Nanninga, 2001). The divisome's functions are integrated in the regulatory network of the overall cell division processes; consequently alterations to the divisome will alter other cell division-related events. In addition to its constriction of the septum, the divisome has another critical function: it positions a specialized penicillin-binding protein required for synthesis of the wall material at the new cell poles.

FtsZ, the first component of the septum ring, is required for the targeting of other proteins to the cell division site. It can be found in virtually all eubacteria, in most *Archaea* and in organelles of many eukaryotes (Beech et al., 2000; Kiessling et al., 2000; Margolin, 2000). *In vitro* FtsZ is able to form various types of filaments (Bramhill and Thompson, 1994; Erickson et al., 1996; Romberg et al., 2001). The structure of FtsZ is similar to that of tubulin (Lowe and Amos, 1998). The ring

Table 1. Proteins located at the septum of cell division and associated with the formation of the septal ring[1]

Genome	*Treponema pallidum*[2]	*Treponema denticola*[3]	*Borrelia burgdorferi*[4]	*Leptospira interrogans* serovar *Lai*[5]	*Escherichia coli*[6]	*Bacillus subtilis*[7]
Gene Name						
ftsZ	Yes	Yes	Yes	Yes (chr I)	Yes	Yes
ftsA	Yes	Yes	Yes	Yes (chr I)	Yes	Yes
zipA	No	No	No	No	Yes	No
ftsE	No	No	No	No	Yes	Yes
ftsX	No	No	No	No	Yes	Yes
ftsK	Yes	Yes	Yes	Yes (chr I)	Yes	No
ftsQ	Yes	Yes	No	Yes (chr I)	Yes	No
ftsL	No	Yes	No	No	Yes	Yes
ygbQ	No	No	No	No	Yes	No
ftsW	Yes	Yes	Yes	Yes (chr I)	Yes	Yes
ftsI	Yes	Yes	Yes	Yes (chr I)	Yes	Yes
ftsN	No	No	No	No	Yes	No
AmiC	No	No	No	Yes (chr I)	Yes	No
divIB	No	No	Yes	No	No	Yes
divIC	No	No	No	No	No	Yes

[1]The protein list is based on published literature (Errington et al., 2003; Schmidt et al., 2004), and the following Web databases: the STD sequence database for *T. pallidum* (www.stdgen.lanl.gov), the Baylor College of Medicine database for *T. denticola* (www.hgsc.bcm.tmc.edu/microbial/Tdenticola/), and from the National Center for Biotechnology Information (www.ncbi.nlm.nih.gov).
[2]Fraser et al. 1998.
[3]Seshadra et al. 2004.
[4]Fraser et al. 1997
[5]Ren et al. 2003.
[6]Blattner et al. 1997.
[7]Kunst et al. 1997.
(chr I): gene located on chromosome I.

seems to be highly dynamic, with the exchange between the cytosol and the filament closely apposed to the inner surface of the cytoplasmic membrane (Stricker et al., 2002). The mechanism of constriction has not yet been resolved, and models need to be tested *in vivo* and *in vitro* (Bramhill, 1997; Gonzalez et al., 2003; Romberg and Levin, 2003). However, the closure of the gap between the two new cells does not require a complete closure of the FtsZ ring (Addinall and Lutkenhaus, 1996).

The proteins associated with the FtsZ ring are listed in Table 1. In *E. coli*, the pathway for assembly is linear: FtsZ → [ZipA, FtsA] → [FtsE, FtsX] →FtsK → FtsQ → [FtsL, YgbQ] →FtsW →FtsI →FtsN →AmiC (Bernhardt and de Boer, 2003; Errington et al., 2003; Schmidt et al., 2004). FtsA is in contact with the FtsZ monomers, and their ratio is important for correct cytokinesis (Dewar et al., 1992). The primary sequence and the crystal structure both show similarities to

Table 2. Presence of *mreB* and *mreB*-like (*mbl*, *mreBH*) genes in different bacterial genomes, including spirochete genomes[1]

Genome	*Treponema pallidum*	*Treponema denticola*	*Borrelia burgdorferi*[2]	*Leptospira interrogans* serovar *Lai*	*Escherichia coli*	*Bacillus subtilis*
Gene Name						
mreB	Yes	Yes	Yes (chr)[3]	Yes (chr I)[4]	Yes	Yes
mbl	No	No	No	No	No	Yes
mreBH	No	No	No	No	No	Yes

[1]The data were collected from the STD sequence database for *T. pallidum* (www.stdgen.lanl.gov), the Baylor College of Medicine database for *T. denticola* (www.hgsc.bcm.tmc.edu/microbial/Tdenticola/), and the database of the National Center for Biotechnology Information (www.ncbi.nlm.nih.gov).
[2]*B. burgdorferi* has a gene annotated *mreB*-2, which is a true homologue of *rodA*.
[3](chr): present on the chromosome.
[4](chr I): present on chromosome I.

actin (Bork et al., 1992; Van den Ent et al., 2001); however, *in vitro* polymerization of FtsA has not been achieved. FtsA is a member of the actin/Hsp70/sugar kinase ATPase superfamily, like another protein present in spirochetes, MreB (see below) (Table 2) (Bork et al., 1992). ZipA in *E. coli* is suspected to be the anchor to the membrane of the FtsZ ring. The *zipA* gene is not present in any spirochete genome available for searches. Spirochetes may use an alternative protein, like EzrA or ZapA, to perform such a function (Gueiros-Filho and Losick, 2002; Levin et al., 2001). With the exception of FtsI, none of the other proteins yet has a precisely delineated function. All are transmembrane proteins. FtsK is present in all spirochetes (see Table 1). FtsQ is present in *Treponema*. *B. burgdorferi* most probably uses DivIB, the homologue of FtsQ in *B. subtilis* (Harry et al., 1994). *Leptospira interrogans*, has no homologues for these two proteins. FtsL and its functional homologues DivC and YgbQ are absent from most spirochetes except *T. denticola*. FtsW is present in all spirochetes. FtsW is part of the SEDS family of proteins (shape, elongation, division and sporulation) that includes RodA and SpoVE. The latter has a specialized role in sporulation in *B. subtilis* (Henriques et al., 1998; Ikeda et al., 1989). Septation requires a dedicated penicillin-binding protein (FtsI or PBP-3 in *E. coli*), which is also present in all spirochetes. FtsN is a poorly conserved protein, with clear homologues in enteric bacteria and in *Haemophilus* species. (Dai et al., 1996).

In the case of FtsN and other proteins involved in cell division, the absence of a detectable homologue may be due to a poor conservation of the protein, the functional replacement by another protein, or compensatory mutations that eliminate the requirement for such a protein. Poor sequence similarity has complicated the naming and functional assignments of certain genes (Limberger et al., 1999). Detailed analysis of *in vivo* mechanisms is required to decipher such issues. Although the overall mechanism of cell division is similar, specific differences can be observed among spirochetes. However, insufficient genomic information is available to understand the roots of those differences.

8. Cell shape and MreB, an actin homologue

The shape of a bacterium is a stable characteristic that has been used extensively as a taxonomic criterion. For the spirochetes, cell shape is a crucial component of their mode of propagation. The helical nature of most spirochetal cells is evident from microscopic observations. The maintenance of such a shape and the presence of periplasmic flagellar filaments together form the basis of the ability of spirochetes to penetrate dense media (Klitorinos et al., 1993; Ruby and Charon, 1998), or cell monolayers and tissues, as demonstrated *in vitro* (Lux et al., 2001; Thomas et al., 1988).

In straight-rod-shaped bacteria like *E. coli* and *B. subtilis*, a number of mutants with spherical morphology have been isolated. The mutations are known to affect various cell processes including their association with the peptidoglycan layer (*pbpA*, *rodA*) (Henriques et al., 1998; Spratt, 1975; Tamaki et al., 1980), their participation in teichoic acid synthesis (*rodC*) (Honeyman and Stewart, 1989); or their direct involvement in cell shape control (*mreB*, and *mbl*) (Abhayawardhane and Stewart, 1995; Levin et al., 1992; Wachi et al., 1987). MreB (present in all spirochetes; Table 2) and Mbl (MreB-like) proteins in *B. subtilis* have been shown to form filamentous helical structures in the cytoplasm of the bacteria. A second homologous gene to *mreB* in *B. subtilis* is *mreBH*, that can be disrupted with no obvious phenotypic consequence (Carballido-Lopez and Errington, 2003b). In *B. subtilis,* MreB is essential for viability (Abhayawardhane and Stewart, 1995; Jones et al., 2001). In *E. coli*, the inactivation of the *mreB* gene results in spherical cells (Wachi et al., 1987). In both organisms, the filaments extend for the length of the cell (Jones et al., 2001; Shih et al., 2003). With a few exceptions, MreB orthologues are present only in species with rod- or helical-shaped cells (Daniel and Errington, 2003) (Table 2); they are usually absent in species with spherical-shaped cells (Daniel and Errington, 2003).

The first approach to an understanding of the mechanisms beyond filament formation that maintain the cell shape was to create mutations of MreB in *B. subtilis* (Jones et al., 2001; Wachi et al., 1987). The filaments control the cell length and the width of the bacterium. The first capability is linked to the ability of the cell to divide in such a way to produce identical daughter cells. The second capability is also tightly controlled (Sharpe et al., 1998; Woldringh et al., 1985). The crystallization of the MreB protein from *Thermotoga maritima* brought a new perspective on the system (Van den Ent et al., 2001). MreB and actin have highly similar three-dimensional folds. The crystallization revealed an actin-like strand at atomic resolution, thus elucidation the monomer-monomer interaction. *In vitro*, MreB forms a large fibrous spiral in the presence of ATP (Van den Ent et al., 2001); this supports the ability of the protein to form helical filaments, *in vivo*, as detected by fluorescence microscopy (Jones et al., 2001). The structures formed by Mbl (MreB-like) filaments are highly dynamic; they re-model continuously during cell elongation (Carballido-Lopez and Errington, 2003a). The Mbl filaments probably constitute only one of the factors influencing cell shape. Another factor is the formation of the peptidoglycan layer. Mbl has a direct influence on the formation of new peptidoglycan (Carballido-Lopez and Errington, 2003a; Daniel and Errington, 2003). In *E. coli*, the alteration of cell shape in mutant forms of the MreB protein was seen to influence cell-division processes, including chromosome segregation (Kruse et al., 2003). Those observations underline the necessity of cell-shape maintenance to other complex functional processes of the cell.

In *Treponema* species, whole-genome analysis has demonstrated the presence of the genes that regulate the cell shape in other bacterial models, including MreB and RodA. As yet, however, functional evidence needed to confirm the functions of these genes is lacking in *Treponema*. As indicated earlier in this chapter, one striking difference between *B. burgdorferi* and the *Treponema* species is that, in the absence of flagella, *B. burgdorferi* becomes a straight rod shaped bacteria (Motaleb et al., 2000; Sadziene et al., 1991). The presence of the flagellar filament in *B. burgdorferi* is directly related to the maintenance of the helical shape (Motaleb et al., 2000). In contrast, in *Treponema* spp., the absence of a flagellar filament does not alter the overall cell morphology (Izard et al., 1999; Limberger et al., 1999). Consequently, other factors must be involved, working together with MreB and associated proteins, to maintain the helical nature of the cell.

9. Hyperstructure, filaments and compartmentalization

The conceptual difference between ultrastructure (cytoskeleton and flagella) and hyperstructure resides in the fact that hyperstructures are not visible by microscopy. However, with the advances of microscopic techniques, including the design of new fluorophores and tagging systems, this difference may soon disappear. A hyperstructure occupies a large portion of the cell's volume and contains many of the cell's constituents, and has a readily assigned function (Norris et al., 1999). Such a structure has many components and requires a flow of building blocks (amino acids, nucleotides, etc.), proteins (enzymes, monomers of protein complexes, etc.), and other components (tRNAs, precursors, etc.). Hyperstructures are activity-based, and they form only when their constituents are actively engaged in carrying out their function. Examples include; the lipid preference of the replisome and the environment surrounding the multi-protein complex (30-40 nm in diameter) associated with the glycolysis (Baker and Bell, 1998; Dowhan, 1997; Lemon and Grossman, 1998; Mileykovskaya et al., 1998; Mowbray and Moses, 1976; Velot et al., 1997). As we begin to understand the dynamic aspect of the cytoskeleton, the concept of dynamic compartmentalization has to be integrated in the temporal studies of bacterial cell biology.

In *Treponema*, the cytoplasmic filament ribbon, the MreB helical filament, the flagellar basal body and filament are components of hyperstructures. Their volume occupancy and their requirement for helper proteins to assist in their function, create and define volumes within the cell where activities are primarily related to the filaments. Moreover, functions associated with formation, anchoring, maintenance, and severing require another set of proteins that are most probably in close proximity to the filaments for a short period of time. Those structures and associated components define, on the temporal scale, volumes that can be transient but create a specific type of compartmentalization. The confluence of genetic, comparative genomic, biochemical, comparative proteomic and biophysical findings will provide a clearer picture of the dynamics of cell architecture.

Acknowledgements

We thank Linda L. Slivienski-Gebhardt for her extensive scientific support of our work on *Treponema*, William A. Samsonoff and Konrad Wicher for their continuous contributions towards treponemal research, and Andrea Carpenter for the illustration in Figure 4. This work was supported by Public Health Research Service grant AI34354 from the National Institutes of Health. The electron microscopy work was made possible by the NIH NCRR Biomedical Research Technology Program Grant RR01219, and the Wadsworth Center's Core Facility for Electron Microscopy. JI was also supported in part by a basic research grant from Health Research Incorporated, and the John W. Hein fellowship.

References

Abhayawardhane, Y., and Stewart, G.C. (1995). *Bacillus subtilis* possesses a second determinant with extensive sequence similarity to the *Escherichia coli mreB* morphogene. J. Bacteriol. *177*, 765-773.

Addinall, S.G., and Lutkenhaus, J. (1996). FtsZ-spirals and -arcs determine the shape of the invaginating septa in some mutants of *Escherichia coli*. Mol. Microbiol. *22*, 231-237.

Baker, T.A., and Bell, S.P. (1998). Polymerases and the replisome: machines within machines. Cell *92*, 295-305.

Baumeister, W. (2002). Electron tomography: towards visualizing the molecular organization of the cytoplasm. Current Opin. Struct. Biol. *12*, 679-684.

Beech, P.L., Nheu, T., Schultz, T., Herbert, S., Lithgow, T., Gilson, P.R., and McFadden, G.I. (2000). Mitochondrial FtsZ in a chromophyte alga. Science *287*, 1276-1279.

Bermudes, D., Chase, D., and Margulis, L. (1988). Morphology as a basis for taxonomy of large spirochetes symbiotic in wood-eating cockroaches and termites: *pillotina* gen. nov., nom. rev.; *Pillotina calotermitidis* sp. nov., nom. rev.; *Diplocalyx* gen. nov., nom. rev.; *Diplocalyx calotermitidis* sp. nov., nom. rev.; *Hollandina* gen. nov., nom. rev.; *Hollandina pterotermitidis* sp. nov., nom. rev.; and *Clevelandina reticulitermitidis* gen. nov., sp. nov. Int. J. Syst. Bacteriol. *38*, 291-302.

Bermudes, D., Fracek, S.P., Jr., Laursen, R.A., Margulis, L., Obar, R., and Tzertzinis, G. (1987). Tubulin-like protein from *Spirochaeta bajacaliforniensis*. Ann. N. Y. Acad. Sci. *503*, 515-527.

Bernhardt, T.G., and de Boer, P.A. (2003). The *Escherichia coli* amidase AmiC is a periplasmic septal ring component exported via the twin-arginine transport pathway. Mol. Microbiol. *48*, 1171-1182.

Bi, E.F., and Lutkenhaus, J. (1991). FtsZ ring structure associated with division in *Escherichia coli*. Nature *354*, 161-164.

Bork, P., Sander, C., and Valencia, A. (1992). An ATPase domain common to prokaryotic cell cycle proteins, sugar kinases, actin, and hsp70 heat shock proteins. Proc. Natl. Acad. Sci. USA *89*, 7290-7294.

Bramhill, D. (1997). Bacterial cell division. Annu. Rev. Cell Dev. Biol. *13*, 395-424.

Bramhill, D., and Thompson, C.M. (1994). GTP-dependent polymerization of *Escherichia coli* FtsZ protein to form tubules. Proc. Natl. Acad. Sci. USA *91*, 5813-5817.

Caimano, M.J., Bourell, K.W., Bannister, T.D., Cox, D.L., and Radolf, J.D. (1999). The *Treponema denticola* major sheath protein is predominantly periplasmic and has only limited surface exposure. Infect. Immun. *67*, 4072-4083.

Canale-Parola, E. (1984a). Motility and chemotaxis of spirochetes. Annu. Rev. Microbiol. *32*, 69-99.

Canale-Parola, E. (1984b). The Spirochetes. In Bergey's Manual of Systematic Bacteriology, N. R. Krieg, and J. G. Holt, eds. (Baltimore, MD, Williams and Wilkins), pp. 38-70.

Carballido-Lopez, R., and Errington, J. (2003a). The bacterial cytoskeleton: *in vivo* dynamics of the actin-like protein Mbl of *Bacillus subtilis*. Dev. Cell *4*, 19-28.

Carballido-Lopez, R., and Errington, J. (2003b). A dynamic bacterial cytoskeleton. Trends Cell. Biol. *13*, 577-583.

Charon, N.C., Goldstein, S.F., Curci, K., and Limberger, R.J. (1991). The bent-end morphology of *Treponema phagedenis* is associated with short, left-handed, periplasmic flagella. J. Bacteriol. *173*, 4820-4826.

Charon, N.W., and Goldstein, S.F. (2002). Genetics of motility and chemotaxis of a fascinating group of bacteria: the spirochetes. Annu. Rev. Genet. *36*, 47-73.

Charon, N.W., Goldstein, S.F., Block, S.M., Kurci, K., Ruby, J.D., Kreiling, J.A., and Limberger, R.J. (1992a). Morphology and dynamics of protruding spirochete periplasmic flagella. J. Bacteriol. *174*, 832-840.

Charon, N.W., Greenberg, E.P., Koopman, M.B.H., and Limberger, R.J. (1992b). Spirochete chemotaxis, motility, and the structure of the spirochetal periplasmic flagella. Res. Microbiol. *143*, 597-603.

Chi, B., Chauhan, S., and Kuramitsu, H. (1999). Development of a system for expressing heterologous genes in the oral spirochete *Treponema denticola* and its use in expression of the *Treponema pallidum flaA* gene. Infect. Immun. *67*, 3653-3656.

Chi, B., Limberger, R.J., and Kuramitsu, H.K. (2002). Complementation of a *Treponema denticola flgE* mutant with a novel coumermycin A1-resistant *T. denticola* shuttle vector system. Infect. Immun. *70*, 2233-2237.

Cox, C.D. (1972). Shape of *Treponema pallidum*. J. Bacteriol. *109*, 943-944.

Crowther, R.A., De Rosier, D.J., and Klug, A. (1970). The reconstruction of a three-dimensional structure from the projections and its application to electron microscopy. Proc. R. Soc. London *A317*, 319-340.

Dai, K., Xu, Y., and Lutkenhaus, J. (1996). Topological characterization of the essential *Escherichia coli* cell division protein FtsN. J. Bacteriol. *178*, 1328-1334.

Daniel, R.A., and Errington, J. (2003). Control of cell morphogenesis in bacteria: two distinct ways to make a rod-shaped cell. Cell *113*, 767-776.

Dewar, S.J., Begg, K.J., and Donachie, W.D. (1992). Inhibition of cell division initiation by an imbalance in the ratio of FtsA to FtsZ. J. Bacteriol. *174*, 6314-6316.

Dowhan, W. (1997). Molecular basis for membrane phospholipid diversity: why are there so many lipids? Annu. Rev. Biochem. *66*, 199-232.

Eipert, S.R., and Black, S.H. (1979). Characterization of the cytoplasmic fibrils of *Treponema refringens* (Nichols). Arch. Microbiol. *120*, 205-214.

Erickson, H.P., Taylor, D.W., Taylor, K.A., and Bramhill, D. (1996). Bacterial cell division protein FtsZ assembles into protofilament sheets and minirings, structural homologs of tubulin polymers. Proc. Natl. Acad. Sci. USA *93*, 519-523.

Errington, J., Daniel, R.A., and Scheffers, D.J. (2003). Cytokinesis in bacteria. Microbiol. Mol. Biol. Rev. *67*, 52-65.

Frank, J., Radermacher, M., Penczek, P., Zhu, J., Li, Y., Ladjadj, M., and Leith, A. (1996). SPIDER and WEB: processing and visualization of images in 3D electron microscopy and related fields. J. Struct. Biol. *116*, 190-199.

Fraser, C.M., Norris, S.J., Weinstock, G.M., White, O., Sutton, G.G., Dodson, R., Gwinn, M., Hickey, E.K., Clayton, R., Ketchum, K.A., et al. (1998). Complete genome sequence of *Treponema pallidum*, the syphilis spirochete. Science *281*, 375-388.

Glaser, P., Sharpe, M.E., Raether, B., Perego, M., Ohlsen, K., and Errington, J. (1997). Dynamic, mitotic-like behavior of a bacterial protein required for accurate chromosome partitioning. Genes Dev. *11*, 1160-1168.

Gonzalez, J.M., Jimenez, M., Velez, M., Mingorance, J., Andreu, J.M., Vicente, M., and Rivas, G. (2003). Essential cell division protein FtsZ assembles into one monomer-thick ribbons under conditions resembling the crowded intracellular environment. J. Biol. Chem. *278*, 37664-37671.

Graves, M.C., and Rabinowitz, J.C. (1986). *In vivo* and *in vitro* transcription of the *Clostridium pasteurianum* ferredoxin gene. Evidence for "extended" promoter elements in Gram-positive organisms. J. Biol. Chem. *261*, 11409-11415.

Grimm, R., Singh, H., Rachel, R., Typke, D., Zillig, W., and Baumeister, W. (1998). Electron tomography of ice-embedded prokaryotic cells. Biophys. J. *74*, 1031-1042.

Gueiros-Filho, F.J., and Losick, R. (2002). A widely conserved bacterial cell division protein that promotes assembly of the tubulin-like protein FtsZ. Genes Dev. *16*, 2544-2556.

Haake, D.A. (2000). Spirochaetal lipoproteins and pathogenesis. Microbiology *146 (Pt 7)*, 1491-1504.

Hampp, E.G. (1951). Further studies on the significance of spirochetal granules. J. Bacteriol. *62*, 347-349.

Harry, E.J., Partridge, S.R., Weiss, A.S., and Wake, R.G. (1994). Conservation of the 168 divIB gene in *Bacillus subtilis* W23 and *B. licheniformis*, and evidence for homology to *ftsQ* of *Escherichia coli*. Gene *147*, 85-89.

Hawley, D.K., and McClure, W.R. (1983). Compilation and analysis of *Escherichia coli* promoter DNA sequences. Nucleic Acids Res. *11*, 2237-2255.

Henriques, A.O., Glaser, P., Piggot, P.J., and Moran, C.P., Jr. (1998). Control of cell shape and elongation by the *rodA* gene in *Bacillus subtilis*. Mol. Microbiol. *28*, 235-247.

Hollande, A., Gharagozlou, I., and Grasse, P. (1967). Morphologie infrastructurale de *Pillotina calotermitidis* nov. gen.. nov. sp. Spirochaetale de l'intestin de Coloterines praecox. C R Acad. Sci. Hebd. Seances. Acad. Sci D *265*, 1309-1312.

Holt, J.G., Krieg, N.R., Sneath, P.H.A., Staley, J.T., and Williams, S.T. (1994). The Spirochetes. In Bergey's Manual of Determinative Bacteriology (Baltimore, Williams and Wilkins), pp. 27-21.

Holt, S.C. (1978). Anatomy and chemistry of spirochetes. Microbiol. Rev. *42*, 114-160.

Honeyman, A.L., and Stewart, G.C. (1989). The nucleotide sequence of the *rodC* operon of *Bacillus subtilis*. Mol. Microbiol. *3*, 1257-1268.

Hovind-Hougen, K. (1972). Further observations on the ultrastructure of *Treponema pallidum* Nichols. Acta. Pathol. Scand. Sect. B *80*, 297-304.

Hovind-Hougen, K. (1974). The ultrastructure of cultivable treponemes. I. *Treponema phagedenis*, *Treponema vincentii*, and *Treponema refringens*. Acta. Pathol. Scand. Sect. B *82*, 329-344.

Hovind-Hougen, K. (1976). Determination by means of electron microscopy of morphological criteria of value for classification of some spirochetes, in particular treponemes. Acta. Pathol. Scand. Sect. B *255*, 1-41.

Hovind-Hougen, K. (1979). *Leptospiraceae*, a new family to include *Leptospira* Noguchi 1917 and *Leptonema* gen. nov. Int. J. Syst. Bacteriol. *29*, 245-251.

Hovind-Hougen, K., and Birch-Andersen, A. (1971). Electron microscopy of endoflagella and microtubules in *Treponema reiter*. Acta. Pathol. Microbiol. Scand. *79*, 37-50.

Hovind-Hougen, K., Birch-Andersen, A., and Jensen, H.J. (1976). Ultrastructure of cells of *Treponema pertenue* obtained from experimentally infected hamsters. Acta. Pathol. Scand. Sect. B *84*, 101-108.

Hughes, R., Olander, H.J., and Williams, C.B. (1975). Swine dysentery: pathogenicity of *Treponema hyodysenteriae*. Am. J. Vet. Res. *36*, 971-977.

Ikeda, M., Sato, T., Wachi, M., Jung, H.K., Ishino, F., Kobayashi, Y., and Matsuhashi, M. (1989). Structural similarity among *Escherichia coli* FtsW and RodA proteins and *Bacillus subtilis* SpoVE protein, which function in cell division, cell elongation, and spore formation, respectively. J. Bacteriol. *171*, 6375-6378.

Indest, K.J., Ramamoorthy, R., and Philipp, M.T. (2000). Transcriptional regulation in spirochetes. J. Mol. Microbiol. Biotechnol. *2*, 473-481.

Izard, J., McEwen, B.F., Barnard, R.M., Portuese, T., Samsonoff, W.A., and Limberger, R.J. (2004). Tomographic reconstruction of treponemal cytoplasmic filaments reveals novel bridging and anchoring components. Mol. Microbiol. *51*, 609-618.

Izard, J., Samsonoff, W.A., Kinoshita, M.B., and Limberger, R.J. (1999). Genetic and structural analysis of the cytoplasmic filaments of wild-type and flagellar filament mutant of *Treponema phagedenis*. J. Bacteriol. *181*, 6739-6746.

Izard, J., Samsonoff, W.A., and Limberger, R.J. (2001). A cytoplasmic filament-deficient mutant of *Treponema denticola* has pleiotropic defects. J. Bacteriol. *183*, 1078-1084.

Jensen, R.B., and Shapiro, L. (1999). The *Caulobacter crescentus smc* gene is required for cell cycle progression and chromosome segregation. Proc. Natl. Acad. Sci. USA *96*, 10661-10666.

Jones, L.J., Carballido-Lopez, R., and Errington, J. (2001). Control of cell shape in bacteria: helical, actin-like filaments in *Bacillus subtilis*. Cell *104*, 913-922.

Kesavalu, L., Falk, C.W., Davis, K.J., Steffen, M.J., Xu, X., Holt, S.C., and Ebersole, J.L. (2002). Biological characterization of lipopolysaccharide from *Treponema pectinovorum*. Infect. Immun. *70*, 211-217.

Kiessling, J., Kruse, S., Rensing, S.A., Harter, K., Decker, E.L., and Reski, R. (2000). Visualization of a cytoskeleton-like FtsZ network in chloroplasts. J. Cell Biol. *151*, 945-950.

Klitorinos, A., Noble, P., Siboo, R., and Chan, E.C. (1993). Viscosity-dependent locomotion of oral spirochetes. Oral Microbiol. Immunol. *8*, 242-244.

Kruse, T., Moller-Jensen, J., Lobner-Olesen, A., and Gerdes, K. (2003). Dysfunctional MreB inhibits chromosome segregation in *Escherichia coli*. EMBO J. *22*, 5283-5292.

Lemon, K.P., and Grossman, A.D. (1998). Localization of bacterial DNA polymerase: evidence for a factory model of replication. Science *282*, 1516-1519.

Levin, P.A., Margolis, P.S., Setlow, P., Losick, R., and Sun, D. (1992). Identification of *Bacillus subtilis* genes for septum placement and shape determination. J. Bacteriol. *174*, 6717-6728.

Levin, P.A., Schwartz, R.L., and Grossman, A.D. (2001). Polymer stability plays an important role in the positional regulation of FtsZ. J. Bacteriol. *183*, 5449-5452.

Li, C., Motaleb, A., Sal, M., Goldstein, S.F., and Charon, N.W. (2000). Spirochete periplasmic flagella and motility. J. Mol. Microbiol. Biotechnol. *2*, 345-354.

Limberger, R.J. (2004). The periplasmic flagellum of spirochetes. J. Mol. Microbiol. Biotechnol. *7*, 30-40.

Limberger, R.J., and Charon, N.W. (1986). Antiserum to the 33,000-dalton periplasmic-flagellum protein of "*Treponema phagedenis*" reacts with other treponemes and *Spirochaeta aurantia*. J. Bacteriol. *168*, 1030-1032.

Limberger, R.J., Slivienski, L.L., El-Afandi, M.C., and Dantuono, L.A. (1996). Organization, transcription, and expression of the 5′ region of the *fla* operon of *Treponema phagedenis* and *Treponema pallidum*. J. Bacteriol. *178*, 4628-4634.

Limberger, R.J., Slivienski, L.L., Izard, J., and Samsonoff, W.A. (1999). Insertional inactivation of *Treponema denticola tap1* results in a nonmotile mutant with elongated flagellar hooks. J. Bacteriol. *181*, 3743-3750.

Listgarten, M.A., and Socransky, S.S. (1964). Electron microscopy of axial fibrils, outer envelope, and cell division of certain oral spirochetes. J. Bacteriol. *88*, 1087-1103.

Lowe, J., and Amos, L.A. (1998). Crystal structure of the bacterial cell-division protein FtsZ. Nature *391*, 203-206.

Lux, R., Miller, J.N., Park, N.H., and Shi, W. (2001). Motility and chemotaxis in tissue penetration of oral epithelial cell layers by *Treponema denticola*. Infect. Immun. *69*, 6276-6283.

Margolin, W. (2000). Organelle division: Self-assembling GTPase caught in the middle. Curr. Biol. *10*, 328-330.

Martinez-Morales, F., Schobert, M., Lopez-Lara, I.M., and Geiger, O. (2003). Pathways for phosphatidylcholine biosynthesis in bacteria. Microbiology *149*, 3461-3471.

Masuda, K., and Kawata, T. (1989). Isolation and characterization of cytoplasmic fibrils from treponemes. Microbiol. Immunol. *33*, 619-630.

McEwen, B.F., and Frank, J. (2001). Electron tomographic and other approaches for imaging molecular machines. Current Opin. Neurobiol. *11*, 594-600.

McEwen, B.F., and Marko, M. (1999). Three-dimensional transmission electron microscopy and its application to mitosis research. Methods Cell Biol. *61*, 81-111.

Mileykovskaya, E., Sun, Q., Margolin, W., and Dowhan, W. (1998). Localization and function of early cell division proteins in filamentous *Escherichia coli* cells lacking phosphatidylethano lamine. J. Bacteriol. *180*, 4252-4257.

Moran, C.P., Jr., Lang, N., LeGrice, S.F., Lee, G., Stephens, M., Sonenshein, A.L., Pero, J., and Losick, R. (1982). Nucleotide sequences that signal the initiation of transcription and translation in *Bacillus subtilis*. Mol. Gen. Genet. *186*, 339-346.

Morioka, H., Ozasa, H., Kishida, T., Yokota, Y., and Suganuma, A. (1979). The peptidoglycan layer of the Reiter treponeme as revealed by thin sectioning and freeze-etching techniques. J. Electron Microsc. (Tokyo) *28*, 189-192.

Motaleb, M.A., Corum, L., Bono, J.L., Elias, A.F., Rosa, P., Samuels, D.S., and Charon, N.W. (2000). *Borrelia burgdorferi* periplasmic flagella have both skeletal and motility functions. Proc. Natl. Acad. Sci. USA *97*, 10899-10904.

Mowbray, J., and Moses, V. (1976). The tentative identification in *Escherichia coli* of a multienzyme complex with glycolytic activity. Eur. J. Biochem. *66*, 25-36.

Nanninga, N. (2001). Cytokinesis in prokaryotes and eukaryotes: common principles and different solutions. Microbiol. Mol. Biol. Rev. *65*, 319-333.

Nauman, R.K., Holt, S.C., and Cox, C.D. (1969). Purification, ultrastructure, and composition of axial filaments from *Leptospira*. J. Bacteriol. *98*, 264-280.

Niki, H., and Hiraga, S. (1998). Polar localization of the replication origin and terminus in *Escherichia coli* nucleoids during chromosome partitioning. Genes Dev. *12*, 1036-1045.

Norris, V., Alexandre, S., Bouligand, Y., Cellier, D., Demarty, M., Grehan, G., Gouesbet, G., Guespin, J., Insinna, E., Le Sceller, L., et al. (1999). Hypothesis: hyperstructures regulate bacterial structure and the cell cycle. Biochimie *81*, 915-920.

Ovcinnikov, N.M., and Delektorskij, V.V. (1970). *Treponema pertenue* under the electron microscope. Br. J. Vener. Dis. *46*, 349-379.

Paster, B.J., Dewhirst, F.E., Weisburg, W.G., Tordoff, L.A., Fraser, G.J., Hespell, R.B., Stanton, T.B., Zablen, L., Mandelco, L., and Woese, C.R. (1991). Phylogenetic analysis of the spirochetes. J. Bacteriol. *173*, 6101-6109.

Penn, C.W., Bailey, M.J., and Cockayne, A. (1985). The axial filament antigen of *Treponema pallidum*. Immunology *54*, 635-641.

Penn, C.W., and Luke, C.J. (1992). Bacterial flagellar diversity and significance in pathogenesis. FEMS Microbiol. Lett. *79*, 331-336.

Ritchie, A.E., and Ellinghausen, H.C. (1965). Electron microscopy of leptospires. I. Anatomical features of *Leptospira pomona*. J. Bacteriol. *89*, 223-233.

Romberg, L., and Levin, P.A. (2003). Assembly dynamics of the bacterial cell division protein FTSZ: poised at the edge of stability. Annu. Rev. Microbiol. *57*, 125-154.

Romberg, L., Simon, M., and Erickson, H.P. (2001). Polymerization of Ftsz, a bacterial homolog of tubulin. is assembly cooperative? J. Biol. Chem. *276*, 11743-11753.

Ruby, J.D., and Charon, N.W. (1998). Effect of temperature and viscosity on the motility of the spirochete *Treponema denticola*. FEMS Microbiol. Lett. *169*, 251-254.

Ruby, J.D., Li, H., Kuramitsu, H., Norris, S.J., Goldstein, S.F., Buttle, K.F., and Charon, N.W. (1997). Relationship of *Treponema denticola* periplasmic flagella to irregular cell morphology. J. Bacteriol. *179*, 1628-1635.

Ryter, A., and Pillot, J. (1963). Etude au microscope electronique de la structure interne du *Treponeme reiter*. Ann. Inst. Pasteur *104*, 496-501.

Sadziene, A., Thomas, D.D., Bundoc, V.G., Holt, S.C., and Barbour, A.G. (1991). A flagella-less mutant of *Borrelia burgdorferi:* structural, molecular and *in vitro* functional characterization. J. Clin. Invest. *88*, 82-92.

Schmidt, K.L., Peterson, N.D., Kustusch, R.J., Wissel, M.C., Graham, B., Phillips, G.J., and Weiss, D.S. (2004). A predicted ABC transporter, FtsEX, is needed for cell division in *Escherichia coli*. J. Bacteriol. *186*, 785-793.

Seshadri, R., Myers, G.S., Tettelin, H., Eisen, J.A., Heidelberg, J.F., Dodson, R.J., Davidsen, T.M., DeBoy, R.T., Fouts, D.E., Haft, D.H., et al. (2004). Comparison of the genome of the oral pathogen *Treponema denticola* with other spirochete genomes. Proc. Natl. Acad. Sci. USA *101*, 5646-5651.

Sharpe, M.E., and Errington, J. (1999). Upheaval in the bacterial nucleoid. An active chromosome segregation mechanism. Trends Genet. *15*, 70-74.

Sharpe, M.E., Hauser, P.M., Sharpe, R.G., and Errington, J. (1998). *Bacillus subtilis* cell cycle as studied by fluorescence microscopy: constancy of cell length at initiation of DNA replication and evidence for active nucleoid partitioning. J. Bacteriol. *180*, 547-555.

Shih, Y.L., Le, T., and Rothfield, L. (2003). Division site selection in *Escherichia coli* involves dynamic redistribution of Min proteins within coiled structures that extend between the two cell poles. Proc. Natl. Acad. Sci. USA *100*, 7865-78670.

Slivienski-Gebhardt, L.L., Samsonoff, W.A., Izard, J., and Limberger, R.J. (2004). Development of a novel chloramphenicol resistance expression plasmid used for genetic complementation of a *fliG* deletion mutant in *Treponema denticola*. Infect. Immun. *72*, 5493-5497.

Smajs, D., McKevitt, M., Howell, J.K., Norris, S.J., Cai, W.W., Palzkill, T., and Weinstock, G.M. (2005). Transcriptome of *Treponema pallidum*: gene expression profile during experimental rabbit infection. J. Bacteriol. *187*, 1866-1874.

Smibert, R.M. (1984). Genus III. *Treponema* Schaudin 1905, 1728 [AL]. In Bergey's manual of systematic bacteriology, vol. 1, N. R. Krieg, and J. G. Holt, eds. (Baltimore, MD, The Williams and Wilkins Co.), pp. 49-57.

Spratt, B.G. (1975). Distinct penicillin binding proteins involved in the division, elongation, and shape of *Escherichia coli* K12. Proc. Natl. Acad. Sci. USA. *72*, 2999-3003.

Stepan, D.E., and Johnson, R.C. (1981). Helical conformation of *Treponema pallidum* (Nichols strain), *Treponema paraluis-cuniculi, Treponema denticola, Borrelia turicatae,* and unidentified oral spirochetes. Infect. Immun. *32*, 937-940.

Stricker, J., Maddox, P., Salmon, E.D., and Erickson, H.P. (2002). Rapid assembly dynamics of the *Escherichia coli* FtsZ-ring demonstrated by fluorescence recovery after photobleaching. Proc. Natl. Acad. Sci. USA *99*, 3171-3175.

Tamaki, S., Matsuzawa, H., and Matsuhashi, M. (1980). Cluster of *mrdA* and *mrdB* genes responsible for the rod shape and mecillinam sensitivity of *Escherichia coli*. J. Bacteriol. *141*, 52-57.

Thomas, D.D., Navab, M., Haake, D.A., Fogelman, A.M., Miller, J.N., and Lovett, M.A. (1988). *Treponema pallidum* invades intracellular junctions of endothelial cell monolayers. Proc. Natl. Acad. Sci. USA *8*, 3608-3612.

Umemoto, T., Ota, T., Sagawa, H., Kato, K., Takada, H., Tsujimoto, M., Kawasaki, A., Ogawa, T., Harada, K., and Kotani, S. (1981). Chemical and biological properties of a peptidoglycan isolated from *Treponema pallidum* kazan. Infect. Immun. *31*, 767-774.

Van den Ent, F., Amos, L.A., and Lowe, J. (2001). Prokaryotic origin of the actin cytoskeleton. Nature *413*, 39-44.

Velot, C., Mixon, M.B., Teige, M., and Srere, P.A. (1997). Model of a quinary structure between Krebs TCA cycle enzymes: a model for the metabolon. Biochemistry *36*, 14271-14276.

Wachi, M., Doi, M., Tamaki, S., Park, W., Nakajima-Iijima, S., and Matsuhashi, M. (1987). Mutant isolation and molecular cloning of *mre* genes, which determine cell shape, sensitivity to mecillinam, and amount of penicillin-binding proteins in *Escherichia coli*. J. Bacteriol. *169*, 4935-4940.

Wang, Q., Ko, K.S., Kapus, A., McCulloch, C.A., and Ellen, R.P. (2001). A spirochete surface protein uncouples store-operated calcium channels in fibroblasts: a novel cytotoxic mechanism. J. Biol. Chem. *276*, 23056-23064.

Webb, C.D., Graumann, P.L., Kahana, J.A., Teleman, A.A., Silver, P.A., and Losick, R. (1998). Use of time-lapse microscopy to visualize rapid movement of the replication origin region of the chromosome during the cell cycle in *Bacillus subtilis*. Mol. Microbiol. *28*, 883-892.

Woldringh, C.L., Valkendurg, J.A., Pas, E., Taschner, P.E., Huls, P., and Wientjes, F.B. (1985). Physiological and geometrical conditions for cell division in *Escherichia coli*. Ann. Inst. Pasteur. Microbiol. *136A*, 131-138.

Yanagihara, Y., Kamisango, K., Yasuda, S., Kobayashi, S., Mifuchi, I., Azuma, I., Yamamura, Y., and Johnson, R.C. (1984). Chemical compositions of cell walls and polysaccharide fractions of spirochetes. Microbiol. Immunol. *28*, 535-544.

You, Y., Elmore, S., Colton, L.L., Mackenzie, C., Stoops, J.K., Weinstock, G.M., and Norris, S.J. (1996). Characterization of the cytoplasmic filament protein gene (*cfpA*) of *Treponema pallidum* subsp. *pallidum*. J. Bacteriol. *178*, 3177-3187.

Zemper, E.D., and Black, S.H. (1978). Morphology of freeze-etched *Treponema refringens* (Nichols). Arch. Microbiol. *117*, 227-238.

4

Metabolism of the *Treponema*

David L. Cox and Justin D. Radolf*

Abstract

The metabolism of the three most extensively studied treponemal species, *Treponema pallidum*, *T. denticola*, and *T. phagedenis*, is presented and discussed. The genomes of *T. pallidum* and *T. denticola* have been sequenced and annotated, while that of *T. phagedenis* has not. The availability of these two genomic sequences allows us not only to explore potential metabolic pathways within the respective organisms but also to assess the accuracy of pre-genomic studies. In the case of *T. pallidum*, performance and interpretation of pre-genomic metabolic investigations were greatly complicated by the unavoidable presence of rabbit testicular contaminants in treponemal suspensions. A tabular comparison of key metabolic pathways and enzymes of *T. pallidum* and *T. denticola* (with the gene and Enzyme Commission {EC} numbers) is also presented as a reference tool. The cultivation methods for all three treponemes are discussed with emphasis on the numerous, but unsuccessful, attempts to propagate *T. pallidum* continuously *in vitro*. Finally, discoveries from genomic studies are addressed in the context of solving the cultivation problem of *T. pallidum.* Although metabolically *T. pallidum* and *T. denticola* share many common pathways, they are still quite diverse in their genomic composition. *T. denticola* has almost 2000 genes that are absent in *T. pallidum*; many of these are predicted to encode proteins with physiological functions. Three areas where the latter appears to be lacking when compared to *T. denticola* are (1) DNA repair and housekeeping, (2) oxygen metabolism and stress, and (3) transport capabilities.

1. Introduction

Soon after the discovery of the syphilis spirochete, *Treponema pallidum*, in 1905 by Schaudinn and Hoffman (1904-05; 1905), attempts were made to cultivate this obligate human pathogen. With no knowledge of this fastidious organism's metabolism, this proved to be an impossible task. However, certain assumptions were made, and many were appropriate. The first of these involved

*For correspondence email dlc6@cdc.gov

energy production by the spirochete. Scientists could make the assumption that *T. pallidum*, being a human pathogen, would use glucose, readily present in blood and most tissue fluids, as at least one major source of energy. Attempts at cultivation of the syphilis spirochete were made by many and some even claimed to have succeeded (Schereschewsky and Hoffman, 1909; Noguchi 1911; 1912; Kast 1929), but the only treponemes cultivated by these investigators were non-pathogens; the virulent syphilis treponeme continued to elude cultivation. For more than 75 years, efforts to elucidate the metabolism of the syphilis spirochete were impeded by a terrible "catch-22": if one cannot cultivate the organism, how can one study its metabolism; and if one knows very little about its metabolic requirements, how can it be cultivated? Despite the demonstration of limited multiplication in 1980 (Fieldsteel et al., 1981), continuous *in vitro* cultivation has not been accomplished. Nonetheless, much has been learned about *T. pallidum* metabolism through experimental and genomic studies.

Additional sections of the chapter will be devoted primarily to two cultivable treponemes, *Treponema denticola* and *Treponema phagedenis*, which have been studied extensively over the years. Of special interest is the comparison of the genome and metabolism of *T. denticola*, a commensal treponeme, with that of the obligate human pathogen *T. pallidum*. The availability of the *T. pallidum* and *T. denticola* genomes should enable us to determine eventually which genes and metabolic capabilities present in *T. denticola*, but absent in *T. pallidum*, enable the former to replicate *in vitro* while the latter is incapable of sustained replication under similar conditions.

2. The Metabolism of *Treponema pallidum*

2.1. Carbohydrate energy sources

Glucose or pyruvate is essential in maintaining the viability of *T. pallidum*. Many of the early cultivation media contained high concentrations of serum or blood products which contained glucose (c.100 mg per dl). For example, Boak and Miller (1954) described a medium that contained 50% rabbit serum to replace Nelson's medium (Nelson and Steinman, 1948); their formulation contained serum ultra filtrate, pyruvate, and crystalline bovine serum albumin. Weber (1960) also noted that increased survival of treponemes was directly correlated to moderate concentrations (15%) of ox serum ultra filtrate. However, the first detailed, definitive study on potential carbon sources utilized by *T. pallidum* was reported by Nichols and Baseman (1975). By using radiolabeled compounds, they demonstrated that glucose, but not galactose, ribose, and fructose could be utilized. Likewise, with respect to short chain organic acids, pyruvate and formate were readily catabolized while acetate, succinate, citrate, lactate or α-ketoglutarate were not. Thus, it was demonstrated that the organism had the metabolic machinery to degrade glucose and pyruvate, but the exact pathways were not elucidated. Schiller and Cox (1977) later determined that glucose was catabolized by the Embden-Meyerhoff-Parnas and pentose phosphate pathways, producing lactate, acetate, and CO_2 as the final end products. Following that study, Barbieri and Cox (1979) reported that pyruvate was broken down to acetyl phosphate and CO_2 via a pyruvate decarboxylase activity. The presence of phosphotransacetylase, acetate kinase, and phosphoenolpyruvate carboxykinase activities in cell-free extracts of *T. pallidum* was also reported, and later pyruvate decarboxylase was demonstrated to be inhibited by high O_2 concentrations (Barberi 1981).

One important question that remained unanswered was whether *T. pallidum* has a functional Krebs cycle and electron transport system. Studies by Cox and Barber (1974) and Lysko and Cox (1977) had previously supported their presence by demonstrating that *T. pallidum* extracted from rabbit testicular tissue consumed O_2 as rapidly as the aerobic spirochete *Leptospira interrogans.* In addition, oxygen consumption was inhibited by cyanide, azide, and other electron transport uncouplers. Yet this finding presented a paradox because for decades syphilis researchers recognized that *T. pallidum* required reducing agents in the medium to protect it from atmospheric concentrations of oxygen and that, without reducing agents, it died rapidly under aerobic conditions. In 1976, Baseman et al. (1976) extended earlier studies by showing inhibition of *T. pallidum* metabolism by carbon monoxide and high levels of cyanide, suggesting that the spirochete might have only a partial electron transport system. Lysko and Cox (1977; 1978) extended these studies by ostensibly demonstrating the presence of cytochromes in treponemal extracts via spectrum analysis and by showing erroneously that oxygen consumption and energy production were coupled to a NADH dehydrogenase activity that was coupled to a putative electron transport chain.

The complete sequence of the genome of *T. pallidum*, published in 1997 (Fraser et al., 1997), has greatly clarified our understanding of the metabolic capability of *T. pallidum* (summarized in Tables 1-9). All of the genes encoding for enzymes of the Embden-Meyerhoff-Parnas pathway are present. As expected, genes for the fermentative enzymes (lactate, alcohol, and glycerate dehydrogenases) associated with the regeneration of oxidized NAD+ from pyruvate, the end-product of glycolysis, are absent (Table 1). The gene for the enzyme phosphoenolpyruvate carboxykinase (TP0122) that converts phosphoenolpyruvate (PEP) to oxalacetate (OAA) is present. It is presumed that PEP can be converted to OAA and that OAA can then be transaminated to aspartic acid at the expense of glutamine or glutamic acid. All of the genes for the enzymes of the pentose phosphate pathway are present except transaldolase (EC 2.2.1.2). Furthermore, only two enzymes of the Entner-Doudoroff pathway are present, glucose 6-phosphate dehydrogenase (TP0477-8), which is also present in the pentose phosphate pathway, and 2-dehydro-3-deoxy-phosphogluconate aldolase (TP0568). Lastly, none of the genes encoding Krebs cycle enzymes or heme-proteins involved in electron transport was identified (Table 1). The genomic sequence confirmed the report by Barbieri and Cox (1979; 1979b) that the terminal reaction of glucose catabolism in *T. pallidum* is the acetate kinase reaction (TP0476). Thus *T. pallidum* must obtain all of its ATP from substrate level phosphorylation reactions. In retrospect, it is now quite apparent that the discrepancies between pre- and post-genomic studies probably were due to contamination of the treponemal suspensions and extracts with minute amounts of testicular tissue components.

Separate from the genome project, Benoit et al. (2001) examined the effect of metals such as Mn^{+2} on the regulation of 3–phosphoglycerate mutase (Pgm). Because Pgm is a key enzyme in the Embden-Meyerhoff-Parnass pathway, its regulation could be very critical for energy production in *T. pallidum.* It was

Table 1. Carbohydrate metabolism

	Genome Designation		
	TP	**TDE**	**Enz Com #**
Glycolysis (Embden-Meyerhoff-Parnass)			
hexokinase	0505	2469	EC 2.7.1.1
phosphoglucoisomerase	0475	2573	EC 5.3.1.9
phosphofructokinase	0108	0064	EC 2.7.1.11
fructose-bisphosphate aldolase	0662	0340	EC 4.1.2.13
triose phosphate isomerase	0537	1236	EC 5.3.1.1
Gly-3-PO4 dehydrogenase	0844	1488	EC 1.2.1.12
phosphoglycerokinase	0538	1715	EC 2.7.2.3
phosphoglyceromutase	0168	1697	EC 5.4.2.1
enolase	0817	0949	EC 4.2.1.11
pyruvate kinase	0746	2535	EC 2.7.1.40
pyruvate-ferredoxin oxidoreductase	–[1]	2198	EC 1.2.7.1
pyruvate oxidoreductase	0939	–	EC 1.2.7.1
phosphoenolpyruvate carboxykinase	0122	–	EC 4.1.1.38
phosphoenolpyruvate-protein phosphotransferase	0575	0084	EC 2.7.3.9
phosphoenolpyruvate synthase, putative	–	0349	EC 2.7.9.2
lactate DH	–	0351	EC 1.1.1.27
alcohol DH	–	0051	EC 1.1.1.1
glycerate DH	–	1616	EC 1.1.1.29
phosphoglucomutase (Gl-6-PO4 --> Gl-1-PO4)	0413	–	EC 5.4.2.6
acetate kinase	0476	0933	EC 2.7.2.1
Pentose Phosphate Pathway			
Glucose 6-PO4 DH	0477-8	–	EC 1.1.1.49
lactonase[2]	0478??	–	EC 3.1.1.17
6-phosphonogluconate DH	0331	–	EC 1.1.1.44
phosphopentose epimerase	0945	2597	EC 5.1.3.1
phosphopentose isomerase	0616	0352	EC 5.3.1.6
Transketolase	0560	1308	EC 2.2.1.1
transaldolase	–	–	EC 2.2.1.2
phosphoglucoisomerase	0475	2573	EC 5.3.1.9
fructose-bisphosphate aldolase	0662	0340	EC 4.1.2.13
Entner- Duodoroff Pathway			
Glucose 6-PO4 DH	0477-8	–	EC 1.1.1.49
6-phosphogluconolactonase	–	–	EC 3.1.1.31
phosphogluconate dehydrase	–	–	EC 4.2.1.12
2-dehydro-3-deoxy-phosphogluconate aldolase	0568	–	EC 4.1.2.14

Table 1. Continued

	Genome Designation		
	TP	TDE	Enz Com #
Kreb's Cycle			
citrate synthase	–	–	EC 2.3.3.1
aconitase	–	–	EC 4.2.1.3
isocitrate DH	–	–	EC 1.1.1.41
a-ketoglutarate DH	–	–	EC 1.2.4.2
succinyl CoA synthase	–	–	EC 6.2.1.4
succinate DH	–	–	EC 1.3.5.1
fumarate hydrase	–	–	EC 4.2.1.2
malate dehydrogenase	–	2446	EC 1.1.1.37
glutamate dehydrogenase	–	0997	EC 1.4.1.2
Glyoxylate Shunt			
malate dehydrogenase	–	2446	EC 1.1.1.37
citrate synthase	–	–	EC 2.3.3.1
malate synthase	–	–	EC 2.3.3.9
isocitrate lyase	–	–	EC 4.1.3.1
aconitase hydratase	–	–	EC 4.2.1.3

[1] No homologous gene present.
[2] putative activity because Tp0477 is Glucose 6-PO4 DH

discovered that Pgm did not require Mn^{+2} for activity but was heat-labile, losing 50% of its activity in 10 min at 37°C. It was speculated that the thermal lability of the enzyme could be responsible for the syphilis spirochete's well-recognized temperature sensitivity.

2.2. Amino acid and protein metabolism

The first modern study on protein synthesis in *T. pallidum* was reported in 1974 by Baseman and Hayes (1974). Using C^{14} labeled amino acids, they demonstrated that maximum protein synthesis occurred at 34° F, a pH of 7.6, and a dO_2 concentration of 10 -20%; this exceptionally high level of dO_2 later was proven to be erroneous (see below). Furthermore, incorporation of radiolabel was linear over a 24-hour period, even though motility of the treponemes was decreasing. The measured anabolic activity was not inhibited by cycloheximide, a potent eukaryotic protein synthesis inhibitor. They also demonstrated the importance of serum to stimulate treponemal protein synthesis; in the absence of serum, little synthesis was observed. In addition to total protein synthesis, their rates of incorporation of 14 specific amino acids into protein were tested. The amino acids fell into three distinct groups: serine and valine (Group 1); glycine, aspartic acid, threonine, leucine, isoleucine, and alanine (Group 2); and arginine, phenylalanine, tyrosine, glutamic acid, lysine and

Table 2. Amino acid metabolism and proteases

	Genome Designation		
	TP	**TDE**	**Enz Com #**
Transaminase			
Aspartate aminotransferase	0223		EC 2.6.1.1
V type	–[1]	0103	EC 2.6.1.X
V type	–	0679	EC 2.6.1.X
V type	–	2052	EC 2.6.1.X
V type	–	2460	EC 2.6.1.X
DegT/DnrJ/EryC1/StrS family	–	1426	EC 2.6.1.X
DegT/DnrJ/EryC1/StrS family	–	1590	EC 2.6.1.X
Glycine Utilization			
glycine reductase selenoproteins	–	0078-9	EC 1.21.4.2
glycine reductase selenoproteins	–	0239-40	EC 1.21.4.2
glycine reductase selenoproteins	–	0745	EC 1.21.4.2
glycine reductase selenoproteins	–	2119-20	EC 1.21.4.2
glycine cleavage system	–	1624-7	EC 1.4.4.2
Proline Biosynthesis			
glutamate-5-semialdehyde dehydrogenase	0350	–	EC 1.2.1.41
glutamate 5-kinase	0351	–	EC 2.7.2.11
1-pyrroline-5-carboxylate reductase	0797	–	EC 1.5.1.2
Tryptophan Utilization			
tryptophanase	–	0251	EC 4.1.99.1
Phenylalanine Utilization			
tyrosine phenol lyase	–	1118	EC 4.1.99.2
Ribose Activation			
adenine phosphoribosyltransferase	1039	1684	EC 2.4.2.7
Arginine Utilization			
ornithine carbamoyltransferase	–	0929	EC 2.1.3.3
arginase	–	–	EC 3.5.3.1
argininosuccinate lyase	–	–	EC 4.3.2.1
argininosuccinate synthase	–	–	EC 6.3.4.5
arginine deiminase	–	0451	EC 3.5.3.6
Shikimate and Chorismate Biosynthesis			
3-deoxy-7-phosphoheptulonate synthase	–	2569	EC 2.5.1.54
3-dehydroquinate synthase	–	0104	EC 4.2.3.4
3-dehydroquinate dehydratase	–	1640	EC 4.2.1.10

Table 2, Continued

	Genome Designation		
	TP	TDE	Enz Com #
shikimate 5-dehydrogenase	–	1640	EC 1.1.1.25
shikimate kinase	–	2447	EC 2.7.1.71
3-phosphoshikimate 1-carboxyvinyltransferase	–	1686	EC 2.5.1.19
chorismate synthase	–	1132	EC 4.2.3.5
isochorismate synthase	–	–	EC 5.4.4.2
Proteases (some putative)			
general protease	–	0546	NA[2]
general protease	–	0773	NA[2]
general protease	–	0841	NA[2]
general protease	–	0997	NA[2]
general protease	–	1343	NA[2]
general protease	–	1966	NA[2]
Hemolysins (cystalysin)	–	1669	NA[2]

[1] No homologous gene present.
[2]No EC number assigned.

histidine (Group 3). Amino acids in Group 1 were incorporated about four times more rapidly than those of Group 3 and about twice as rapidly as those in Group 2. In 1977, Baseman and Hayes (1977) followed this study by publishing the protein profile of *T. pallidum* incubated with tritiated amino acids, comparing it with that obtained with *T. phagedenis*. This comparison demonstrated that *T. pallidum* possessed considerable anabolic competence with respect to protein synthesis even though there was a rapid loss of motility over the 24 hour observation period.

A year later, Sandok and Jenkin (1978) published important observations on protein synthesis carried out over an extended period of incubation in a medium designed to sustain treponemal motility and viability. Their rationale for reinvestigating treponemal protein synthesis was that rapidly degenerating treponemes might exhibit abnormal metabolism and, therefore, short-term experiments could yield erroneous results. The study demonstrated a linear uptake of amino acids over a 96-hour period that was inhibited by erythromycin. However, erythromycin did not affect the motility of the treponemes during this same period. These results differed from those of Baseman and Hayes (1977) who were unable to detect protein synthesis under anaerobic conditions. Sandok and Jenkin did, however, confirm the observation that motility does not directly parallel treponemal biosynthetic activity. Methods to refine and improve treponemal protein labeling with [^{35}S]-methionine were later reported by both Thornburg and Baseman (1983) and Stamm and Bassford (1985). In the latter study, the efficiency of methionine incorporation was reported to be much higher than reported in previous studies. Lastly, after Fieldsteel et al. (1981) had achieved successful *in vitro* replication of *T. pallidum*, Cox (unpublished) reproduced the Baseman and Hayes study. In these co-cultures with Sf1Ep cells, the treponemes were monitored over a 12-14 day period during which they were dividing. There was a lag of approximately 2-3 days before any radiolabel could be detected in the treponemal fraction (presumably reflecting the 100-fold lower concentration of treponemes in the cultures vs. the suspensions used earlier). Thereafter, uptake of ^{14}C-amino acids was linear until treponemal numbers had peaked; the rates of uptake for each of the amino acids were very similar to those reported earlier (Baseman and Hayes, 1977).

The genome of *T. pallidum* revealed more precisely the spirochete's limited biosynthetic capacity for amino acids (Table 2). It does possess the ability to transaminate oxaloacetic acid (OAA) to form aspartic acid via aspartate aminotransferase (TP0223). Aspartate can be transaminated to make asparagine by asparagine synthetase (TP0556). There are three genes encoding for proline biosynthesis (Tp0350, TP0351, and TP0797, Table 2) which encode glutamate-5-semialdehyde dehydrogenase, glutamate 5-kinase, and pyrroline-5-carboxylate reductase, respectively. Furthermore, Gherardini et al. (1990) had previously demonstrated this capability by complementing a null mutation in *E. coli* for *proC* which encodes a 1-pyrroline-5-carboxylate reductase (TP0797). Thus, *T. pallidum* can synthesize proline from glutamate or glutamine but cannot synthesize arginine, citrulline or ornithine, three other members of the glutamic acid family of amino acids. The genomic sequence also revealed that *T. pallidum* had the genetic competency to synthesize serine and glycine from 3-phosphogylcerate, but not the genes necessary to encode for enzymes needed to synthesize cysteine (Fraser et al., 1998). However upon detailed inspection, *T. pallidum* appears to lack the enzymes phosphoserine phosphatase (EC 3.1.3.3) and phosphoserine aminotransferase (EC 2.6.1.52). *T. pallidum* also lacks the genes necessary for the biosynthesis of the pyruvate family of amino acids (alanine, valine, and leucine). Lastly, it does not have the genes necessary for the biosynthesis of any of the aromatic amino acids (tryptophan, histidine, or phenylalanine).

2.3. Fatty acid and lipid metabolism

It has long been recognized that *T. pallidum* requires an exogenous source of fatty acids and lipids for viability. As previously mentioned, one common ingredient in most of the cultivation efforts was the inclusion of serum or other blood products (Nelson and Steinman, 1948; Boak and Miller, 1954; Weber 1960; Chalmers and Taylor-Robinson, 1979). Albumin, which is known to function in the transportation of fatty acids and some serum lipids, is the most abundant protein in serum. Nelson and Diesendruck (1951) actually demonstrated that oleic and butyric acids increased the survival of *T. pallidum in vitro*. During their metabolic studies, Nichols and Baseman (1975) did not detect the release of CO_2 during the degradation of radiolabeled oleic or palmitic acids. This was confirmed by Schiller and Cox (1977), who also reported the absence of enzymes associated with β-oxidation. These results suggested that *T. pallidum* has a fatty acid metabolism similar to that of nonpathogenic treponemes, which are unable to catabolize fatty acids either for energy or for use as carbon skeletons.

Several studies have examined the detailed fatty acid composition of *T. pallidum*. Livermore and Johnson (1974) determined that the lipid composition of *T. pallidum* was 41% phospholipids, 27% glycolipids, and 22% neutral lipids. Upon further

Table 3. Lipid metabolism

	Genome Designation		
	TP	TDE	Enz Com #
β-oxidation of FAs			
Fatty acyl-CoA DH	–[1]	–[1]	EC 1.3.99.3
enoyl-CoA hydratase	–	–	EC 4.2.1.17
3-Hydroxyacyl-CoA DH	–	–	EC 1.1.1.35
β-ketothiolase	–	–	EC 2.3.1.16
Anabolic Metabolism			
long chain fatty-acid--CoA ligase	0145	0942	EC 6.2.1.3
long chain fatty-acid--CoA ligase	0812	–	EC 6.2.1.3
lysophosphatidic acid acyltransferase	0361	1708	EC 3.1.1.3
lysophosphatidic acid acyltransferase	–	2299	EC 3.1.1.3
biotin--acetyl-CoA-carboxylase ligase	0357	–	EC 6.3.4.15
sn-1,2-diacylglycerol cholinephosphotransferase, putative	0671	–	EC 2.7.8.2
prolipoprotein diacylglyceryl transferase	0852	1399	EC 3.4.23.36
prolipoprotein diacylglyceryl transferase, putative	–	0030	EC 3.4.23.36
lipoyltransferase and lipoate-protein ligase family protein	–	1551	EC 3.4.23.36
lipoyltransferase and lipoate-protein ligase family protein	–	2647	EC 3.4.23.36

[1] No homologous gene present.

analysis, the presence of phosphatidylcholine (PC), phosphatidylinisitol, and lysophosphatidylcholine was detected. Using treponemes purified on Hypaque gradients, Matthews et al. (1979; 1979b) repeated this study and found that 68% of the lipid composition was phospholipids, while 32% was neutral lipids; no glycolipids were detected. The phospholipids were primarily phosphatidylcholine (PC), cardiolipin and sphingomyelin. In addition, phosphatidylinisitol/serine, phosphatidylethanolamine, and lysophosphatidylcholine also were detected in small quantities. Fatty acid analysis revealed that the most prevalent fatty acid was palmitate, followed by stearate, oleate and linoleate. They also noted that the fatty acid composition of *T. pallidum* was very similar to that of the infected testis, reinforcing the concept that *T. pallidum* was incapable of fatty acid synthesis. Belisle et al. (1994) subsequently investigated the lipid composition of *T. pallidum* and its ability to incorporate fatty acids into lipids. Fatty acids profiles of *T. pallidum* were similar to those reported by Matthews et al., (1979) as were the relative concentrations of fatty acids (palmitate > stearate > oleate > linoleate). However, only trace amounts of linoleic acid were detected. Treponemes labeled with either tritiated palmitate or oleate contained tritiated labeled phosphatidyl glycerol, phosphatidyl choline, phosphatidyl ethanolamine, phosphatidyl serine, phosphatidyl inisitol, and cardiolipin.

The genomic sequence confirmed the prediction from pre-genomic studies that *T. pallidum* lacks the ability to synthesize fatty acids and cannot carry out the β-oxidation of fatty acids (Table 3). *T. pallidum* contains two enzymes associated with the cleavage and utilization of pre-formed fatty acids. These are long-chain-fatty-

Table 4. Nucleic acid metabolism

	Genome Designation		
	TP	**TDE**	**Enz Com #**
Ribose Activation			
adenine phosphoribosyltransferase	1039	1684	EC 2.4.2.7
Pyrimidine Biosynthesis			
carbamate kinase	–[1]	2476	EC 2.7.2.2
carbamoyl-phosphate synthase (glutamine)	–	–	EC 6.3.5.5
aspartate carbamoyltransferase	–	–	EC 2.1.3.2
dihydroorotase	–	2107	EC 3.5.2.3
dihydroorotate dehydrogenase/oxidoreductase, FAD	–	0982	EC 1.3.1.14
orotate phosphoribosyltransferase	–	1001	EC 2.4.2.10
orotidine-5'-phosphate decarboxylase	–	2110	EC 4.1.1.23
Purine Biosynthesis			
phosphoribosylamine—glycine ligase	–	1267	EC 6.3.4.13
phosphoribosylglycinamide formyltransferase	0695	1897	EC 2.1.2.2
phosphoribosylformylglycinamidine synthase	–	1279,1405	EC 6.3.5.3
phosphoribosylformylglycinamidine cyclo-ligase	–	1266	EC 6.3.3.1
phosphoribosylaminoimidazole carboxylase	–	0687	EC 4.1.1.21
phosphoribosylaminoimidazolesuccinocarboxamide synthase	–	0686	EC 6.3.2.6
adenylosuccinate lyase	–	1499	EC 4.3.2.2
phosphoribosylaminoimidazolecarboxamide formyltransferase	–	0992?	EC 2.1.2.3
IMP cyclohydrolase	–	1896	EC 3.5.4.10
amidophosphoribosyltransferase	–	2407	EC 2.4.2.14
adenine phosphoribosyltransferase	1039	1684	EC 2.4.2.7
Nucleotide Pool Enzymes			
nucleoside-diphosphate kinase	1010	–	EC 2.7.4.6
adenylate kinase	0595	1112	EC 2.7.4.10
cytidylate kinase	0279	1087	EC 2.7.4.14
thymidylate kinase	0354	2599	EC 2.7.4.9
uridine kinase	0667	2289	EC 2.7.1.48
uridylate kinase	0099	–	NA[2]
thymidine kinase	–	–	EC 2.7.1.21

[1] No homologous gene present.
[2]No EC number assigned.

acid-CoA ligase (TP0145) and biotin--acetyl-CoA-carboxylase ligase (TP0357). The presence of these enzymes confirmed previous observations that the fatty acid profile of *T. pallidum* closely resembles that of host tissue (Matthews 1979; Belisle et al., 1994), and confirms *T. pallidum's* dependence upon a host cell or a supplemental source of fatty acids for growth. Kent et al. (2004) recently examined the genomes of *T. pallidum* and *T. denticola* for PC biosynthesis pathways and found that they both contained a putative gene *licCA* which encodes for a fusion protein containing choline kinase and CTP:phosphocholine cytidylyltransferase activities. Both enzymatic functions were demonstrated in *T. denticola* (see below), and it was suggested that TP0671 was a *licCA* ortholog in *T. pallidum*. The enzymatic function of the *T. pallidum* gene product has yet to be confirmed.

2.4. Nucleic acid metabolism

In 1978, Nichols and Baseman (1978) investigated the ability of *T. pallidum* to synthesize nucleic acids from radiolabeled precursors. RNA synthesis was monitored by uridine incorporation and was linear for the first 12 hours of a 48-hour incubation period. The treponemes remained motile over the entire incubation. In addition, treponemal RNA biosynthesis was inhibited by actinomycin D. On further investigation using pulse-labeling, it was determined that most of the incorporation occurred during the first 3 hours and rapidly decreased thereafter, demonstrating a rapid shut down of RNA synthesis long before the loss of motility. This observation again reinforced the concept that motility is a poor indicator of metabolic activity. In other studies on RNA synthesis, Leschine and Canale-Parola (1980) determined that many treponemal RNA polymerases were resistant to rifampin and that this antibiotic could be included in media used for the isolation of *T. pallidum* from infected tissues.

In a complementary study, Baseman et al. (1979) monitored DNA synthesis in *T. pallidum* and reported that the organism did not take up tritiated thymidine and lacked a thymidine kinase (EC 2.7.1.21). However, DNA was synthesized from tritiated uridine, demonstrating the treponeme's capacity for nucleotide biosynthesis. In parallel studies, *T. phagedenis* behaved in exactly the same manner. They also witnessed the same phenomenon as previously described for RNA synthesis, namely that motility and DNA biosynthesis were totally independent of each other. The rate of DNA replication was also explored in this study by measuring DNA polymerase activity in treponemal extracts. They speculated that the low levels of DNA polymerase found in *T. pallidum* extracts when compared to *T. phagedenis* and other typical bacteria correlated with *T. pallidum*'s longer generation time (33-35 hours).

One relatively recent interesting finding by Rhodes et al. (2000) was that the DNA polymerase I (*PolA*) of *T. pallidum* is unique. The cysteine content of the protein is 2.4% (24 residues) compared to an average of <0.05% (0-2 residues) for the ortholog in most other bacteria. This unique composition suggests that a highly reduced redox state may be required for native enzymatic activity. *PolA* could be expressed in *E. coli* but not in an active form that could complement *polA*-deficient *E. coli* mutants. The redox properties of this enzyme suggest additional explanations for the oxygen sensitivity of *T. pallidum* and why its chromosomal DNA is exquisitely sensitive to oxygen radicals.

Genomics validates the previous studies (Baseman et al., 1979) that *T. pallidum* lacks a thymidine kinase (EC 2.7.1.21) but can synthesize DNA from uridine nucleotides (Table 4). Both uridine kinase and uridylate kinase (TP0667 and TP0099) are present as well as a thymidylate kinase (TP0354). *T. pallidum* possesses the necessary enzymes to synthesize DNA, *polA* (TP0105), and three

Table 5. Peptidoglycan synthesis

	Genome Designation		
	TP	TDE	Enz Com #
UDP-N-acetylglucosamine 1-carboxyvinyltransferase	0029	641	EC 2.5.1.7
UDP-N-acetylmuramate dehydrogenase	0090	–[1]	EC 1.1.1.158
UDP-N-acetylmuramate--alanine ligase	0341	1085	EC 6.3.2.8
UDP-N-acetylmuramoylalanine—D-glutamate ligase	0903	370	EC 6.3.2.9
UDP-N-acetylmuramoylalanyl-D-glutamate—2,6-diaminopimelate ligase	0933	835	EC 6.3.2.13
UDP-N-acetylmuramoyl-L-alanyl-D-glutamate—L-lysine ligase	—	—	EC 6.3.2.7
UDP-N-acetylmuramoyl-tripeptide—D-alanyl-D-alanine ligase	0386	1200	EC 6.3.2.10
phospho-N-acetylmuramoyl-pentapeptide-transferase	0345	2249	EC 2.7.8.13
undecaprenyldiphospho-muramoylpentapeptide b-N-acetylglucosaminyltransferase	0523	1974	EC 2.4.1.227
peptidoglycan transglycosylase	0043	2149	EC 2.4.1.129
peptidoglycan transglycosylase	0623	2280	EC 2.4.1.129
undecaprenyl-diphosphatase	—	2344	EC 3.6.1.27
octaprenyl-diphosphate synthase	0683	—	EC 2.5.1.11

[1] No homologous gene present.

polymerase III genes (TP0069, TP0002, and TP0521). In addition a gyrase (TP0005), originally reported by Stamm et al. (1997) and a topoisomerase (TP0394) are also present. What appears to be minimal is the number of DNA recombination and repair (rec) genes; only recN (TP0442), recG (TP0687), recJ (TP0704), and sms (TP1022) are present as well as a gene for recA (TP0692) which was not in agreement with Steiner et al. (1984) previous predictions, but was suggested by Stamm et al. (1991). In addition, there are several helicase genes for putative DNA (TP0058, TP0162, TP0380, and TP1028) and RNA repair (TP0770).

2.5. Miscellaneous Nutrients

T. pallidum has traditionally been recovered from infected tissues using solutions whose base includes physiological saline and serum. Concentrations of serum as low as 10-20% should provide essential vitamins and minerals. Few studies have concentrated on identifying these requirements or determining the minimal concentrations necessary for treponemal survival. One study by Doak et al.. (1959) suggested that physiological concentrations of sodium might be toxic for *T. pallidum* and that it should be replaced with potassium, a putative "growth

Table 6. Enzymes involved in maintenance of redox state

	Genome Designation		
	TP	TDE	Enz Com #
NADH oxidase	0921	0096	EC 1.11.1.1
superoxide dismutase	–[1]	—	EC 1.15.1.1
catalase	—	—	EC 1.11.1.6
peroxidase, glutathione	—	1729	EC 1.11.1.9
neelaredoxin	0823	1754	EC 1.15.1.2
alkyl hydroperoxide reductase	0509	0011	EC 1.11.1.15
Thioredoxin	0100	0238	NA[2]
Thioredoxin	0919	0744	NA[2]
Thioredoxin reductase	0814	0743	EC 1.8.1.9
bacteroferretin**	1038	—	NA[2]
ferredoxin	—	1736	NA[2]
flavodredoxin	—	0249	NA[2]
FAD- oxoreductase	—	0134	EC 1.2.7.7
FAD- oxoreductase	—	0868	EC 1.2.7.7
FAD- oxoreductase	—	2643	EC 1.2.7.7
rubrerythin	—	0434	EC 1.11.1.7
oxidoreductase	—	0685	NA[2]
nitroreductase	—	0708	NA[2]
nitroreductase	—	1363	NA[2]

[1] No homologous gene present.
[2] No EC number assigned.

factor". This concept was based upon a study by Eagle (1956) which demonstrated that sodium ions could inhibit anaerobic glycolysis. They observed the optimum survival of *T. pallidum* at sodium and potassium ion concentrations of 0.29 M and 0.01M, respectively.

Weber (1960) investigated the effects of several cations on the survival of *T. pallidum*, including sodium, potassium, and ammonium chloride. It was demonstrated that sodium was essential and could not be replaced by potassium; however, sodium could be replaced by ammonium ions. It also was demonstrated that phosphate was required for long-term survival and vitamin supplements could stimulate treponemal survival.

Since many of these nutrients can come from serum or testicular extracts, quantitation has been difficult because *T. pallidum* does not survive for long periods, and definitely does not grow, in serum-free media. One study, mentioned previously, investigated which fraction of serum was responsible for providing those factors (Norris and Edmondson, 1986). Essential nutrients for treponemal survival and growth were determined to be present in the protein fraction more than the soluble ultra-filtrate fraction. However, these studies were performed in tissue culture medium which is relatively rich in vitamins and other nutrients.

Table 7. DNA housekeeping and repair

	Genome Designation		
	TP	TDE	Enz Com #
DNA polymerase I	0105	1637	EC 2.7.7.7
DNA polymerase III -alpha	0669	2786	EC 2.7.7.7
DNA polymerase III-beta	0002	0231	EC 2.7.7.7
DNA polymerase III -gamma	0521	2586	EC 2.7.7.7
DNA polymerase IV	–[1]	0963	EC 2.7.7.7
DNA gyrase	0005	0002	EC 5.99.1.3
DNA gyrase	—	0295	EC 5.99.1.3
DNA helicase	0380	1990	NA[2]
DNA helicase dnaB	0058	1662	NA[2]
DNA helicase UvrD	1028	0085	NA[2]
DNA helicase UvrD		2335	NA[2]
DNA helicase ruvB	0162	1914	NA[2]
RNA helicase	0770	2109	NA[2]
DNA topoisomerase	0394	1208	EC 5.99.1.2
Repair- Rec A	0692	0872	NA[2]
Repair- Rec F	—	0230	NA[2]
Repair- Rec Q	—	0326	NA[2]
Repair- Rec Q family	—	0435	NA[2]
Repair- Rec N	0442	1301	NA[2]
Repair- Rec J	0704	2190	EC 3.1.-. -
Repair- Rec G	0687	2579	EC 3.6.1.-
sms	1022	—	NA[2]
Other repair damage	—	0973	NA[2]
Protein J	—	0506	NA[2]
Protein J	—	1573	NA[2]
DNA modification (13)	—	0099	EC 3.1.21.3
DNA modification	—	0299	EC 3.1.21.3
DNA modification	—	0366	EC 3.1.21.3
DNA modification	—	0368	EC 3.1.21.3
DNA modification	—	0369	EC 3.1.21.3
DNA modification	—	0706	EC 3.1.21.3
DNA modification	—	0909	EC 3.1.21.3
DNA modification	—	1309	EC 3.1.21.3
DNA modification	—	1310	EC 3.1.21.3
DNA modification	—	2176	EC 3.1.21.3
DNA modification	—	2744	EC 3.1.21.3
DNA modification	—	2746	EC 3.1.21.3
DNA modification	—	2747	EC 3.1.21.3
DNA binding Proteins (12)	—	0033	NA[2]
DNA binding Proteins	—	0079	NA[2]
DNA binding Proteins	—	0127	NA[2]
DNA binding Proteins	—	0440	NA[2]

Table 7, Continued

	Genome Designation		
	TP	TDE	Enz Com #
DNA binding Proteins	—	0496	NA[2]
DNA binding Proteins	—	0608	NA[2]
DNA binding Proteins	—	1115	NA[2]
DNA binding Proteins	—	1170	NA[2]
DNA binding Proteins	—	1305	NA[2]
DNA binding Proteins	—	1610	NA[2]
DNA binding Proteins	—	1647	NA[2]
DNA binding Proteins	—	1709	NA[2]
DNA binding Proteins	—	1912	NA[2]
Prevent Host Death	—	0225	NA[2]
Prevent Host Death	—	1455	NA[2]
Stress Response (4)	—	0012	NA[2]
Stress Response	—	0354	NA[2]
Stress Response	—	0655	NA[2]
Stress Response	—	1210	NA[2]

[1] No homologous gene present.
[2]No EC number assigned.

2.6. Putative synthesis and secretion of extracellular products

A major controversy in the late 1970s and early 1980s centered around the ability of *T. pallidum* to synthesize mucopolyscaccharide (glycosaminoglycans) material found in many infected testis extracts. Several publications by Fitzgerald and co-workers (1978; 1979; 1979b; 1979c) suggested that *T. pallidum* synthesized this material and they demonstrated its presence in tissue cultures of normal rabbit testis cells. The material could be stained with ruthenium red and was degraded by both bovine and *Streptomyces* hyaluronidase. It was suggested that this material may function as a capsule, providing protection against host defenses, and also exerting immunosuppressive activities. In further studies, it was also claimed that suspensions of viable *T. pallidum* exhibited hyaluronidase activity that was not present in heat-inactivated or filtered suspensions. Furthermore, *T. pertenue* suspensions had similar activities, but that cultivable *T. denticola* and *T. vincentii* did not. It was proposed that this putative treponemal hyaluronidase altered capillary permeability, facilitating dissemination of spirochetes and invasion of tissues (Fitzgerald and Johnson, 1979; Fitzgerald et al., 1982; Fitzgerald, 1987). Subsequent studies disputed the origin of both the mucopolysaccharide and the hyaluronidase. Tissue cultures of Sf1Ep cells containing replicating *T. pallidum* were stained with ruthenium red and observed by electron microscopy (Konishi 1984). Only very light staining was observed, and this could have been due to the agar overlay employed in the fixation process; definitive mucopolysaccharide production was not detected over a 12-day period. Two additional studies provided more convincing evidence that the material was of host and not treponemal origin. First, Strugnell et al. (1986) were not able to detect the uptake of radiolabeled sulfate, a precursor of glycosaminoglycans. The uptake and incorporation of another

Table 8. Secretory pathway and secreted products

	Genome Designation	
	TP	TDE
SecA	0379	1898
SecB	–[1]	—
SecD	0410	0681
SecE	0235	2427
SecF	0411	0680
SecG	—	1238
SecY	0208	0787
MFP	—	2284
YajC	—	0682
TolC	—	—
hlyB (8)	—	0028-0029
hlyB (8)	—	0185
hlyB (8)	—	0234
hlyB (8)	—	0315
hlyB (8)	—	1061-2
hlyB (8)	—	2000
VirB	—	—
YscC	—	—
MATE efflux	—	0333
MtrF	—	2328
Macrolide efflux	—	2654
Bacteriocin	—	0416
Bacteriocin	—	0422
Bacteriocin	—	0424
Bacteriocin	—	0425
Bacteriocin	—	0719-20
Bacteriocin	—	2430-1
Proteases[2]		

[1] No homologous gene present.

[2]Proteases and hemolysins are listed in Table 2.

precursor, glucosamine, into an unknown molecule was detected, but that product was not degraded by specific glycosaminoglycanases. These authors concluded that *T. pallidum* does not synthesize a capsule containing glycosaminoglycans. In the second study (Strugnell et al., 1988), the same group demonstrated by monitoring the uptake of radiolabeled sulfate and glucosamine by infected testicular tissue that the host tissue synthesized molecules resembling glycosaminoglycans. Uninfected testicular tissue or infected tissue treated for 2 days with penicillin incorporated lower amounts of precursors than infected testis allowed to progress to the peak of orchitis, suggesting that synthesis of glycosaminoglycans was a host response to active infection. An analysis of the *T. pallidum* genome does not reveal the presence of an ortholog of hyaluronidase. Moreover, *T. pallidum* has only secA, D, E, F, and Y genes (Table 8) and appears to lacks the machinery required to secrete proteins extracellularly.

2.7. Membrane transporters

Metabolically, a major surprise of the genome study was the large number (57) of genes which encode 18 putative transport systems (Table 9A, B) (Fraser et al., 1998; Weinstock et al., 1998) (Figure 1). The predicted substrates of these transporters include amino acids, carbohydrates, and cations. In hindsight, this makes perfect biological sense for an organism that appears to be an obligate parasite. Why synthesize a vital nutrient that may require a whole pathway of enzymes, if the nutrient is readily available from the immediate environment?

Predicted carbohydrate transporters include those for ribose, galactose, glucose, glycerol, malate, succinate, and fumarate (Table 9B). Deka et al. (2004) demonstrated that the lipoprotein Tp38 (TP0684) was a periplasmic sugar binding protein. The protein was subsequently cloned in *E. coli* and the recombinant protein tested for its specific sugar binding properties using intrinsic florescence microscopy. They demonstrated that Tp38 bound D-glucose and to a lesser extent D-galactose, but not D-fucose, D-ribose, L-glucose, or L-galactose. This established that Tp38 was a primary glucose receptor for *T. pallidu*m and may be involved in glucose-mediated chemotaxis. One transporter missing from *T. pallidum* as a result of examination of the genome is phosphoenolpyruvate phosphotransferase for carbohydrate uptake. This activity may be replaced by one of the many V type ATPases (TP0426-30; 0527-33) for which the predicted substrate is unknown. Also missing is a specific transporter for inorganic phosphate, although this could possibly be replaced by the transporter for glycerol 3-phosphate (TP0256). Some of the amino acid transporters include alanine, glycine, glutamate, aspartate, and neutral (aliphatic) amino acids. Lastly, there is also a transporter for the uptake of thiamine or thiamine pyrophosphate (TP0142-0144) and another for spermidine /or putrescine (TP0652-5). The uptake of thiamine pyrophosphate may be a second mechanism by which *T. pallidum* could acquire inorganic phosphate.

Cation transporters include those for K^+, Cu^{+2}, Mg^{+2} and trace metals such as Zn^{+2}, Mn^{2+}, and possibly iron (see Chapter 5, Gherardini et al.). In 1997, Hardam et al.. (1997) described a six gene transport-related or *tro* operon consisting of four genes for an ABC transporter, one for a DtxR-like repressor protein. and the last for the glycolytic enzyme phosphoglyceromutase (Pgm). Radolf, Norgard and co-

Table 9. Transporters and permeases

A. Transport /permease systems shared by *T. pallidum and T. denticola*

Substrate	Type	TP	TDE
thiamine	ABC	0142-0144	0143-5
Zn, Mn, ?Fe	ABC	0163-7	1223-6
galactose/ Mg	ABC	0585,0684	2217
Mg		0863, 2718	0774, 0917
metal	ABC	TP0034	TDE2226
metal	ABC	TP0035-6	TDE2561-2

B. Transport /permease systems found only in *T. pallidum*

Substrate	Type	TP
Na/ Cl		0023
Ala, Na		0998
K		0140
K		0513
Me++	ABC	0034-36
Co	ABC	0227
Mg		0774
Mg		0917
cation		1036
sugar	ABC	0074-76
sugar	ABC	0737
sugar	ABC	0786
sugar	ABC	0804
sugar	ABC	0838
sugar	ABC	0881
rib-gal	ABC	0300
rib-gal	ABC	0321-3
Me-Gal	ABC	0545
Me-Gal	ABC	0684-6
carnitine		0106
Amino Acid	ABC	0119-20
Amino Acid	ABC	0265
Amino Acid	ABC	0308-9
Glu-Asp		0555
glut		0934
Sperm-putresc	ABC	0652-5
Oligpept	ABC	0585
dicarboxylate		0958
Transporter	ABC	0581
Transporter	ABC	0611
Transporter	ABC	0786
Transporter	ABC	0964

Table 9, Continued

C. Transport /permease systems ifound only in *T. denticola*				
Substrate	**Type**	**TDE**	**Substrate**	**Type**
transporter (106)	ABC	0044-5	sugar	
(Unknown)	ABC	0182-4	galactoside (3)	ABC
(Unknown)	ABC	0243-5	permease (5)	
(Unknown)	ABC	0255	permease	
(Unknown)	ABC	0273-4	permease	
(Unknown)	ABC	0282-3	dicarboxylate	
(Unknown)	ABC	0286-7	Amino Acid (16)	
(Unknown)	ABC	0316	Amino Acid	
(Unknown)	ABC	0359-61	Amino Acid	
(Unknown)	ABC	0364	Amino Acid	
(Unknown)	ABC	0372-3	Amino Acid	
(Unknown)	ABC	0375	Amino Acid	ABC
(Unknown)	ABC	0385-6	Amino Acid	
(Unknown)	ABC	0460	Amino Acid	ABC
(Unknown)	ABC	0475	Amino Acid	ABC
(Unknown)	ABC	0487	cysTW	ABC
(Unknown)	ABC	0535	oligopeptide (33)	ABC
(Unknown)	ABC	0625	oligopeptide	ABC
(Unknown)	ABC	0644	oligopeptide	ABC
(Unknown)	ABC	0657-8	oligopeptide	ABC
(Unknown)	ABC	0691	oligopeptide	ABC
(Unknown)	ABC	0694	oligopeptide	ABC
(Unknown)	ABC	0702	oligopeptide	ABC
(Unknown)	ABC	0798	Iron (16)	ABC
(Unknown)	ABC	0812-3	Iron	ABC
(Unknown)	ABC	0818-9	Iron	ABC
(Unknown)	ABC	0892-3	Iron	
(Unknown)	ABC	0896-7	Iron	ABC
(Unknown)	ABC	0899-900	Iron	ABC
(Unknown)	ABC	0902	cation	

workers (Deka et al. 1999; Lee et al., 1999) elucidated the 3-dimensional structure of the Tro binding protein and identified the bound metal as zinc. Hazlett et al., (2003) further characterized the regulation of the operon and the activities of the transporter and determined that (1) TroA has identical affinities for Zn^{+2} and Mn^{+2}; (2) the Tro transporter expressed in *E. coli* imported Zn^{+2}, Mn^{+2}, and possibly Fe^{+3}; (3) TroR represses transporter expression in *E. coli* at significantly lower concentrations of Zn^{+2} than of Mn^{+2}; and (4) TroR-mediated repression causes a much greater down-regulation of the transporter genes than of phosphoglycerol mutase, *pgm,* and therefore, glycolysis would not be abrogated when *T. pallidum* encounters high Zn^{+2} levels. In addition, there are four other genes for metal

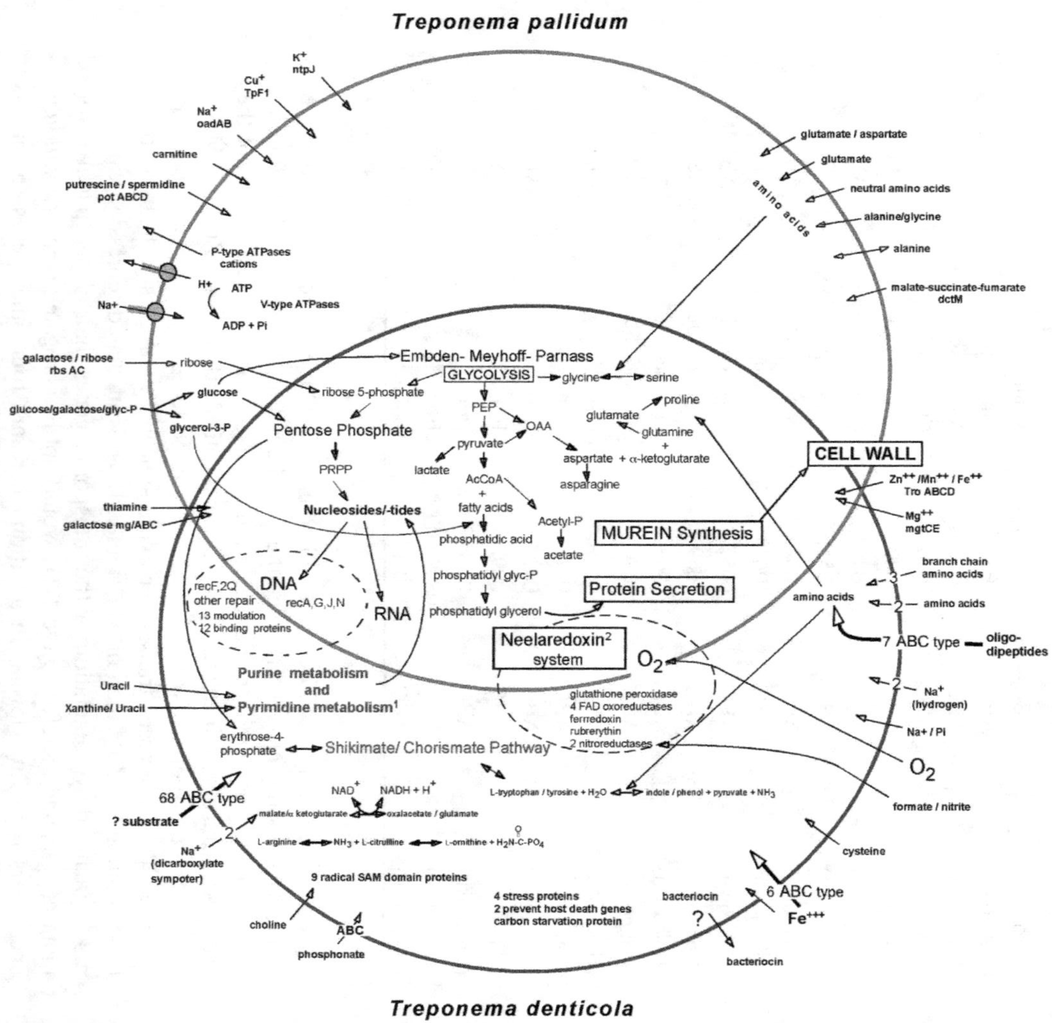

Figure 1. The metabolism of *T. pallidum* and *T. denticola*. The top oval represents the metabolism of *T. pallidum*, and the bottom oval that of *T. denticola*. The overlapping region represents pathways and mechanisms shared by these two pathogens, while those outside it represent each organism's unique metabolic capabilities. (Modified from Fraser et al. (1998) with permission).

transport. Three of these, TP0034-36, comprise another ABC transport system with homology to the Tro transporter and hypothesized to be Zn-specific. Finally, TP1036 is a putative cation transporting ATPase (P-type). These alternate transporters probably complement the tro operon and may be critical for metal transport under environmental conditions in which expression of the tro operon is down regulated by its repressor, TroR (Posey et al., 1999; Hazlett et al., 2003). The alternate metal transporters could also function in transporting other metals like Co^{+2} or Cu^{+2}.

2.8. Oxygen utilization and protective mechanisms

The sensitivity of *T. pallidum* to atmospheric oxygen was recognized shortly after its discovery (Noguchi 1911; Noguchi 1912), and for many years, the bacterium was thought to be an anaerobe (Brewer 1940; Kast 1940). Furthermore, it also was recognized that *T. pallidum* had a requirement for sulfhydryl compounds. Cysteine, thioglycollate, and glutathione, as well as the sulfhydryl compounds mercaptosuccinate and homocysteine, were common components of many media in attempts to cultivate the organism (Nelson and Steinman, 1948; Nelson and Mayer, 1949; Nelson and Diesendruck, 1951; Weber 1960). The putative roles of these sulfhydryl compounds were two-fold: (1) they were required for essential metabolic functions, and (2) they lowered the redox potential (E_H) of the medium, thus reducing oxygen toxicity (Metzger 1966). In fact, Metzger and Smogor (1966) determined that the optimal survival of *T. pallidum* occurred at a pH between 7.2 and 7.3 and an E_H between –230 and –240 mV.

Although a chemically-reduced medium protected treponemal suspensions for a few days, serious problems arose from the oxidation of these agents. Weber (1960) demonstrated this phenomenon using Nelson's medium and Metzger (1966) proved that rapid oxidation accelerated treponemal death. Thus, new methods were sought that did not depend heavily on oxygen-sensitive reducing agents.

Sandok (1976) and Fitzgerald et al. (1975) reported the effects of tissue culture cells on maintaining the E_H and treponemal survival *in vitro*. A combination of tissue culture cells and sulfhydryl compounds was studied and determined to provide superior protection (Fitzgerald et al., 1977) than either alone. Furthermore, the addition of a synthetic reducing agent dithiothreitol (DTT) was found to be beneficial in that DTT maintained the cysteine and glutathione in a reduced state and was not readily oxidized by molecular oxygen (Fitzgerald et al., 1977).

Investigators next sought to determine the optimal oxygen concentration for *in vitro* cultivation. Using tissue cultures on vertical (oxygen) gradients, Fieldsteel et al. (1977) determined that optimal growth of *T. pallidum* occurred between 1.5 and 2.0% dissolved oxygen (dO_2), thereby demonstrating the microaerophilic nature of *T. pallidum*. Norris et al., (1978) compared the influence of dO_2 on treponemal growth in the presence and absence of sulfhydryl agents. At 0% dO_2, *T. pallidum* survived best without DTT; however, at 3% dO_2 the presence of DTT increased survival suggesting a microaerophilic metabolism, which was later confirmed by Graves and Billington (1979). Chalmers and Taylor-Robinson (1979) determined that DTT and cysteine helped to prolong survival at 3% dO_2, while thioglycollate and glutathione were detrimental.

Toxic oxygen radicals can damage a variety of cellular structures including genomic DNA. Steiner, Wong, and Graves (1983) demonstrated that *T. pallidum* was as sensitive as anaerobes to superoxide and peroxide. Further investigation (Steiner et al. 1984) revealed that the treponeme's DNA suffered rapid single strand breaks upon exposure to peroxide at a rate 10-100 times greater than repair-proficient *E. coli* K12. *T. pallidum*, in comparison, showed little or no repair *in vitro* and they suggested a deficiency in DNA repair and housekeeping enzymes such as *RecA*l. However, *recA* was later reported to be present by Stamm et al., (1991).

Most aerobes possess two critical enzymes to protect from the toxicity of oxygen radicals, superoxide dismutase (SODase) and either catalase or peroxidase, (Hewitt and Morris, 1975; Morris 1976) and most anaerobes lack one or both of these enzymes. Lysko and Cox (1978) reported that treponemal suspensions had catalase activity and Austin et al. (1981) examined *T. pallidum* for the presence of catalase, peroxidase, and SODase and noted that *T. pallidum* suspensions contained activities for catalase and SODase, but not peroxidase. However, the putative superoxide dismutase had properties consistent with the copper-zinc eukaryotic SODases present in rabbit tissues. Furthermore, Steiner et al. (1983; 1984) demonstrated that the catalase activity in *T. pallidum* extracts was from rabbits. The absence of these two enzymes appeared to explain the spirochete's exquisite sensitivity to atmospheric oxygen while, at the same time, raising a metabolic paradox: How does *T. pallidum* exist as a microaerophile without them?

Genomic studies of *T. pallidum* confirmed the lack of SODase and catalase (Table 6). However, the genes for several other oxygen protective enzymes were discovered, and a scheme for oxygen detoxification has been proposed (Hazlett et al., 2002) (See also Chapter 5, Gherardini et al.). Neelaredoxin (TP0823), a superoxide reductase which reduces superoxide to peroxide, has been demonstrated in *T. pallidum* (Jovanovic et al., 2000). Alkyl hydroperoxide reductase C (AhpC, TP0509) then reduces the peroxide to water and thus substitutes for catalase activity. The regeneration of neelaredoxin is accomplished by its reduction with rubredoxin (TP0991), which in turn is reduced by NADH oxidase (TP0921). In fact, NADH oxidase is the only enzyme in *T. pallidum* known to require molecular oxygen and could be the primary reason for the treponeme's microaerophilic nature. AhpC is regenerated by its reduction with thioredoxin (TP0919) and thioredoxin is regenerated by the action of thioredoxin reductase (TP0814). In addition, Hazlett et al. (2002) proposed that NADH oxidase itself provides some limited protection by the direct conversion of molecular O_2 to water, thus lowering the potential for superoxide and peroxide generation. While these mechanisms offer an alternative pathway for protection from toxic oxygen radicals, they are probably less efficient than the traditional one composed of SODase and catalase, thus providing a potential explanation for the increased sensitivity of *T. pallidum* to atmospheric levels of molecular oxygen.

2.9. *In vitro* cultivation of *T. pallidum*

One of the long-term goals of syphilis research has been sustained cultivation of the spirochete *in vitro*. In the first 15 years following the discovery of *T. pallidum* in 1905, there were 78 reports on attempts for successful cultivation. Attempts declined, however, over the next 46 years, with only 29 studies reported from 1920 through 1940 and 20 from 1940 through 1966 (Fitzgerald 1981). This progressive decline was probably the result of the enormous difficulties encountered in cultivation attempts. This situation was compounded by the complete lack of knowledge and understanding of treponemal metabolism, the development of serological tests for syphilis, and the successful treatment of the disease with penicillin and other antibiotics.

The mid-1960's and early 1970's witnessed a resurgence of interest in cultivating *T. pallidum*. Many of the studies performed during this time sought to prolong viability *in vitro* as opposed to achieving continuous propagation. Wright's study was one of the first to co-cultivate tissue culture cells and treponemes (Wright, 1962). Using the medium of Weber (1960), she investigated not only rabbit testis cells, but also cells from rabbit liver and kidney, human amnion and embryonic liver, as well as HeLa cells. None of the cultures was successful, but she did note that organisms attached to host cells appeared to survive better and retained motility longer than those in the medium. Over a decade later, Fitzgerald et al. (1975) demonstrated that human cervical carcinoma cells significantly prolonged treponemal survival *in vitro*, no replication was observed and treponemes rarely survived longer than one week. In 1977, Fieldsteel et al., (1977), experimenting with twelve cell lines in a vertical oxygen gradient system with basal reduced medium (BRM), reported the prolonged survival (>50% motile) of *T. pallidum* for up to 21 days. A follow-up study determined that cotton-tail rabbit epithelium (Sf1Ep) cells were the best cellular substrate for treponemal survival (Fieldsteel et al., 1979). Although a 3- to 5-fold increase in numbers of *T. pallidum* was observed, replication of spirochetes in this system could not be proven unequivocally.

The Fieldsteel laboratory soon made another important discovery. When testing different lots of fetal bovine serum (FBS), they noticed some lots were toxic to the organisms (Fieldsteel et al., 1981). Thus, they began to screen serum batches for their ability to support treponemal viability. Simultaneously, several studies revealed the importance of reduced sulfhydryl compounds for treponemal survival (Fitzgerald et al., 1980) in the presence of oxygen. Studies prior to 1970 routinely used reducing agents such as thioglycollate, glutathione, and cysteine to maintain a reduced environment and oxygen tension low. However in 1964, Cleland developed an extremely potent reducing agent, dithiothreitol (DTT) with two terminal sulfhydryl groups on the 4-carbon sugar threitol that was resistant to oxidation by molecular oxygen. DTT became very important in replacing other reducing agents like thioglycollate, glutathione, and cysteine that were rapidly oxidized in the presence of molecular oxygen and produced toxic by-products. Several independent discoveries culminated, therefore, in the final development of the first successful cultivation system: (1) the best cell line (Sf1Ep) to use as host cells (Fitzgerald et al., 1977; Fieldsteel et al., 1979); (2) determination of the optimum dO_2 concentration; (3) DTT needed to be included as the sole reducing agent with the removal of cysteine and glutathione; and (4) the correct batch of FBS selected. When all these findings were utilized in a tissue culture system, *T. pallidum* cells increased in number about 20-50 fold over a 7 to 10 day period while >90% remained motile (Fieldsteel et al., 1981).

In light of the many false claims for successful cultivation, it was imperative that Fieldsteel's results be verified independently; the success of this cultivation system was quickly corroborated by Norris (1982). Subsequently, Norris and Edmondson (1986) investigated the effects of pH, E_H, dissolved oxygen and glucose levels in the cultures, as well as various methods for serial passage. As the cultures grew, the pH and glucose levels fell and the reduced sulfhydryl level and dissolved O_2 content remained relatively unchanged, while the E_H increased slightly. They concluded that replacing individual components, such as glucose or reducing agents, as they were consumed would do little to enhance growth or survival and that the only way to maintain the culture was through serial passage. They experimented with several passage techniques including: (1) using unattached organisms present in the medium; (2) mechanical disruption of the monolayers with a rubber policeman, (3) chemical disruption using trypsin and/or EDTA. None of the passage methods was successful; after 9 to 15 days the cultures stopped growing. Nevertheless, they concluded that the cessation of growth in the serial passages was not due to any mechanical or biological damage during the manipulations, but to inherent "starvation" of the organisms for some critical growth factor. However, this explanation has yet to be proven.

Although the previous studies utilized several methods of passage, all involved some level of disruption except passage of unattached organisms. Regardless, one immediate concern that has dominated passage studies was the fragility of *T. pallidum in vitro*, and thus methods were sought to eliminate any manipulation of the treponemes during serial passage in order to resolve the above two issues. One solution was to grow the cells on polystyrene or agarose beads in suspension cultures. Cultivated cells and treponemes could easily be transferred from flask to flask with minimal disruption. In 1988, Riley and Cox (1988) reported the procedures for successful microcarrier cultivation in Wheaton stir flasks. Growth was almost equal to that in static tissue cultures and serial passage was easily performed. However, subsequent passages failed to replicate continuously. Further attempts to enhance replication during primary cultivation were published by Cox et al., (1990). In the original cultivation system, vessels were gassed with 5% O_2 and remained sealed for the entire cultivation period (14-21 days). One concern was that as the cells and treponemes grew consuming the atmospheric O_2, decreased O_2 levels could eventually inhibit growth; thus cultivation under a constant O_2 atmosphere could enhance and /or prolong growth. Along with changing the oxygen source from a fixed to a constant supply, manipulation of the reducing agents and antioxidants was concurrently investigated. The optimum constant oxygen concentration was demonstrated to be 3%. Furthermore, the addition of 4 antioxidants (mannitol, histidine, superoxide dismutase, and catalase) enhanced treponemal growth at this optimum and at higher oxygen concentrations. The average growth in these cultures was increased from 30 - 40 fold to between 50 and 100 fold. This modified medium was called TpCM (*T. pallidum* cultivation medium). A detailed description of the current cultivation system and some variations, can be found in Volume 236 of Methods of Enzymology (Cox 1994).

3. The cultivable treponemes, *T. denticola* and *T. phagedenis*

3.1. Carbohydrate energy sources

Both *T. denticola* and *T. phagedenis* can catabolize glucose via the Embden-Meyerhoff-Parnass pathway. However, *T. denticola* does not utilize glucose as its primary energy source but instead catabolizes amino acids. This preference for amino acids instead of sugars as a primary carbon source correlates with the high levels of protease produced tissue degradation that occurs in of periodontal disease (see below). The primary end product of glucose fermentation in *T. phagedenis* is acetic acid with traces of lactic acid. Traces of butyric and proprionic acids, and ethanol, butanol, and propanol are also detected in cultures of *T. phagedenis*. In addition to glucose, *T. phagedenis* can also ferment lactose, fructose, sucrose, mannitol, mannose, and ocasionally galactose and trehalose (Smibert 1973; Smibert 1973b).

The genome of *T. phagedenis* has not been sequenced therefore only the genome of *T. denticola* can be discussed with reference to the accuracy of previous metabolic studies (Seshadri 2004). Table 1 lists the metabolic enzymes involved in carbohydrate catabolism. As expected, genes for all the enzymes of the Embden-Meyerhoff-Parnass pathway are present. In addition, several enzymes for the regeneration of NAD^+ by the reduction of pyruvate were also present: lactate, glycerate, and alcohol dehydrogenase. The catabolic enzymes for the Entner-Doudoroff pathway are totally absent. Furthermore, only five of the pentose phosphate pathway are present; genes for the first three enzymes are lacking. As expected, genes for a complete Krebs cycle are absent but genes encoding two of the enzymes (malate and glutamate dehydrogenase) are present (Table 1). These enzymes are probably utilized in the degradation of amino acids. Therefore, as in *T. pallidum,* energy production in *T. denticola* is limited to substrate level phosphorylation of ADP to make ATP and other high-energy compounds.

3.2. Amino acid and protein metabolism

The amino acid composition of 14 strains of cultivable treponemes were published by Moss (1971). The most prevalent amino acids were alanine, leucine and aspartic acid, each with around 10-12% of the total amino acid composition. These were followed by glycine, phenylalanine, glutamic acid, and lysine, each with 8 – 10 %. Although there were minor differences in amino acid composition between strains, no qualitative or large quantitative differences that might distinguish between them was observed.

As cited above, *T. denticola* preferentially ferments amino acids rather than glucose as a source of energy. *T. denticola* primarily utilizes serine, alanine, valine, and cysteine as an energy source (Smibert 1973; Canale-Parola 1977). Blakemore and Canale-Parola (1976) reported extensive studies on the arginine requirements of *T. denticola*. Additional evidence of amino acid fermentation comes from the presence of several selenium-dependent glycine reductases (TDE0078-9; TDE0239-40; TDE0745; and TDE2119-20), enzymes previously thought to be restricted to Gram-positive bacteria (Rother et al., 2001). The origin of free amino acids needed for catabolism may be a result of several proteolytic enzymes (Table 2). Uitto et al. (1988) reported a chymotrypsin-like protease present in *T. denticola* and suggested that it may play a role in the invasiveness of the organism because of its ability to hydrolyze serum proteins such as transferrin, fibrinogen, serum albumin and IgA and IgG. This may partly explain why albumin is required by *T. denticola* for growth (Van Horn 1983). In addition, Mikx et al. (1997) demonstrated that the proteolytic capability of *T. denticola* ATCC 33520 is dependent upon the rate of growth and pH; greater proteolytic activity occurred at higher growth rates and at a lower pH.

Chu et al. (1994; 1995) discovered a novel protein in *T. denticola*, cystalysin (TDE1669), which participated in the destruction of red blood cells and exhibited hemooxidative and hemolytic activities. It was later demonstrated that cystalysin was actually a cysteine desulfhydrase which degraded cysteine to produce H_2S, pyruvate and ammonia (Chu et al., 1997). The pyruvate from the breakdown of cysteine could be utilized for ATP production via the Embden-Meyerhoff-Parnass pathway. Furthermore, cystalysin was shown to be present in several strains of *T. denticola* but not in other oral spirochetes, *T. pectinovorum, T. vincentii*, or *T. socranskii*. This enzyme appears to play a critical role both in the metabolism and the pathogenesis of *T. denticola*. Recently, Chu et al. (2002) began an investigation of the metabolism of the tripeptide glutathione in *T. denticola*. Cystalysin was demonstrated to be incapable of degrading glutathione; cysteine was its sole substrate for catalysis. However, they proposed a pathway for glutathione metabolism. In 1997, Makinen and Makinen (1997) isolated a γ-glutamyltransferase (GGT, TDE0444) supposedly from the outer envelope of *T. denticola,* which Chu et al. (2002) maintain could serve as the first enzyme in the degradation of glutathione to produce glutamate and the dipeptide, Cys-Gly. This dipeptide could then be broken down by cysteinyl glycinase to produce glycine and cysteine, the latter being the substrate for cystalysin. They proposed that the metabolism of glutathione could serve multiple biologic purposes that are critical for the survival of *T. denticola*. First the H_2S produced may be critical for virulence as H_2S plays a major role in hemolysis and hemoxidation. Second, pyruvate, another byproduct of cysteine degradation, has been shown to enhance the growth of *T. denticola* (Chu et al., 2002). Lastly, eukaryotic cells usually have high levels of glutathione (up to 4 mM) for macromolecule synthesis, transport, and enzyme regulation, so it is an abundant source for a pathogenic spirochete to use for energy and supply of amino acids. TDE0444 was later cloned and its enzyme activities characterized (Chu et al., 2003). Of note, PSORT analysis of the amino acid sequence predicts that the enzyme is located in the cytoplasmic, not the outer, membrane of *T. denticola*.

Genomic evidence also supports the metabolic data that *T. denticola* preferentially utilizes amino acids as both carbon and energy sources (Seshadri 2004) (Table 2). First, there are 18 genes encoding proteases including a serine protease (TDE0762), a glycoprotease (TDE1468), a prtB protease (TDE0346), a ptrB protease (TDE2140), three genes for CAAX amino terminal proteases (TDE0275, TDE0716, and TDE1870), a putative zinc metalloprotease (TDE 2341), and five genes encoding an ATP-dependent Clp protease (TDE1672, TDE1673, TDE2124, TDE2327, and TDE2388). Second, there are at least 49 genes encoding amino acid and peptide permeases (Table 9C) for the transport of protease-digestion products into the *T. denticola* cytoplasm. Thirty-three of these genes code for ABC type permeases for oligo- and di-peptide transport and 16 more code amino acid transport. Lastly, in comparison, the genome of *T. pallidum,* a spirochete whose metabolism prefers to obtain energy and carbon from sugar fermentation, contains only five genes for amino acid transport.

In contrast to *T. denticola*, *T. phagedenis* uses arginine, histidine, threonine, serine, and glutamate as energy sources. This explains why Steinman et al., (1953) reported that the amino acids arginine, cysteine, and phenylalanine were required components of their growth medium. Furthermore, Steinman et al. (1952) identified thirteen essential amino acids required for cultivation and two additional amino acids (serine, tyrosine) which greatly stimulated growth. This was confirmed by Barban (1954). Barban (1956) later reported that *T. phagedenis* could transaminate alanine, arginine, aspartate, histidine, and threonine to glutamate and could transaminate glyoxylate to glycine. Further studies demonstrated that *T. phagedenis* is capable of synthesizing the amino acids glycine, serine, alanine, aspartate, glutamate, proline and ornithine (Allen 1971). However, phenylalanine, arginine, lysine, leucine, isoleucine, valine, threonine, and histidine could not be synthesized and exogenous sources were required for growth.

In addition to malate and glutamate dehydrogenase, which can be utilized for the degradation of the aspartic acid and glutamic acid families of amino acids, there are also genes for several enzymes that could be used to degrade aromatic amino acids and arginine. These include tryptophanase (TDE0251), tyrosine phenol lyase (TDE1118), arginine deiminase (TDE0451). and ornithine carbamoyltransferase (TDE0929).

3.3. Fatty acid and lipid metabolism

Early studies by Oyama (1953) demonstrated that *T. phagedenis* requires crystallized serum albumin, which was suspected to function as a carrier for essential lipids, and detoxify essential lipids. When closely examined, most of the lipid requirements can be satisfied with supplemental oleic acid. Esters of various fatty acids proved less effective in supporting growth and alcohols of those fatty acids were completely ineffective. The optimal ratio of fatty acid to protein was determined to be 4:1.

Most treponemes require either short- or long-chained FAs for growth suggesting they lack the pathways to synthesize them (Meyer and Meyer, 1971; Livermore and Johnson, 1970). Meyer and Meyer (1971) studied lipid metabolism in both the FA requiring *T. phagedenis* and the free-living spirochete *T. zuelzerae*. *T. zuelzerae* can synthesize unsaturated fatty acids as well as some simple monoenoic FAs. Not only is *T. phagedenis* unable to synthesize FAs, but it also cannot modify the chain length or desaturate FAs (Livermore and Johnson, 1970). Thus as expected, both *T. denticola* and *T. phagedenis* require exogenous sources of fatty acids with albumin serving as a fatty acid carrier. *T. phagedenis* can be cultivated with either 6- or 8-octadecanoic, linoleic, linolenic, elaidic, ricinoleic, methyl or ethyl ester of oleic, 8- or 9-octadecynoic acid, or sodium oleate. Tween 80 and TEM-4T also can be used to meet the organism's fatty acid requirements. When individual fatty acids were studied, oleic gave higher yields than stearic, palmitic, or linoleic acids (Power and Pelczar, 1959). The best mixture of FAs was an equimolar ratio of stearic, palmitic, and oleic acids or the lipid substitute TEM-4T (later shown to contain predominantly these three fatty acids).

Also as expected, there were no genes encoding enzymes required for the synthesis or degradation of fatty acids (Table 3). However genes that code for enzymes to acylate and transacylate proteins were present. As mentioned above, Kent et al. (2004) discovered that the *T. denticola* genome contained a putative *licCA* (TDE1260) gene for PC biosynthesis. The predicted choline kinase and CTP: phosphocholine cytidylyltransferase activities were demonstrated to be functional by cloning the gene and demonstrating those activities in a recombinant protein.

In addition, the disruption *of licCA* expression prevented conversion of labeled precursors to phosphocholine and PC, and eliminated all detectable PC from this organism. Finally, this study revealed that *T. denticola* uses a CDP-choline pathway for PC biosynthesis, which has not previously been reported for bacteria.

Prokaryotes have only been shown to make PC by either the phospholipid N-methyltransferase pathway (Kaneshiro and Law, 1964) or the phosphatidylcholine synthase pathway; the CDP-choline pathway for PC biosynthesis is one of the two pathways used by eukaryotic cells.

3.4. Nucleic acid metabolism

Early studies on the cultivation of the Reiter treponeme revealed that supplemental purines were not required for growth (Steinman et al., 1952), but the addition of adenine to the medium did stimulate growth. However, a source of pyrimidine was essential for growth; either uracil or cytosine was able to satisfy this nutritional requirement. (Rathlev and Pfau 1964). Surprisingly, *T. denticola* contains all of the genes encoding enzymes required for the biosythesis of purines and most of those for pyrimidines (Seshadri 2004) (Table 4). However, genes encoding carbamylphosphate synthetase and aspartate transcarbamylase, the first 2 enzymes in the pyrimidine pathway, are absent. However, *T. denticola* possesses a gene for carbamate kinase (TDE2476) which catalyzes a reaction similar to carbamylphosphate synthetase. However, the latter is an anabolic enzyme and carbamate kinase is normally used for catabolic degradation. This suggests that *T. denticola* may also be using nucleotides as a source of energy and nitrogen.

3.5. Miscellaneous nutrients

3.5.1. Vitamin requirements

Early in the analysis of growth requirements for *T. phagedenis*, it was discovered that the organism requires a mixture of B vitamins (Steinman 1950). Analysis of individual B vitamins revealed the critical compound was pantothenic acid (B5). Further studies (Steinman et al., 1954) revealed that nicotinic acid and biotin were also essential; the optimal concentrations were 100 and 1 pg/ml respectively. Three additional vitamins (choline, riboflavin and thiamin) were found to stimulate growth, and supplemental thiamin also increased the cellular yield.

3.5.2. Transporters and trace metal requirements

The genome of *T. denticola* revealed seven (33 genes) ABC-type uptake systems for oligopeptides and ten (16 genes) amino acid transport systems (Seshadri 2004) (Table 9C) that support its catabolism of amino acids for energy production. In contrast, there are few (4) known genes that code for sugar uptake. Incredibly, there are 106 genes that code for 68 different ABC transport systems whose substrates have not yet been identified. This surely reflects the organism's ability to efficiently transport a variety of potential nutrients in its niche and to effectively compete with a myriad of other oral bacteria for limited nutrients. In addition, there are five (16 genes) ABC-type uptake systems and one non-ABC type system for iron uptake. *T. denticola* also possesses the genes for a Tro operon (TDE1223-6) similar to that found in *T. pallidum*. It seems unusual that *T. denticola* would have so many systems for iron transport since, like *T. pallidum,* it lacks the cytochromes of electron transport. Lastly, three of the transport systems were probable bacteriocin secreting systems (TDE 0719-0720; TDE0425-0426; and TDE2430-2431) that the spirochete may use for competing with other flora for nutrients (Table 8).

3.6. Oxygen utilization and protective mechanisms

T. denticola and *T. phagedenis* are best cultivated under anaerobic conditions, however anaerobiosis is not an absolute requirement. *T. denticola* is especially sensitive to oxygen (Loesche 1969), whereas *T. phagedenis* appears to be more aerotolerant. While *T. phagedenis* can be cultivated in tall test tubes with small air spaces above the media, *T. denticola* cultures grow much better if the air in the head space is replaced with nitrogen or CO_2 (Smibert 1971) and the E_H is below -125 mV. Kawata (1967) reported the presence of cytochromes in *T. phagedenis*, but this has never been confirmed and is now presumed to be an error. Syad et al. (Syad 1993) also reported low and variable activities of a peroxidase, catalase and superoxide dismutase in *T. denticola*. Caldwell and Marquis (1999) reported that *T. denticola* utilized molecular O_2 at a rate of 0.46 μmoles/hr/mg of cell protein. They found no heme-containing proteins (*i.e.*, cytochromes) but noted that the spirochete has very strong NADH oxidase (TDE0096) activity. Further investigation revealed the presence of a glutathione peroxidase (TDE1729) and four oxoreducatases (TDE0134; 0685; 0868; and 2643). NADH oxidase activity was dependent upon the presence of flavin and was not diminished by fluoride, a known inhibitor of heme-based oxidases. These findings suggested that, although *T. denticola* lives predominantly in an anaerobic environment, it is capable of utilizing low levels of oxygen and protecting itself from oxygen radicals.

As expected, there are no genes for SODase, or catalase in *T. denticola* (Table 6); however, as mentioned above a gene encoding glutathione peroxidase is present (TDE1729), which would reduce peroxide to water at the expense of glutathione. In addition, as in *T. pallidum*, genes are present for neelaredoxin, alkyl hydroperoxide reductase, NADH oxidase, thioredoxin, and thioredoxin reductase. It is presumed, therefore, that *T. denticola* possesses an oxygen radical detoxification pathway similar to that proposed by Hazlett (2002) for *T. pallidum*. In addition, there is also a gene encoding a rubrerythin (TDE0434) that typically has peroxidase activity (Weinberg et al., 2004) and two genes for nitroreductases (TDE0708; 1363).

3.7. Cultivation of *T. denticola* and *T. phagedenis*

Because *T. denticola and T. phagedenis* are relatively easy to cultivate compared to *T. pallidum*, less emphasis will be given to their propagation. One of the earliest investigations into the nutritional requirements for the cultivation of *T. denticola* was reported by Hardy and Munro (1966). Their basal medium was one part Brain Heart Infusion (BHI) and three parts Spirolate broth supplemented with thioglyocollate and rabbit serum. Socransky and Humersak (1967) discovered that the serum could be replaced by α-globulin. A more modern description of the medium and methods for the cultivation of *T. denticola* in PTY medium has been described by Mikx et al. (1992) In addition, Wyss (1992) developed a new chemically defined medium without serum named OMIZ-W1 which not only supports the growth of *T. denticola* but many other oral spirochetes including *T. vincentii*. The high concentrations of ascorbic acid and ammonium ions distinguish this medium from others like PTY. Some strains of oral spirochetes required serum (FCS), but type strains of the above three species did not.

An early investigation by Barban (1954) describes the cultivation of the treponeme in BHI, thiotone peptone, thioglycollate and 10% ox serum. Current methods for the cultivation of *T. phagedenis* were originally described by Wachter and Johnson (1976).

3.8. Genome comparison for metabolic functions

When one compares the genomes of *T. pallidum* and *T. denticola* with respect to their metabolic capabilities, one obvious difference must be considered: *T. denticola* is a cultivable spirochete while *T. pallidum* is not. One also must consider their natural niches. *T. denticola* lives in the gingiva where it co-habits with many other bacterial species; from a microbiological standpoint its environment is dynamic and extremely complex. Variations in pH, periodic absence of nutrients, competition with other microflora, and accumulation of toxic waste products from bacterial metabolism are the norm. In contrast, *T. pallidum* survives in an otherwise sterile, homeostatic environment. The host readily supplies a complement of nutrients (glucose, amino acids, fatty acid, nucleotides, vitamins, etc.), maintains a narrow pH range, and removes all toxic metabolic waste products. Thus, one would expect to see major differences in the number and variety of metabolic genes for survival and growth. In fact when the genomes are compared, *T. denticola* has almost 2000 genes not found or shared in the *T. pallidum* genome (Seshadri 2004). The small size of *T. pallidum*'s genome compared to other published spirochetal genomes suggests reductive evolution that was most likely a result of adaptation to a host tissue environment. This limited genome (about 40% the size of that of *T. denticola*) may directly contribute to its difficulty in surviving and replicating *in vitro*. To no surprise, many of the biosynthetic pathways are absent, most notably those for amino acids, fatty acids, nucleic acid bases, and vitamins. Although these end products can be supplemented in media designed for *in vitro* growth, as discussed below, none of these components satisfy the metabolic requirements of the organism to permit continuous serial passage.

3.9. Catabolic energy producing pathways

The enzymes associated with glycolysis and the Embden-Meyerhoff-Parnass pathway are listed in Table 1. The EC (Enzyme Commission) number of each enzyme is listed as a unique identifier along with the gene number for each protein in both *Treponema* species. The major differences are that *T. pallidum* has a pyruvate oxidoreductase (TP0939), whereas *T. denticola* has a ferreredoxin oxidoreductase (TDE2198) to perform that function. *T. pallidum* also has a phosphoenolpyruvate carboxykinase (TP0122) that can reversibly convert PEP + CO_2 to OAA (oxaloacetic acid) and *T. denticola* does not. Instead, *T. denticola* has a putative phosphoenolpyruvate synthase (TDE0349) that catalyzes a similar, but irreversible reaction. *T. denticola* also possesses both lactate (TDE0351), glycerate (TDE1616), and alcohol dehydrogenases (TDE0051), to regenerate oxidized NAD^+ for glycolysis, all of which are lacking in *T. pallidum*.

Neither *T. pallidum* nor *T. denticola* has the full complement of genes required for the pentose phosphate pathway (Table 1). *T. pallidum* possesses all of the genes necessary for this pathway except transaldolase. *T. denticola* lacks that enzyme as well as the first three enzymes in the pathway namely glucose 6-phosphate dehydrogenase, lactonase, and 6-phosphonogluconate dehydrogenase. From genomic analysis, it is uncertain whether TP0478 is a putative gene for lactonase; however, the preceding gene, TP0477, shows strong homology for glucose 6-phosphate dehydrogenase, the enzyme immediately preceding it in the pathway. It seems reasonable to propose, therefore, that TP0478 has lactonase-like activity.

Only *T. pallidum* has genes encoding enzymes for the Entner-Doudoroff Pathway (Table 1), and it possesses only two: the first enzyme in the pathway glucose 6-phosphate dehydrogenase (TP0477-8) and the last, 2-dehydro-3-deoxy-phosphogluconate aldolase (TP0568). Both organisms lack the remaining enzymes.

The enzymes associated with the Krebs or TCA cycle and the glyoxylate shunt are also listed in Table 1. *T. pallidum* is lacking all of the genes for these two pathways. *T. denticola* has both malate (TDE2446) and glutamate dehydrogenase (TDE0997) but, like *T. pallidum,* lacks all the others.

Neither *T. pallidum* nor *T. denticola* possesses any of the four enzymes required for the oxidation of fatty acids (Table 3). This explains their dependence on exogenous source of fatty acids (usually supplied from serum), and their respective fatty acid content profiles being similar to the host (Matthews et al., 1979) or that of the culture medium (Belisle et al., 1994).

3.10. Anabolic pathways

Both organisms are restricted by the lack of complete pathways for the biosynthesis of some key metabolites. As mentioned previously, neither can synthesize nor modify fatty acids owing to the lack of genes of the β-oxidation pathway and enzymes necessary for the desaturation of fatty acid hydrocarbon chains. However both possess a fatty acid CoA ligase (TP0145; TDE 0942) that results in the utilization of exogenous fatty acids for lipid synthesis. In addition, genes are present for the synthesis of phospholipids (Table 3).

With respect to amino acid biosynthesis, *T. pallidum* lacks all of the genes necessary for tryptophan, phenylalanine, and arginine biosynthesis (Table 2). However, *T. denticola* possesses all but one enzyme of the pathway for chorismate, a precursor for aromatic amino acid biosynthesis. The one enzyme lacking in this pathway is isochorismate synthase, but it does have a gene for chorismate synthase (TDE1132). In addition, *T. denticola* possesses genes for tryptophanase (TDE0251), tyrosine phenol lyase (TDE1118), and two of the enzymes for arginine biosynthesis, ornithine transcarbamylase (TDE0929), and arginine deaminase (TDE0451). However, all of these enzymes are thought to be more important in amino acid degradation for energy production than for biosynthesis, especially when many other genes in the biosynthetic pathways are lacking.

With one exception, TP0695, the genes for purine and pyrimidine biosynthesis are totally lacking in *T. pallidum* (Table 4). TP0695 has homology with other bacterial genes coding for phosphoribosylglycinamide formyltransferase (EC 2.1.2.2), which may allow a mechanism for purine base conversion, but not synthesis. However, *T. denticola* possesses all of the genes for purine biosynthesis, and all but two genes for pyrimidine biosynthesis. What is surprising about the missing enzymes, carbamylphosphate synthetase (CPSase; EC 6.3.4.16) and aspartate transcarbamylase (ATCase; EC 2.1.3.2), is that they are the first two in the pathway and both catalyze reversible reactions. However, *T. denticola* does possess carbamate kinase (TDE2476) that is similar in function to CPSase and can transphosphorylate ADP to ATP. With the pyrimidine enzymes it does possess, *T. denticola* can break down pyrimidines into carbamylaspartate. However, it appears to be missing ATCase, the enzyme necessary to break carbamylaspartate into aspartic acid and carbamylphosphate upon which carbamate kinase could utilize for energy production (ATP). In addition, aspartic acid would add to the pool of amino acids being fermented for energy. It seems paradoxical that an organism scavenging nutrients from host tissue and other organisms in gingival plaque would be lacking ATCase. Most likely one of the many hypothetical proteins will eventually be discovered to carry out this function. Lastly, both organisms have thioredoxin reductase necessary for the production of thioredoxin that is critical for the interconversion of purines and pyrimidines.

Because both *T. pallidum* and *T. denticola* synthesize intact cell walls, they must have a full complement of enzymes necessary for peptidoglycan synthesis (Table 5). However, they both lack one enzyme, UDP-N-acetylmuramoyl-L-alanyl-D-glutamate—L-lysine ligase (EC 6.3.2.7), which is found in most other bacteria. This function must be substituted by some other enzymatic activity.

Both species have mechanisms to detoxify oxygen radicals (Table 6). Even though both lack the more classical and common detoxification enzymes, SODase and catalase, both organisms possess genes encoding alternate pathways for detoxification of oxygen radicals via neelaredoxin (TP0823; TDE1754), NADH oxidase (TP0921; TDE0096), alkyl hydroperoxide reductase (TP0509; TDE0011), and thioredoxin reductase (TP0814; TDE0743). As mentioned earlier, one major difference between the two genomes is the presence of nine oxidation-reduction enzymes found exclusively in *T. denticola*: four flavin enzymes, a ferredoxin, a rubrerythin, an uncharacterized oxidoreductase, and two nitroreductases. In addition, *T. denticola* contains a gene for glutathione peroxidase (TDE1729). This enzyme could play a critical role in *T. denticola*'s ability to respond to toxic oxygen radicals because the spirochete has no pathway for glutathione biosynthesis and the gene encoding glutathione synthase (EC 6.3.2.3) is absent. Again, these additional redox proteins may be required by *T. denticola* because of its apparent increased utilization of iron which is known to catalize more oxygen radicals than other transition metals. The increased number of redox proteins along with increased iron utilization probably reflects a balance in the metabolism of *T. denticola*.

Another critical difference between the two genomes is the larger number of DNA replication, repair, and housekeeping genes in *T. denticola* (Table 7). *T. pallidum* has 9 genes for these functions whereas *T. denticola* has five times that. Of special interest are three putative "prevent host death genes" (TDE0255; TDE0455; TDE1954), one encoding a carbon starvation protein (TDE0012), one stress gene (TDE0354), 13 DNA modification/recombination genes, and 14 DNA binding genes of unknown function (Table 7).

Lastly, *T. denticola* possesses a considerable number of transport systems not found in *T. pallidum* (Table 9C). For example, there are 106 unidentified ABC transporters in *T. denticola* whose substrates have not yet been identified compared to only four in *T. pallidum*. In addition, another major difference is that *T. denticola* appears to have many more genes (16) for iron transport than *T. pallidum*. Much of the transport capability of *T. denticola* appears to be involved in amino acid and oligopeptide incorporation, supporting the observation that it derives most of its energy from amino acid fermentation (Smibert 1973; Canale-Parola 1977).

4. Conclusions

Two critical areas of future research on the two species of *Treponema* are: (1) the determination of virulence factors of both *T. denticola* and *T. pallidum*, and (2) the unrestricted ability to cultivate *T. pallidum*. Because this chapter deals with metabolism, the latter is the most relevant to discuss here.

There are basically two explanations for the inability to cultivate *T. pallidum*. The first and more historical explanation is that some nutritional factor is still missing or limited. As mentioned above, it appears that none of the substrates (sugars, amino acids, nucleotides, fatty acids, vitamins, minerals, and antioxidants) for the numerous transport systems revealed in the genome completely satisfied the growth requirements of *T. pallidum*. As almost every nutrient tested fails to produce continuous growth *in vitro*, is there some exotic chemical that been overlooked as a potential growth requirement? Erythritol as a growth requirement for the pathogen *Brucella abortus* is an example (Meyer 1966; Sperry and Robertson, 1975). Another possibility is that a labile nutrient rapidly disintegrates over time in the tissue cultures. This possibility appears remote because refeeding the cultures with fresh TpCM does not significantly prolong treponemal viability. Even with the most gentle methods, serial passage of treponemes into new uninfected cultures of Sf1Ep cells also does not significantly prolong their viability. Lastly, even if a critical missing nutrient is supplied *in vitro*, does *T. pallidum* have the capability to transport the molecule across its membranes? If not, then how would it acquire the same nutrient or its precursors *in vivo*?

The second explanation is that some condition or stress exists in the current *in vitro* system that inhibits the continual growth of *T. pallidum* and the organism is incapable of overcoming this stress because of its limited genome. The evidence to support this hypothesis becomes apparent when as discussed earlier the genes for stress response, DNA repair and housekeeping are examined. *T. pallidum* has only one-fifth the genes for this vital response than does *T. denticola* (Table 8). Besides damage to the genome, stress could also interfere with other metabolic functions including the transport of critical nutrients. Thus, *T. pallidum* in a stressful environment might be incapable of incorporating vital nutrients even when they are supplied. *In vivo*, it is presumed that the host tissue provides protection from this stress via a homeostatic environment that has not been fully replicated in the tissue culture cultivation system.

To further frustrate research efforts, this fragile and restricted spirochete also limits our ability to utilize mutagenesis techniques to study its metabolism. Nevertheless, some molecular tools are still at our disposal. Real time RT-PCR can be used to examine gene expression under various conditions and to compare those levels of expression under ideal growth conditions (*in vivo*) and under the current cultivation system (potential stress) to reveal the differences. If "environmental stress" is the explanation for the demise of *T. pallidum* cultures, then the stress first needs to be identified. Towards that goal, the expression of two stress related genes, *rpoN* and *rpoS*, have been compared between treponemes cultivated in rabbits and in tissue cultures. Their level of expression appears to be elevated *in vitro* (unpublished data). In addition, the expression of several of the oxygen protection enzymes also appear elevated, suggesting the organisms may be under oxidative stress. The effects of this presumed stress on the overall metabolism of the treponemes has yet to be determined. Another powerful molecular tool is microarray analysis, which can be used to assess the whole genome for differences in gene expression under various *in vitro* and *in vivo* conditions that can be studied in greater detail using real time RT-PCR.

In addition, we need more information about the metabolism of *T. pallidum* based on the information published in the genome. One unknown factor that may hold several answers are the numerous "hypothetical proteins" which remain unidentified. Over 30% of the genes in *T. pallidum* have yet to be matched with any known protein, and another 17% are conserved hypothetical proteins whose function has yet to be verified. Thus, a huge gap in our knowledge of the organism's overall metabolic capability still remains. We assume, because annotated genes were lacking for some metabolic functions, that *T. pallidum* may be metabolically deficient in certain areas. However, the absence of a homolog does not necessarily mean that the organism lacks that particular metabolic function which may be performed by an enzyme with little overall homology to other known proteins. One molecular tool to identify potential functions of a "hypothetical protein" is to use computer programs to predict its 3D structure from the primary sequence; those predictions can be used to screen proteomic databases containing the structure of known proteins whose functions are well characterized. To confirm the functionality of any proteins identified, they can be cloned, purified, and their properties fully characterized. Finally, libraries of the genes for these "hypothetical proteins" can be tested to determine whether they complement *E. coli* "knockouts" such as demonstrated for the *proC* gene (Gherardini et al., 1990). Only when these unknown proteins are identified can we begin to understand the complete physiology of *T. pallidum,* critical knowledge that will be required to overcome the persistent limits of cultivation.

Acknowledgment

Supported by grants AI26756 and AI38894 to JDR from the National Institutes of Health.

References

Allen, S., Johnson, R.C., and Peterson, D. (1971). Metabolism of Common Substrates by the reiter strain of *Treponema pallidum*. Infect. Immun. *3*(6), 727-734.

Austin, F. E., Barbieri, J. T., Corin, R. E., Grigas, K. E., and Cox, C. D. (1981). Distribution of superoxide dismutase, catalase, and peroxidase activities among *Treponema pallidum* and other spiorochetes. Infect. Immun. *33*, 372-379.

Barban, S. (1954). Studies on the metabolism of the *Treponemata*. I. Amino acid metabolism. J. Bacteriol. *68*(4), 493-497.

Barban, S. (1956). Studies on the metabolism of the *Treponemata*. II. Transamination in the Reiter treponeme. J. Bacteriol. *71*(3), 274-7.

Barberi, J. T. and Cox, C.D. (1979). Glucose incorporation by *Treponema pallidum*. Infect. Immun. *24*, 291-293.

Barberi, J. T. and Cox, C.D. (1979). Pyruvate oxidation by *Treponema pallidum*. Infect. Immun. *25*, 157-163.

Barberi, J. T. and Cox, C.D. (1981). Influence of oxygen on respiration and glucose catabolism by *Treponema pallidum*. Infect. Immun. *31*, 992-997.

Baseman, J. B. and Hayes, N.S. (1974). Protein synthesis by *Treponema pallidum* extracted from infected rabbit tissue. Infect. Immun. *10*, 1350-1355.

Baseman, J. B. and Hayes, N.S. (1977). Anabolic potential of virulent *Treponema pallidum*. Infect. Immun. *18*, 857-859.

Baseman, J.B., J.C. Nichols, and Hayes, N.S. (1976). Virulent *Treponema pallidum*, aerobe or anaerobe. Infect. Immun. *13*, 704-711.

Baseman, J.B., and Nichols, J. C. (1979). Capacity of virulent *Treponema pallidum* (Nichols) for deoxyribonucleic acid synthesis. Infect. Immun. *23*(2), 392-397.

Belisle, J.T., Brandt ME, Radolf, J.D., and Norgard, M.V. (1994). Fatty acids of *Treponema pallidum* and *Borrelia burgdorferi* lipoproteins. J. Bacteriol. *176*, 2151-2157.

Benoit, S., J. Posey, Chenoweth, M.R., and Gherardini, F.C. (2001). *Treponema pallidum* 3-phosphoglycerate mutase is a heat-labile enzyme that may limit the maximum growth temperature for the spirochete. J. Bacteriol *183*(16), 4702-4708.

Blakemore, R. P. and Canale-Parola, E. (1976). Arginine catabolism by *Treponema denticola*. J. Bacteriol *128*(2), 616-622.

Boak, M. and Miller, J. N. (1954). A simple medium for maintaining the viability of *Treponema pallidum* in the *Treponema pallidum* immobilization test. Am. J. Syphilis, Gonorrhea, and Venereal Diseases *38*(5), 429-33.

Brewer, J. H. (1940). Clear liquid mediums for anaerobic cultivation of anaerobes. JAMA *115*, 598-600.

Caldwell, C. E., and Marquis R. E. (1999). oxygen metabolism by *Treponema denticola*. Oral Microbiology and Immunology *14*, 66-72.

Canale-Parola, E. (1977). Physiology and evolution of spirochetes. Bact. Rev. *41*(1), 181-204.

Chalmers, S. K. and Taylor-Robinson, D. (1979). The effect of reducing and other agents on the motility of *Treponema pallidum* in an acellular culture medium. J. Gen. Microbiol. *114*, 443-447.

Chu, L., Burgum, A, Kolodrubetz, and Holt, S.C. (1995). The 46-kilodalton-hemolysin gene from *Treponema denticola* encodes a novel hemolysin homologous to aminotransferases. Infect. Immun. *63*, 4448-4455.

Chu, L., Dong, Z., Xu, X., Cochran, D.and J. Ebersole. (2002). Role of glutathione metabolism od *Treponema denticola* in bacterial growth and virulence expression. Infect. Immun. *70*(3), 1113-1120.

Chu, L., Ebersole, J. L., Kurzben, G. P. and Holt, S.C. (1997). Cystalysin, a 46-kilodalton cysteine desulfhydrase from *Treponema denticola* with hemolytic and hemoxidative activities. Infect. Immun. *65*, 3231-3238.

Chu, L., Kennell, W., and Holt, S.C. (1994). Characterization of hemolysis and hemoxidation activities by *Treponema denticola*. Microb. Pathog. *16*, 197-212.

Chu, L., Xu, X., Dong, Z., Cappelli, D., and Ebersole, J. (2003). Role for recombinant g-glutamyltransferase from *Treponema denticola* in glutathione metabolism. Infect. Immun. *71*(1), 335-342.

Cox, C. D. and Barber, M.K.(1974). Oxygen uptake by *Treponema pallidum*. Infect Immun *10*, 123-127.

Cox, D.L. (1994). Culture of *Treponema pallidum*. In Methods of Enzymology. Volume 236. Pages 390-404. Elsevier Science. New York.

Cox, D. L., Riley, B., Chang, P.,Sayahtaheri, S.,Tassell, S., and Hevelone, J. (1990). Effects of molecular oxygen, oxidation-reduction potential, and antioxidants upon *in vitro* replication of *Treponema pallidum* subsp. *pallidum*. Appl. Environ. Microbiol. *56*, 3063-3072.

Deka, R., Goldberg, M., Hagman, K., and Norgard, M.V. (2004). The Tp38 (TpMglB-2) lipoprotein binds glucose in a manner consistent with receptor function in *Treponema pallidum*. J. Bacteriol. *186*(8), 2303-2308.

Deka, R., Lee, YH., Hagman, L.K., Shevchenko, D., Lingwood, C., Hasemann, C., Norgard, M.V., and Radolf, J. D. (1999). Physiochemical evidednce that *Treponema pallidum* TroA is a zinc-containing metalloprotein that lacks porin-like structure. J. Bacteriol. *181*(14), 4420-3.

Doak, G. O., Freedman, L. D. and Clark, J. W. (1959). Ionic requirements of *Treponema pallidum*. I. Sodium and potassium. J. Bacteriol *77*, 322-327.

Eagle, H. (1956). The salt requirements of mammalian cells in tissue culture. Arch. Biochem. and Biophys. *61*(2), 356-66.

Fieldsteel, A. H., Becker, F. A., and Stout, J. G. (1977). Prolonged survival of virulent *T. pallidum* (Nichols strains) in cell-free and tissue culture systems. Infect. Immun. *18*, 173-182.

Fieldsteel, A. H., Cox, D. L., and Moeckli, R. A. (1981). Cultivation of virulent *T. pallidum* in tissue culture. Infect. Immun. *32*, 908-915.

Fieldsteel, A. H., Stout, J. G., and Becker, F. A. (1979). Comparative behavior of virulent strains of *Treponema pallidum* and *Treponema pertenue* in Gradient Cultures of Various Mammalian Cells. Infect. Immun. *24*, 337-345.

Fieldsteel, A. H., Stout, J. G., and Becker, F. A. (1981). Role of serum in survival of *Treponema pallidum* in tissue cultures. Infect. Immun. *17*, 28-32.

Fitzgerald, T. J. (1981). In vitro cultivation of *Treponema pallidum*. Bulletin of the WHO *59*(5), 787-912.

Fitzgerald, T. J. (1987). Activation of the classical and alternate pathways of complement by *Treponema pallidum* subsp. *pallidum* and *Treponema vincentii*. Infect. Immun. *55*, 2066-2073.

Fitzgerald, T. J. and Johnson, R.C. (1979). Mucopolysaccharidase of *Treponema pallidum*. Infect. Immun. *24*, 261-268.

Fitzgerald, T. J. and Johnson, R.C. (1979). Surface mucoploysaccharides of *Treponema pallidum*. Infect. Immun. *24*, 244-251.

Fitzgerald, T. J., Johnson, R.C., and Ritzi, D. M. (1979). Relationship of *Treponema pallidum* to acidic mucopolysaccharides. Infect. Immun. *24*, 252-260.

Fitzgerald, T. J., Johnson, R.C., Sykes, J. A., and Miller, J. N. (1977). Interaction of *Treponema pallidum* (Nichols Strain) with cultures mammalian cells, Effects of oxygen, reducing agents, serum supplements, and different cell types. Infect. Immun. *15*, 444-52.

Fitzgerald, T. J., Johnson, R.C., and Wolff, E.T. (1978). Mucopolysaccharide material resulting from the interaction of *Treponema pallidum* (Nichols strain) with cultures mammalian cells. Infect. Immun. *22*, 575-584.

Fitzgerald, T. J., Johnson, R.C., and Wolff, E.T. (1980). Sulfhydryl oxidation using procedures and experimental conditions commonly used for *Treponema pallidum*. Brit. J. Vener. Dis. *56*, 129-136.

Fitzgerald, T. J., Miller, J. N., and Sykes, J. A. (1975). *Treponema pallidum* (Nichols strain) in tissue cultures, cellular attachment, entry, and survival. Infect. Immun. *11*, 1133-1140.

Fitzgerald, T. J., Repesh, L.A., and Oakes, S.G. (1982). Morphological destruction of cultured cells by the attachment of *Treponema pallidum*. Brit. J. Vener. Dis. *58*, 1-11.

Fraser, C. M., Casjens, S., Huang, W.M.,Sutton, G.G., Clayton, R., Lathigra, R., White, O., Ketchum, K.A., Dodson, R. et al., (1997). Genomic sequence of a Lyme disease spirochaete, *Borrelia burgdorferi*. Nature *390*(6660), 580-6.

Fraser, C., Norris, SJ., Weinstock, GM.,White, O., Sutton, G.G., Dodson, R., Gwinn, M., Hickey, E.K., Clayton, R., et al. (1998). Complete genome sequence of *Treponema pallidum*, the syphilis spirochete. Science *281*(5375), 375-88.

Gherardini, F.C, Hobbs, M,, Stamm, L.V., and Bassford, Jr., P.J. (1990). Complementation of an *Escherichia coli* proC Mutation by a Gene Cloned from *Treponema pallidum*. J. Bacteriol. *172*(5), 2996-3002.

Graves, S. R. and Billington, T. (1979). Optimum concentration of dissolved oxygen for the survival of virulent *Treponema pallidum* under conditions of low oxidation-reduction potential. Brit. J. Vener. Dis. *55*, 387-393.

Hardham, J., Stamm, L.V., Porcella, S.F., Frye, J.G., Barnes, N.Y., Howell, J.K., Mueller, S.L., Radolf, J.D., Weinstock, G.M., and Norris, S.J. (1997). Identification and transcriptional analysis of a *Treponema pallidum* operon encoding a putative ABC transport system, an iron-activated repressor protein homolog, and a glycolytic pathway enzyme homolog. Gene *197*, 47-64.

Hardy, P. and Munro, C. (1966). Nutritional requirements of anaerobic spirochetes. I. Demonstration of isobutyrate and bicarbonate as growth factors for a strain of *Treponema microdentium*. J. Bacteriol. *91*(1), 27-32.

Hazlett, K., Cox, D.L., Sikkink, R.A., Auch'ere, F., Rusnak, F., and Radolf, J.D. (2002). Contribution of neelaredoxin to oxygen tolerance by *Treponema pallidum*., In Methods in Enzymology, Volume 353. pages 140-56. Elsevier Science. New York.

Hazlett, K.R., Rusnak, F.Kehres, D.G., Bearden, S.W., La Vake, C.J. , La Vake, M.E., Maguire, M.E., Perry, R.D., and Radolf, J.D. (2003). The *Treponema pallidum tro* operon encodes a multiple metal transporter, a zinc-dependent transcriptional repressor, and a semi-autonomously expressed phosphoglycerate mutase. J. Biol. Chem. *278*, 20687-94.

Hewitt, J. and Morris, J. G. (1975). Superoxide dismutase in some obligately anaerobic bacteria. FEBS Letters *50*(3), 315-318.

Jovanovic, T., C. Ascenso, C., Hazlett, K.R., Sikkink, R., Krebs, C., Litwiller, R., Benson, L.M., Moura, I., Moura, J.J., Radolf, J.D., Huynh, B.H., Naylor, S., and Rusnak, F. (2000). Neelaredoxin, an iron-binding protein from the syphilis spirochete, *Treponema pallidum*, is a superoxide reductase. J. Biol. Chem. *275*(37), 28439-28448.

Kaneshiro, T. and Law, J. H. (1964). Phosphatidylcholine synthesis in *Agrobacterium tumefaciens* I. Purification and properties of a phosphatidylethanolamine N-methyltransferase. J. Biol. Chem. *239*, 1705–1713.

Kast, C. C., and Kolmer, J.A. (1929). Concerning the cultivation of *Spirochaeta pallida*. Am. J. Syphilis *13*, 419-453.

Kast, C. C., and Kolmer, J.A. (1940). Methods for the isolation and cultivation of treponemes, with special reference to culture media. Am. J. Syphilis *6*, 671-683.

Kawata, T. (1967). Presence of cytochromes in the Reiter Treponeme. J. Gen. Appl. Microbiol. *13*, 405-6.

Kent, C., Gee, P., Lee, S. Y., Bian, X., and Fenno, J. C. (2004). A CDP-choline pathway for phosphatidylcholine biosynthesis in *Treponema denticola*. Molecular Microbiology *51*(2), 471–481.

Konishi, H., Yoshii, Z., and Cox, D.L. (1984). Electron microscopy of *Treponema pallidum* (Nichols) cultivated in tissue cultures of Sf1Ep cells. Infect. Immun. *53*, 32-37.

Lee, Y., Deka, R., Norgard, M. V., Radolf, J. D., and Hasemann, C. (1999). *Treponema pallidum TroA, is a periplasmic zinc-binding protein with a helical backbone*. Nature Structural Biology *6*(7), 628-633.

Leschine, S. and Canale-Parola, E. (1980). Rifampin as a selective agent for isolation of oral spirochetes. J. Clin. Microbiol. *12*(6), 792-795.

Livermore, B. and Johnson, R.C. (1970). Isolation and characterization of a glycolipid from Treponema pallidum, Kazan 5. Biochemica Biophysica Acta *210*(2), 315-318.

Livermore, B. and Johnson, R.C. (1974). Lipids of the *Spirochaetales*, comparison of the lipids of several members of the genera *Spirochaeta*, *Treponema*, and *Leptospira*. J. Bacteriol. *120*(3), 1268-1273.

Loesche, W. (1969). Oxygen sensitivity of various anaerobic bacteria. Appl. Microbiol. *18*(5), 723-727.

Lysko, P. G. and Cox, C.D. (1977). Terminal electron transport in *T. pallidum*. Infect. Immun. *16*, 885-890.

Lysco, P. and Cox, C.D. (1978). Respiration and oxidative phosphorylation in *T. pallidum*. Infect. Immun. *21*, 462-473.

Makinen, P. L. and Makinen, K.K. (1997). g-Glutamyltransferase from the outer cell envelope of *T. denticola* ATCC35405. Infect. Immun. *65*, 685-691.

Matthews, H. M., Jenkin, H.M., Crilly, K., and Sandok, P.L. (1979). Effects of fatty acids on motility retention by *Treponema pallidum in vitro*. Infect. Immun. *19*(3), 814-21.

Matthews, H.M.,Yang, T.K., and Jenkin, H.M. (1979b). Unique lipid composition of *Treponema pallidum* (Nichols virulent strain). Infect. Immun. *24*(3), 713-719.

Metzger, M. and Smogor, W. (1966). Study of the effect of pH and Eh values of the Nelson-Diesendruck medium on the survival of virulent *T. pallidum*. Archivum Immunologiae et Therapiae Experimentalis. *14*, 445-453.

Meyer, H. and Meyer, F. (1971). Lipid metabolism in the parasitic and free-living spirochetes *Treponema pallidum* (Reiter) and *Treponema zuelzerae*. Biochimica et Biophysica Acta *231*(1), 93-106.

Meyer, M. (1966). Metabolic characterization of the genus *Brucella*. V. Relationship of strain oxidation rate of i-erythritol to strain virulence for guinea pigs. J. Bacteriol. *92*(3), 584-548.

Mikx, F., Jacobs, F., and Satumalay, C. (1992). Cell-bound peptidase activities of Treponema denticola ATCC 33520 in continuous culture. J. Gen. Micro. *138*(9), 1837-1842.

Mikx, F. H. (1997). Environmental effects on the growth and proteolysis of *Treponema denticola* ATCC 33520. Oral Microbiology and Immunology. *12*, 249-253.

Morris, J. (1976). Fifth Stenhous-Williams memorial lecture. Oxygen and the obligate anaerobe. J. Appl. Bacteriol. *40*(3), 229-244.

Moss, C. W., Thomas, M. L., and Lambert, M. A. (1971). Amino acid composition of treponemes. Brit. J. Vener. Dis. *47*, 165-168.

Nelson, R. A. and Diesendruck, J.A. (1951). Studies on treponemal immobilizing antibodies in syphilis. I. Techniques of measurement and factors influencing immobilization. J. Bacteriol. *66*, 667-685.

Nelson, R. A. and Mayer, M.M. (1949). Immobilization of *Treponema pallidum in vitro* by antibody produced ib syphilitic infection. J. Exp. Med. *89*, 369-393.

Nelson, R. A. and Steinman, H.G. (1948). Factors affecting the survival of *Treponema pallidum in vitro*. Am. J. of Syph., Gonor., and Ven. Dis. *36*, 588.

Nichols, J. C. and Baseman, J.B. (1975). Carbon sources utilized by virulent *Treponema pallidum*. Infect. Immun. *12*, 1044-1050.

Nichols, J. C. and Baseman, J.B. (1978). Ribosomal ribonucleic acid synthesis by virulent *Treponema pallidum*. Infect. Immun. *19*, 854-860.

Noguchi, H. (1911). A method for the pure cultivation of the pathogen *Treponema pallidum* (*Spirochaeta pallida*). J. Exper. Med. *14*, 99-108.

Noguchi, H. (1912). A method for cultivating *Treponema pallidum* in fluid media. J. Exper. Med. *16*, 211-215.

Norris, S. J. (1982). In vitro cultivation of *T. pallidum*, independent conformation. Infect. Immun. *36*, 437-439.

Norris, S. J. and Edmondson, D.G. (1986). Factors affecting the multiplication and subculture of *Treponema pallidum* subsp. *pallidum* in a tissue culture system. Infect. Immun. *53*, 534-539.

Norris, S. J. Miller, J. N., Sykes, J. A., and Fitzgerald, T. J. (1978). Influence of oxygen tension, sulfhydryl compounds, and serum on the motility and virulence of *Treponema pallidum* (Nichols strain) in a cell-free system. Infect. Immun. *22*, 689-697.

Oyama, V., Steinman, H.G., and Eagle, H. (1953). The nutritional requirements of *Treponemata*. V. A detoxified lipide as the essential growth factor supplied by crystallized serum albumin. J. Bacteriol. *65*, 609-16.

Posey, J., Hardham, J.M., Norris, S.J., and Gherardini, F.C (1999). Characterization of a manganese-dependent regulatory protein, TroR, from *Treponema pallidum*. PNAS *96*(10), 10887-92.

Power, D. and Pelczar, M.J. (1959). Growth enhancement of the Reiter treponeme by fatty acids. J. Bacteriol. *77*, 789-93.

Rhodes, B., Liu, H., Johnson, S., George, R., and Steiner, B. (2000). Molecular cloning of a gene (polA) coding for an unusual DNA polymerase from *Treponema pallidum*. J. Med. Microbiol. *49*, 657-667.

Riley, B. S. and Cox, D.L. (1988). Cultivation of cottontail rabbit epidermal (Sf1Ep) cells on microcarrier beads and their use for suspension cultivation of *T. pallidum* subsp. *pallidum*. Appl. and Environ. Microbiol. *54*, 2862-2865.

Rother, M., Bock, A., and Wyss, C. (2001). Selenium-dependent growth of *Treponema denticola*, evidence for a clostridial-type glycine reductase. Arch. Microbiol. *177*, 113-116.

Sandok, P.L., Knight, S.T., and Jenkin, J.H.. (1976). Examination of various cell culture techniques for co-incubation of virulent *Treponema pallidum* (Nichols I strain) under anaerobic conditions. J. Clin. Microbiol. *4*(4), 360-371.

Sandok P.L., and Jenkin, J.H. (1978). Radiolabeling of *Treponema pallidum* (Nichols virulent strain) *in vitro* with precursors for protein and RNA biosynthesis. Infect. Immun. *22*(1), 22-28.

Schereschewsky, F. and Hoffman, E. (1909). Vorlaufiger Bericht uber dasVorkommen von Spirochoeten in syphilischen Krankheitsprodukten und bei Papillomen. Deutsche Medizinische Wochenschrift *35*, 835.

Schiller, N. L. and Cox, C.D. (1977). Catabolism of glucose and fatty acids by virulent *Treponema pallidum*. Infect. Immun. *16*, 60-68.

Schaudinn, F. and Hoffman, E. (1904-05). Vorlaüufiger Bericht über des Vorkommen von Spirochäten in Syphilitischen Krankheitsprodukten und bei Papillomen. Arb. Gesundh amte (Berl.) *22*, 527-534.

Schaudinn, F. and Hoffman, E. (1905). Über Spirochaetenbefunde im Lynphdrasensaft syphilitischer. Dtsch. Med. Wochenschr. *31*, 711-714.

Seshadri R, Meyers, G.S., Tettelin, H., Eisen, J.A., Heidelberg. J.F., Dodson, R.J., Davidsen, T.M., DeBoy, R.T., et al. (2004). Comparison of the genome of the oral pathogen *Treponema denticola* with other spirochete genomes. PNAS *101*(15), 5646-5651.

Smibert, R. (1971). The Isolation, Cultivation, and Characterization of Treponemes. WHO VDT Res. *71*, 242-252.

Smibert, R. (1974). The Spirochaetales in CRC Handbook of Microbiology.pages152-181. CRC Press. Cleveland OH

Smibert, R. (1973). Spirochaetales -A Review. CRC Critical Reviews in Microbiology *2*, 491-552.

Socransky, S. and Hubersak, C. (1967). Replacement of ascitic fluid or rabbit serum requirement of *Treponema dentium* by alpha-globulin. J. Bacteriol. *94*(5), 1795-1796.

Sperry, J. and Robertson, D. (1975). Erythritol catabolism by *Brucella abortus*. J. Bacteriol. *121*(3), 619-30.

Stamm, L., Greene, S.,and Barnes, N.Y. (1997). Identification and characterization of the gyrB gene from *Treponema pallidum* subsp. *pallidum*. FEMS Micro. Let. *153*(1), 129-134.

Stamm, L., Parrish, E., and Gherardini, F.C. (1991). Cloning of the recA gene from a free-living leptospire and distribution of RecA-like protein among spirochetes. J Antimicrobiol. Chemother. *29*(4), 447-53.

Stamm, L. V. and Bassford Jr., P. J. (1985). Cellular and extracellular protein antigens of *T. pallidum* synthesized during *in vitro* incubation of freshly extracted organisms. Infect. Immun. *47*, 799-807.

Steiner, B. M., Wong, G. H. W., Drummond, L., and Graves, S. R. (1983). The role of respiratory protection on increased survival of *Treponema pallidum* (Nichols) when co-cultivated with mammalian cells *in vitro*. Brit. J. Vener. Dis. *29*, 1595-1600.

Steiner, B. M., Wong, G. H. W., and Graves, S. R. (1984). Susceptability of *Treponema pallidum* to the toxic products of oxygen reduction and the non-treponemal nature of its catalase. Br. J. Vener. Dis. *60*, 14-22.

Steiner, B. M., Wong, G. H. W., Sutrave, P., and Graves, S. R. (1983). Oxygen toxicity in *Treponema pallidum*, deoxyribonucleic acid single-stranded breakage induced by low doses of hydrogen peroxide. Can. J. Microbiol. *30*, 1467-1476.

Steinman, H., and Eagle, H, (1950). Nutritional requirements of *Treponemata*. II. Pantothenic acid, glutamine, and phenylalanine as additional growth-promoting factors for the Reiter treponeme. J. Bacteriol. *60*, 57-68.

Steinman, H., H. Eagle, and Oyama, V.I. (1952). The nutritional requirements of *Treponemata*. III. A defined medium for cultivation of the Reiter treponeme. J. Bacteriol. *64*, 265-9.

Steinman, H., H. Eagle, and Oyama, V.I. (1953). Nutritional requirements of *Treponemata*. IV. The total nitrogen requirements of the Reiter Treponeme. J. Biol. Chem. *200*, 775-85.

Steinman, H., V. Oyama, V.I. and Schulze, HO. (1954). The nutritional requirements of *Treponemata*. VI. The total vitamin requirements of the Reiter treponeme. J. Bacteriol. *67*(5), 597-602.

Strugnell, R., Handley, C.J., Drummond, L.P., and Faine, S. (1986). Characterization of the proteoglycans synthesized by rabbit testis in response to infection by *Treponema pallidum*. Am. J. Path. *124*(2), 216-225.

Strugnell, R., Kent, T., Handley, C.J., and Faine, S. (1988). Experimental syphilitic orchitis. Relationship between *Treponema pallidum* infection and testis synthesis of proteoglycans. Am. J. Path. *133*(1), 110-117.

Syad, S. A., Makinen K. K., Makinin P.L., Chen C.Y., and Muhammad Z. (1993). Proteolytic and oxidoreductase activity of *Treponema denticola* ATCC 35405 grown in an aerobic and anaerobic gaseous environment. Res. Microbiol. *144*, 317-326.

Thornburg, R. and Baseman, J.B. (1983). Comparison of major protein antigens and protein profiles of *Treponema pallidum* and *Treponema pertenue*. Infect. Immun. *42*(2), 623-627.

Uitto, V., Grenier, D., Chan, EC., and McBride, BC. (1988). Isolation of a chymotrypsinlike enzyme from *Treponema denticola*. Infect. Immun. *56*, 2717-2722.

Van Horn, K.G. and Smibert, R.M. (1983). Albumin requirement of *Treponema denticola* and *Treponema vincentii*. Can. J. Microbiol. *29*(9), 1141-1148.

Wachter, M., and Johnson, R.C. (1976). Treponeme outer envelope, chemical analysis. Proceedings of the Society for Experimental Biology and Medicine. *151*(1), 97-100.

Weber, M. M. (1953). Factors influencing the *in vitro* survival and the virulent strain of *Treponema pallidum*. Baltamore, MD, Johns Hopkins University School of Hygeine and Public Health.

Weber, M. M. (1960). Factors influencing the *in vitro* survival of Treponema pallidum. Am. J. Hygeine *71*, 401-417.

Weinberg, M., Jenney Jr, F.E., Cui,, X., andAdams, M.W. (2004). Rubrerythrin from the hyperthermophilic archaeon *Pyrococcus furiosus* is a rubredoxin-dependent, iron-containing peroxidase. J. Bacteriol. *186*(23), 7888-7895.

Weinstock, G.M., Hardham, J.M., McLeod, M.P., Sodergren, E.J., and Norris, S.J. (1998). The genome of *Treponema pallidum*, new light on the agent of syphilis. FEMS Microbiology Reviews *22*(4), 323-332.

Wright, M. (1962). Exploratory studies in tissue culture of *Treponema pallidum*. Washington, DC, Exerpta Media Foundation.

Wyss, C. (1992). Growth of *Porphyromonas gingivalis, Treponema denticola, T. pectinovorum, T. socranskii*, and *T. vincentii* in a chemically defined medium. J. Clin. Microbiol. *30*(9), 2225-2229.

5

Metal Utilization and Oxidative Stress

Frank C. Gherardini, Julie A. Boylan and Paul J. Brett*

Abstract

Spirochetes living in oxygen-rich environments or when challenged by host immune cells are exposed to reactive oxygen (ROS) and reactive nitrogen (RNS) species. These can harm polyunsaturated lipids, cysteinyl residues, iron-sulfur clusters and DNA (Fenton Reaction) leading to inhibition of growth and, if the damage is too extensive, cell death. As is the case with other bacteria, spirochetes harbor genes encoding proteins such as superoxide dismutases (SOD), superoxide reductase (SOR), alkylhydroperoxide reductases (AhpR), peroxidases and small molecular weight thiols (*i.e.*, thioredoxin, glutathione, CoA) to protect cellular components and maintain intracellular oxidation/reduction potential (redox). While defending against an oxidative challenge, cells must also maintain sufficient, biochemically relevant intracellular concentrations of the metals which constitute a major ROS/RNS target. To maintain this balance, bacterial cells coordinately regulate metal transport and defense enzymes using oxidative stress response regulators and metal-dependent repressors. Within the *Spirochetales*, different systems have evolved to deal with the interplay between the requirements for metals and oxidative challenge. In this chapter, we compare the more "conventional" mechanisms seen in *Treponema denticola* and *Leptospira interrogans* with the more limited systems seen in *Treponema pallidum* and *Borrelia burgdorferi*.

1. Introduction

One imposing challenge faced by living organisms is that of dealing with incomplete reduction products of oxygen. This assault primarily comes from reactions involving endogenous metalloproteins and flavoproteins as a consequence of exposure to or metabolism of oxygen. Pathogenic bacteria face a similar metabolic challenge but also are exposed to exogenous reactive oxygen (ROS) and/or nitrogen species (RNS) generated by host immune cells. These compounds constitute a formidable challenge that must be overcome for these organisms to successfully survive,

*For correspondence email fgherardini@niaid.nih.gov

colonize and cause disease in mammalian hosts. This holds true for pathogenic spirochetes particularly members of *Leptospira*, *Treponema* and *Borrelia*.

The intracellular targets of ROS and NOS include polyunsaturated lipids in bacteria that contain them (Beermann et al., 2000; Ben-Menachem et al., 2003; Imlay, 2003; Tasaka et al., 1996), cysteinyl residues, iron-sulfur clusters (Imlay, 2003) and DNA as a result of Fe^{2+} oxidation by H_2O_2 generating $OH\bullet^-$ (Fenton Reaction) which subsequently causes oxidative DNA damage (Imlay, 2003). Several recent studies suggest that DNA may be the major target of ROS and RNS in bacteria when intracellular "free" iron concentrations increase significantly (Keyer and Imlay, 1996) and H_2O_2 concentrations exceed 0.2 -1 μM (Seaver and Imlay, 2001). This poses an interesting "dilemma" for most pathogenic bacteria which must accumulate significant levels of iron for growth then defend against cellular damage resulting from its reactivity with ROS/RNS. Pathogenic spirochetes have evolved different "strategies" to deal with this. For example, *Treponema denticola* and *Leptospira interrogans* harbor numerous Fe-containing metalloproteins and inactivate ROS using enzymes, such as superoxide reductase (SOR), AhpR, and glutathione peroxidases that are commonly found in other bacteria (see Table 3) (Nascimento et al., 2004; Ren et al., 2003; Seshadri et al., 2004). Similarly, the *Treponema pallidum* genome contains similar defense enzymes and the genes encoding Fe metalloproteins (Fraser et al., 1998). Thus all of these spirochetes are capable of accumulating intracellular "free" Fe at concentrations sufficient to generate $OH\bullet^-$ suggesting that DNA may be a main target of ROS in these bacterium. In contrast, *Borrelia burgdorferi* contains no intracellular iron (Posey and Gherardini, 2000) and few ROS scavenging enzymes suggesting that it is innately more resistant to ROS and the primary targets in this bacteria may be cysteinyl residues and polyunsaturated lipids. Clearly, metal metabolism (particularly Fe) and oxidative stress response are interrelated and therefore, are addressed jointly in this chapter. Additionally, we will use *T. denticola* as a model for traditional metal and oxidative stress systems described in other bacteria (and *L. interrogans*) and for comparison with those more limited systems described in *T. pallidum* and *B. burgdorferi*.

2. Metal utilization

2.1. Introduction

Nearly one third of all structurally characterized proteins are metalloproteins (Butler, 1998). It is not surprising, therefore, that certain metals are essential for the survival of all known life forms. Metals are biologically significant due to their ability to function as catalysts, activate biochemical processes as well as adopt structural roles. Paradoxically, however, when present in excess these elements can be extremely toxic (Finney and O'Halloran, 2003; Rosen, 2002). Thus both eukaryotic and prokaryotic cells must possess specialized mechanisms for regulating the uptake and efflux of metals across biological membranes (Hall, 2002; Rosen, 2002). Traditionally, essential metals have included the alkali metals potassium (K) and sodium (Na), the alkaline earth metals calcium (Ca) and magnesium (Mg), and the transition metals copper (Cu), iron (Fe), manganese (Mn) and zinc (Zn). Other transition metals which have less well defined or more obscure biological functions include cobalt (Co), molybdenum (Mo) and nickel

(Ni). Metals such as iron are particularly important to biological systems because of their ability to adopt multiple redox and spin states under physiological conditions. Such properties make these metals extremely versatile prosthetic components for incorporation into proteins that serve as biocatalysts or electron carriers.

Metals are important for many biological processes including respiration, nitrogen fixation, methanogenesis, nucleic acid synthesis, oxygen transport, oxygen tolerance, proteolysis and hydrogen production/consumption since the essential steps in these processes are catalyzed by metalloenzymes (Andrews et al., 2003; Butler, 1998; Klebba, 2003). There are two main classes of metal-requiring enzymes that can be distinguished from one another based upon the strengths of their cation-protein interactions. Metalloenzymes contain tightly bound metals, most commonly the transition metals Co, Cu, Fe, Mn or Zn, whereas, metal-activated enzymes loosely bind metal ions such as the alkali and alkaline earth metals Ca, K, Na and Mg. Metals are known to participate in the catalytic process in three ways: i) by binding to substrates to orient them properly for a chemical reaction, ii) by mediating redox reactions through reversible changes in the metals oxidation state and iii) by electrostatically stabilizing or shielding negative charges.

Although dogma dictates that all living organisms require the use of iron for survival, a number of studies have indicated that there may be exceptions to this rule. Lactobacilli have been identified as iron-independent organisms (Archibald, 1983; Bruyneel et al., 1990). In particular, *Lactobacillus plantarum* has been shown to contain only one to two iron atoms per cell, a number that is considered to be far too low to be of any biological significance. It has been suggested, therefore, that the ability of these organisms to survive in the absence of iron is consistent with their ability to grow in highly iron-limited environments such as milk (Weinberg, 1997). More recently, analysis of the genome sequence of *B. burgdorferi* indicated that it has few orthologs of known iron-requiring proteins (Table 1) (Fraser et al., 1997). In addition, studies demonstrated that severe iron limitation had no effect on growth in defined media and that the iron content of the organism was less than 10 iron atoms per cell (Posey and Gherardini, 2000). In order to circumvent host-imposed iron limitations, it appears that this spirochete has eliminated the need for iron by reducing the number of genes that encode iron-dependent proteins and has opted to utilize cations other than iron as cofactors for the remaining metalloproteins. While simplistic in approach, the ability to eliminate cellular iron requirements is only likely to be feasible for organisms like *B. burgdorferi* that do not require iron-dependent metalloenzymes because of their ability to scavenge essential biosynthetic intermediates and growth factors from their environmental niches (Brown and Holden, 2002).

T. pallidum, like *B. burgdorferi*, is an obligate parasite with limited biosynthetic capabilities. It has thus been proposed that this pathogen may have a limited requirement for iron. While analyses of the *T. pallidum* genome do reveal the absence of genes predicted to encode heme-containing proteins and few Fe-S or non-Fe-S proteins, orthologs of iron-dependent proteins have been identified (Table 1) (Fraser et al., 1998; Hazlett et al., 2003). In fact, a number studies have recently confirmed the importance of iron as a cofactor in some of these proteins (Table 1) (Jovanovic et al., 2000). Thus, while *T. pallidum* may possess a diminished requirement for

Table 1. Metalloprotein orthologs identified in *T. pallidums*, *T. denticola* and *B. burgdorferi*

Protein	*T. pallidum* Gene Designation	*T. denticola* Gene Designation	*B. burgdorferi* Gene Designation
Iron-dependent enzymes			
Superoxide reductase (Neelaredoxin)	TP0823	TDE1754	--*
Rubredoxin	TP0991	TDE1052	--
Peptide Deformylase	TP0757	TDE1645	See Zn enzymes[1]
Ribonucleotide-diphosphate reductase (α/β)	TP1008/ TP0053	TDE0341/ TDE0335	-- [2]
Pyruvate dehydrogenase	TP0939	TDE0665 TDE2198	--
Bacterioferrin (DPS)	TP1038	--	--
(Dpr)	--	--	BB0690[3]
Manganese-dependent enzymes			
Methionine aminopeptidase	TP0842	TDE1733	BB0105
Aminopeptidase P	TP0569	TDE1482	BB0067
Phosphoenolpyruvate carboxykinase	TP0122	--	--
Superoxide dismutase	--	--	BB0153
Enolase	TP0817	TDE0949	BB0337
Adenine deaminase	--	--	BBK17
Zinc-dependent enzymes			
Aminopeptidase	TP0112	TDE2337	BB0069
DNA Topoisomerase I	TP0394	TDE1208	BB0828
DNA primase	TP0492	TDE1345	BB0710
Fructose-biphosphate aldolase	TP0662	-	BB0445
RNA polymerasae β′	TP0242	TDE2420	BB0388
Adenosine deaminase	TP0045	TDE0188	--
5′-nucleotidase	TP0104	TDE2257	--
Threonyl-tRNA synthetase	TP0837	TDE1090	BB0720
Methionyl-tRNA synthetase	TP0798	TDE1063	BB0587
Isoleucyl-tRNA synthetase	TP0452	TDE2663	BB0833
Zinc protease	TP0600	TDE2341	BB0118
α-Amylase	TP0147	TDE1395	--
Carboxypeptidases	TP0221 TP0800	TDE0633 TDE0979	BB0605 BB0582
[1]Peptide Deformylase	--	--	BB0065

* -- Means that no orthologs has been identified.

[1] Purified enzyme from *B. burgdorferi* is Zn-dependent.

[2] *B. burgdorferi* has no homolog and no detectable ribonucleotide reductase activity.

[3] The enzyme purified from *B. burgdorferi* contains no detectable metal.

iron in comparison to other well characterized bacterial pathogens, mounting evidence indicates that this organism, unlike *B. burgdorferi,* is likely to require iron for growth. More recently, the genome of the oral spirochete, *T. denticola*, has been sequenced and annotated. Genome analyses indicate that this spirochete has a variety of iron transport systems as well as a number of orthologs of well characterized iron-dependent enzymes (Table 1). Based upon these findings, it has been predicted that iron plays an important role in the physiology of this bacterium (Seshadri et al., 2004).

2.2. Acquisition from host sources

The ability of bacterial pathogens such as *T. pallidum*, *T. denticola* and *B. burgdorferi* to acquire essential elements from their host during an infection is critical for their survival (Blencowe and Morby, 2003; Cornelissen, 2003; Mulrooney and Hausinger, 2003; Outten and O'Halloran, 2001; Rensing and Grass, 2003). Although host tissues, fluids and secretions are seemingly rich in nutrients, elements such as iron are generally not freely available for use by pathogenic bacteria. In a human host, under aerobic conditions at neutral pH, the concentration of free iron is approximately 10^{-9} M because of the tendency of the metal to form insoluble hydroxy-aquo polymers (Braun, 2003; Ratledge and Dover, 2000). In tissues and extracellular fluids the concentration of soluble iron is further reduced due to the siderophilin, transferrin (Wooldridge and Williams, 1993). At mucosal surfaces, iron availability is similarly limited due to the activity of the glycoprotein, lactoferrin (Wooldridge and Williams, 1993). Interestingly, one of the major features of the acute phase response during periods of stress, trauma, inflammation or infection is to restrict serum concentrations of iron even further (Gaetke et al., 1997; Goldblum et al., 1987). Therefore, it has been estimated that the concentration of free iron in physiological fluids and tissues during an active infection would be $<10^{-18}$ M which is significantly lower than that required to sustain the growth of most pathogens (10^{-6} to 10^{-7} M) (Andrews et al., 2003; Bullen, 1981; Ferguson and Deisenhofer, 2004; Weinberg, 1978). The capacity of the host to limit Fe availability provides a powerful mechanism by which the growth of pathogenic microbes can be effectively limited. In many respects, the ability of host factors to restrict metal availability can be viewed as an important extension of the innate immune response. At present, the mechanisms by which pathogenic spirochetes obtain metals from tissues and fluids remains relatively undefined. The ability of these organisms to replicate and cause disease in their mammalian hosts, however, indicates that they must possess the means to compete for and acquire metals from host sources.

Bacterial pathogens have evolved both passive and active mechanisms for acquiring metals from their host environments (Ferguson and Deisenhofer, 2004). Passive mechanisms are associated with high rates of transport and low metal-binding affinities while active mechanisms show slower transport kinetics with higher binding affinities (Nikaido and Saier, 1992). With the exception of Fe, Gram-negative bacteria appear to acquire cations using outer membrane porins (Blencowe and Morby, 2003; Rensing and Grass, 2003). These proteins form cylindrical, fluid-filled channels in the outer membrane of cells allowing for the transport of small, hydrophilic solutes such as divalent metals into the periplasmic space (Jap and Walian, 1996). Since porin-based mechanisms are passive, cations transported in this way must be present at higher concentrations outside the bacterial cell. Potential host sources of these cations include ceruloplasmin, albumin, α2-macroglobulin and β1-globulin as well as other host serum proteins (de la Calle Guntinas et al., 2002). Although a number of putative outer membrane proteins have been identified in the genomes of *T. denticola*, *T. pallidum* and *B. burgdorferi*, their involvement in divalent cation transport remains to be determined.

Due to the extremely low availability of free Fe in physiological fluids and tissues, bacterial pathogens have evolved high-affinity active uptake systems to acquire Fe from their host environments (Brown and Holden, 2002; Cornelissen, 2003; Faraldo-Gomez and Sansom, 2003). Such high-affinity systems are capable of scavenging iron from transferrin and lactoferrin, heme-containing proteins including hemoglobin, hemopexin and haptoglobin as well as heme and hemin (Brown and Holden, 2002). In general, bacterial Fe uptake systems are siderophore- or receptor-mediated. Although these mechanisms are not mutually exclusive, mucosal pathogens appear to favor receptor-mediated strategies (Cornelissen, 2003; Ratledge and Dover, 2000). Analysis of the *T. pallidum* genome indicates that it lacks genes required for the synthesis, secretion or uptake of siderophores. Based upon these findings, it seems unlikely that siderophore-mediated uptake is an important means by which *T. pallidum* acquires Fe from a host. Results from a number of studies indicate that *T. pallidum* obtains Fe via direct binding of lactoferrin to cell-surface receptors. Three proteins with relative masses of 45, 40 and 34 kDa have been partially purified and shown bind lactoferrin. Interestingly, the mobility of the 34-kDa protein was affected by phospholipase D suggesting that this antigen may be a lipoprotein (Staggs et al., 1994). Transferrin may also serve as an iron source for *T. pallidum*. Although the *T. pallidum* genome does not contain orthologs for previously characterized transferrin binding proteins, a study demonstrated that the bacterium is capable of acquiring Fe from transferrin by an undefined mechanism (Alderete et al., 1988). Taken together, these data suggest that on a mucosal surface, *T. pallidum* utilizes lactoferrin as an Fe source while in deep tissues, it acquires Fe from sources such as transferrin. Further studies will be needed, however, in order determine the exact mechanisms by which *T. pallidum* acquires iron during an infection (Staggs et al., 1994).

Studies have also shown that *T. denticola* strains do not appear to produce siderophores (Chu et al., 1994; Scott et al., 1993). These findings are consistent with recent genome analyses indicating the absence of siderophore biosynthetic genes in the genome of the organism. Mounting evidence suggests that it may acquire iron via direct interactions with hemin. It has been proposed that the ability of *T. denticola* to express proteases, hemolysins and hemagglutinans may aid in this process by liberating hemin from subgingival tissues (Karunakaran and Holt, 1994; Scott et al., 1996). Under both iron rich and limiting conditions, *T. denticola* strains have been shown to express at least three hemin binding proteins with predicted masses of 47, 44 and 43 kDa (Chu et al., 1994; Scott et al., 1993). The 47-kDa antigen, which appears to be constitutively expressed, has been purified from *T. denticola* ATCC 35405 as well as *Treponema vincentii* and *Treponema socranskii* (Scott et al., 1993) and has the ability to bind hemin. The genes encoding the 44- and 43-kDa antigens from *T. denticola* ATCC 35404 have been cloned and sequenced (Xu et al., 2001).

They have been designated *hbpA* and *hbpB*, respectively, and sequence analysis of the predicted gene products indicated that they are 49% identical. However, no significant homology was found to other proteins in the protein databases nor were hemin-binding motifs identified suggesting that these proteins may represent a novel means by which *T. denticola* can acquire hemin-bound iron (Xu et al., 2001). Based upon the identification of at least two lactoferrin binding proteins, it has also been speculated that *T. denticola* may be capable of acquiring Fe from the lactoferrin found in saliva (Scott et al., 1996). Recently, a study supporting this hypothesis demonstrated that growth defects associated with *T. denticola hbpA* and/or *hbpB* mutants grown in Fe-limited media could be overcome by the addition of lactoferrin to the media. The addition of hemin or inorganic Fe only partially restored the mutant growth phenotype (Xu and Kolodrubetz, 2002). Transferrin does not appear to be utilized by *T. denticola* as an Fe source (Staggs et al., 1994).

2.3. Metal transport

Very little is known about the mechanisms by which metals are transported across the cell envelopes of spirochetes. Like other organisms, however, membrane transport systems presumably regulate the exchange of metals between the outer membrane, periplasmic space, the cytoplasmic membrane and the cytoplasm (Ferguson and Deisenhofer, 2004). In order to accomplish this, bacteria typically utilize three types of transport systems: i) ion channels (or porins) which function as highly selective pores that open and close in response to specific stimuli allowing ions to rapidly flow down a gradient, ii) primary active transporters, or ATP-binding cassette (ABC) transporters that directly utilize the free energy released from ATP hydrolysis to pump metals against a concentration gradient, and iii) secondary active transporters that couple electrochemical proton gradients to substrate uniport, symport or antiport across the cytoplasmic membrane (Ferguson and Deisenhofer, 2004). Most cations appear to diffuse passively into the periplasmic space via outer membrane porins following which they are actively pumped into the cytoplasm by specific transport systems (Jap and Walian, 1996). Iron, in contrast, must be actively transported through each layer of the cell envelope. The stepwise transport of Fe across cell envelopes has discrete energy requirements and involves both primary and secondary active transport systems. Since electrical potential gradients cannot be maintained across outer membrane and nucleoside triphosphates are not available in the periplasmic space, Fe transport depends upon the ability of cells to generate electrochemical gradients across cytoplasmic membranes (Bradbeer, 1993; Ferguson and Deisenhofer, 2004).

Analysis of the *B. burgdorferi* genome demonstrates the presence of very few putative metal transporters. Interestingly, none of those that have been identified are predicted to participate in the transport of transition metals (Table 2). The transport mechanisms by which zinc and other metals that are essential for the survival of *B. burgdorferi* remain to be identified. In contrast, the *T. pallidum* genome harbors a number of ORFs predicted to encode metal-transport related proteins (Table 2). Other than the Tro (transport related operon) system, very little is known about the metal specificity and/or significance of many of these proteins. The *tro* operon encodes six proteins including TroA, TroB, TroC, TroD, TroR and Pgm (Figure 1),

Table 2. Putative metal-transport orthologs identified in the genomes of *B. burgdorferi* and *T. pallidum*

Gene ID	Putative Function	Protein		Metal Specificity
B. burgdorferi				
BB0380	Cation transport protein	MgtE		Magnesium
BB0724	Cation transport protein	NtpJ		Potassium
BB0164	Cation exchange protein	-		Sodium/Calcium
BB0447	Cation antiporter	NapA		Sodium
BB0637	Cation antiporter	NhaC-1		Sodium
BB0638	Cation antiporter	NhaC-2		Sodium
T. pallidum				
TP0227*	ATP-binding protein	-		Cobalt
TP1036*	Cation-transporting ATPase, P-type	-		Copper
TP0972*	High-affinity permease	-		Iron
TP0163*	Periplasmic binding protein	TroA	ABC transporter	Manganese, Zinc
TP0164*	ATP-binding protein	TroB		
TP0165*	Permease protein	TroC		
TP0166*	Permease protein	TroD		
TP0024*	Cation transport protein	KtrA		Potassium
TP0139*	Cation transport protein	KtrA		Potassium
TP0140*	Cation transport protein	KtrB		Potassium
TP0513*	Cation transport protein	TrkA		Potassium
TP0774*	Cation transport protein	MgtC		Magnesium
TP0917*	Cation transport protein	MgtE		Magnesium
TP0056	Oxaloacetate decarboxylase - subunit α	OadA		Sodium
TP0057	Oxaloacetate decarboxylase - subunit β	OadB		Sodium
TP0034*	Periplasmic binding protein	ZnuA	ABC transporter	Zinc
TP0035*	ATP-binding protein	ZnuB		
TP0036*	Permease protein	ZnuC		

* orthologs identified in the genome of *T. denticola*

the first 5 of which are coordinately expressed from a σ^{70}-like promoter upstream of *troA* (Hardham et al., 1997). Four of these proteins (TroA-TroD) have been shown to be homologous to members of the ABC family of bacterial transport proteins. TroA, formerly known as Tromp1, has significant homology to bacterial periplasmic binding proteins and a family of streptococcal binding proteins and adhesins (Tasaka et al., 1996). TroB is homologous to the ATP-binding component, while TroC and TroD are similar to the membrane components of ABC transporters. TroR is homologous to a number of Gram-positive iron-activated repressor proteins

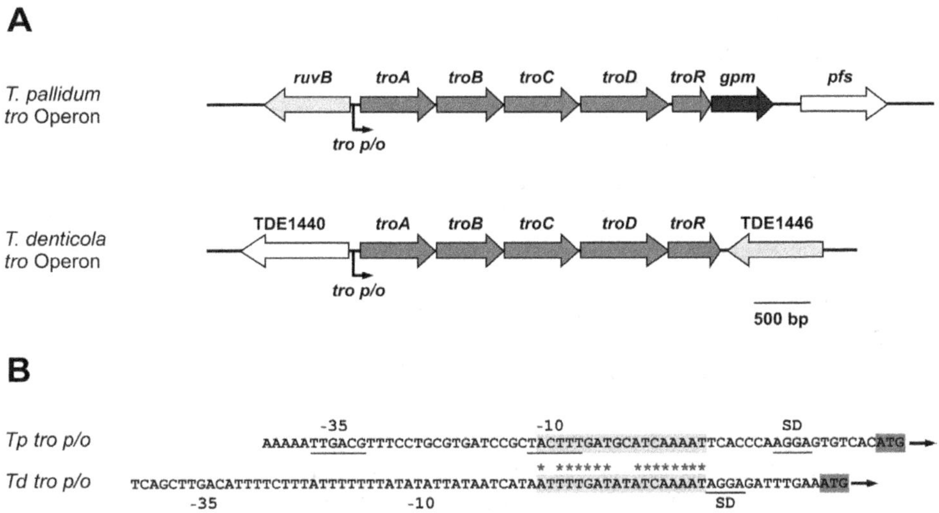

Figure 1. A schematic representation of the gene organization and regulatory region of the Transport Related Operons (Tro) from *T. pallidum* (top) and *T. denticola* (bottom). Panel A. The genes *troABCD* encode the Zn-Mn ABC transport system, *troR* encodes the metal-dependent repressor and *pgm* encodes a phosphate-dependent phosphoglycerate mutase. Panel B. The shaded area denotes the TroR binding site upstream of the *Tro* operon.

including DtxR and SirR. Pgm has been shown to encode for the glycolytic pathway enzyme, phosphoglycerate mutase (Hardham et al., 1997).

Originally, TroA was identified as a putative surface exposed outer membrane porin, however, subsequent studies have now confirmed that the protein is a periplasmic metal binding protein (Akins et al., 1997; Blanco et al., 1995; Deka et al., 1999). Lee et al. have solved a crystal structure of recombinant TroA complexed with zinc. The results of this study demonstrated that the protein was similar to other periplasmic solute-binding proteins in that it possessed two globular domains that interacted with one another to form a metal-binding pocket (Lee et al., 1999). More recently, Hazlett et al. have shown that purified TroA has equal affinity for Zn (1.27 ± 0.4 μM) and Mn (0.95 ± 0.2 μM) and that the expression of the *troABCD* genes in *E. coli* facilitated the transport of Zn, Mn and Fe into a variety of transport-deficient mutants (Hazlett et al., 2003). The results of the various studies indicate that the *T. pallidum* Tro transport system is likely to facilitate the uptake of multiple metal species and serve as an important means by which the spirochete acquires essential metals from host tissues. The *T. pallidum* genome also contains the genes encoding two additional transition metal transport systems. TP0034 (*znuA*), TP0035 (*znuB*) and TP0036 (*znuC*) encode proteins with homology to high affinity Zn ABC transport systems in Gram-positive bacteria (*i.e., Clostridium perfringens, Bacillus subtilis, Enterococcus faecalis* and *Streptococcus mutans,* www.tigr.org/tigrscripts/CMR2/GenePage.spl?locus=TP0034) while TP0972 has significant homology to a group of Gram-negative high affinity Fe permeases (Table 2). Taken together, these data suggest that *T. pallidum* has high affinity transport systems for Fe (TP0972), Mn (TroABCD) and Zn (ZnuABC and TroABCD).

Consistent with the predicted metabolic diversity of *T. denticola* in comparison to *T. pallidum* and *B. burgdorferi*, analysis of the genome indicates the presence of a significant number of putative metal-transport related loci. Besides harboring the genes encoding homologs for *T. pallidum* ZnuABC and TP0972 (Table 2), *T. denticola* appears to possess eight ABC-type transporters specifically devoted to the transport of Fe suggesting the importance of this metal to this spirochete (Seshadri et al., 2004). For a comprehensive list of putative metal transport related loci expressed by *T. denticola* refer to the following web link: http://www.tigr.org/tigr-scripts/CMR2/gene_table.spl?db=gtd. Of particular interest is the presence of a Tro ABC-transport system similar to that identified in *T. pallidum*. Comparisons of the two *tro* operons from *T. pallidum* and *T. denticola* demonstrates that the genetic organization of *troA-troR* is almost identical in both species except the *T. denticola pgm* is not physically linked at the end of the Tro operon as is the case in *T. pallidum*. Also, there is a high degree of sequence identity between the various gene products in the operons at the amino acid level (Figure 1). Further analyses have demonstrated that residues predicted to be important for the binding of metals to the *T. denticola* TroA protein are identical to those facilitating the same function in the *T. pallidum* orthologs (Brett et al., unpublished data). Interestingly, these preliminary findings suggest that the transport functions of the two Tro operons may be very similar (*i.e.*, zinc/manganese transport). At the present, the function(s) of the Tro operon in *T. dentiocola* sp. has not been determined experimentally.

2.4. Regulation of metal uptake and metabolism

Metals are important for the survival of all living organisms. Since elevated concentrations of both essential and non-essential metals can be toxic to cells, their uptake is often regulated by metal-dependent repressors (Glasfeld et al., 2003; Outten and O'Halloran, 2001; Patzer and Hantke, 2001). These proteins serve as a direct link between intracellular metal levels and the expression of appropriate uptake, efflux and storage systems. Typically, metal cations bind reversibly to a metal-sensing site altering the conformation of the repressor to effect changes in gene expression. The affinity of metal ions for the regulatory site serves to set the intracellular concentration of metals within the cell. The selectivity of the site is essential to ensure that other metals do not interfere with these homeostasis mechanisms (Guedon and Helmann, 2003). In addition to mediating physiological responses to environmental levels of metals, metal-dependent repressors modulate the virulence phenotypes of several important human pathogens (Guedon and Helmann, 2003; Jakubovics and Jenkinson, 2001). DxtR and Fur represent two distinct families of metal-dependent repressors. In many Gram-negative and Gram-positive bacteria, genes encoding for Fe uptake are negatively regulated by Fur (Patzer and Hantke, 2001). As intracellular concentrations of ferric iron increase, the Fe-loaded form of Fur binds to palindromic sequences within Fur-regulated promoters and represses transcription of downstream genes. Conversely, as intracellular levels of Fe decrease, the regulated genes become derepressed since the *apo*-form of Fur has a much lower affinity for cognate DNA operators than does the iron-bound form. In Gram-positive species with high GC content, DtxR-like proteins similarly regulate the expression of genes involved in both Fe uptake and virulence (Boyd et al., 1990).

Previous studies by Hardham et al. identified an open reading frame whose product, TroR, was predicted to function as a metal-dependent transcriptional regulator. TroR was shown to be homologous to members of the DtxR family of iron-dependent repressor proteins including DtxR, IdeR, SirR and DesR. Interestingly, unlike many of these known repressor proteins, TroR, was found to lack the characteristic C-terminal or SH3-like domain. Although analyses of the *tro* promoter region failed to identify a DtxR-like operator motif, a region with potential stem-loop structure overlapping the –10 motif was found. Comparison of this nucleotide sequence with other known operators, such as those for Fur and Fnr, demonstrated a significant level of identity (Hardham et al., 1997). Using DNase I footprinting, Posey et al. demonstrated that purified TroR could bind to and protect the *tro* promoter associated region of dyad symmetry previously identified by Hardham et al. indicating that the region was likely a TroR operator motif (Posey et al., 1999). Gel mobility-shift assays were also used to demonstrate the ability of specific metals to activate the binding of the purified TroR to DNA fragments containing the *tro* promoter region (Posey et al., 1999). The results of this analysis indicated that the only metal cation that promoted binding of TroR to its target sequence was Mn. Zn, Fe, Ni or Co had no effect on TroR binding. These findings are unique in that many other repressors belonging to the same family of proteins can be activated *in vitro* by a variety of metal cation effectors. Based upon the results of this study, it was proposed that the *T. pallidum* TroR regulated the expression of the *tro* operon in a Mn-dependent manner (Posey et al., 1999).

More recently, Hazlett et al. have challenged this hypothesis. By utilizing *E. coli* as a surrogate host for expression of the entire *T. pallidum tro* operon, they demonstrated that TroR-mediated repression of the *tro* genes responded to much lower external concentrations of Zn than Mn. Based upon these results, Hazlett et al. have proposed that while TroABCD are capable of transporting Zn, Mn or Fe, the operon is regulated primarily by Zn (Hazlett et al., 2003) which raises some interesting conundrums. In other well characterized bacterial systems, metal-dependent uptake systems are regulated by the metal they transport in order to maintain toxic metal ions within acceptable limits. In addition to the *tro* operon, *T. pallidum* harbors the genes encoding a putative high-affinity ABC Zn transport system (Table 2, TP0034, TP0035 and TP0036: Znu operon) and a putative high-affinity Fe uptake permease (Table 2, TP0972). Analyses of the DNA sequences upstream of the Znu operon and TP0972 failed to identify a TroR binding sequence (Figure 1, panel B) suggesting that these uptake systems are not regulated by TroR. Additionally, no other putative metal-dependent transcriptional regulators have been identified in the genome. Therefore, it is difficult to understand how *T. pallidum* maintains intracellular Zn and Fe homeostasis without these putative uptake systems being tightly regulated. It is possible that novel metal-dependent regulators remain to be identified, that TroR does regulate these transport systems by recognizing a different target sequence or that small RNAs play a role in metal-dependent regulation (Masse and Gottesman, 2002) in *T. pallidum*. Interestingly, the DNA sequence immediately upstream of the *T. denticola tro* operon contains a *T. pallidum* TroR binding sequence (Figure 1, panel B). Preliminary analyses indicate that both Tro proteins recognize the target sequence from either spirochete in a metal-dependent fashion (Brett et al., unpublished data). Thus more detailed analyses of TroR from *T. denticola* may help define the exact biochemical function of this important regulator.

3. Oxidative stress response

3.1. Sources of Reactive oxygen species in Spirochetes

Oxidative stress can be functionally defined as an excess of prooxidants in the cell. This stress occurs when the rate of generation and/or accumulation of reactive oxygen species exceeds the detoxification capacity of the cell. Active oxygen species are a consequence of life in an oxygen-rich environment and are produced as a by-product of normal aerobic metabolism (*e.g.*, respiration or fatty acid oxidation) or when O_2 oxidizes electron donating redox enzymes. In addition, ROS are used as cytotoxic weapons of the immune system and inflammatory response against invading microorganisms. The key reactive species are superoxide radicals ($O_2\bullet^-$), hydrogen peroxide (H_2O_2), hydroxyl radicals ($OH\bullet^-$), and singlet oxygen, which result from the sequential univalent reductions of molecular oxygen and nitric oxide (NO).

3.1.1. Superoxide radicals

Originally, $O_2\bullet^-$ was thought to be generated by the partial reduction of oxygen by cytochrome oxidase during oxygen metabolism. However, this does not seem to be the case. Superoxide radicals are generated when a single electron is transferred from a univalent donor to molecular oxygen and, *in vivo,* the most likely donors are reduced flavins functioning as cofactors in respiratory flavoproteins (*e.g.*, respiratory dehydrogenases). In exponentially growing *E. coli*, Imlay and Fridovich (1991) have shown that both $O_2\bullet^-$ and H_2O_2 are generated by the autoxidation of components of the respiratory chain (NADH dehydrogenase II, succinate dehydrogenase, and D-lactate dehydrogenase). Fumarate reductase, a terminal oxidase that is induced during anaerobic growth, reacts rapidly with oxygen and can be a major source of $O_2\bullet^-$ *in vivo* when cells are exposed to increased levels of O_2 (Imlay, 1995). The production of superoxide radicals can also occur by the autoxidation of other cellular components including ubiquinols, catechols, and thiols (Imlay, 2003). Quinone compounds, both natural cellular constituents and exogenous sources, can undergo a cycle of reduction/oxidation generating $O_2\bullet^-$. Finally, reduced transition metals, both free and in complexed form, can potentially donate a single electron to molecular O_2 to generate $O_2\bullet^-$. This reaction is readily reversible and depends upon the concentrations of reduced and oxidized metals, and of O_2 and $O_2\bullet^-$ (Farr and Kogoma, 1991).

Two key factors are most likely to mediate the levels of $O_2\bullet^-$ generated in the cytosol of the pathogenic spirochetes: the exposure to O_2 and the availability of univalent electron donors. For example, *T. denticola* colonizes anaerobic dental gingival pockets (Socransky et al., 1998) and may only be transiently exposed to O_2 concentrations high enough to generate significant levels of intracellular $O_2\bullet^-$ (Marquis, 1995). *B. burgdorferi*, during its infective cycle, is exposed to anaerobic and microaerobic conditions during passage through its arthropod vector (unpublished data) and microaerobic conditions in its mammalian hosts.

Table 3. Putative oxidative stress genes common to *T. pallidum*, *T. denticola* and *B. burgdorferi*

Oxidative Stress Protein	*T. pallidum* Nichols	*T. denticola* ATCC 35405	*B. burgdorferi* B31
Superoxide dismutase	--*	--	BB0153
ahpC	TP0509 (peroxiredoxin)	TDE0011	--
ahpF (NADH dehydrogenase)	--	TDE0153	--
Thioredoxin	TP0919	TDE0238 TDE0744	BB0061
Thioredoxin reductase	TP0814	TDE0743	BB0515
CoEnzyme A disulfide reductase	--	--	BB0728
Dps	TP1038	--	--
Dpr	--	--	BB0690
Fur	--	--	--
BosR	--	--	BB0647
TroR	TP0167	TDE1222	--
Sulfoxide reductase	TP0678 TP0633	TDE0709	BB0340
Superoxide reductase/ desulfoferrodoxin	TP0823	TDE1754	--
Glutathione peroxidase	--	TDE1729	--
NADH oxidase	TP0921	TDE0096 TDE2205 TDE0868	--
***In vivo* sources of superoxide**			
NADH dehydrogenase	--	TDE0153	BB0728
Lactate dehydrogenase	TP0037	TDE0351	BB0087
Fumarate reductase	--	Hydratase TDE1521 TDE1520	--
Other flavoproteins	TP0921	TDE1065 TDE0357 TDE1078	BB0728

* -- = no ortholog

T. pallidum survives on mucosal surfaces, skin and disseminated sites where O_2 concentrations vary from aerobic to microaerobic. It seems clear that *T. pallidum* and *B. burgdorferi*, to a lesser extent *T. denticola*, encounter O_2 at levels that could potentially generate biological relevant levels of cytosolic $O_2{\bullet}^-$.

Genome analyses of *T. pallidum*, *T. denticola*, and *B. burgdorferi* indicate that the central metabolism in these spirochetes relies on fermenting sugars to lactate or acetate via the Emden-Myerhof pathway (Fraser et al., 1997; Fraser et al., 1998; Seshadri et al., 2004). *T. pallidum* and *B. burgdorferi* are very metabolically limited and have no TCA cycle or respiratory chain. However, they do harbor the genes encoding putative flavin containing lactate dehydrogenases, NADH oxidases and a few other dehydrogenases and oxidases (Table 3, section titled *In vivo* sources of superoxide) which could potential serve as electron donors for reducing O_2 to $O_2{\bullet}^-$. *T. denticola* also has no respiratory chain but harbors several potential flavin containing enzymes (Table 3). In addition, *T. denticola* harbors the genes encoding the alpha and beta subunits of a putative fumarase although this enzyme has not been characterized and its potential for generating $O_2\bullet^-$ *in vivo* remains to be determined. Therefore, the extent to which superoxide radicals are produced during normal growth is unknown but is unlikely to contribute significantly to oxidative damage in these bacteria.

3.1.2. Hydrogen peroxide

Hydrogen peroxide, unlike $O_2\bullet^-$, readily diffuses across biological membranes so that its intracellular concentration is influenced by external as well as internal sources. Intracellular sources of peroxides include non-enzymatic, superoxide dismutase (SOD)-catalyzed dismutation, or superoxide reductase (SOR) reduction of superoxide. Since $O_2\bullet^-$ will oxidize transition metals such as iron (particularly iron-sulfur clusters) or reduced flavins, or is enzymatically converted to H_2O_2, it is highly likely that as the steady-state concentration of $O_2\bullet^-$ rises within a cell, the steady-state concentrations of H_2O_2 and $OH\bullet^-$ will also rise. In *E. coli*, the intracellular concentration of H_2O_2 (endogenous origin) is maintained at <20 nM by alkylhydroperoxide reductase (AphR) and catalase, and when concentrations approach 100 nM, DNA damage is detected (Seaver and Imlay, 2001). As concentrations reach 2 μM H_2O_2, cell growth was affected. However, as previously mentioned, it seems unlikely that under normal conditions, intracellular concentrations of $O_2\bullet^-$ in spirochetes reach levels that would dramatically increase H_2O_2 to levels that would cause cellular damage. It is much more likely that external H_2O_2 that is encountered by *T. pallidum*, *T. denticola*, and *B. burgdorferi* at inflammation sites (*e.g.*, chancres, rashes, inflamed dental gingival, etc.) is the major source of H_2O_2 challenging these pathogenic spirochetes.

3.2. Cellular targets of ROS

The biological targets for ROS are DNA, RNA, proteins, and lipids. Small molecules, including polyphenols, catecholamine, ascorbate, and thiols, can rapidly react with $O_2\bullet^-$ and thus, contribute to the propagation of radical chain reactions. Superoxide can inactivate enzymes like catalase or glutathione peroxidase and proteins containing iron-sulfur clusters (*e.g.*, aconitase) are highly sensitive to attack by $O_2\bullet^-$. Superoxide, however, cannot traverse biological membranes except in its protonated state. It does not react very rapidly with most lipids or DNA. The deleterious effects of $O_2\bullet^-$ are mainly related to its propensity to be converted to more potent oxidants like H_2O_2 or $OH\bullet^-$. Hydrogen peroxide efficiently oxidizes protein-thiols, and is likely to inactive enzymes that rely upon active-site cysteine residues for catalytic activity. However, the oxidation of cysteine residues and subsequent inactivation of thiol containing proteins may occur during oxidative stress in bacteria, but does not appear to be a major target of ROS (Imlay, 2002).

The most damaging effects of H_2O_2 result from its interactions with reduced iron (and to a lesser extent, copper) centers in proteins. The initial oxidation reaction causes the release of Fe^{3+} from Fe-S proteins contributing to the "free" Fe pool and generates $OH\bullet^-$ via the Fenton reaction (Imlay, 2003) (Figure 2). $OH\bullet^-$ reacts with most biomolecules and because it is so reactive, will react with them within a cell at diffusion-limited rates. The average diffusion distance is only a

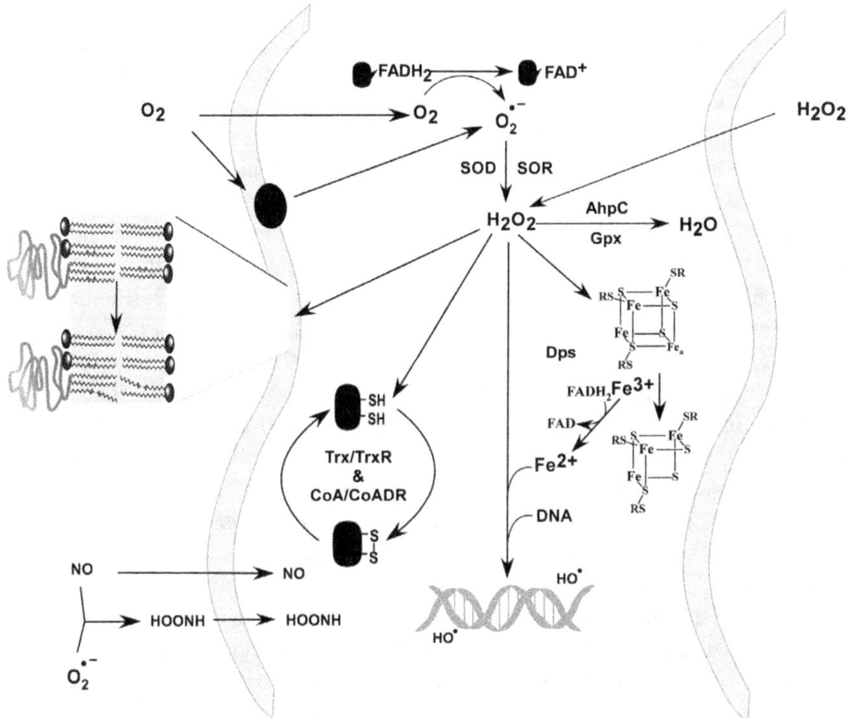

Figure 2. A schematic representation of oxidative stress in spirochetes. The arrows into the cell represent the potential sources of reactive oxygen species (ROS) (designated $O_2\bullet^-$, H_2O_2) or reactive nitrogen species (NOS) (designated NO, HOONH). The reaction of O_2 or $O_2\bullet^-$ with flavins ($FADH_2$) or from external sources (upper right, H_2O_2) probably represents the most significant challenge to *T. pallidum*, *T. denticola*, or *B. burgdorferi*. The arrow in the upper left represents the incomplete reduction of O_2 via a respiratory chain which is not likely to occur in these pathogenic spirochetes. Nitric oxide (NO), $O_2\bullet^-$, and peroxinitrites (HOONH) produced by the host represent a significant challenge during the infection process.

The arrows inside the cell represent the potential targets for ROS/NOS including Fe-S centers (designated: Fe^{3+}) and DNA via the Fenton reaction (designated: $Fe^{2+} + H_2O_2$ and DNA). These reactions do not occur in *B. burgdorferi* and may be limited in *T. pallidum* but would be a significant challenge to the nucleic acids of *T. denticola*. The ROS-dependent oxidation of reduced cysteines in proteins (designated S-S) could inhibit enzymatic activity and, like other organisms, spirochetes contain small molecular weight thiols (*e.g.*, thioredoxin, CoA) to protect enzyme function and maintain intracellular redox. The reaction within the membrane (magnified inset) represents lipid peroxidation which could potentially occur in *B. burgdorferi, T. pallidum* and *T. denticola*.

Defense enzymes against ROS are designated as: SOD for superoxide dismutase (in *B. burgdorferi*), SOR for superoxide reductase (in *T. pallidum* and *T. denticola*), AhpC for alkylhydroperoxide reductase (in *T. pallidum* and *T. denticola*), Gpx for glutathione peroxidase (in *T. denticola*), Dps for DNA/Fe binding protein (in *T. pallidum*), Trx/TrxR for thioredoxin and thioredoxin reductase (in *T. pallidum*, *T. denticola* and *B. burgdorferi*), and CoA/CoADR for coenzyme A and coenzyme A disulfide reductase (in *B. burgdorferi,* BB0728, Boylan et al., 2006).

few nanometers, and thus, its effect on any given biomolecule will depend largely upon proximity to the target. Because Fe can localize along the phosphodiester backbone of nucleic acid, DNA is a major target of HO•⁻ Active species can attack both the base and sugar moieties producing single and double strand breaks in the backbone and crosslinks to other molecules. Since some of the base damage can result in miscoding, lesions formed by endogenous oxidants may be a significant or even preponderant source of spontaneous mutagenesis in aerobically growing cells.

The amount of DNA damage that results from Fenton chemistry in spirochetes probably varies greatly and depends upon the metal metabolism of the bacteria. The extreme case is *B. burgdorferi* which has been shown to not require Fe for growth and has intracellular Fe concentrations estimated to be <10 atoms per cell (Posey and Gherardini, 2000). At those levels, it is unlikely that *B. burgdorferi* DNA is a target for ROS. In support of this, growth of *B. burgdorferi* in the presence of high concentrations of peroxides (50 mM H_2O_2 or *t*-butyl peroxide) has little to no effect on the DNA mutation rate (Boylan and Gherardini, unpublished data). *T. pallidum* harbors Fe-containing proteins (Table 1) but there are far fewer than the numbers observed in bacteria with larger genomes. In *E. coli*, it has been estimated that 5-100 nM "free" Fe is required to generate enough OH•⁻ to cause a biologically significant number of DNA lesions (Imlay, 2003). With a limited number of Fe-S proteins in its genome, it seems quite possible that "free" Fe concentrations in ROS challenged *T. pallidum* may not reach those levels (Posey et al., 1999; Radolf et al., 1999). This suggests that DNA may not be a major target of ROS in *T. pallidum*. Conversely, it has been shown that *T. denticola* requires significant amount of Fe for growth and its genome harbors several genes encoding Fe-S proteins (Table 1). When exposed to oxygen or challenged by exogenous ROS, it seems likely that the *T. denticola* chromosome would be a major target of OH•⁻.

Lipids are a major target during oxidative stress in eukaryotes. Free radicals can attack polyunsaturated fatty acids (*e.g.*, linoleic acid) in membranes and initiate lipid peroxidation. A primary effect of this is a decrease in membrane fluidity which affects the properties of the membrane and alters the function of membrane-associated proteins. Almost all reactions of singlet oxygen with biomolecules are additions of the oxygen to conjugated bonds. Typically, peroxides are the initial products. Singlet oxygen can add to olefins to produce allylic hydroperoxides and a shift in the double bond. Many of the sensitizers which produce singlet oxygen are embedded within the membrane and it is possible that singlet oxygen is capable of initiating lipid peroxidation. There is also growing evidence that implicates iron-oxygen complexes, called perferryl radicals, in the initiation of lipid peroxidation. Once lipid peroxides have formed they react with adjacent polyunsaturated lipids causing an amplification of the damage. Lipid peroxides degraded into a variety of products including aldehydes which can subsequently damage membrane proteins. Unlike reactive free radicals, aldehydes are rather long-lived and can therefore diffuse from the site of origin and attack targets which are distant from the initial reaction.

Are bacterial membranes targets of ROS? Until recently, it was assumed that bacterial lipids were not subject to oxidative damage observed in eukaryotic cells. Only certain polyunsaturated lipids, such as linoleic and linolenic acid, are subject to attack (Gutteridge and Halliwell, 1990) and it is clear that most bacteria do not synthesize or contain these types of lipids in their membranes. One notable exception is the photosynthetic bacteria which synthesize and incorporate significant levels of linoleic acid in their membrane (Tasaka et al., 1996). Since their photosynthetic and respiratory systems generate incomplete reduction products of O_2, it seems likely that these bacteria would be susceptible to lipid peroxidation. Interestingly, linoleic acid is incorporated into the membranes of some pathogenic bacteria including *B. burgdorferi* and *T. pallidum.* Since these bacteria cannot synthesize their own lipids, they must scavenge them so that their membrane composition reflects the host's lipid profile or that of their growth medium (Barbour, 1984; Fraser et al., 1997; Matthews et al., 1979). Approximately 10% of the total lipids in *B. burgdorferi* is linoleic acid and 3-5% is linolenic acid (unpublished data) (Beermann et al., 2000; Ben-Menachem et al., 2003) suggesting that their membranes could undergo peroxidation when exposed to external ROS. This seems to be the case since significant levels of aldehydes were measured in membranes isolated from *B. burgdorferi* cells treated with purified lipoxigenase (Treglown and Gherardini, unpublished data). It is possible that *T. denticola* and *T. pallidum* membrane lipids are also able to be oxidized by ROS since these bacteria also contain measurable levels of linoleic acid [~5% (Yotis et al., 1993) and ~ 10% (Matthews et al., 1979) respectively]. In *B. burgdorferi,* membrane lipids may possibly be a more significant target for ROS than DNA.

Finally, mechanisms of NO toxicity are complex and may include its direct reactions with heme, non-heme iron, and copper proteins. Nitric oxide blocks bacterial respiration *in vitro* by binding the heme and/or copper sites of cytochrome oxidases and inactivation of aconitase. Reactions between NO and $O_2{\bullet}^-$ produce another oxidant, peroxynitrite, that rapidly oxidizes dehydratase clusters, similar to $O_2{\bullet}^-$, and oxidizes protein thiols, like H_2O_2. All of these ROS are deleterious to the cell, since they lead to a loss of function of membranes and proteins, and block DNA replication or cause mutations. Clearly, for pathogenic spirochetes, these compounds represent as significant a challenge as H_2O_2 during certain stages of their respective diseases.

3.3. Cellular defenses

3.3.1. Superoxide dismutases and reductases

Cellular defenses against the damaging effects of ROS involve both enzymatic and nonenzymatic components. The enzymatic components may directly scavenge ROS or may act by producing the nonenzymatic antioxidants. There are five categories of enzymes that provide the bulk of protection against deleterious reactions involving ROS in bacteria: Superoxide dismutases (SODs)/superoxide reductases (SORs), hydroperoxidase, catalases, AhpR, Dps/Dpr proteins. Superoxide dismutases, ubiquitous among aerobic and most aerotolerant organisms, provide a defense against oxidative stress by catalyzing the dismutation of superoxide into oxygen and hydrogen peroxide through alternate oxidation and reduction of their catalytic metal ions (McCord and Fridovich, 1969; McCord et al., 1971). SODs react with superoxide with a rate constant close to the diffusion limit ($k= 2 \times 10^9\ M^{-1}\ s^{-1}$ at pH 7.4). To date, four types of superoxide dismutase have been identified: manganese (MnSODs), iron (FeSODs), copper/zinc (Cu/ZnSODs), and a recently identified nickel containing form of the enzyme (NiSOD) from *Streptomyces* (Youn et al., 1996a; 1996b) and cyanobacteria (Palenik et al., 2003). The MnSOD was the first identified form and is encoded by *sodA*, whose transcription is controlled by mechanisms that depend on metal ions (Touati, 1983). The overall dismutation is carried out in two steps:

$$Ez_{ox} + O_2\bullet^- + H^+ \leftrightarrow Ez_{red}(H^+) + O_2 \quad (1a)$$

$$Ez_{red}(H^+) + O_2\bullet^- + H^+ \leftrightarrow Ez_{ox} + H_2O_2 \quad (1b)$$

MnSODs occur in eukaryotic mitochondria and the cytoplasm of prokaryotes. The activity of MnSOD is sensitive to external oxidative stress and is called the inducible form. In *E. coli,* it is absent under anaerobic conditions, but is rapidly produced when oxygen concentrations increase. In contrast, the FeSOD is constitutively expressed (Touati, 1991) in the cytosols of prokaryotes, primitive eukaryotes, and some green plants. MnSOD and FeSOD are homologous homodimeric or homotetrameric proteins with a two-domain primarily α-helical fold (Stallings et al., 1985; 1983). Three histidine residues, one aspartic acid, and a solvent molecule in trigonal bipyramidal geometry coordinate the Fe/Mn ion. Austin et al. (Austin et al., 1981; Whitehouse et al., 1997) cloned and characterized a gene encoding a SOD from *B. burgdorferi*. This gene had 68% similarity to the *sodA* from *Haemophilus influenzae* and the partially purified enzyme had superoxide dismutase activity. Based on homology, the metal ligand associated with the SOD is most likely manganese. *T. pallidum* and *T. denticola* genomes do not contain genes encoding SOD orthologs (Table 3) (Fraser et al., 1998; Seshadri et al., 2004).

Anaerobic and some microaerophilic bacteria which lack enzymes like SOD and catalase have varying degrees of tolerance for O_2. What was particularly perplexing was how these bacteria tolerated O_2 and its inevitable reactive reduction products without these enzymes. The discovery of superoxide reductases (SOR) and the characterization of their biochemical function has led to a better understanding of how these bacteria defend against ROS (Jenney et al., 1999; Lombard et al., 2000). SORs catalyze only one of the two reactions of SODs, *i.e.* the reduction of superoxide to hydrogen peroxide.

$$O_2\bullet^- + 1e^- + 2H^+ \rightarrow H_2O_2$$

With a few notable exceptions, most SORs are poor SODs but rather function as oxidoreductases, deriving the reducing equivalents for superoxide reduction ultimately from reduced pyridine nucleotides via intermediate electron carriers. Like SOD, SOR is physically configured so its reaction with superoxide is diffusion limited. Two types of SOR have been described (Jenney et al., 1999; Lombard et al., 2000). The first is a small protein found in anaerobic sulfate-reducing bacteria called desulfoferrodoxin (Dfx). Dfx is a homodimer organized in two domains, each with a specific mononuclear iron site. Center I contains a mononuclear ferric iron coordinated by four cysteines in a distorted rubredoxin-type center. Center II has an oxygen-stable ferrous iron that has been shown to be the active site for the SOR activity (Lombard et al., 2000). The second type has been isolated from the anaerobic archaea, *Pyrococcus furiosus*. The protein exhibited strong homologies to neelaredoxin (Nlr), a small protein containing a single mononuclear center, from sulfate-reducing bacteria (Chen et al., 1994). Based on the three-dimensional structure of the *P. furiosus* SOR, the unique mononuclear iron center is similar to the second domain of Dfx. In *P. furiosus*, an electron-transferring chain, including NADH, NADH rubredoxin oxidoreductase, and rubredoxin, was proposed to provide the electrons necessary for the reaction (Jenney et al., 1999).

Both *T. pallidum* and *T. denticola* possess an ortholog (TP0823 and TDE1754, respectively) for neelaredoxin (Lombard et al., 2000). Like neelaredoxin from *P. furiosus*, the *T. pallidum* protein is postulated to function as a SOR, accepting electrons from NAD(P)H, via rubredoxin, to reduce superoxide to hydrogen peroxide. Indeed, Niviere et al. (2001) have demonstrated that the *T. pallidum* neelaredoxin exhibits superoxide reductase activity. Recently, Auchere et al. (2004) have demonstrated the existence of a superoxide-mediated electron transfer between rubredoxin and neelaredoxin. Also, it has been shown that the recombinant *D. vulgaris* rubredoxin can serve as an electron donor for the *T. pallidum* Nlr reduction (Hazlett et al., 2002). These findings suggest that *T. pallidum* does utilize Nlr and rubredoxin to reduce superoxide. Because *T. pallidum* does not possess a homolog for the rubredoxin oxidoreductase, the electron donor to rubredoxin is yet to be determined. Although *T. denticola* does possess orthologs for both neelaredoxin and rubredoxin, these enzymes have not been characterized.

3.3.2. Catalases, peroxidases and alkylhydroperoxide reductases

The product of SOD or SOR is H_2O_2, which is itself a reactive species. The function of SOD would not appear to benefit the cells unless, (i) superoxide is more toxic than H_2O_2, (ii) H_2O_2 is removed very rapidly, or (iii) SOD directs superoxide to H_2O_2. Although, the relative toxicities of H_2O_2 and superoxides *in vivo* are not clear, it is true that cellular catalases destroy H_2O_2 with remarkable rapidity. The turnover number for typical catalase is about 10^9 molecules of H_2O_2 disproportionated per active site per second (at 1M H_2O_2). The electron source is from H_2O_2 itself, so the reaction does not require an exogenous reducing source. Catalases therefore provide protection even in energy-depleted cells. Peroxidases are also capable of destroying H_2O_2. Peroxidases require NADH or NADPH as an electron source. Under conditions where reducing power is limited, the protective role or peroxidase is likely to be small.

However, *T. pallidum*, *T. denticola*, and *B. burgdorferi* all lack the genes encoding these enzymes. A potential mechanism by which *T. pallidum* and *T. denticola* remove peroxides may involve the putative alkyl hydroperoxide reductase (AhpC) homologs identified in their respective genomes (TP0509 and TDE0011). In aerobic bacteria, the role in protecting cells against organic peroxides is exerted in part by the peroxide-scavenging enzyme AhpR. It is a heterodimer consisting of two components; a small 22,000-Da protein, AhpC, the enzymatic subunit and a larger 57,000-Da flavoprotein, AhpF. These two proteins act together; AhpF uses NADH or NADPH as electron donor to AhpC, which reduces physiological lipid peroxides such as linoleic acid hydroperoxide and thymine hydroperoxide and nonphysiological alkyl hydroperoxides to their respective nontoxic alcohol forms. In addition to alkyl hydroperoxide reductase activity, the *S. typhimurium* AhpC/F complex has been shown to reduce H_2O_2 to water at the expense of NADPH. No *ahpF* homolog as been identified within the *T. pallidum* and *T. denticola* genomes. However, both contain homologs for thioredoxin reductase (TP0814 and TDE0743) and NADH oxidase (TP0921, TDE2205, TDE0096, and TDE0868) which have been shown to function as AhpF in the reduction of AhpC in *H. pylori* (Baker et al., 2001). The enzyme(s) by which *B. burgdorferi* removes peroxides has not been identified.

Glutathione (GSH) is an important antioxidant and is synthesized by glutathione synthetase. The high steady-state levels of glutathione in bacteria such as *E. coli* maintain a strong reducing environment in the cell. GSH will react with H_2O_2, superoxide, or HOO^- to form a stable glutathione radical (GS·). One of the most important functions of GSH is to reduce disulfide bridges caused by oxidative stress in proteins. However, GSH synthesis enzymes are not found in genomes of *B. burgdorferi*, *T. denticola* or *T. pallidum*. Instead, these bacteria contain the small protein thioredoxin (BB0061, TDE0238, TDE0744, and TP0919) which can be used as a reductant in many cellular reactions. Similar to GSH, thioredoxin proteins act as disulfide oxidoreductases and are electron donors for thioredoxin peroxidases. Oxidized thioredoxin is reduced by the NADPH-dependent thioredoxin reductase. In general thioredoxins are better disulfide bond reductants than glutareductants.

Additionally, some organisms produce alternative low molecular weight thiols that participate in unique thiol/disulfide redox systems. Mycobacteria and Actinomycetes synthesize a small glycothiol called mycothiol in high concentrations, suggesting a role for mycothiol analogous to glutathione. It was recently shown that *Staphylococcus aureus* and *Pyrococcus horikoshii* and *P. furiosus* produce coenzyme A as the predominant thiol and possess a disulfide reductase responsible for maintaining CoA in its reduced form (delCardayre et al., 1998; Harris et al., 2005). *B. burgdorferi* genome analysis has revealed the presence of an ortholog of the *S. aureus* coenzyme A-disulfide reductase (BB0728), and recently it has been reported that BB0728 does indeed function as an NADH-dependent coenzyme A disulfide reductase (Boylan et al., 2006). This suggests a role for BB0728 in intracellular redox.

3.3.3. Regulation

The most extensively studied organisms for understanding the genetic regulation in response to reactive oxygen/nitrogen species have been the enteric bacteria *E. coli* and *Salmonella typhimurium* and *Bacillus subtilis*. These bacteria can activate the expression of some 30-40 proteins in response to sublethal levels of H_2O_2 and another 40 in response to $O_2^{\bullet-}$ generating agents. The key regulators of these adaptive responses are the SoxRS and OxyR transcription factors of *E. coli* and the peroxide-sensitive repressor PerR from *Bacillus subtilis*. The expression of the H_2O_2-inducible activities is regulated by the OxyR transcription factor, which is directly sensitive to oxidation, forming an intramolecular disulfide bond upon oxidation. It is this oxidized form that is capable of activating transcription of genes in the oxidative stress response. Positive regulation of genes involved in the glutaredoxin system results in the reduction of OxyR, returning the regulator to an inactive, thiol-containing form (Zheng et al., 1998). The SoxR protein contains a redox active Fe-S cluster that regulates DNA binding. Once oxidized, SoxR induces transcription of *soxS,* and SoxS activates transcription of genes in response to $O_2{\bullet}^-$ (Hidalgo et al., 1997). The PerR protein is a metalloprotein which binds one or more copies of a conserved, 14-bp inverted repeat to repress expression of target genes. The oxidation of the Zn^{2+}-Fe^{2+} form of PerR (active) by H_2O_2 abolishes binding, causing derepression of target genes (Herbig and Helmann, 2001).

B. burgdorferi, *T. denticola* and *T. pallidum* do not contain a SoxRS, OxyR, or PerR ortholog but *B. burgdorferi* does contains a protein that responds to ROS. The *Borrelia* oxidative stress regulator, BosR, acts as a transcriptional activator of oxidative stress genes (Boylan et al., 2003; Seshu et al., 2004). Although there is no apparent amino acid homology (>15%), BosR appears to be functionally similar to OxyR. BosR is capable of binding DNA under both oxidizing and reducing conditions, requires Zn^{2+} for efficient binding, and has an unusually large (45 bp to 50 bp) target sequences based upon DNase I footprint analysis. BosR binds target sequences upstream of *sodA* (BB0153), the gene encoding the neutrophil activating protein, *napA*, (BB0690, a Dpr homolog) and *bosR* (BB0647) (see Table 3). *E. coli* reporter assays performed on strains harboring an inducible BosR and p*napAP/O-lacZ* indicated that BosR activates transcription under oxidizing conditions (the presence of *t*-butyl peroxide) but it is not known if reduced BosR is able to activate transcription. More recently, Katona et al. (2004) using a His-tagged BorR purified with a Zn affinity column, demonstrated that purified protein binds to target sequences upstream of a gene encoding a hydrolase (BB0646), *bosR* (BB0647) and *napA* (BB0690). This binding is greatly decreased by either the exposure to oxidants or the addition of Zn which suggests that BosR functions as a repressor to regulate the oxidative stress response of *B. burgdorferi.* It should be noted that Seshu et al. (2004) demonstrated that a single amino acid change in the N-terminal end of BosR (R39K) dramatically affected the promoter affinity and activity of BosR so it is not clear if the N-terminal modification introduced by Katona et al. affected protein activity (Katona et al., 2004). The oxidative stress regulators of *T. denticola* and *T. pallidum* have yet to be identified.

References

Akins, D. R., Robinson, E., Shevchenko, D., Elkins, C., Cox, D. L., and Radolf, J. D. (1997). Tromp1, a putative rare outer membrane protein, is anchored by an uncleaved signal sequence to the *Treponema pallidum* cytoplasmic membrane. J. Bacteriol. *179*, 5076-5086.

Alderete, J. F., Peterson, K. M., and Baseman, J. B. (1988). Affinities of *Treponema pallidum* for human lactoferrin and transferrin. Genitourin Med. *64*, 359-363.

Andrews, S. C., Robinson, A. K., and Rodriguez-Quinones, F. (2003). Bacterial iron homeostasis. FEMS Microbiol. Rev. *27*, 215-237.

Archibald, F. S. (1983). *Lactobacillus plantarum*, an organism not requiring iron. FEMS Microbiology Letters *19*, 29-32.

Auchere, F., Sikkink, R., Cordas, C., Raleiras, P., Tavares, P., Moura, I., and Moura, J. J. (2004). Overexpression and purification of *Treponema pallidum* rubredoxin; kinetic evidence for a superoxide-mediated electron transfer with the superoxide reductase neelaredoxin. J. Biol. Inorg Chem. *9*, 839-849.

Austin, F. E., Barbieri, J. T., Corin, R. E., Grigas, K. E., and Cox, C. D. (1981). Distribution of superoxide dismutase, catalase, and peroxidase activities among *Treponema pallidum* and other spirochetes. Infect. Immun. *33*, 372-379.

Baker, L. M., Raudonikiene, A., Hoffman, P. S., and Poole, L. B. (2001). Essential thioredoxin-dependent peroxiredoxin system from *Helicobacter pylori*: genetic and kinetic characterization. J. Bacteriol. *183*, 1961-1973.

Barbour, A. G. (1984). Isolation and cultivation of Lyme disease spirochetes. Yale Journal of Biology and Medicine *57*, 521-525.

Beermann, C., Lochnit, G., Geyer, R., Groscurth, P., and Filgueira, L. (2000). The lipid component of lipoproteins from *Borrelia burgdorferi*: structural analysis, antigenicity, and presentation via human dendritic cells. Biochem. Biophys. Res. Comm. *267*, 897-905.

Ben-Menachem, G., Kubler- Kielb, J., Coxon, B., Yergey, A., and Schneerson, R. (2003). A newly discovered cholesteryl galactoside from *Borrelia burgdorferi*. PNAS *100*, 7913-7918.

Blanco, D. R., Champion, C. I., Exner, M. M., Erdjument-Bromage, H., Hancock, R. E., Tempst, P., Miller, J. N., and Lovett, M. A. (1995). Porin activity and sequence analysis of a 31-kilodalton *Treponema pallidum* subsp. *pallidum* rare outer membrane protein (Tromp1). J. Bacteriol. *177*, 3556-3562.

Blencowe, D. K., and Morby, A. P. (2003). Zn(II) metabolism in prokaryotes. FEMS Microbiol. Rev. *27*, 291-311.

Boyd, J., Oza, M. N., and Murphy, J. R. (1990). Molecular cloning and DNA sequence analysis of a diphtheria tox iron-dependent regulatory element (dtxR) from *Corynebacterium diphtheriae*. Proc. Natl. Acad. Sci. USA. *87*, 5968-5972.

Boylan, J., Hummel, C.S., Benoit, S., Garcia-Lara, J., Treglown-Downey, J., Crane, E.J. and Gherardini, F. (2006). *Borrelia burgdorferi* bb0728 encodes a coenzyme A disulphide reductase whose function suggests a role in intracellular redox and its oxidative stress response. Mol. Microbiol. *59*, 475-486.

Boylan, J. A., Posey, J. E., and Gherardini, F. C. (2003). Borrelia oxidative stress response regulator, BosR: a distinctive Zn-dependent transcriptional activator. Proc. Natl. Acad. Sci. USA. *100*, 11684-11689.

Bradbeer, C. (1993). The proton motive force drives the outer membrane transport of cobalamin in *Escherichia coli*. J. Bacteriol. *175*, 3146-3150.

Braun, V. (2003). Iron uptake by *Escherichia coli*. Front Biosci *8*, s1409-1421.

Brown, J. S., and Holden, D. W. (2002). Iron acquisition by Gram-positive bacterial pathogens. Microbes Infect. *4*, 1149-1156.

Bruyneel, B., Vande Woestyne, M., and Verstraete, W. (1990). Iron complexation as a tool to direct mixed clostridia-lactobacilli fermentations. J. Appl Bacteriol. *69*, 80-85.

Bullen, J. J. (1981). The significance of iron in infection. Rev. Infect. Dis. *3*, 1127-1138.

Butler, A. (1998). Acquisition and utilization of transition metal ions by marine organisms. Science *281*, 207-210.

Chen, L., Sharma, P., Le Gall, J., Mariano, A. M., Teixeira, M., and Xavier, A. V. (1994). A blue non-heme iron protein from *Desulfovibrio gigas*. Eur. J. Biochem. *226*, 613-618.

Chu, L., Kennell, W., and Holt, S. C. (1994). Characterization of hemolysis and hemoxidation activities by *Treponema denticola*. Microb Pathog *16*, 183-195.

Cornelissen, C. N. (2003). Transferrin-iron uptake by Gram-negative bacteria. Front Biosci *8*, d836-847.

de la Calle Guntinas, M. B., Bordin, G., and Rodriguez, A. R. (2002). Identification, characterization and determination of metal-binding proteins by liquid chromatography. A review. Anal. Bioanal Chem. *374*, 369-378.

Deka, R. K., Lee, Y. H., Hagman, K. E., Shevchenko, D., Lingwood, C. A., Hasemann, C. A., Norgard, M. V., and Radolf, J. D. (1999). Physicochemical evidence that *Treponema pallidum* TroA is a zinc-containing metalloprotein that lacks porin-like structure. J. Bacteriol. *181*, 4420-4423.

delCardayre, S. B., Stock, K. P., Newton, G. L., Fahey, R. C., and Davies, J. E. (1998). Coenzyme A disulfide reductase, the primary low molecular weight disulfide reductase from *Staphylococcus aureus*. Purification and characterization of the native enzyme. J. Biol. Chem. *273*, 5744-5751.

Faraldo-Gomez, J. D., and Sansom, M. S. (2003). Acquisition of siderophores in gram-negative bacteria. Nat. Rev. Mol. Cell Biol. *4*, 105-116.

Farr, S. B., and Kogoma, T. (1991). Oxidative stress responses in *Escherichia coli* and *Salmonella typhimurium*. Microbiol. Rev. *55*, 561-585.

Ferguson, A. D., and Deisenhofer, J. (2004). Metal import through microbial membranes. Cell *116*, 15-24.

Finney, L. A., and O'Halloran, T. V. (2003). Transition metal speciation in the cell: insights from the chemistry of metal ion receptors. Science *300*, 931-936.

Fraser, C. M., Casjens, S., Huang, W. M., Sutton, G. G., Clayton, R., Lathigra, R., White, O., Ketchum, K. A., Dodson, R., Hickey, E. K., et al. (1997). Genomic sequence of a Lyme disease spirochaete, Borrelia burgdorferi. Nature *390*, 580-586.

Fraser, C. M., Norris, S. J., Weinstock, G. M., White, O., Sutton, G. G., Dodson, R., Gwinn, M., Hickey, E. K., Clayton, R., Ketchum, K. A., et al. (1998). Complete genome sequence of *Treponema pallidum*, the syphilis spirochete. Science *281*, 375-388.

Gaetke, L. M., McClain, C. J., Talwalkar, R. T., and Shedlofsky, S. I. (1997). Effects of endotoxin on zinc metabolism in human volunteers. Amer. J. Physiol *272*, E952-956.

Glasfeld, A., Guedon, E., Helmann, J. D., and Brennan, R. G. (2003). Structure of the manganese-bound manganese transport regulator of *Bacillus subtilis*. Nat. Struct Biol. *10*, 652-657.

Goldblum, S. E., Cohen, D. A., Jay, M., and McClain, C. J. (1987). Interleukin 1-induced depression of iron and zinc: role of granulocytes and lactoferrin. Amer. J. Physiol *252*, E27-32.

Guedon, E., and Helmann, J. D. (2003). Origins of metal ion selectivity in the DtxR/MntR family of metalloregulators. Mol. Microbiol. *48*, 495-506.

Gutteridge, J. M., and Halliwell, B. (1990). The measurement and mechanism of lipid peroxidation in biological systems. Trends Biochem. Sci. *15*, 129-135.

Hall, J. L. (2002). Cellular mechanisms for heavy metal detoxification and tolerance. J. Exp. Bot *53*, 1-11.

Hardham, J. M., Stamm, L. V., Porcella, S. F., Frye, J. G., Barnes, N. Y., Howell, J. K., Mueller, S. L., Radolf, J. D., Weinstock, G. M., and Norris, S. J. (1997). Identification and transcriptional analysis of a *Treponema pallidum* operon encoding a putative ABC transport system, an iron-activated repressor protein homolog, and a glycolytic pathway enzyme homolog. Gene *197*, 47-64.

Harris, D. R., Ward, D. E., Feasel, J. M., Lancaster, K. M., Murphy, R. D., Mallet, T. C., and Crane, E. J., 3rd (2005). Discovery and characterization of a Coenzyme A disulfide reductase from *Pyrococcus horikoshii*. Implications for this disulfide metabolism of anaerobic hyperthermophiles. Febs J. *272*, 1189-1200.

Hazlett, K. R., Cox, D. L., Sikkink, R. A., Auch'ere, F., Rusnak, F., and Radolf, J. D. (2002). Contribution of neelaredoxin to oxygen tolerance by *Treponema pallidum*. Methods Enzymol *353*, 140-156.

Hazlett, K. R., Rusnak, F., Kehres, D. G., Bearden, S. W., La Vake, C. J., La Vake, M. E., Maguire, M. E., Perry, R. D., and Radolf, J. D. (2003). The *Treponema pallidum tro* operon encodes a multiple metal transporter, a zinc-dependent transcriptional repressor, and a semi-autonomously expressed phosphoglycerate mutase. J. Biol. Chem. *278*, 20687-20694.

Herbig, A. F., and Helmann, J. D. (2001). Roles of metal ions and hydrogen peroxide in modulating the interaction of the *Bacillus subtilis* PerR peroxide regulon repressor with operator DNA. Mol. Microbiol. *41*, 849-859.

Hidalgo, E., Ding, H., and Demple, B. (1997). Redox signal transduction via iron-sulfur clusters in the SoxR transcription activator. Trends Biochem. Sci. *22*, 207-210.

Imlay, J. A. (1995). A metabolic enzyme that rapidly produces superoxide, fumarate reductase of *Escherichia coli*. J. Biol. Chem. *270*, 19767-19777.

Imlay, J. A. (2002). How oxygen damages microbes: oxygen tolerance and obligate anaerobiosis. Adv. Microb Physiol *46*, 111-153.

Imlay, J. A. (2003). Pathways of oxidative damage. Annu. Rev. Microbiol. *57*, 395-418.

Imlay, J. A., and Fridovich, I. (1991). Assay of metabolic superoxide production in *Escherichia coli*. J. Biol. Chem. *266*, 6957-6965.

Jakubovics, N. S., and Jenkinson, H. F. (2001). Out of the iron age: new insights into the critical role of manganese homeostasis in bacteria. Microbiology *147*, 1709-1718.

Jap, B. K., and Walian, P. J. (1996). Structure and functional mechanism of porins. Physiol Rev. *76*, 1073-1088.

Jenney, F. E., Jr., Verhagen, M. F., Cui, X., and Adams, M. W. (1999). Anaerobic microbes: oxygen detoxification without superoxide dismutase. Science *286*, 306-309.

Jovanovic, T., Ascenso, C., Hazlett, K. R., Sikkink, R., Krebs, C., Litwiller, R., Benson, L. M., Moura, I., Moura, J. J., Radolf, J. D., et al. (2000). Neelaredoxin, an iron-binding protein from the syphilis spirochete, *Treponema pallidum*, is a superoxide reductase. J. Biol. Chem. *275*, 28439-28448.

Karunakaran, T., and Holt, S. C. (1994). Cloning and expression of hemolysin genes from Treponema denticola strains ATCC 35404 (TD-4) and human clinical isolate GM-1 in Escherichia coli. Microb Pathog *16*, 337-348.

Katona, L. I., Tokarz, R., Kuhlow, C. J., Benach, J., and Benach, J. L. (2004). The fur homologue in *Borrelia burgdorferi*. J. Bacteriol. *186*, 6443-6456.

Keyer, K., and Imlay, J. A. (1996). Superoxide accelerates DNA damage by elevating free-iron levels. Proc. Natl. Acad. Sci. USA. *93*, 13635-13640.

Klebba, P. E. (2003). Three paradoxes of ferric enterobactin uptake. Front Biosci *8*, s1422-1436.

Lee, Y. H., Deka, R. K., Norgard, M. V., Radolf, J. D., and Hasemann, C. A. (1999). *Treponema pallidum* TroA is a periplasmic zinc-binding protein with a helical backbone. Nat. Struct Biol. *6*, 628-633.

Lombard, M., Touati, D., Fontecave, M., and Niviere, V. (2000). Superoxide reductase as a unique defense system against superoxide stress in the microaerophile *Treponema pallidum*. J. Biol. Chem. *275*, 27021-27026.

Marquis, R. E. (1995). Oxygen metabolism, oxidative stress and acid-base physiology of dental plaque biofilms. J. Ind Microbiol. *15*, 198-207.

Masse, E., and Gottesman, S. (2002). A small RNA regulates the expression of genes involved in iron metabolism in *Escherichia coli*. Proc. Natl. Acad. Sci. USA. *99*, 4620-4625.

Matthews, H. M., Yang, T. K., and Jenkin, H. M. (1979). Unique lipid composition of *Treponema pallidum* (Nichols virulent strain). Infect. Immun. *24*, 713-719.

McCord, J. M., and Fridovich, I. (1969). The utility of superoxide dismutase in studying free radical reactions. I. Radicals generated by the interaction of sulfite, dimethyl sulfoxide, and oxygen. J. Biol. Chem. *244*, 6056-6063.

McCord, J. M., Keele, B. B., Jr., and Fridovich, I. (1971). An enzyme-based theory of obligate anaerobiosis: the physiological function of superoxide dismutase. Proc. Natl. Acad. Sci. USA. *68*, 1024-1027.

Mulrooney, S. B., and Hausinger, R. P. (2003). Nickel uptake and utilization by microorganisms. FEMS Microbiol. Rev. *27*, 239-261.

Nascimento, A. L., Verjovski-Almeida, S., Van Sluys, M. A., Monteiro-Vitorello, C. B., Camargo, L. E., Digiampietri, L. A., Harstkeerl, R. A., Ho, P. L., Marques, M. V., Oliveira, M. C., et al. (2004). Genome features of *Leptospira interrogans* serovar *Copenhageni*. Braz J. Med. Biol. Res. *37*, 459-477.

Nikaido, H., and Saier, M. H., Jr. (1992). Transport proteins in bacteria: common themes in their design. Science *258*, 936-942.

Niviere, V., Lombard, M., Fontecave, M., and Houee-Levin, C. (2001). Pulse radiolysis studies on superoxide reductase from *Treponema pallidum*. FEBS Lett. *497*, 171-173.

Outten, C. E., and O'Halloran, T. V. (2001). Femtomolar sensitivity of metalloregulatory proteins controlling zinc homeostasis. Science *292*, 2488-2492.

Palenik, B., Brahamsha, B., Larimer, F. W., Land, M., Hauser, L., Chain, P., Lamerdin, J., Regala, W., Allen, E. E., McCarren, J., et al. (2003). The genome of a motile marine *Synechococcus*. Nature *424*, 1037-1042.

Patzer, S. I., and Hantke, K. (2001). Dual repression by Fe(2+)-Fur and Mn(2+)-MntR of the mntH gene, encoding an NRAMP-like Mn(2+) transporter in *Escherichia coli*. J. Bacteriol. *183*, 4806-4813.

Posey, J. E., and Gherardini, F. C. (2000). Lack of a role for iron in the Lyme disease pathogen. Science *288*, 1651-1653.

Posey, J. E., Hardham, J. M., Norris, S. J., and Gherardini, F. C. (1999). Characterization of a manganese-dependent regulatory protein, TroR, from *Treponema pallidum*. Proc. Natl. Acad. Sci. USA. *96*, 10887-10892.

Radolf, J. D., Steiner, B., and Shevchenko, D. (1999). *Treponema pallidum*: doing a remarkable job with what it's got. Trends Microbiol. *7*, 7-9.

Ratledge, C., and Dover, L. G. (2000). Iron metabolism in pathogenic bacteria. Annu. Rev. Microbiol. *54*, 881-941.

Ren, S. X., Fu, G., Jiang, X. G., Zeng, R., Miao, Y. G., Xu, H., Zhang, Y. X., Xiong, H., Lu, G., Lu, L. F., et al. (2003). Unique physiological and pathogenic features of *Leptospira interrogans* revealed by whole-genome sequencing. Nature *422*, 888-893.

Rensing, C., and Grass, G. (2003). *Escherichia coli* mechanisms of copper homeostasis in a changing environment. FEMS Microbiol. Rev. *27*, 197-213.

Rosen, B. P. (2002). Transport and detoxification systems for transition metals, heavy metals and metalloids in eukaryotic and prokaryotic microbes. Comp Biochem. Physiol A Mol. Integr Physiol *133*, 689-693.

Scott, D., Siboo, I. R., Chan, E. C., Klitorinos, A., and Siboo, R. (1993). Binding of hemin and congo red by oral hemolytic spirochetes. Oral Microbiol. Immunol. *8*, 245-250.

Scott, D., Siboo, R., Chan, E. C., and Siboo, R. (1996). An extracellular enzyme with hyaluronidase and chondroitinase activities from some oral anaerobic spirochaetes. Microbiology *142 (Pt 9)*, 2567-2576.

Seaver, L. C., and Imlay, J. A. (2001). Hydrogen peroxide fluxes and compartmentalization inside growing *Escherichia coli*. J. Bacteriol. *183*, 7182-7189.

Seshadri, R., Myers, G. S., Tettelin, H., Eisen, J. A., Heidelberg, J. F., Dodson, R. J., Davidsen, T. M., DeBoy, R. T., Fouts, D. E., Haft, D. H., et al. (2004). Comparison of the genome of the oral pathogen *Treponema denticola* with other spirochete genomes. Proc. Natl. Acad. Sci. USA. *101*, 5646-5651.

Seshu, J., Boylan, J. A., Hyde, J. A., Swingle, K. L., Gherardini, F. C., and Skare, J. T. (2004). A conservative amino acid change alters the function of BosR, the redox regulator of *Borrelia burgdorferi*. Mol. Microbiol. *54*, 1352-1363.

Socransky, S. S., Haffajee, A. D., Cugini, M. A., Smith, C., and Kent, R. L., Jr. (1998). Microbial complexes in subgingival plaque. J. Clin. Periodontol *25*, 134-144.

Staggs, T. M., Greer, M. K., Baseman, J. B., Holt, S. C., and Tryon, V. V. (1994). Identification of lactoferrin-binding proteins from *Treponema pallidum* subspecies *pallidum* and *Treponema denticola*. Mol. Microbiol. *12*, 613-619.

Stallings, W. C., Pattridge, K. A., Strong, R. K., and Ludwig, M. L. (1985). The structure of manganese superoxide dismutase from *Thermus thermophilus* HB8 at 2.4-A resolution. J. Biol. Chem. *260*, 16424-16432.

Stallings, W. C., Powers, T. B., Pattridge, K. A., Fee, J. A., and Ludwig, M. L. (1983). Iron superoxide dismutase from *Escherichia coli* at 3.1-A resolution: a structure unlike that of copper/zinc protein at both monomer and dimer levels. Proc. Natl. Acad. Sci. USA. *80*, 3884-3888.

Tasaka, Y., Gombos, Z., Nishiyama, Y., Mohanty, P., Ohba, T., Ohki, K., and Murata, N. (1996). Targeted mutagenesis of acyl-lipid desaturases in *Synechocystis*: evidence for the important roles of polyunsaturated membrane lipids in growth, respiration and photosynthesis. Embo J. *15*, 6416-6425.

Touati, D. (1983). Cloning and mapping of the manganese superoxide dismutase gene (*sodA*) of *Escherichia coli* K-12. J. Bacteriol. *155*, 1078-1087.

Touati, D. (1991). The molecular genetics of superoxide dismutase in *E. coli*. Free Radic Res. Commun. *12-13 Pt 1*, 379-382.

Weinberg, E. D. (1978). Iron and infection. Microbiol. Rev. *42*, 45-66.

Weinberg, E. D. (1997). The Lactobacillus anomaly: total iron abstinence. Perspect Biol. Med. *40*, 578-583.

Whitehouse, C. A., Williams, L. R., and Austin, F. E. (1997). Identification of superoxide dismutase activity in *Borrelia burgdorferi*. Infect. Immun. *65*, 4865-4868.

Wooldridge, K. G., and Williams, P. H. (1993). Iron uptake mechanisms of pathogenic bacteria. FEMS Microbiol. Rev. *12*, 325-348.

Xu, X., Holt, S. C., and Kolodrubetz, D. (2001). Cloning and expression of two novel hemin binding protein genes from *Treponema denticola*. Infect. Immun. *69*, 4465-4472.

Xu, X., and Kolodrubetz, D. (2002). Construction and analysis of hemin binding protein mutants in the oral pathogen *Treponema denticola*. Res. Microbiol. *153*, 569-577.

Yotis, W., Keene, J., Hoerman, K., and Simonson, L. G. (1993). Fatty acid profiles of the outer membrane of ATCC strains 35405, 35404 and 33521 of *Treponema denticola*. J. Periodontal Res. *28*, 387-395.

Youn, H. D., Kim, E. J., Roe, J. H., Hah, Y. C., and Kang, S. O. (1996a). A novel nickel-containing superoxide dismutase from *Streptomyces* spp. Biochem. J. *318 (Pt 3)*, 889-896.

Youn, H. D., Youn, H., Lee, J. W., Yim, Y. I., Lee, J. K., Hah, Y. C., and Kang, S. O. (1996b). Unique isozymes of superoxide dismutase in *Streptomyces griseus*. Arch. Biochem. Biophys *334*, 341-348.

Zheng, M., Aslund, F., and Storz, G. (1998). Activation of the OxyR transcription factor by reversible disulfide bond formation. Science *279*, 1718-1721.

6

The Beguiling Motility of the Genus *Treponema*

Nyles W. Charon, Chunhao Li, and Stuart F. Goldstein*

Abstract

Understanding the motility of *Treponema* species is still in its infancy. An analysis of motility mutants indicate that the periplasmic flagella, which are situated between the outer membrane sheath and cell cylinder, are directly involved in the motility in both *Treponema denticola* and *Treponema phagedenis*. Furthermore, not only do these organelles rotate in a manner similar to the flagella of rod-shaped bacteria, but they also have a skeletal function, *i.e.*, they influence the shape of the cell of both species. A model of how these two species swim states the following: during cell displacement, the periplasmic flagella rotate in one direction (counter-clockwise) as viewed from behind the cell, and the right-handed cell cylinder rolls in the opposite direction (clockwise). Because a gel-like medium is essential for translocation of both these spirochetes, this rolling of the cell cylinder is essential for cell displacement. In contrast, almost nothing is known about the dynamics of motility of *Treponema pallidum*.

The organelle for motility, the periplasmic flagellum, has been studied in some detail in these three species. Its structure is unique, as a protein sheath composed of FlaA surrounds a core of three different FlaB proteins, FlaB1, FlaB3, and FlaB3. Although the function of each protein is unclear, recent results with a related spirochete species suggest that some of these proteins have an overlap in function. Specifically, in *Brachyspira hyodysenteriae*, mutants that lack FlaB1 or FlaB2 are still motile, but mutants that lack both are completely non-motile.

Genomic analysis of *T. denticola* and *T. pallidum* reveals that both species have all the necessary genes for flagella-based motility. Although sequence analysis indicates that their motility genes are well conserved between these species, their gene order is quite different. Furthermore, several lines of evidence indicate that motility gene regulation is controlled at both the transcriptional and translational

*For correspondence email ncharon@hsc.wvu.edu

levels. Finally, because a large number of genes in both species are dedicated to motility, this cell function likely plays an important role in its survival and disease progression in the host. However, there is no experimental evidence as yet to support this contention.

1. Introduction

"Because of the narrow width of *Treponema pallidum*, the ordinary light microscope does not permit sufficient resolution to visualize the organism. However, this can be accomplished by the use of darkfield microscopy. *T. pallidum* can be identified by its morphologic characteristics and by its characteristic motions. These motions consist of a slow forward and backward movement, rotation about the long axis like a corkscrew, and a slight bending, twisting or undulation from side to side..."

The above description of *Treponema pallidum* motility, as observed by darkfield microscopy, was written over 35 year ago in a treatise on syphilis written by the United States Public Health Service (U.S.Department of Health, 1968a). We know now that approximately five percent of its genome is involved in motility and chemotaxis (Fraser et al., 1998). With so many genes involved, and their notable expense to the cell with respect to energy consumption when expressed, these functions are obviously vital to the survival of the spirochete in the infected host (Armitage, 1999). As discussed in this chapter, we have begun to understand some aspects of the motility of other *Treponema* species, but we are still quite in the dark with respect to a molecular understanding of motility and chemotaxis of *T. pallidum*. The description above from the USPHS is still accurate for *T. pallidum* even in the genomic era. Here we review what is known about bacterial motility in general, present recent information on the motility of the genus *Treponema*, and point out new avenues of research.

2. General features of flagellated bacterial motility

Organisms such as *Escherichia coli* and *Salmonella enterica* serovar Typhimurium serve as model systems in helping to decipher spirochete motility (Macnab, 1996; Berry and Armitage, 1999; Berg, 2003). These species swim via the rotation of externally located helical flagellar filaments. A rotary motor at the base of each flagellum powers this rotation, which is driven by a proton gradient.

The flagellar structure is complex, with a rotary motor apparatus beginning in the cytoplasm and extending through the cytoplasmic membrane, peptidoglycan wall layer, outer membrane, and finally the helical flagellar filament that contacts the ambient medium. In the region below the flagellum, a ring structure (C-ring complex) composed of several different proteins (FliG, FliM, FliN) is attached to another ring (MS ring, encoded by *fliF*). This C-ring structure and the MS ring serve as the rotary part of the flagellar motor. MotA and MotB proteins are embedded in the cytoplasmic membrane and cell wall around the MS ring and function as the stator, or nonrotating part of the motor. The P and L rings serve as bushings and localize at the peptidoglycan layer and outer-membrane layers, respectively. The MS, P, and L rings surround a central rod, which in turn is attached to the hook, an external structure approximately 55 nm long, composed of a polymer of the FlgE protein. The hook serves as a flexible coupling to the flagellar filament, a helically shaped structure that is many μm long. This filament, composed of a polymer of the

Table 1. Components of motility apparatus

Structure	Proteins
Motor rotor and switch	FliG, FliM, FliN (C ring complex); FliF (MS ring)
Motor stator	MotA, MotB
P ring	FlgI
L ring	FlgH
Rod	FlgB, FlgC, FlgF (motor shaft) FlgG (cap of rod) FliE (rod-MS ring interface?)
Hook	FlgE FlgD (hook formation) FliK (hook length control) FlgK, FlgL (hook-filament interface)
Filament	FliC [FlaA, FlaB in spirochetes] FliD (filament cap)
Basal transport apparatus	FlhA, FlhB, FliH, FliI, FliO, FliP, FliQ, FliR

FliC protein, is a hollow tubular quasi-rigid polymorphic structure that can have either left- or right-handed configurations, depending on the direction it rotates (a right-handed helix rotates clockwise [CW], and a left handed helix rotates counter-clockwise [CCW] while moving along the helix away from the observer). At the distal tip of the flagellar filament is the FliD protein, which serves as the flagellar cap. This is the site where new monomers that travel through the hollow filament attach to the growing filament. Associated with the flagellar apparatus near the C-ring complex are several different proteins involved in subunit transport and flagellar assembly. External proteins of the hook-basal body complex are secreted by a Type III secretion pathway through the basal body. The only exceptions are some of the proteins that comprise the L and P rings, which are secreted by the SecA-dependent pathway. The bacterium constructs the architecturally complex flagellum using constant feedback mechanisms that govern its assembly at the growing tip (Macnab, 2000; Karlinsey et al., 2000). The proteins and their approximate location are summarized in Table 1.

Most externally flagellated bacteria have two different swim modalities: runs and tumbles (Berg, 2003; Macnab, 1996; Armitage, 1999; Berg, 1975). For example, *E. coli*, which is peritrichous, "runs" (translational motility) when the flagella rotate CCW (as a frame of reference, the flagella are viewed from the end of the flagella toward their insertion into the cell). During a run, the flagella form a rotating bundle behind the cell and propel the cell forward. When one or more of the flagella reverse direction and rotate CW, the cell tumbles (Macnab, 1996; Turner et al., 2000). During the tumble, some of the filaments leave the flagellar

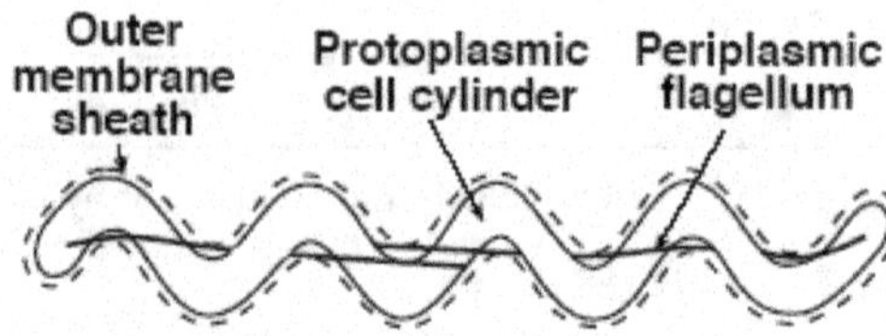

Figure 1. Schematic diagram of a spirochete (Printed with permission from the Annual Review of Genetics, Charon and Goldstein, 2002).

bundle and the cell body reorients. The cell then runs in a new direction when the flagella revert to the CCW direction. In some species, such as *Rhodobacter sphaeroides* and *Sinorhizobium meliloti*, the flagella stop or slow down instead of changing their direction of rotation (Armitage, 1999).

3. General features of spirochete motility

3.1. Structural overview

The general characteristics of spirochete motility have been described in several older and recent reviews (Canale-Parola, 1978; Goldstein and Charon, 1988; Charon et al., 1992b; Li et al., 2000b; Lux et al., 2000; Charon and Goldstein, 2002; Limberger, 2004). Here we summarize the general features of motility of spirochetes as related to structure and genetics, with a special emphasis on the genus *Treponema*. The reader is referred to the accompanying chapters by Ellen, Limberger , Lux and Shi.

Most spirochetes are helically shaped (Holt, 1978), but some species have a flat sinusoidal or meandering waveform (Goldstein et al., 1996; 1994; Charon and Goldstein, 2002; Holt et al., 1994). In addition to a typical bacterial plasma membrane surrounded by a cell wall containing peptidoglycan, referred to as the protoplasmic cell cylinder (PC), they have an outer lipid bilayer membrane, referred to as either an outer membrane sheath, outer membrane, or outer sheath (OS, Figure 1). The space between the protoplasmic cell cylinder and the outer membrane sheath is referred to as the periplasm or periplasmic space. The spirochetes have flagella that are similar in many respects to the external flagella of rod-shaped bacteria. However, the spirochetes are unique in that their flagella are located within the periplasm and, hence, are referred to as periplasmic flagella (PFs). Each PF is subterminally attached to only one end of the cell cylinder and extends toward the opposite end. Spirochete species vary with respect to size, number of PFs, and whether the PFs overlap in the center of the cell. *Cristispira*, for example, are 0.5-3 μm wide, 30-180 μm long, and have over 100 PFs attached at each cell end, while the Leptospiraceae (which include *Leptospira* and *Leptonema* sp.) are approximately 0.1 μm in diameter, 10-20 μm long, and have only one PF at each end (Figure 1).

3.2. Role of the periplasmic flagellum in spirochete motility

The PF is clearly the organelle of motility. Early on, the analysis of chemically induced and spontaneously occurring mutants and their revertants pointed towards this conclusion (Paster and Canale-Parola, 1980; Charon et al., 1992a; Li et al.,

2000b), but recent targeted mutagenesis studies have been conclusive: Mutations that inhibited the synthesis of PFs resulted in nonmotility. These mutations include *fliG2*, *flgE*, *fliF*, *flaB* [periplasmic flagellar filament protein] for *Borrelia burgdorferi* (Motaleb et al., 2000; M.A. Motaleb, M. Sal, C. Li, N.W. Charon, unpublished data); *flgE, fliG, tap1* [hook assembly protein (Li et al., 1996; Limberger et al., 1999)] in *Treponema denticola*; *fliG*, *flaB1-flaB2* double mutant in *Brachyspira hyodysenteriae* (C. Li and N. W. Charon, unpublished data, see below); and *flaB* in *L. biflexa* (Picardeau et al., 2001). In some cases, complementation of the above mutations restored the wild-type phenotype (Sartakova et al., 2001; Chi et al., 2002).

Although it has not been directly proven that the PFs rotate within the outer membrane sheath, several lines of evidence are strongly suggestive. Protrusions surrounded by an outer membrane sheath are often seen in stationary phase cells of many spirochete species, and at a high frequency throughout all growth phases in certain mutants of *Treponema phagedenis* (Charon et al., 1992a). These protrusions are clearly PFs, as they have the same helix pitch and helix diameter of purified PFs, and are not present in mutants lacking PFs. Most importantly, these protruding PFs have been shown to rotate using both darkfield and differential interference contrast light microscopy (Charon et al., 1992a; Goldstein et al., 1994). In addition, as discussed below, many of the motions observed in swimming spirochetes are best explained by rotation of the internally located PFs. Finally, both structural and genetic analyses indicate that the PFs are very similar to their flagellar counterparts in other bacteria (Canale-Parola, 1984; Li et al., 2000b). This obviously suggests that they function in a comparable manner.

3.3. Skeletal Function of PFs

Sorting out the factors that dictate cell morphology is essential in understanding spirochete models of motility (see section D, below). Spirochetes exhibit a range of shapes. They can be wave-like or helical. Often the proximal ends of the cells of certain species have a different shape then the central cell region. The first hint that the PFs in spirochetes had a skeletal function came from the early studies on *Leptonema* (formerly *Leptospira*) *illini* (Hovind Hougen, 1979; Bromley and Charon, 1979). All members of the Leptospiraceae have right-handed helical cell bodies, and have one short PF attached near each cell end that does not overlap in the central part of the cell (Carleton et al., 1979; Bromley and Charon, 1979). The swimming Leptospiraceae have regions proximal to and including the ends that are either spiral-shaped or hook-shaped (Figure 2, Inada et al., 1916; Goldstein and Charon, 1988). These hook and spiral shapes are superimposed on the helically shaped cell cylinder. A given end can change from one shape to the other. When non-motile mutants were isolated, one class had ends that were no longer hook or spiral shaped; their ends were straight. When the PFs were examined in these mutants, these organelles no longer coiled; compared to the flagella of other bacteria and other spirochete species, this coiling is unique to the Leptospiraceae. The PFs of the mutants were straight (Bromley and Charon, 1979). These results indicated that the PFs caused the ends of the cells to influence the shape of the cell in the domain where they resided.

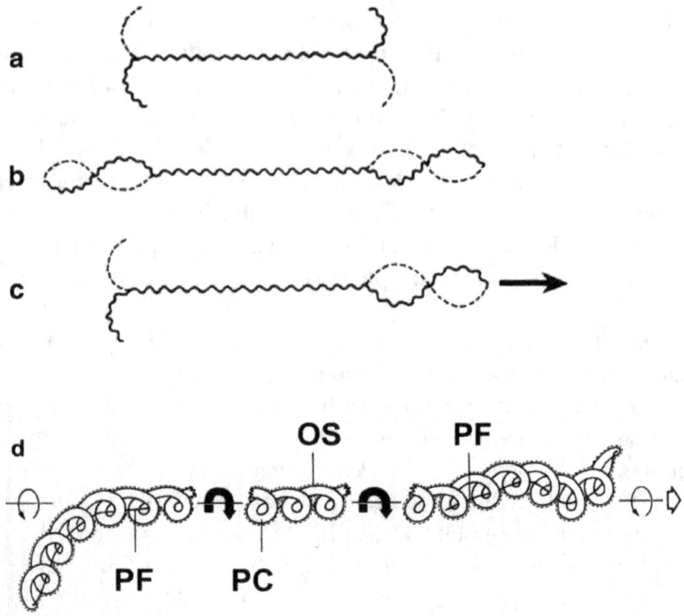

Figure 2. Swimming cells of Leptospiraceae (*Leptonema illini*) illustrating the hook- and spiral-shaped ends during swimming. Non-translating cells have both ends being either hook-shaped (a) or spiral-shaped (b). Translating cells have their anterior end spiral shaped, and their posterior end hook-shaped (c) (Figure from Goldstein and Charon, 1990). Proposed model of Leptospiraceae motility (d) (Berg et al., 1978; Figure from Goldstein and Charon, 1988, with permission from Wiley-Liss, Inc. a subsidiary of John Wiley and Sons, Inc.). An observer behind the cell sees the spiral- and hook-shaped ends gyrate CCW (thin arrows) because of the rotation of the PFs in that direction, while the PC rotates CW (wide arrow; Figure from Goldstein and Charon, 1988). The *Treponema* and *Borrelia* species are hypothesized to swim using a variation of this model.

A similar relationship between short PFs and altered cell end shape was found in *T. phagedenis* Kazan 5 (Figure 3, Charon et al., 1991). *T. phagedenis* is approximately 14-15.5 μm long. It has two flagellar bundles containing approximately 4-8 PFs subterminally attached at each end that fail to overlap in the center of the cell. The cell body is a right-handed helix in the central part of the cell where there are no PFs. However, the ends of the cells have bent-shaped ends, which are often left-handed. Mutants that do not synthesize the PFs lack the bent ends. Furthermore, these bent ends and the PFs are approximately the same length (2.4 - 2.5 μm). These results indicate a close relationship between the formation of the bent shaped ends and the presence of the PFs and indicate a causal relationship.

There is a very complex relationship between the PFs and the bent-shaped ends of *T. phagedenis* (Charon et al., 1991). Purified PFs are left-handed with a helix diameter of 0.36 μm and a pitch of 1.26 μm. The bent ends are left-handed helices with a helix diameter of 0.56 μm and a pitch of 1.85 μm. Thus, the shape of the bent ends differs significantly from those of the right-handed cell body and of the isolated PFs. Together, these results indicate a complex mechanical interaction between the cell body and the PFs. This complex interaction is amplified by the observation that certain mutants of *T. phagedenis* have a cell cylinder that is more

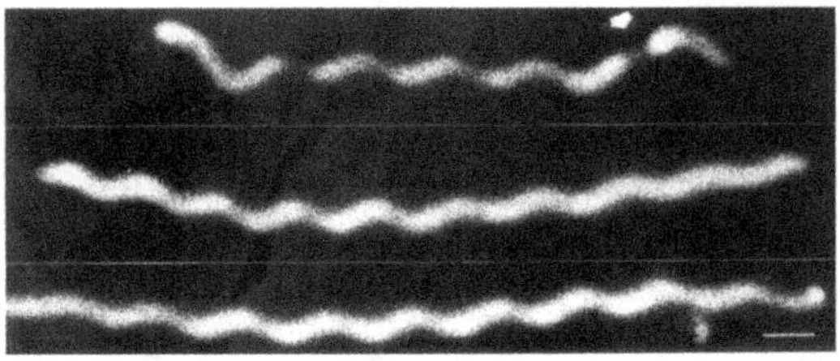

Figure 3. Darkfield micrograph of *T. phagedenis* Kazan 5 (top) illustrating the right-handed helix in the central part of the cell and the bent ends. The plane of focus is below the axis of the cell. Arrow indicates a non-helical bend in a bent-end. The bent-ends are not evident a PF-deficient mutants T-40 (middle) and T-55 (Bottom). (From Charon et al. 1991, with permission from the Journal of Bacteriology).

helical than that of the wild-type while others are rod- shaped. Both of these mutants protrude their PFs at a frequency greater than the wild type (Charon et al., 1992a). For example, the rod-shaped cell body mutant protrudes its PFs at greater than 90% throughout all growth phases compared to a maximum of 25% of wild-type cells in stationary phase. These results indicate that there is a critical shape of the cell cylinder for an essential fit with the PFs to form the bent-shaped ends; otherwise, the PFs protrude.

The complex interaction between the PFs and the cell cylinder is quite marked in the oral *T. denticola* (Figure 4, Ruby et al., 1997). *T . denticola* has two PFs at each end that are long enough to overlap in the center of the cell. The cell bodies, as seen in PF-less mutants, are right-handed helices (Ruby et al., 1997). Isolated PFs are left-handed helices, with a helix diameter of about 0.26 μm and a pitch of 0.78 μm. Swimming cells often show a mixture of cell morphologies. Specifically, some cells are right-handed. However, translating cells often have an irregular, twisted morphology that can contain both planar and right-handed helical regions. Such irregularly shaped cells were absent in the PF deficient mutants. These irregularly shaped cells were also absent when cells were treated with detergent to remove the outer membrane sheath. These results suggest that, in a certain fraction of swimming cells, the PFs interact with the cell cylinder, and most likely with the PFs at the opposite ends of the cells, to cause an irregular shape. In addition, the outer membrane sheath is hypothesized to be necessary to keep the PFs and cell cylinder in close juxtaposition to one another to bring about this irregularity.

The most remarkable example of the PFs having a skeletal function is found with *B. burgdorferi* (Motaleb et al., 2000; Sartakova et al., 2001). This spirochete is

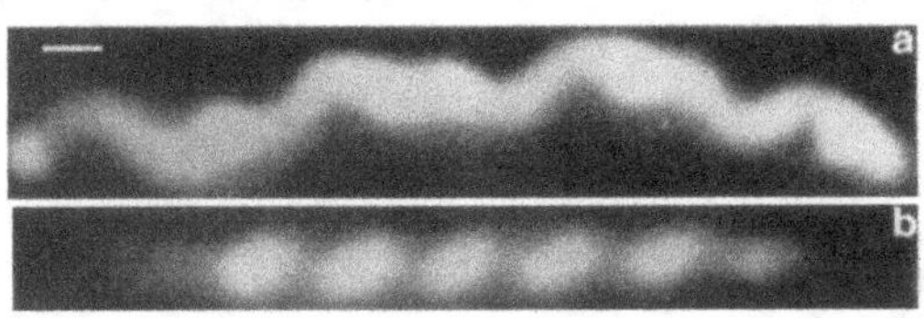

Figure 4. (a) Wild-type *T. denticola* exhibiting the typical irregular morphology. (b) Flagellar hook mutant HL51 exhibiting a right handed helical morphology. Bar, 1.0 mμ. (Ruby et al., 1997, printed with permission from the Journal of Bacteriology).

significantly larger than most pathogenic bacteria (10-20 μm long, 0.33 μm wide), and has 7 to 11 PFs attached to each end of the cell. These filaments overlap in the center of the cell and form a continuous bundle in the periplasmic space (Goldstein et al., 1996). These organisms differ from all the other spirochetes so far mentioned in that they are not helically shaped; they have a flat-wave morphology (Goldstein et al., 1996; Goldstein et al., 1994). Remarkably, insertion mutations in the gene that encodes the major PF filament protein FlaB (analogous to FliC of *E. coli* and *S. enterica*) result in cells that are rod-shaped; these results suggest that the PFs affect the morphology of *B. burgdorferi* in quite a striking manner.

We know very little about whether the PFs influence the shape of *T. pallidum. T. pallidum* has approximately two to four PFs subterminally attached at each end overlapping in the center of the cell (Hovind-Hougen, 1976; Norris et al., 2001; Holt, 1978). As with other spirochete species, the speed of *T. pallidum* increases when suspended in a gel-like medium (Van Weelden et al., 1990). Several studies suggest that the intact cell swims with a planar rather than helical wave form (Cox, 1972; Sequeira, 1956). This observation suggests that *T. pallidum* could be similar to *B. burgdorferi* with respect to structure and motility. If this similarity completely holds for *T. pallidum*, then we predict that putative PF mutants of *T. pallidum* would be rod- shaped. Because of the inherent difficulty in isolating PF deficient mutants of this uncultivable spirochete, this question remains to be answered. On the other hand, results with scanning electron microscopy suggest that the predominant form of *T. pallidum* is left-handed, with approximately twenty five percent of the cells being flat waves (Stephan and Johnson, 1981). These results amplify one of the problems with analyzing the three dimensional structure of spirochetes; in some species, such as *T. pallidum, T. denticola*, and *B. burgdorferi*, the cell cylinders are so flexible that more than one morphological shape is evident on microscopic analysis.

In summary, *T. phagedenis* and *T. denticola* have right-handed cell cylinders. This right-handed morphology is independent of the presence of the PFs. However, the final shapes of the cells are markedly influenced by the PFs in the domains where these organelles reside. In cells with relatively short PFs, such as *T. phagedenis*, regions at and near the ends of the cells show this alteration of cell shape. In species that have overlapping PFs such as *T. denticola,* the PFs influence the shape along the entire cell. In *T. denticola*, such cells often appear irregular. To draw comparisons to other spirochete species, *T. phagedenis* is most analogous to the Leptospiraceae, whereas *T. denticola* is most similar to *B. burgdorferi*. Note, however, that *B. burgdorferi* differs from *T. denticola* by having a rod-shaped cell cylinder and a flat wave swimming form. With respect to *T. pallidum*, there is not a sufficient amount of data to determine how (or if) the PFs influence its morphology.

3.4. How do we think *Treponema* swims?

Our understanding of the mechanics of *Treponema* motility is still in its infancy. The indication that PFs have a skeletal function in *T. phagedenis* and *T. denticola* is significant. As discussed below, this information has allowed us to draw inferences from the well studied *Leptospiraceae* model system and apply this knowledge to these organisms.

In a translating cell of the *Leptospiraceae*, the posterior end is hook-shaped, and the anterior end is spiral-shaped (Inada et al., 1916; Goldstein and Charon, 1988) (Figure 2). When the cells reverse direction, the ends of the cell concomitantly change. In addition, non-translating cells are also seen; the ends of these cells are either both spiral-shaped or both hook-shaped (Berg et al., 1978; Goldstein and Charon, 1988; Charon et al., 1984). An analysis of motility mutants of *L. illini*, along with an analysis of swimming *Leptospiraceae* in both low viscosity (water like) and highly viscous gel-like media, led to a model for their motility (Figure 2, Berg et al., 1978). The model states that the PFs rotate between the outer membrane sheath and the right-handed protoplasmic cylinder. A given PF is more rigid than the cell cylinder, causing the end of the cell to conform somewhat to the shape of its flagellum. As viewed from the center of the cell towards one of the cell ends, CCW rotation of a PF results in the flagellum at that end forming a left-handed helix with a pitch and diameter larger than those of the cell body. This form is referred to as the spiral shaped end. On the other hand, rotation of a PF in the opposite CW direction results in the flagellum and cell body at that end having a planar hook-shaped morphology. When the PFs are rotating in the same direction (CCW or CW), the cell does not translate (*i.e.*, show displacement in a given direction). The cell is either spiral-spiral-shaped or hook-hook-shaped.

In the model for *Leptospiraceae*, forward thrust results from two rotations. The model can be conveniently described from the viewpoint of an observer located behind a swimming cell (Goldstein and Charon, 1988; Berg et al., 1978). First, the left-handed helical anterior PF and the posterior hook-shaped PF both rotate CCW. Second, the torque of the rotating PFs produces a CW counter-rotation of the right-handed helical cell body. Two modes of thrust are proposed. CCW rotation of the anterior PF causes that end of the cell to gyrate (to bend in a circular manner without necessarily rotating); its rotation generates a backward moving spiral-wave. This wave is left-handed (Goldstein et al., 1994), and it generates sufficient thrust for the cell to translate forward in a low viscosity medium. Concomitantly, CW rolling of the cell cylinder gives the organism the unique capacity to bore through gel-like media and connective tissue (Goldstein et al., 1994). Note that most bacteria slow down or stop in such media, whereas spirochete species speed up (Charon and Goldstein, 2002; Berg and Turner, 1979). The torque due to the roll of the cell cylinder is counter-balanced by the gyrations of the cell ends. Several lines of evidence support this model including an analysis of tethered cells, high-speed cinematography, and mutant cell analysis (Charon et al., 1984; Bromley and Charon, 1979; Goldstein and Charon, 1990; Picardeau et al., 2001). Furthermore, genomic analysis shows that the *Leptospiraceae* have the full complement of genes necessary for PF rotation (Ren et al., 2003).

A similar approach used for the *Leptospiraceae* has been initiated in analyzing the motility of the more complex *T. phagedenis* (Charon et al., 1991). As already mentioned, this spirochete has a right handed cell cylinder with 4-8 short PFs attached subterminally at each end. These organelles cause the regions near the cell ends to be left-handed or irregularly shaped. CCW rotation of the PFs, as viewed from behind the cell, results in a gyration of the bent-shaped ends. This gyration does not yield sufficient thrust for swimming in a low viscosity medium as it does for the *Leptospiraceae*; the gyration is either too slow or not large enough. Yet, because the cell cylinder is right-handed, *T. phagedenis* can translate in a highly viscous gel-like medium such as one percent methylcellulose with little slippage. The torque generated by the CCW rotation of the PFs (and gyrating ends) results in a CW counter-rotation of the cell body. It is the CW counter-rotation of the cell cylinder that is responsible for this translation. Thus, the prediction is that in translating *T. phagedenis*, the ends gyrate CCW as viewed from the back of the cell and the cell rolls CW to balance this motion. This CW roll of the right-handed cell cylinder promotes forward motion. Further experimentation is necessary to rigorously test this model for *T. phagedenis*.

T. denticola is considerably more complex in structure, and the dynamics of its motility are not understood in detail. As mentioned, *T. denticola* has two PFs subterminally attached at each end that are long enough to overlap and interact with one another in the center of the cell with those attached to the other end (Hovind-Hougen, 1976; Listgarten and Socransky, 1964; Charon et al., 1992a). The two morphological types of *T. denticola* previously described, *i.e.* the helical form and the irregular shaped form, fail to translate in a low viscosity medium. However, in a gel-like medium with high macroscopic viscosity, the cells translate quite efficiently, especially if the temperature of the ambient medium is raised from 23°C to 35°C. Raising the temperature from room temperature to 35°C is quite critical, as their velocity increases close to five-fold (4 to 20 μm /sec, Ruby and Charon, 1998). Purified PFs of *T. denticola* are left-handed. Thus, our working hypothesis is that with respect to *T. denticola*, CCW rotation of the PFs, as viewed from behind the cell, generates backward moving helical or irregularly shaped gyrating waves. These gyrating waves do not have sufficient thrust for noticeable translation in a low viscosity medium. However, the CCW rotation of the PFs produces a CW counter-rotation of the cell cylinder. This rolling (or rotation) of the cell cylinder has been observed in *T. denticola* in a gel like medium; they swim with little or no slippage. Because *T. denticola* translates quite efficiently in a gel-like medium, we propose that CW rolling of the cell cylinder is its main means of thrust. As with all spirochete species studied, non-translating forms are seen. These forms occur when the PFs rotate in the same direction as viewed from the center of the cells. In contrast to the *Leptospiraceae* and *T. phagedenis*, when this occurs in *T. denticola*, the cell bodies can become quite distorted and twisted. This type of form is referred to as a flex. We predict that a flex occurs when the bundle of PFs of each end interact, intertwine, and rotate against one another. (Try taking two left-handed wires, hold one in each hand, and rotate each CCW as viewed from your elbow to the end of each wire: They wind around each other). This is predicted to occur when the bundles rotate CCW. When they both rotate CW, we predict an elongated non-translating form.

The discussion above assumes that both PF bundles actively rotate during swimming. As an alternative model, perhaps one bundle rotates while the other is stopped, and cell reversals occur when motor rotation switches from one bundle to the other. One prediction of this hypothesis is that the bundles at a given end should show frequent stops. To test this hypothesis, we analyzed spirochete cells tethered to a glass surface. *T. phagedenis* gave the most straightforward results. As mentioned, its PFs are quite short and influence the shape of the cell in the domains where they reside. We found that the terminal regions gyrated due to the rotation of the PFs (Charon et al., 1991). Moreover, both of the terminal regions constantly gyrated and often changed direction. In some cells, one end constantly gyrated in a specific direction, and the other end switched direction quite frequently (N.W. Charon, unpublished data). These results indicate that, for translation in *T. phagedenis*, both bundles of PFs rotate. A similar conclusion was reached with the *Leptospiraceae* and *B. burgdorferi* (Charon et al., 1984; Li et al., 2002).

We know very little about the motility of *T. pallidum*. Phylogenetic analysis of the 16S ribosomal RNA gene of *T. pallidum* indicates that it clusters most closely with *T. denticola* and *T. phagedenis* (Paster and Dewhirst, 2000). Genomic analysis indicates that *T. pallidum* is genetically quite similar to *T. denticola* with respect to conservation of motility genes (Fraser et al., 1998; Seshadri et al., 2004; Stamm and Bergen, 1999). We, therefore, suspect that the mechanics of motility of *T. pallidum* is quite similar to that of either or both of these species. Because the PFs of *T. palllidum* are long and overlap in the cell center, it is most likely that its motility will be similar to that of *T. denticola*. However, the *T. denticola* cell body is quite helical (Ruby et al., 1997), and *T. pallidum* has been reported to swim both as a helix and as a flat wave (DeLamater et al., 1951; Cox, 1972). Both rotary and undulatory swimming patterns have been described (DeLamater et al., 1951). If *T. pallidum* does primarily swim as a flat-wave, then its motility may be most similar to that of *B. burgdorferi*. Another model, based on observations of *Spirochaeta aurantia*, is also conceivable for *T. palllidum* (Berg, 1976). The model states that the outer membrane sheath rotates or flows around the cell body in one direction and the protoplasmic cell cylinder rotates in the opposite direction. Forward thrust is obtained by rolling of the cell cylinder in one direction and rotation of the sheath in the opposite direction. In analyzing *T. pallidum* motility, one concern is that it is essential that the spirochetes be freshly isolated and examined in an optimal atmospheric and temperature environment; otherwise, *T. palllidum* cells become sluggish and distorted. This brief discussion on *T. pallidum* leads us to the obvious conclusion that some of the very basic experiments related to motility are yet to be done on this most important *Treponema* species.

3.5. Structure and composition of PFs

The PFs of several spirochete genera (*Leptospira*, *Treponema*, *Brachyspira*, and *Spirochaeta*) possess some features that distinguish them from the flagella of other bacteria (Figure 5, Li et al., 2000b; Charon et al., 1992b; Charon and Goldstein, 2002). Because of a unique protein sheath surrounding the filament core, the flagellar filaments of spirochetes are thicker than those of other bacteria, *e.g.*, 25 nm in diameter for *B. hyodysenteriae* compared to 20 nm for *E. coli* and *S. enterica* (Macnab, 1996; Li et al., 2000a). This sheath is composed of a protein designated FlaA. Its sequence is well conserved among the spirochetes. The PF core is composed of a family of FlaB proteins. FlaB proteins show sequence similarity to flagellin proteins of other bacteria, especially at the N- and C-terminal regions. Most spirochetal species have three different FlaB proteins (FlaB1, FlaB2, and FlaB3), each encoded by a separate gene. In addition, there is extensive sequence similarity among the FlaB proteins within a given species and also among species

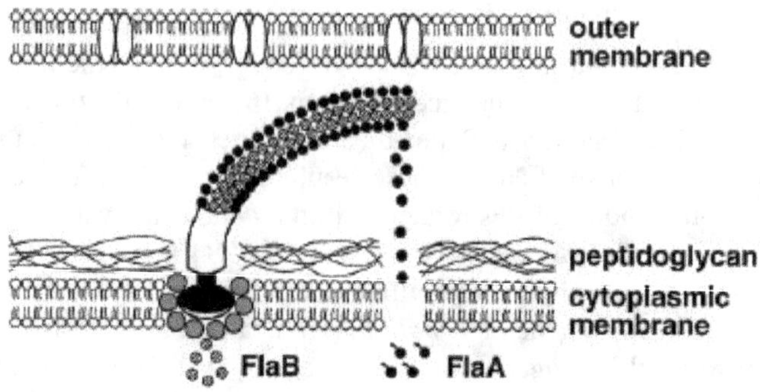

Figure 5. Structure of a spirochete PF. In several spirochete species, the basal bodies lack the L and P rings found in *E. coli* and *S. enterica* (Brahamsha and Greenberg, 1988; Hovind-Hougen, 1976; Nauman et al., 1969). FlaA and FlaB are secreted to the periplasmic space by SecA and type III secretion systems, respectively. This complex periplasmic flagellar structure is found in *Treponema* sp., *Spirochaeta aurantia*, *Leptospiraceae* sp., and *Brachyspira hyodysenteriae* (Charon et al., 1992b; Li et al., 2000b). (Figure from Charon and Goldstein, 2002, printed with permission from the Annual Review of Genetics).

of different genera. The FlaA protein is secreted into the periplasmic space by a SecA-dependent system, whereas FlaB is secreted by a Type III secretion system. The mechanism of assembly of the sheath around the core is not known, but the confining environment of the periplasmic space likely plays a critical role. The PF composition of *B. burgdorferi* is somewhat different. It has only one FlaB protein, and there is considerably more FlaB than FlaA. The location of FlaA in *B. burgdorferi* is uncertain, but evidence suggests that it is closely associated with the periplasmic flagellar filament (Ge et al., 1998; Motaleb et al., 2004).

B. hyodysenteriae has been used as a model system to analyze spirochete PF structure. Besides having PFs that are structurally and antigenically similar to those of treponemes, this spirochete species can be genetically manipulated in the laboratory using a generalized transducing phage-like particle (Humphrey et al., 1997). Recent results indicate that the interaction of FlaA with the FlaB subunits affects flagellar filament shape in a manner that has not been observed in any other flagellar system (Li et al., 2000a). Wild-type cells of *B. hyodysenteriae* have PFs that are left-handed with a defined shape, as measured by helix pitch and helix diameter. The approximate ratios of flagellin proteins to one another are the following: FlaA : FlaB1 : FlaB2 : FlaB3 of 1 : 0.42 : 0.22 : 0.45. Thus, there is approximately one FlaA per FlaB family subunit. Targeted mutagenesis of each of the genes encoding *flaA*, *flaB1*, *flaB2*, and *flaB3* resulted in mutant cells that were still motile, and also possessed PFs that retained their left-handed configurations. However, comparison of mutant and wild-type purified PF shapes yielded surprising results. Each of the *flaB* mutants had PFs that differed slightly in shape from those of the wild-type. However, the PFs from the *flaA* mutants displayed major decreases in both helix pitch and helix diameter compared to the wild type. These results suggest that FlaA markedly impacts the shape of the PFs. Because bacterial flagella are often quasi-rigid and undergo helical transformations, perhaps the sheath helps stabilize the FlaB helical core into one of these configurations for optimal thrust as it rotates between the outer membrane sheath and cell cylinder.

The analysis of double mutants of *B. hyodysenteriae* indicates a remarkably complex interaction between the FlaA and FlaB subunits. The location of each FlaB protein along the PF is unknown, but two different models could account for their arrangement (Rosey et al., 1995). One model proposes that FlaB1, FlaB2, and FlaB3 reside in specific regions of the flagellum in a manner analogous to the multiple flagellar subunits of *Caulobacter crescentus* (Driks et al., 1989). For example, perhaps FlaB3 is adjacent to the hook, FlaB2 is in the central region, and FlaB1 is at the distal region. However, a recent analysis of double mutants support another model in which the different FlaB proteins are distributed along the entire length of the PF (C. Li and N.W. Charon, unpublished data). In this analysis, the PFs from the double mutants *flaA-flaB1*, *flaA-flaB2*, and *flaA-flaB3* were found to be left-handed. Surprisingly, their helix pitch and diameter were more drastically different from the wild-type and the single *flaA* mutants. These results suggest that in the absence of FlaA, the inhibition of synthesis of one FlaB protein dramatically influences the shape of the entire PF. Such a result would not be expected if a given FlaB protein resided in only one domain of the PF. These results also suggest that FlaA interacts with each FlaB protein to affect PF shape.

The role of the different flagellin proteins in motility is unclear. Although single mutants of *flaB1*, *flaB2*, *flaB3*, or *flaA* of *B. hyodysenteriae* were motile, observations of swimming cells and swarm-agar-plate assays indicate that all these mutants swam less efficiently than the wild type (Kennedy et al., 1997; Rosey et al., 1995; Li et al., 2000a). In addition, the *flaA-flaB1*, *flaA-flaB2*, and *flaA-flaB3* double mutants were also motile, but their swarming deficiency was even more pronounced than was that of the single mutants (C. Li and N.W. Charon, unpublished data). The analyses of double *flaB* mutants were intriguing. Whereas the single *flaB* mutants and the double *flaB1–flaB3* mutant were still motile and had PFs, the *flaB1-flaB2* double mutant was completely nonmotile and lacked intact PFs. These results indicate that the FlaB1 and FlaB2 overlap in function. Clearly, to fully understand the function of the multiple flagellin proteins, we need a more thorough analysis of the structure and the dynamics of motility wild-type and derived mutants of *B. hyodysenteriae*.

3.6. Role of cytoplasmic filaments in motility

Treponema and *Leptonema* are the only spirochete genera shown to have cytoplasmic filaments (Hovind-Hougen, 1976; Limberger, 2004). These filaments are approximately 7.75 nm wide, and they are anchored to the cytoplasmic membrane. Tomographic reconstruction in *T. phagedenis* indicate that these filaments form a bundle that extends from one end of the cell to the other (Izard et al., 1999). Early observations suggested that these filaments were anchored to the cytoplasmic side of the basal bodies of the PFs, and could conceivably affect motility (Hovind-Hougen, 1976). To better understand their role in cell function, mutants that lack these filaments in *T. denticola* were isolated and characterized (Izard et al., 1999; Izard et al., 2001). These mutants had pleiotropic effects with altered DNA segregation, motility, and cell length. Cells often grew in chains. However, when short cells were examined, they were motile. Furthermore, close examination of their location in the cell did not confirm attachment to the PF basal bodies (Izard et al., 1999). These results indicate that the cytoplasmic filaments are not necessary for motility. It is still conceivable, however, that they could play a role in coordinating rotation of the bundles at each end of the cell, and perhaps in chemotaxis.

3.7. Organization and function of motility genes

Complete sequencing of the *T. pallidum* and *T. denticola* genomes has yielded important information about motility (Fraser et al., 1998; Seshadri et al., 2004). We find that all the genes necessary for flagella-based motility are present in *T. pallidum* and *T. denticola*. *T. pallidum* lacks the homolog that encodes the basal body P ring found in the Enterobacteriacae, whereas *T. denticola* evidently has this gene. Both contain multiple genes that show high identity to *flaA* (two for *T. pallidum* and three for *T. denticola*), and for *flaB* (three each). Neither contains the Class I regulatory genes (*flhC, flhD*) that modulate the synthesis of turning on (or off) flagellar genes found in other bacteria (Chilcott and Hughes, 2000). The lack of these genes in these two *Treponema* species is in agreement with PFs consistently being present in cells under all conditions examined. With one exception, both *T. pallidum* and *T. denticola* have only one copy of the many genes that encode the basal body and C-ring complex. The exception is the switch protein FliG--both organisms have two homologs of *fliG*. Both species also have *fliY* homologs, which encode a phosphatase associated with basal bodies first found in *Bacillus* species (Szurmant et al., 2004; Szurmant and Ordal, 2004).

In comparing the gene organization of the motility gene clusters between the two species, no consistent pattern emerges. Individual motility genes have the highest sequence identity among the *Treponema* species and with other spirochete species. Furthermore, *T. denticola* has four large motility gene clusters, and *T. pallidum* only has two. Some of the large gene clusters have been shown to function as operons, and their transcription is apparently initiated by a σ28 transcription factor (Heinzerling et al., 1997; Limberger et al., 1996). *T. denticola* and *T. pallidum* each have several genes of unknown function that map within the motility gene clusters. Because of their locations, these genes are likely to be involved in motility.

3.8. Control of synthesis of motility-related proteins

Unique aspects of motility and chemotaxis gene expression in spirochetes are beginning to emerge. In most bacteria studied, there is cascade control of motility gene expression (Chilcott and Hughes, 2000). For example, in *E. coli* and *S. enterica*, the expression of *flhC* and *flhD* (class 1) initiates transcription of several motility genes (class 2). Two of the class 2 genes play a critical role in regulating class 3 gene transcription. Specifically, *fliA* encodes σ28, which binds to RNA polymerase and promotes transcription of class 3 genes such as FliC. The other regulatory gene is *flgM*, which encodes the anti-σ28 factor FlgM. FlgM is secreted into the medium when the hook basal body is complete. Secretion of FlgM insures that the initiation of class 3 gene transcription is timed to the assembly of the initial flagellar filament structure. To our knowledge, *B. burgdorferi* is the only bacterial species that lacks transcriptional cascade control of motility gene expression by alternative sigma factors (Charon and Goldstein, 2002; Li et al., 2000b). *T. pallidum, S. aurantia,*

T. denticola, *T. phagedenis,* and *B. hyodysenteriae* have many motility promoter sequences homologous to σ28 recognition sequences (Li et al., 2000b; Champion et al., 1990; Limberger et al., 1992; Fraser et al., 1998; Seshadri et al., 2004; Stamm and Bergen, 1999). Thus, at first glance, it is likely that flagellar gene regulation is similar in these spirochete species to that found in other bacteria. However, a *T. denticola fliK* mutant was found to lack PFs and failed to make PF proteins but still synthesized wild type levels of PF mRNA (Limberger et al., 1999). These results suggest that there is a strong likelihood that translational control is involved in PF synthesis similar to that recently reported for *B. burgdorferi* (Motaleb et al., 2004).

There are many unknown factors concerning motility gene regulation in spirochetes. For example, the PFs are inserted into the cell cylinder subterminally at each end. Observations of dividing cells indicate that before the time of division, new PFs emerge in both daughter cells proximal to the division plane (Holt, 1978; Izard et al., 1999). However, the factors that promote synchronization and localization of basal body synthesis with cell division are unknown. Furthermore, little is known about the factors that control PF length and number, although we know that PF length is tightly controlled in some species (Charon et al., 1991). Finally, DNA microarray analysis of many bacterial species indicates that motility genes are differentially regulated *in vitro* as compared to *in vivo*. Both the environmental factors within the host and the molecular details within the spirochete that promote this differential regulation of gene expression have yet to be elucidated for the treponemes.

3.9. Control of flagellar motor direction

The swimming behavior of spirochetes is more complicated than that of other bacteria as a consequence of their complex geometry. Spirochetes have three motility modes: two modes of runs (*i.e.*, translation through the medium with either end leading), and one mode of nontranslational motility (Fosnaugh and Greenberg, 1988; Charon et al., 1984; Goldstein et al., 1994; Bilwes et al., 1999). During runs, as seen be a viewer looking into the cell along the motor shaft, the anterior bundle of PFs rotates CCW and the posterior bundle rotates CW. Thus, in contrast to bacteria such as *E. coli* and *S. enterica,* the PF bundles at the two ends of the cell rotate in opposite directions, *i.e.*, asymmetrically. Reversals occur when both bundles change their direction of rotation. In the third (nontranslational) mode, the bundles rotate in the same direction: both CW or both CCW (in a sense, there are two non-translational modes, one with both bundles of PFs going CW, and the other CCW). How spirochetes coordinate the direction of rotation of their two PF bundles for directed movement during chemotaxis is not known (Charon and Goldstein, 2002) (see Chapter 7, Lux and Shi).

3.10. Motility and virulence

Deciphering the roles of motility and chemotaxis of spirochetes in the disease process is at the early stages (see Chapter 7, Lux and Shi, and Chapter 15, Ellen; Lux et al., 2000; Barbour and Hayes, 1986). Many spirochete researchers have long believed that these processes are major virulence factors. This belief is supported by the fact that approximately 5 per cent of the genes for *T. pallidum* are putative chemotaxis and motility genes. In addition, several *Treponema* species are known to be highly invasive in host tissues. This is particularly true for *T. pallidum* and the other treponemes associated digital dermatitis (U.S.Department of Health, 1968b; Moter et al., 1998; Stamm and Bergen, 1999). (See Chapter 17, Stamm and Trott). In addition, quorum sensing, the ability for bacteria to sense cell density and regulate gene expression, is a key step in the regulation of virulence genes and biofilm formation in other bacterial species (Fuqua et al., 2001). Treponemes achieve high densities when attached to cells and tissues (Soames and Davies, 1975; Holt, 1978). In some species, motility plays an important role in biofilm formation (O'Toole et al., 2000). Recent results suggest that biofilm formation of a non-motile mutant of *T. denticola* is reduced, suggesting that motility may be important for that species (Vesey and Kuramitsu, 2004). Clearly, sorting out the the relationships between motility, tissue invasion, quorum sensing, and biofilm formation are intriguing areas for future research on the treponemes.

Acknowledgements

We thank R. Limberger, M. Motaleb, M. Sal, L. Stamm, J. Vuksanovic, and C. Wolgemuth for their helpful suggestions. This research was supported by U.S. Public Health Service grants AI29743, DE12046, and by USDA grant 95-37204-2132 to NWC; U.S.P.H.S. grant AR050656 to C. Li, and National Science Foundation Grant DMS 0201063 to SFG.

References

Armitage, J.P. (1999). Bacterial tactic responses. Adv. Microb. Physiol *41*, 229-289.

Barbour, A.G. and Hayes, S.F. (1986). Biology of *Borrelia* species. Microbiol. Rev. *50*, 381-400.

Berg, H.C. (1975). Chemotaxis in bacteria. Annu. Rev. Biophys. Bioeng. *4*, 119-136.

Berg, H.C. (1976). How spirochetes may swim. J. Theor. Biol. *56*, 269-273.

Berg, H.C. (2003). The rotary motor of bacterial flagella. Annu. Rev. Biochem. *72*, 19-54.

Berg, H.C., Bromley, D.B., and Charon, N.W. (1978). Leptospiral motility. Symp. Soc. Gen. Microbiol. *28*, 285-294.

Berg, H.C. and Turner, L. (1979). Movement of microorganisms in viscous environments. Nature (London) *278*, 349-351.

Berry, R.M. and Armitage, J.P. (1999). The bacterial flagellar motor. Adv. Microb. Physiol *41*, 291-337.

Bilwes, A.M., Alex, L.A., Crane, B.R., and Simon, M.I. (1999). Structure of CheA, a signal-transducing histidine kinase. Cell *96*, 131-141.

Brahamsha, B. and Greenberg, E.P. (1988). A biochemical and cytological analysis of the complex periplasmic flagella from *Spirochaeta aurantia*. J. Bacteriol. *170*, 4023-4032.

Bromley, D.B. and Charon, N.W. (1979). Axial filament involvement in the motility of *Leptospira interrogans*. J. Bacteriol. *137*, 1406-1412.

Canale-Parola, E. (1978). Motility and chemotaxis of spirochetes. Annu. Rev. Microbiol. *32*, 69-99.

Canale-Parola, E. (1984). The Spirochetes. In Bergey's Manual of Systematic Bacteriology, N.R.Krieg and J.G.Holt, eds. (Baltimore: Williams and Wilkins), pp. 38-70.

Carleton, O., Charon, N.W., Allender, P., and O'Brien, S. (1979). Helix handedness of *Leptospira interrogans* as determined by scanning electron microscopy. J. Bacteriol. *137*, 1413-1416.

Champion, C.I., Miller, J.N., Lovett, M.A., and Blanco, D.R. (1990). Cloning, sequencing, and expression of two class B endoflagellar genes of *Treponema pallidum* subsp. *pallidum* encoding the 34.5- and 31.0- kilodalton proteins. Infect. Immun. *58*, 1697-1704.

Charon, N.W., Daughtry, G.R., McCuskey, R.S., and Franz, G.N. (1984). Microcinematographic analysis of tethered *Leptospira illini*. J. Bacteriol. *160*, 1067-1073.

Charon, N.W. and Goldstein, S.F. (2002). The genetics of motility and chemotaxis of a fascinating group of bacteria: the spirochetes. Annu. Rev. Genet. *36*, 47-53.

Charon, N.W., Goldstein, S.F., Block, S.M., Curci, K., Ruby, J.D., Kreiling, J.A., and Limberger, R.J. (1992a). Morphology and dynamics of protruding spirochete periplasmic flagella. J. Bacteriol. *174*, 832-840.

Charon, N.W., Goldstein, S.F., Curci, K., and Limberger, R.J. (1991). The bent-end morphology of *Treponema phagedenis* is associated with short, left-handed periplasmic flagella. J. Bacteriol. *173*, 4820-4826.

Charon, N.W., Greenberg, E.P., Koopman, M.B.H., and Limberger, R.J. (1992b). Spirochete chemotaxis, motility, and the structure of the spirochetal periplasmic flagella. Res. Microbiol. *143*, 597-603.

Chi, B., Limberger, R.J., and Kuramitsu, H.K. (2002). Complementation of a *Treponema denticola flgE* mutant with a novel coumermycin A1-resistant *T. denticola* shuttle vector system. Infect. Immun. *70*, 2233-2237.

Chilcott, G.S. and Hughes, K.T. (2000). Coupling of flagellar gene expression to flagellar assembly in *Salmonella enterica* serovar Typhimurium and *Escherichia coli*. Microbiol. Mol. Biol. Rev. *64*, 694-708.

Cox, C.D. (1972). Shape of *Treponema pallidum*. J. Bacteriol. *109*, 943-944.

DeLamater, E.D., Haanes, M., and Wiggall, R.H. (1951). Studies on the life cycle of spirochetes. V. The life cycle of the Nichols nonpathogenic *Treponema pallidum* in culture. Am. J. Syph. , Gonor. and Ven. Dis. *35*, 164-187.

Driks, A., Bryan, R., Shapiro, L., and DeRosier, D.J. (1989). The organization of the *Caulobacter crescentus* flagellar filament. J. Mol. Biol. *206*, 627-636.

Fosnaugh, K. and Greenberg, E.P. (1988). Motility and chemotaxis of *Spirochaeta aurantia*: Computer-assisted motion analysis. J. Bacteriol. *170*, 1678-1774.

Fraser, C.M., Norris, S.J., Weinstock, C.M., White, O., Sutton, G.G., Dodson, R., Gwinn, M., Hickey, E.K., Clayton, R., Ketchum, K.A., Sodergren, E., Hardham, J.M., McLeod, M.P., et al. (1998). Complete genome sequence of *Treponema pallidum*, the syphilis spirochete. Science *281*, 375-388.

Fuqua, C., Parsek, M.R., and Greenberg, E.P. (2001). Regulation of gene expression by cell-to-cell communication: acyl- homoserine lactone quorum sensing. Annu. Rev. Genet. *35*, 439-468.

Ge, Y., Li, C., Corum, L., Slaughter, C.A., and Charon, N.W. (1998). Structure and expression of the FlaA periplasmic flagellar protein of *Borrelia burgdorferi*. J Bacteriol. *180*, 2418-2425.

Goldstein, S.F., Buttle, K.F., and Charon, N.W. (1996). Structural analysis of *Leptospiraceae* and *Borrelia burgdorferi* by high-voltage electron microscopy. J. Bacteriol. *178*, 6539-6545.

Goldstein, S.F. and Charon, N.W. (1988). The motility of the spirochete *Leptospira*. Cell Motil. Cytoskel. *9*, 101-110.

Goldstein, S.F. and Charon, N.W. (1990). Multiple exposure photographic analysis of a motile spirochete. Proc. Natl. Acad. Sci. U. S. A. *87*, 4895-4899.

Goldstein, S.F., Charon, N.W., and Kreiling, J.A. (1994). *Borrelia burgdorferi* swims with a planar waveform similar to that of eukaryotic flagella. Proc. Natl. Acad. Sci. U. S. A. *91*, 3433-3437.

Heinzerling, H.F., Olivares, M., and Burne, R.A. (1997). Genetic and transcriptional analysis of *flgB* flagellar operon constituents in the oral spirochete *Treponema denticola* and their heterologous expression in enteric bacteria. Infect. Immun *65*, 2041-2051.

Holt, J.G., Krieg, N.R., Sneath, P.H.A., Staley, J.T., and Williams, S.T. (1994). The Spirochetes. In Bergey's Manual of Determinative Bacteriology, Lippincott Williams and Wilkins), pp. 27-31.

Holt, S.C. (1978). Anatomy and chemistry of spirochetes. Microbiol. Rev. *42*, 114-160.

Hovind Hougen, K. (1979). Leptospiraceae: A new family to include *Leptospira* Noguchi 1917 and *Leptospira* gen. nov. Int. J. Syst. Bacteriol. *29*, 245-251.

Hovind-Hougen, K. (1976). Determination by means of electron microscopy of morphological criteria of value for classification of some spirochetes, in particular treponemes. Acta. Pathol. Microbiol. Scand. [B] (Suppl.) *255*, 1-41.

Humphrey, S.B., Stanton, T.B., Jensen, N.S., and Zuerner, R.L. (1997). Purification and characterization of VSH-1, a generalized transducing bacteriophage of *Serpulina hyodysenteriae*. J. Bacteriol. *179*, 323-329.

Inada, R., Ido, Y., Hoki, R., Kaneko, R., and Ito, H. (1916). The etiology, mode of infection, and specific therapy of Weil's disease (spirochaetosis icterhaemorrhagica). J. Exp. Med. *23*, 377-402.

Izard, J., Samsonoff, W.A., Kinoshita, M.B., and Limberger, R.J. (1999). Genetic and structural analyses of cytoplasmic filaments of wild-type *Treponema phagedenis* and a flagellar filament-deficient mutant. J Bacteriol. *181*, 6739-6746.

Izard, J., Samsonoff, W.A., and Limberger, R.J. (2001). Cytoplasmic filament-deficient mutant of *Treponema denticola* has pleiotropic defects. J. Bacteriol. *183*, 1078-1084.

Karlinsey, J.E., Tanaka, S., Bettenworth, V., Yamaguchi, S., Boos, W., Aizawa, S.I., and Hughes, K.T. (2000). Completion of the hook-basal body complex of the *Salmonella typhimurium* flagellum is coupled to FlgM secretion and *fliC* transcription. Mol. Microbiol. *37*, 1220-1231.

Kennedy, M.J., Rosey, E.L., and Yancey, R.J., Jr. (1997). Characterization of *flaA*$^-$ and *flaB*$^-$ mutants of *Serpulina hyodysenteriae*: both flagellin subunits, FlaA and FlaB, are necessary for full motility and intestinal colonization. FEMS Mirobiol. let. *153*, 119-128.

Li, C., Bakker, R.G., Motaleb, M.A., Sartakova, M.L., Cabello, F.C., and Charon, N.W. (2002). Asymmetrical flagellar rotation in *Borrelia burgdorferi* nonchemotactic mutants. Proc. Natl. Acad. Sci. U. S. A. *99*, 6169-6174.

Li, C., Corum, L., Morgan, D., Rosey, E.L., Stanton, T.B., and Charon, N.W. (2000a). The spirochete FlaA periplasmic flagellar sheath protein impacts flagellar helicity. J. Bacteriol. *182*, 6698-6706.

Li, C., Motaleb, M.A., Sal, M., Goldstein, S.F., and Charon, N.W. (2000b). Spirochete periplasmic flagella and motility. J. Mol. Microbiol. Biotechnol. 2, 345-354.

Li, H., Ruby, J., Charon, N., and Kuramitsu, H. (1996). Gene inactivation in the oral spirochete *Treponema denticola:* construction of a *flgE* mutant. J. Bacteriol. *178*, 3664-3667.

Limberger, R.J. (2004). The periplasmic flagellum of spirochetes. J. Mol. Microbiol. Biotechnol. *7*, 30-40.

Limberger, R.J., Slivienski, L.L., El-Afandi, M.C.T., and Dantuono, L.A. (1996). Organization, transcription, and expression of the 5' region of the *fla* operon of *Treponema phagedenis* and *Treponema pallidum*. J. Bacteriol. *178*, 4628-4634.

Limberger, R.J., Slivienski, L.L., Izard, J., and Samsonoff, W.A. (1999). Insertional inactivation of *Treponema denticola tap1* results in a nonmotile mutant with elongated flagellar hooks. J Bacteriol. *181*, 3743-3750.

Limberger, R.J., Slivienski, L.L., Yelton, D.B., and Charon, N.W. (1992). Molecular genetic analysis of a class B periplasmic flagellum gene of *Treponema phagedenis*. J. Bacteriol. *174*, 6404-6410.

Listgarten, M.A. and Socransky, S.S. (1964). Electron microscopy of axial fibrils, outer envolope, and cell division of certain oral spirochetes. J. Bacteriol. *88*, 1087-1103.

Lux, R., Moter, A., and Shi, W. (2000). Chemotaxis in pathogenic spirochetes: directed movement toward targeting tissues? J Mol. Microbiol. Biotechnol. 2, 355-364.

Macnab, R.M. (1996). Flagella and Motility. In *Escherichia coli* and *Salmonella.* Cellular and Molecular Biology., F.C.Neidhardt, R.Curtiss III, J.L.Ingraham, E.C.C.Lin, K.B.Low, B.Magasanik, W.S.Reznikoff, M.Riley, M.Schaechter, and H.E.Umbarger, eds. (Washington, D.C.: ASM Press), pp. 123-145.

Macnab, R.M. (2000). Microbiology. Action at a distance--bacterial flagellar assembly. Science *290*, 2086-2087.

Motaleb, M.A., Corum, L., Bono, J.L., Elias, A.F., Rosa, P., Samuels, D.S., and Charon, N.W. (2000). *Borrelia burgdorferi* periplasmic flagella have both skeletal and motility functions. Proc. Natl. Acad. Sci. U. S. A *97*, 10899-10904.

Motaleb, M.A., Sal, M.S., and Charon, N.W. (2004). The decrease in FlaA observed in a *flaB* mutant of *Borrelia burgdorferi* occurs posttranscriptionally. J. Bacteriol. *186*, 3703-3711.

Moter, A., Leist, G., Rudolph, R., Schrank, K., Choi, B.K., Wagner, M., and Gobel, U.B. (1998). Fluorescence in situ hybridization shows spatial distribution of as yet uncultured treponemes in biopsies from digital dermatitis lesions. Microbiol. *144*, 2459-2467.

Nauman, R.K., Holt, S.C., and Cox, C.D. (1969). Purification, ultrastructure, and composition of axial filaments from *Leptospira*. J. Bacteriol. *98*, 264-280.

Norris, S.J., Cox, D.L., and Weinstock, G.M. (2001). Biology of *Treponema pallidum*: correlation of functional activities with genome sequence data. J. Mol. Microbiol. Biotechnol. *3*, 37-62.

O'Toole, G., Kaplan, H.B., and Kolter, R. (2000). Biofilm formation as microbial development. Annu. Rev. Microbiol. *54*, 49-79.

Paster, B.J. and Canale-Parola, E. (1980). Involvement of periplasmic fibrils in motility of spirochetes. J. Bacteriol. *141*, 359-364.

Paster, B.J. and Dewhirst, F.E. (2000). Phylogenetic foundation of spirochetes. J. Mol. Microbiol. Biotechnol. *2*, 341-344.

Picardeau, M., Brenot, A., and Saint Girons, I. (2001). First evidence for gene replacement in *Leptospira* spp. Inactivation of *L. biflexa flaB* results in non-motile mutants deficient in endoflagella. Mol. Microbiol. *40*, 189-199.

Ren, S.X., Fu, G., Jiang, X.G., Zeng, R., Miao, Y.G., Xu, H., Zhang, Y.X., Xiong, H., Lu, G., Lu, L.F., Jiang, H.Q., Jia, J., Tu, Y.F., Jiang, J.X., Gu, W.Y., Zhang, Y.Q., Cai, Z., Sheng, H.H., et al. (2003). Unique physiological and pathogenic features of *Leptospira interrogans* revealed by whole-genome sequencing. Nature *422*, 888-893.

Rosey, E.L., Kennedy, M.J., Petrella, D.K., Ulrich, R.G., and Yancey Jr., R.J. (1995). Inactivation of *Serpulina hyodysenteriae flaA1* and *flaB1* periplasmic flagellar genes by electroporation-mediated allelic exchange. J. Bacteriol. *177*, 5959-5970.

Ruby, J.D. and Charon, N.W. (1998). Effect of temperature and viscosity on the motility of the spirochete *Treponema denticola*. FEMS Microbiol. Lett. *169*, 251-254.

Ruby, J.D., Li, H., Kuramitsu, H., Norris, S.J., Goldstein, S.F., Buttle, K.F., and Charon, N.W. (1997). Relationship of *Treponema denticola* periplasmic flagella to irregular cell morphology. J. Bacteriol. *179*, 1628-1635.

Sartakova, M.L., Dobrikova, E.Y., Motaleb, M.A., Godfrey, H.P., Charon, N.W., and Cabello, F.C. (2001). Complementation of a nonmotile *flaB* mutant of *Borrelia burgdorferi* by chromosomal integration of a plasmid containing a wild-type *flaB* allele. J Bacteriol. *183*, 6558-6564.

Sequeira, P.J.L. (1956). The morphology of *Treponema pallidum*. Lancet *271*, 749-and figure.

Seshadri, R., Myers, G.S., Tettelin, H., Eisen, J.A., Heidelberg, J.F., Dodson, R.J., Davidsen, T.M., DeBoy, R.T., Fouts, D.E., Haft, D.H., Selengut, J., Ren, Q., Brinkac, L.M., Madupu, R., et al. (2004). Comparison of the genome of the oral pathogen *Treponema denticola* with other spirochete genomes. Proc. Natl. Acad. Sci. U. S. A *101*, 5646-5651.

Soames, J.V. and Davies, R.M. (1975). The structure of subgingival plaque in a beagle dog. J. Periodont. *9*, 333-341.

Stamm, L.V. and Bergen, H.L. (1999). Molecular characterization of a flagellar (fla) operon in the oral spirochete *Treponema denticola* ATCC 35405. FEMS Microbiol. Lett. *179*, 31-36.

Stephan, D.E. and Johnson, R.C. (1981). Helical conformation of *Treponema pallidum* (Nichohs strain), *Treponema paraluis-cuniculi, Treponema denticola, Borrelia turicatae* and unidentified oral spirochetes. Infect. Immun *32*, 937-940.

Szurmant, H., Muff, T.J., and Ordal, G.W. (2004). *Bacillus subtilis* CheC and FliY are members of a novel class of CheY-P-hydrolyzing proteins in the chemotactic signal transduction cascade. J. Biol. Chem. *279*, 21787-21792.

Szurmant, H. and Ordal, G.W. (2004). Diversity in chemotaxis mechanisms among the bacteria and archaea. Microbiol. Mol. Biol. Rev. *68*, 301-319.

Turner, L., Ryu, W.S., and Berg, H.C. (2000). Real-time imaging of fluorescent flagellar filaments. J. Bacteriol. *182*, 2793-2801.

U.S.Department of Health, E.a.W.P.H.S. (1968a). Clinical Diagnosis of Syphilis. In Syphilis, A Synopsis, (Washington, DC: U.S. Government Printing Office), pp. 41-44.

U.S.Department of Health, E.a.W.P.H.S. (1968b). Pathophysiology of Syphilis. In Syphilis, A Synopsis, (Washington, DC: U.S. Government Printing Office), pp. 17-40.

Van Weelden, G.J., Charon, N.W., and Norris, S.J. (1990). Viscosity and the motility of *Treponema phagedenis* and *Treponema pallidum*. Abstr. 90th Annu. Meet. Am. Soc. Microbiol. I76, p.211.

Vesey, P.M. and Kuramitsu, H.K. (2004). Genetic analysis of *Treponema denticola* ATCC 35405 biofilm formation. Microbiol. *150*, 2401-2407.

7

Chemotaxis Signaling Systems in Spirochetes: Their Role in Directed Cell Movement and Pathogenesis

Renate Lux and Wenyuan Shi*

Abstract

Chemotaxis enables motile bacteria including spirochetes to monitor their environment and change their motility according to the stimuli perceived. In pathogenic species, chemotaxis and motility have emerged as important virulence factors. Host invasion appears to be greatly affected when these features are missing. The general mechanism of the basic chemotaxis pathway and the proteins involved are highly conserved throughout motile bacteria and archaea. Spirochetes, such as *Treponema ssp.*, however, appear to contain a few interesting variations in the form of novel fusion proteins and specific signal co-ordination problems related to their unique morphology. In addition to the chemotaxis pathway and the role of chemotaxis in spirochete pathogenesis, with a special focus on treponemes, phylogenetic aspects of the individual chemotaxis proteins are examined.

1. Chemotaxis in spirochetes – the early work

The distinct morphology and swimming behavior of spirochetes have intrigued researchers ever since this interesting group of bacteria was first observed several centuries ago (van Leeuwenhoek, 1683). Since Antony van Leeuwenhoek was examining plaque samples, he was likely looking at *Treponema ssp.* the genus that comprises the majority if not all of the oral spirochetes.

Spirochetes are characterized by helicial or flat-wave shaped long cell bodies, as well as by the periplasmic location of their flagella. Depending on the species, more than a hundred or only one or two of these periplasmic flagella are inserted subterminally at either cell pole (Canale-Parola, 1984; Holt, 1978). They can overlap in the middle of the cell or be so short that they barely extend to a third

*For correspondence email lux@ucla.edu

of the cell body. Rotation of the flagella around the cytoplamic cylinder generates thrust (Berg, 1976; Charon et al., 1992), and results in a rotational or snake-like movement of the bacterium (Canale-Parola, 1978; Goldstein et al., 1994). The motility pattern of spirochetes includes translational movements (runs) with either cell pole being able to assume the role of the leading end. Switching of the leading end is referred to as a reversal. Uncoordinated movement of both cell ends results in a non-translational twisting motion also known as flexing. Current models for spirochete motility (see the accompanying motility chapter of this book by Charon and Goldstein for details) imply that flagella motors at both cell ends need to rotate in opposite directions in order to achieve translational movement, a very special coordination problem typically not encountered by externally flagellated bacteria.

The duration of the distinct spirochete movements is modulated by the presence or absence of environmental stimuli (Fosnaugh and Greenberg, 1988; Greenberg et al., 1985), implying the regulation of motility via a sophisticated chemotaxis signal transduction system. Pioneering work by Greenberg, Canole-Parola, and coworkers provided the first direct measurements of spirochete chemotaxis using the freshwater mud-dwelling spirochete *Spirochaeta aurantia* as their model organism. Even though genome information or genetic tools for manipulation of spirochetes were not available at the time, these researchers were able to obtain a comprehensive analysis of the basic chemotactic behavior of *S. aurantia* by adapting various chemotaxis assays including swarm-plates, capillary assays (Adler, 1973; Greenberg and Canale-Parola, 1977) and video microscopy (Fosnaugh and Greenberg, 1988). This bacterium was found to perceive a variety of carbohydrates but not amino acids as attractants (Greenberg and Canale-Parola, 1977) whereas acids, short-chained alcohols, sulfide and high oxygen concentrations acted as repellents (Kaempf and Greenberg, 1990). Distinct changes in the motility pattern were observed upon the addition of chemoeffectors: an attractant response resulted in prolonged translocation in one direction, whereas repellents caused a significant increase in the cellular reversal rate and extensive flexing (Fosnaugh and Greenberg, 1988; Greenberg et al., 1985). The most compelling discovery, however, includes the change in membrane potential ($\Delta\psi$) during chemotaxis responses of *S. aurantia*, a characteristic not commonly found in other bacterial species (Goulbourne and Greenberg, 1981). Attractant carbohydrates triggered a transient membrane depolarization, whereas non-chemoeffector substances had no effect on the membrane potential. This finding was supported by the fact that disruption of $\Delta\psi$ by certain chemicals or application of a voltage clamp specifically abolished chemotaxis responses without influencing motility (Goulbourne and Greenberg, 1983a; Goulbourne and Greenberg, 1983b). ΔpH, in contrast, did not play a role in mediating chemotaxis. Mutant strains with an aberrant membrane potential were generally defective in chemotaxis (Fosnaugh and Greenberg, 1989). The membrane potential was discussed to be required for rapid transmission of the chemosensory information from one cell end to the other or even be involved in coordinating flagella rotation at both cell poles.

These early studies clearly established the fact that spirochete motility is guided by a chemotaxis system. Due to the lack of effective genetic tools, the molecular mechanism for spirochete chemotaxis remained a mystery for a long time, despite the detailed understanding of chemotaxis signal transduction systems in other

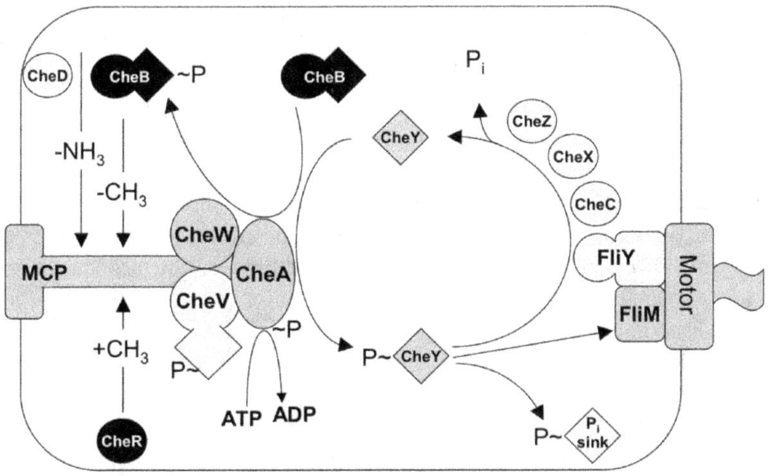

Figure 1. General chemotaxis pathway in flagellated bacteria (adapted with permission from Szurmant and Ordal, 2004). Shown are all the components known to participate in chemotaxis signal transduction and the reactions involved. Grey symbols indicate proteins that are present in every chemotaxis pathway studied so far (MCP, CheA, CheW, CheY, FliM and flagellar motor proteins). Black symbols represents proteins present in most organisms, whereas light grey ones represent more specialized chemotaxis components of individual groups of bacteria. For further details see text.

bacteria on a molecular level. Recent genetic and genomic studies of pathogenic spirochetes have significantly improved the situation. In this review, we will first give a general overview on the chemotaxis signaling transduction systems found in other bacteria, followed by a detailed analysis of the treponemal chemotaxis genes/ proteins that includes other spirochete species for a more complete picture. The potential role of chemotaxis in *Treponema* pathogenesis will also be discussed.

2. An overview of chemotaxis proteins and chemotaxis signal transduction pathways in other bacteria

Bacterial movements are typically directed by sophisticated chemotaxis signal transduction systems that are conserved throughout bacteria and archaea. The detailed examination of a growing number of organisms revealed a multitude of complex variations for the chemotaxis signal transduction pathway (for recent reviews see: (Lux and Shi, 2004; Szurmant and Ordal, 2004) that seem to share a basic excitation pathway comprised of four major chemotaxis proteins: the chemoreceptors or methyl-accepting chemotaxis proteins (MCPs), CheA, CheY and CheW (Figure 1). In brief, the generally transmembrane MCPs constantly monitor the environment and form a tight complex with the central histidine kinase CheA mediated by the coupling factor CheW (Gegner et al., 1992; Schuster et al., 1993). Changes in the concentration of environmental stimuli trigger conformational changes in the CheA-interacting portion of the MCP (Chervitz et al., 1995; Liu et al., 1997; Ottemann et al., 1999). These conformational changes regulate the ATP-dependent autophosphorylation activity of CheA according to the quality of the perceived stimulus (Borkovich et al., 1989). As a consequence, phosphotransfer reactions from CheA to its cognate response regulator CheY (Hess et al., 1988; 1987; Wylie et al., 1988), a small molecule that in its phosphorylated form interacts with the motor (Welch et al., 1993), can greatly vary depending on the nature of the environmental cue. Since binding of CheY~P to the flagella motor modulates the direction or duration of flagella rotation and triggers changes in the motility pattern of the bacterium, a sensible control of CheY phosphorylation levels is crucial to achieve bacterial chemotaxis. A small protein such as CheY that can freely diffuse throughout the cytoplasm enables most bacterial species (whose cell bodies are usually only a few microns long) to easily synchronize all motors. This feature becomes especially important for chemotactic movement of species containing more than one flagellum.

Interestingly, bacteria have developed a variety of different mechanisms that enable adaptation to environmental stimuli, even though all species studied so far employ the highly conserved pathway described above for excitation responses (Figure 1). Processes involved in adaptation target both the chemosensory signal input and the motor control output part of the chemotaxis signal transduction chain. Most bacteria contain several of the proteins required for adaptation but none of the species examined so far appears to implement all of the known adaptation systems. A common mechanism utilized by the majority of bacteria resets CheA activity to pre-stimulus levels by desensitizing the chemosensors to the presence of a stimulus. This is achieved via methylation and demethylation of the cytoplasmic portion of the MCPs by the antagonists CheR, a methyltransferase, and CheB, an enzyme with methylesterase and deamidase activities (Goy et al., 1977; Kehry et al., 1983; Toews et al., 1979). The deamidase function of CheB is required for chemoreceptor maturation, an activity that in numerous bacteria including the well-studied *B. subtilis* is performed by a separate enzyme, CheD (Kristich and Ordal, 2002). In addition to this methylation-dependent adaptation mechanism, *B. subtilis* and several other species utilize a methylation-independent pathway involving CheV (Karatan et al., 2001). This CheW-CheY fusion protein is able to substitute for lack of CheW and couple CheA to the chemoreceptors (Rosario et al., 1994). In species, such as *Helicobacter pylori*, that are lacking any apparent CheB and CheR homologs, this methylation-independent pathway appears to be the only adaptation mechanism on the chemoreceptor end (Pittman et al., 2001).

In contrast to this almost universal theme for adaptation on the input side, a variety of different mechanisms are involved in reducing phosphorylation levels of the motor-interacting CheY to enhance adaptation at the output end. In the enterobacteria that have represented the bacterial chemotaxis paradigm for decades, the CheY-specific phosphatase CheZ rapidly hydrolyzes CheY~P (Bren et al., 1996; Hess et al., 1987). Analysis of the increasing number of genomes available for a large selection of different species, however, revealed that CheZ is a very specialized enzyme only found in β- and γ-proteobacteria (Szurmant and Ordal, 2004). The majority of species including spirochetes appear to employ a recently described novel class of CheC/FliY/CheX (CYX) phosphatases (Kirby et al., 2001; Szurmant et al., 2004). Two members of this new family of CheY~P hydrolyzing enzymes, CheC and FliY, have been examined in detail in *B. subtilis* (Szurmant et al., 2003; Szurmant et al., 2004). Recently, CheY phosphatase activity was also demonstrated for CheC and CheX of *Thermotoga maritima* (Park et al., 2004).

α-proteobacteria including *Sinorhizobium meliloti* and *Rhodobacter sphaeroides* favor the "phosphate sink" mechanism (Porter and Armitage, 2002; Sourjik and Schmitt, 1998). This means of adaptation decreases phosphorylation levels of the motor-interacting CheY by transferring phospho-groups to other CheY homologues that are unable to bind to the motor switch.

3. The chemotaxis proteins of spirochetes

Greenberg and Kathariou observed two decades ago that certain spirochetal proteins were methylated/demethylated upon addition or removal of chemoattractants, indicating the presence of chemosensors that are functionally similar to MCPs (Kathariou and Greenberg, 1983). Later studies in various spirochete species identified homologues to the central chemotaxis proteins, CheA, CheW, and CheY (Ge and Charon, 1997b; Greene and Stamm, 1999; Greene et al., 1997; Trueba et al., 1997), as well as MCPs (Greene and Stamm, 1997; Greene and Stamm, 1998; Hagman et al., 1997; Kataoka et al., 1997; Li et al., 1999), and motor switch proteins (Ge and Charon, 1997a; Stamm and Bergen, 1999), implying that this interesting group of bacteria follows the general chemotaxis theme realized in all other flagellated species studied so far. This hypothesis is further supported by the complete sets of chemotaxis genes found in several pathogenic spirochetes whose whole genome sequences are now available (see below).

The first genome of a spirochete, the Lyme disease agent *Borrelia burgdorferi*, was completed in 1997 (Fraser et al., 1997), followed by other important pathogenic spirochetes including the syphilis agent *Treponema pallidum* (Fraser et al., 1998), the periodontitis-associated *T. denticola* (Seshadri et al., 2004), two different serovars of the leptospirosis agent *L. interrogans* (Nascimento et al., 2004a; 2004b), and *B. garinii* (Glockner et al., 2004), a close relative of *B. burgdorferi*. The chemotaxis proteins of *B. burgdorferi* and *B. garinii* are virtually identically. A closer look into these genomes reveals that, similar to other motile bacteria, a considerable portion of the genomic information is devoted to chemotaxis and motility functions. The importance of chemotaxis and motility becomes most apparent in the minimalistic chromosomes of the two sequenced *Borrelia ssp.* and *T. pallidum* that are only about 1 Mb in size but still contain complete sets of chemotaxis (Table 1) and flagellar genes. Both organisms devote 5–6 % of their genomes to enable directed motility while lacking numerous functions necessary for autotrophic growth. In the larger genomes of *T. denticola* (~2.7 Mb) and the two serovars of *L. interrogans* (~4.8 Mb) the proportion of the chromosome dedicated to genes encoding motility and chemotaxis functions is equivalent to other motile bacteria with a comparable genome size. In this context, it is interesting to note that the motility and chemotaxis related genes are among the only about 300 ORFs found in *L. interrogans* that contain homologues in other spirochete species (Nascimento et al., 2004b).

Since the motility and chemotaxis behaviors of spirochetes produce unique phenotypes among motile microorganism, the proteins involved in these characteristic features would be expected to exhibit close homology between the different spirochetes species. A more detailed analysis of the chemotaxis components encoded by the individual genomes, however, yielded surprising results. Most strikingly, all chemotaxis proteins present in *L. interrogans* find their

Table 1. Spirochete chemotaxis genes

Gene	Protein function	*B. burgdorferi*/ *B. garinii*	*L. interrogans* [a]	*T. denticola*	*T. pallidum*
mcp	chemoreceptor	5	12 (11) [b]	20	4
cheW	coupling factor connecting CheA to the chemoreceptors	3 [c]	3	2 [c]	2 [c]
cheA	cognate histidine kinase for CheY and usually CheB	2	2 [d]	1	1
cheY	cognate response regulator for CheA	3	2	1	1
cheB	methyl esterase, CheR antagonist, usually contains cognate response regulator domain for CheA	2	3	1	1
cheR	methyl transferase	3 [c]	2	2 [c]	2 [c]
cheD	chemoreceptor deamidation (maturation)	1	2	-	-
cheC	phosphatase, likely CheY~P specific	-	1 [d]	-	-
cheX	putative CheC and FliY-like phosphatase	1	1	1	1
fliY	phosphatase, likely CheY~P specific	- [e]	1	1	1

a), the chemotaxis proteins encoded by the genomes of the genomes of the two different serovars of *L. interrogans* whose complete genome sequence is available are mostly identical and are therefore not listed separately

b), the genome of *L. interrogans* serovar Lai contains an additional *mcp* not present in the closely related *L. interrogans* serovar Fiocruz

c), *B. burgdorferi, B. garinii, T. denticola* and *T. pallidum* encode a copy of CheW in their general chemotaxis operons that contains a C-terminal CheR-like domain

d), CheA1 of both *L. interrogans* serovars contains a C-terminal CheC-like domain

e), the genome of *B. burgdorferi* and *B. garinii* do not contain *fliY* but *fliN* instead closest homologues not within the other spirochete species but mainly among the firmicutes, including *Bacillus* and *Clostridium ssp.*, and a variety of representatives of the different subdivisions of the proteobacteria. Also, several chemotaxis components in the more closely related *B. burdorferi*, *T. denticola* and *T. palllidum* produce very good alignments with each other, while overall homologies to related proteins in other species are relatively poor. Another interesting feature is the occurrence of unusual protein fusions such as the CheW-CheR fusions of *B. burdorferi*, *T. denticola* and *T. palllidum* and the CheA-CheC fusion protein found in *L. interrogans*. In the numerous organisms being sequenced, these proteins are always found as separate components within the respective chemotaxis pathways. Described below are detailed analyses and comparisons of the individual chemotaxis proteins present in the different spirochete genomes including functional, phylogenetic and evolutionary aspects of spirochete chemotaxis systems.

3.1. The MCPs

The different spirochetes vary greatly in the number of putative MCPs identified in their respective genomes (Table 1). Like numerous other bacteria the sequenced *Borrelia ssp.* and *T. pallidum* only contain five and four chemoreceptors, respectively (Fraser et al., 1997; 1998). With 11 and 12 MCPs for the two sequenced serotypes, *L. interrogans* has significantly more (Nascimento et al., 2004a; 2004b). The most putative chemosensor encoding ORFs are present in the *T. denticola* genome (Seshadri et al., 2004); this oral pathogen possesses as many as 20 chemoreceptors. Two of these MCPs, DmcA and DmcB, have been characterized via gene inactivation and mutant analysis (Kataoka et al., 1997; Li et al., 1999). Recent yeast two-hybrid based protein interaction analyses revealed that the majority of the 20 MCPs encoded by the *T. denticola* genome including DmcA and DmcB interact with CheW and are therefore likely to be functional in chemotaxis (Hower et al., 2004). A role for having numerous chemoreceptors has been discussed for *L. interrogans* as an adaptation to its lifecycle (Nascimento et al., 2004a). Since the leptospirosis agent shuttles between mammalian hosts rich in nutrition and the relatively poor environment of freshwater sources, it is frequently challenged with very divergent living conditions. The assumption that the number of MCPs reflects the complexity of the ecological niche of a bacterium may hold true for *T. denticola*. This oral pathogen resides within the host tissue and the subgingival plaque of periodontal pockets, and frequently encounters hundreds of different bacterial species as well as immune and tissue responses.

Between species, alignment of the different spirochete chemoreceptors exhibited the expected high homology at the HCD (highly conserved domain) of the CheA and CheW interacting C-terminal portion (MA domain) (Figure 2A). The N-terminal regions that determine substrate specificity, in contrast, did not display any significant homology between species for the majority of MCPs. Since most of the chemoreceptor-encoding ORFs are scattered about the chromosome, rather than being organized with other chemotaxis genes, this lack of N-terminal homology could indicate that some of the chemoreceptors were acquired through horizontal gene transfer, significantly modified during evolution to accommodate environmental diversity, or both. Within species, however, clusters of closely

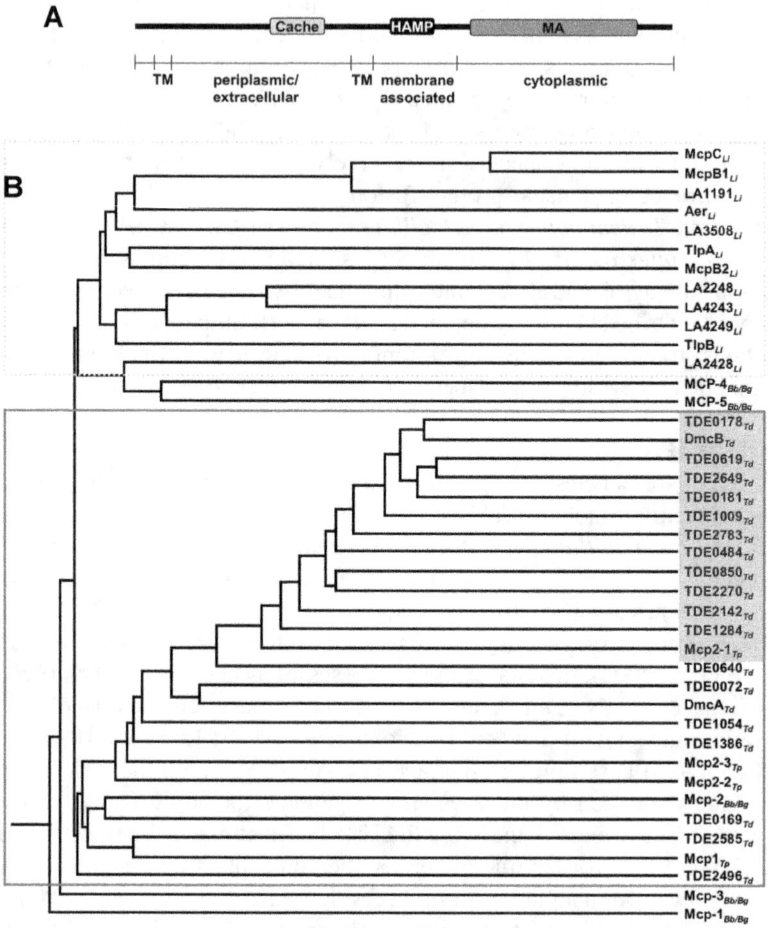

Figure 2. (A) MCP domain organization. The MA domain (grey) constitutes the signature domain of chemosensors that is present in all MCPs, and HAMP (black) domains are typical for a variety of sensory proteins including MCPs. The N-terminal sensory domain that is generally exposed to the outside via two transmembrane sections (TM) can contain Cache domains (grey) in certain types of MCPs. (B) Phylogenetic tree of the putative chemosensors encoded by the genomes of *B. borrelia* (*Bb*)/ *B. garinii* (*Bg*), *L. interrogans* (*Li*), *T. denticola* (*Td*) and *T. pallidum* (*Tp*). The tree was generated using the ClustalW and T-coffee online sequence comparison programs. Two major

related chemoreceptors are found especially in *T. denticola* (Figure 2B), implying that gene duplication may have been involved in the evolution of its numerous MCPs.

Interestingly, in 10 of the 20 MCPs encoded by *T. denticola,* the N-terminal 250 amino acids exhibit a striking homology to PrtB (Arakawa and Kuramitsu, 1994), one of the numerous proteases identified for *T. denticola* (Figure 2C). Mcp2-1$_{Tp}$ of the closely related species *T. pallidum* falls into the same category. Some of these "protease-like" MCPs are encoded by operons that contain additional proteases and

C

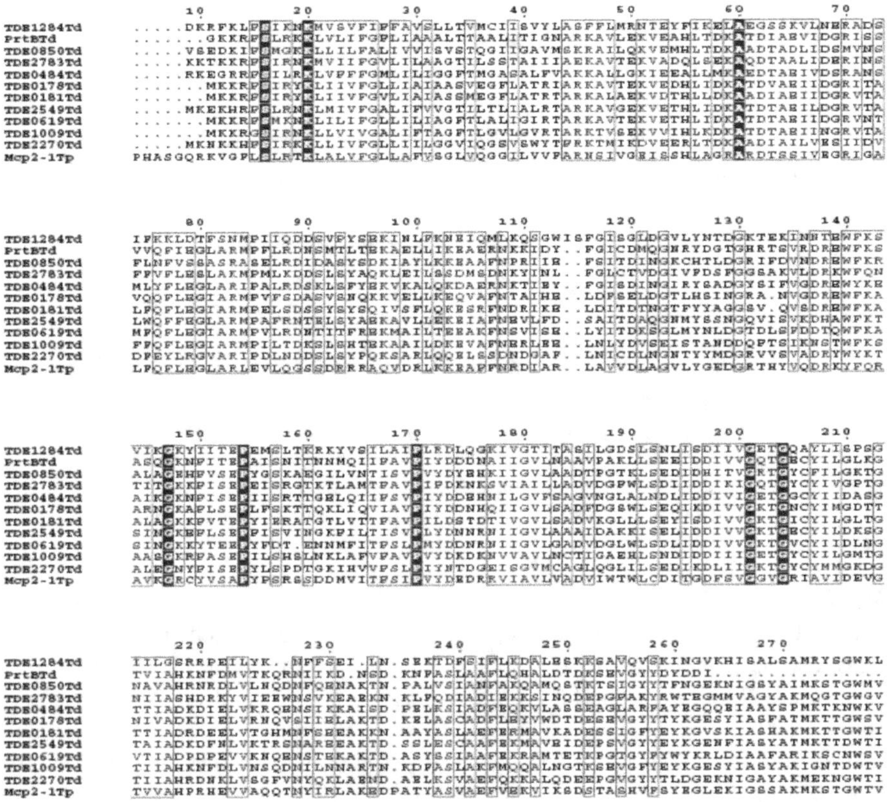

clusters are found, one of whom contains all *L. interrogans* MCPs (boxed in light gray – dotted line), wheras the other one is mainly comprised of *T. denticola* and *T. pallidum* MCPs (boxed in dark gray). A subgroup of MCPs in the *T. denticola* and *T. pallidum* cluster exhibits N-terminal homology to PrtB, a protease of *T. denticola* (shaded in gray). (C) Alignment of the N-termini of PrtB-like MCPs with PrtB. Identical amino acids are shaded in black, conserved amino acids are boxed.

peptidases, indicating a potential link of these MCPs to virulence-related events. Intriguingly, similar MCPs are present in the genomes of pathogenic *Bacillus ssp.* including *B. anthracis* (Read et al., 2003), *B. cereus* (Rasko et al., 2004), *B. licheniformis* (Rey et al., 2004), and *B. thuringensis* (www.genome.jp/kegg/kegg2.html). The remaining *T. denticola* MCPs are most homologous to the MCPs of *T. pallidum.* The chemoreceptors encoded by the genomes of the two different serovars of *L. interrogans* group together, whereas those found in the *Borrelia ssp.* do not appear to cluster with any particular spirochete.

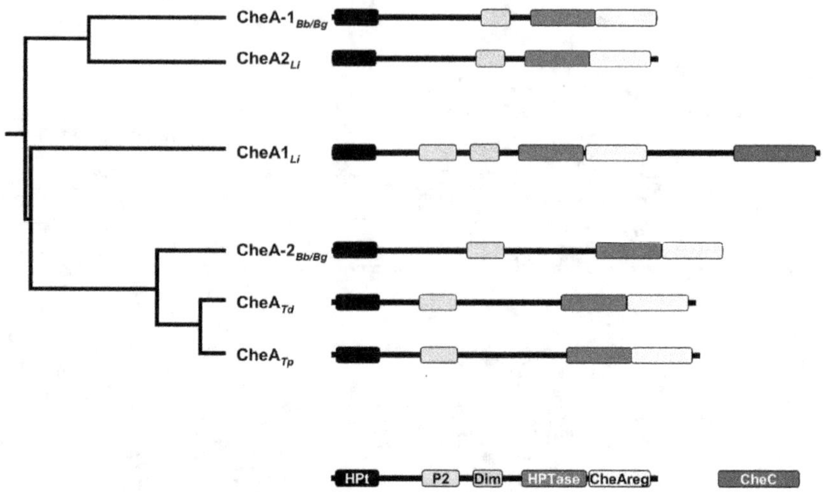

Figure 3. Phylogenetic tree and domain structure of the CheAs encoded by the genomes of *B. borrelia*/ *B. garinii*, *L. interrogans*, *T. denticola* and *T. pallidum*. The tree was generated using the ClustalW and T-coffee online sequence comparison programs. The domain representation was adapted from the BLASTp domain analysis. For abbreviations of bacterial names see legend Figure 2.

3.2. CheA

The histidine kinase CheA comprises the centerpiece of chemotactic signal transduction and appears to be (almost) ubiquitous among motile species. The genomes of *T. denticola* and *T. pallidum* encode one copy of *cheA* (Greene and Stamm, 1999; Greene et al., 1997), whereas the *Borrelia ssp.* and *L. interrogans* each contain two copies of *cheA*. Since these pathogenic spirochetes frequently change their host environment, in contrast to the *Treponema* species that are generally transmitted between the same type of mammalian host, a correlation between the host diversity and the number of general chemotaxis operons has been discussed (Li et al., 2002; Nascimento et al., 2004a). All spirochete CheAs contain the signature domains of histidine kinases, the HPT (Histidine Phosphotransfer domain) and HATPase (Histidine kinase-like ATPase) domains, in an order typical for CheA kinases (Figure 3) (Dutta et al., 1999). As expected, they also display significant homologies to the CheW interacting C-terminal CheW-like domain typically present in CheA. Additionally, CheA-2$_{Bb/Bg}$, CheA1$_{Li}$ and the treponemal CheAs include a domain homologous to the CheY and CheB binding P2 domain present in many bacterial species that is not apparent in CheA-1$_{Bb/Bg}$ and CheA2$_{Li}$. With exception of CheA1$_{Li}$, however, none of the spirochete CheAs contains a clear dimerization domain. Surprisingly, domain analysis of CheA1$_{Li}$, revealed the presence of an additional domain that exhibits significant homology to CheC, a member of the CYX family of CheY phosphatases (Kirby et al., 2001; Szurmant et al., 2004), fused to its C-terminal CheW-like domain. This is especially interesting since CheC of *B. subtilis* has been demonstrated to interact with CheA (Kirby et al., 2001), and interacting domains that are present as separate proteins in one species can be found as fusion proteins in others (Marcotte and Marcotte, 2002; Marcotte et al., 1999).

Phylogenetic analysis shows that CheA-$2_{Bb/Bg}$ and the CheAs encoded by *T. denticola* and *T. pallidum* are closely related to each other, but, surprisingly, neither one of the two *cheA* copies present in the genome of *L. interrogans* is included in this group (Figure 3). CheA2_{Li} clusters with CheA-$1_{Bb/Bg}$, whereas CheA1_{Li} comprises a completely independent branch of the tree. Further homology searches with other species reveal that CheA-$2_{Bb/Bg}$, CheA$_{Td}$ and CheA$_{Tp}$ find their closest full-length homologues within the *Clostridia* including *Clostridium acetobutylicum*, *Moorella thermoacetica*, *Thermoanaerobacter tengcongensis* among others and interestingly *Archeae* of the genus *Pyrococcus*. The second CheA copy found in the *Borrelia ssp.*, CheA-$1_{Bb/Bg}$, and CheA2 of *L. interrogans* exhibit considerable homology to each other but find their best matches within the various subgroups of proteobacteria. It has been discussed earlier for *B. burgdorferi* that this copy of CheA may have been acquired via horizontal gene transfer (Li et al., 2002). A putative role of CheA-1_{Bb} during transmission or dissemination in one of its hosts has been suggested. Further experimental evidence to support such a role, however, remains to be generated. Interestingly, CheA1_{Li} does not demonstrate any close relationship to other spirochete CheAs including the second CheA copy of *L. interrogans*, CheA2_{Li}. Even though CheA1_{Li} is most homologous to a variety of *Firmicutes* including *Bacillales* and *Clostridia*, it is only distantly related to CheA-$2_{Bb/Bg}$, CheA$_{Td}$ or CheA$_{Tp}$ that find their best matches within the same clostridia species. This could indicate that firmicutes and spirochetes share a common ancestor and subsequently adapted their respective CheAs according to their individual requirements. Since the *Leptospiraceae* that contain *L. interrogans*, and the *Spirochaetaceae* that include *B. burgdorferi*, *B. garinii*, *T. denticola* and *T. palllidum* comprise independent families within the order *Spirochaetales*, divergent evolution of the "archetype spirochete" may have led to these very different CheAs found in present day spirochete families. Alternatively, copies of *cheA* acquired during horizontal gene transfer could have proven more proficient for the lifestyle of the respective spirochete thus eventually resulting in loss of the original *cheA* copy.

Compared to CheAs present in most species that are about 600 to 700 amino acids in length, CheA-$2_{Bb/Bg}$, CheA$_{Td}$ and CheA$_{Tp}$ are unusually long, ranging from 801 amino acids (CheA$_{Td}$) to 864 amino acids (CheA-$2_{Bb/Bg}$). With the exception of the 798 amino acid long CheA of *C. tetani* that possesses two adjacent P2 domains, "long" CheAs in most other species including *Campylobacter jejuni*, *Helicobacter ssp.*, or *Wollinella succinogenes* contain extra CheY-like response regulators at their C-termini. Even though no additional domains with homologies to known motifs are found in CheA-$2_{Bb/Bg}$, CheA$_{Td}$ and CheA$_{Tp}$, the possibility that a novel domain is hidden within these additional 100 to 200 amino acids cannot be ruled out at this point.

3.3. CheW

CheW is the coupling factor that connects CheA to the chemoreceptors. The dependency of CheA on CheW is underscored by the finding that *cheA* and *cheW* are localized in the same (putative) operon in the majority of species, including the spirochetes, whose genomes have been sequenced,. The *cheWs* that are organized in one operon with *cheA-$2_{Bb/Bg}$*, *cheA$_{Td}$* and *cheA$_{Tp}$* have an interesting feature that is so far unique among the hundreds of species for which complete or partial

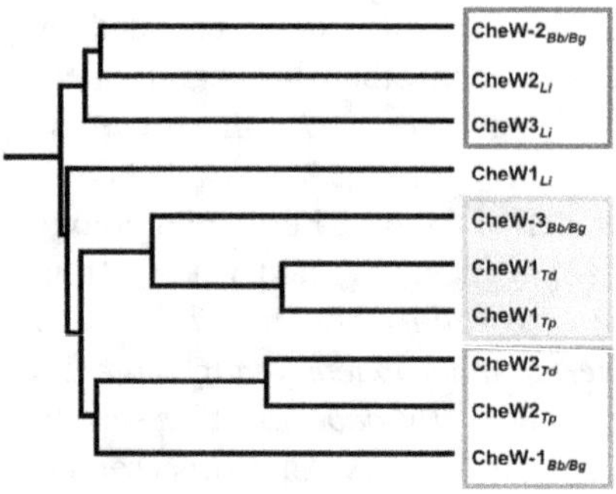

Figure 4. Phylogenetic tree of the putative CheWs encoded by the genomes of *B. borrelia*/ *B. garinii*, *L. interrogans*, *T. denticola* and *T. pallidum*. The tree was generated using the ClustalW and T-coffee online sequence comparison programs. Three main clusters were identified: CheWs most similar to various proteobacteria (dark gray box), CheWs present in the *cheAWXY* operon of *B. borrelia*, *T. denticola* and *T. pallidum* (medium gray box), and CheWs not organized with chemotaxis genes (light gray box). For abbreviations of bacterial names see legend Figure 2.

genome sequence information is available: In addition to their N-terminal CheW portions that exhibit significant homology to the CheWs present in proteobacteria, these proteins contain C-terminal CheR-like domains. Interestingly, *L. interrogans* whose two copies of CheA are not closely related to the above mentioned CheAs of the sequenced *Borrelia ssp.*, *T. denticola* or *T. pallidum* does not possess this distinctive CheW-CheR fusion protein (Figures 3 and 4). In agreement with the homology results for CheA-$1_{Bb/Bg}$ and CheA2_{Li}, the accompanying CheWs, CheW-$2_{Bb/Bg}$ and CheW3_{Li}, are related to each other (Figure 4) as well as with representatives of various subdivisions of proteobacteria.

Intriguingly, all spirochetes contain exactly one more copy of CheW than CheA. These additional CheWs cluster together for the *Borrelia ssp.*, *T. denticola* and *T. pallidum* and exhibit significant homologies with CheWs present in clostridia and some proteobacterial species. The extra CheW of *L. interrogans*, CheW2_{Li}, however, is closely related to CheW-2_{BbBg} and most homologous to CheWs in various proteobacteria. Since most of these additional *cheWs* – with the exception of *cheW-1*$_{Bb/Bg}$ that is organized with possible flagellum related genes – are localized in putative operons encoding genes completely unrelated to chemotaxis and motility, their potential involvement in chemotaxis-guided cellular translocation is purely hypothetical. Current yeast-two hybrid analysis revealed that CheW2_{Td} is able to interact with several if not all of the MCPs of *T. denticola* but not CheA, implying a role in chemotaxis for this additional coupling factor that is likely distinct from CheW1_{Td} (Renate Lux and Dawn Hower, unpublished).

3.4. CheY

In many bacterial chemotaxis systems, *cheY* and their respective CheA encoding genes are often localized in the same operon. Accordingly, the *cheYs* found in the different spirochete chemotaxis operons exhibit the same relationship among each other as the *cheAs* and *cheWs* with which they co-localize. Like numerous other bacteria, spirochetes contain extra copies of CheY-like response regulator encoding genes that are not organized with other genes involved in chemotaxis and motility related features. Even though additional *cheY* copies are known to mediate the phosphate sink mechanism employed for adaptation to chemotactic stimuli in some organisms, CheY-like response regulators are quite commonly found in bacterial genomes including non-motile species. Therefore, these extra CheYs could comprise a more general means of regulating cellular processes that is not limited to motor control.

3.5. CheX and CheC

CheX was first discovered in spirochetes when one of the chemotaxis operons of *B. burgdorferi* was being sequenced and found to contain a novel ORF localized between *cheW* and *cheY* (Ge and Charon, 1997b) whose putative gene product did not exhibit any significant homologies to known chemotaxis proteins. Later work revealed that CheX comprises a member of the CYX family of CheY-phosphatases that also includes CheC and FliY (Kirby et al., 2001; Park et al., 2004; Szurmant et al., 2004). Soon after its discovery in the Lyme disease spirochete, *cheX* was found in *T. pallidum* and expression studies confirmed this novel gene as part of the *cheAWXY* operon (Greene et al., 1997) that is also present in *T. denticola* (Greene and Stamm, 1999). Surprisingly, the *cheX* homologue found in the *L. interrogans* genomes (Nascimento et al., 2004a; 2004b) is not encoded within a chemotaxis related operon but in a separate locus together with hypothetical proteins unique to this species. In addition to CheX, *L. interrogans* contains a CheC that is fused to the C-terminus of CheA1$_{Li}$. Since similar CheA-CheC fusions have not yet been found in any other species, it may represent an event that occurred only in *L. interrogans* to facilitate the likely interaction between these two proteins.

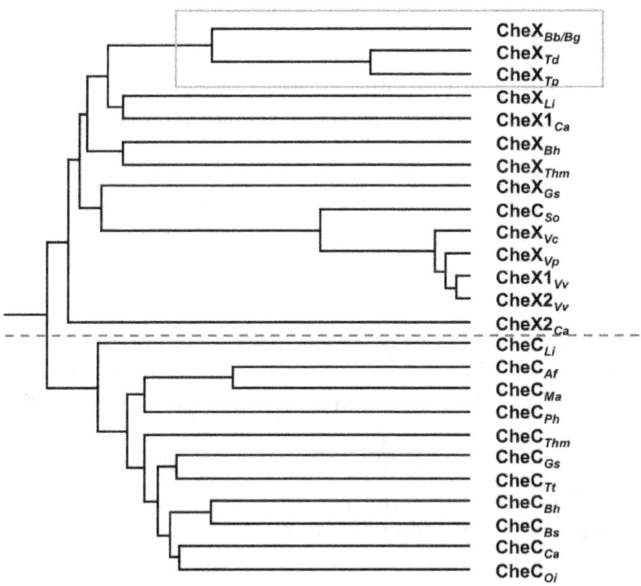

Figure 5. Phylogenetic tree of the CheXs and CheCs encoded by the genomes of CheX-containing bacteria. The tree was generated using the ClustalW and T-coffee online sequence comparison programs. As expected, the CheCs that are organized within chemotaxis operons cluster together (below dotted lline). With the exception of those in spirochetes (light gray box), CheXs do not co-localize with chemotaxis-related functions and form a separate group (above dotted line). The following species were included in the analysis: *Archaeoglobus fulgidus* (*Af*), *Bacillus halodurans* (*Bh*), *B. subtilis* (*Bs*), *Clostridium acetobutylicum* (*Ca*), *Geobacter sulfurreducens* (*Gs*), *Methanosarcina acetivorans* (*Ma*), *Oceanobacillus iheyensis* (*Oi*), *Pyrococcus horikoshii* (*Ph*), *Shewanella oneidenis* (*So*), *Thermotoga maritime* (*Thm*), *Thermoanaerobacter tengcongensis* (*Tt*), *Vibrio cholerae* (*Vc*), *V. parahaemolyticus* (*Vp*) and *V. vulnificus* (*Vv*). Spirochete species are abbreviated according to legend Figure 2.

With hundreds of completed and partial genomes available, CheX and CheC homologues are found in numerous organisms, covering a wide spectrum of species (Figure 5). Similar to *L. interrogans*, many species including *C. acetobutylicum* (Nolling et al., 2001), *Geobacter sulfurreducens* (Methe et al., 2003) and *T. maritima* (Nelson et al., 1999) among others contain both CheX and CheC encoding genes. CheC is exclusively found in connection with other chemotaxis genes (www.genome.jp/kegg/kegg2.html). With the exception of the closely related *B. burgdorferi*, *B. garinii*, *T. denticola* and *T. pallidum*, however, none of the other species encode their copies of *cheX* in chemotaxis related operons (Figure 5). The *cheX* gene can be present as single gene operon or co-localize with genes that encode a variety of very diverse cellular function indicating a role for CheX that is not limited to chemotaxis. Intriguingly, despite being part of a putative operon encoding cobalamin transport related functions, one of the two copies of CheX present in *C. acetobutylicum* is fused to a C-terminal CheY-like response regulator supporting the idea of a functional interaction between these two proteins.

A recent study provided a major breakthrough in the understanding of CheX and CheC function by solving their crystal structures, identifying their putative active centers and characterizing interesting differences between these proteins (Park et al., 2004). CheC contains two active centers whereas CheX only has one and requires dimerization to achieve the same symmetry. This dimerization was independently observed using a yeast two-hybrid approach (Sim et al., 2005). The detailed examination of structural features present in this novel class of CheY phosphatases will prove very useful for future genetic and biochemical studies and enhance further analysis of CYX phosphatase function in the spirochete chemotaxis pathway. The CheY phosphatase activity of CheX in spirochetes was recently confirmed for *B. borrelia* in an exciting study by Motaleb et al. (2005).

3.6. FliY

FliY has been first described in *B. subtilis* as a novel motor switch protein (Bischoff and Ordal, 1992) and was later found to exhibit CheY-specific phosphatase activity (Szurmant et al., 2003). This two domain protein is basically a fusion protein between an N-terminal CheC/CheX like portion that contains the CheY-binding region, and a C-terminal FliN domain (Kirby et al., 2001). Since FliN is an integral part of the cytoplasmic portion of the motor (Garza et al., 1995; Khan et al., 1992); see also accompanying chapter by Charon et al.) and most species containing FliY do not possess an extra copy of FliN (www.genome.jp/kegg/kegg2.html), the FliN domain of FliY is likely to assume FliN function in the flagellar motor. *L. interrogans*, *T. denticola* and *T. pallidum* contain *fliY* rather than *fliN* in their motor and flagellar functions encoding operons. In the genome of *L. interrogans*, however, an additional copy of *fliN* appears to be present. Interestingly, *B. burgdorferi* and *B. garinii* lack a FliY homologue and encode instead a FliN within their flagellar operons (Ge and Charon, 1997a). Since these spirochetes are missing any further apparent CheY-specific phosphatases with the exception of CheX, the control of CheY phosphorylation levels at the output end of their chemotaxis signal transduction chains is likely to rely on CheX function alone.

3.7. CheR and CheB

In the genomes of *T. denticola* and *T. pallidum*, one set of *cheR* and *cheB* is in the same operon, emphasizing their functional relatedness (Strong et al., 2003). Co-localization of these genes is typical for the majority of bacterial species (www.genome.jp/kegg/kegg2.html) including other spirochete genomes. Both *Treponema* species only possess a single copy of *cheB* whereas *B. burgdorferi*, *B. garnii* and *L. interrogans* encode additional copies of this gene, but not *cheR*, in their chemotaxis operons that are most similar to the ones found in proteobacteria.

All spirochete genomes contain extra copies of *cheR*. The most interesting ones are present in the representatives of the family *Spirochaetaceae* (*B. burgdorferi*, *B. garinii*, *T. denticola*, and *T. pallidum*) in the form of the C-terminal fusion to CheW mentioned earlier. This CheR does not produce any significant matches with CheRs from other species despite containing the signature motifs of CheR-like methyltransferases. Fusion proteins between CheW and CheR are not found in other bacterial species with genome information currently available (www.genome.jp/kegg/kegg2.html,www.ncbi.nlm.nih.gov/sutils/genom_table.cgi).

It could therefore be interesting to determine the function of this CheR-like domain connected to the CheW of this particular branch of spirochetes. A role of CheR-$1_{Bb/Bg}$ and CheR2_{Li}, additional *cheR* copies present in the genomes of the sequenced *Borrelia ssp.* and *L. interrogans*, respectively, in chemotaxis is questionable: both *cheRs* do not co-localize with other chemotaxis function encoding genes and both organisms contain CheR/ CheB pairs likely to perform chemoreceptor modification.

3.8. CheD

CheD is a protein involved in receptor maturation in some bacterial species, enhances CheC acitivity and has been studied in detail in *B. subtilis* (Kristich and Ordal, 2002). Homologues to *cheD* are absent in both *T. denticola* and *T. pallidum*, indicating that in contrast to CheC, CheX function does not require the presence of CheD. The genomes of the two sequenced *Borrelia ssp.* and *L. interrogans* contain putative *cheDs*. Since *cheD1*$_{Li}$, one of the two *cheD* homologues present in the *L. interrogans* genome, is localized in an operon encoding other general chemotaxis components, its involvement in chemotaxis is likely. The genomic environment of *cheD2*$_{Li}$ and *cheD*$_{Bb/Bg}$ does not encode any chemotaxis or motility related functions, thus leaving confirmation of their true function to further experimental examination.

3.9. Chemotaxis mutant analysis

In addition to the above genomic evidence, recent genetic studies employing targeted gene inactivation analysis have confirmed that several of these putative chemotaxis genes are indeed required for spirochete chemotaxis. Lack of CheA, CheW or CheY in *T. denticola* as well as CheA-1, CheW-3 or CheY-2 in *B. burgdorferi* results in non-reversing phenotypes that likely correspond to the smooth swimming phenotypes reported for mutants deficient in these key chemotaxis proteins in externally flagellated species (Charon and Goldstein, 2002; Li et al., 2002; Lux et al., 2002). Accordingly, inactivation of the two *T. denticola* MCPs that have been further analyzed so far, DmcA and DmcB, triggers specific chemotaxis defects

(Kataoka et al., 1997; Li et al., 1999) and reduced tissue penetration efficiencies (Lux et al., 2001). As already suggested by the characteristics of certain *S. aurantia* proteins that were assumed to be MCPs (Kathariou and Greenberg, 1983), both *T. denticola* MCPs undergo methylation (Kataoka et al., 1997; Li et al., 1999) similar to their homologues in other bacterial species (Kirby et al., 1999; Kleene et al., 1979; Toews et al., 1979; Zimmer et al., 2000). Since all spirochete genomes contain at least one pair of CheR and CheB homologues, and receptor methylation has been demonstrated, it appears likely that adaptation to environmental stimuli on the chemosensor level follows known mechanisms. Furthermore, lack of CheX in *B. burgdorferi* and *T. denticola* causes distinct motility and chemotaxis deficient phenotypes, confirming an important role of the CheY -phosphatase CheX in chemotaxis (Charon and Goldstein, 2002).

4. Model for chemotaxis signal transduction pathways in spirochetes

The information available from spirochete genome and mutant analysis in combination with in depth knowledge available for chemotaxis protein function in other organisms suggests the following model for spirochete chemotaxis (Figure 6): In compliance with the central signaling pathway employed by all bacterial chemotaxis systems studied so far, MCPs communicate the quality of environmental cues to the autokinase CheA that is linked to these chemosensors via the coupling protein CheW. The chemotactic signal is translated into the corresponding changes in CheA autophosphorylation activity to modulate CheY phosphorylation levels and hence the motor response. Adaptation to chemoeffectors on the chemoreceptor (input) side is achieved through methylation and demethylation events catalyzed by CheR and CheB, respectively. On the motor (output) side, reduction of CheY phosphorylation levels follows the mechanisms employed by *B. subtilis* and *T. maritima* via the action of CheY-specific phosphatases of the CYX family, CheX, CheC and FliY. Since CheC, the CheX homologue of *B. subtilis* (Kirby et al., 2001), and CheX of *T. denticola* (Sim et al., 2005) were demonstrated to interact with CheA, a feedback communication to monitor the CheA activity level appears likely. The different spirochete species seem to utilize separate mechanisms for receptor maturation, a process that involves deamidation of specific glutamine residues to generate the glutamate residues required for CheR-mediated methylation. Lacking other apparent proteins to perform this reaction, both *T. denticola* and *T. pallidum* are likely to employ the CheB catalyzed reactions present in a variety of other species. Since *B. burgdorferi*, *B. garinii* and *L. interrogans* contain copies of *cheD* homologues in their genomes, they possibly utilize CheD for chemoreceptor deamidation similar to the reaction identified in *B. subtilis*.

Despite this standard approach to chemotactic signaling, the spirochete chemotaxis pathway contains some interesting interspecies variations among the different spirochetes as well as features not present in other bacterial species: The CheW homologue encoded by the *cheAWXY* operons of both sequenced *Borrelia ssp.*, *T. denticola* and *T. pallidum* is fused to a C-terminal CheR domain whose role is currently completely unknown. This unique CheW-CheR fusion, however, is missing in *L. interrogans*, thus restricting its elusive function so far to one branch of spirochetes. Also, the corresponding CheAs are longer than the average CheAs

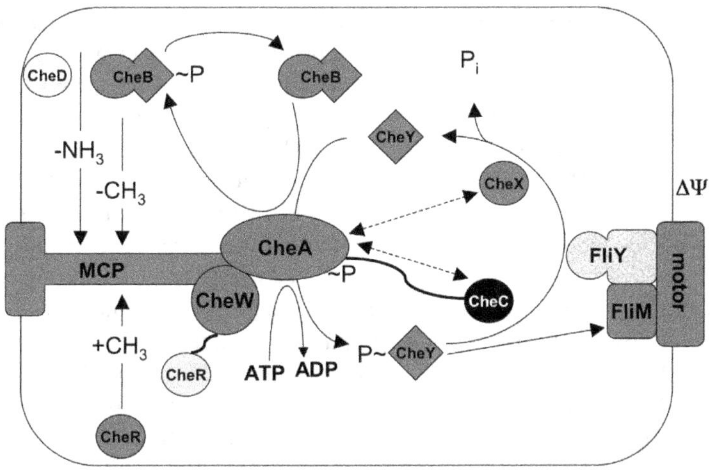

Figure 6. Putative chemotaxis pathway in spirochetes. MCPs, CheW, CheA, CheY, CheX, CheR, CheB, FliM, and the flagellar motor are shown in dark gray to indicate that they are found in all spirochetes. FliY is shown with diagonal lines to indicate that it is found only in the genomes of *L. interrogans*, *T. denticola* and *T. pallidum*. The CheR portion of the CheW-CheR fusion that is exclusively found in *B. borrelia*, *B. garinii*, *T. denticola* and *T. pallidum* is in the lightest shade of gray. CheD which might function in chemotaxis of *B. burgdorferi* and *L. interrogans* is stippled, whereas CheC of *L. interrogans* is black.

and could potentially contain an additional domain for novel interactions. Another remarkable aspect is the diversity of combinations in which CheY-phosphatases of the CYX family appear to be employed among spirochetes: Unlike the other spirochete species, *B. burgdorferi* and *B. garinii* do not contain the motor associated FliY but seemingly rely on CheX action alone to dephosphorylate CheY. This is especially peculiar since both *Borrelia ssp.* possess two complete general chemotaxis operons including separate copies of CheY. *T. denticola* and *T. pallidum* contain CheX as well as FliY, while the *L. interrogans* genome encodes all three known members of the CYX family of CheY phosphatases. Given that CheC is fused to $CheA1_{Li}$, one of the two CheAs present in the leptospirosis agent, CheX could specifically remove phosphogroups from the corresponding CheY. FliY and CheX could either be specific to the CheY encoded by the second chemotaxis operon or generally control CheY~P levels within the cell.

Another problem not encountered by the majority of bacterial species is the coordination of the direction of flagellar rotation at both cell poles in spirochetes whose cell body can be up to 250 μm long (Canale-Parola, 1984). CheY diffusion is too slow to enable fast signal coordination over such long distances (Segall et al., 1985), therefore rapid communication between the cell ends could involve the changes in membrane potential observed during spirochete chemotaxis responses (Goulbourne and Greenberg, 1981). CheX was suggested to probably play a role in this process, since *B. burgdorferi* strains lacking this protein appear to loose motility co-ordination (Charon and Goldstein, 2002). This hypothesis still that cannot be ruled out at this point even though CheY phosphatase activity was recently confirmed for CheX of *T. maritima* (Park et al., 2004).

5. Implication for chemotaxis in spirochete pathogenesis

Neglected for a long-time, chemotaxis is now being recognized as a virulence factor in an increasing number of organisms including the oral spirochete *T. denticola* (Andermann et al., 2002; Butler and Camilli, 2004; Dons et al., 2004; Foynes et al., 2000; Lux et al., 2001; O'Toole et al., 1996; Parke et al., 1987; Senesi et al., 2002; Yao et al., 1997). Oral spirochetes are implicated in periodontal disease and their ability to efficiently migrate through the epithelial barrier into the deeper layers of the connective tissue is thought to contribute greatly to the progression of the disease. Transmigration experiments demonstrated that oral spirochete isolated from plaque and several cultivable oral *Treponema* species including *T. denticola* can actively overcome epithelial and endothelial layers in tissue models (Peters et al., 1999; Riviere et al., 1991). Interestingly, only strains of the *T. denticola* serovars "A", "B" and "D" but not "C" were found to be invasive.

Recently, mutants of *T. denticola* defective in chemotaxis were found to be significantly impaired or completely unable to penetrate epithelial tissue layers *in vitro* (Lux et al., 2001). Closely related *Treponema* species that cause digital dermatitis, an infection of the soft tissue of cow feet whose etiology resembles periodontitis, were found to specifically accumulate around certain cells within the invaded tissue (Moter and Gobel, 2000). Chemotaxis towards substances released from damaged cells would be an obvious explanation for this phenomenon of cell targeting. In agreement with this hypothesis, the chemotaxis response of *T. denticola* include a variety of small molecules including sugars and amino acids that are likely metabolites to be leaked from injured tissue cells (Mayo et al., 1990). Serum and albumin are also found among the chemoattractants identified for this oral spirochete (Umemoto et al., 2001). Since oral spirochetes typically reside in inflamed periodontal pockets that are prone to bleeding, the chemotaxis response to serum could facilitate identification of preferred colonization sites.

The syphilis spirochete *T. pallidum* also uses motility to disseminate throughout its host and several studies demonstrated its ability to efficiently penetrate through endothelial and epithelial tissue barriers (Riviere et al., 1989; Thomas et al., 1989; 1988). Migration appeared to be directional from the epithelial side of the tissue into the connective tissue but not vice versa (Riviere et al., 1989). Tissue penetration was enhanced by the presence of the potential attractant serum on the exit side of the tissue model system, indicating a role for chemotaxis as a driving force for targeted movement (Riviere et al., 1989). Recent microarray experiments revealed that expression of the genes encoding the general chemotaxis proteins CheA, CheX and CheY, as well as one of the chemosensors, MCP2-1, is significantly upregulated (Smajs et al., 2005), similar to findings in *B. burgdorferi* (Revel et al., 2002; Tokarz et al., 2004). Even though lack of *in vitro* growth and genetic tools currently prevent the construction of mutant strains in *T. pallidum* to confirm the importance of chemotaxis for pathogenesis, the circumstantial evidence described above strongly suggests the implication of chemotaxis in the etiology of this syphilis spirochete.

The idea that chemotaxis is an important factor in navigating the host tissue is supported by findings for other pathogenic spirochetes that include differential chemotaxis gene expression during its life cycle for *B. burgdorferi* (Revel et al., 2002; Tokarz et al., 2004) as well as chemotaxis response towards substances likely to be present in their host environment for *B. burgdorferi* (Shi et al., 1998; Shih et al., 2002), *Brachyspira hyodysenteriae*, the causative agent of swine dysenteriae (Kennedy et al., 1988; Kennedy and Yancey, 1996; Milner and Sellwood, 1994), *B. pilocilosi* which triggers colonic spirochetosis in various mammals including humans (Witters and Duhamel, 1999), and *L. interrogans* (Hiramune et al., 1990; Trueba et al., 2004; Yuri et al., 1993).

In conclusion, treponemes are members of a fascinating group of bacteria, the spirochetes that intrigue with their very distinct morphology and chemotaxis-guided motility behavior. The completion of several spirochetal genomes revealed numerous interesting features, such as the unusual chemotaxis protein fusions, CheW-CheR or CheA-CheD, and chemosensors with potential protease function in their sensory domains. Recent advances, especially through the development of genetic gene inactivation and complementation techniques but also microarrays and other genetic tools, have confirmed the assumed role of several chemotaxis genes motility control and implicated an important role in pathogenesis. All of this taken together is a very promising start and will provide a solid basis for further studies exploring the importance of chemotaxis-guided motility in the pathogenic events of spirochetes.

Acknowledgement

This work has been supported by 2 RO1 GM54666 to Wenyuan Shi

References

Adler, J. (1973). A method for measuring chemotaxis and use of the method to determine optimum conditions for chemotaxis by *Escherichia coli*. J. Gen. Microbiol. *74*, 77-91.

Andermann, T. M., Chen, Y. T., and Ottemann, K. M. (2002). Two predicted chemoreceptors of *Helicobacter pylori* promote stomach infection. Infect. Immun. *70*, 5877-5881.

Arakawa, S., and Kuramitsu, H. K. (1994). Cloning and sequence analysis of a chymotrypsinlike protease from *Treponema denticola*. Infect. Immun. *62*, 3424-3433.

Berg, H. C. (1976). How spirochetes may swim. J. Theor Biol. *56*, 269-273.

Bischoff, D. S., and Ordal, G. W. (1992). Identification and characterization of FliY, a novel component of the *Bacillus subtilis* flagellar switch complex. Mol. Microbiol. *6*, 2715-2723.

Borkovich, K. A., Kaplan, N., Hess, J. F., and Simon, M. I. (1989). Transmembrane signal transduction in bacterial chemotaxis involves ligand-dependent activation of phosphate group transfer. Proc. Natl. Acad. Sci. USA. *86*, 1208-1212.

Bren, A., Welch, M., Blat, Y., and Eisenbach, M. (1996). Signal termination in bacterial chemotaxis: CheZ mediates dephosphorylation of free rather than switch-bound CheY. Proc. Natl. Acad. Sci. USA. *93*, 10090-10093.

Butler, S. M., and Camilli, A. (2004). Both chemotaxis and net motility greatly influence the infectivity of *Vibrio cholerae*. Proc. Natl. Acad. Sci. USA. *101*, 5018-5023.

Canale-Parola, E. (1978). Motility and chemotaxis of spirochetes. Annu. Rev. Microbiol. *32*, 69-99.

Canale-Parola, E. (1984). The spirochetes. In Bergey's Manual of Systematic Bacteriology, N. R. Krieg, and J. G. Holt, eds. (Baltimore, Williams and Wilkins), pp. 38-70.

Charon, N. W., and Goldstein, S. F. (2002). Genetics of motility and chemotaxis of a fascinating group of bacteria: the spirochetes. Annu. Rev. Genet. *36*, 47-73.

Charon, N. W., Greenberg, E. P., Koopman, M. B., and Limberger, R. J. (1992). Spirochete chemotaxis, motility, and the structure of the spirochetal periplasmic flagella. Res. Microbiol. *143*, 597-603.

Chervitz, S. A., Lin, C. M., and Falke, J. J. (1995). Transmembrane signaling by the aspartate receptor: engineered disulfides reveal static regions of the subunit interface. Biochem. *34*, 9722-9733.

Dons, L., Eriksson, E., Jin, Y., Rottenberg, M. E., Kristensson, K., Larsen, C. N., Bresciani, J., and Olsen, J. E. (2004). Role of flagellin and the two-component CheA/CheY system of *Listeria monocytogenes* in host cell invasion and virulence. Infect. Immun. *72*, 3237-3244.

Dutta, R., Qin, L., and Inouye, M. (1999). Histidine kinases: diversity of domain organization. Mol. Microbiol. *34*, 633-640.

Fosnaugh, K., and Greenberg, E. P. (1988). Motility and chemotaxis of *Spirochaeta aurantia*: computer-assisted motion analysis. J. Bacteriol. *170*, 1768-1774.

Fosnaugh, K., and Greenberg, E. P. (1989). Chemotaxis mutants of *Spirochaeta aurantia*. J. Bacteriol. *171*, 606-611.

Foynes, S., Dorrell, N., Ward, S. J., Stabler, R. A., McColm, A. A., Rycroft, A. N., and Wren, B. W. (2000). *Helicobacter pylori* possesses two CheY response regulators and a histidine kinase sensor, CheA, which are essential for chemotaxis and colonization of the gastric mucosa. Infect. Immun. *68*, 2016-2023.

Fraser, C. M., Casjens, S., Huang, W. M., Sutton, G. G., Clayton, R., Lathigra, R., White, O., Ketchum, K. A., Dodson, R., Hickey, E. K., et al. (1997). Genomic sequence of a Lyme disease spirochaete, *Borrelia burgdorferi*. Nature *390*, 580-586.

Fraser, C. M., Norris, S. J., Weinstock, G. M., White, O., Sutton, G. G., Dodson, R., Gwinn, M., Hickey, E. K., Clayton, R., Ketchum, K. A., et al. (1998). Complete genome sequence of *Treponema pallidum*, the syphilis spirochete. Science *281*, 375-388.

Garza, A. G., Harris-Haller, L. W., Stoebner, R. A., and Manson, M. D. (1995). Motility protein interactions in the bacterial flagellar motor. Proc. Natl. Acad. Sci. USA. *92*, 1970-1974.

Ge, Y., and Charon, N. W. (1997a). Identification of a large motility operon in *Borrelia burgdorferi* by semi-random PCR chromosome walking. Gene *189*, 195-201.

Ge, Y., and Charon, N. W. (1997b). Molecular characterization of a flagellar/chemotaxis operon in the spirochete *Borrelia burgdorferi*. FEMS Microbiol. Lett. *153*, 425-431.

Gegner, J. A., Graham, D. R., Roth, A. F., and Dahlquist, F. W. (1992). Assembly of an MCP receptor, CheW, and kinase CheA complex in the bacterial chemotaxis signal transduction pathway. Cell *70*, 975-982.

Glockner, G., Lehmann, R., Romualdi, A., Pradella, S., Schulte-Spechtel, U., Schilhabel, M., Wilske, B., Suhnel, J., and Platzer, M. (2004). Comparative analysis of the *Borrelia garinii* genome. Nucleic Acids Res. *32*, 6038-6046.

Goldstein, S. F., Charon, N. W., and Kreiling, J. A. (1994). *Borrelia burgdorferi* swims with a planar waveform similar to that of eukaryotic flagella. Proc. Natl. Acad. Sci. USA. *91*, 3433-3437.

Goulbourne, E. A., Jr., and Greenberg, E. P. (1981). Chemotaxis of *Spirochaeta aurantia*: involvement of membrane potential in chemosensory signal transduction. J. Bacteriol. *148*, 837-844.

Goulbourne, E. A., Jr., and Greenberg, E. P. (1983a). Inhibition of *Spirochaeta aurantia* chemotaxis by neurotoxins. J. Bacteriol. *155*, 1443-1445.

Goulbourne, E. A., Jr., and Greenberg, E. P. (1983b). A voltage clamp inhibits chemotaxis of *Spirochaeta aurantia*. J. Bacteriol. *153*, 916-920.

Goy, M. F., Springer, M. S., and Adler, J. (1977). Sensory transduction in *Escherichia coli*: role of a protein methylation reaction in sensory adaptation. Proc. Natl. Acad. Sci. USA. *74*, 4964-4968.

Greenberg, E. P., Brahamsha, B., and Fosnaugh, K. (1985). The motile behavior of *Spirochaeta aurantia*: a twist to chemosensorytransduction in bacteria. In Sensing and Response in Microorgansims, M. Eisenbach, and M. Balaban, eds. (Amsterdam, Netherlands, Elsevier Science Publishers), pp. 107-117.

Greenberg, E. P., and Canale-Parola, E. (1977). Chemotaxis in *Spirochaeta aurantia*. J. Bacteriol. *130*, 485-494.

Greene, S. R., and Stamm, L. V. (1997). Identification, sequence, and expression of *Treponema denticola mcpA*, a putative chemoreceptor gene. FEMS Microbiol. Lett. *157*, 245-249.

Greene, S. R., and Stamm, L. V. (1998). Molecular characterization of *Treponema pallidum mcp2*, a putative chemotaxis protein gene. Infect. Immun. *66*, 2999-3002.

Greene, S. R., and Stamm, L. V. (1999). Molecular characterization of a chemotaxis operon in the oral spirochete, *Treponema denticola*. Gene *232*, 59-68.

Greene, S. R., Stamm, L. V., Hardham, J. M., Young, N. R., and Frye, J. G. (1997). Identification, sequences, and expression of *Treponema pallidum* chemotaxis genes. DNA Seq *7*, 267-284.

Hagman, K. E., Porcella, S. F., Popova, T. G., and Norgard, M. V. (1997). Evidence for a methyl-accepting chemotaxis protein gene (*mcp1*) that encodes a putative sensory transducer in virulent *Treponema pallidum*. Infect. Immun. *65*, 1701-1709.

Hess, J. F., Oosawa, K., Kaplan, N., and Simon, M. I. (1988). Phosphorylation of three proteins in the signaling pathway of bacterial chemotaxis. Cell *53*, 79-87.

Hess, J. F., Oosawa, K., Matsumura, P., and Simon, M. I. (1987). Protein phosphorylation is involved in bacterial chemotaxis. Proc. Natl. Acad. Sci. USA. *84*, 7609-7613.

Hiramune, T., Shiraiwa, C., Kikuchi, N., and Yanagawa, R. (1990). A basic study of chemotaxis of leptospiras. Zentralbl Veterinarmed B *37*, 749-752.

Holt, S. C. (1978). Anatomy and chemistry of spirochetes. Microbiol. Rev. *42*, 114-160.

Hower, D. A., Lopez, R., Lux, R., and Shi, W. (2004). Generation of a yeast two-hybrid genomic library for *Treponema denticola*. J. Dent Res. *83*.

Kaempf, C., and Greenberg, E. P. (1990). Negative chemotaxis in *Spirochaeta aurantia*. Curr. Microbiol. *21*, 187-192.

Karatan, E., Saulmon, M. M., Bunn, M. W., and Ordal, G. W. (2001). Phosphorylation of the response regulator CheV is required for adaptation to attractants during *Bacillus subtilis* chemotaxis. J. Biol. Chem. *276*, 43618-43626.

Kataoka, M., Li, H., Arakawa, S., and Kuramitsu, H. (1997). Characterization of a methyl-accepting chemotaxis protein gene, *dmcA*, from the oral spirochete *Treponema denticola*. Infect. Immun. *65*, 4011-4016.

Kathariou, S., and Greenberg, E. P. (1983). Chemoattractants elicit methylation of specific polypeptides in *Spirochaeta aurantia*. J. Bacteriol. *156*, 95-100.

Kehry, M. R., Bond, M. W., Hunkapiller, M. W., and Dahlquist, F. W. (1983). Enzymatic deamidation of methyl-accepting chemotaxis proteins in *Escherichia coli* catalyzed by the *cheB* gene product. Proc. Natl. Acad. Sci. USA. *80*, 3599-3603.

Kennedy, M. J., Rosnick, D. K., Ulrich, R. G., and Yancey, R. J., Jr. (1988). Association of *Treponema hyodysenteriae* with porcine intestinal mucosa. J. Gen. Microbiol. *134 (Pt 6)*, 1565-1576.

Kennedy, M. J., and Yancey, R. J., Jr. (1996). Motility and chemotaxis in *Serpulina hyodysenteriae*. Vet. Microbiol. *49*, 21-30.

Khan, I. H., Reese, T. S., and Khan, S. (1992). The cytoplasmic component of the bacterial flagellar motor. Proc. Natl. Acad. Sci. USA. *89*, 5956-5960.

Kirby, J. R., Kristich, C. J., Saulmon, M. M., Zimmer, M. A., Garrity, L. F., Zhulin, I. B., and Ordal, G. W. (2001). CheC is related to the family of flagellar switch proteins and acts independently from CheD to control chemotaxis in *Bacillus subtilis*. Mol. Microbiol. *42*, 573-585.

Kirby, J. R., Saulmon, M. M., Kristich, C. J., and Ordal, G. W. (1999). CheY-dependent methylation of the asparagine receptor, McpB, during chemotaxis in *Bacillus subtilis*. J. Biol. Chem. *274*, 11092-11100.

Kleene, S. J., Hobson, A. C., and Adler, J. (1979). Attractants and repellents influence methylation and demethylation of methyl-accepting chemotaxis proteins in an extract of *Escherichia coli*. Proc. Natl. Acad. Sci. USA. *76*, 6309-6313.

Kristich, C. J., and Ordal, G. W. (2002). *Bacillus subtilis* CheD is a chemoreceptor modification enzyme required for chemotaxis. J. Biol. Chem. *277*, 25356-25362.

Li, C., Bakker, R. G., Motaleb, M. A., Sartakova, M. L., Cabello, F. C., and Charon, N. W. (2002). Asymmetrical flagellar rotation in *Borrelia burgdorferi* nonchemotactic mutants. Proc. Natl. Acad. Sci. USA. *99*, 6169-6174.

Li, H., Arakawa, S., Deng, Q. D., and Kuramitsu, H. (1999). Characterization of a novel methyl-accepting chemotaxis gene, *dmcB*, from the oral spirochete *Treponema denticola*. Infect. Immun. *67*, 694-699.

Liu, Y., Levit, M., Lurz, R., Surette, M. G., and Stock, J. B. (1997). Receptor-mediated protein kinase activation and the mechanism of transmembrane signaling in bacterial chemotaxis. Embo J. *16*, 7231-7240.

Lux, R., Miller, J. N., Park, N. H., and Shi, W. (2001). Motility and chemotaxis in tissue penetration of oral epithelial cell layers by *Treponema denticola*. Infect. Immun. *69*, 6276-6283.

Lux, R., and Shi, W. (2004). Chemotaxis-guided movements in bacteria. Crit. Rev. Oral Biol. Med. *15*, 207-220.

Lux, R., Sim, J. H., Tsai, J. P., and Shi, W. (2002). Construction and characterization of a *cheA* mutant of *Treponema denticola*. J. Bacteriol. *184*, 3130-3134.

Marcotte, C. J., and Marcotte, E. M. (2002). Predicting functional linkages from gene fusions with confidence. Appl Bioinformatics *1*, 93-100.

Marcotte, E. M., Pellegrini, M., Ng, H. L., Rice, D. W., Yeates, T. O., and Eisenberg, D. (1999). Detecting protein function and protein-protein interactions from genome sequences. Science *285*, 751-753.

Mayo, J. A., Blake, A., and Donze, D. (1990). Chemotaxis by *Treponema denticola*. J. Dent Res. *69*, 382.

Methe, B. A., Nelson, K. E., Eisen, J. A., Paulsen, I. T., Nelson, W., Heidelberg, J. F., Wu, D., Wu, M., Ward, N., Beanan, M. J., et al. (2003). Genome of *Geobacter sulfurreducens*: metal reduction in subsurface environments. Science *302*, 1967-1969.

Milner, J. A., and Sellwood, R. (1994). Chemotactic response to mucin by *Serpulina hyodysenteriae* and other porcine spirochetes: potential role in intestinal colonization. Infect. Immun. *62*, 4095-4099.

Motaleb, M. A., Miller, M. R., Li, C., Bakker, R. G., Goldstein, S. F., Silversmith, R. E., Bourret, R. B., and Charon, N. W. (2005). CheX is a CheY-P phosphatase essential for *Borrelia burgdorferi* chemotaxis. J. Bacteriol. *187*, 7963-7969.

Moter, A., and Gobel, U. B. (2000). Fluorescence *in situ* hybridization (FISH) for direct visualization of microorganisms. J. Microbiol. Methods *41*, 85-112.

Nascimento, A. L., Ko, A. I., Martins, E. A., Monteiro-Vitorello, C. B., Ho, P. L., Haake, D. A., Verjovski-Almeida, S., Hartskeerl, R. A., Marques, M. V., Oliveira, M. C., et al. (2004a). Comparative genomics of two *Leptospira interrogans* serovars reveals novel insights into physiology and pathogenesis. J. Bacteriol. *186*, 2164-2172.

Nascimento, A. L., Verjovski-Almeida, S., Van Sluys, M. A., Monteiro-Vitorello, C. B., Camargo, L. E., Digiampietri, L. A., Harstkeerl, R. A., Ho, P. L., Marques, M. V., Oliveira, M. C., et al. (2004b). Genome features of *Leptospira interrogans* serovar *Copenhageni*. Braz J. Med. Biol. Res. *37*, 459-477.

Nelson, K. E., Clayton, R. A., Gill, S. R., Gwinn, M. L., Dodson, R. J., Haft, D. H., Hickey, E. K., Peterson, J. D., Nelson, W. C., Ketchum, K. A., et al. (1999). Evidence for lateral gene transfer between Archaea and bacteria from genome sequence of *Thermotoga maritima*. Nature *399*, 323-329.

Nolling, J., Breton, G., Omelchenko, M. V., Makarova, K. S., Zeng, Q., Gibson, R., Lee, H. M., Dubois, J., Qiu, D., Hitti, J., et al. (2001). Genome sequence and comparative analysis of the solvent-producing bacterium *Clostridium acetobutylicum*. J. Bacteriol. *183*, 4823-4838.

O'Toole, R., Milton, D. L., and Wolf-Watz, H. (1996). Chemotactic motility is required for invasion of the host by the fish pathogen *Vibrio anguillarum*. Mol. Microbiol. *19*, 625-637.

Ottemann, K. M., Xiao, W., Shin, Y. K., and Koshland, D. E., Jr. (1999). A piston model for transmembrane signaling of the aspartate receptor. Science *285*, 1751-1754.

Park, S. Y., Chao, X., Gonzalez-Bonet, G., Beel, B. D., Bilwes, A. M., and Crane, B. R. (2004). Structure and function of an unusual family of protein phosphatases; the bacterial chemotaxis proteins CheC and CheX. Mol. Cell *16*, 563-574.

Parke, D., Ornston, L. N., and Nester, E. W. (1987). Chemotaxis to plant phenolic inducers of virulence genes is constitutively expressed in the absence of the Ti plasmid in *Agrobacterium tumefaciens*. J. Bacteriol. *169*, 5336-5338.

Peters, S. R., Valdez, M., Riviere, G., and Thomas, D. D. (1999). Adherence to and penetration through endothelial cells by oral treponemes. Oral Microbiol. Immunol. *14*, 379-383.'

Pittman, M. S., Goodwin, M., and Kelly, D. J. (2001). Chemotaxis in the human gastric pathogen *Helicobacter pylori*: different roles for CheW and the three CheV paralogues, and evidence for CheV2 phosphorylation. Microbiology *147*, 2493-2504.

Porter, S. L., and Armitage, J. P. (2002). Phosphotransfer in *Rhodobacter sphaeroides* chemotaxis. J. Mol. Biol. *324*, 35-45.

Rasko, D. A., Ravel, J., Okstad, O. A., Helgason, E., Cer, R. Z., Jiang, L., Shores, K. A., Fouts, D. E., Tourasse, N. J., Angiuoli, S. V., et al. (2004). The genome sequence of *Bacillus cereus* ATCC 10987 reveals metabolic adaptations and a large plasmid related to *Bacillus anthracis* pXO1. Nucleic Acids Res. *32*, 977-988.

Read, T. D., Peterson, S. N., Tourasse, N., Baillie, L. W., Paulsen, I. T., Nelson, K. E., Tettelin, H., Fouts, D. E., Eisen, J. A., Gill, S. R., et al. (2003). The genome sequence of *Bacillus anthracis* Ames and comparison to closely related bacteria. Nature *423*, 81-86.

Revel, A. T., Talaat, A. M., and Norgard, M. V. (2002). DNA microarray analysis of differential gene expression in *Borrelia burgdorferi*, the Lyme disease spirochete. Proc. Natl. Acad. Sci. USA. *99*, 1562-1567.

Rey, M. W., Ramaiya, P., Nelson, B. A., Brody-Karpin, S. D., Zaretsky, E. J., Tang, M., Lopez de Leon, A., Xiang, H., Gusti, V., Clausen, I. G., et al. (2004). Complete genome sequence of the industrial bacterium *Bacillus licheniformis* and comparisons with closely related *Bacillus* species. Genome Biol. *5*, R77.

Riviere, G. R., Thomas, D. D., and Cobb, C. M. (1989). *In vitro* model of *Treponema pallidum* invasiveness. Infect. Immun. *57*, 2267-2271.

Riviere, G. R., Weisz, K. S., Adams, D. F., and Thomas, D. D. (1991). Pathogen-related oral spirochetes from dental plaque are invasive. Infect. Immun. *59*, 3377-3380.

Rosario, M. M., Fredrick, K. L., Ordal, G. W., and Helmann, J. D. (1994). Chemotaxis in *Bacillus subtilis* requires either of two functionally redundant CheW homologs. J. Bacteriol. *176*, 2736-2739.

Schuster, S. C., Swanson, R. V., Alex, L. A., Bourret, R. B., and Simon, M. I. (1993). Assembly and function of a quaternary signal transduction complex monitored by surface plasmon resonance. Nature *365*, 343-347.

Segall, J. E., Ishihara, A., and Berg, H. C. (1985). Chemotactic signaling in filamentous cells of *Escherichia coli*. J. Bacteriol. *161*, 51-59.

Senesi, S., Celandroni, F., Salvetti, S., Beecher, D. J., Wong, A. C., and Ghelardi, E. (2002). Swarming motility in *Bacillus cereus* and characterization of a *fliY* mutant impaired in swarm cell differentiation. Microbiology *148*, 1785-1794.

Seshadri, R., Myers, G. S., Tettelin, H., Eisen, J. A., Heidelberg, J. F., Dodson, R. J., Davidsen, T. M., DeBoy, R. T., Fouts, D. E., Haft, D. H., et al. (2004). Comparison of the genome of the oral pathogen *Treponema denticola* with other spirochete genomes. Proc. Natl. Acad. Sci. USA. *101*, 5646-5651.

Shi, W., Yang, Z., Geng, Y., Wolinsky, L. E., and Lovett, M. A. (1998). Chemotaxis in *Borrelia burgdorferi*. J. Bacteriol. *180*, 231-235.

Shih, C. M., Chao, L. L., and Yu, C. P. (2002). Chemotactic migration of the Lyme disease spirochete (*Borrelia burgdorferi*) to salivary gland extracts of vector ticks. Amer. J. Trop Med. Hyg *66*, 616-621.

Sim, J. H., Shi, W., and Lux, R. (2005). Protein-protein interactions in the chemotaxis signaling pathway of *Treponema denticola*. Microbiology 151:1801-1807.

Smajs, D., McKevitt, M., Howell, J. K., Norris, S. J., Cai, W. W., Palzkill, T., and Weinstock, G. M. (2005). Transcriptome of *Treponema pallidum*: gene expression profile during experimental rabbit infection. J. Bacteriol. *187*, 1866-1874.

Sourjik, V., and Schmitt, R. (1998). Phosphotransfer between CheA, CheY1, and CheY2 in the chemotaxis signal transduction chain of *Rhizobium meliloti*. Biochemistry *37*, 2327-2335.

Stamm, L. V., and Bergen, H. L. (1999). Molecular characterization of a flagellar (*fla*) operon in the oral spirochete *Treponema denticola* ATCC 35405. FEMS Microbiol. Lett. *179*, 31-36.

Strong, M., Mallick, P., Pellegrini, M., Thompson, M. J., and Eisenberg, D. (2003). Inference of protein function and protein linkages in *Mycobacterium tuberculosis* based on prokaryotic genome organization: a combined computational approach. Genome Biol. *4*, R59.

Szurmant, H., Bunn, M. W., Cannistraro, V. J., and Ordal, G. W. (2003). *Bacillus subtilis* hydrolyzes CheY-P at the location of its action, the flagellar switch. J. Biol. Chem. *278*, 48611-48616.

Szurmant, H., Muff, T. J., and Ordal, G. W. (2004). *Bacillus subtilis* CheC and FliY are members of a novel class of CheY-P-hydrolyzing proteins in the chemotactic signal transduction cascade. J. Biol. Chem. *279*, 21787-21792.

Szurmant, H., and Ordal, G. W. (2004). Diversity in chemotaxis mechanisms among the bacteria and archaea. Microbiol. Mol. Biol. Rev. *68*, 301-319.

Thomas, D. D., Fogelman, A. M., Miller, J. N., and Lovett, M. A. (1989). Interactions of Treponema pallidum with endothelial cell monolayers. Eur. J. Epidemiol. *5*, 15-21.

Thomas, D. D., Navab, M., Haake, D. A., Fogelman, A. M., Miller, J. N., and Lovett, M. A. (1988). Treponema pallidum invades intercellular junctions of endothelial cell monolayers. Proc. Natl. Acad. Sci. USA. *85*, 3608-3612.

Toews, M. L., Goy, M. F., Springer, M. S., and Adler, J. (1979). Attractants and repellents control demethylation of methylated chemotaxis proteins in *Escherichia coli*. Proc. Natl. Acad. Sci. USA. *76*, 5544-5548.

Tokarz, R., Anderton, J. M., Katona, L. I., and Benach, J. L. (2004). Combined effects of blood and temperature shift on *Borrelia burgdorferi* gene expression as determined by whole genome DNA array. Infect. Immun. *72*, 5419-5432.

Trueba, G., Zapata, S., Madrid, K., Cullen, P., and Haake, D. (2004). Cell aggregation: a mechanism of pathogenic Leptospira to survive in fresh water. Int. Microbiol. *7*, 35-40.

Trueba, G. A., Old, I. G., Saint Girons, I., and Johnson, R. C. (1997). A *cheA cheW* operon in *Borrelia burgdorferi*, the agent of Lyme disease. Res. Microbiol. *148*, 191-200.

Umemoto, T., Jinno, T., Taiji, Y., and Ogawa, T. (2001). Chemotaxis of oral treponemes toward sera and albumin of rabbit. Microbiol. Immunol. *45*, 571-577.

van Leeuwenhoek, A. (1683). Microscopical observations about animals in the scurf of the teeth. Phil Trans R Soc. Lond *14*, 568-574.

Welch, M., Oosawa, K., Aizawa, S., and Eisenbach, M. (1993). Phosphorylation-dependent binding of a signal molecule to the flagellar switch of bacteria. Proc. Natl. Acad. Sci. USA. *90*, 8787-8791.

Witters, N. A., and Duhamel, G. E. (1999). Motility-regulated mucin association of *Serpulina pilosicoli*, the agent of colonic spirochetosis of humans and animals. Adv. Exp. Med. Biol. *473*, 199-205.

Wylie, D., Stock, A., Wong, C. Y., and Stock, J. (1988). Sensory transduction in bacterial chemotaxis involves phosphotransfer between Che proteins. Biochem. Biophys Res. Commun. *151*, 891-896.

Yao, R., Burr, D. H., and Guerry, P. (1997). CheY-mediated modulation of *Campylobacter jejuni* virulence. Mol. Microbiol. *23*, 1021-1031.

Yuri, K., Takamoto, Y., Okada, M., Hiramune, T., Kikuchi, N., and Yanagawa, R. (1993). Chemotaxis of leptospires to hemoglobin in relation to virulence. Infect. Immun. *61*, 2270-2272.

Zimmer, M. A., Tiu, J., Collins, M. A., and Ordal, G. W. (2000). Selective methylation changes on the *Bacillus subtilis* chemotaxis receptor McpB promote adaptation. J. Biol. Chem. *275*, 24264-24272.

8

Genetic Manipulation of Cultivable Treponemes

*Howard K. Kuramitsu**

Abstract

Until relatively recently, genetic characterization of treponemes was limited to cloning treponemal genes into nonspirochete hosts such as *Escherichia coli*. However, the recent development of transformation systems for a few cultivable spirochetes, particularly the oral spirochete *Treponema denticola*, has now enabled the construction of monospecific mutants in these organisms. In addition, construction of novel shuttle plasmids has resulted in the expression of heterologous genes in these transformable strains. This will now allow for the expression and characterization of genes from noncultivable treponemes in selected cultivable treponemes. Thus, the development of these genetic tools should lead to more detailed characterization of virulence factors in both cultivable and noncultivable treponemes.

1. Introduction

The introduction of recombinant DNA techniques has remarkably enhanced our ability to dissect the virulence properties of important microbial pathogens. This revolutionary technique led to the development of a "molecular Koch's principles" approach for defining the pathogenicity of an organism (Falkow, 1988). Unfortunately, for those studying the virulence of pathogenic treponemes this approach was limited since many of these organisms could not be cultivated in the laboratory. This limitation was particularly a handicap in the investigation of *Treponema pallidum* and the multiple species of oral treponemes associated with periodontal diseases. In addition, procedures to genetically manipulate these organisms (construct isogenic mutants, express homologous or heterologous genes) which could be cultivated (*Treponema denticola, T. phagedenis, etc.*) were

*For correspondence email kuramits@buffalo.edu

also not available. However, the recent development of techniques to genetically manipulate treponemes has now improved our ability to better understand the physiology and virulence of some of these organisms

2. Isolation and characterization of treponeme genes

The introduction of recombinant DNA techniques in the late 1970's to investigate the genetics of bacterial pathogens (So et al., 1976) allowed for the examination of potential virulence factors from organisms even if some of these were difficult to grow and examine in the laboratory. This approach was initially applied to the study of treponemes in the early 1980's when *T. pallidum* antigens were cloned into *Escherichia coli* (Stamm and Bassford, 1982; Walfield et al., 1982; Norgard and Miller, 1983). Since that time, a number of genes from this organism have been isolated in *E. coli* and partially characterized. Many of these genes code for antigenic molecules and may be relevant for the pathogenicity of the organism. In addition, expression of these genes could be readily detected with appropriate antisera. More recently, the complete sequencing of the *T. pallidum* genome (Fraser et al., 1998) has now allowed for the isolation and characterization of many of the genes from this treponeme. However, functional investigations of these genes with these approaches are more problematic because of the absence of techniques to construct mutants in these organisms. Nevertheless, biochemical characterization of *T. pallidum* genes expressed in *E. coli* has yielded valuable information regarding the functional aspects of some of the genes. However, morphological differences between the treponemes and the *E. coli* host utilized to express these genes may limit some functional studies. This limitation may be overcome in some cases by the expression of *T. pallidum* genes in a suitable spirochete host as described later in this chapter.

It is likely that the greatest variety of treponemes associated with human diseases are present in the oral cavity. A number of investigations have demonstrated that the predominant group of bacteria associated with human periodontitis are spirochetes (Linde et al., 1980). Among these, several have been isolated and characterized including *T. denticola, T. socranskii, T. vincentii, T. medium, T. maltophilum,* and *T. lecithinolyticum* (see Paster and Dewhirst, Chapter 1) all of which can be grown in the laboratory under suitable conditions. However, it is likely that a number of periodontitis-associated spirochetes still remain to be cultivated and identified.

Among the oral treponemes associated with periodontitis, *T. denticola* has received the greatest attention since a number of studies have directly associated this species with periodontitis (Simonson et al., 1988; Sela, 2001) and the organism can be readily grown in the laboratory. Beginning with the isolation of the *tdp*A gene coding for a presumptive surface protein of *T. denticola* (Miyamoto et al., 1991), a number of genes from this species have been isolated and characterized in *E. coli*. Among these are genes coding for a major outer sheath protein, Msp (Haapasalo et al., 1992), metabolic enzymes (Ishihara et al., 1992), the extracellular serine protease dentilisin (Ishihara et al., 1996), flagellar-associated genes (Li et al., 1996), a cytoplasmic filament protein (Izard et al., 2001) as well as chemotaxis proteins (Greene and Stamm, 1999; Kataoka et al., 1997; Lux et al., 2002).

In addition to *T. pallidum* and *T. denticola,* genes from *T. phagedenis* (Limberger et al., 1992), *T. socranskii* (Heuner et al., 2001), *T. maltophilum* (Heuner et al.,

2000) as well as *T. lecithinolyticum* (Park et al., 2002) have also been isolated. This list of genes isolated from treponemes will undoubtedly increase over the next few years as more of these organisms are isolated and characterized. However, prior to the introduction of gene inactivation systems in these organisms it was not possible to identify the functions of the cloned genes in detail. For example, although the *T. denticola tdpA* gene was originally characterized as coding for a putative surface-associated antigenic lipoprotein (Miyamoto *et.al*, 1991), the absence of studies with a *tdpA* mutant has precluded any conclusions regarding the function of this gene. Likewise, the isolation and preliminary characterization of several genes from *T. pallidum* has not allowed for a detailed investigation into the function of these genes since mutants of these organisms cannot be constructed and grown in the laboratory. For example, one of these genes, *tplrr*, apparently codes for an extracytoplasmic protein containing tandem leucine-rich repeats typically involved in protein-protein interactions (Shevchenko et al., 1997). Nevertheless, the function of the protein in the physiology or virulence of these organisms still remains unclear. However, the function of some of these genes has been suggested based upon homology with genes of known function.

2.1. Genetic transformation of treponemes

The availability of treponeme genes cloned into *E. coli* allowed for the biochemical characterization of the gene products. However, the inability to utilize the isolated genes as a means to construct isogenic mutants made it difficult to unequivocally confirm the function of the genes. Therefore, attempts were initially made to develop gene transfer systems for the cultivable *T. denticola*. The presence of naturally occurring cryptic plasmids in at least one strain of this organism, ATCC 33520 (Ivic et al., 1991), suggested the possibility that natural gene transfer via either conjugation or transformation occurred in this strain. Since electroporation of the spirochetes *Borrelia burgdoferi* (Samuels et al., 1994) and *Serpula hyodysenteriae* (ter Huurne et al., 1992) had been previously demonstrated, electroporetic transformation of *T. denticola* 33520 was attempted with the broad-host range shuttle plasmid pKT210 (Li and Kuramitsu, 1996). The successful isolation of chloramphenicol resistant transformants harboring the plasmid following transformation indicated that this procedure could be used to introduce DNA into this oral treponeme and further indicated the compatibility of pTD1 and pKT210 in strain 33520. However, the presence of the naturally occurring plasmid of strain 33520, pTD1, was not necessary for transformation since plasmid pKT210 could also be introduced into strain ATCC 35405 which does not contain a naturally occurring plasmid.

In order to determine if transformation of *T. denticola* via electroporation could be utilized to construct mutants of this organism, inactivation of the flagellar hook protein gene, *flg*E, was attempted (Li et al., 1996) since this would lead to a readily identifiable nonmotile phenotype. Since the utilization of the chloramphenicol resistance gene contained on the shuttle plasmid pKT210 in *T. denticola* led to the development of relatively high levels of chloramphenicol resistance, another readily selectable marker was sought. Utilizing an erythromycin resistance cassette containing both the *ermF* and *ermAM* genes originally constructed for use in the oral bacterium *Porphyromonas gingivalis* (Fletcher et.al., 1995), it was possible to

Table 1. *T. denticola* mutants constructed

Genes	Alterations	Reference
flgE	Motility	Li et.al., 1996
dmcA	Chemotaxis	Kataoka et.al., 1997
dmcB	Chemotaxis	Li et.al., 1999
prtP	Protease	Ishihara et.al. 1998
msp	Adherence	Fenno et.al. 1998
cfpA	Cell division	Izard et.al., 2001
cheA	Chemotaxis	Lux et.al., 2002
opdB	Protease	Fenno et.al. 2001
tap1	Motility	Limberger et.al. 1999
prcA	Protease	Lee et.al., 2002
hbpA	Iron uptake	Xu and Kolodubretz, 2002
oppA	Protein binding	Fenno *et.al.*, 2000

insertionally inactive the *flgE* gene on the chromosome of strain 35405 following electroporation and homologous recombination. As predicted, the resultant transformant, HL51, was nonmotile, did not swarm on agarose plates and also was defective in chemotaxis. In addition, electron micrographs of the mutant confirmed the absence of visible flagella. Additional studies demonstrated that the *ermF* gene was responsible for erythromycin resistance in *T. denticola* and the *ermAM* gene allowed for subsequent identification of these gene fragments in *E. coli*. In addition, the *erm* cassette utilized for these studies did not induce major polar effects on the genes immediately downstream from the inactivation sites in the spirochete (Limberger et.al, 1999). Inactivation of the *flgE* gene was also documented in strain 33520 with this procedure (Li et al., 1996).

The successful development of a system to construct isogenic mutants in *T. denticola* led to the functional characterization of a number of genes of this organism. In addition to the *flgE* gene, a large variety of mutants have now been constructed (Table 1) and have allowed for the identification of potential virulence factors in these organisms. These developments together with the use of a murine model (Ishihara et al., 1998) has further suggested a role for some of these genes in pathogenicity.

Several studies have suggested that the Msp protein of *T. denticola* is an important virulence factor of this organism (see Ellen, Chapter 15). The construction of a *msp* mutant (Fenno et al., 1998) has supported such a hypothesis. For example, the utilization of such a mutant has suggested a role for Msp in disrupting host cell structural integrity. Likewise, construction of mutants of the major serine proteinase of strain 35405, dentilisin, suggested that a defect in this mutant leads to pleiotrophic effects in the spirochete (Ishihara et al., 1998). The proteinase mutant K1 is not only defective in cleaving host cell proteins but also is attenuated in tissue penetration (Chi et al., 2003), altered in the formation of synergistic biofilms with *P. gingivalis* (Ikegami and Kuramitsu, unpub. results), attenuated in the induction of host cell responses (Ellen et al., 2000), is less virulent in a mouse abscess model (Ishihara et al., 1998), as well displaying abnormal oligomerization of the Msp (Ishihara et al., 1998). These multiple properties illustrate that, like any mutant constructed, precautions need to be taken in terms of directly associating a mutant phenotype with specific gene products. The successful utilization of electroporation for genetic manipulation of *T. denticola* has further suggested that this approach may be useful for gene transfer in other cultivable treponemes. This has been confirmed for at least one other spirochete, *T. phagedenis* (see below). Therefore, it is likely that this procedure could be used to construct specific mutants in this species as well in some other transformable treponemes.

3. Development of treponeme shuttle vectors

The ability to introduce genes into treponemes suggested that it might be possible to develop shuttle vectors for the expression of homologous or heterologous genes in these organisms. The former would be valuable for complementing genetic defects while the latter might allow for the characterization of genes from uncultivable treponemes. For *T. denticola,* the broad-host range plasmid pKT210 did not appear to be a suitable candidate for this approach since some instability of the plasmid in the oral treponeme was noted (Li and Kuramitsu, 1996). Therefore, it appeared reasonable to modify one of the two cryptic plasmids previously identified in oral treponemes as potential shuttle plasmids. One of these plasmids, pTD1, is present in the ATCC strain 33520 and had been sequenced (MacDougall et al., 1992). Sequencing of the plasmid revealed two open reading frames, one of which appeared to correspond to a replication region. However, attempts to construct shuttle plasmids with DNA fragments from pTD1 were unsuccessful (MacDougall, person. commun.). Therefore, the 4.2 kb plasmid pTS1 that was originally identified in a natural presumptive isolate of *T. denticola,* U9b, (Chan et al., 1996) was investigated as a potential component of a treponeme shuttle vector. Utilizing a DNA fragment of pTS1, the *ermFermAM* cassette under the control of the promoter region of the *T. denticola prtB* gene (Arakawa and Kuramitsu, 1994), and the *E. coli* plasmid pKMOZ19, it was possible to construct a 7.7 kb shuttle plasmid, pKMR4PE. This plasmid could be electroporated into strain 33520 to produce Erm^r transformants (Chi et al., 1999). Interestingly, the plasmid could not be introduced into strain 35405 despite repeated attempts.

Plasmid pKMR4PE was then utilized to express the gene coding for the flagellar protein FlaA from *T. pallidum* in strain 33520 (Chi et al., 1999). Expression of this gene was detected using anti-FlaA sera but appeared to be relatively weak. This may result from the distinct G+C ratios of *T. pallidum* (52.7%) compared to *T. denticola* (38%). Nevertheless, these results indicated that it is possible to express genes from noncultivable treponemes in *T. denticola* for possible characterization. Therefore, the utilization of *T. denticola* and other transformable treponemes as surrogate organisms for the expression of heterologous genes appears to be a promising strategy for characterizing genes from noncultivable treponemes.

One strategy required for a complete genetic analysis of a gene is to complement mutants of the gene with homologous or heterologous genes. Since all of the mutants constructed in *T. denticola* involved the utilization of the Erm^r cassette as a selectable marker, it was necessary to develop other markers for complementation analysis. Toward this end, a spontaneous mutant altered in the *T. denticola gyr*B gene was isolated for use as a coumermycin resistant (Cou^r) marker (Chi et al., 2002). A gene fragment containing the Cou^r *gyrB* gene was then used to construct the shuttle plasmid pKMCou (Chi et al., 2002). Introduction of the *T. denticola flgE* gene and electroporation of the resulting plasmid into the *flgE* mutant of strain 33520 resulted in Cou^r Erm^r transformants. Complementation of the chromosomal *flgE* mutation was demonstrated by the restoration of swarming on agarose plates and by the detection of FlgE by Western blot analysis with anti-*T. pallidum* FlgE sera (the sera detected both *T. denticola* and *T. pallidum* FlgE proteins). In addition, very recently a novel shuttle plasmid based upon the pKMR4PE shuttle plasmid expressing a chloramphenicol resistance gene from a *Staphylococcus aureus* strain has been constructed and utilized to complement a *fliG* mutant of *T. denticola* (R. Limberger, person. commun.). The utilization of this plasmid appears to yield lower background levels compared to plasmid pKMCou. Therefore, it is now possible to complement mutations in *T. denticola* with appropriate shuttle plasmids. Furthermore, three selectable markers (Erm^r , Cou^r , and Cm^r) are currently available for such studies.

It would also be beneficial to develop novel shuttle vectors containing other replication regions to increase the copy number of these plasmids in transformable treponemes. However, only two plasmid replication regions have been identified so far in these organisms and only one these, derived from pTS1, has been successfully utilized to construct shuttle plasmids as described earlier. It may be possible to identify broad-host range plasmids for such use. It will also be important to develop inducible expression systems for use with treponeme shuttle vectors. In this regard, limited information is currently available regarding the regulation of gene expression in these organisms. Thus, many potential approaches are available to improve the expression of exogenous genes in transformable treponemes.

In order to expand the utilization of the above-mentioned genetic approaches to other cultivable treponemes, attempts are currently in progress to transform several other treponeme species. Utilizing shuttle plasmid pKMR4PE it has been possible to express the *lrrA* gene from *T. denticola* 35405 in *T. phagedenis* Kazan 5 (Ikegami and Kuramitsu, unpub. results). The utilization of monospecific anti-LrrA antibody confirmed the expression of this protein in the latter strain. These results were not surprising given the close phylogenetic relationship between *T. phagedenis* and *T. denticola* (Dewhirst et al., 2000). Attempts are currently in progress in this laboratory to examine the introduction of the pTS1-based shuttle plasmids into other oral treponeme such as *T. socranskii*, one of whose members apparently also harbors plasmid pTS1 (Chan et al., 1996).

The shuttle plasmids have also been recently utilized to express nonspirochete genes in *T. denticola* (Girons et al., 2000). Both the *E. coli lacZ* gene as well as a gene for the green fluorescent protein (GFP) have been expressed in strain 33520. β-Galactosidase activity was detected both by direct enzyme assays as well as by the typical blue color of colonies grown on X-gal containing agarose plates (Figure 1). Therefore, gene fusions with *E. coli lacZ* under the control of *T. denticola* promoters contained on either the shuttle plasmids or on the chromosome should be useful for examining the regulation of gene expression in these organisms. Furthermore, GFP expression from a shuttle plasmid in strain 33520 was detected under fluorescent

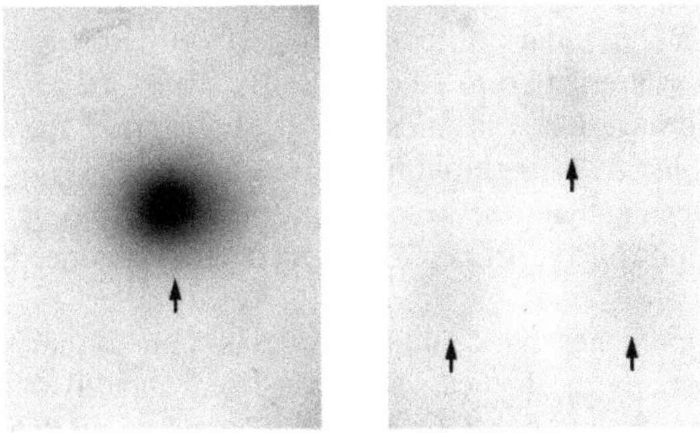

Figure 1. Expression of β-galactosidase activity in *T. denticola* 33520. Left, strain 33520 harboring the shuttle plasmid expressing β-galactosidase; right, strain 33520 containing the shuttle plasmid pKMR4PE. Arrows indicate presence of colonies on TYGVS-agarose-X-gal plates.

microscopy. However, the fluorescence was relatively weak compared to expression from other organisms and also appeared to be unstable after prolonged culturing of the transformed strains. Therefore, additional studies employing perhaps other GFP derivatives may be necessary to both increase the fluorescence of these strains and stably maintain the constructs. Nevertheless, these preliminary results suggest that it may be possible to examine gene regulation in the transformable treponemes using gene fusion technology.

4. Future prospects for the genetic manipulation of treponemes

Since the genetic manipulation of treponemes is still in the early stages of development, it is likely that additional strategies will soon be forthcoming to expand our repertoire of genetic tools for the study of these organisms. For example, the relatively low frequency of transformation of *T. denticola* by electroporation (Li and Kuramitsu, 1996) has precluded the utilization of transposon mutagenesis for construction of mutant clone banks. Attempts to increase the transformation rates by heating the organism (to partially inactivate restriction enzymes in the treponemes) have led to only a moderate increase in the recovery of transformants (Chi et al., 2002). In this regard, it may be possible to take advantage of the genome sequencing of *T. denticola* 35405 (TIGR data base) and identify potential restriction endonuclease genes in this organism. The existence of such genes was suggested by earlier observations that shuttle plasmids isolated from *T. denticola* transformants subsequently transformed the treponeme at significantly higher frequencies compared to the same plasmid isolated from *E. coli* (Li and Kuramitsu, 1996). These endonuclease genes could be systematically inactivated to determine if such mutants exhibit higher rates of transformation.

Recently, the cryptic plasmid pTS1 has been fully sequenced (Chauhan and Kuramitsu, 2004). Two potential ORFs were identified on the plasmid: one coding for a suggested replication region (which was used to construct the shuttle plasmid pKMR4PE) and another coding for a putative mobilization gene. This latter possibility suggests that plasmid pTS1 may be capable of conjugation between treponemes. The latter ORF also shared low homology with a potential ORF identified on plasmid pTD1 from strain 33520. In addition, the demonstration that a plasmid similar to, if not identical with, pTS1 is present in a human oral isolate of *T. socranskii* (Chan et al., 1996) is compatible with such transfer and may also suggest intra- and intergeneric conjugal transfer of plasmids within subgingival plaque. Therefore, it may be possible to develop conjugative systems for gene transfer in some treponemes as an alternative to electroporative transformation. In addition, it was also demonstrated that plasmids can be transferred from *T. denticola* to *Streptococcus gordonii* in *in vitro* biofilms via transformation (Wang and Kuramitsu, 2002) further suggesting that genetic exchange may occur in dental plaque. However, transfer of plasmids into *T. denticola* was not detected but needs further evaluation.

One of the more promising approaches that could be used with transformable treponemes is to utilize these systems to characterize genes from uncultivable spirochetes. The expression of the *T. pallidum flaA* gene in *T. denticola* 33520 (Chi et al.,1999) suggests that this system could be used to complement mutants with defects in homologous genes in the latter organism. In this manner, the characterization of genes from uncultivable treponemes may be possible. However, it is not yet clear if all genes from *T. pallidum* or other uncultivable treponemes can be expressed in *T. denticola*. The relatively weak expression of the *T. pallidum* FlaA protein in *T. denticola* suggests that this may be a problem for certain genes. Furthermore, proper localization of the gene products could also be a limitation in this approach. Alternatively, as other treponeme hosts for gene cloning become available such as *T. phagedenis,* these may prove to be more suitable for specific genes. Clearly, much work still needs to be done in exploiting these gene transfer systems in this regard. Nevertheless, the availability of the *T. pallidum* genome data base together with systems to genetically manipulate cultivable treponemes suggests that this combination may prove useful in characterizing genes in the former organism. This approach may prove to be less problematic compared to examining the phenotypic expression of treponeme genes in non-treponeme hosts such as *E. coli*.

Another potential use of the treponeme gene transfer systems could be to characterize genes from the uncultivable and not yet identified treponemes associated with diseased tissue. The sequencing of 16S rRNA genes from dental plaque samples has identified over fifty different species of treponemes most of which have not yet been cultivated in the laboratory (Dewhirst et al., 2000). Since some of these organisms may turn out to be uncultivable with available growth media, the utilization of metagenomic approaches (Schloss and Handelsman, 2003) may be applicable to eventually characterize these organisms and their potential virulence genes. In this approach, large random fragments of chromosomal DNA (>50 kb) isolated directly from plaque samples could be cloned in *E. coli* using baculovirus vectors. Treponeme-specific probes (based upon the sequences of available treponeme virulence-associated genes) or functional screening approaches could then be utilized to identify clones containing DNA fragments from these organisms. Following sequencing of the fragments, individual genes of interest could then be subcloned into *T. denticola* or *T. phagedenis* for further characterization. Similar approaches could also be applied with other diseased tissue which might be suspected to involve treponeme infection (Choi et al., 1997).

Clearly, there is much potential in utilizing genetic manipulation of cultivable treponemes as a means of further defining the virulence of these organisms as well as pathogenic factors in uncultivable treponemes. Whether or not such promise is fulfilled will depend primarily upon the ingenuity of spirochete researchers.

Acknowledgments

The author wishes to thank R. Limberger for sharing unpublished information. Data generated in the author's laboratory was supported in part by National Institutes of Health grant DE09821.

References

Arakawa, S., and Kuramitsu, H.K. (1994). Cloning and sequence analysis of a chymotrypsinlike protease from *Treponema denticola*. Infect. Immun. *62*, 3424-3433.

Chan, E.C., Klitorinos, A., Gharbia, S., Caudry, S.D., Rahal, M.D., and Siboo, R. (1996). Characterization of a 4.2 kb plasmid isolated from periodontopathic spirochetes. Oral Microbiol. Immunol. *11*, 365-368.

Chauhan, S., and Kuramitsu, H.K. (2004). Sequence analysis of plasmid pTS1 from oral spirochetes. Plasmid *51*, 61-65.

Chi, B., Chauhan, S., and Kuramitsu, H. (1999). Development of a system for expressing heterologous genes in the oral spirochete *Treponema denticola* and its use in expression of the *Treponema pallidum flaA* gene. Infect. Immun. *67*, 3653-3656.

Chi, B., Limberger, R.J., and Kuramitsu, H.K. (2002). Complementation of a *Treponema denticola flgE* mutant with a novel coumermycin A1-resistant *T. denticola* shuttle vector system. Infect. Immun. *70*, 2233-2237.

Chi, B., Qi, M., and Kuramitsu, H.K. (2003). Role of dentilisin in *Treponema denticola* epithelial layer penetration. Res. in Microbiol. *154*, 637-643.

Choi, B.K., Nattermann, H., Grund, S., Haider, W., and Gobel, U.B. (1997). Spirochetes from digital dermatitis lesions in cattle are closely related to treponemes associated with human periodontitis. Int. J. Sys. Bacteriol. *47*, 175-181.

Dewhirst, F.E., Tanner, M.A., Ericson, R.E., Lau, C.N., Levanos, V.A., Boches, S.K., Galvin, J.L., and Paster, B.J. (2000). The diversity of periodontal spirochetes by 16S rRNA analysis. Oral Microbiol. Immunol. *15*, 196-202.

Ellen, R.P., Ko, K.S., Lo, C.M., Grove, D.A., and Ishihara, K. (2000). Insertional inactivation of the *prtP* gene of *Treponema denticola* confirms dentilisin's disruption of epithelial junctions. J. Mol. Microbiol. Biotechnol. *2*, 581-586.

Falkow, S. (1988). Molecular Koch's postulates applied to microbial pathogenicity. Rev. Infect. Diseases 10 (Suppl. 2), S274-S276.

Fenno, J.C., Wong, G.W., Hannam, P.M., and McBride, B.C. (1998). Mutagenesis of outer membrane virulence determinants of the oral spirochete *Treponema denticola*. FEMS Microbiol. Lett. *163*, 209-215.

Fenno, J.C., Lee, S.Y., Bayer, C.H., and Ning, Y. (2001). The *opdB* locus encodes the trypsin-like peptidase activity of *Treponema denticola*. Infect. Immun. *69*, 6193-6200.

Fenno, J.C., Tamura, M., Hannam, P.M., Wong, G.W., Chan, R.A., and McBride, B.C. (2000). Identification of a *Treponema denticola* OppA homologue that binds host proteins present in the subgingival environment. Infect. Immun. *68*, 1884-1892.

Fletcher, H.M., Schenkein, H.A., Morgan, R.M., Bailey, K.A., Berry, C.R., and Marcina, F.L. (1995). Virulence of a *Porphyromonas gingivalis* W83 mutant defective in the *prtH* gene. Infect. Immun. *63*, 1521-1528.

Fraser, C.M., Norris, S.J., Weinstock, G.M., White, O., Sutton, G.G., Dodson R. et al., (1998). Complete genome sequence of *Treponema pallidum*, the syphilis spirochete. Science *281*, 375-388.

Girons, I.S., Chi, B., and Kuramitsu, H. (2000). Development of shuttle vectors for spirochetes. J. Mol. Microbiol. Biotechnol. *2*, 443-445.

Greene, S.R., and Stamm, L.V. (1999). Molecular characterization of a chemotaxis operon in the oral spirochete, *Treponema denticola*. Gene *232*, 59-68.

Haapasalo, M., Muller, K.-H., Uitto, V.-J., Leung, W.K. and McBride, B.C. (1992). Characterization, cloning, and binding properties of the major 53-kilodalton *Treponema denticola* surface antigen. Infect. Immun. *60*, 2058-2065.

Heuner, K., Grosse, K., Schade, R., and Gobel, U.B. (2000). A flagellar gene cluster from the oral spirochete *Treponema maltophilum*. Microbiol. *146*, 497-507.

Heuner, K., Bergman, I., Heckenbach, K., and Gobel, U.B. (2001). Proteolytic activity among various oral *Treponema* species and cloning of a *prtP*-like gene from *Treponema socranskii* subsp. *socranskii*. FEMS Microbiol. Lett. *201*, 169-176.

Ishihara, K., Ishihara, M., Takazoe, I., and Okuda, K. (1992). Cloning and expression of the aspartate transcarbamylase gene from *Treponema denticola*. Appl. Environ. Microbiol. *58*, 3399-3403.

Ishihara, K., Miura, T., Kuramitsu, H.K., and Okuda, K. (1996). Characterization of the *Treponema denticola prtP* gene encoding a prolyl-phenylalanine-specific protease (dentilisin). Infect. Immun. *64*, 5178-5186.

Ishihara, K., Kuramitsu, H.K., Miura, T., and Okuda, K. (1998). Dentilisin activity affects the organization of the outer sheath of *Treponema denticola*. J. Bacteriol. *180*, 3837-3844.

Ivic, A., MacDougall, J., Russell, R.R.B., and Penn, C.W. (1991). Isolation and characterization of a plasmid from *Treponema denticola*. FEMS Microbiol. Lett. *78*, 189-194.

Izard, J., Samsonoff, W.A., and Limberger, R.J. (2001). Cytoplasmic-filament deficient mutant of *Treponema denticola* has pleiotropic defects. J. Bacteriol. *183*, 1078-1084.

Kataoka, M., Li, H., Arakawa, S., and Kuramitsu, H. (1997). Characterization of a methyl-accepting chemotaxis protein gene, *dmcA*, from the oral spirochete *Treponema denticola*. Infect. Immun. *65*, 4011-4016.

Lee, S.Y., Bian, X.L., Wong, G.W., Hannum, P.M., McBride, B.C., and Fenno, J.C. (2002). Cleavage of *Treponema denticola* PrcA polypeptide to yield protease complex-associated proteins Prca1 and Prca2 is dependent on PrtP. J. Bacteriol. *184*, 3864-3870.

Li, H., and Kuramitsu, H.K. (1996). Development of a gene transfer system in *Treponema denticola* by electroporation. Oral Microbiol. Immunol. *11*, 161-165.

Li, H., Ruby, J., Charon, N., and Kuramitsu, H. (1996). Gene inactivation in the oral spirochete *Treponema denticola*, construction of a *flgE* mutant. J. Bacteriol. *178*, 3664-3667.

Li, H., Arakawa, S., Deng, Q.-D., and Kuramitsu, H. (1999). Characterization of a novel methyl-accepting chemotaxis gene, *dmcB*, from the oral spirochete *Treponema denticola*. Infect. Immun. *67*, 694-699.

Limberger, R.J., Slivienski, L.L., Yelton, D.B., and Charon, N.W. (1992). Molecular genetic analysis of a class B periplasmic flagellum gene of *Treponema phagedenis*. J. Bacteriol. *174*, 6404-6410.

Limberger, R.J., Slivienski, L.L., Izard, J., and Samsonoff, W.A. (1999). Insertional inactivation of *Treponema denticola tap1* results in a nonmotile mutant with elongated flagellar hooks. J. Bacteriol. *181*, 3743-3750.

Lindhe, J.E., Liljenberg, B., and Listgarten, M.A. (1980). Some microbiological and histopathological features of periodontal disease in man. J. Periodontol. *51*, 264-269.

Lux, R., Sim J.-H., Tsai, J.P. and Shi, W. (2002). Construction and characterization of a *cheA* mutant of *Treponema denticola*. J. Bacteriol. *184*, 3130-3134.

MacDougall, J., Margarita, D., and Saint Girons, I. (1997). Homology of a plasmid from the spirochete *Treponema denticola* with single-stranded DNA plasmids. J. Bacteriol. *174*, 2724-2728.

Miyamoto, M., Noji, S., Kokeguchi, S., Kato, K., Kurihara, H., Murayama, Y., and Taniguchi, S. (1991). Molecular cloning and sequence analysis of antigen gene *tdpA* of *Treponema denticola*. Infect. Immun. *59*, 1941-1947.

Norgard, M.V., and Miller, J.N. (1983). Cloning and expression of *Treponema pallidum* (Nichols) antigen genes in *Escherichia coli*. Infect. Immun. *42*, 435-445.

Park, K.K., Heuner, K., Gobel, U.B., Yoo, Y.J., Kim, C.K., and Choi, B.K. (2002). Cloning and characterization of a major surface protein (MspTL) of *Treponema lecithinolyticum* associated with rapidly progressing periodontitis. FEMS Microbiol. Lett. *207*, 185-192.

Samuels, D.S., Mach, K.E., and Garon, C.F. (1994). Genetic transformation of the lyme disease agent *Borrelia burgdorferi* with coumarin-resistant *gyrB*. J. Bacteriol. *176*, 6045-6049.

Schloss, P.D., and Handelsman, J. (2003). Biotechnological prospects from metagenomics. Curr. Opin. Biotechnol. *14*, 303-310.

Sela, M. (2001). Role of *Treponema denticola* in periodontal diseases. Crit. Rev. Oral Biol. Med. *12*, 399-413.

Shevchenko, D.V., Akins, D.R., Robinson, E., Li, M., Popova, T.G., Cox, D.L., and Radolf, J.D. (1997). Molecular characterization and cellular localization of TpLRR, a processed leucine-rich repeat protein of *Treponema pallidum*, the syphilis spirochete. J. Bacteriol. *179*, 3188-3195.

Simonson, L.G., Goodman, C.H., Bial, J.J., and Morton, H.E. (1988). Quantitative relationship of *Treponema denticola* to severity of periodontal disease. Infect. Immun. *56*, 726-728.

So, M., Boyer, H.W., Betlach, M., and Falkow, S. (1976). Molecular cloning of an *Esherichia coli* plasmid determinant that encodes for the production of heat-stable enterotoxin. J. Bacteriol. *128*, 463-472.

Stamm, L.V., and Bassford, P.J. Jr. (1982). Cloning and expression of *Treponema pallidum* protein antigens in *Escherichia coli*. DNA *1*, 329-333.

ter Huurne, A.A.H., van Houten, M., Muir, S., Kusters, J.G., van der Zeijst, B., and Gaastra, W. (1992). Inactivation of a *Serpula* (*Treponema*) *hyodysenteriae* hemolysin gene by homologous recombination, importance of the hemolysin in the pathogenesis of *S. hyodysenteriae* in mice. FEMS Microbiol. Lett. *92*, 109-114.

Walfield, A.M., Hanff, P.A., and Lovett, M.A. (1982). Expression of *Treponema pallidum* antigens in *Escherichia coli*. Science *216*, 522-523.

Wang, B.Y., Chi, B., and Kuramitsu, H.K. (2002). Genetic exchange between *Treponema denticola* and *Streptococcus gordonii* in biofilms. Oral Microbiol. Immunol. *17*, 108-112.

Xu, X., and Kolodrubetz, D. (2002). Construction and analysis of hemin binding protein mutants in the oral pathogen *Treponema denticola*. Res. in Microbiol. *153*, 569-577.

9

Historical Evidence of Syphilis and Other Treponemes

By Laura J. McGough and Emily Erbelding*

Abstract

This chapter explains the methodological and theoretical issues that historians face in researching the history of syphilis and other treponemes. Differences in disease definitions make simple comparisons between past and present difficult. On the other hand, a wealth of newly tapped historical material regarding mortality, demography, and population movements makes it possible to reach some conclusions about the early history of the late fifteenth century epidemic. Wartime played a critical role in producing the syphilis, or French disease, epidemic in the Italian peninsula, which apparently was recognized as a new disease, not just a clinical misdiagnosis of leprosy. Evidence does not support the claim that the disease was unusually virulent during this period. The disease's association with sexuality, later understood as sexual transmission as contagion theory became accepted, complicated descriptions of the disease. Prejudices about the sexual behaviors of different races, ethnic groups, or social classes often contributed to the confusion between venereal syphilis and other treponemal diseases. A collaborative research project between scientists and historians focusing on an ecological history of disease could yield more definitive conclusions about the history of treponemal transmission.

1. Methodological problems in the history of diseases

During the last twenty years, it has become increasingly difficult to bridge the disciplinary divide between history and microbiology. We cannot even agree on the words themselves, as microbiologists prefer the modern term "syphilis" while many historians now opt for the term the "French disease" or the "pox" to more accurately reflect the diagnostic categories used at the time (Arrizabalaga et al, 1997). How did this difference in disciplinary approaches develop? Historians were once comfortable writing the history of diseases according to their modern

*For correspondence email mcgoughlj@msn.com

names. Current aversion to using modern diagnostic categories arises from historians' deep respect for the changes brought about by the laboratory revolution of the early twentieth century (Cunningham, 2002), in addition to an appreciation for how scientific knowledge is always partially shaped by the intellectual worlds and institutional cultures in which scientific knowledge is produced (Rosenberg, 1992).

During the fifteenth century, diseases were identified according to a set of symptoms, not an underlying, specific causative agent. The way physicians and patients understand disease before and after the laboratory revolution is fundamentally different: it therefore makes little sense to trace the history of a category that did not exist, like trying to trace the history of "teenage angst" or the "mid-life crisis" before these concepts existed. To trace these concepts backward in time would be to create and invent their histories, not uncover them. Furthermore, to try to understand past therapeutic practices, patients' experiences of illness, or the physician-patient relationship without understanding how different diseases were defined and conceptualized is a nearly impossible task. Because historians rely on written records, language is central to our analysis. Paradoxically, as historians have grown to appreciate the impact of the laboratory revolution on the history of medicine, it has become more difficult for us to communicate with scientists interested in the history of specific, modern diseases, which are built on different definitions of what a disease is (Wilson, 2000). What follows is a modest attempt to bridge this divide: to explain to an audience of scientists how historians evaluate evidence about syphilis, or the French disease, and what conclusions we can reach.

Furthermore, this chapter will also try to explain some of the recent developments in the field of history, as a new generation of historians has tackled previously unexamined archival documents and primary sources. Broad surveys of the history of syphilis, such as Quetel's excellent study, are still useful for their overall picture, but need to be supplemented and revised in light of new research (Quetel, 1990). This chapter suggests that a collaborative approach between scientists and historians towards an ecological history of disease could yield more definitive results than work in isolation, with historians providing crucial information about large-scale population migration and human behavior and with scientists providing information about specific pathogens. Although it is still difficult to pinpoint the origins of modern venereal syphilis, a social, ecological approach to the history of disease can nonetheless provides clues about the combination of human and biological conditions that lead to disease epidemics.

2. The first European syphilis epidemic and disease origins

The first European "syphilis" - that is, French disease - epidemic occurred in the context of wartime, during the French invasion of Italy in 1494 and the Spanish defense of Naples. Although the French army retreated, it did not do so quickly. At the Battle of Fornovo in 1495, the French army faced the combined forces of the Holy Roman Empire, the Spanish, several Italian city-states, including the papacy and the republics of Milan and Venice, with their mercenary soldiers hired from throughout Europe. The other major European powers remained to vie for control of the Italian peninsula. The French returned in 1508 in a war of constantly shifting alliances between the major powers of Europe and the Italian city-states, which shifted alliances among themselves. Meanwhile, these armies continued to rely on mercenary soldiers (Mallet, 1974). The Italian Wars' long duration (until 1530) meant that foreign soldiers remained on Italian soil long enough to form short- and long-term sexual relationships with local women, not to mention war's usual impact on rape and prostitution. Venetian soldiers, who often came from the Holy Roman Empire, were billeted in towns and intermarried with the local population (Mallet, 1984). Furthermore, the disruptions of war brought famine and epidemics of a variety of diseases, including numerous "fevers," and, of course, "syphilis," more commonly referred to as the French disease during the period. It is important to underscore the significance of this war, which included the violent sacking of the city of Rome in 1527, as the context in which this epidemic took place. Too much of the literature on the historical origins of syphilis have described only the initial battle, not the subsequent war. The creation of new sexual networks (through consensual or forced, temporary or long-term sexual relations) as a result of the war and occupation helps explain why an epidemic emerged on Italian soil, in addition to the fact (as explained below) that the Italian city-states were the only part of Europe that possessed a rudimentary disease surveillance system. In order to trace the ultimate origins of disease, it is useful to start with the epidemic itself and consider various hypotheses about why an epidemic occurred at this time on the Italian peninsula.

One hypothesis is that the syphilis epidemic may have occurred during the 1490s due to changes in sexual behavior or hygiene independent of the impact of war (Meyer et al. 2002). Archival evidence does not support the claim that the 1490s were a particular time of change in sexual behavior independent of the impact of the war, nor a time of widespread adoption of urban hygiene measures. Depending on the city-state, efforts to prevent plague such as quarantine had been implemented periodically since the mid-fifteenth century and continued to be implemented throughout the sixteenth century (Palmer, 1978). Widespread changes in sexual behavior usually occur slowly. There is no evidence of a noticeable shift in behavior during the 1490s, at least from what can be learned from a study of Venice's criminal records of fornication, rape, sodomy (defined as oral and anal sex), and adultery (Ruggiero, 1985).

Generally speaking, continuity in the numbers and types of crime suggests, but does not prove, an overall continuity in behavior. Changes in the numbers and types of crimes recorded do not necessarily suggest broad changes in behavior, however. A new government may decide to prosecute sexual crimes more vigorously in response to political pressure, religious revival, or some other cause. Such was the case in Florence, where the Dominican friar Girolamo Savonarola enjoyed a brief period of enormous political influence following the French invasion. For Savonarola, the French invasion was evidence that his campaign against sin and immorality was well-founded: God had sent an avenger in the form of the French army to persuade Florentines to abandon their "immoral ways," which included card-playing and reading books written by "pagan authors," such as the Greek and Roman classics. For approximately two years, Florentines agreed and joined in the repression of vice, supplying books to be burned in massive bonfires (Weinstein,

1970). Public attention focused on sexual crimes as well, including prostitution and sodomy, but it is not possible to assert that these behaviors had increased simply because Florentines had become more preoccupied with them. The political situation had clearly changed; sexual behavior probably had not.

A second hypothesis is that treponemal disease already existed in Europe, either in the form of yaws, which evolved into venereal syphilis, or was already present in northern Europe and then spread to the rest of Europe during the Italian Wars. In 2000, paleopathologists reported signs of syphilis on skeletons in Hull, England; the skeletons apparently dated from between 1300 and 1450, although the dating is inexact. But this evidence is also controversial in that it is difficult to prove definitively that syphilis, rather than any other treponemal disease, caused the bone deformations. Perhaps recent work in genome sequencing will be able to clarify which treponemal disease is the causative agent (Zimmer, 2001). Although this work has not yet proved the existence of modern venereal syphilis in Europe prior to the epidemic during the Italian Wars, it nonetheless remains a potentially viable hypothesis.

A third hypothesis, and probably most familiar to readers, is that syphilis existed in the New World and was carried to Naples via Columbus. This argument has enjoyed favor for several centuries at least partially because of the enormous symbolic weight that the New World presented to Europeans. Often represented in female form as an object of potential military and sexual conquest, the New World became associated with "unbridled" female sexuality and, by extension, with disease (Montrose, 1991). It is important always to analyze stories of disease origins within the cultural, political, and social context of the time, since disease origin stories are usually trying to explain more than just the biological origins of disease. Origin stories also attempt to explain why a particular region suffers from disease and draw on wider cultural perceptions of scapegoats and blame (Foa, 1990).

Paleopathologists have turned to archaeological evidence in order to investigate disease origins in an attempt to find evidence that is less influenced by changing cultural patterns. Bone evidence from archaeological sites—although less influenced by culture than written evidence—nonetheless requires careful interpretation. Scientists have used bone evidence from modern syphilis patients in order to identify evidence of disease in centuries-old bones. Scientists are therefore assuming that modern syphilis, named the "Great Imitator" due to its highly variable clinical presentations in the modern era, exhibited the same pattern of bony destruction many centuries ago that it does today: the assumption may or may not be true. With this important limitation, however, it is worth presenting the archaeological evidence

The hypothesis that syphilis originated in the New World and traveled to the Old World is based on the following: New World bone evidence shows signs of damage similar to that caused by modern-day congenital or late syphilis, while no bone evidence from the Old World prior to the late fifteenth century shows similar signs of damage. To make this argument, it is necessary to distinguish the impact of syphilis on skeletons from the impact of the other treponemal diseases of pinta, yaws, and endemic (nonvenereal) syphilis. Congenital infection is most likely to occur in the setting of venereal syphilis, not nonvenereal treponematoses, if the disease manifestations of centuries ago were similar to modern-day infections. In the situation of venereal transmission, however, a woman would be most likely to acquire treponemal infection early in her reproductive life, when she was also most likely to become pregnant. In contrast, skeletal remains that do not show specific features of congenital syphilis, but show only general inflammatory changes of treponemal infection may indicate transmission due to exposures earlier in childhood (nonvenereal transmission).

Some changes in the skeletal structure or in permanent teeth, when available, are thought to be very specific for the pathologic effects of syphilis on the developing skeleton early in life, suggestive of treponemal infection acquired at birth. These specific pathologic features include bowing of the tibia (saber shins), naso-palatine destruction, characteristic morphologic abnormalities of the permanent teeth, especially the central incisors (Hutchinson's teeth) and 6-year molars (mulberry molars) that have not been described in other processes. Remains excavated from New World populations show some of these specific changes, as well as a high prevalence of other changes consistent with treponematosis. The paleopathologic record arising from the Old World populations whose existence dated prior to the late fifteenth-century does not record the types of changes suggestive of congenital syphilis; the bony abnormalities of inflammation that exist on these remains may be attributable to any number of pathogens acquired later in life, including endemic treponemotosis, leprosy, and other bacterial pathogens (Baker and Armelagos, 1988).

The New World origins of syphilis hypothesis has usually included the suggestion that the disease was brought back by Columbus's sailors. At least one team of researchers has claimed to find skeletal evidence of syphilis at the exact place where Columbus's crew landed, the present-day Dominican Republic (Rothschild et al., 2000). Although many of Columbus's sailors did become sick and some researchers have argued that the illness must have been syphilis based on symptoms described as "buboes," Spanish physicians did not refer to a new disease, or mention the pox or the French disease by name, until the early sixteenth century, after the epidemic had broken out on the Italian peninsula (Arrizabalaga at al., 1997). It is not clear whether Columbus's ships transported enough infected sailors, who in turn defended Naples from the French invasion, to launch an epidemic.

Furthermore, being able to identify diseases from bone evidence is only part of the problem in determining the geographical origins of syphilis. Finding archaeological evidence of syphilis in England or northern Europe (assuming that it can be verified through genome sequencing) prior to 1494 does not mean that syphilis was widespread through the entire European continent at the same time. Research needs to be much more geographically focused and make greater use of historical records of large-scale population movements. The problem is that there has been less archaeological work done on bone evidence from the Italian peninsula prior to the French invasion, although some evidence (a single skeleton) has been found from the sixteenth century (Fornaciari, 1989).

The ultimate geographical origins of the late fifteenth-century syphilis epidemic, from the New World via Columbus and his sailors, or from northern Europe, are difficult to ascertain from present evidence, but we can establish that the disease was probably new to a very localized region, specifically where the first reports of an epidemic originated. Mortality records, which had been carefully maintained in Milan since 1452 to track plague epidemics, do provide evidence that the disease was new to the Italian peninsula and constituted an "epidemic." Deaths attributed to *morbus gallicus* occur after 1503: this disease was a new diagnostic category. It is not possible to determine the date of the first death attributed to *morbus gallicus,* since the records between 1495 and 1503 are missing. Furthermore, because diagnosis of *morbus gallicus* was dependent on external symptoms such as ulcers on the genitals and skin rather than on serological evidence, these mortality records do not provide reliable indications of overall morbidity or mortality. The data do show, however, that Milanese physicians had diagnosed leprosy as a cause of death since 1452, and that *morbus gallicus* represented a new disease category since 1503 (Carmichael, 1991). Therefore, the syphilis epidemic was not merely an artifact of reporting errors based on inability to distinguish between the two diseases, as has been suggested (Baker and Armelagos, 1988). The Italian city-states were the first to develop permanent public health offices and track causes of death. Consequently, it is not possible to evaluate records from England or any other part of Europe during this period to evaluate whether the disease was new to that region. In northern and central Italy, the disease was new.

Although a lot of attention has been focused on the geographical origins of disease, it is by no means the only, or even the most, interesting question to ask about this period. What combination of social, political, and biological conditions fostered the widespread transmission of disease? How did the disease come to be associated with sexual transmission and what were the implications of that association? For historians, the French disease has become an excellent case study of how physicians and the wider public undergo a process of intellectually and culturally assimilating a new disease. The newness of the disease was disorienting. The psychological difficulties of coping with an unknown disease led some writers, medical and non-medical, to draw parallels between syphilis and diseases already known to Europeans, such as leprosy. In art as well as literature, Europeans drew on the rich symbolic legacy of leprosy, a frightening but familiar disease, in their representations of syphilis (Foa, 1990). Symbolic association of syphilis and leprosy was not the same as clinical misdiagnosis, however, although occasional misdiagnosis may have occurred (Carmichael, 1991).

The new disease was typically diagnosed as the "French disease," often with the Latin name "*morbus gallicus*" rather than the term syphilis. Although the word syphilis was first used in 1530 by Girolamo Fracastoro in his epic poem, the word syphilis was seldom used clinically. Even the name "the French disease" perfectly illustrates the problems presented in interpreting the historical evidence. Named after the invader, the French disease evoked a dual set of intertwined reactions: to wartime and ultimate military defeat, as well as to the disease itself. As already mentioned, the 1490s were a period of intense religious fervor, as popular preachers such as Girolamo Savonarola anticipated the second coming of Christ in the year 1500 and God's punishment of sin. It was easy to see the French army, and subsequent disease, famine, and destruction, as divine punishment. With skin lesions figuring prominently as one of the symptoms of the new French disease, the public was quick to see parallels with the afflictions of the Biblical Job, whose faith was tested by God with boils on his skin. Although the disease was not consistently associated with sexual activity until 1500 (and linked to sexual transmission until 1530), it could still represent God's punishment, much as Job, an innocent man, was tested by God (Arrizabalaga et al., 1997).

The view that syphilis was unusually virulent during its first one or two decades in Europe is not well supported by the evidence. Medical and lay writers alike provide conflicting reports, some arguing that the disease was seldom fatal and others insisting that many had died (Arrizabalaga et al., 1997). The early authors were not trying to write neutral scientific accounts of a disease, but to draw attention to, or alternatively maintain calm in the face of, yet another problem during a time of crisis. It is difficult to know whether reports of high mortality were influenced by co-morbidities, or by a desire to bring public attention to a new problem, or by simple panic.

By the mid-sixteenth century, the French disease had become endemic throughout Europe. Precise estimates of prevalence are not possible from the historical evidence. Nonetheless, substantial evidence supports the theory that the disease was widespread, based on three different kinds of records. First, the thriving medical marketplace in cures, which targeted both genders and every social class and income level, suggests that the disease was common enough to support the sale of books describing cures and a wide variety of medical practitioners, trained and untrained. Books, advertisements, and applications for licenses to sell medications all provide evidence of the medical marketplace which specialized in pox remedies. Second, institutional records of poorhouses and hospitals, which were overwhelmed by syphilis patients, suggest that the disease was fairly common. For example, approximately 20 percent of all hospital patients in London during the seventeenth century were being treated for syphilis (Siena, 2004). Finally mortality records, although they underestimate prevalence, show a widespread distribution of the disease across neighborhoods, occupational groups, and a roughly even distribution between men and women.

3. Sexual transmission and problems of diagnosis: A continuing historical theme

Despite evidence that the disease was widespread, both lay and medical writers associated the disease with groups of suspect morality. It is important to distinguish between two concepts: the association of the French disease with sexuality, and the idea of sexual transmission of a communicable agent. The former idea does not necessarily involve the idea of contagion. Based in Galenic theory, the idea that imbalances in a body's humors produced disease could explain why some people who engaged in frequent sexual intercourse could experience a disease without having acquired that disease from someone else. Leprosy, for example, was often associated with sexuality, but not necessarily with sexual transmission (Allen, 2000). By the mid-sixteenth century, however, medical opinion coalesced around the idea that the French disease was primarily, but not exclusively, transmitted through sexual contact. The idea of sexual transmission depended on widespread acceptance of the idea of contagion, an idea propagated (but not invented) by Girolamo Fracastoro during the 1540s (Nutton, 1983). The theory of contagion differed significantly from nineteenth-century germ theory, in that "seeds" of disease were considered transmissible by air, contact with clothing or other objects, and so forth. Furthermore, this theory of contagion coexisted other theories about how the climate and bad air, or miasmas, influenced disease, until the late nineteenth century, when modern germ theory became influential.

By 1530, university-educated physicians had reached a consensus (with occasional dissent) that the French disease spread primarily through sexual contact (Temkin, 1977). In his book on the French disease, which was first published in 1527, the physician Niccolò Massa (d. 1569) thought it was necessary to remind readers that French disease patients did not acquire the disease through sexual contact alone. Although most people contracted the disease through sexual contact, he explained, a few patients he had treated that year had not, among them three children. The oldest child, at age 11, had surpassed the age where he could have acquired the disease from breast milk, according to Massa; it was possible, he argued, to acquire the disease simply by sharing a bed with a person who had open sores (Massa, 1566). Massa did not criticize the general understanding that the French disease was primarily acquired through coitus, but instead simply reminded other physicians not to assume that all French disease patients were sexually experienced. It is possible, however, that Massa was trying to protect patients from the stigma of a sexually acquired disease.

Because Renaissance physicians regarded sexual activity as the mode of transmission of the French disease in the overwhelming majority of cases, their perception of the disease itself was closely linked to their attitudes towards sexuality. Renaissance physicians' advice on prevention invoked a double standard of morality that found ways for single men to enjoy sex and avoid infection, while holding women responsible for infection. Gabriele Falloppio's lectures on the French disease instructed young medical students to clean their genitals after intercourse with a prostitute. "I would seem to have achieved nothing unless I taught you how someone who sees a most beautiful 'siren' and sleeps with her although she is infected can avoid the French disease" (quoted by Schleiner, 1994). Italian writers consistently applied a sexual double standard, which allowed sexual freedom to men but demanded repentance and moral reform of the women who served these men. By contrast, English Protestant writers of the same period condemned the practice of prostitution itself. William Clowes (the Elder, d. 1604), for example, refrained from offering advice on how to wash the genitals to avoid infection, because he wanted men to avoid the sin that led them to infection. Clowes' idea of prevention was moral reform (Schleiner, 1994).

Whether known as the pox, the French or Neapolitan disease, or syphilis, the disease had acquired a sense of stigma and became a widespread symbol of moral corruption in art and literature (Qualtiere and Slights, 2003). During the eighteenth century, for example, the nobility and the clergy came under scrutiny as privileged, debauched persons who carried syphilis. Voltaire's popular satirical novel *Candide* traced the genealogy of syphilis through homosexual relations between a Jesuit priest and one of Christopher Columbus' shipmates and cross-class relations between a noblewoman and her page. Voltaire thereby used syphilis as a literary weapon to discredit the clergy as sodomites and noblewomen as sexually voracious with unnatural appetites. Pornography was a popular weapon in the political warfare of eighteenth-century Europe, making it crucially important to distinguish satirical representations of syphilis and sexuality from actual behavior (Hunt 1993; Maza 1993). Shame and stigma made syphilis a powerful metaphor in society and sometimes confounded physicians' ability to diagnose the disease.

During the nineteenth century, for example, suspicion settled firmly on the lower classes, who began to crowd Europe's new industrial cities, and also people of different races, especially Africans (Engelstein, 1986). The idea that the lower-classes and Africans were violent, hyper-sexual people provided justification for upper-class rule at home and European rule abroad. This racist and upper-class ideology undermined physicians' ability to diagnose disease accurately. Convinced that workers and Africans were sexually profligate, they mistook yaws and endemic syphilis for venereal syphilis (Setel et al., 1999; Jochelson, 2003). Even when a few public health officials as individuals were relatively free of prejudice, they often operated in political environments and under public health laws that were discriminatory. In colonial Rhodesia, for example, Andrew Fleming, head of the public health office, tried to convince the white public that rates of syphilis and gonorrhea were higher among British troops than among African troops. Nonetheless, the Public Health Act of 1924 was introduced, which allowed colonial officers to examine forcibly any African, disinfect their clothing, and even destroy their homes (McCulloch, 1999). In imperial Russia, public health officials may have contributed to the spread of disease by forcibly examining groups of prostitutes and using the same speculum for examination (Bernstein, 1995).

Children presented a vexing problem in tracing the source of venereal disease. When confronted with widespread evidence of gonorrhea in 5 to 9 year-old white, middle-class American girls during the early twentieth century, physicians were reluctant to acknowledge or even investigate the role of incest. Unable to believe that middle-class fathers were capable of molesting their own daughters, physicians revised their description of the etiology of childhood venereal diseases and concluded that toilet seats were the culprit (Sacco, 2002).

4. The hidden epidemic

As the title, *The Hidden Epidemic*, of the 1997 Institute of Medicine's report about current sexually transmitted disease morbidity suggests, syphilis and other STDs represent a "hidden epidemic," largely because the American public is uncomfortable with diseases that are sexually transmitted. This statement about the present could apply equally well to many past societies and profoundly different cultures, in Europe, Africa, and the United States. Widespread population-based surveillance did not occur until the twentieth century when the possibility of serological testing (via the Wassermann test, available in 1906), a reasonable therapy (salvarsan), and the motive for testing coincided with the outbreak of the First World War. Syphilis and gonorrhea accounted for a significant loss of military manpower, thereby making these diseases an issue of urgent public concern. Hospital admissions for military personnel in 1910 had reached the rate of 196.99 per 1000, making the surgeon general worried about the effect of wartime increase in VD rates on American military capacity (Brandt, 1987).

Selective Service records for World War II provide some indication of how widespread syphilis was prior to the availability of penicillin. In 1943, Otis

Anderson, Senior Surgeon of the Division of Venereal Diseases, reported on the prevalence of syphilis among the first million selectees and volunteers given routine serologic blood tests. The overall rate of seroreactivity for the United States was 45.3 cases per 1,000, with substantial regional and racial variations. In Savannah, Georgia, for example, the rate was 35.4 per 1,000 among whites and 430.5 per 1,000 among blacks (records of the U.S. Public Health Service). After the war, however, these numbers were contested on the grounds that the Wassermann test was not sufficiently specific and was recording a number of false-positives, perhaps as many as half of all those who tested positive in low-prevalence populations (Löwy, 1993). Nonetheless, even if the numbers are halved, the evidence suggests that the disease was relatively common before the massive scale-up of public health efforts to control venereal diseases. During the 1940s, public health efforts included education, screening, contact tracing, the opening of residential Rapid Treatment Centers for civilians where counseling and job training were offered, in addition to the widespread availability of penicillin to civilians after 1945. This comprehensive approach to venereal disease control was continued until 1953 in the United States, when the Rapid Treatment Centers offering free treatment and counseling were closed (National Archives, Southeast Region, East Point, GA).

5. Conclusion: Suggestions for collaborating across disciplines

It would be useful for historians, physical anthropologists, and microbiologists to work together to formulate methods for studying the dynamic relationship between syphilis and human populations—towards an ecological study of disease. Scientists can provide concrete information about pathogens, disease etiology, and biological conditions that facilitate disease transmission, while historians can provide evidence about the behavior and activities of the host population. Scientists might be surprised to learn how rich the records of past human activities are, even for the fifteenth and sixteenth centuries. For example, demographic studies of fifteenth-century Florence are possible based on a tax assessment carried out by the Florentine government in 1427-29 (Herlihy and Klapish-Zuber, 1985); it is also possible to learn about the household structure and number, gender, and ages of the majority of the residents of Venice from sixteenth-century surveys (Chojnacka, 2000; 2001). Particularly crucial is evidence of large-scale population mobility or displacement, since infectious disease outbreaks can often be traced to human movements. Historians cannot, however, identify pathogens with any precision from the historical records. The expertise of scientists is required for the identification of pathogens. It will be useful if current tools (such as nucleic acid amplification techniques) can be applied to the paleopathologic record to further our understanding of Old and New World treponemal transmission. Scientists can also contribute information about the underlying biological conditions that facilitate or retard the spread of specific diseases.

To facilitate a collaboration between historians and scientists, however, it is also important to explore a few methodological and theoretical points of potential conflict. First, historians avoid trying to diagnose famous individuals from history based on a description of their symptoms by contemporaries. The potential for errors in "retrospective diagnosis" is enormous. Gonorrhea and syphilis were not even recognized as distinct diseases prior to the nineteenth century. More importantly, descriptions of diseases are shaped by cultural expectations and symbolic associations that change over time, so the disease could be a wide range of possibilities. Even if a methodology for diagnosing disease at the individual level could be developed, historians would still insist on placing that individual, and his or her disease-experience, in a wider context. An individual's behavior cannot be attributed to the presence of disease alone, but must be considered in conjunction with a wide variety of other influences.

Second, historians are primarily interested in the impact of the pathogen and its resulting epidemic on the course of human events and behavior. The interaction between human beings and pathogens encompasses a broad variety of topics: scientists' efforts to understand disease, demographic and economic impact of disease, public health efforts to prevent and control disease outbreaks, and cultural and artistic responses to disease, to name just a few.

Finally, historians regard scientific inquiry itself as partially a product of its historical context. The way that scientific research is organized, the questions that are regarded as most salient, the metaphors that scientists often use to describe or understand what they are studying, the underlying concepts in defining particular diseases: all these things are influenced by the particular historical context in which scientific research is conducted. To argue that science is influenced by its context is not to deny the claim that scientific inquiry produces knowledge. On the contrary, as Thomas Kuhn's work has shown in the case of astronomy, scientific knowledge is produced within different paradigms (Kuhn, 1962). But it may be more useful to acknowledge, rather than ignore, the role that cultural assumptions play in medical and scientific inquiry. If the assumptions are stated, they can be investigated and studied. With an acknowledgement of the contributions of each discipline, a collaborative effort between scientists and historians could be fruitful and yield information that is more definitive than speculative about the dynamic relationship between human beings and treponemal diseases.

References

Allen, P.L. (2000). The Wages of Sin: Sex and Disease, Past and Present. (Chicago: University of Chicago Press).

Arrizabalaga, J., Henderson, J., and French, R. (1997). The Great Pox: The French Disease in Renaissance Europe. (New Haven and London: Yale University Press).

Baker, B.J., and Armelagos, G.J. (1988). The Origin and antiquity of syphilis: Paleopathological diagnosis and interpretation. Current Anthropology *29*, 703-737.

Bernstein, L. (1995). Sonia's Daughters: Prostitutes and their regulation in Imperial Russia. (Berkeley: University of California Press).

Brandt, A.M. (1987). No Magic Bullet: A Social History of Venereal Disease in the United States since 1880. (New York, Oxford: Oxford University Press).

Carmichael, A. (1991). Syphilis and the Columbian Exchange: Was the Disease Really New? In The Great Maritime Discoveries and World Health, M.G. Marques and J. Cule, eds. (Lisbon: Ecola Nacional de Saúde Pública)), pp. 187-200.

Chojnacka, M. (2000). Women, men, and residential patterns in early modern Venice. J.Family History *25*, 6-25.

Chojnacka, M. (2001). Working Women of Early Modern Venice. (Baltimore and London: Johns Hopkins University Press).

Cunningham, A. (2002). Transforming Plague: The laboratory and the identity of infectious disease. In The laboratory revolution in medicine, A. Cunningham and P. Williams, eds. (Cambridge, UK: Cambridge University Press), pp. 209-44.

Engelstein, L. (1986). Syphilis, actual and historical: cultural geography of a disease. Rev. Infectious Diseases. *8*, 1036-1048.

Foa, A. (1990). The New and the Old: The Spread of Syphilis (1494-1530). In Sex and Gender in Historical Perspective, E. Muir and G. Ruggiero, eds. (Baltimore: Johns Hopkins University Press).

Fornaciari, G., Castagna, M., Tognetti, A., Tornaboni, D., and Bruno, J. (1989). Syphilis in Italian renaissance mummy. The Lancet *334*, 614.

Herlihy, D. and Klapisch-Zuber, C. (1985). Tuscans and their Families: A Study of the Florentine Catasto of 1427. (New Haven: Yale University Press).

Hunt, L., ed. (1993). The Invention of Pornography, 1500-1800. (Boston: Zone Books).

Institute of Medicine. (1997). The Hidden Epidemic: Confronting Sexually Transmitted Diseases. (The National Academies Press: Washington, DC).

Kuhn, T. (1962). The Structure of Scientific Revolutions. (Chicago: University of Chicago Press).

Löwy, I. (1993). Testing for a sexually transmissible disease, 1907-1970: the history of the Wassermann reaction. In AIDS and contemporary history, V. Berridge and P. Strong, eds. (New York, Cambridge, UK: Cambridge University Press), pp. 74-92.

Mallett, M. (1974). Mercenaries and their Masters: Warfare in Renaissance Italy. (Totoaw, NJ: Rowman and Littlefield).

Mallett, M. (1984). The Military Organization of a Renaissance State: Venice c. 1400 to 1617. (Cambridge, UK: Cambridge University Press).

Massa, N. (1566). Il libro del mal francese. (Venice: G. Zilletti).

Maza, S. (1993). Private Lives and Public Affairs: The Causes Célèbres of Prerevolutionary France. (Berkeley: University of California Press).

Montrose, L. (1991). The work of gender in the discourse of discovery. Representations *33*, Special Issue, 1-41.

McCulloch, J. (1999). The Management of venereal disease in a settler society: Colonial Zimbabwe, 1900-30. In Histories of Sexually Transmitted Diseases and HIV/AIDS in Sub-Saharan Africa, P.W. Setel, M. Lewis, and M. Lyons, eds. (Westport, CT: Greenwood Press), pp. 195-216.

Nutton, V. (1983). The seeds of disease: An explanation of contagion and infection from the Greeks to the renaissance. Medical History *27*, 1-34.

Palmer, R. (1978). The Control of Plague in Venice and Northern Italy, 1348-1600. Ph.D. Thesis, University of Kent.

Qualtiere, L.F. and Sights, W.W.E. (2003) Contagion and blame in early modern England: The Case of the French Pox. Literature and Medicine *22*, 1-23.

Quetel, C. (1990; original French edition, 1986). History of Syphilis, J. Braddock and B. Pike, trans. (Baltimore: Johns Hopkins University Press).

Rosenberg, C.E. (1992). Framing disease: Illness, society, and history. In Framing Disease: Studies in Cultural History, C.E. Rosenberg and J. Golden, eds.. (New Brunswick, NJ: Rutgers University Press).

Rothschild, B.M. and Heathcote, G.M. (1993). Characterization of the skeletal manifestations of the treponemal disease yaws as a population phenomenon. Clinical Infectious Diseases 17, 198-203.

Rothschild, B.M., Calderon, F. L., Coppa, A., and Rothschild, C. (2000). First European exposure to syphilis: The Dominican Republic at the time of columbian contact. Clin. Infect. Dis. *31*, 936-41.

Ruggiero, G. (1985). The Boundaries of Eros: Sex Crime and Sexuality in Renaissance Venice. (New York, Oxford: Oxford University Press.)

Sacco, L. (2002). Sanitized for your protection: Medical discourse and the denial of incest in the United States, 1890-1940. J. Women's History *14*, 80-104.

Schleiner, W. (1994). Moral attitudes towards syphilis and its prevention in the renaissance. Bulletin of the History of Medicine *68*, 389-410.

Siena, Kevin. (2004). Venereal Disease, Hospitals, and the Urban Poor: London's "Foul Wards," 1600-1800. (Rochester: University of Rochester Press).

Weinstein, D. (1970). Savonarola and Florence: Prophecy and Patriotism in the Renaissance. (Princeton, NJ: Princeton University Press).

Wilson, A. (2002). On the history of disease-concepts: The case of pleurisy. Hist. Sci., 38, 271-319.

Zimmer, C. (2001). "Can genes solve the syphilis mystery?" Science *292*, 1091.

10

Pathogenesis of Syphilis

Justin D. Radolf, Karsten R. O. Hazlett, and Sheila A. Lukehart*

Abstract

Venereal syphilis is caused by *Treponema pallidum,* a noncultivatable, microaerophilic spirochete and obligate human pathogen. *T. pallidum* utilizes glycolysis for energy production, has extremely limited biosynthetic capacity, and possesses limited tolerance for environmental stress. Transmission of *T. pallidum* occurs in a high percentage of individuals exposed to primary or secondary syphilitic lesions. *T. pallidum* is extremely invasive. Invasion at the local site occurs shortly after inoculation, while hematogenous dissemination occurs well before the appearance of the primary lesion (chancre). Penetration of the blood-brain barrier by *T. pallidum* is common during early syphilis and, in untreated individuals, sets the stage for the subsequent development of neurological complications (neurosyphilis). Clinical manifestations result from the inflammatory responses elicited by spirochetes and spirochetal constituents (e.g., lipoproteins) locally and systemically. Clearance of spirochetes from sites of infection is accomplished by activated macrophages most likely in concert with opsonic antibodies. The limited surface antigenicity of *T. pallidum* is thought to be a major factor in the spirochete's ability to establish persistent infection; selection of *tprK* sequence variants also may contribute to immune evasion. As with other spirochetal pathogens, motility is thought to be critical for treponemal invasiveness. The lack of orthologs for well characterized virulence determinants in the *T. pallidum* genome emphasizes the uniqueness of the bacterium's parasitic strategies. Genetic approaches for identifying treponemal virulence determinants are beginning to emerge.

1. Introduction

Venereal syphilis, an infectious disease which can involve virtually any organ system and structure within the human body, has long been recognized for the dilemmas it poses in everyday clinical practice. In the AIDS era, these challenges have intensified as clinicians grapple with the impact of HIV co-infection on the

*For correspondence email JRadolf@up.uchc.edu

already complex clinical course of syphilis and its responsiveness to established therapeutic regimens. Researchers studying syphilis pathogenesis face a challenge that is at least as daunting, namely, to explain how a fragile bacterium with a minimalist genome and miniscule metabolic capabilities (Fraser et al., 1998; Norris et al., 2001) can be so destructive and persistent within its obligate human host. To further complicate their task, basic syphilologists must contend with the fact that *Treponema pallidum* subspecies *pallidum* (*T. pallidum*), the etiologic agent of venereal syphilis, is a most recalcitrant bacterium to study. The syphilis spirochete has defied all attempts at continuous cultivation in artificial medium (see Chapter 4, Cox and Radolf; Cox, 1994; Norris et al., 2001), and, therefore, cannot be manipulated genetically using *in vitro* selection techniques. *T. pallidum* generally does not cause productive infection in mice (Gueft and Rosahn, 1948; Turner and Hollander, 1957; Willcox and Guthe, 1966), confounding efforts to address fundamental immunological issues that have long been resolved with other pathogens. Nevertheless, syphilis researchers over the years have shown impressive resourcefulness in circumventing these obstacles. Most recently, molecular biology and genomics have added versatile and powerful tools to this armamentarium (Fraser et al., 1998; Norris et al., 1998). Thus, while many aspects of syphilitic infection remain beyond our understanding, much has been learned in the one hundred years since the discovery of *Spirochaeta pallidum* by Schaudinn and Hoffman (1905). Our current understanding of syphilis pathogenesis is an amalgam of observations, some dating back to the early twentieth century, made from clinical and epidemiological investigations, autopsy studies, studies with experimental animal models, and *in vitro* investigations. This review is an attempt to summarize and integrate this eclectic body of information, providing a useful conceptual framework for a new generation of molecularly-oriented syphilologists who will enlist their talents in the campaign to eliminate a disease that has afflicted humans for centuries (see Chapter 9, McGough and Erbelding). It is divided into three sections. The first describes the pathogenic treponemes, among which is the agent of venereal syphilis. The second outlines the natural history of syphilis, juxtaposing salient microbiological and immunological events and processes with the clinical course of the disease. The final section builds upon the rich history of experimental syphilis research to present a genomics-era perspective of the syphilis spirochete's virulence properties and determinants.

2. Etiologic agent

Four members of the genus *Treponema* cause the infectious disorders collectively referred to as the "treponematoses": venereal syphilis, endemic syphilis, yaws, and pinta. Syphilis is the only one of the four that is routinely transmitted by sexual contact and by vertical transmission from a pregnant woman to fetus *in utero* (Antal et al., 2002). Based on the high degree of cross-hybridization of two genomes (Miao and Fieldsteel, 1980), the agents of venereal syphilis, endemic syphilis, and yaws, originally considered separate species, were re-classified in the 1980s as subspecies of *Treponema pallidum* (subsp. *pallidum, endemicum*, and *pertenue*, respectively). *T. carateum*, which has not been reproducibly passed in laboratory animals, maintained its status as a distinct treponemal species. The discovery of genetic differences among treponemal strains and subspecies, while often minor (i.e., point mutations) and not clearly segregating along subspecies lines (Noordhoek et al., 1990; Centurion-Lara et al., 1998; Stamm et al., 1998; Cameron et al., 1999; Centurion-Lara et al., 2000a; 2000b), has helped to put to rest the idea, championed initially by Hollander and Turner (1954) and subsequently by Hudson (1965), that the treponematoses are caused by a single pathogenic strain and that differences in disease expression are due to environmental factors, primarily temperature. On the other hand, the genetic similarities of these organisms raise an interesting conundrum, implying that variation in a small proportion of the treponemal genome gives rise to the discordances in tissue tropisms, modes of transmission, levels of virulence and host range in experimental animals displayed by the *T. pallidum* subspecies and *T. carateum* (Hollander and Turner, 1954; Turner and Hollander, 1957; Antal et al., 2002).

All subspecies of *T. pallidum* are obligate mammalian parasites. The syphilis spirochete has resisted numerous attempts at continuous cultivation on artificial medium (see Chapter 4, Cox and Radolf) and, for this reason, must be passaged in animals, usually by serial intratesticular inoculation of rabbits. Rabbits are also the animal of choice for studying syphilitic infection; they are highly susceptible to the syphilis spirochete and they develop manifestations that clinically and histopathologically resemble those in human disease (Turner and Hollander, 1957; Willcox and Guthe, 1966; Schell et al., 1983). Chimpanzees are the most susceptible of the non-human primates to *T. pallidum* infection, and they develop manifestations that closely resemble those of acquired human syphilis (Turner and Hollander, 1957; Willcox and Guthe, 1966). Mice are susceptible to *T. pallidum* but fail to develop pathology even in the face of persistent infection and, thus, have not been looked upon favorably as an experimental model for studying disease pathogenesis (Gueft and Rosahn, 1948; Turner and Hollander, 1957; Willcox and Guthe, 1966). A report in 1980 by Klein and co-workers (Klein et al., 1980) that genetic resistance to infection in mice could be overcome by irradiation has not, to our knowledge, been reproduced. Although the *T. pallidum* subspecies are typically considered to be strictly human pathogens outside of the laboratory, Fribourg-Blanc and Mollaret (1963), provided serological evidence for treponemal infection in African primates, primarily baboons, and they subsequently isolated treponemes from several monkeys with reactive serologies by hamster inoculation (Fribourg-Blanc and Mollaret, 1969). They postulated that *T. pertenue* is the cause of the simian treponematosis based on its geographical overlap with yaws, serological cross-reactivity, and the resemblance of lesions elicited in hamsters by the simian isolates with those produced by known strains of *T. pertenue*.

The syphilis spirochete is the only one of the four pathogenic treponemes whose metabolic capabilities, nutritional requirements, and resistance to environmental stress has been examined systematically (see Chapter 4, Cox and Radolf). *T. pallidum* is a slow-growing, fastidious microorganism that poorly tolerates desiccation, wide fluctuations in temperature, and elevated oxygen tensions (Turner et al., 1941; Turner et al., 1948; Fitzgerald et al., 1977b; Fieldsteel et al., 1982; Norris and Edmondson, 1986; Cox et al., 1990; Norris et al., 2001). These traits undoubtedly explain why efficient transmission of treponemal subspecies requires direct person-to-person contact, venereal or non-venereal. Magnuson and co-workers (1948) showed in classic studies that *T. pallidum* has a remarkably slow dividing time of 30 to 33 hours in infected rabbits. The dividing time of the bacterium is even slower (38 to 40 h) in the *in vitro* tissue culture systems in which limited replication has been achieved (Fieldsteel et al., 1981; Norris, 1982).

Because of the unavoidable presence of rabbit testicular contaminants, pre-genomics era studies of *T. pallidum*'s biosynthetic and metabolic capabilities yielded confusing and sometimes erroneous results (Cox, 1983). For example, it was concluded that *T. pallidum* is capable of oxidative phosphorylation (Lysko and Cox, 1978), while appearing to lack a functional Krebs cycle (Schiller and Cox, 1977). The genomic sequence revealed that the syphilis spirochete relies upon glycolysis for energy production, that it lacks both the tricarboxylic acid cycle and oxidative phosphorylation pathways, and that it is unable to synthesize fatty acids, enzyme cofactors, and most amino acids (Fraser et al., 1998). To compensate for its loss of biosynthetic capability, *T. pallidum* maintains a repertoire of transport systems predicted to have broad specificity for host-derived nutrients (Fraser et al., 1998). One must presume that, beginning with a free-living proto-spirochete progenitor (Paster and Dewhirst, 2000), the spirochete has undergone reductive evolution, jettisoning genes encoding "redundant" metabolic functions provided by the mammalian host, while acquiring new genes needed for its invasive parasitic lifestyle (Ochman and Moran, 2001). Of particular interest from this standpoint is a recent microarray analysis by Smajs et al. (2005) indicating that virtually every gene in the bacterium is expressed during testicular infection of rabbits. This gene expression pattern contrasts markedly with that of *Borrelia burgdorferi*, the Lyme disease spirochete, which undergoes a complex pattern of largely plasmid-based differential gene expression as it cycles between arthropod and mammalian hosts (Revel et al., 2002; Brooks et al., 2003; Narasimhan et al., 2003; Tokarz et al., 2004). Unlike *B. burgdorferi* which must adapt to the mammalian host during tick feeding and prior to transmission, *T. pallidum* can be viewed as always being mammalian host-adapted. While being perpetually adapted to humans may severely limit its host-range possibilities, it may serve to foster person-to-person transmission.

It has long been known that increased ambient temperature decreases growth and survival of *T. pallidum*. Hollander and Turner (1954) reported that the optimal temperature for multiplication of treponemes *in vivo* is 35°C to 37°C, while Fieldsteel et al. (1982) observed that optimal replication during co-culture with rabbit Sf1Ep cells occurs within a slightly lower temperature range, 33°C to 35°C. Consistent with the replication data are studies demonstrating that 34°C is the optimal temperature for incorporation of radiolabeled amino acids *in vitro* and that protein synthesis essentially stops at 42°C (Baseman and Hayes, 1974; Stamm et al., 1991). The inability of the spirochete to mount a heat shock response was proposed as an explanation for its temperature sensitivity (Stamm et al., 1991). The thermal instability of the glycolytic enzyme 3-phosphoglycerate mutase (Benoit et al., 2001) also may explain why treponemal growth is temperature-restricted.

Though once considered an anaerobe, *T. pallidum* takes up oxygen and replicates best *in vitro* at ambient oxygen concentrations of 3% to 5% (Cox and Barber, 1974; Fieldsteel et al., 1982; Cover et al., 1982; Cox et al., 1990; Norris et al., 2001) and, thus, is currently classified as a microaerophile. Using the genomic sequence in concert with biochemical evidence demonstrating that *T. pallidum* expresses a neelaredoxin ortholog with superoxidase reductase activity (Jovanovíc et al., 2000; Lombard et al., 2000), Hazlett et al. (2002) identified *in silico* the novel pathways which *T. pallidum* might utilize to detoxify reactive oxygen intermediates. They also pointed out the apparent contradiction between the syphilis spirochete's susceptibility to oxygen toxicity *in vitro* and its presumed frequent exposure *in vivo* to environments with high redox potentials. Interestingly, in their recent microarray analysis, Smajs et al. (2005) observed that a number of anti-oxidant genes appear to be expressed at relatively high levels, suggesting that the spirochete experiences considerable redox stress during infection in the rabbit.

Because of its double-membrane ultrastructure (Johnson et al., 1973; Hovind-Hougen, 1983), *T. pallidum* has often been analogized to enteric Gram-negative bacteria, such as *Escherichia coli*. The molecular architecture and composition of the *T. pallidum* outer membrane, however, differ markedly from those of Gram-negative bacteria. In addition to lacking lipopolysaccharide, the potent proinflammatory glycolipid found in the outer membranes of Gram-negative (Young et al., 1982; Fraser et al., 1998), multiple lines of evidence indicate that the treponemal outer membrane contains a paucity of integral membrane proteins and that the syphilis spirochete's dominant immunogens are lipid-modified and cytoplasmic membrane-associated (Radolf, 1995; Blanco et al., 1997; Norris et al., 2001). Despite extensive efforts over the past 25 years and the availability of the genomic sequence (Radolf, 1995; Fraser et al., 1998), unambiguous identification of *T. pallidum* outer membrane proteins has eluded syphilis researchers (see Chapter 11, Cameron). Not surprisingly, the *T. pallidum* subspecies are morphologically and ultrastructurally indistinguishable (Hovind-Hougen, 1976; Engelkens et al., 1991; Walker et al., 1989).

3. Natural history of syphilis

3.1. Microbiological events

3.1.1. Transmission

Venereal syphilis is acquired when treponemes are inoculated onto mucosa or skin, usually in the genital or anorectal areas, most commonly during sexual intercourse. The importance of oral sex as a means of transmission, recognized for decades (Stokes et al., 1944; Mindel et al., 1989a), has received renewed emphasis in a recent analysis of a syphilis outbreak among men who have sex with men (Centers for Disease Control and Prevention, 2004). In contrast to venereal syphilis, the agents of yaws, pinta, and endemic syphilis are transmitted by direct, non-venereal contact with open lesions during childhood (Antal et al., 2002).

Individuals with venereal syphilis are most infectious during the primary and secondary stages of disease (including secondary relapses of early latency) when moist, mucocutaneous lesions are present. Persons with early syphilis may be infectious, however, even if they lack open lesions; microscopic abrasions created during sexual activity are theoretically capable of transmitting infection to a sex partner. Estimates of the risk of sexual transmission derived from partner notification or prophylactic treatment studies vary from 10% to 80% (Alexander et al., 1949; Moore et al., 1963; Schroeter et al., 1971; Schober et al., 1983; Singh and Romanowski, 1999). A transmission rate of 30% derived by Schroeter et al (Schroeter et al., 1971) is a commonly quoted figure. Studies spanning the antibiotic era found that primary and secondary syphilis were equally transmissible (Alexander et al., 1949; Schober et al., 1983). Nonsexual contact with mucocutaneous lesions, mainly through kissing, also can transmit disease on rare occasions (Stokes et al., 1944). Transfusion-acquired syphilis has ceased to be of epidemiologic importance in developed countries (Schmidt, 2001).

Vertical transmission of syphilis has been recognized for several centuries (Radolf et al., 1999). The risk of congenital syphilis ranges from thirty to eighty percent in pregnant females with primary or secondary syphilis and decreases as the interval between infection and pregnancy lengthens (Paley, 1937; Ingraham, 1951; Schulz et al., 1987). A recent prospective study from Parkland Memorial Hospital (Dallas, Texas) reported that 16 of 24 women with untreated early syphilis had fetuses with congenital syphilis or amniotic fluid infected with *T. pallidum* (Hollier et al., 2001). The finding of spirochetes in the placenta and umbilical cord, along with histopathologic changes consistent with syphilis (Qureshi et al., 1993; Sheffield et al., 2002), supports transplacental invasion as the major route of fetal infection. Alternatively, Wendel and co-workers (Nathan et al., 1993; 1997; Wendel et al., 1989; Wendel, Jr. et al., 1991) proposed that treponemes gain access to the fetal circulation by first traversing the fetal membranes and infecting amniotic fluid. Less commonly, perinatal infection may occur when a neonate is exposed to an infectious lesion or infected maternal blood during passage through the birth canal (Stokes et al., 1944; Dorfman and Glaser, 1990).

3.1.2. Early invasion and dissemination

Animal inoculation studies early in the twentieth century revealed *T. pallidum's* impressive capacity to disseminate systemically shortly after infection (Turner et al., 1969; Stokes et al., 1944). As stated by Brown and Pearce (1920), "for practical purposes, there is probably no appreciable time during which a syphilitic infection can be regarded as confined to the focus of entry." In 1906-09, Neisser showed that treponemes were detectable in bone marrow, spleen, and testes of apes 48 hours following intradermal inoculation (Stokes et al., 1944). In 1920, Brown and Pearce (1920) recovered *T. pallidum* from inguinal lymph nodes and blood of rabbits with in 48 h after intratesticular inoculation, well before the onset of orchitis. Raiziss and Severac (1935) found that the interval between intratesticular inoculation and spirochetemia in rabbits may be as short as five minutes; in retrospect, one wonders (as did Turner and Hollander [1957]) whether these striking results were due to inadvertent intravascular injection.

Early and widespread dissemination of spirochetes appears to be the rule as well during acquired human syphilis. Uhlenhuth and Mulzer (1913a; 1913b) showed that an astonishing 82% of patients with primary syphilis, including persons with non-reactive serological tests, harbor treponemes in their blood; though of uncertain statistical significance, this figure exceeded the 71% isolation rate for secondary syphilis patients. Stokes et al. (1944) considered infection of the vascular system to be a universal feature of syphilitic infection and he used the term "vasculotropism" to describe the ubiquitous vasculitis and perivasculitis that result from spirochetal dissemination during early disease. Numerous reports in the literature of the early twentieth century documented transmission of infection following the transfusion of blood from individuals with incubating syphilis, providing further clinical evidence that spirochetemia precedes disease onset (Post and Cooney, 1933;

Raiziss and Severac, 1935). These cases are essentially equivalent to modern day descriptions of congenital syphilis in infants born to recently infected mothers with nonreactive syphilis serologies (Sanchez et al., 1991; Dorfman and Glaser, 1990).

Whereas central nervous system (CNS) invasion is unusual following intradermal or intratesticular inoculation of rabbits and even non-human primates (Willcox and Guthe, 1966; Turner and Hollander, 1957), *T. pallidum* regularly breaches the blood-brain barrier in humans with early syphilis (Merritt et al., 1946; Lukehart et al., 1988; Hook, III, 1991; Rolfs et al., 1997; Golden et al., 2003). In 1912, Nichols and Hough (1913) accomplished the first successful isolation of *T. pallidum* in rabbits when they cultivated the organism from the CSF of a patient experiencing neurorelapse following salvarsan therapy for secondary syphilis. In addition to being the first demonstration of early CNS invasion by *T. pallidum*, their findings suggested that the CNS may serve as a sanctuary from which infection could re-emerge following inadequate therapy. Parenthetically, the strain isolated by Nichols and Hough subsequently became the type strain of *T. pallidum* and was the source of DNA for the genomic sequencing project (Fraser et al., 1998). Based on the human inoculation studies conducted by Magnuson and co-workers (1956) in the 1950s and a laboratory accident in 1976 (Fitzgerald et al., 1976), the Nichols strain is presumed to have retained its virulence for humans despite decades of intratesticular passage in rabbits.

Abnormalities in the cerebrospinal fluids (CSFs) of early syphilis patients were first recorded in the early 1900s (Wile and Stokes, 1914), and it was subsequently found that CSF abnormalities occur in up to one-third of early syphilis patients, peaking between twelve and fifteen months post-infection (Mills, 1927; Moore and Hopkins, 1930; Merritt et al., 1946). Infectivity testing and longitudinal studies conducted in the pre-antibiotic era demonstrated convincingly that CSF abnormalities result from CNS invasion by *T. pallidum* and that they portend a substantial risk for progression to symptomatic neurosyphilis if they persist beyond the first two years of infection (Moore and Hopkins, 1930; Merritt et al., 1946; Hook, III, 1991). Recent studies have confirmed both the high prevalence of CNS invasion by *T. pallidum* during early syphilis and the frequent occurrence of spinal fluid abnormalities in neurologically asymptomatic patients with early disease (Lukehart et al., 1988; Rolfs et al., 1997; Marra et al., 2004). It is extremely important to note that normal CSF parameters at the time of lumbar puncture do not rule out the presence of treponemes. In a classic study published in 1924, Chesney and Kemp (Chesney and Kemp, 1924) isolated *T. pallidum* from 4 of 34 untreated early syphilis patients with normal spinal fluids, and comparable results have been obtained in contemporary, prospective studies using RIT and/or PCR analysis (Lukehart et al., 1988; Rolfs et al., 1997). "Normal" CSF values presumably indicate that treponemes have recently gained access to the CNS and have yet to elicit an inflammatory response.

Over the years, there has been considerable debate over whether some *T. pallidum* strains are neurotropic (Stokes et al., 1944; Merritt et al., 1946). Previously, no suitable experimental animal model existed for comparing the ability of clinical isolates to cross the blood-brain barrier. Tantalo et al. (2005), however, recently showed that differences in neuroinvasive capacity, as determined by CSF pleocytosis and detection of *T. pallidum* by RT-PCR, could be identified among six *T. pallidum* strains inoculated intravenously into rabbits. Consistent with its history, the Nichols strain was found to be neuroinvasive, although it was less so than at least two other "street strains." It will be quite interesting to see if genetic correlates for the neuroinvasive phenotype are identified in future studies.

3.2. Role of the inflammatory response

In their remarkably comprehensive textbook published more than six decades ago, Stokes et al. (1944) emphasized the importance of the inflammatory response to the development of disease manifestations, lesion resolution, and protective immunity in venereal syphilis. Indeed, one forms the impression that they conceptualized early syphilis as a "race" between the spirochete's capacity for replication, invasion and dissemination and the ability of host defenses to "catch up" with the spirochete in order to contain infection and promote healing. In their classic monograph, Turner and Hollander (1957) described the massive infiltration of lymphocytes, epithelioid cells (macrophages) and heterophiles (neutrophils), along with the subsequent necrotic reaction, elicited in rabbits following intradermal inoculation with *T. pallidum*. These perspectives clearly laid the groundwork for the current consensus among syphilologists that lesion development and lesion resolution are direct consequences of the inflammatory processes triggered by treponemes and treponemal constituents within infected tissues (Norris, 1988; Lukehart, 1992; Salazar et al., 2002). Toxigenic molecules elaborated by the bacterium, if they exist, are now thought to play no more than a subsidiary role in disease production (Fraser et al., 1998; Norris et al., 2001).

Sell, Lukehart, and coworkers deserve the lion's share of the credit for placing our understanding of syphilis pathogenesis on a contemporary immunological footing. In carefully performed time-course studies, they delineated three distinct phases following primary testicular or intradermal infection of rabbits. During the proliferative and inductive phases, a rapid increase in spirochete numbers is followed by infiltration of T cells (Lukehart et al., 1980a; Baker-Zander and Sell, 1980; Sell et al., 1980b). Parallel experiments demonstrated that hyperplasia of T cells in lymphoid tissues of infected rabbits and the appearance of *T. pallidum*-specific T cells in spleen and lymph nodes correlate temporally with infiltration of T cells into the site of primary infection (Sell et al., 1980a; Baker-Zander and Sell, 1980; Lukehart et al., 1980b; Baker-Zander et al., 1988; Arroll et al., 1999). The reactive phase is characterized by an influx of macrophages that rapidly eliminate spirochetes by phagocytosis (Lukehart et al., 1980a; Sell et al., 1980b; Baker-Zander and Sell, 1980; Sell et al., 1982). Clearance of spirochetes *in vivo* temporally corresponds to the appearance of serum opsonic antibodies, and companion *ex vivo* studies showed that opsonic antibodies are required for phagocytosis and killing of treponemes by macrophages (Lukehart and Miller, 1978; Alder et al., 1990; Baker-Zander and Lukehart, 1992). Activation of macrophages by T cell-derived cytokines (most likely IFN-γ) was found to enhance bacterial clearance (Lukehart, 1992). Interestingly, studies with a non-human primate model indicate that local production of IFN-γ is likely to be involved in the clearance of spirochetes from the CNS (Marra et al., 1998). Although comparable human studies have not been performed, it is reasonable to extrapolate from this body of work to human disease. Indeed, as would be predicted, immunohistochemical and RT-PCR analyses of biopsy specimens from humans have demonstrated that early syphilis lesions contain large numbers of macrophages and T cells expressing a Th1 cytokine profile (Bjerke et al., 1981; Tosca et al., 1988; Van Voorhis et al., 1996a; McBroom et al., 1999).

The past several years have witnessed dramatic advances in our understanding of how innate immune mechanisms are mobilized during bacterial infections (Aderem and Ulevitch, 2000; Medzhitov, 2001; Sieling and Modlin, 2002). Four lines of evidence have established an etiological link between innate immunity and syphilis pathogenesis: (i) *T. pallidum* contains an abundance of lipid-modified proteins (Chamberlain et al., 1989; Schouls et al., 1989); (ii) these lipoproteins, as well as synthetic lipoprotein analogs (lipopeptides) based upon their N-termini, are potent proinflammatory agonists *in vitro* (Radolf et al., 1991; Norgard et al., 1996; Radolf et al., 1995a); (iii) lipoproteins/lipopeptides stimulate cells via interactions with CD14, Toll-like receptor 1 (TLR1), and TLR2 expressed by monocytes/macrophages and dendritic cells (Lien et al., 1999; Sellati et al., 1998; Brightbill et al., 1999; Wooten et al., 1998; Hirschfeld et al., 1999; Alexopoulou et al., 2002); and (iv) intradermal injection of synthetic treponemal lipopeptides into rabbits (Norgard et al., 1995) or humans (Sellati et al., 2001) elicits cellular infiltrates similar to those observed in early syphilis lesions. It is now widely accepted, based upon these findings, that treponemal lipoproteins are major proinflammatory agonists during syphilitic infection.

Using admittedly broad brushstrokes, it is now possible to paint a picture of syphilis pathogenesis in which innate and adaptive immune responses to *T. pallidum* function coordinately throughout the disease process. According to a recently proposed schema (Salazar et al., 2002), treponemal lipoproteins maintain a proinflammatory milieu at sites of infection via pre-programmed, innate mechanisms that drive the recruitment of immune cells from peripheral blood into tissues. This stereotypical "lipoprotein effect" also can be invoked to help explain the common histopathological features observed in all types and stages of syphilis, congenital as well as acquired (Stokes et al., 1944). Superimposed on the proinflammatory backdrop created by the lipoproteins is an adaptive immune response that becomes increasingly robust as treponemal antigens are processed and presented to naïve T cells within draining lymph nodes. Neo-sensitized T cells enter the circulation and are recruited to sites of infection where they participate in a secondary immune response, releasing cytokines that activate macrophages and promote spirochetal clearance, almost certainly in conjunction with opsonizing antibodies. The molecular architecture of the *T. pallidum* outer membrane is conjectured to be one of the pathogen's principal means of counteracting these defenses (see below and Chapter 11, Cameron). The lack of surface-exposed lipoproteins (Radolf, 1995; Blanco et al., 1997) enables spirochetes to minimize direct interactions with pattern recognition receptors on the surfaces of macrophages and dendritic cells, thereby delaying or dampening the "danger signals" that trigger innate mechanisms. The spirochete's poor surface antigenicity also limits the effectiveness of the *T. pallidum*-specific antibodies that are subsequently produced in high titers, as evidenced by the fact that phagocytosis and killing of treponemes by macrophages proceeds slowly and requires relatively high concentrations of opsonic antibodies (Alder et al., 1990; Baker-Zander and Lukehart, 1992). Failure to eradicate local and systemic treponemal infection is responsible for the latent or persistent stage which may last for the remainder of the individual's lifetime.

3.3. Course of the disease

Two principal objectives of syphilis pathogenesis research are to explain individual steps in the disease process and specific disease manifestations in cellular and molecular terms. Pathogenesis researchers also want to characterize spirochetal and host factors responsible for the extraordinary diversity of disease at the individual and population levels. None of these goals could be achieved without a sound understanding of how syphilitic infection "plays out" over time. While numerous investigations have contributed to our knowledge of the clinical events that transpire in patients with syphilis, the clinical studies collectively known as "The Oslo Study of Untreated Syphilis," stand out because of the panoramic view they provided of the entire disease process. Convinced that the mercurial compounds available in the late nineteenth century for the treatment of syphilis were more toxic than therapeutic, Boeck (University of Oslo) opted to hospitalize nearly 2000 patients with primary and secondary syphilis between the years 1890 and 1910, withholding therapy while maintaining meticulous records on each person from one to 12 months (average of 3.6 months) until all traces of the disease had disappeared (Clark and Danbolt, 1955). In 1925, Bruusgaard, Boeck's immediate successor, initiated a follow-up of 443 persons (309 living and 164 dead) from the original cohort with intervals of infection ranging from 3 to 40 years (Clark and Danbolt, 1955). Prompted by an internationally prominent group of spirochetologists who recognized the potentially biased nature of Bruusgaard's follow-up data, in 1948, Gjestland, Bruusgaard's successor, began a painstaking re-examination of the available records for the 1,404 patients who lived near Oslo (Gjestland, 1955). Because Boeck assembled his cohort prior to the availability of contemporary syphilis diagnostics, patients had been diagnosed exclusively by clinical criteria. Examination of hospital records as part of the Gjestland re-examination confirmed the diagnosis of syphilis in the great majority of the patients. The monumental re-analysis of the original Boeck cohort has become the single greatest source of prognostic information about syphilis available to clinicians and a pillar of pathogenesis research. One of its principal findings, that approximately one-third of untreated patients will experience complications of late disease, lent statistical validity to the well established perception of syphilis as a remarkably persistent and pernicious infectious disorder. At the same time, the re-examination also corroborated Bruusgaard's conclusion that most untreated patients do not experience serious complications.

Some general points about staging are in order before discussing individual stages. The need to stage patients with suspected or proven syphilis is a time-honored principle of clinical syphilology (Stokes et al., 1944). Yet, despite its clinical and public health utility (Centers for Disease Control and Prevention, 2002; Lukehart, 2004), the practice has the potential to obscure the fact that syphilis tends to unfold in a waxing and waning manner, rather than in discrete steps. Moreover, as one might expect with an infection which becomes systemic almost from the very outset but induces a slowly evolving local and systemic immunologic response, a clear demarcation between primary and secondary stages does not always exist (Chapel, 1980; Hira et al., 1987; Rompalo et al., 2001). Confusion about staging also arises in clinical settings because specific manifestations may not be confined to a single stage of disease. These overlaps undoubtedly reflect the limited tissue response patterns that treponemes can elicit. For example, genital ulcers can occur as secondary lesions and lesions resembling chancres can occur in non-genital regions (Hira et al., 1987; Rademacher and Radolf, 1996). Perhaps most confusing of all to the general practitioner from the standpoint of staging is neurosyphilis. Neurological complications of syphilis can occur in early as well as late syphilis (Merritt et al., 1946; Hook, III, 1991; Golden et al., 2003), while, as noted earlier, CSF abnormalities due to CNS invasion by *T. pallidum* are commonplace in early syphilis (Wile and Stokes, 1914; Moore and Hopkins, 1930; Lukehart et al., 1988; Rolfs et al., 1997; Marra et al., 2004).

3.3.1. Incubating syphilis

The incubation period of syphilis is said to range from 9 to 90 days, and an asymptomatic patient is presumed to have incubating syphilis if exposure is known to have occurred within the previous 90 days. As noted above, dissemination of spirochetes from the site of inoculation and draining lymphatics is well underway during this pre-symptomatic stage.

3.3.2. Primary syphilis

Syphilis commences clinically when spirochetes replicating at the site of inoculation induce a local inflammatory response sufficient to generate a papule which subsequently ulcerates, forming a chancre, the defining lesion of primary syphilis. In their human inoculation studies conducted at Sing Sing Prison, Magnuson and co-workers (1956) observed mean incubation periods of 24 days for normal volunteers inoculated intracutaneously with either 10 and 100 spirochetes of the Nichols strain. The similarity between this figure and the three- to four-week mean incubation period observed clinically (Garnett et al., 1997; Lukehart, 2004) suggests that inoculating doses often are in the vicinity of the ID_{50} of 57 organisms determined for the Nichols strain (Magnuson et al., 1956). The classic histologic picture of primary syphilis consists of proliferating small vessels (neovascularization) surrounded by spirochetes, macrophages, lymphocytes, and variable numbers of plasma cells (Azar et al., 1970; Lukehart, 2004). Immunocytochemical studies have revealed that the dermal lymphocytes are mixtures of CD4 and CD8 cells (Tosca et al., 1988; Van Voorhis et al., 1996b). The reduction of blood flow due to small vessel vasculitis (capillary endothelial swelling and hyperplasia and thrombosis) is thought to be directly responsible for tissue necrosis and ulceration (Norris, 1988; Azar et al., 1970). Electron microscopic examination of biopsy specimens has revealed that spirochetes are predominantly, if not exclusively, extracellular, interdigitating with collagen bundles in the dermis and often in proximity to blood vessels (Drusin et al., 1969; Azar et al., 1970; Sykes and Kalan, 1975). Unlike the exquisitely painful genital ulcers caused by *Haemophilus ducreyi* (chancroid) and herpes simplex virus (herpes genitalis), the classical chancre is often painless

(Stokes et al., 1944). Spirochetal infiltration of cutaneous sensory nerves may explain this curious phenomenon (Sell et al., 1985; Sell and Salman, 1992). In untreated, immunocompetent individuals, spontaneous healing of chancres occurs within three to eight weeks and, as in rabbits with experimental syphilis (Lukehart et al., 1980a; 1980b; Baker-Zander and Sell, 1980; Baker-Zander et al., 1993), is presumed to coincide with the influx of sensitized T cells and macrophages and the concurrent generation of *T. pallidum*-specific opsonizing antibodies.

3.3.3. Secondary syphilis

Manifestations of secondary syphilis usually begin 4 to 10 weeks after the chancre has appeared (Lukehart, 2004; Golden et al., 2003). Infectivity testing established many years ago that the vast majority of complications of secondary syphilis result from hematogenous dissemination of spirochetes (Turner et al., 1969). Circulating immune complexes have been demonstrated in early syphilis (Jorizzo et al., 1986; Baughn et al., 1986) but, with the exception of the rare glomerulopathies occasionally observed in secondary syphilis patients (Hunte et al., 1993; Walker et al., 1984), are no longer thought to be central to disease pathogenesis (McBroom et al., 1999). Although mucocutaneous lesions (syphilids) are far and away the most common manifestation of disseminated disease, secondary syphilis can affect virtually any organ system (Mindel et al., 1989b; Lukehart, 2004; Golden et al., 2003) It is generally believed that the predominance of skin lesions, as well as their often characteristic distribution on palms and soles, during secondary syphilis reflects the spirochete's predilection for temperatures below core, as discussed earlier. Diffuse nontender lymphadenopathy, present in up to 75% of patients, (Chapel, 1980; Hira et al., 1987; Mindel et al., 1989b), is almost certainly indicative of the intense systemic degree of T cell activation which has been well documented in the rabbit experimental model (Pavia et al., 1977b; Lukehart et al., 1980b). Constitutional symptoms, another indicator of systemic infection, are common but usually mild (Chapel, 1980; Hira et al., 1987; Stokes et al., 1944; Lukehart, 2004). The rarity of sepsis-like symptomatology during the disseminated stage of the disease distinguishes the spirochetemia of secondary syphilis from bacteremia with more "conventional" pathogens and presumably reflects, at least in part, the spirochete's lack of lipopolysaccharide (Fraser et al., 1998; Young et al., 1982).

Cutaneous histologic patterns in secondary syphilis have been extensively characterized. Infiltrates consisting of variable proportions of lymphocytes, macrophages, neutrophils, and plasma cells, often perivascular in distribution, are most intense in the papillary dermis but often extend into the reticular dermis (Abell et al., 1975; Jeerapaet and Ackerman, 1973; Jordaan, 1988; Pandhi et al., 1995). Consistent with the Th1 cytokine profile revealed by RT-PCR analysis of skin biopsies (Van Voorhis et al., 1996a), frank granulomatous reactions also are well recognized (Abell et al., 1975; Pandhi et al., 1995; Kahn and Gordon, 1971). Follicles and sweat glands are frequently sleeved with inflammatory cells; severe follicular involvement presumably is the cause of syphilitic alopecia. Histologic abnormalities in the epithelium include exocytosis, necrosis of keratinocytes, mild to moderate acanthosis, and varying degrees of hyperkeratosis. Immunohistochemical analysis has revealed that keratinocytes overlying dermal infiltrates frequently express HLA-DR, a marker of cellular activation (McBroom et al., 1999). One of the great mysteries of syphilology is how these dermal and epidermal reaction patterns are translated into the remarkably diverse types of rashes observed in syphilis skin lesions (Stokes et al., 1944; Lukehart, 2004).

Syphilologists have long been intrigued by the paradoxical observation that disseminated infection becomes clinically evident, and indeed, may be rampant, at a time when the chancre is resolving or healed (Norris, 1988; Lukehart, 1992; Musher and Knox, 1983). In the 1970s, this enigma was attributed to a putative immunosuppressive effect of systemic *T. pallidum* infection. This concept was based on reports (i) that lymphocytes from infected rabbits (Maret et al., 1980; Pavia et al., 1976; 1977a; Wicher and Wicher, 1977a) or humans (Musher et al., 1975; Levene et al., 1971) display decreased blastogenic responses to mitogens or treponemal antigens and (ii) that rabbit (Pavia et al., 1977b; Wicher and Wicher, 1977b; Ware et al., 1979) and human (Levene et al., 1969; From et al., 1976; Friedmann and Turk, 1978) syphilitic sera or plasma inhibit blastogenic responses. As already noted, by the 1980s, the notion that an active immune response occurs during syphilitic infection had gained sway in the field (Sell and Norris, 1983; Norris, 1988; Lukehart, 1992). Accordingly, Norris (1988) speculated that the inability to prevent widespread multiplication of *T. pallidum* at the time of chancre resolution represents a delay in the transition from local to systemic immunity. Lukehart (1992) subsequently proposed that lesion formation doesn't occur until organisms at a particular site reach a threshold density or "critical antigenic mass" capable of triggering local inflammatory processes that result in the recruitment of circulating innate and newly sensitized T cells. According to this latter scenario, which is currently the most attractive, the "delay" in the evolution of secondary lesions would reflect the small numbers of spirochetes believed to be implanted at a metastatic site (e.g., skin), the bacterium's extraordinarily slow rate of replication, and its inherent ability to avoid triggering local danger signals on resident innate immune cells.

3.3.4. Jarisch-Herxheimer reaction

Within hours following treatment for primary or secondary syphilis, as many as one-quarter of patients experience an acute febrile reaction consisting of arthralgias, myalgias, headache, and increased prominence of lesions (Lukehart, 2004). This phenomenon, called the Jarisch-Herxheimer (J-H) reaction, has long been thought to be an acute inflammatory response induced by the sudden release of spirochetal products (Bryceson, 1976; Beutler and Munford, 1996). In the early 1980s, before it had been definitively shown that *T. pallidum* lacks lipopolysaccharide, Young et al. (Young et al., 1982) reported that endotoxin was unlikely to be the inciting molecule. Subsequently, Beutler and Munford (1996) proposed that the J-H reaction is due to the release of spirochetal lipoproteins. The *in vitro* observation that treponemal lysates have a much greater capacity to activate monocytes than comparable numbers of intact organisms may be the *in vitro* correlate for this clinical phenomenon (Sellati et al., 1999). A central role for TNFα in the pathogenesis of the J-H response is supported by a controlled, prospective study from Ethiopia in which the reaction was aborted or attenuated when antibodies directed against TNFα were administered prior to antimicrobial therapy for louse-borne relapsing fever caused by the spirochete *B. hermsi* (Fekade et al., 1996). Like *T. pallidum*, *B. hermsii* also lacks LPS (Hardy, Jr. and Levin, 1983; Takayama et al., 1987) but contains abundant proinflammatory lipoproteins (Scragg et al., 2000).

3.3.5. Latency

Mucocutaneous lesions of secondary syphilis usually resolve spontaneously in approximately three to twelve weeks, leading to the "asymptomatic" (or "post-symptomatic") stage referred to as latency. During early years of syphilis infection, patients experience recurrences of spirochetemia, which need not be clinically evident, as well as secondary relapses. One of the most important findings from the Oslo Study is that nearly 25% of patients will experience secondary relapses and, of these, approximately one-quarter will have multiple relapses (Clark and Danbolt, 1955; Gjestland, 1955). Although 90% of relapses occur during the first year of infection, they can occur as late as five years into the illness. Improved immunological control eventually causes reduction in spirochetemia(s), a decline in total spirochetal burden, and an end to infectiousness. It does not, however, necessarily result in the total eradication of the disease. Treponemes can persist at local sites throughout latency, inducing a smoldering inflammatory state that sets the stage for recrudescent or tertiary disease years later. In the pre-antibiotic era, persistent infection was common. A cross-sectional review of all autopsies conducted at Yale University from 1917 to 1941 by Rosahn (1947) found that 30 percent of the syphilitic patients autopsied had late anatomic lesions (mostly cardiovascular) and that continued reactivity of serological tests was not a prerequisite for autopsy evidence of late disease. The mechanisms by which *T. pallidum* persists in the face of the host's efforts at eradication during the long latency period have been the subject of much speculation but are still a mystery. Residence within intracellular niches has been proposed based upon electron micrographs of infected rabbit tissues and biopsy specimens (Ovcinnikov and Delektorskij, 1972; Lauderdale and Goldman, 1972; Sykes and Miller, 1971; Sykes et al., 1974). Persistence in the extracellular milieu is the alternative and currently favored possibility, though without much precedent among bacterial pathogens. Consistent with this notion was the demonstration by Lukehart and coworkers (Lukehart et al., 1992) of a subpopulation of phagocytosis-resistant spirochetes after the majority of bacteria were cleared from primary testicular infections.

3.3.6. Tertiary syphilis

More than 30 percent of the patients in the Oslo Study developed tertiary syphilis, with a calculated excess mortality of approximately 20 percent (Clark and Danbolt, 1955; Gjestland, 1955). Gummatous syphilis was the most common form of tertiary disease, followed by cardiovascular syphilis and neurosyphilis. In the post-antibiotic era, all forms of tertiary syphilis have decreased markedly in incidence, and there are no reliable figures on the relative frequency of gummatous, cardiovascular, and neurological late complications. There is an impression, however, that during the past several decades neurosyphilis has become the most commonly encountered late complication (Swartz et al., 1999). As discussed below, involvement of the CNS is a major concern in the AIDS era.

Benign tertiary or gummatous syphilis is a proliferative or destructive granulomatous process usually of skin, bone, or mucous membranes, but occasionally involving viscera or CNS (Stokes et al., 1944; Lukehart, 2004). Because treponemes are difficult to visualize microscopically in gummas, it has long been thought that they arise from extreme "hypersensitivity" reactions to minute numbers of organisms in foci of persistent infection (Stokes et al., 1944; Swartz et al., 1999). The importance of cellular hypersensitivity phenomena is supported by the observation in the Sing Sing Study that two patients with treated late syphilis developed gumma following intradermal challenge with the Nichols strain of *T. pallidum* (Magnuson et al., 1956). While an intense cell-mediated response is certainly consistent with the histopathologic abnormalities, it is not clear how the host would become hypersensitized to *T. pallidum* at a stage of disease when spirochetal burdens are thought to be at a nadir. One possible explanation, suggested by an immunofluorescent analysis of a gumma revealing substantial numbers of organisms not detected by silver stain (Handsfield et al., 1983), is that a diminution in local surveillance mechanisms, perhaps as a result of aging, allows increased spirochetal replication; the increased local spirochetal burden would then "boost" systemic and local cellular responses, provoking the intense local reaction.

Aortitis is the *sine qua non* of cardiovascular syphilis (Jackman and Radolf, 1989). Organisms lodge within the vasa vasora, the nutrient blood vessels of the aortic adventitia, initiating inflammatory changes which eventually cause obliterative endarteritis, adventitial scarring, and patchy medial necrosis with destruction of elastic fibers. The resultant weakening of the aortic wall leads to dilatation of the aortic root and aneurysm formation over time. The predilection for involvement of the proximal aorta has been attributed to the larger number of *vasa vasora* in the ascending and transverse arches (Kampmeier and Morgan, 1952). An intriguing alternative explanation is suggested by the finding from the Oslo Study that cardiovascular syphilis did not develop in patients infected before the age of fifteen (Clark and Danbolt, 1955). Pre-existing atheromatous disease might predispose to treponemal invasion of central arterial structures by disrupting endothelial barriers and exposing extracellular matrix components to which treponemes can adhere (Cameron, 2003; Cameron et al., 2004).

Neurosyphilis is traditionally divided into asymptomatic, meningeal, meningovascular, and parenchymatous syndromes; a temporal progression of syndromes, along with distinctive pathological abnormalities, is well recognized (Merritt et al., 1946; Hook, III, 1991; Golden et al., 2003). In the absence of treatment, asymptomatic neurosyphilis either resolves spontaneously or progresses to symptomatic disease (Moore and Hopkins, 1930). Syphilitic meningitis, which usually occurs within the first two years of infection, results from vasculitis of the small vessels lining the meninges. Meningovascular syphilis typically develops five to ten years post-infection and denotes the various stroke syndromes caused by a proliferative endarteritis of small to medium-sized arteries that supply blood to CNS parenchyma. The parenchymatous forms of neurosyphilis, generalized paresis and tabes dorsalis, are the last in the temporal progression, and involve neuronal loss, demyelination, and astrocyte proliferation with gliosis. Nothing is known as to how persistent *T. pallidum* infection gives rise to these tissue reactions.

3.3.7. Syphilis and HIV infection

A statistical association between HIV infection and syphilis became evident early in the AIDS epidemic (Jaffe et al., 1983). Since then, numerous studies have documented the high rate of HIV coinfection among syphilis patients (Hutchinson et al., 1991; Ansell et al., 1994; Dickerson et al., 1997; Steen and Dallabetta, 2004). Although initially thought to merely reflect similar risk factors for acquisition or transmission of the two infections, it has become apparent over the years that there are complex epidemiologic and biologic relationships between these disorders (Greenblatt et al., 1988; Wasserheit, 1992; Fleming and Wasserheit, 1999; Rottingen et al., 2001; Galvin and Cohen, 2004). From a pathogenesis standpoint, the interactions between these two sexually transmitted infections can be separated into two distinct, though interrelated, aspects: (i) the effect of syphilitic genital ulcers on the bidirectional transmission of HIV and (ii) the effect of HIV infection on the natural history and manifestations of syphilis.

Chancres theoretically can facilitate HIV transmission either by increasing susceptibility or infectiousness (Galvin and Cohen, 2004). The former involves creating a local environment that favors the acquisition of virus, whereas the latter involves augmenting viral load in blood, genital fluids, and/or lesion exudates. A reasonable body of evidence, admittedly mostly indirect, suggests that chancres can do both. In addition to disrupting normal epithelia or mucosal barriers, early syphilis lesions contain infiltrates enriched in activated lymphocytes and macrophages, potential targets and donors for HIV (Greenblatt et al., 1988; Tosca et al., 1988; Van Voorhis et al., 1996a; 1996b; McBroom et al., 1999). *In vitro* and *in vivo* studies have demonstrated that viable treponemes, treponemal lipoproteins, and treponemal lipoprotein analogs (lipopeptides) have the capacity to induce enhanced expression of CCR5, the coreceptor for sexually transmitted, macrophage-trophic (M-trophic) strains of HIV, on the surfaces of macrophages and dendritic cells (Sellati et al., 2000; 2001), the principal targets of HIV infection at mucosal sites (Moore et al., 2004). The biological relevance of these findings is supported by the demonstration that the increased surface expression of CCR5 induced by these agonists promoted entry of an M-tropic, but not a T-tropic, HIV strain into peripheral blood monocytes (Sellati et al., 2000). Interestingly, lipooligosaccharide from *Haemophilus ducreyi*, the agent of chancroid, does not induce increased expression of CCR5 by macrophages (Humphreys et al., 2002), suggesting that this response is lipoprotein-specific. With respect to increased infectiousness, HIV has been recovered from genital ulcers, including chancres (Gadkari et al., 1998; Kreiss et al., 1989; Erbelding and Rompalo, 2004). *T. pallidum* and treponemal lipoproteins/lipopeptides have been shown to induce HIV gene expression and increased viral replication in tissue culture models of HIV infection, and these effects correlated with increased translocation and binding of nuclear factor (NF)-κB to the promoter region of the HIV provirus long-terminal repeat (Theus et al., 1998).

There is considerable controversy as to whether HIV infection worsens the manifestations of early syphilis and/or accelerates the course of the disease. Since the late 1980s, a number of case reports and small series have documented ophthalmologic and neurological complications, as well as unusual or highly destructive non-neurological syphilis syndromes, in HIV-infected patients (Johns et al., 1987; Radolf and Kaplan, 1988; Lukehart et al., 1988; Musher et al., 1990;

Marra, 1992; Horowitz et al., 1994; Kastner et al., 1994; Shalaby et al., 1997). The prospective study by Lukehart and co-workers (Lukehart et al., 1988) was immensely influential because it was the first in the AIDS era to use infectivity testing to demonstrate the persistence of treponemes in the CSF of HIV-infected patients. The sheer volume of cases led many authorities to conclude that infection with HIV poses a profoundly increased risk of complications, particularly neurologic, during active syphilis (Musher et al., 1990; Marra, 1992). To the surprise of many, a multi-center prospective study sponsored by the Centers for Disease Control and Prevention (the Syphilis and HIV Study) found that HIV infection had only a minimal effect on early syphilis (Rolfs et al., 1997; Rompalo et al., 2001). HIV-infected patients with primary syphilis tended to present with more genital ulcers, and genital ulcers were present more frequently in HIV-infected patients with secondary syphilis. Manifestations of disseminated infection, including neurological complications, however, were not worsened by concomitant HIV infection. These results are generally consistent with the modest effect of retrovirus-induced immunodeficiency on treponemal clearance in a rhesus macaque model of syphilis and simian immunodeficiency virus coinfection (Marra et al., 1992).

Unfortunately, there is reason to believe that the Syphilis and HIV Study may not have provided definitive answers to the questions it sought to address. It did not include patients that required hospitalization, so it may have overlooked a subpopulation of coinfected individuals critical for examining the thesis that HIV infection exacerbates early syphilis. The original recruitment goals were not met and, the study probably was under-powered. Lastly, few of the HIV-infected patients had advanced HIV disease. On the other hand, the study clearly established that early syphilis in persons with severely depressed CD4 counts may be no more severe than that which occurs in immunocompetent individuals. Thus, it appears as if a distinction must be made between syphilis and those HIV-related opportunistic infections for which a predictable relationship exists between absolute CD4 count and susceptibility to infection (Rompalo et al., 2001; Kaplan et al., 2002). An interesting concordance between the study by Lukehart et al. (1988) and the Syphilis and HIV Study (Rolfs et al., 1997) also should be noted. Neither group of investigators observed a correlation between HIV infection and isolation or detection of *T. pallidum* in CSF. Thus, the bacterium's inherent invasiveness, not the integrity of host defenses, appears to be the primary determinant of CNS invasion.

3.3.8. Congenital syphilis

Rates of congenital syphilis closely parallel the rates of primary and secondary syphilis in young women (Radolf et al., 1999). In 1938, Beck and Dailey (1938) proposed that the prominence of the Langerhans cell layer of the cytotrophoblast during early pregnancy created a barrier against transplancental infection during the first 18 weeks of gestation. This theory was disproved, however, by the detection of spirochetes in 9 and 10 week abortuses (Benirschke, 1974; Harter and Benirschke, 1976) and by the demonstration of spirochetes in amniotic fluid as early as 17 weeks of pregnancy (Nathan et al., 1997). It is now believed that the fetus' inability to mount a characteristic histopathologic response to *T. pallidum* explains the belated recognition of transplacental syphilis during the first trimester.

The fundamental histologic lesion of congenital syphilis is an obliterative endarteritis typically consisting of mononuclear and plasma cell infiltrates surrounding blood vessels with intimal hyperplasia of larger vessels and swollen, hyperplastic endothelial cells. Fibrosis and granulomatous changes also are frequently found in congenitally infected tissues (Fraser, 1920; McCord, 1929; Hoffmann, 1936; Stowens, 1959). Inflammatory changes in the placenta may be as striking as those in the fetus and are under-utilized for confirming the diagnosis of congenital syphilis (Sheffield et al., 2002). The infected placenta is grossly large, thick, and pale. Typical microscopic abnormalities in the placenta, the "syphilis triad", consist of enlarged hypercellular villi, proliferative vascular changes, and acute or chronic villitis (McCord, 1934; Russell and Altshuler, 1974; Sheffield et al., 2002; Radolf et al., 1999). An inflammatory process within the cord matrix, termed funisitis, is well recognized (Fojaco et al., 1989; Jacques and Qureshi, 1992; Schwartz et al., 1995; Radolf et al., 1999). In a recent study from Bolivia, visualization of *T. pallidum* by immunohistochemical staining of cords with funisitis was strongly associated with marked inflammation and necrosis (Guarner et al., 2000).

Two years of age is used to demarcate early from late congenital syphilis. Because of its probable hematogenous origin, early congenital syphilis is analogous to the secondary stage of acquired infection and can involve almost any fetal organ, with liver, kidneys, bone, pancreas, spleen, lungs, heart, and brain the most frequently afflicted (Oppenheimer and Hardy, 1971; Radolf et al., 1999). As with acquired syphilis, *T. pallidum* has a remarkable propensity to invade the CNS of infected neonates. In a recent prospective study, *T. pallidum* was isolated from the CSFs of 22% of infants with untreated congenital syphilis and 60% of infants with abnormalities on physical examination consistent with congenital syphilis (Michelow et al., 2002). Interestingly, as is usually the case with early acquired syphilis in adults, clinically evident CNS involvement was not seen, implying that CNS invasion mainly poses a risk for the child's subsequent neurological development (Michelow et al., 2002). Late congenital syphilis corresponds to tertiary syphilis in the adult. The so-called stigmata of late congenital syphilis are developmental abnormalities caused by chronic inflammation; the best known are Hutchinson's teeth, interstitial keratitis, saddle nose deformity, frontal bossing, and saber shins (Radolf et al., 1999).

4. Virulence properties of *T. pallidum*

As noted at the outset, the search for the genes and gene products that enable *T. pallidum* to fulfill its complex parasitic program continues to be an extraordinarily difficult task given the lack of *in vitro* cultivation systems and the consequent inability to genetically manipulate the organism. Indeed, it is quite accurate to state that no *T. pallidum* molecule or constituent has been identified as a virulence determinant in the classical sense of the term (Finlay and Falkow, 1989). On the other hand, as discussed below, investigators have had a fair degree of success in defining some virulence properties of the bacterium even if these comprise at present somewhat disconnected pieces of a very intricate puzzle. The availability of the genomic sequence has markedly facilitated the search for candidate virulence determinants though, somewhat disappointingly, orthologs for virulence determinants that have been well characterized in other bacterial pathogens have not been identified. As will be discussed at the end of the chapter, bioinformatics strategies for defining virulence-related genes are beginning to emerge.

4.1. Adherence

It is generally well accepted that *T. pallidum* must attach to epithelial cells and extracellular matrix components within the dermis in order to gain the "foothold" needed to establish infection. The diversity of cell types to which *T. pallidum* binds (Baseman and Alderete, 1983; Fitzgerald, 1983; Thomas et al., 1985a; 1985b; 1985c) is consistent with the spirochete's broad tissue tropisms *in vivo*. The observation that syphilitic serum blocks the attachment of spirochetes to cells (Fitzgerald, 1983) has been interpreted as indicating that the treponemal constituents mediating attachment to tissue culture cells are expressed during infection and, presumably, function in a similar capacity *in vivo*. Failure to observe attachment of cultivatable treponemes to tissue culture cells was, at one time, cited as evidence that *T. pallidum* possesses virulence determinants lacking in "nonpathogenic" treponemes (Fitzgerald, 1983). This idea, however, is no longer tenable in light of *in vitro* attachment studies performed using oral treponemes (Baehni et al., 1992; De Filippo et al., 1995; Walker et al., 1999). Micrographs showing *T. pallidum* attached to cell surfaces by its tip generated speculation that the ends of the bacterium are specialized attachment organelles (Hayes et al., 1977; Fitzgerald, 1983). This viewpoint is not universally accepted, however, given that spirochetes also frequently appear to be attached to cells along variable portions of the cell cylinder. Konishi et al. (1986) proposed that *T. pallidum* can attach anywhere along its length but that the point of attachment slides towards the tip as a result of the bacterium's motility and the fluidity of its outer membrane (OM). Complicating interpretation of these electron microscopic data is the possibility that dehydration and fixation during processing for electron microscopy might alter the spatial relationships between bacteria and cells. An important observation to emerge from cell attachment studies is that cultivation with tissue culture cells prolongs the viability of *T. pallidum* and enables limited replication of the bacterium (Fitzgerald, 1983; Jenkin and Sandok, 1983). Cells derived from rabbit epithelium (Sf1ep) proved superior to other cell types for supporting replication of treponemes and, consequently, came to be used routinely in the co-cultivation systems developed in the early 1980s (Fieldsteel et al., 1981; Norris, 1982; Cox et al., 1984).

Binding studies have suggested that fibronectin and laminin are of particular importance for treponemal cytadherance (Peterson et al., 1983; Fitzgerald et al., 1984; Thomas et al., 1985b; Steiner et al., 1987; Baughn, 1987). Identifying the surface receptor(s) that mediate these interactions, however, has been challenging. Affinity-based studies in the 1980s identified several putative fibronectin-binding proteins whose molecular identities remain unknown (Peterson et al., 1983; Thomas et al., 1985c). The availability of the *T. pallidum* genomic sequence has breathed new life into the search for these receptors. Cameron (2003) recently confirmed that freshly harvested, motile treponemes bind fibronectin, collagen I and laminin. To identify the receptors mediating these activities, she used the PSORT algorithm to screen the *T. pallidum* genomic sequence *in silico* for candidate outer membrane proteins. Ten candidates then were expressed as recombinant polypeptides, purified, and assessed for their ability to bind to ECM components. One protein, TP0751, attached to laminin in a saturable, dose-dependent manner and has been partially characterized (Cameron, 2003). Two treponemal proteins, TP0155 and TP0483, demonstrated specific attachment to fibronectin, blocked adherence to fibronectin-coated slides, and bound fibronectin-producing cell lines (Cameron et al., 2004).

4.2. Infectivity and invasiveness

As noted earlier, clinical and experimental studies established early in the twentieth century that *T. pallidum* is highly infectious and invasive. Magnuson and colleagues (Magnuson et al., 1948; Magnuson et al., 1956) deserve much of the credit for perfecting the techniques for quantitatively assessing these hallmark virulence traits. Using titrated doses of the Nichols strain, they established that ten spirochetes or fewer inoculated intradermally will routinely infect rabbits and humans and that the ID_{50} of the Nichols strain for humans and rabbits are highly similar, 57 and 10 -23 organisms, respectively. Turner and Hollander (1957) and subsequently Izzat et al. (1970) demonstrated that inoculum size correlates directly with lesion severity and inversely with time to lesion development; each ten-fold decrease in inoculum size results in a 3 to 4 day increase in incubation period. Unfortunately, because of the heavy reliance upon the use of the Nichols isolate through the years, there is a paucity of information regarding the relative virulence of *T. pallidum* street strains. Using lesion severity following intradermal inoculation as an indicator of tissue spirochete burden, Turner and Hollander (1957) demonstrated modest differences in virulence among *T. pallidum* strains, with the Nichols isolate being one of the most virulent tested. They also made the intriguing observation that *T. pallidum* isolates were generally better than *T. pertenue* isolates at producing dermal lesions in rabbits, whereas the converse was true in hamsters.

In the early 1930s, Mahoney and Bryant (1933; 1934) demonstrated that *T. pallidum* can rapidly traverse intact genital mucosa in rabbits and could be detected in the underlying dermis within 2 to 3 h of exposure. In the 1980s, *in vitro* models were used to demonstrate the spirochete's invasiveness. *T. pallidum*, but not nonpathogenic treponemes, efficiently penetrates cultured vascular endothelial cells via intercellular junctions (Thomas et al., 1988). Electron microscopic visualization of treponemes traversing endothelial cells in biopsy specimens (Azar et al., 1970) suggests that this model mimics the spirochete's actual mode of penetrating vascular structures during hematogenous dissemination. Riviere and Thomas (1989) took this model one step further by using abdominal walls excised from mice to demonstrate the syphilis spirochete's capacity to traverse complex anatomical barriers analogous to those invaded during actual disease. Based on the widely accepted notion that motility is essential for treponemal invasiveness, investigators in the pre-genomics era extensively characterized the spirochete's organelles for motility, the periplasmic flagella (Isaacs et al., 1989; Charon and Goldstein, 2002; Champion et al., 1990; Norris et al., 1988). The presumptive importance of chemotaxis and motility for the syphilis spirochete's parasitic lifestyle is underscored by the finding that 5-6% of the syphilis spirochete's genome consists of ORFs with putative chemotaxis- and motility-related functions (Fraser et al., 1998). Results obtained in animal models and tissue culture systems with targeted motility and chemotaxis mutants in *Serpulina hyodysenteriae* (Rosey et al., 1996) and *T. denticola* (Lux et al., 2001) lead one to predict that deletion of the *T. pallidum* orthologs also would result in attenuated virulence. Comprehensive reviews of treponemal motility and chemotaxis can be found in the chapters by Charon and Lux.

4.3. Immunoevasiveness

Essential for the ability of *T. pallidum* to establish persistent infection is its capacity to evade host clearance mechanisms, a property we have termed "immunoevasiveness" (Radolf, 1994). Over the years, much thought and energy have been devoted to understanding how *T. pallidum* manages to accomplish this impressive feat, and there has been substantial evolution of our thinking on the subject during the past two decades. As noted earlier, suppression of host defenses was once a favored explanation for this phenomenon. In a similar vein, molecular mimicry was in vogue at one time (Marchitto et al., 1986; Baughn, 1987; Baughn et al., 1996), but this concept, centering about a small number of treponemal proteins, is equally at odds with the robust immunological responses now known to be elicited by virulent treponemes during acute infection. Consequently, during the past decade or so, attention has shifted towards understanding how the syphilis spirochete, or, perhaps a subset within a spirochetal population, survives the host's immunological onslaught during the acute and chronic stages of infection. Because *T. pallidum* has come to be regarded as an extracellular pathogen, much of this work has focused on the spirochete's resistance to humoral defenses. The validity of this line of investigation is bolstered by the simple observation that early syphilis patients characteristically have virulent treponemes in blood despite high titers of anti-treponemal antibodies directed against numerous *T. pallidum* antigens (Hanff et al., 1982; Baker-Zander et al., 1985). Nevertheless, there is no way to ascertain whether the immune evasion mechanisms operative during acute infection are those employed during latency. This facet of syphilis research needs fresh experimental approaches; particularly lacking is a means of studying the antigenic composition and gene expression profiles of spirochetes in different stages of disease.

It has been recognized for more than fifty years that motile treponemes resist binding the specific anti-treponemal antibodies in syphilitic serum. In 1949, Nelson and Mayer (1949) demonstrated the extraordinarily slow time course required for the immobilization of *T. pallidum* by antibody and complement. Hardy and Nell (1957) later showed that organisms could be agglutinated with immune sera only after prolonged storage ("aging") at 4°C, a process now believed to result in the loss of outer membrane integrity and exposure of subsurface antigenic targets. Investigators subsequently used a variety of approaches, including immunoelectron and immunofluorescence microscopy (Hovind-Hougen et al., 1979; Penn and Rhodes, 1982; Radolf et al., 1986), radioimmunoprecipitation (Stamm et al., 1987), and radioimmunoassay (Cox et al., 1992), to demonstrate that binding of antibody to the surface of intact treponemes is below the threshold of detectability by these assays. The poor antigenicity of motile treponemes formerly was attributed to the presence of a coat of serum proteins and/or mucopolysaccharides that render surface antigens inaccessible to antibodies (Christiansen, 1963; Alderete and Baseman, 1979; Fitzgerald and Johnson, 1979). However, beginning in the early 1980s, skepticism about the existence of this protective layer began to develop. Some investigators noted that such an external coat or capsule is inconsistent with the large amount of internal structure visible when freshly harvested, negatively stained treponemes are examined by whole-mount transmission electron microscopy (Hovind-Hougen, 1983; Radolf et al., 1986; Cox et al., 1992). Furthermore, serum protein constituents of the putative outer coat were not detected when freshly harvested treponemes were collected onto low-protein binding polycarbonate filters before being probed with specific antibodies in a highly sensitive radioimmunoassay format (Cox et al., 1992).

The difficulty in detecting the putative outer coat, coupled with the finding that the *T. pallidum* outer membrane is much more fragile than its Gram-negative counterparts (Cox et al., 1992), led Radolf and his co-workers to consider the possibility that the OM comprises a barrier to antibody binding because it contains a paucity of integral proteins. Over the years, complementary approaches have been used to garner support for this "outer membrane" hypothesis. One strategy was to prove that the major membrane immunogens of *T. pallidum* are sub-surface and, therefore, inaccessible to antibodies as long as the integrity of the outer membrane is maintained (Cox et al., 1992; 1995; Radolf et al., 1995b). Instrumental in this regard has been an immunofluorescence assay in which treponemes are encapsulated in porous agarose gel beads (gel microdroplets) to minimize damage to outer membranes during repeated washing steps (Cox et al., 1995). The discovery that the treponeme's membrane immunogens are lipoproteins (Chamberlain et al., 1989; Schouls et al., 1989; Fraser et al., 1998) has led to the notion that they are hydrophilic polypeptides anchored to the periplasmic leaflet of the cytoplasmic membrane by their N-terminal lipid head groups (Cox et al., 1992; Radolf, 1995). Failure to appreciate the lability of the outer membrane during routine experimental manipulation most likely explains why, in earlier studies (Jones et al., 1984; Norris and Sell, 1984; Hindersson et al., 1986), these essentially periplasmic proteins were inadvertently localized to the treponemal surface.

Freeze-fracture electron microscopy, a technique that visualizes integral membrane proteins within lipid bilayers as intramembranous particles (IMPs), has played a pivotal role in the formulation of our model of *T. pallidum* molecular architecture (Radolf, 1995; Blanco et al., 1997). In 1989, two groups independently reported that the density of IMPs in freeze-fractured *T. pallidum* OMs is two to three orders of magnitude less than in that of *Escherichia coli* (Radolf et al., 1989; Walker et al., 1989). These low-density particles were designated *T. pallidum* rare outer membrane proteins (Tromps) (Blanco et al., 1990; 1997). Using freeze-etching, a variant of freeze-fracture, it was demonstrated that the particles actually protrude from the spirochetal surface, furthering the case that they represent proteins with one or more transmembrane domains (Bourell et al., 1994; Radolf et al., 1989). Based on the uniformity of particle size, it was proposed that the OM particles represent a single protein or protein oligomer (Radolf et al., 1989; Radolf, 1995). Obviously, this notion is a major departure from contemporary views about the diversity of proteins in "conventional" bacterial outer membranes (Nikaido, 2003). Inasmuch as unrelated integral membrane proteins can form morphologically indistinguishable IMPs, it is possible, in retrospect, that the freeze-fracture and freeze-etching data were over-interpreted. As discussed in the chapter by Cameron, "mining" of the *T. pallidum* genome, in concert with other strategies for functionally characterizing treponemal open reading frames (ORFs), has resulted in the identification of a number of candidate Tromps (Cameron, 2003).

As noted already, efforts to directly detect the binding of antibodies to intact treponemes revealed that the *T. pallidum* surface behaves as if it were antigenically inert (Fehniger et al., 1986; Hovind-Hougen et al., 1979; Penn and Rhodes, 1982; Radolf et al., 1986; Stamm et al., 1987). On the other hand, biological assays utilizing live treponemes have consistently shown that syphilitic serum does, in fact, contain antibodies directed against treponemal surface constituents. Numerous investigators have shown that syphilitic sera can inhibit cytadherence (Fitzgerald et al., 1977a), promote phagocytosis (Lukehart and Miller, 1978; Alder et al., 1990; Baker-Zander and Lukehart, 1992; Lukehart et al., 1992; Shevchenko et al., 1999), and mediate complement-dependent treponemicidal activity (Nelson, Jr. and Mayer, 1949; Bishop and Miller, 1976a; 1976b). Blanco et al. (1990) extended these findings by demonstrating aggregation of OM particles following incubation of *T. pallidum* with rabbit syphilitic serum; the prolonged incubation periods (8 h to 16 h) required for this phenomenon led them to postulate that aggregation of rare OM proteins was the rate-limiting step for complement-mediated killing of *T. pallidum*. Thus, while antibodies appear to be generated against Tromps, the biological activities of these antibodies are likely to be constrained by the small number of available targets. Nevertheless, the fact that "functional" antibodies directed against *T. pallidum* surface constituents are generated during infection has profound implications; an OM of static antigenic composition, despite its intrinsically poor antigenicity, may not suffice for evading host humoral responses.

The *T. pallidum* genome encodes a 12-membered polymorphic gene family with sequence homology to the major surface protein (MSP) of *T. denticola* (Fraser et al., 1998), a multi-functional protein reported to have porin-like activity, to bind laminin and fibronectin, to form hexagonal OM arrays, and to disrupt the actin cytoskeletons of cultured gingival fibrobasts (Haapasalo et al., 1992; Egli et al., 1993; Fenno et al., 1996; Fenno et al., 1997; Amin et al., 2004). The discovery of the *T. pallidum* repeat (Tpr) family has helped to rejuvenate the quest for the syphilis spirochete's OM proteins and raised the possibility that it comprises a previously undetected system for antigenic variation. Of the several Tprs with a predicted OM location, TprK, thus far, has attracted the most attention. In 1999, Centurion-Lara and colleagues (1999) reported that (i) TprK is the predominantly expressed member of the Tpr family during experimental infection; (ii) anti-TprK antibodies have *in vitro* opsonic activity for *T. pallidum*; and (iii) animals immunized with recombinant TprK show substantial protection against intradermal challenge with *T. pallidum*. Although Hazlett et al. (2001) have presented evidence that TprK is actually periplasmic, reports of TprK sequence diversity (Stamm and Bergen, 2000; Centurion-Lara et al., 2000a; LaFond et al., 2003) have bolstered the idea that it is a surface-exposed OM protein. Based on genomic sequence and Southern hybridization data showing that *tprK* is a single-copy gene (Fraser et al., 1998; Stamm and Bergen, 2000), it was deduced that PCR amplification of multiple *tprK* alleles from *T. pallidum* clinical isolates indicates the existence of genetically different subpopulations within treponemal isolates (Stamm and Bergen, 2000; Centurion-Lara et al., 2000a). This line of thought was extended by demonstrating heterogeneity of *tprK* sequences amplified directly from syphilitic chancres of three patients and by the observation that antigenic variants accumulate during successive passages of a clinical isolate (LaFond et al., 2003). For unclear reasons, infection with the Nichols strain appears to give rise to much less *tprK* sequence variation than does infection with isolates obtained more recently (Stamm and Bergen, 2000; Centurion-Lara et al., 2000a). In an elegant recent study, Centurion-Lara et al. (2004) described the progression of TprK variants during infection of rabbits with a clonal *T. pallidum* ("Chicago C") strain. They proposed that sequence diversity arises from a segmented gene conversion mechanism in which donor sequences adjacent to the *tprD* gene are copied and inserted into seven discrete variable (V) regions of the recipient *tprK* gene. Additional studies by the Seattle group have shown that antibodies to TprK generated during infection are specifically targeted to the V regions (Morgan et al., 2003; Morgan et al., 2002), suggesting that host immune pressures are involved in the selection of TprK variants. Other members of the Tpr protein family appear to be subject to much more limited degrees of sequence variation (Centurion-Lara et al., 2000b; Sun et al., 2004).

4.4. Mining the genome-the final frontier

Identification of virulence determinants was one of the primary rationales for sequencing of the *T. pallidum* genome (Fraser et al., 1998; Norris et al., 1998). Unfortunately, this aspect of the sequencing project has yielded a limited dividend to date. Only one ORF in the genome, TP0516 (*mviN*), was annotated by TIGR as a virulence determinant. While *mviN* has orthologs in a number of bacterial genomes, including those of pathogens, as best we can tell, the evidence for its having a definite virulence-related function is scant. Five ORFs are annotated as hemolysins. One of these (TP1037) has sequence similarity to hemolysin III of *Bacillus cereus* (Baida and Kuzmin, 1995), while the other four paralogous genes (TP0027, TP0028, TP0649, and TP0936) share significant sequence similarity with the *tlyC* hemolysin from *Serpulina hyodysenteriae* (ter Huurne et al., 1994). Hemolysis is not a clinical feature of syphilis so the significance of these homologies is hard to discern. An alternative possibility is that these ORFs encode cytotoxins, but this idea is also hard to reconcile with our current conception of syphilis as being the product of immune-mediated processes driven by the presence of treponemes. *T. pallidum* is cytotoxic *in vitro* for neuroblasts and some other cell types but only at extraordinarily high multiplicities of infection (Fitzgerald and Repesh, 1983). *In silico* topological analysis brings to light additional problems with the annotation of these genes as hemolysins. Three (TP0027, TP0028, and TP1037) are predicted by PSORT to be polytopic cytoplasmic membrane proteins which are unlikely to be secreted into the extracellular milieu. TP0649 is predicted to be cytoplasmic and would require some type of secretory apparatus, which was not identified in *T. pallidum*, in order to become extracellular. TP0936 is predicted by PSORT to be periplasmic, but visual inspection of its N-terminus revealed that it has an extremely poor signal peptide and is probably also cytoplasmic. In short, one can question the validity of these designations on both pathogenesis and molecular grounds.

It would be premature, and almost certainly incorrect, however, to conclude that *T. pallidum* lacks virulence determinants. Of the 1041 predicted ORFs in the genome, predicted biological roles could be assigned to only 55% when the *T. pallidum* genome was originally annotated (Fraser et al., 1998). Given the spirochete's evolutionary "decision" to dispense with large portions of its metabolic capacity in exchange for near-total reliance on the human host for nutritional support, one can speculate that a substantial proportion of the treponemal genome is available to participate in the infectious process. Lacking targeted or generalized mutagenesis systems, how do we go about identifying novel virulence-related genes? Recently, Garbom et al. (2004) described a bioinformatics strategy which enabled them to identify potential virulence-associated genes (*vag* genes) by comparing the *T. pallidum* genome with those of several other bacterial pathogens; mutagenesis of the corresponding orthologs in *Yersinia pestis* confirmed that a number were involved in disease production. On closer examination, however, this work cannot be considered an unqualified success. While the stated objective of their strategy was to screen ORFs of unknown function for virulence-associated factors, nearly all of the *T. pallidum vag* genes have homologies with ORFs in other bacteria that have been annotated as physiologic or metabolic enzymes and thus, are unlikely to be virulence-related factors in the traditional sense. The reason for the discrepancy between the authors' objective and their results was not addressed but probably reflects the fact that the *T. pallidum* genome was sequenced and annotated when bioinformatics was in its infancy. If correct, then a re-annotation of the *T. pallidum* genome, exploiting the wealth of bioinformatics currently available, will be needed in order to fully implement their (or any other) comparative genomics strategy. The more serious limitation of comparative genomics approaches, however, is that they can be used only to study *T. pallidum* genes that have orthologs in other bacteria. Without being able to generate knockouts for selected genes, it will be very difficult to identify virulence-related *T. pallidum*-specific genes.

5. Future directions

Winston Churchill once described Soviet Russia as "a riddle wrapped in a mystery inside an enigma." Nearly one hundred years after its discovery, one might argue that the same phrase aptly describes *T. pallidum.* While such a viewpoint may be more nihilistic than is warranted, it is safe to assume that the syphilis spirochete is not going to give up its secrets without a struggle. Nevertheless, the goals of contemporary syphilis pathogenesis research are clear even if the paths to achieve them are not. Efforts to cultivate the bacterium need to be redoubled; nearly two decades have passed since this work was conducted in earnest. Pending this singular and long-awaited breakthrough, there must be even greater reliance upon genetically malleable surrogate systems. Advances in the genetic manipulation of *B. burgdorferi* (Rosa et al., 2005) and *Treponema denticola* (see Chapter 8, Kuramitsu) indicate that it may soon be possible to study *T. pallidum* orthologs in organisms that are phylogenetically "closer to home." Because the surface of the bacterium is its interface with the host, OM proteins need to be definitively identified and characterized. The methodologies and model systems required for this work are largely in hand and can be further optimized. Latency is another area that needs renewed attention. Assuming that latency in humans and rabbits is similar, use of the rabbit model, in conjunction with PCR-based techniques for detection of spirochetes and spirochetal transcripts, should provide novel insights into this stage of disease. As mentioned earlier, microarray technology has recently been applied to syphilis research; the time is now ripe for efforts to elucidate how the spirochete orchestrates its minimalist genome to make a living *in vivo*. Such studies may provide physiological information critical for solving the *in vitro* cultivation problem as well as identifying virulence-related functions of *T. pallidum*-specific genes. In the past, much emphasis has been placed on vaccine development. This may be an example of the proverbial "putting the cart before the horse." It is possible that we need to understand much more about how *T. pallidum* interacts with and survives within its human host before we can develop effective strategies for curtailing these interactions.

Acknowledgment

Supported by grants AI-26756 and AI-38894 (JDR), AI-10573 (KROH), and AI -34616, AI- 63940, AI-42143 (SAL) from the National Institutes of Health

References

Abell, E., Marks, R., and Jones, E.W. (1975). Secondary syphilis: a clinico-pathological review. Br. J. Dermatol. *93*, 53-61.

Aderem, A. and Ulevitch, R.J. (2000). Toll-like receptors in the induction of the innate immune response. Nature *406*, 782-787.

Alder, J.D., Friess, L., Tengowski, M., and Schell, R.F. (1990). Phagocytosis of opsonized *Treponema pallidum* subsp. *pallidum* proceeds slowly. Infect. Immun. *58*, 1167-1173.

Alderete, J.F. and Baseman, J.B. (1979). Surface-associated host proteins on virulent *Treponema pallidum*. Infect. Immun. *26*, 1048-1056.

Alexander, L.J., Schoch, A.G., and Mantooth, W.B. (1949). Abortive treatment of syphilis. Results obtained in the incubation, primary, and secondary stages of syphilis. Am. J. Syph. Gon. Vener. Dis. *33*, 429-436.

Alexopoulou, L., Thomas, V., Schnare, M., Lobet, Y., Anguita, J., Schoen, R.T., Medzhitov, R., Fikrig, E., and Flavell, R.A. (2002). Hyporesponsiveness to vaccination with *Borrelia burgdorferi* OspA in humans and in TLR1- and TLR2-deficient mice. Nat. Med. *8*, 878-884

Amin, M., Ho, A.C., Lin, J.Y., Batista da Silva, A.P., Glogauer, M., and Ellen, R.P. (2004). Induction of *de novo* subcortical actin filament assembly by *Treponema denticola* major outer sheath protein. Infect. Immun. *72*, 3650-3654.

Ansell, D.A., Hu, T.C., Straus, M., Cohen, M., and Sherer, R. (1994). HIV and syphilis seroprevalence among clients with sexually transmitted diseases attending a walk-in clinic at Cook County Hospital. Sex Transm. Dis. *21*, 93-96.

Antal, G.M., Lukehart, S.A., and Meheus, A.Z. (2002). The endemic treponematoses. Microbes. Infect. *4*, 83-94.

Arroll, T.W., Centurion-Lara, A., Lukehart, S.A., and Van Voorhis, W.C. (1999). T-Cell responses to *Treponema pallidum* subsp. *pallidum* antigens during the course of experimental syphilis infection. Infect. Immun. *67*, 4757-4763.

Azar, H.A., Pham, T.D., and Kurban, A.K. (1970). An electron microscopic study of a syphilitic chancre. Engulfment of *Treponema pallidum* by plasma cells. Arch. Pathol. *90*, 143-150.

Baehni, P.C., Song, M., McCulloch, C.A., and Ellen, R.P. (1992). *Treponema denticola* induces actin rearrangement and detachment of human gingival fibroblasts. Infect. Immun. *60*, 3360-3368.

Baida, G.E. and Kuzmin, N.P. (1995). Cloning and primary structure of a new hemolysin gene from *Bacillus cereus*. Biochim. Biophys. Acta *1264*, 151-154.

Baker-Zander, S. and Sell, S. (1980). A histopathologic and immunologic study of the course of syphilis in the experimentally infected rabbit. Demonstration of long-lasting cellular immunity. Am. J. Pathol. *101*, 387-413.

Baker-Zander, S.A., Fohn, M.J., and Lukehart, S.A. (1988). Development of cellular immunity to individual soluble antigens of *Treponema pallidum* during experimental syphilis. J. Immunol. *141*, 4363-4369.

Baker-Zander, S.A., Hook, E.W., III, Bonin, P., Handsfield, H.H., and Lukehart, S.A. (1985).

Antigens of *Treponema pallidum* recognized by IgG and IgM antibodies during syphilis in humans. J. Infect. Dis. *151*, 264-272.

Baker-Zander, S.A. and Lukehart, S.A. (1992). Macrophage-mediated killing of opsonized *Treponema pallidum*. J. Infect. Dis. *165*, 69-74.

Baker-Zander, S.A., Shaffer, J.M., and Lukehart, S.A. (1993). Characterization of the serum requirement for macrophage-mediated killing of *Treponema pallidum* ssp. *pallidum*: relationship to the development of opsonizing antibodies. FEMS Immunol. Med. Microbiol. *6*, 273-279.

Baseman, J.B. and Alderete, J.F. (1983). The parasitic strategies of *Treponema pallidum*. In The Pathogenesis and Immunology of Treponemal Infection, R.F.Schell and D.M.Musher, eds. (New York: Marcel Dekker, Inc.), pp. 229-240.

Baseman, J.B. and Hayes, N.S. (1974). Protein synthesis by *Treponema pallidum* extracted from infected rabbit tissue. Infect. Immun. *10*, 1350-1355.

Baughn, R.E. (1987). Role of fibronectin in the pathogenesis of syphilis. Rev. Infect. Dis. *9 Suppl.4*, S372-S385.

Baughn, R.E., Jiang, A., Abraham, R., Ottmers, V., and Musher, D.M. (1996). Molecular mimicry between an immunodominant amino acid motif on the 47-kDa lipoprotein of *Treponema pallidum* (Tpp47) and multiple repeats of analogous sequences in fibronectin. J. Immunol. *157*, 720-731.

Baughn, R.E., McNeely, M.C., Jorizzo, J.L., and Musher, D.M. (1986). Characterization of the antigenic determinants and host components in immune complexes from patients with secondary syphilis. J. Immunol. *136*, 1406-1414.

Beck, A.C. and Dailey, W.T. (1938). Syphilis in pregnancy. Pub. Am. Assoc. Adv. Sci. *6*, 101.

Benirschke, K. (1974). Syphilis--the placenta and the fetus. Am. J. Dis. Child. *128*, 142-143.

Benoit, S., Posey, J.E., Chenoweth, M.R., and Gherardini, F.C. (2001). *Treponema pallidum* 3-phosphoglycerate mutase is a heat-labile enzyme that may limit the maximum growth temperature for the spirochete. J. Bacteriol. *183*, 4702-4708.

Beutler, B. and Munford, R.S. (1996). Tumor necrosis factor and the Jarisch-Herxheimer reaction. N. Engl. J. Med. *335*, 347-348.

Bishop, N.H. and Miller, J.N. (1976a). Humoral immunity in experimental syphilis. I. The demonstration of resistance conferred by passive immunization. J. Immunol. *117*, 191-196.

Bishop, N.H. and Miller, J.N. (1976b). Humoral immunity in experimental syphilis. II. The relationship of neutralizing factors in immune sera to acquired resistance. J. Immunol. *117*, 197-207.

Bjerke, J.R., Krogh, H.K., and Matre, R. (1981). *In situ* identification of mononuclear cells in cutaneous infiltrates in discoid lupus erythematosus, sarcoidosis and secondary syphilis. Acta Dermatologica Venereologica *61*, 371-380.

Blanco, D.R., Miller, J.N., and Lovett, M.A. (1997). Surface antigens of the syphilis spirochete and their potential as virulence determinants. Emer. Infect. Dis. *3*, 11-20.

Blanco, D.R., Walker, E.M., Haake, D.A., Champion, C.I., Miller, J.N., and Lovett, M.A. (1990). Complement activation limits the rate of *in vitro* treponemicidal activity and correlates with antibody-mediated aggregation of *Treponema pallidum* rare outer membrane protein. J. Immunol. *144*, 1914-1921.

Bourell, K.W., Schulz, W., Norgard, M.V., and Radolf, J.D. (1994). *Treponema pallidum* rare outer membrane proteins: analysis of mobility by freeze-fracture electron microscopy. J. Bacteriol. *176*, 1598-1608.

Brightbill, H.D., Libraty, D.H., Krutzik, S.R., Yang, R.B., Belisle, J.T., Bleharski, J.R., Maitland, M., Norgard, M.V., Plevy, S.E., Smale, S.T., Brennan, P.J., Bloom, B.R., Godowski, P.J., and Modlin, R.L. (1999). Host defense mechanisms triggered by microbial lipoproteins through toll-like receptors. Science *285*, 732-736.

Brooks, C. and Akins, D. R. Microarray-based genetic analysis of *Borrelia burgdorferi* in a mammalian host-adapted state. 2002.

Brooks, C.S., Hefty, P.S., Jolliff, S.E., and Akins, D.R. (2003). Global analysis of *Borrelia burgdorferi* genes regulated by mammalian host-specific signals. Infect Immun. *71*, 3371-3383.

Brown, W.H. and Pearce, L. (1920). A note on the dissemination of *Spirochaeta pallida* from the primary focus of infection. Arch. Derm. Syph. *2*, 470-472.

Bryceson, A.D. (1976). Clinical pathology of the Jarisch-Herxheimer reaction. J. Infect. Dis. *133*, 696-704.

Cameron, C.E. (2003). Identification of a *Treponema pallidum* laminin-binding protein. Infect. Immun. *71*, 2525-2533.

Cameron, C.E., Brown, E.L., Kuroiwa, J.M., Schnapp, L.M., and Brouwer, N.L. (2004). *Treponema pallidum* fibronectin-binding proteins. J. Bacteriol. *186*, 7019-7022.

Cameron, C.E., Castro, C., Lukehart, S.A., and Van Voorhis, W.C. (1999). Sequence conservation of glycerophosphodiester phosphodiesterase among *Treponema pallidum* strains. Infect. Immun. *67*, 3168-3170.

Centers for Disease Control and Prevention (2002). Sexually Transmitted Diseases Treatment Guidelines. Morbid. Mortal. Weekly Rep. *51, RR-6*, 1-84.

Centers for Disease Control and Prevention (2004). Transmission of primary and secondary syphilis by oral sex--Chicago, Illinois, 1998-2002. MMWR Morb. Mortal. Wkly. Rep. *53*, 966-968.

Centurion-Lara, A., Castro, C., Barrett, L., Cameron, C., Mostowfi, M., Van Voorhis, W.C., and Lukehart, S.A. (1999). *Treponema pallidum* major sheath protein homologue TprK is a target of opsonic antibody and the protective immune response. J. Exp. Med. *189*, 647-656.

Centurion-Lara, A., Castro, C., Castillo, R., Shaffer, J.M., Van Voorhis, W.C., and Lukehart, S.A. (1998). The flanking region sequences of the 15-kDa lipoprotein gene differentiate pathogenic treponemes. J. Infect. Dis. *177*, 1036-1040.

Centurion-Lara, A., Godornes, C., Castro, C., Van Voorhis, W.C., and Lukehart, S.A. (2000a). The *tprK* gene is heterogeneous among *Treponema pallidum* strains and has multiple alleles. Infect. Immun. *68*, 824-831.

Centurion-Lara, A., LaFond, R.E., Hevner, K., Godornes, C., Molini, B.J., Van Voorhis, W.C., and Lukehart, S.A. (2004). Gene conversion: a mechanism for generation of heterogeneity in the *tprK* gene of *Treponema pallidum* during infection. Mol. Microbiol. *52*, 1579-1596.

Centurion-Lara, A., Sun, E.S., Barrett, L.K., Castro, C., Lukehart, S.A., and Van Voorhis, W.C. (2000b). Multiple alleles of *Treponema pallidum* repeat gene D in *Treponema pallidum* isolates. J. Bacteriol. *182*, 2332-2335.

Chamberlain, N.R., Brandt, M.E., Erwin, A.L., Radolf, J.D., and Norgard, M.V. (1989). Major integral membrane protein immunogens of *Treponema pallidum* are proteolipids. Infect. Immun. *57*, 2872-2877.

Champion, C.I., Miller, J.N., Lovett, M.A., and Blanco, D.R. (1990). Cloning, sequencing, and expression of two class B endoflagellar genes of *Treponema pallidum* subsp. *pallidum* encoding the 34.5 and 31.0 kilodalton proteins. Infect. Immun. *58*, 1697-1704.

Chapel, T.A. (1980). The signs and symptoms of secondary syphilis. Sex. Transm. Dis. *7*, 161-164.

Charon, N.W. and Goldstein, S.F. (2002). Genetics of motility and chemotaxis of a fascinating group of bacteria: The Spirochetes. Annu. Rev. Genet. *36*, 47-73.

Chesney, A.M. and Kemp, J.E. (1924). Incidence of *Spirochaeta pallidum* in cerebrospinal fluid during early stages of syphilis. JAMA *22*, 1725-1728.

Christiansen, S. (1963). Protective layer covering pathogenic treponemata. Lancet *1*, 423-425.

Clark, E.G. and Danbolt, N. (1955). The Oslo study of the natural history of untreated syphilis. An epidemiologic investigation based on a restudy of the Boeck-Bruusgaard material. J. Chronic Dis. *2*, 311-344.

Cover, W.H., Norris, S.J., and Miller, J.N. (1982). The microaerophilic nature of *Treponema pallidum*: enhanced survival and incorporation of tritiated adenine under microaerobic conditions in the presence or absence of reducing compounds. Sex. Transm. Dis. *7*, 1-8.

Cox, C.D. (1983). Metabolic activities. In Pathogenesis and Immunology of Treponemal Infection, R.F.Schell and D.M.Musher, eds. (New York: Marcel Dekker, Inc.), pp. 57-70.

Cox, C.D. and Barber, M.K. (1974). Oxygen uptake by *Treponema pallidum*. Infect. Immun. *10*, 123-127.

Cox, D.L. (1994). Culture of *Treponema pallidum*. Methods Enzymol. *236*, 390-405.

Cox, D.L., Akins, D.R., Porcella, S.F., Norgard, M.V., and Radolf, J.D. (1995). *Treponema pallidum* in gel microdroplets: a novel strategy for investigation of treponemal molecular architecture. Mol. Microbiol. *15*, 1151-1164.

Cox, D.L., Chang, P., McDowall, A., and Radolf, J.D. (1992). The outer membrane, not a coat of host proteins, limits the antigenicity of virulent *Treponema pallidum*. Infect. Immun. *60*, 1076-1083.

Cox, D.L., Moeckli, R.A., and Fieldsteel, A.H. (1984). Cultivation of pathogenic treponema in tissue cultures of SflEp cells. In Vitro *20*, 879-883.

Cox, D.L., Riley, B., Chang, P., Sayahtaheri, S., Tassell, S., and Hevelone, J. (1990). Effects of molecular oxygen, oxidation-reduction potential, and antioxidants upon *in vitro* replication of *Treponema pallidum* subsp. *pallidum*. Appl. Environ. Microbiol. *56*, 3063-3072.

De Filippo, A.B., Ellen, R.P., and McCulloch, C.A. (1995). Induction of cytoskeletal rearrangements and loss of volume regulation in epithelial cells by *Treponema denticola*. Arch. Oral Biol. *40*, 199-207.

Dickerson, M.C., Johnston, J., Delea, T.E., White, A., and Andrews, E. (1997). The causal role for genital ulcer disease as a risk factor for transmission of human immunodeficiency virus. An application of the Bradford Hill criteria. Sex. Transm. Dis. *23*, 429-440.

Dorfman, D.H. and Glaser, J.H. (1990). Congenital syphilis presenting in infants after the newborn period. N. Engl. J. Med. *323*, 1299-1302.

Drusin, L.M., Rouiller, G.C., and Chapman, G.B. (1969). Electron microscopy of *Treponema pallidum* occurring in a human primary lesion. J. Bacteriol. *97*, 951-955.

Egli, C., Leung, W.K., Muller, K.H., Hancock, R.E., and McBride, B.C. (1993). Pore-forming properties of the major 53-kilodalton surface antigen from the outer sheath of *Treponema denticola*. Infect. Immun. *61*, 1694-1699.

Engelkens, H.J., Vuzevski, V.D., ten Kate, F.J., van der Heul, P., Van der Sluis, J.J., and Stolz, E. (1991). Ultrastructural aspects of infection with *Treponema pallidum* subspecies *pertenue* (Pariaman strain). Genitourin. Med. *67*, 403-407.

Erbelding, E. and Rompalo, A. (2004). Changing epidemiology of syphilis and its persistent relationship with HIV. Curr. Infect. Dis. Rep. *6*, 135-140.

Fehniger, T.E., Radolf, J.D., Walfield, A.M., Cunningham, T.M., Miller, J.N., and Lovett, M.A. (1986). Native surface location of a recombinant 38-kDa *Treponema pallidum* antigen isolated from the *Escherichia coli* outer membrane. Infect. Immun. *52*, 586-592.

Fekade, D., Knox, K., Hussein, K., Melka, A., Lalloo, D.G., Coxon, R.E., and Warrell, D.A. (1996). Prevention of Jarisch-Herxheimer reactions by treatment with antibodies against tumor necrosis factor α. N. Engl. J. Med. *335*, 311-315.

Fenno, J.C., Muller, K.H., and McBride, B.C. (1996). Sequence analysis, expression, and binding activity of recombinant major outer sheath protein (Msp) of *Treponema denticola*. J. Bacteriol. *178*, 2489-2497.

Fenno, J.C., Wong, G.W., Hannam, P.M., Muller, K.H., Leung, W.K., and McBride, B.C. (1997). Conservation of *msp*, the gene encoding the major outer membrane protein of oral *Treponema* spp. J. Bacteriol. *179*, 1082-1089.

Fieldsteel, A.H., Cox, D.L., and Moeckli, R.A. (1981). Cultivation of virulent *Treponema pallidum* in tissue culture. Infect. Immun. *32*, 908-915.

Fieldsteel, A.H., Cox, D.L., and Moeckli, R.A. (1982). Further studies on replication of virulent *Treponema pallidum* in tissue cultures of Sf1Ep cells. Infect. Immun. *35*, 449-455.

Finlay, B.B. and Falkow, S. (1989). Common themes in microbial pathogenicity. Microbiol. Rev. *53*, 210-230.

Fitzgerald, T.J. (1983). Attachment of treponemes to cell surfaces. In Pathogenesis and Immunology of Treponemal Infection, R.F.Schell and D.M.Musher, eds. (New York: Marcel Dekker, Inc.), pp. 195-228.

Fitzgerald, T.J. and Johnson, R.C. (1979). Surface mucopolysaccharides of *Treponema pallidum*. Infect. Immun. *24*, 244-251.

Fitzgerald, T.J., Johnson, R.C., Miller, J.N., and Sykes, J.A. (1977a). Characterization of the attachment of *Treponema pallidum* (Nichols strain) to cultured mammalian cells and the potential relationship of attachment to pathogenicity. Infect. Immun. *18*, 467-478.

Fitzgerald, T.J., Johnson, R.C., and Smith, M. (1976). Accidental laboratory infection with *Treponema pallidum*, Nichols strain. J. Am. Vener. Dis. Assoc. *3*, 76-78.

Fitzgerald, T.J., Johnson, R.C., Sykes, J.A., and Miller, J.N. (1977b). Interaction of *Treponema pallidum* (Nichols strain) with cultured mammalian cells: effects of oxygen, reducing agents, serum supplements, and different cell types. Infect. Immun. *15*, 444-452.

Fitzgerald, T.J., Repesh, A., Blanco, D.R., and Miller, J.N. (1984). Attachment of *Treponema pallidum* to fibronectin, laminin, collagen IV, and collagen I and blockage of attachment by immune rabbit IgG. Br. J. Vener. Dis. *60*, 357-363.

Fitzgerald, T.J. and Repesh, L.A. (1983). Toxic activities of *Treponema pallidum*. In Pathogenesis and Immunology of Treponemal Infection, R.F.Schell and D.M.Musher, eds. (New York: Marcell Dekker, Inc.), pp. 173-193.

Fleming, D.T. and Wasserheit, J.N. (1999). From epidemiological synergy to public health policy and practice: the contribution of other sexually transmitted diseases to sexual transmission of HIV infection. Sex Transm. Infect. *75*, 3-17.

Fojaco, R.M., Hensley, G.T., and Moskowitz, L. (1989). Congenital syphilis and necrotizing funisitis. JAMA *261*, 1788-1790.

Fraser, C.M., Norris, S.J., Weinstock, G.M., White, O., Sutton, G.G., Dodson, R., Gwinn, M., Hickey, E.K., Clayton, R., Ketchum, K.A., Sodergren, E., Hardham, J.M., McLeod, M.P., et al. (1998). Complete genome sequence of *Treponema pallidum*, the syphilis spirochete. Science *281*, 375-388.

Fraser, J.F. (1920). The pathology of congenital syphilis. Acta Dermatol Syph *38*, 491.

Fribourg-Blanc, A. and Mollaret, H.H. (1963). Confirmation sérologique et microscopique de la tréponémose du cynocéphale de Guinée. Bulletin de la Société Pathologie Exotique *59*, 54-59.

Fribourg-Blanc, A. and Mollaret, H.H. (1969). Natural treponematosis of the African primate. Primates in Medicine *3*, 113-121.

Friedmann, P.S. and Turk, J.L. (1978). The role of cell-mediated immune mechanisms in syphilis in Ethiopia. Clin. Exp. Immunol. *31*, 59-65.

From, E., Thestrup-Pedersen, K., and Thulin, H. (1976). Reactivity of lymphocytes from patients with syphilis towards *T. pallidum* antigen in the leucocyte migration and lymphocyte transformation tests. Br. J. Vener. Dis. *52*, 224-229.

Gadkari, D.A., Quinn, T.C., Gangakhedkar, R.R., Mehendale, S.M., Divekar, A.D., Risbud, A.R., Chan-Tack, K., Shepherd, M., Gaydos, C., and Bollinger, R.C. (1998). HIV-1 DNA shedding in genital ulcers and its associated risk factors in Pune, India. J. Acquir. Immune. Defic. Syndr. Hum. Retrovirol. *18*, 277-281.

Galvin, S.R. and Cohen, M.S. (2004). The role of sexually transmitted diseases in HIV transmission. Nat. Rev. Microbiol. *2*, 33-42.

Garbom, S., Forsberg, A., Wolf-Watz, H., and Kihlberg, B.M. (2004). Identification of novel virulence-associated genes via genome analysis of hypothetical genes. Infect. Immun. *72*, 1333-1340.

Garnett, G.P., Aral, S.O., Hoyle, D.V., Cates, W., Jr., and Anderson, R.M. (1997). The natural history of syphilis. Implications for the transmission dynamics and control of infection. Sex Transm. Dis. *24*, 185-200.

Gjestland, T. (1955). The Oslo study of untreated syphilis; an epidemiologic investigation of the natural course of the syphilitic infection based upon a re-study of the Boeck-Bruusgaard material. Acta Derm Venereol. *35(Suppl 34)*, 3-368

Golden, M.R., Marra, C.M., and Holmes, K.K. (2003). Update on syphilis: resurgence of an old problem. JAMA *290*, 1510-1514.

Greenblatt, R.M., Lukehart, S.A., Plummer, F.A., Quinn, T.C., Critchlow, C.W., Ashley, R.L., D'Costa, L.J., Ndinya-Achola, J.O., Corey, L., Ronald, A.R., and Holmes, K.K. (1988). Genital ulceration as a risk factor for human immunodeficiency virus infection. AIDS *2*, 47-50.

Guarner, J., Southwick, K., Greer, P., Bartlett, J., Santander, A., Blanco, S., Pope, V., Levine, W., and Zaki, S. (2000). Testing umbilical cords for funisitis due to *Treponema pallidum* infection, Bolivia. Emerg. Infect. Dis. *6*, 487-492.

Gueft, B. and Rosahn, P.D. (1948). Experimental mouse syphilis, a critical review of the literature. Am. J. Syph. Gonor. Vener. Dis. *32*, 59-88.

Haapasalo, M., Muller, K.H., Uitto, V.J., Leung, W.K., and McBride, B.C. (1992). Characterization, cloning, and binding properties of the major 53-kilodalton *Treponema denticola* surface antigen. Infect. Immun. *60*, 2058-2065.

Handsfield, H.H., Lukehart, S.A., Sell, S., Norris, S.J., and Holmes, K.K. (1983). Demonstration of *Treponema pallidum* in a cutaneous gumma by indirect immunofluorescence. Arch. Dermatol. *119*, 677-680.

Hanff, P.A., Fehniger, T.E., Miller, J.N., and Lovett, M.A. (1982). Humoral immune response in human syphilis to polypeptides of *Treponema pallidum*. J. Immunol. *129*, 1287-1291.

Hardy, P.H., Jr. and Levin, J. (1983). Lack of endotoxin in *Borrelia hispanica* and *Treponema pallidum*. Proc. Soc. Exp. Biol. Med. *174*, 47-52.

Hardy, P.H., Jr. and Nell, E.E. (1957). Study of the antigenic structure of *Treponema pallidum* by specific agglutination. Am. J. Hyg. *66*, 160-172.

Harter, C. and Benirschke, K. (1976). Fetal syphilis in the first trimester. Am. J. Obstetr. Gynecol. *124*, 705-711.

Hayes, N.S., Muse, K.E., Collier, A.M., and Baseman, J.B. (1977). Parasitism by virulent *Treponema pallidum* of host cell surfaces. Infect. Immun. *17*, 174-186.

Hazlett, K.R.O., Cox, D.L., Sikkink, R.A., Auch'ere, F., Rusnak, F., and Radolf, J.D. (2002). Contribution of neelaredoxin to oxygen tolerance by *Treponema pallidum*. Methods Enzymol. *353*, 140-156.

Hazlett, K.R.O., Sellati, T.J., Nguyen, T.T., Cox, D.L., Clawson, M.L., Caimano, M.J., and Radolf, J.D. (2001). The TprK protein of *Treponema pallidum* is periplasmic and is not a target of opsonic antibody or protective immunity. J. Exp. Med. *193*, 1015-1026.

Hindersson, P., Cockayne, A., Schouls, L.M., and Van Embden, J.D.A. (1986). Immunochemical characterization and purification of *Treponema pallidum* antigen TpD expressed by *Escherichia coli* K12. Sex. Transm. Dis. *13*, 237-244.

Hira, S.K., Patel, J.S., Bnhart, S.G., Chilikima, K., and Mooney, N. (1987). Clinical manifestations of secondary syphilis. Int. J. Dermatol. *26*, 103-107.

Hirschfeld, M., Kirschning, C.J., Schwandner, R., Wesche, H., Weis, J.H., Wooten, R.M., and Weis, J.J. (1999). Inflammatory signaling by *Borrelia burgdorferi* lipoproteins is mediated by toll-like receptor 2. J. Immunol. *163*, 2382-2386.

Hoffmann, E. (1936). Congenital syphilis in light of 30 year's investigation of the spirochete and 25 year's experience with salvarsan. J. Pediatr. *9*, 569.

Hollander, D.H. and Turner, T.B. (1954). The role of temperature in experimental treponemal infection. Am. J. Syph. Gon. Vener. Dis. *38*, 489-505.

Hollier, L.M., Harstad, T.W., Sanchez, P.J., Twickler, D.M., and Wendel, G.D., Jr. (2001). Fetal syphilis: clinical and laboratory characteristics. Obstet. Gynecol. *97*, 947-953.

Hook, E.W., III (1991). Central nervous system syphilis. In Infections of the Central Nervous System, W.M.Scheld, R.J.Whitley, and D.T.Durack, eds. (New York: Raven Press, Ltd.), pp. 639-656.

Horowitz, H.W., Valsamis, M.P., Wicher, V., Abbruscato, F., Larsen, S.A., Wormser, G.P., and Wicher, K. (1994). Brief report: cerebral syphilitic gumma confirmed by the polymerase chain reaction in a man with human immunodeficiency virus infection. N. Engl. J. Med. *331*, 1488-1491.

Hovind-Hougen, K. (1976). Determination by means of electron microscopy of morphological criteria of value for classification of some spirochetes, in particular treponemes. Acta Pathol. Microbiol. Scand. Suppl. *P 1-41.*, -41.

Hovind-Hougen, K. (1983). Morphology. In Pathogenesis and Immunology of Treponemal Infection, R.F.Schell and D.M.Musher, eds. (New York: Marcel Dekker), pp. 3-28.

Hovind-Hougen, K., Birch Andersen, A., and Nielsen, H.A. (1979). Electron microscopy of treponemes subjected to the *Treponema pallidum* immobilization (TPI) test. II. Immunoelectron microscopy. Acta Pathol. Microbiol. Scand. *87C*, 263-268.

Hudson, E.H. (1965). Treponematosis in perspective. Bull World Health Organ. *32*, 735-748

Humphreys, T.L., Schnizlein-Bick, C.T., Katz, B.P., Baldridge, L.A., Hood, A.F., Hromas, R.A., and Spinola, S.M. (2002). Evolution of the cutaneous immune response to experimental

Haemophilus ducreyi infection and its relevance to HIV-1 acquisition. J. Immunol. *169*, 6316-6323.

Hunte, W., al-Ghraoui, F., and Cohen, R.J. (1993). Secondary syphilis and the nephrotic syndrome. J. Am. Soc. Nephrol. *3*, 1351-1355.

Hutchinson, C.M., Rompalo, A.M., Reichart, C.A., and Hook, E.W., III (1991). Characteristics of patients with syphilis attending Baltimore STD clinics. Multiple high-risk subgroups and interactions with human immunodeficiency virus infection. Arch. Intern. Med. *151*, 511-516.

Ingraham, N.R. (1951). The value of penicillin alone in the prevention and treatment of congenital syphilis. Acta Dermatologica Venereologica *31 (suppl 24)*, 60.

Isaacs, R.D., Hanke, J.H., Guzman-Verduzco, L.-M., Newport, G., Agabian, N., Norgard, M.V., Lukehart, S.A., and Radolf, J.D. (1989). Molecular cloning and DNA sequence analysis of the 37-kilodalton endoflagellar sheath protein of *Treponema pallidum*. Infect. Immun. *57*, 3403-3411.

Izzat, N.N., Knox, J.M., Werth, J.A., and Dacres, W.G. (1970). Evolution of syphilitic chancres with virulent *Treponema pallidum* in the rabbit. Br. J. Vener. Dis. *47*, 67-72.

Jackman, J.D. and Radolf, J.D. (1989). Cardiovascular syphilis. Am. J. Med. *87*, 425-433.

Jacques, S.M. and Qureshi, F. (1992). Necrotizing funisitis: a study of 45 cases. Human Pathol. *23*, 1278-1283.

Jaffe, H.W., Choi, K., Thomas, P.A., Haverkos, H.W., Auerbach, D.M., Guinan, M.E., Rogers, M.F., Spira, T.J., Darrow, W.W., Kramer, M.A., Friedman, S.M., Monroe, J.M., Friedman Kien, A.E., Laubenstein, L.J., Marmor, M., Safai, B., Dritz, S.K., Crispi, S.J., Fannin, S.L., Orkwis, J.P., Kelter, A., Rushing, W.R., Thacker, S.B., and Curran, J.W. (1983). National case-control study of Kaposi's sarcoma and *Pneumocystis carinii* pneumonia in homosexual men: Part 1. Epidemiologic results. Ann. Intern. Med. *99*, 145-151.

Jeerapaet, P. and Ackerman, A.B. (1973). Histologic patterns of secondary syphilis. Arch. Dermatol. *107*, 373-377.

Jenkin, H.M. and Sandok, P.L. (1983). *In vitro* cultivation of *Treponema pallidum*. In Pathogenesis and Immunology of Treponemal Infection, R.F.Schell and D.M.Musher, eds. (New York: Marcel Dekker, Inc.), pp. 71-98.

Johns, D.R., Tierney, M., and Felsentein, D. (1987). Alteration in the natural history of neurosyphilis by concurrent infection with the human immunodeficiency virus. N. Engl. J. Med. *316*, 1569-1572.

Johnson, R.C., Ritzi, D.M., and Livermore, B.P. (1973). Outer envelope of virulent *Treponema pallidum*. Infect. Immun. *8*, 291-295.

Jones, S.A., Marchitto, K.S., Miller, J.N., and Norgard, M.V. (1984). Monoclonal antibody with hemagglutination, immobilization, and neutralization activities defines an immunodominant, 47, 000 mol wt, surface-exposed immunogen of *Treponema pallidum* (Nichols). J. Exp. Med. *160*, 1404-1420.

Jordaan, H.F. (1988). Secondary syphilis. A clinicopathological study. Am. J. Dermatopathol. *10*, 399-409.

Jorizzo, J.L., McNeely, M.C., Baughn, R.E., Solomon, A.R., Cavallo, T., and Smith, E.B. (1986). The role of circulating immune complexes in human secondary syphilis. J. Infect. Dis. *153*, 1014-1022.

Jovanovíc, T., Ascenso, C., Hazlett, K.R.O., Sikkink, R., Krebs, C., Litwiller, R., Benson, L.M., Moura, I., Moura, J.J.G., Radolf, J.D., Huynh, B.H., and Rusnak, F. (2000). Neelaredoxin, an iron-binding protein from the syphilis spirochete, *Treponema pallidum*, is a superoxide reductase. J. Biol. Chem. *275*, 28439-28448.

Kahn, L.B. and Gordon, W. (1971). Sarcoid-like granulomas in secondary syphilis. A clinical and histopathologic study of five cases. Arch. Pathol. *92*, 334-337.

Kampmeier, R.H. and Morgan, H.J. (1952). The specific treatment of syphilitic aortitis. Circulation *5*, 771-778.

Kaplan, J.E., Masur, H., and Holmes, K.K. (2002). Guidelines for preventing opportunistic infections among HIV-infected persons--2002. Recommendations of the U.S. Public Health Service and the Infectious Diseases Society of America. MMWR Recomm. Rep. *51*, 1-52.

Kastner, R.J., Malone, J.L., and Decker, C.F. (1994). Syphilitic osteitis in a patient with secondary syphilis and concurrent human immunodeficiency virus infection. Clin. Infect. Dis. *18*, 250-252.

Klein, J.R., Monjan, A.A., Hardy, P.H., Jr., and Cole, G.A. (1980). Abrogation of genetically controlled resistance of mice to *Treponema pallidum* by irradiation. Nature *283*, 572-574.

Konishi, H., Yoshii, Z., and Cox, D.L. (1986). Electron microscopy of *Treponema pallidum* (Nichols) cultivated in tissue cultures of Sf1Ep cells. Infect. Immun. *53*, 32-37.

Kreiss, J.K., Coombs, R., Plummer, F., Holmes, K.K., Nikora, B., Cameron, W., Ngugi, E., Ndinya-Achola, J.O., and Corey, L. (1989). Isolation of human immunodeficiency virus from genital ulcers in Nairobi prostitutes. J. Infect. Dis. *160*, 380-384.

LaFond, R.E., Centurion-Lara, A., Godornes, C., Rompalo, A.M., Van Voorhis, W.C., and Lukehart, S.A. (2003). Sequence diversity of *Treponema pallidum* subsp. *pallidum tprK* in human syphilis lesions and rabbit-propagated isolates. J. Bacteriol. *185*, 6262-6268.

Lauderdale, V. and Goldman, J.N. (1972). Serial ultrathin sectioning demonstrating the intracellularity of *T. pallidum*. An electron microscopic study. Br. J. Vener. Dis. *48*, 87-96.

Levene, G.M., Turk, J.L., Wright, D.J., and Grimble, A.G. (1969). Reduced lymphocyte transformation due to a plasma factor in patients with active syphilis. Lancet *2*, 246-247.

Levene, G.M., Wright, D.J., and Turk, J.L. (1971). Cell-mediated immunity and lymphocyte transformation in syphilis. Proc. R. Soc. Med. *64*, 426-428.

Lien, E., Sellati, T.J., Yoshimura, A., Flo, T.H., Rawadi, G., Finberg, R.W., Carroll, J.D., Espevik, T., Ingalls, R.R., Radolf, J.D., and Golenbock, D.T. (1999). Toll-like receptor 2 functions as a pattern recognition receptor for diverse bacterial products. J. Biol. Chem. *274*, 33419-33425.

Lombard, M., Touati, D., Fontecave, M., and Niviere, V. (2000). Superoxide reductase as a unique defense system against superoxide stress in the microaerophile *Treponema pallidum*. J. Biol. Chem. *275*, 27021-27026.

Lukehart, S.A. (1992). Immunology and pathogenesis of syphilis. In Advances in Host Defense Mechanisms. Sexually Transmitted Diseases, T.C.Quinn, ed. (New York: Raven Press), pp. 141-163.

Lukehart, S.A. (2004). Syphilis. In Harrison's Principles of Internal Medicine, E.Brauwald, A.S.Faucci, S.L.Hauser, D.L.Longo, and J.L.Jameson, eds. (New York: McGraw Hill), pp. 977-985.

Lukehart, S.A., Baker-Zander, S.A., Cheri Lloyd, R.M., and Sell, S. (1980a). Characterization of lymphocyte responsiveness in early experimental syphilis. II. Nature of cellular infiltration and *Treponema pallidum* distribution in testicular lesions. J. Immunol. *124*, 461-467.

Lukehart, S.A., Baker-Zander, S.A., and Sell, S. (1980b). Characterization of lymphocyte responsiveness in early experimental syphilis. I. *In vitro* response to mitogens and *Treponema pallidum* antigens. J. Immunol. *124*, 454-460.

Lukehart, S.A., Hook, E.W., III, Baker-Zander, S.A., Collier, A.C., Critchlow, C.W., and Handsfield, H.H. (1988). Invasion of the central nervous system by *Treponema pallidum*: implications for diagnosis and treatment. Ann. Intern. Med. *109*, 855-862.

Lukehart, S.A. and Miller, J.N. (1978). Demonstration of the *in vitro* phagocytosis of *Treponema pallidum* by rabbit peritoneal macrophages. J. Immunol. *121*, 2014-2024.

Lukehart, S.A., Shaffer, J.M., and Baker-Zander, S.A. (1992). A subpopulation of *Treponema pallidum* is resistant to phagocytosis: possible mechanisms of persistence. J. Infect. Dis. *166*, 1449-1453.

Lux, R., Miller, J.N., Park, N.H., and Shi, W. (2001). Motility and chemotaxis in tissue penetration of oral epithelial cell layers by *Treponema denticola*. Infect. Immun. *69*, 6276-6283.

Lysko, P.G. and Cox, C.D. (1978). Respiration and oxidative phosphorylation in *Treponema pallidum*. Infect. Immun. *21*, 462-473.

Magnuson, H.J., Eagle, H., and Fleischman, R. (1948). The minimal infectious inoculum of *Spirochaeta pallida* (Nichols Strain), and a consideration of its rate of multiplication *in vivo*. Am. J. Syph. Gon. Vener. Dis. *32*, 1-19.

Magnuson, H.J., Thomas, E.W., Olansky, S., Kaplan, B.I., de Mello, L., and Cutler, J.C. (1956). Inoculation syphilis in human volunteers. Medicine *35*, 33-82.

Mahoney, J.F. and Bryant, K.K. (1933). Contact infection of rabbits in experimental syphilis. Am. J. Syph. *17*, 188-193.

Mahoney, J.F. and Bryant, K.K. (1934). Time element in penetration of genital mucosa by *Treponema pallidum*. J. Vener. Dis. Inf. *15*, 1-5.

Marchitto, K.S., Kindt, T.J., and Norgard, M.V. (1986). Monoclonal antibodies directed against major histocompatibility complex antigens bind to the surface of *Treponema pallidum* isolated from infected rabbits or humans. Cell. Immunol. *101*, 633-642.

Maret, S.M., Baseman, J.B., and Folds, J.D. (1980). Cell-mediated immunity in *Treponema pallidum* infected rabbits: in vitro response of splenic and lymph node lymphocytes to mitogens and specific antigens. Clin. Exp. Immunol. *39*, 38-43.

Marra, C.M. (1992). Syphilis and human immunodeficiency virus infection. Semin. Neurol. *12*, 43-50.

Marra, C.M., Castro, C.D., Kuller, L., Dukes, A.C., Centurion-Lara, A., Morton, W.R., and Lukehart, S.A. (1998). Mechanisms of clearance of *Treponema pallidum* from the CSF in a nonhuman primate model. Neurol. *51*, 957-961.

Marra, C.M., Handsfield, H.H., Kuller, L., Morton, W.R., and Lukehart, S.A. (1992). Alterations in the course of experimental syphilis associated with concurrent simian immunodeficiency virus infection. J. Infect. Dis. *165*, 1020-1025.

Marra, C.M., Maxwell, C.L., Smith, S.L., Lukehart, S.A., Rompalo, A.M., Eaton, M., Stoner, B.P., Augenbraun, M., Barker, D.E., Corbett, J.J., Zajackowski, M., Raines, C., Nerad, J., Kee, R., and Barnett, S.H. (2004). Cerebrospinal fluid abnormalities in patients with syphilis: association with clinical and laboratory features. J. Infect. Dis. *189*, 369-376.

McBroom, R.L., Styles, A.R., Chiu, M.J., Clegg, C., Cockerell, C.J., and Radolf, J.D. (1999). Secondary syphilis in persons infected with and not infected with HIV-1: a comparative immunohistologic study. Am. J. Dermatopathol. *21*, 432-441.

McCord, J.R. (1929). A study of two hundred autopsies made on syphilitic fetuses. Am. J. Obstetr. Gynecol. *18*, 597-601.

McCord, J.R. (1934). Syphilis of the placenta. The histologic examination of 1, 085 placentas of mothers with strongly positive Wasserman reactions. Am. J. Obstetr. Gynecol. *28*, 743.

Medzhitov, R. (2001). Toll-like receptors and innate immunity. Nature Rev. Immunol. *1*, 135-145.

Merritt, H.H., Adams, R.D., and Solomon, H.C. (1946). Neurosyphilis. (New York: Oxford University Press).

Miao, R.M. and Fieldsteel, A.H. (1980). Genetic relationship between *Treponema pallidum* and *Treponema pertenue*, two noncultivable human pathogens. J. Bacteriol. *141*, 427-429.

Michelow, I.C., Wendel, G.D., Jr., Norgard, M.V., Zeray, F., Leos, N.K., Alsaadi, R., and Sanchez, P.J. (2002). Central nervous system infection in congenital syphilis. N. Engl. J. Med. *346*, 1792-1798.

Mills, C.H. (1927). Routine examination of the cerebrospinal fluid in syphilis: its value in regard to more accurate knowledge, prognosis, and treatment. Br. Med. J. 2, 527-532.

Mindel, A., Tovey, S.J., Timmins, D.J., and Williams, P. (1989a). Primary and secondary syphilis, 20 years' experience. 2. Clinical features. Genitourin. Med. *65*, 1-3.

Mindel, A., Tovey, S.J., and Williams, P. (1989b). Primary and secondary syphilis, 20 years' experience 3: diagnosis, treatment, and follow-up. Genitourin. Med. *65*, 239-243.

Moore, J.E. and Hopkins, H.H. (1930). Asymptomatic neurosyphilis. VI. the prognosis of early and late asymptomatic neurosyphilis. JAMA *95*, 1637-1641.

Moore, J.P., Kitchen, S.G., Pugach, P., and Zack, J.A. (2004). The CCR5 and CXCR4 coreceptors--central to understanding the transmission and pathogenesis of human immunodeficiency virus type 1 infection. AIDS Res. Hum. Retroviruses *20*, 111-126.

Moore, M.B.Jr.P.E.V., Knox, J.M., and ELGIN, L.W. (1963). Epidemiologic treatment of contacts to infectious syphilis. Public Health Rep. *78*, 966-970.

Morgan, C.A., Lukehart, S.A., and Van Voorhis, W.C. (2003). Protection against syphilis correlates with specificity of antibodies to the variable regions of *Treponema pallidum* repeat protein K. Infect. Immun. *71*, 5605-5612.

Morgan, C.A., Molini, B.J., Lukehart, S.A., and Van Voorhis, W.C. (2002). Segregation of B and T cell epitopes of *Treponema pallidum* repeat protein K to variable and conserved regions during experimental syphilis infection. J. Immunol. *169*, 952-957.

Musher, D.M., Hamill, R.J., and Baughn, R.E. (1990). Effect of human immunodeficiency virus (HIV) infection on the course of syphilis and on the response to treatment. Ann. Intern. Med. *113*, 872-881.

Musher, D.M. and Knox, J.M. (1983). Syphilis and Yaws. In Pathogenesis and Immunology of Treponemal Infection, R.F.Schell and D.M.Musher, eds. (New York: Marcel Dekker), pp. 101-120.

Musher, D.M., Schell, R.F., Jones, R.H., and Jones, A.M. (1975). Lymphocyte transformation in syphilis: an *in vitro* correlate of immune suppression in vivo? Infect. Immun. *11*, 1261-1264.

Narasimhan, S., Camaino, M.J., Liang, F.T., Santiago, F., Laskowski, M., Philipp, M.T., Pachner, A.R., Radolf, J.D., and Fikrig, E. (2003). *Borrelia burgdorferi* transcriptome in the central nervous system of non-human primates. Proc. Natl. Acad. Sci. U. S. A *100*, 15953-15958.

Nathan, L., Bohman, V.R., Sánchez, P.J., Leos, N.K., Twickler, D.M., and Wendel, G.D. (1997). *In utero* infection with *Treponema pallidum* in early pregnancy. Prenat. Diagn. *17*, 119-123.

Nathan, L., Twickler, D.M., Peters, M.T., Sánchez, P.J., and Wendel, G.D., Jr. (1993). Fetal syphilis: correlation of sonographic findings and rabbit infectivity testing of amniotic fluid. J. Ultrasound Med. *12*, 97-101.

Nelson, R.A., Jr. and Mayer, M.M. (1949). Immobilization of *Treponema pallidum in vitro* by antibody produced in syphilitic infection. J. Exp. Med. *89*, 369-383.

Nichols, H.J. and Hough, H.H. (1913). Demonstration of *Spirochaeta pallida* in the cerebrospinal fluid. From a patient with nervous relapse following the use of salvarsan. JAMA *40*, 108-110.

Nikaido, H. (2003). Molecular basis of bacterial outer membrane permeability revisited. Microbiol. Mol. Biol. Rev. *67*, 593-656.

Noordhoek, G.T., Wieles, B., Van der Sluis, J.J., and Van Embden, J.D. (1990). Polymerase chain reaction and synthetic DNA probes: a means of distinguishing the causative agents of syphilis and yaws? Infect. Immun. *58*, 2011-2013.

Norgard, M.V., Arndt, L.L., Akins, D.R., Curetty, L.L., Harrich, D.A., and Radolf, J.D. (1996). Activation of human monocytic cells by *Treponema pallidum* and *Borrelia burgdorferi* lipoproteins and synthetic lipopeptides proceeds via a pathway distinct from that of lipopolysaccharide but involves the transcriptional activator NF-□B. Infect. Immun. *64*, 3845-3852.

Norgard, M.V., Riley, B.S., Richardson, J.A., and Radolf, J.D. (1995). Dermal inflammation elicited by synthetic analogs of *Treponema pallidum* and *Borrelia burgdorferi* lipoproteins. Infect. Immun. *63*, 1507-1515.

Norris, S.J. (1982). *In vitro* cultivation of *Treponema pallidum*: independent confirmation. Infect. Immun. *36*, 437-439.

Norris, S.J. (1988). Syphilis. In Immunology of Sexually Transmitted Diseases, D.J.M.Wright, ed. (Boston: Kluwer Academic Publishers), pp. 1-31.

Norris, S.J., Charon, N.W., Cook, R.G., Fuentes, M.D., and Limberger, R.J. (1988). Antigenic relatedness and N-terminal sequence homology define two classes of periplasmic flagellar proteins of *Treponema pallidum* subsp. *pallidum* and *Treponema phagedenis*. J. Bacteriol. *170*, 4072-4082.

Norris, S.J., Cox, D.L., and Weinstock, G.M. (2001). Biology of *Treponema pallidum*: correlation of functional activities activities with genome sequence data. J. Mol. Microbiol. Biotechnol. *3*, 37-62.

Norris, S.J. and Edmondson, D.G. (1986). Factors affecting the multiplication and subculture of *Treponema pallidum* subsp. *pallidum* in a tissue culture system. Infect. Immun. *53*, 534-539.

Norris, S.J., Fraser, C.M., and Weinstock, G.M. (1998). Illuminating the agent of syphilis: the *Treponema pallidum* genome project. Electrophoresis *19*, 551-553.

Norris, S.J. and Sell, S. (1984). Antigenic complexity of *Treponema pallidum*: antigenicity and surface localization of major polypeptides. J. Immunol. *133*, 2686-2692.

Ochman, H. and Moran, N.A. (2001). Genes lost and genes found: evolution of bacterial pathogenesis and symbiosis. Science *292*, 1096-1099.

Oppenheimer, E.H. and Hardy, J.B. (1971). Congenital syphilis in the newborn infant: clinical and pathological observations in recent cases. Johns Hopkins Med. J. *129*, 63-82.

Ovcinnikov, N.M. and Delektorskij, V.V. (1972). Electron microscopy of phagocytosis in syphilis and yaws. Br. J. Vener. Dis. *48*, 227-248.

Paley, S.S. (1937). Syphilis in pregnancy. N. Y. State J. Med. *37*, 585-590.

Pandhi, R.K., Singh, N., and Ramam, M. (1995). Secondary syphilis: a clinicopathologic study. Int. J. Dermatol. *34*, 240-243.

Paster, B.J. and Dewhirst, F.E. (2000). Phylogenetic foundation of spirochetes. J. Mol. Microbiol. Biotechnol. *2*, 341-344.

Pavia, C.S., Baseman, J.B., and Folds, J.D. (1977a). Selective response of lymphocytes from *Treponema pallidum*- infected rabbits to mitogens and *Treponema reiteri*. Infect. Immun. *15*, 417-422.

Pavia, C.S., Folds, J.D., and Baseman, J.B. (1976). Depression of lymphocyte response to concanavalin A in rabbits infected with *Treponema pallidum* (Nichols strain). Infect. Immun. *14*, 320-322.

Pavia, C.S., Folds, J.D., and Baseman, J.B. (1977b). Selective *in vitro* response of thymus-derived lymphocytes from *Treponema pallidum*-infected rabbits. Infect. Immun. *18*, 603-611.

Penn, C.W. and Rhodes, J.G. (1982). Surface-associated antigens of *Treponema pallidum* concealed by an inert outer layer. Immunology *46*, 9-16.

Peterson, K.M., Baseman, J.B., and Alderete, J.F. (1983). *Treponema pallidum* receptor binding proteins interact with fibronectin. J. Exp. Med. *157*, 1958-1970.

Post, C.D. and Cooney, G.C. (1933). Accidental transmission of syphilis by blood transfusion. JAMA *100*, 258-259.

Qureshi, F., Jacques, S.M., and Reyes, M.P. (1993). Placental histopathology in syphilis. Human Pathol. *24*, 779-784.

Rademacher, S.E. and Radolf, J.D. (1996). Prominent osseous and unusual dermatologic manifestations of early syphilis in two patients with discordant serological statuses for human immunodeficiency virus infection. Clin. Infect. Dis. *23*, 462-467.

Radolf, J.D. (1994). Role of outer membrane architecture in immune evasion by *Treponema pallidum* and *Borrelia burgdorferi*. Trends Microbiol. 2, 307-311.

Radolf, J.D. (1995). *Treponema pallidum* and the quest for outer membrane proteins. Mol. Microbiol. *16*, 1067-1073.

Radolf, J.D., Arndt, L.L., Akins, D.R., Curetty, L.L., Levi, M.E., Shen, Y., Davis, L.S., and Norgard, M.V. (1995a). *Treponema pallidum* and *Borrelia burgdorferi* lipoproteins and synthetic lipopeptides activate monocytes/macrophages. J. Immunol. *154*, 2866-2877.

Radolf, J.D., Fehniger, T.E., Silverblatt, F.J., Miller, J.N., and Lovett, M.A. (1986). The surface of virulent *Treponema pallidum*: Resistance to antibody binding in the absence of complement and surface association of recombinant antigen 4D. Infect. Immun. *52*, 579-585.

Radolf, J.D. and Kaplan, R.P. (1988). Unusual manifestations of secondary syphilis and abnormal humoral response to *Treponema pallidum* antigens in a homosexual man with asymptomatic human immunodeficiency virus infection. J. Am. Acad. Dermatol. *18 (part 2)*, 423-428.

Radolf, J.D., Norgard, M.V., Brandt, M.E., Isaacs, R.D., Thompson, P.A., and Beutler, B. (1991). Lipoproteins of *Borrelia burgdorferi* and *Treponema pallidum* activate cachectin/tumor necrosis factor synthesis: analysis using a CAT reporter construct. J. Immunol. *147*, 1968-1974.

Radolf, J.D., Norgard, M.V., and Schulz, W.W. (1989). Outer membrane ultrastructure explains the limited antigenicity of virulent *Treponema pallidum*. Proc. Natl. Acad. Sci. U. S. A *86*, 2051-2055.

Radolf, J.D., Robinson, E.J., Bourell, K.W., Akins, D.R., Porcella, S.F., Weigel, L.M., Jones, J.D., and Norgard, M.V. (1995b). Characterization of outer membranes isolated from *Treponema pallidum*, the syphilis spirochete. Infect. Immun. *63*, 4244-4252.

Radolf, J.D., Sanchez, P.J., Schulz, K.F., and Murphy, F.K. (1999). Congenital Syphilis. In Sexually Transmitted Diseases, K.K.Holmes, P.F.Sparling, P.-A.Mardh, S.M.Lemon, W.E.Stamm, and P.Piot, eds. (New York: McGraw Hill), pp. 1165-1189.

Raiziss, G.W. and Severac, M. (1935). Rapidity with which *Spirochaeta pallida* invades the blood stream. Arch. Derm. Syph. *35*, 1101-1109.

Revel, A.T., Talaat, A.M., and Norgard, M.V. (2002). DNA microarray analysis of differential gene expression in *Borrelia burgdorferi*, the Lyme disease spirochete. Proc. Natl. Acad. Sci. U. S. A *99*, 1562-1567.

Riviere, G.R., Thomas, D.D., and Cobb, C.M. (1989). *In vitro* model of *Treponema pallidum* invasiveness. Infect. Immun. *57*, 2267-2271.

Rolfs, R.T., Joesoef, M.R., Hendershot, E.F., Rompalo, A.M., Augenbraun, M.H., Chiu, M., Bolan, G., Johnson, S.C., French, P., Steen, E., Radolf, J.D., and Larsen, S. (1997). A randomized trial of enhanced therapy for early syphilis in patients with and without human immunodeficiency virus infection. The Syphilis and HIV Study Group. N. Engl. J. Med. *337*, 307-314.

Rompalo, A.M., Joesoef, M.R., O'Donnell, J.A., Augenbraun, M., Brady, W., Radolf, J.D., Johnson, R., and Rolfs, R.T. (2001). Clinical manifestations of early syphilis by HIV status and gender: results of the syphilis and HIV study. Sex Transm. Dis. *28*, 158-165.

Rosa, P.A., Tilly, K., and Stewart, P.E. (2005). The burgeoning molecular genetics of the Lyme disease spirochaete. Nat. Rev. Microbiol. *3*, 129-143.

Rosahn, P.D. (1947). Autopsy studies in syphilis. J. Vener. Dis. Inf. *(suppl. 21)*, 1-67.

Rosey, E.L., Kennedy, M.J., and Yancey, R.J., Jr. (1996). Dual *flaA1 flaB1* mutant of *Serpulina hyodysenteriae* expressing periplasmic flagella is severely attenuated in a murine model of swine dysentery. Infect. Immun. *64*, 4154-4162.

Rottingen, J.A., Cameron, D.W., and Garnett, G.P. (2001). A systematic review of the epidemiologic interactions between classic sexually transmitted diseases and HIV: how much really is known? Sex Transm. Dis. *28*, 579-597.

Russell, P. and Altshuler, G. (1974). Placental abnormalities of congenital syphilis. Am. J. Dis. Child. *128*, 160-163.

Salazar, J., Hazlett, K.R.O., and Radolf, J. (2002). The immune response to infection with *Treponema pallidum*, the stealth pathogen. Microbes. Infect. *4*, 1133.

Sanchez, P.J., Wendel, G.D., and Norgard, M.V. (1991). Congenital syphilis associated with negative results of maternal serologic tests at delivery. Am. J. Dis. Child *145*, 967-969.

Schaudinn, F.N. and Hoffman, E. (1905). Volaufiger Bericht uber das Vorkommen von Spirochaeten in syphilitischen Krankheits produckten un bei Papillomen. Arbeiten K Gesundheits *22*, 527-534.

Schell, R.F., Chan, J.K., and LeFrock, J.L. (1983). T-cell-mediated resistance. In Pathogenesis and Immunology of Treponemal Infection, R.F.Schell and D.M.Musher, eds. (New York: Marcel Dekker, Inc.), pp. 331-348.

Schiller, N.L. and Cox, C.D. (1977). Catabolism of glucose and fatty acids by virulent *Treponema pallidum*. Infect. Immun. *16*, 60-68.

Schmidt, P.J. (2001). Syphilis, a disease of direct transfusion. Transfusion *41*, 1069-1071.

Schober, P.C., Gabriel, G., White, P., Felton, W.F., and Thin, R.N. (1983). How infectious is syphilis? Br. J. Vener. Dis. *59*, 217-219.

Schouls, L.M., Mout, R., Dekker, J., and Van Embden, J.D.A. (1989). Characterization of lipid-modified immunogenic proteins of *Treponema pallidum* expressed in *Escherichia coli*. Microb. Pathog. *7*, 175-188.

Schroeter, A.L., Turner, R.H., Lucas, J.B., and Brown, W.J. (1971). Therapy for incubating syphilis. Effectiveness of gonorrhea treatment. JAMA *218*, 711-713.

Schulz, K.F., Cates, W., Jr., and O'Mara, P.R. (1987). Pregnancy loss, infant death, and suffering: legacy of syphilis and gonorrhoea in Africa. Genitourin. Med. *63*, 320-325.

Schwartz, D.A., Larsen, S.A., Beck-Sague, C., Fears, M., and Rice, R.J. (1995). Pathology of the umbilical cord in congenital syphilis: analysis of 25 specimens using histochemistry and immunofluorescent antibody to *Treponema pallidum*. Hum. Pathol. *26*, 784-791.

Scragg, I.G., Kwiakowski, D., Vidal, V., Reason, A., Paxton, T., Dell, A., and Morris, H. (2000). Structural characterization of the inflammatory moiety of a variable major lipoprotein of *Borrelia recurrentis*. J. Biol. Chem. *276*, 937-941.

Sell, S., Baker-Zander, S., and Powell, H.C. (1982). Experimental syphilitic orchitis in rabbits: ultrastructural appearance of *Treponema pallidum* during phagocytosis and dissolution by macrophages *in vivo*. Lab. Invest. *46*, 355-364.

Sell, S., Baker-Zander, S.A., and Lloyd, R.M. (1980a). T-cell hyperplasia of lymphoid tissues of rabbits infected with Treponema pallidum: evidence for a vigorous immune response. Sex Transm. Dis. *7*, 74-84.

Sell, S., Gamboa, D., Baker-Zander, S.A., Lukehart, S.A., and Miller, J.N. (1980b). Host response to *Treponema pallidum* in intradermally-infected rabbits: evidence for persistence of infection at local and distant sites. J. Invest Dermatol. *75*, 470-475.

Sell, S. and Norris, S.J. (1983). The biology, pathology, and immunology of syphilis. Int. Rev. Exp. Pathol. *24*, 203-276.

Sell, S., and Salman, J. (1992). Demonstration of *Treponema pallidum* in axons of cutaneous nerves in experimental chancres of rabbits. Sex Transm Dis. *19*, 1-6

Sell, S., Salman, J., and Norris, S.J. (1985). Reinfection of chancre-immune rabbits with *Treponema pallidum*. I. Light and immunofluorescence studies. Am. J. Pathol. *118*, 248-255.

Sellati, T.J., Bouis, D.A., Caimano, M.J., Feulner, J.A., Ayers, C., Lien, E., and Radolf, J.D. (1999). Activation of human monocytic cells by *Borrelia burgdorferi* and *Treponema pallidum* is facilitated by CD14 and correlates with surface exposure of spirochetal lipoproteins. J. Immunol. *163*, 2049-2056.

Sellati, T.J., Bouis, D.A., Kitchens, R.L., Darveau, R.P., Pugin, J., Ulevitch, R.J., Gangloff, S.C., Goyert, S.M., Norgard, M.V., and Radolf, J.D. (1998). *Treponema pallidum* and *Borrelia burgdorferi* lipoproteins and synthetic lipopeptides activate monocytic cells via a CD14-dependent pathway distinct from that used by lipopolysaccharide. J. Immunol. *160*, 5455-5464.

Sellati, T.J., Waldrop, S.L., Salazar, J.C., Bergstresser, P.R., Picker, L.J., and Radolf, J.D. (2001). The cutaneous response in humans to *Treponema pallidum* lipoprotein analogues involves cellular elements of both innate and adaptive immunity. J. Immunol. *166*, 4131-4140.

Sellati, T.J., Wilkinson, D.A., Sheffield, J.S., Koup, R.A., Radolf, J.D., and Norgard, M.V. (2000). Virulent *Treponema pallidum*, lipoprotein, and synthetic lipopeptides induce CCR5 on human monocytes and enhance their susceptibility to infection by human immunodeficiency virus type 1. J. Infect. Dis. *181*, 283-293.

Shalaby, I.A., Dunn, J.P., Semba, R.D., and Jabs, D.A. (1997). Syphilitic uveitis in human immunodeficiency virus-infected patients. Arch. Ophthalmol. *115*, 469-473.

Sheffield, J.S., Sanchez, P.J., Wendel, G.D., Jr., Fong, D.W., Margraf, L.R., Zeray, F., McIntire, D.D., and Barton, R.B. (2002). Placental histopathology of congenital syphilis. Obstet. Gynecol. *100*, 126-133.

Shevchenko, D.V., Sellati, T.J., Cox, D.L., Shevchenko, O.V., Robinson, E.J., and Radolf, J.D. (1999). Membrane topology and cellular location of the *Treponema pallidum* glycerophosphodiester phosphodiesterase (GlpQ) ortholog. Infect. Immun. *67*, 2266-2276.

Sieling, P.A. and Modlin, R.L. (2002). Toll-like receptors: mammalian "taste receptors" for a smorgasbord of microbial invaders. Curr. Opin. Microbiol. *5*, 70-75.

Singh, A.E. and Romanowski, B. (1999). Syphilis: review with emphasis on clinical, epidemiologic, and some biologic features. Clin. Microbiol. Rev. *12*, 187-209.

Smajs, D., McKevitt, M., Howell, J.K., Norris, S.J., Cai, W.W., Palzkill, T., and Weinstock, G.M. (2005). Transcriptome of *Treponema pallidum*: gene expression profile during experimental rabbit infection. J. Bacteriol. *187*, 1866-1874.

Stamm, L.V. and Bergen, H.L. (2000). The sequence-variable, single-copy *tprK* gene of *Treponema pallidum* Nichols strain UNC and street strain 14 encodes heterogeneous TprK proteins. Infect. Immun. *68*, 6482-6486.

Stamm, L.V., Gherardini, F.C., Parrish, E.A., and Moomaw, C.R. (1991). Heat shock response of spirochetes. Infect. Immun. *59*, 1572-1575.

Stamm, L.V., Greene, S.R., Bergen, H.L., Hardham, J.M., and Barnes, N.Y. (1998). Identification and sequence analysis of *Treponema pallidum tprJ*, a member of a polymorphic multigene family. FEMS Microbiol. Lett. *169*, 155-163.

Stamm, L.V., Hodinka, R.L., Wyrick, P.B., and Bassford, P.J., Jr. (1987). Changes in the cell surface properties of *Treponema pallidum* that occur during the *in vitro* incubation of freshly extracted organisms. Infect. Immun. *55*, 2255-2261.

Steen, R. and Dallabetta, G. (2004). Genital ulcer disease control and HIV prevention. J. Clin. Virol. *29*, 143-151.

Steiner, B.M., Sell, S., and Schell, R.F. (1987). *Treponema pallidum* attachment to surface and matrix proteins of cultured rabbit epithelial cells. J. Infect. Dis. *155*, 742-748.

Stokes, J.H., Beerman, H., and Ingraham, N.R. (1944). Modern Clinical Syphilology. (Philadelphia: W.B. Saunders).

Stowens, D. (1959). Pediatric Pathology. (Baltimore: Williams & Wilkins).

Sun, E.S., Molini, B.J., Barrett, L.K., Centurion-Lara, A., Lukehart, S.A., and Van Voorhis, W.C. (2004). Subfamily I *Treponema pallidum* repeat protein family: sequence variation and immunity. Microbes. Infect. *6*, 725-737.

Swartz, M.N., Healy, B.P., and Musher, D.M. (1999). Late Syphilis. In Sexually Transmitted Diseases, K.K.Holmes, P.F.Sparling, P.-A.Mardh, S.M.Lemon, W.E.Stamm, P.Piot, and J.N.Wasserheit, eds. (New York: McGraw-Hill), pp. 487-509.

Sykes, J.A. and Kalan, J. (1975). Intracellular *Treponema pallidum* in cells of a syphilitic lesion of the uterine cervix. Am. J. Obstet. Gynecol. *122*, 361-367.

Sykes, J.A. and Miller, J.N. (1971). Intracellular location of *Treponema pallidum* (Nichols strain) in the rabbit testis. Infect. Immun. *4*, 307-314.

Sykes, J.A., Miller, J.N., and Kalan, A.J. (1974). *Treponema pallidum* within cells of a primary chancre from a human female. Br. J. Vener. Dis. *50*, 40-44.

Takayama, K., Rothenberg, R.J., and Barbour, A.G. (1987). Absence of lipopolysaccharide in the Lyme disease spirochete, *Borrelia burgdorferi*. Infect. Immun. *55*, 2311-2313.

Tantalo, L.C., Lukehart, S.A., and Marra, C.M. (2005). *Treponema pallidum* strain-specific differences in neuroinvasion and clinical phenotype in a rabbit model. J. Infect. Dis. *191*, 75-80.

ter Huurne, A.A., Muir, S., van Houten, M., van der Zeijst, B.A., and Gaastra, W. (1994). Characterization of three putative *Serpulina hyodysenteriae* hemolysins. Microb. Pathog. *16*, 269-282.

Theus, S.A., Harrich, D.A., Gaynor, R., Radolf, J.D., and Norgard, M.V. (1998). *Treponema pallidum*, treponemal lipoproteins, and synthetic lipoprotein analogs (lipopeptides) induce human immunodeficiency virus-1 gene expression in monocyte/macrophages via nuclear translocation of NF-kB. J. Infect. Dis. *177*, 941-950.

Thomas, D.D., Baseman, J.B., and Alderete, J.F. (1985a). Fibronectin mediates *Treponema pallidum* cytadherence through recognition of fibronectin cell-binding domain. J. Exp. Med. *161*, 514-525.

Thomas, D.D., Baseman, J.B., and Alderete, J.F. (1985b). Fibronectin tetrapeptide is target for syphilis spirochete cytadherence. J. Exp. Med. *162*, 1715-1719.

Thomas, D.D., Baseman, J.B., and Alderete, J.F. (1985c). Putative *Treponema pallidum* cytadhesins share a common functional domain. Infect. Immun. *49*, 833-835.

Thomas, D.D., Navab, M., Haake, D.A., Fogelman, A.M., Miller, J.N., and Lovett, M.A. (1988). *Treponema pallidum* invades intercellular junctions of endothelial cell monolayers. Proc. Natl. Acad. Sci. U. S. A *85*, 3608-3612.

Tokarz, R., Anderton, J.M., Katona, L.I., and Benach, J.L. (2004). Combined effects of blood and temperature shift on *Borrelia burgdorferi* gene expression as determined by whole genome DNA array. Infect. Immun. *72*, 5419-5432.

Tosca, A., Lehou, J., Hatjivasiliou, M., Varelzidis, A., and Stratigos, J.D. (1988). Infiltrate of syphilitic lesions before and after treatment. Genitourin. Med. *64*, 289-293.

Turner, T.B., Bauer, J.A., and Kluth, F.C. (1941). The viability of the spirochetes of syphilis and yaws in dessicated blood serum. Am. J. Med. Sci. *202*, 416-423.

Turner, T.B., Hardy, P.H., and Newman, B. (1969). Infectivity tests in syphilis. Br. J. Vener. Dis. *45*, 183-196.

Turner, T.B. and Hollander, D.H. (1957). Biology of the Treponematoses. (Geneva: World Health Organziation).

Turner, T.B., Kluth, F.C., McLeod, C., and Winsor, C.P. (1948). Protective antibodies in the serum of syphilitic patients. Am. J. Hyg. *48*, 173-181.

Uhlenhuth, P. and Mulzer, P. (1913a). Weitere mitteilungen über die infectiosität des blutes und anderer körperflüssigkeiten syphilitischer menschen für das kaninchen. Berl. klin. Wchnschr. *50*, 769-775.

Uhlenhuth, P. and Mulzer, P. (1913b). Weitere mitteilungen über ergebnisse der experimetnellen syphilisforschung. Berl. klin. Wchnschr. *50*, 2031-2035.

Van Voorhis, W.C., Barrett, L.K., Koelle, D.M., Nasio, J.M., Plummer, F.A., and Lukehart, S.A. (1996a). Primary and secondary syphilis lesions contain mRNA for Th1 cytokines. J. Infect. Dis. *173*, 491-495.

Van Voorhis, W.C., Barrett, L.K., Nasio, J.M., Plummer, F.A., and Lukehart, S.A. (1996b). Lesions of primary and secondary syphilis contain activated cytolytic T cells. Infect. Immun. *64*, 1048-1050.

Walker, E.M., Zampighi, G.A., Blanco, D.R., Miller, J.N., and Lovett, M.A. (1989). Demonstration of rare protein in the outer membrane of *Treponema pallidum* subsp. *pallidum* by freeze-fracture analysis. J. Bacteriol. *171*, 5005-5011.

Walker, P.D., Deeves, E.C., Sahba, G., Wallin, J.D., and O'Neill, W.M., Jr. (1984). Rapidly progressive glomerulonephritis in a patient with syphilis. Identification of antitreponemal antibody and treponemal antigen in renal tissue. Am. J. Med. *76*, 1106-1112.

Walker, S.G., Ebersole, J.L., and Holt, S.C. (1999). Studies on the binding of *Treponema pectinovorum* to HEp-2 epithelial cells. Oral Microbiol. Immunol. *14*, 165-171.

Ware, J.L., Folds, J.D., and Baseman, J.B. (1979). Serum of rabbits infected with *Treponema pallidum* (Nichols) inhibits *in vitro* transformation of normal rabbit lymphocytes. Cell Immunol. *42*, 363-372.

Wasserheit, J.N. (1992). Epidemiological synergy. Interrelationships between human immunodeficiency virus infection and other sexually transmitted diseases. Sex. Transm. Dis. *19*, 61-77.

Wendel, G.D., Maberry, M.C., Christmas, J.T., Goldberg, M.S., and Norgard, M.V. (1989). Examination of amniotic fluid in diagnosing congenital syphilis with fetal death. Obstetr. Gynecol. *74*, 967-970.

Wendel, G.D., Jr., Sánchez, P.J., Peters, M.T., Harstad, T.W., Potter, L.L., and Norgard, M.V. (1991). Identification of *Treponema pallidum* in amniotic fluid and fetal blood from pregnancies complicated by congenital syphilis. Obstetr. Gynecol. *78*, 890-895.

Wicher, V. and Wicher, K. (1977a). *In vitro* cell response of *Treponema pallidum*-infected rabbits. I. Lymphocyte transformation. Clin. Exp. Immunol. *29*, 480-486.

Wicher, V. and Wicher, K. (1977b). In vitro cell response of Treponema pallidum-infected rabbits. II. Inhibition of lymphocyte response to phytohaemagglutinin by serum of T. pallidum-infected rabbits. Clin. Exp. Immunol. *29*, 487-495.

Wile, U.J. and Stokes, J.H. (1914). A study of the spinal fluid with reference to involvement of the nervous system in secondary syphilis. J. Cut. Dis. *32*, 607-623.

Willcox, R.R. and Guthe, T. (1966). *Treponema pallidum.* A bibliographical review of the morphology, culture and survival of *T. pallidum* and associated organisms. Bull. World Health Organ *35*, 1-169.

Wooten, R.M., Morrison, T.B., Weis, J.H., Wright, S.D., Thieringer, R., and Weis, J.J. (1998). The role of CD14 in signaling mediated by outer membrane lipoproteins of *Borrelia burgdorferi*. J. Immunol. *160*, 5485-5492.

Young, E.J., Weingarten, N.M., Baughn, R.E., and Duncan, W.C. (1982). Studies on the pathogenesis of the Jarisch-Herxheimer reaction: development of an animal model and evidence against a role for classical endotoxin. J. Infect. Dis. *146*, 606-615.

11

T. pallidum Outer Membrane and Outer Membrane Proteins

*Caroline E. Cameron**

Abstract

The spirochetal bacterium that causes syphilis, *Treponema pallidum* subsp. *pallidum*, has been termed the "stealth pathogen" due to its remarkable ability to evade the host immune response and remain quiescent for extended periods of time. This capability is due, at least in part, to the unusual outer membrane that surrounds the bacterium and interfaces with the host. This chapter correlates the pathogenic potential of *T. pallidum* with the cellular morphology of the outer membrane, and details the current state of knowledge of the composition of the *T. pallidum* outer membrane and its constituent proteins. Within this chapter the treponemal outer membrane ultrastructure, antigenicity, lipid composition, and permeability are described; *in silico* and experimental techniques that have been used to identify outer membrane proteins are reviewed; and current outer membrane protein candidates that have been identified via these methods are outlined.

1. Introduction

At the end of the nineteenth century the great physician Sir William Osler claimed, 'Know syphilis and all its manifestations and all things clinical will be added unto you.' Syphilis is a multistage disease, with localized, disseminated and chronic phases of infection. The disease is characterized by periods of active clinical infection separated by intervals of latent, asymptomatic infection. Varied and extensive clinical manifestations are observed throughout the body, and if left untreated infection can persist for decades despite the generation of an active host immune response. The diverse clinical symptoms and remarkable prolonged latency that are associated with this disease emphasize the highly invasive/limited toxigenic nature of the spirochetal bacterium that causes syphilis, *Treponema pallidum* subsp. *pallidum*.

*For correspondence email caroc@u.washington.edu

One factor that is central to the impressive "pathogenic success" of *T. pallidum* is the outer membrane that surrounds the bacterium, an unusual cellular structure that has prompted the appropriate designation of this organism as a "stealth" pathogen. Prior to 1973, it was uncertain whether *T. pallidum* possessed an outer membrane, due to difficulties encountered with visualizing this structure using routine experimental methods. In their seminal report Johnson and coworkers prepared ultrathin sections of fixed, virulent treponemes and definitively demonstrated by electron microscopy the presence of an outer membrane in *T. pallidum* (Johnson et al., 1973). Since that time a plethora of investigations have been conducted on the interface of this elusive pathogen that is presented to the host. Despite years of intensive research, many aspects of the surface morphology of this structure are still unknown.

Investigation of the *T. pallidum* outer membrane is worthwhile from both a pathogenesis and a vaccine development perspective. The outer membrane of a bacterial pathogen is the medium that is directly presented to the host and, therefore, the constituent proteins of the outer membrane are responsible for many of the pathogenic mechanisms displayed by the organism. Interrelated with this notion, outer membrane molecules of bacterial pathogens have traditionally comprised the most successful vaccine candidates. Generation of an immune response to a surface-exposed molecule not only promotes killing and elimination of the organism by host defenses, it also helps to prevent disease by inhibiting the molecule's pathogenic function.

The quest for identification of *T. pallidum* outer membrane proteins has been a fascinating and constantly evolving journey that at times has been mired in controversy. Although ultimately this remains a largely unresolved issue, our understanding of the physiology and pathogenesis of *T. pallidum* has been significantly advanced by the rigorous research that has been conducted in this field. Presented herein is a compilation of the current knowledge of the *T. pallidum* outer membrane and its constituent protein components.

2. The *T. pallidum* outer membrane

The cellular architecture of *T. pallidum* is similar to that of conventional Gram-negative bacteria, in that the bacterium possesses a cytoplasmic membrane, a periplasmic space and an outer membrane. However, there are several key features that clearly differentiate *T. pallidum* from other eubacteria. First, analogous to other spirochetes, *T. pallidum* contains internal flagella that lie between the protoplasmic cylinder and the outer sheath (Wile et al., 1942; Jepsen et al., 1968; Jackson and Black, 1971). Second, contrary to conventional Gram-negative bacteria, the *T. pallidum* outer membrane is devoid of lipopolysaccharide, a finding that has been shown through both experimental (Belisle et al., 1994; Hardy, Jr. and Levin, 1983; Bailey et al., 1985; Penn et al., 1985; Radolf and Norgard, 1988; Johnson, 1983) and genomic (Fraser et al., 1998; Weinstock et al., 1998) analyses. Third, for typical Gram-negative bacteria, peptidoglycan underlies the outer membrane. Conversely, in *T. pallidum* the peptidoglycan layer overlies the cytoplasmic membrane (Hovind-Hougen, 1983; Radolf et al., 1989a; Holt, 1978). This unusual cellular morphology of the murein layer is likely responsible for the extremely labile nature of the *T. pallidum* outer membrane, which is easily disrupted by detergent treatment (Cox et al., 1992; Radolf et al., 1988; Cunningham et al., 1988; Penn and Lichfield, 1982) or experimental manipulations such as centrifugation and washing (Cox et al., 1992; Penn et al., 1985; Stamm et al., 1987; Radolf et al., 1986). These experimental limitations have hampered progress in identifying the constituent proteins in the outer membrane of *T. pallidum*.

3. The poorly antigenic nature of the *T. pallidum* outer membrane

The outer membrane delimits the surface of the organism and, as such, is the cellular structure that mediates interaction with host components. Throughout many years of study the *T. pallidum* outer membrane has become widely regarded as a structure with limited antigenic reactivity. Anti-*T. pallidum* antibodies in syphilitic sera bind poorly to intact treponemes, and it is only after treponemal storage or disruption of bacterial integrity that serological reactivity is observed (Fehniger et al., 1986; Cox et al., 1992; Stamm et al., 1987; Hovind-Hougen et al., 1979b; Radolf et al., 1986; Penn and Rhodes, 1982; Hardy, Jr. and Nell, 1957; Penn, 1981; Metzger and Ruczkowska, 1964; Robertson et al., 1982). Killing of *T. pallidum* by specific antibody and complement *in vitro* requires prolonged incubation times. A minimum of four hours of incubation with high-titer antiserum are required to demonstrate any degree of treponeme killing, and extended incubations of 16 hours are required to attain complete killing (Bishop and Miller, 1976; Miller et al., 1963; Nelson, Jr. and Mayer, 1949). In addition, phagocytosis of *T. pallidum* by macrophages proceeds unusually slowly (Lukehart and Miller, 1978; Alder et al., 1990), with one study reporting only 65% of treponemes phagocytized over a 24 hour period (Alder et al., 1990).

As supporting evidence for the poorly antigenic nature of the *T. pallidum* outer membrane, conventional techniques used with other pathogenic organisms to identify surface-localized outer membrane proteins have proven to be unsuccessful with *T. pallidum.* Surface radioiodination experiments conducted under conditions where bacterial integrity has been maintained demonstrated there is little incorporation of radiolabel into surface-localized proteins (Penn et al., 1985; Shevchenko et al., 1999; Norris and Sell, 1984). Further, attempts at detecting surface-bound antibodies by immunofluorescence (Deacon et al., 1957; Metzger and Ruczkowska, 1964; Penn, 1981) or immunoelectron microscopy (Fehniger et al., 1986; Radolf et al., 1986; Hovind-Hougen et al., 1979a) have been ineffectual.

The reported poor antigenicity and limited reactivity of the *T. pallidum* outer membrane was hypothesized to result from the presence of an outer layer masking the treponemal surface (Christiansen, 1963; Fitzgerald et al., 1977; Zeigler et al., 1976). This layer was postulated to consist of mucopolysaccharide (Fitzgerald and Johnson, 1979; Fitzgerald et al., 1985) or host components such as fibronectin and albumin bound to the treponemal surface, forming an outer coat (Penn et al., 1985; Alderete and Baseman, 1979; Fitzgerald and Johnson, 1979; Fitzgerald et al., 1975; Peterson et al., 1983). Although electron-microscopy (Hovind-Hougen, 1983; Radolf et al., 1986) and radioimmunoassay (Cox et al., 1992) studies have argued against the presence of an extensive outer coat structure on the treponemal surface, these studies do not preclude the possibility that small yet biologically significant

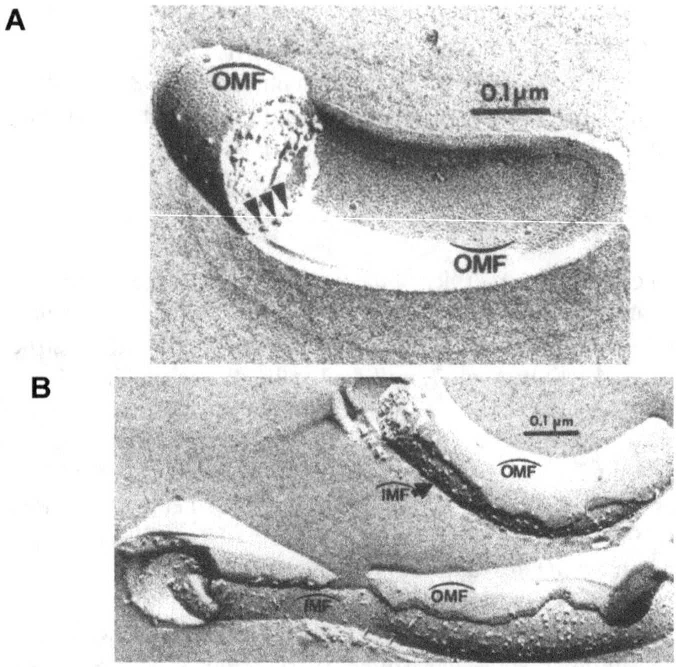

Figure 1. Freeze-fracture electron micrographs. (A) High-magnification view of the outer membrane fracture face (OMF); shown are the convex (⌒) OMF and concave (◡) OMF of *T. pallidum*. Endoflagella are visible in the transverse fracture (indicated by arrowheads). (B) High-magnification view of the convex (⌒) OMF and inner membrane fracture face (IMF) of *T. pallidum*. Reprinted with permission from Walker et al. (1989).

amounts of host components are deposited on the bacterial surface. In support of this, early experiments identified several *T. pallidum* fibronectin-binding proteins (Alderete and Baseman, 1980; Peterson et al., 1987; Peterson et al., 1983; Thomas et al., 1985), and recent investigations have characterized the molecular identity of two treponemal fibronectin-binding proteins (Cameron et al., 2004) and a laminin-binding protein (Cameron, 2003). These data are consistent with the hypothesis that host proteins may be deposited on surface-exposed outer membrane proteins. This feature, in combination with the unique cellular ultrastructure of the *T. pallidum* outer membrane (see below), would thus result in a surface of limited antigenicity being presented to the host.

4. Outer membrane ultrastructure

To investigate the underlying reason for the overall lack of reactivity observed with the outer membrane, freeze-fracture electron microscopy studies were conducted on *T. pallidum*. By the process of freeze-fracture, the two halves of a membrane are separated by the passage of the fracture plane between the apposed lipid bilayers. The smooth fracture face formed by the exposed lipid fatty acid tails are interrupted by integral membrane-spanning proteins which originally spanned the bilayer. These proteins partition with one leaflet of the fractured membrane, and when the fracture faces are obliquely shadowed with metal the protruding protein molecules are observed as intramembranous particles (IMPs) (Shotton and Severs, 1995). Proteins identified by this methodology are exclusively molecules that span the membrane, and thus can be differentiated from lipoproteins that do not span the membrane and therefore do not form IMPs in freeze-fracture analysis (Jones et al., 1995). This line of experimentation was initiated to estimate the overall content of membrane-spanning proteins in the *T. pallidum* outer membrane, and the findings from these studies significantly advanced the understanding of the molecular architecture of the treponemal outer membrane. These studies directly demonstrated that the *T. pallidum* outer membrane is unique among bacterial pathogens in that it contains few integral membrane proteins (Walker et al., 1989; Radolf et al., 1989b). Analyses showed the low IMP densities of the convex (inner leaflet) and concave (outer leaflet) outer membrane fracture faces (Figure 1, Panel A). This is in sharp contrast to the high density of IMPs observed within the convex inner membrane fracture face (Figure 1, Panel B) (Walker et al., 1989). The particle densities in the inner and outer leaflets were calculated to be approximately 70 and 100 particles per μm^2, respectively (Walker et al., 1989). These values are in direct contrast to a conventional Gram-negative organism, such as *E. coli*, which contains 6,000-10,000 and 500-700 particles per μm^2 on the outer and inner leaflets, respectively, of the outer membrane fracture face (Lugtenberg and Van Alphen, 1983). The vast difference in the densities of IMPs in the *T. pallidum* and *E. coli* outer membranes can be observed in Figure 2, which compares the outer leaflets of the outer membranes from each organism. For *T. pallidum*, at least a portion of the particles extend through the outer membrane for exposure on the treponemal surface, as demonstrated by freeze-etch electron microscopy (Radolf et al., 1989b).

Several unusual features of the *T. pallidum* outer membrane were revealed through these studies and deserve further consideration. Upon fracture of a membrane, the proteins partition with one leaflet of the fractured membrane and leave complementary pits in the other leaflet from which they have been removed (Shotton and Severs, 1995). As shown by freeze-fracture analysis of *T. pallidum*,

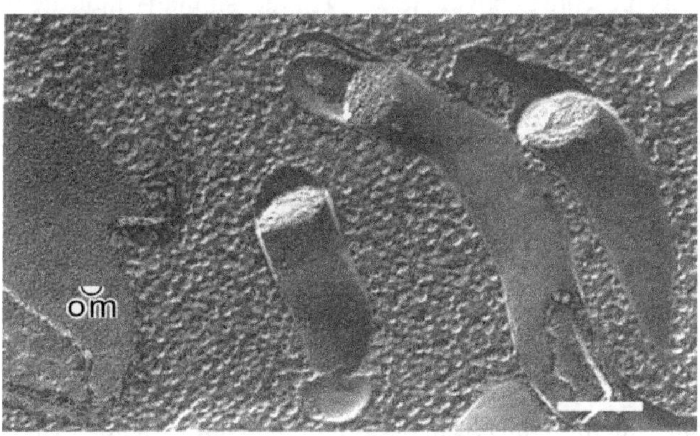

Figure 2. Outer membrane fractures of *E. coli* and *T. pallidum*. The concave outer membrane (om) of *E. coli* shows a uniformly dense distribution of IMPs in contrast to the scarce particles in the fracture faces of the *T. pallidum* outer membrane. Bar = 0.2 µm. Reprinted with permission from Radolf et al. (1989b).

both leaflets of the outer membrane exhibit a greater proportion of particles than pits (Bourell et al., 1994), further emphasizing the atypical organization of proteins present in the outer membrane of this organism. In addition, the sizes of IMPs in both leaflets of the outer membrane are similar (Bourell et al., 1994; Radolf et al., 1989b), compared with a visually heterogenous population of particles present in the cytoplasmic membrane (Bourell et al., 1994). The significance of this observation has not been clearly established, but Radolf and coworkers hypothesized that IMP size uniformity correlates with a limited repertoire of proteins present in the outer membrane (Radolf et al., 1989b).

Another unusual characteristic of the *T. pallidum* outer membrane revealed by freeze-fracture analysis was the similar distribution of IMPs in both leaflets of the outer membrane (Bourell et al., 1994; Radolf et al., 1989b), with the particles being grouped nonrandomly into regions of alternating high and low density. In contrast, the particles present in the cytoplasmic membrane have a random distribution (Bourell et al., 1994). The investigators postulated that the nonrandom distribution of outer membrane IMPs may imply that their translational mobility is restricted. To investigate this prediction, temperature-induced lateral phase separations of *T. pallidum* membrane lipids were performed. The rationale for these studies is that when biological membranes containing mixtures of lipids are cooled slowly, the various lipids partition into fluid and crystalline phases and integral membrane proteins can be visualized as aggregates that are excluded from the crystalline phases of the lipids. Such studies did not cause aggregation of the rare outer membrane proteins (Bourell et al., 1994). This restricted mobility may be caused by interactions between the periplasmic domains of the outer membrane proteins and components of the protoplasmic cylinder, as suggested by Bourell and coworkers.

In independent studies, Blanco et al. used an *in vitro* immobilization assay in conjunction with freeze-fracture analysis to show that *T. pallidum* outer membrane proteins aggregate following an extended 16 hour incubation of treponemes in the presence of immune rabbit serum (Blanco et al., 1990). Several implications can be drawn from this observation. First, although these results seem to be dichotomous with the results presented above, they may in fact similarly indicate that the lateral mobility of the proteins in the outer membrane is restricted, thus resulting in the requirement for extensive incubation times for detection of protein aggregation. Second, the antibody-mediated aggregation demonstrates that the outer membrane proteins have surface-exposed antigenic sites. And lastly, this result suggests that the extremely low density of *T. pallidum* outer membrane proteins necessitates an extended incubation time to allow for cross-linking of these molecules. The slow kinetics of this antibody-mediated protein aggregation directly correlates with the prolonged incubation time required to attain complete killing of *T. pallidum*.

5. Significance of the *T. pallidum* outer membrane ultrastructure

The overwhelming inference from the *T. pallidum* ultrastructural studies is that few membrane-spanning proteins are found in the outer membrane of the organism and, accordingly, are presented to the host. This finding explains previous descriptions of the poorly antigenic nature of the *T. pallidum* outer membrane, including the inability of conventional techniques such as immunofluorescence or immunoelectron microscopy to detect surface-associated proteins. In addition, this observation would also provide a rationale for the limited incorporation of radiolabel upon surface iodination of viable *T. pallidum*, as well as the requirement for extended incubations of *T. pallidum* with antibody preparations before significant levels of treponemicidal or opsonic activities are observed. From a pathogenic perspective, the presentation of a small number of molecules on the *T. pallidum* surface would account for its remarkable ability to successfully evade host defenses and persist for extended periods in spite of the humoral and cellular immune responses that are elicited during infection.

6. Lipoprotein association with the outer membrane

Although the compiled experimental evidence suggests that the *T. pallidum* outer membrane contains a paucity of protein, caution must be exercised with this interpretation. Although lipoproteins are classified as integral membrane proteins, it is the lipid moiety, and not the protein component, that extends into the lipid bilayer. Therefore, these proteins do not form particles upon analysis by freeze-fracture electron microscopy. The lack of detectable binding of antibody to the surface of intact *T. pallidum*, as measured by immunofluorescence (Deacon et al., 1957; Metzger and Ruczkowska, 1964; Penn, 1981) or immunoelectron microscopy (Fehniger et al., 1986; Hovind-Hougen et al., 1979b; Radolf et al., 1986), and the limited incorporation of radioactive label (Alderete and Baseman, 1980; Penn et al., 1985; Norris and Sell, 1984) suggests that, if present, such extrinsic proteins are of low abundance on the outer membrane.

Lipoprotein cellular sorting has been well characterized in *E. coli*, where lipoproteins are found on the periplasmic face of the inner membrane and on both the inner and outer leaflets of the outer membrane. The sorting of lipoproteins to each of these cellular locations is determined by the amino acid residue in the +2 position relative to the lipoprotein signal peptidase cleavage site (Seydel et al., 1999). This residue interacts with the LolA periplasmic carrier protein that mediates translocation of lipoproteins from the inner to the outer membrane (Matsuyama et al., 1995), at which point the LolA-lipoprotein complex interacts with the LolB receptor and the lipoprotein is anchored in the outer membrane (Matsuyama et al., 1997). The lipoprotein sorting pathways appear to differ between *E. coli* and *T. pallidum*, with no clear consensus for the +2 position relative to the peptidase cleavage site and only a partial Lol pathway. Whether this is indicative of a lack of lipoprotein associated with the *T. pallidum* outer membrane, or simply indicates that lipoprotein routing to the outer membrane occurs via a Lol-independent pathway, is currently unknown.

Cox et al. have proposed that the treponemal lipoproteins comprise a 'family' of polypeptides located predominantly (Cox et al., 1995) or exclusively (Cox et al., 1992) within the periplasmic leaflet of the cytoplasmic membrane. However, more recent findings presented by this group (Shevchenko et al., 1999; Hazlett et al., 2005) demonstrated that one lipoprotein, designated p30.5 (Tp0453), was radiolabeled when isolated *T. pallidum* outer membranes and intact treponemes were incubated with the lipophilic reagent [^{125}I]TID, a compound that labels proteins and lipids within phospholipid bilayers. Thus, these experimental results suggest an outer membrane localization for at least this treponemal lipoprotein.

7. Lipid Composition of the *T. pallidum* outer membrane

The outer leaflet of the outer membrane of conventional Gram-negative bacteria consists of lipopolysaccharides (LPS). In contrast, the *T. pallidum* outer membrane is devoid of LPS, a finding that has been shown through direct experimentation (Belisle et al., 1994; Hardy, Jr. and Levin, 1983; Johnson, 1983; Bailey et al., 1985; Penn et al., 1985; Radolf and Norgard, 1988), as well as through the lack of genes encoding LPS biosynthesis enzymes (Fraser et al., 1998; Weinstock et al., 1998). In Gram-negative bacteria, the inner and outer leaflets of the cytoplasmic membrane and the inner leaflet of the outer membrane are composed of phosphatidylethanolamine (70-80%), phosphatidylglycerol, and cardiolipin (Koebnik et al., 2000). By comparison, early investigations demonstrated a higher degree of complexity in the *T. pallidum* cellular lipid profile (Livermore and Johnson, 1974; Vaczi et al., 1966). Specifically, studies performed by Matthews and coworkers (Matthews et al., 1979) determined the total lipid composition of *T. pallidum* to be 32.2% neutral lipids (mainly cholesterol) and 67.8% phospholipids. The composition of phospholipids was determined to be phosphatidylcholine (32.1%), sphingomyelin (14.8%), cardiolipin (13.0%), phosphatidylethanolamine (6.2%), phosphatidylinositol-serine (1.2%) and lysophosphatidylcholine (0.4%). A caveat that must be factored into interpretation of these results is that the treponemes used in these studies were harvested from rabbit testicles, and therefore are at risk of having contaminating host-derived lipids present in the preparations.

In more recent studies, the total cellular lipid composition of *T. pallidum* was determined by metabolic labeling with [^{3}H]palmitate or [^{3}H]oleate, extraction of the cellular lipids and resolution of the lipid profile by thin layer chromatography (Belisle et al., 1994). These studies identified phosphatidylglycerol, phosphatidylcholine, phosphatidylethanolamine, phosphatidylinositol, phosphatidylserine, and cardiolipin as being components of the cellular lipid profile of *T. pallidum*. Insight into the lipid content of the outer membrane also comes from the ability of antibodies to the Venereal Disease Research Laboratory (VDRL) antigen to mediate opsonization and phagocytosis of *T. pallidum* (Baker-Zander et al., 1993b). The predominant antigen present in the VDRL complex is cardiolipin, thus providing independent verification of the presence of this lipid within the *T. pallidum* outer membrane.

Subsequent studies performed on [^{3}H]palmitate-labeled lipids extracted from *T. pallidum* whole cells and isolated outer membranes showed that, with the exception of cardiolipin, similar proportions of phospholipids were found in whole cells and outer membranes (Radolf et al., 1995). If a situation exists similar to that observed in conventional Gram-negative bacteria, apart from the lack of LPS, one can assume the five phospholipids are approximately equally distributed within the inner and outer leaflets of the *T. pallidum* outer membrane.

Lipids present in the outer membrane may serve more than a structural role in *T. pallidum*, functioning as key targets of the protective immune response against infection. Evidence in support of this hypothesis includes the presence of high titers of VDRL antibody during the primary and secondary stages of *T. pallidum* infection, which via treponemal opsonization and phagocytosis may contribute to bacterial clearance. In additional studies, isolation of monoclonal antibodies from mice immunized with purified *T. pallidum* outer membrane vesicles identified a phosphorylcholine-specific antibody with potent complement killing activity. This has led to speculation that phosphorylcholine is in a specific, unique arrangement in the *T. pallidum* outer membrane, and that this phospholipid is surface-exposed, immunogenic and a target of treponemicidal activity (Blanco et al., 2005).

8. Permeability of the *T. pallidum* outer membrane

In typical Gram-negative bacteria, LPS contributes to the low permeability of the outer membrane and increases the resistance of the membrane towards hydrophobic agents (Wiese et al., 1999). Therefore, the lack of LPS in the *T. pallidum* outer membrane suggests an increased permeability to lipophilic compounds (Cox and Radolf, 2001). Further substantiating this prediction, early studies investigating the fatty acid composition of *T. pallidum* indicated that the treponemal cellular lipids contained primarily palmitic and unsaturated fatty acids, with the latter in relatively high proportions. In addition, cyclopropane ring-containing fatty acids were not detected in *T. pallidum.* As high amounts of unsaturated fatty acids and a lack of saturation of double bonds by cyclopropane rings correlates with an increased membrane permeability, the authors surmised the fatty acid composition of the treponemal cellular lipids would lead to an increased membrane permeability (Vaczi et al., 1966). More recent investigations have confirmed and extended these studies, showing that the *T. pallidum* cellular lipid profile consists mainly of the saturated fatty acid palmitate and the unsaturated fatty acids oleate and stearate, with the latter representing a high proportion of the total cellular fatty acids recovered (21.4%) (Belisle et al., 1994). Consistent with these findings, investigations using fluorescent fatty acids probes demonstrated that the *T. pallidum* outer membrane is significantly more permeable to lipophilic compounds than the *E. coli* outer membrane. This result has led to the speculation that *T. pallidum* obtains long-chain fatty acids, which they are unable to synthesize de novo, by direct permeation of their outer membranes (Cox and Radolf, 2001).

The enhanced permeability of the *T. pallidum* outer membrane to lipophilic reagents may represent an overall increased permeability of the outer membrane, compared to conventional Gram-negative bacteria. This may in turn reflect the need to compensate for the metabolically crippled nature of this organism, as well as the unique paucity of proteins present in the outer membrane. Specifically, *T. pallidum* is likely dependent upon the host for a multitude of factors necessary for cell survival during infection. Due to the limited number of proteins predicted to be present on the treponemal surface, it is further likely that *T. pallidum* does not possess as many transporter proteins or porins as are present in Gram-negative bacterial outer membranes. This is exemplified by the fact that *T. pallidum* contains 57 open reading frames sharing sequence similarity with known transport and binding proteins (Fraser et al., 1998), while *E. coli* possesses 281 such open reading frames (Blattner et al., 1997). Although the majority of these represent transport and binding proteins present in the periplasm and cytoplasmic membrane, extrapolation from these numbers, combined with the experimental data compiled to date on the treponemal outer membrane, would suggest a similarly low number of transport proteins exist in the *T. pallidum* outer membrane. This may result in a slow rate of movement across the outer membrane and, accordingly, may contribute to both the inability to culture the bacterium *in vitro* and the slow doubling time *in vivo* (30-33 hours) (Cumberland and Turner, 1949; Magnuson and Eagle, 1948). This represents an unresolved conundrum between *T. pallidum* metabolic and structural biology; the few transporter proteins/porins that are present in the outer membrane, the lipid bilayer itself, or conceivably both, would have to be broadly permissive and allow transport or diffusion of a wide variety of necessary nutrients and compounds. Recently, Hazlett and coworkers have identified a novel *T. pallidum* lipoprotein that may shed light on this conundrum. This protein, Tp0453, is proposed to reside in the treponemal outer membrane and increase membrane permeability, allowing non-selective transport of small hydrophilic molecules such as nutrients (Hazlett et al., 2005).

9. The journey towards *T. pallidum* outer membrane protein identification

Due to the unusual molecular architecture of *T. pallidum*, experimental methods used to isolate outer membranes from Gram-negative bacteria cannot be easily applied to *T. pallidum*. This impediment has hampered progress in identification of outer membrane proteins in *T. pallium*. In particular, the "gold-standard" method of using sucrose density centrifugation to isolate bacterial outer membranes from inner membranes is not easily applied to *T. pallidum*, since this technique relies upon the higher density imparted to the outer membrane by the LPS of Gram-negative bacteria, a structure lacking in *T. pallidum*. To circumvent these complications, investigators have designed creative modifications of conventional methodologies to isolate the *T. pallidum* outer membrane and characterize its constituent proteins. The following is a detailed, historical account of the methods that have been used in the quest for outer membrane proteins.

9.1. Detergent solublization of the *T. pallidum* outer membrane

Detergent extraction techniques have been widely employed to identify the constituents of the *T. pallidum* outer membrane. Extraction of *T. pallidum* with high concentrations of detergent (1%) resulted in recovery of the major spirochetal antigens in the detergent phase, indicating these proteins are integral membrane proteins (Cunningham et al., 1988; Radolf et al., 1988). However, these studies provided little useful information on the constituents of the outer membrane, since high detergent concentrations solubilize all cellular membranes and therefore proteins originating from the inner and outer membranes are unable to be differentiated. Studies using a low concentration (0.04%) of the ionic detergent sodium dodecyl sulfate (SDS) have been used to dissociate the *T. pallidum* outer membrane from the protoplasmic cylinder (Stamm et al., 1987). This approach resulted primarily in isolation of the endoflagellar subunits, and did not identify any convincing candidates for proteins present in the *T. pallidum* outer membrane.

The use of low concentrations of nonionic detergents allows the *T. pallidum* outer membrane to be selectively solubilized, while the peptioglycan-stablized inner membrane remains intact. In particular, the detergent Triton X-114 has the unique advantage of having a low cloud point; detergent preparations warmed above the cloud point can be separated into distinct aqueous and detergent phases, with integral membrane proteins localizing to the heavier detergent phase (Bordier, 1981). This hydrophobicity separation capability, when combined with the selective outer membrane solubilization, should theoretically allow differentiation between periplasmic and outer membrane proteins. Studies have been conducted using 0.02-0.1% of the detergent Triton X-114 (Radolf et al., 1988) and 0.2% of the closely-related nonionic detergent Triton X-100 (Penn et al., 1985). From these investigations several predominant proteins were identified, some of which had been previously identified and many of which were further investigated in independent studies. These included the 47 kDa lipoprotein (TpN47) (Chamberlain et al., 1989b; Chamberlain et al., 1989a), the treponemal membrane protein A (TpN44.5) (Schouls et al., 1989a; Hansen et al., 1985), TpN35 (TmpC) (Schouls et al., 1989b; Hubbard et al., 1991; Van Embden et al., 1983), TpN29-35 (TpD) (Schouls et al., 1991; Swancutt et al., 1990; S1989), TpN17 (Tpp17) (Akins et al., 1993; Chamberlain et al., 1989a) and TpN15 (Purcell et al., 1989; Chamberlain et al., 1989a; Purcell et al., 1990). In these experiments, the observation that the greatest proportion of these proteins did not coextract with the outer membrane using low detergent concentrations led to the assumption that these proteins are associated primarily with the cytoplasmic membrane. In further studies, a *T. pallidum* surface-specific radioimmunoassay (Cox et al., 1992) and immunoelectron microscopy performed on either detergent-treated, metabolically-labeled *T. pallidum* (Radolf et al., 1988) or *T. pallidum* that had been subjected to mechanical disruption (Cox et al., 1992) suggested these major treponemal immunogens are associated with the cytoplasmic membrane, and subsequent investigations showed these molecules are lipoproteins (Chamberlain et al., 1989a; Schouls et al., 1989b; Purcell et al., 1990; Swancutt et al., 1990; Jones et al., 1995).

Overall, these studies indicated that detergent solubilization was not an effective method to definitively identify proteins that are present in the outer membrane, since detergent treatment resulted in nonspecific release of polypeptides from other treponemal structural compartments. However, these investigations were instrumental in defining some of the prominent constituent proteins of the cytoplasmic membrane.

9.2. Outer membrane purification and protein content analyses

In subsequent studies, alternative strategies to isolate the outer membrane were employed that were independent of detergent isolations and instead took advantage of the unusual cellular architecture of *T. pallidum*. The first of these investigations exploited the cellular location of the endoflagella, which have been proposed to physically interact with the *T. pallidum* outer membrane, thus providing the treponeme with its characteristic motility (Berg, 1976). Blanco and coworkers used a hypotonic citrate buffer to depolymerize the endoflagella, thus promoting dissociation of the outer membrane (Blanco et al., 1994). Treponemes were purified from contaminating host material using a Ficoll step gradient, labeled with the chromophore octadecyl rhodamine B chloride to monitor outer membrane release, and immersed in 0.05M citrate buffer (pH3.2). The isolated outer membranes were free of penicillin-binding proteins and TpN47, both markers of the cytoplasmic membrane, and only trace amounts of the periplasmic endoflagella could be detected. In addition, freeze-fracture electron microscopy of the purified membrane vesicles demonstrated rare IMPs similar to those observed in native *T. pallidum* outer membranes. Two-dimensional immunoblot analyses of the isolated membrane vesicles identified five constituent proteins. These included two lipoproteins (TpN44.5 and TpN17) and three additional *T. pallidum* rare outer membrane proteins (TROMPs) of molecular weights 65, 31 and 28 kDa present in approximately equal amounts in the isolated outer membranes (Blanco et al., 1994).

In separate investigations, isopycnic density ultracentrifugation was used to isolate the outer membrane fraction following treponeme plasmolysis in hypertonic sucrose (Radolf et al., 1995; Cox et al., 1995; Shevchenko et al., 1997b; 1999). This method exploited the cellular location of the treponemal murein layer, specifically by expansion of the periplasmic space during plasmolysis and subsequent isolation of the destabilized outer membrane (Radolf et al., 1995). The low specific gravity of the outer membranes (1.05 g/cm^3) allowed for purification away from the protoplasmic cylinders and emphasized the high lipid:protein ratio of this preparation, compared to a specific gravity of 1.02 g/cm^3 for pure phospholipids liposomes (Jones et al., 1995) and 1.20 g/cm^3 for the *E. coli* outer membrane (Osborn et al., 1972). Protein analyses performed on the isolated membranes (Shevchenko et al., 1997b; Radolf et al., 1995; Shevchenko et al., 1999) identified more than 20 proteins, including p30.5 (Tp0453), orthologs of thioredoxin (Tlp), peptidyl prolyl *cis-trans* isomerase's (PpiB), and glycerophosphodiester phosphodiesterase (GlpQ)(Shevchenko et al., 1997b; Stebeck et al., 1997; Cameron et al., 1998; 1999), and the previously identified proteins TpN44.5 (Schouls et al., 1989a), Trompl/TroA (Blanco et al., 1994; Hardham et al., 1997), *T. pallidum* lysine rich repeat protein (TpLRR) (Shevchenko et al., 1997a), basic membrane protein (TpN39) (Dallas et al., 1987) and the 47 kDa (TpN47) and 17 kDa (TpN17) lipoproteins (Chamberlain et al., 1989a).

Many of the proteins identified using these outer membrane purification techniques had previously been shown to be located in the periplasm. Therefore, some of the proteins identified using these methods may actually be components from other cellular compartments that comigrated with outer membranes during the isolation procedure. For this reason, outer membrane purification protocols are not a reliable method for identifying the true residents of the *T. pallidum* outer membrane.

9.3. TnPhoA mutagenesis

Another method used to identify *T. pallidum* proteins that may be present in the outer membrane was that of TnPhoA mutagenesis (Stamm et al., 1998; Hardham et al., 1995; Blanco et al., 1991). This technique creates fusions between foreign genes and the signal sequence-less alkaline phosphatase gene, which encodes an easily assayable enzyme normally located in the periplasm of *E. coli* (Hoffman and Wright, 1985). By fusing portions of *T. pallidum* genes to *phoA* lacking the signal sequence, treponemal proteins containing signal sequences can be identified. However, the limitation of this procedure is that all proteins that are transported to or across the inner membrane are identified, including proteins that ultimately reside in the inner membrane, periplasm, outer membrane, or extracellular space. Therefore, this technique was not able to specifically identify proteins present in the treponemal outer membrane, and no viable outer membrane protein candidates arose from these studies.

9.4. Indirect identification of putative outer membrane protein candidates

As an alternative means to experiments designed to directly identify protein components of the *T. pallidum* outer membrane, several indirect experimental approaches have been utilized. These approaches rely upon a link between the success of an experimental technique being tested and a requirement for an outer membranc locale for that technique to have succeeded. In the first of these experiments, an outer membrane location was inferred for the encoded products of the Tp0155 and Tp0483 open reading frames (Cameron et al., 2004) and the Tp0751 open reading frame (Cameron, 2003) based upon the capability of these proteins to mediate attachment to host extracellular matrix components. In other studies, the capacity of antigen-specific polyclonal sera to mediate opsonophagocytic uptake of *T. pallidum* by macrophages was determined. The primary clearance mechanism responsible for removal of *T. pallidum* from syphilitic chancres is believed to be antibody-mediated treponemal opsonization and subsequent phagocytosis and killing by macrophages (Lukehart and Miller, 1978; Baker-Zander and Lukehart, 1992; Baker-Zander et al., 1993a). The targets of opsonic antibody are presumed to reside on the treponemal surface, therefore allowing opsonic antibody to bind to intact *T. pallidum* and phagocytosis to proceed. These studies showed that the *T. pallidum* proteins Tp92 (corresponding to open reading frame Tp0326) (Cameron et al., 2000) and TprK (corresponding to open reading frame Tp0897) (Centurion-Lara et al., 1999) were the targets of opsonic antibody, thereby strongly suggesting surface exposure. The putative outer membrane location for all of these proteins is discussed in greater detail in the section titled "The current status of *T. pallidum* outer membrane protein candidates".

9.5. Characterization of *T. pallidum* in gel microdroplets

A novel experimental approach for the identification of outer membrane proteins has been the encapsulation of *T. pallidum* in gel microdroplets and subsequent probing of the encased treponemes via immunofluorescence microscopy (Cox et al., 1995). This technique was developed to overcome the experimental limitations associated with the fragile *T. pallidum* outer membrane. By encapsulating virulent treponemes in agarose bead microdroplets, manipulations such as centrifugations and washings can be safely performed without compromising the outer membrane integrity. Maintenance of treponemal structural integrity was confirmed by the lack of reactivity of both a non-membrane permeable DNA-binding dye and an endoflagella-specific monoclonal antibody. Using this technique, antibodies generated against isolated *T. pallidum* outer membranes labeled intact, encapsulated organisms in a pattern consistent with the low density and non-random distribution of outer membrane proteins observed by freeze-fracture electron microscopy. This represents a promising approach to the identification of surface-exposed outer membrane proteins. However, since no known outer membrane protein has been definitively identified via this method, it is presently unknown if this procedure would provide sufficient sensitivity to detect the rare proteins present in the *T. pallidum* outer membrane.

9.6. Summary of results from experimental outer membrane protein identification techniques

As demonstrated by each of the procedures outlined above, determination of the constituent proteins present in the *T. pallidum* outer membrane is not a straightforward process. Contamination of proteins from alternative cellular compartments is a recurrent issue, and therefore distinguishing contaminating proteins from genuine outer membrane proteins and, accordingly, definitively assigning a candidate protein to the outer membrane, are not easy tasks. In addition, the lack of a classic marker for outer membrane location complicates the development of any technique for outer membrane protein identification. Despite these experimental confines and the conflicting data that have been generated in the search for the true residents of the outer membrane, several candidates for putative outer membrane proteins have been identified, as discussed in depth below in the section titled "The current status of *T. pallidum* outer membrane protein candidates."

What can be stated with certainty is that the multitude of experimental and analytical investigations conducted on the outer membrane of *T. pallidum* have collectively shaped our image of this structure. Data compiled from freeze-fracture analyses, detergent fractionation studies, immunological analyses, radioiodination experiments, and outer membrane isolation techniques have led us to the overwhelming conclusion that the *T. pallidum* outer membrane is poorly antigenic and contains a limited number of integral outer membrane proteins. These observations have led to the proposed model that *T. pallidum* possesses a unique molecular architecture (Cox et al., 1992; Radolf, 1994) (Figure 3). In this model, the major membrane immunogens are proposed to be associated predominantly with the cytoplasmic membrane via lipid tails (Cox et al., 1992; Cox et al., 1995; Radolf, 1994), and a limited number of poorly antigenic outer membrane proteins are found on the treponemal surface. The presentation of such a bacterial surface

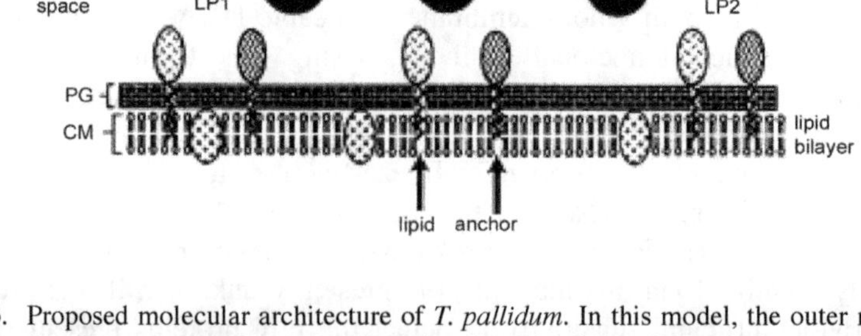

Figure 3. Proposed molecular architecture of *T. pallidum*. In this model, the outer membrane (OM) contains rare transmembrane proteins visualized by freeze-fracture electron microscopy. The major membrane immunogens (designated LP1, LP2, etc.) are anchored via lipids to the periplasmic leaflet of the cytoplasmic membrane (CM) in close association with the peptidoglycan (PG) layer. Endoflagella (Ef) are located in the periplasmic space. Reprinted with permission from Cox et al. (1992).

to the host would contribute, at least in part, to successful bacterial evasion of the host immune response and establishment of chronic infection.

9.7. Genomic analysis

Due to the limitations encountered in the experimental search for *T. pallidum* outer membrane proteins, the genome sequence was eagerly welcomed with the hope that candidate proteins would be revealed through sequence similarity with outer membrane proteins of other microorganisms. Interestingly, no orthologs for highly conserved Gram-negative porins were identified, and overall only a limited number of *T. pallidum* open reading frames exhibited sequence similarity with confirmed bacterial outer membrane proteins. These included the similarity of the *T. pallidum* open reading frame Tp0326 with the D15/Omp85 orthologs present in various microbial pathogens, and the twelve open reading frames encoding the Tpr protein family with the major sheath protein (Msp) of *Treponema denticola.*

Bioinformatic analyses performed on the *T. pallidum* genome have revealed additional potential outer membrane proteins, with the number of identified candidates varying depending on the nature of the analyses performed. Use of the programs TopPred (Claros and von Heijne, 1994) for prediction of membrane-spanning domains, Signal-P (Nielsen et al., 1997) for identification of signal peptides, and BLOCKS (Henikoff et al., 1998) and ProDom (Corpet et al., 1998) for identification of conserved protein domains identified thirteen previously unknown outer membrane protein candidates, not including the members of the Tpr family (Weinstock et al., 1998). Independent analyses performed using PSORT (Nakai and Horton, 1999) for cellular locale prediction, 3D-PSSM (Kelley et al., 2000) for tertiary structure predictions, and PROSITE (http://www2.ebi.ac.uk/ppsearch) for identification of characteristic outer membrane protein motifs revealed a total of twenty-seven outer membrane protein candidates (Cameron, 2003). Of course, it must be realized that identification of open reading frames encoding putative outer membrane proteins is only one step in the discovery process, and direct functional and/or experimental proof of an outer membrane locale is required to verify such predictions.

9.8. The current status of *T. pallidum* outer membrane protein candidates

The intractability of the *T. pallidum* outer membrane to conventional isolation and protein identification techniques has necessitated the design of innovative strategies to decipher the protein profile of this cellular structure. The *T. pallidum* outer membrane has been the focus of intensive research over the last three decades, yet still we understand little about what is present on the surface of this elusive pathogen. One leading factor impeding progress in identifying *T. pallidum* outer membrane proteins is the lack of a definitive "gold-standard" protein by which to access the success of experiments designed to isolate and/or purify the outer membrane and its constituent proteins. This critical deficiency must be kept in mind when interpreting any results obtained from such experimentation.

Despite the technical limitations that confront investigators in the field of *T. pallidum* research, or more accurately as a result of these limitations, several candidate outer membrane proteins have been identified through the use of creative

Table 1. Candidate *T. pallidum* outer membrane proteins

Candidate OM Protein		Route of Identification
Experimental designation	**ORF designation**	
Tromp1/TroA	Tp0163	Outer membrane isolation
Tromp2	Tp0663	Outer membrane isolation
Tromp3	unknown	Outer membrane isolation
Lipoproteins		
Gpd/GlpQ	Tp0257	Outer membrane isolation, differential immunologic screening
p30.5	Tp0453	Outer membrane isolation
ECM-binding proteins		
Tp0155	Tp0155	Genomic analyses, adhesion assays
Tp0483	Tp0483	
Tp0751	Tp0751	
Tpr protein family		
TprF	Tp0316	Genomic analyses
TprI	Tp0620	Genomic analyses
TprK	Tp0897	Genomic analyses, subtraction hybridization, differential immunologic screening
Tp92	Tp0326	Differential immunologic screening

experimental design (Table 1). Included in this section are all proteins that have been reported to be associated with the treponemal outer membrane with convincing experimental data to support this assumption. The candidate proteins may account for all or part of the proteins that are present in the bacterial outer membrane, but the possibility also exists that the true residents of the outer membrane await future discovery.

9.8.1. TROMPs

Three candidate outer membrane proteins were identified by Blanco and coworkers from isolated *T. pallidum* outer membranes (Blanco et al., 1994). The proteins were designated *T. pallidum* rare outer membrane proteins, or TROMPs, and had molecular masses of 65, 31 and 28 kDa. These proteins necessitated very low dilutions (1:25) of immune syphilitic rabbit serum for detection by immunoblot analysis, a finding which is consistent with the predicted rare nature of these proteins. Further investigations have been conducted on these putative outer membrane proteins, as outlined below.

9.8.1.1. Tromp1

The 31 kDa protein isolated from outer membrane preparations, designated Tromp1, corresponds to open reading frame Tp0163 as assigned in the *T. pallidum* genome. Native Tromp1 was shown by the black lipid bilayer assay to have porin activity and a topographical model was proposed in which Tromp1 contains 14 transmembrane domains (Blanco et al., 1995). This porin activity correlated well with the activity observed when the black lipid bilayer assay was used with Triton X-100 detergent-solubilized proteins from purified *T. pallidum* outer membranes (Blanco et al., 1994). Additional studies performed on a recombinant version of Tromp1 (rTromp1) showed similar porin activity upon renaturation and reconstitution into planar lipid bilayers (Blanco et al., 1996; Zhang et al., 1999), as well as porin activity and surface antigenic exposure following expression of rTromp1 in *E. coli* (Blanco et al., 1996). The investigators used MALDI-TOF mass spectrometry to demonstrate native Tromp1 has a cleaved signal peptide (Blanco et al., 1999b), consistent with an outer membrane location. Further, Blanco et al. showed that immunization with *T. pallidum* outer membrane vesicles generated antibodies with treponemicidal activity and induced aggregation of Tromps, including Tromp1, thus suggesting this molecule may be an outer membrane target for bactericidal antibodies (Blanco et al., 1999a).

Despite these intensive investigations, controversy exists surrounding the outer membrane location of this molecule. Independent studies revealed that Tromp1 (designated TroA in these studies) is located in an operon which includes ABC transporter systems, and that Tromp1/TroA possesses 28% sequence identity with periplasmic binding proteins found in such systems (Hardham et al., 1997). These findings led to the hypothesis that Tromp1/TroA is a periplasmic binding protein anchored by an uncleaved signal sequence to the cytoplasmic membrane (Akins et al., 1997). Subsequent investigations performed on recombinant Tromp1/TroA demonstrated that rTromp1/TroA does not have porin activity (Akins et al., 1997; Deka et al., 1999), and that it possesses a crystal structure similar to that of periplasmic ligand-binding proteins (Lee et al., 1999). These investigators also showed rTromp1/TroA was a metalloprotein (Deka et al., 1999) with equal affinity for the metals Mn^{2+} and Zn^{2+} (Hazlett et al., 2003). Separate corroborating investigations have shown that the gene encoding the metal-dependent regulator protein TroR is located in the same operon as *tromp1/troA*, and that TroR represses transcription of the *tro* operon (Posey et al., 1999). These functional findings argue against the presence of this protein in the *T. pallidum* outer membrane, and instead support the designation of Tromp1/TroA as a periplasmic substrate-binding protein involved in the transport of metals across the treponemal cytoplasmic membrane (Deka et al., 1999).

9.8.1.2. Tromp2

The 28 kDa protein isolated from *T. pallidum* outer membranes, designated Tromp2 (Tp0663), has been cloned, sequenced and expressed in *E. coli* (Champion et al., 1997). This molecule shares sequence similarity (26% identity, 40% similarity) with the *T. pallidum* flagellar filament outer layer protein FlaA. However, additional experimental evidence suggests Tromp2 is a putative outer membrane protein. Alkali and high salt treatment of purified outer membranes showed this molecule to be an integral membrane protein, and the predicted secondary structure contains nine transmembrane segments of amphipathic beta-sheets typical of outer membrane proteins. Recombinant Tromp2 showed only occasional porin activity in the black lipid bilayer assay, and therefore a true porin function could not be assigned to this molecule. Following Triton X-114 extraction and phase separation of *T. pallidum*, Tromp2 was detected predominantly in the detergent phase. In addition, in heterologous expression studies rTromp2 fractionated exclusively with the *E. coli* outer membrane and was recognized by anti-Tromp2 antibodies on the surface of *E. coli*. These experimental findings suggest Tromp2 may be a *T. pallidum* membrane-spanning outer membrane protein.

9.8.1.3. Tromp3

The third protein isolated from *T. pallidum* outer membrane preparations, Tromp3, is a 65 kDa protein that has not been further characterized (Blanco et al., 1997). The presence of this protein among outer membrane preparations is variable, thus suggesting this protein is either differentially expressed or represents a contaminating protein from another cellular compartment (Blanco et al., 1999a).

9.8.2. Lipoproteins

The *T. pallidum* lipoprotein glycerophosphodiester phosphodiesterase (Gpd or GlpQ) was identified via screening of a *T. pallidum* genomic expression library (Stebeck et al., 1997; Cameron et al., 1998). This lipoprotein was inferred from experimental analyses to be associated with the treponemal outer membrane; these analyses included a decrease in anti-Gpd/GlpQ reactivity in lysates where the *T. pallidum* outer membrane had been removed by extensive washing, and the demonstration that the lipoprotein binds human immunoglobulins (Stebeck et al., 1997; Cameron et al., 1998). In additional studies, this lipoprotein was shown to be present in outer membrane fractions isolated from *T. pallidum* (Shevchenko et al., 1997b) and the related spirochete *Borrelia hermsii* (Shang et al., 1997), and a homolog of this protein has been shown to be surface exposed in *Haemophilus influenzae* (Ruan et al., 1990). However, Shevchenko et al. (1999) surmised that this lipoprotein is instead associated with the cytoplasmic membrane. This prediction was based upon a lack of incorporation of radiolabel when isolated *T. pallidum* outer membranes and viable treponemes were incubated with ^{125}I[TID], a compound that labels proteins and lipids within phospholipid bilayers. Due to these contradictory experimental findings, the definitive cellular location of this lipoprotein in *T. pallidum* remains to be determined. However, the outer surface lipoproteins of the Lyme disease spirochete *Borrelia burgdorferi* have been shown to be both anchored in the cytoplasmic membrane and associated with the outer membrane (Barbour and Hayes, 1986; Cox et al., 1996; Radolf et al., 1994), and therefore the possibility exists that a subset of lipoproteins present in *T. pallidum* may have a similar dual distribution.

Outer membrane isolation via hypertonic plasmolysis of *T. pallidum* followed by isopycnic density ultracentrifugation resulted in the identification of a 30.5 kDa putative outer membrane protein (Cox et al., 1995; Shevchenko et al., 1997b; Radolf et al., 1995). This protein corresponds to open reading frame Tp0453, as assigned in the *T. pallidum* genome. Subsequent studies have demonstrated that this molecule is a lipoprotein that incorporates radiolabel when isolated *T. pallidum* outer membranes and viable treponemes are incubated with ^{125}I[TID], thus suggesting this lipoprotein is present in the treponemal outer membrane (Shevchenko et al., 1999; Hazlett et al., 2005). Hazlett and coworkers propose

Tp0453 is a novel type of protein that inserts into the periplasmic leaflet of the outer membrane via multiple membrane-inserting, amphipathic α-helices. In this way, Tp0453 is hypothesized to increase the permeability of the outer membrane and allow non-selective transport of nutrients through the outer membrane (Hazlett et al., 2005).

9.8.3. Extracellular matrix (ECM)-binding adhesins

The ECM components fibronectin and laminin are ideal adhesion targets that are used by a multitude of pathogens for colonization of tissues and initiation of infection. Fibronectin is an abundant ECM protein that is secreted by cells as a soluble dimer and is subsequently assembled into insoluble multimeric fibrils at the cell surface. Fibronectin synthesis is upregulated in areas of ulceration, including syphilis chancres formed at the primary site of infection, thereby providing additional targets for treponemal colonization. Similarly, laminin is a large multidomain glycoprotein that is a major constituent of basement membranes, which are condensed, polymer-like aggregates of the extracellular matrix. Laminins are widely distributed throughout the host, and are specifically associated with basement membranes that underlie epithelial and endothelial cell layers and surround muscle cells and peripheral nerves (Timpl and Dziadek, 1986). Attachment to laminin could be an important portal for the initiation of *T. pallidum* infection, via exposure of the epithelial laminin-rich basement membrane following tissue injury. In addition, as has been observed with other disseminating pathogens (Vicentini et al., 1994; McMahon et al., 1995; Yan et al., 1998; Lima et al., 1999), attachment to laminin may be crucial for subsequent treponemal dissemination to distant tissue sites, due to both the close proximity of laminin-rich basement membranes to the vasculature and the fact that treponemes disseminate via the bloodstream.

As an alternative strategy for identifying candidate outer membrane proteins, bioinformatic analyses were performed on the published *T. pallidum* genome to identify open reading frames predicted to encode outer membrane proteins, and specifically candidate bacterial adhesins (Cameron, 2003). Analyses used included PSORT (Nakai and Horton, 1999) for cellular locale prediction, 3D-PSSM (Kelley et al., 2000) for tertiary structure predictions, and PROSITE (http://www2.ebi.ac.uk/ppsearch) for identification of characteristic outer membrane protein motifs. Recombinant proteins were produced from open reading frames identified via these analyses and tested for attachment to the ECM components fibronectin and laminin. These analyses identified the *T. pallidum* open reading frames Tp0155 and Tp0483 as encoding proteins that mediate attachment to the ECM component fibronectin (Cameron et al., 2004), and the open reading frame Tp0751 as encoding a protein that mediates attachment to the ECM component laminin (Cameron, 2003). Subsequent studies performed with Tp0751 have demonstrated that antibodies specific for this treponemal adhesin, as well as synthetic peptides derived from the Tp0751 sequence, prevent attachment of *T. pallidum* to laminin-coated surfaces (Cameron et al., 2005). This direct demonstration of functionality for the protein products encoded by these *T. pallidum* open reading frames is suggestive of their exposure on the treponemal surface. However, as with all other candidate outer membrane proteins in *T. pallidum*, definitive determination of the cellular location of these molecules must await independent confirmation.

9.8.4. The Tpr protein family

The sequencing of the *T. pallidum* genome revealed the presence of a twelve member gene family, *tprA-tprL*, that represents 2% of the organism's total DNA content, a striking proportion when viewed in the context of the small genome size (1,138,006 base pairs) (Fraser et al., 1998). The maintenance of such a large proportion of the genome to this gene family, in an organism that has eliminated numerous metabolic genes during evolution, suggests that the Tpr proteins may play a critical role in survival and virulence. This family of proteins shares sequence similarity with the major sheath protein (Msp) of *Treponema denticola* (Haapasalo et al., 1992; Fenno et al., 1996). Msp has been reported to mediate attachment to host cells and the extracellular matrix, to have porin capabilities, and to form a hexagonal array on the surface of *T. denticola* (Fenno et al., 1996; Haapasalo et al., 1992; Egli et al., 1993), although this structural organization has been debated in independent studies (Caimano et al., 1999).

In addition to the genomic identification of the *tpr* family, individual members of the Tpr protein family have also been identified experimentally via TnPhoA mutagenesis studies (Stamm et al., 1998), subtraction hybridization analysis and differential immunologic screening (Centurion-Lara et al., 1999), and reverse transcriptase PCR and Southern hybridization analyses (Stamm and Bergen, 2000). The twelve-member Tpr family can be divided into three subfamilies according to their amino acid sequence homology: subfamily I consists of Tpr C, D, F and I; subfamily II consists of Tpr E, G and J; and subfamily III consists of Tpr A, B, H, K and L (Centurion-Lara et al., 1999). A separate *tprD* allele, termed *tprD2*, has also been found in 7 of 12 *T. pallidum* subsp. *pallidum* strains (Centurion-Lara et al., 2000). In the examination of different strains of *T. pallidum*, *tpr* sequence heterogeneity has been found in the *tprC*, *D*, *J*, and *K* loci.

Several lines of evidence suggest that at least a portion of the Tpr proteins are located in the outer membrane with surface-exposed domains. First, the paralogous Msp protein of *T. denticola* has been shown to be surface-exposed (Haapasalo et al., 1992; Fenno et al., 1996). Second, Hazlett and coworkers (Hazlett et al., 2001) demonstrated TprK has a cleavable signal sequence, and computer analyses identified putative cleavable signal sequences in the members of Subfamily I and predicted a putative outer membrane location for TprF, I and K. Third, computational structural analyses performed on the predicted Tpr protein sequences identified central hydrophilic variable domains flanked by amphipathic transmembrane helices, suggesting that the variable domains may be surface-exposed. Third, antibodies directed against TprK demonstrate opsonic activity, thus suggesting this molecule is accessible at the surface of the intact organism (Centurion-Lara et al., 1999). And fourth, antibodies that develop during the course of experimental infection are specifically directed towards the variable regions of TprK, implying surface-exposure of these regions (Morgan et al., 2002).

Finally, since protein targets under immunologic pressure tend to undergo antigenic variation, the sequence heterogeneity observed among several members of the Tpr family, including TprK, is strongly supportive of exposure on the treponemal surface. For *tprK*, the sequence heterogeneity is localized in 7 discrete variable regions; the degree of heterogeneity differs among the variable regions, and the heterogeneity is produced by base changes, insertions and deletions, with no frameshifts or stop codons introduced into the sequence (Centurion-Lara et al., 2004). The variable region sequences diversify during infection, suggesting this sequence diversity may play an important role in immune evasion and establishment of persistent infection during chronic syphilis (Centurion-Lara et al., 2004). It is difficult to imagine why such an elaborate antigenic variation system would exist for an internal antigen that was not exposed to the host immune system, further substantiating the hypothesis that TprK is present on the treponemal surface.

Although collectively these studies provide strong evidence for an outer membrane location for TprK, and possibly a subset of other Tpr proteins, independent findings have questioned this cellular location (Hazlett et al., 2001). In this report, the investigators conclude that TprK is entirely periplasmic based upon their observations from four experimental techniques. These included: (a) a lack of immunoreactivity of TprK antibodies with intact treponemes via indirect immunofluorescence analysis of agarose-encapsulated organisms; (b) a lack of opsonophagocytic activity observed with TprK antibodies and intact spirochetes; (c) the intractability of TprK to proteinase K digestion of motile *T. pallidum*; and (d) aqueous phase partitioning of TprK upon Triton X-114 solubilization of *T. pallidum*. Thus, as is the case with the other *T. pallidum* outer membrane protein candidates, the definitive determination of the surface exposure of TprK, and other members of the Tpr family, must await future studies.

9.8.5. Tp92

The Tp92 (Tp0326) molecule was first identified through differential screening of a *T. pallidum* expression library with opsonic antiserum collected from experimentally infected rabbits (Cameron et al., 2000). This method of identification indicated Tp92 was a target of opsonic antibodies, a finding which was later confirmed experimentally. The implication from these studies was that the targets of opsonic antibodies must be present on the outer leaflet of the outer membrane to successfully mediate phagocytosis of intact, viable *T. pallidum*. Thus, the identification of Tp92 as a target of opsonic antibody was suggestive that this molecule is present on the surface of *T. pallidum*.

Homologs of the Tp92 molecule are widely distributed throughout Gram-negative bacteria (Cameron et al., 2000). Although many of these homologs have been identified only through whole genome sequencing of the bacterial species in which they are found, several members of this protein family have been well characterized. These include homologs from *H. influenzae* (Loosmore et al., 1997; Flack et al., 1995), *Haemophilus ducreyi* (Thomas et al., 2001), *Pasteurella multocida* (Ruffolo and Adler, 1996), *Shigella flexneri* (Robb et al., 2001), *Neisseria gonorrhoeae* and *Neisseria meningitidis* (Manning et al., 1998). The Tp92-homologs from *N. gonorrhoeae*, *N. meningitidis*, *H. influenzae* and *P. multocida* have been shown to be present on the surface of the bacteria (Manning et al., 1998; Loosmore et al., 1997; Yang et al., 1998; Ruffolo and Adler, 1996). In addition, immunization with the Tp92-homologs of *H. influenzae* and *P. multocida* elicit protection in animal models of infection (Loosmore et al., 1997; Ruffolo and Adler, 1996). Attempts to disrupt the *tp92*-homolog gene by knockout mutagenesis have been unsuccessful for *H. ducreyi* (Thomas et al., 2001), and no other Tp92-homolog mutants have been reported in other organisms, suggesting this molecule may be required for survival.

Homology analysis places Tp92-homologs in a family of protein-translocating outer membrane porins of Gram-negative bacteria (Yen et al., 2002). Along these lines, two recent reports have investigated the possible function of the Tp92-homolog, Omp85, from *N. meningitidis* (Manning et al., 1998). These studies used a *N. meningitidis* Omp85 knock-out strain with an IPTG-inducible copy of Omp85 to demonstrate that this molecule is essential for viability and has a probable role in outer membrane assembly (Voulhoux et al., 2003; Genevrois et al., 2003). As Omp85 was depleted from *N. meningitidis* after removal of IPTG, bacterial growth was inhibited, lipid structures (Genevrois et al., 2003) and unassembled forms of various outer membrane proteins (Voulhoux et al., 2003) accumulated in the periplasm, and the outer membrane became depleted of phospholipids and lipopolysaccharide (LPS) (Genevrois et al., 2003). Thus, Omp85 is likely to play a role in outer membrane assembly, with one group favoring a role for lipid transport to the outer membrane (Genevrois et al., 2003) and the other group favoring a role in outer membrane protein assembly (Voulhoux et al., 2003). A function in outer membrane assembly could explain why the Tp92-homologs are required for survival in many species of bacteria.

10. Summary

The accurate cellular location of the molecules reported herein remains to be definitively determined. In many cases controversy exists over the association of a particular molecule with the treponemal outer membrane, with convincing supporting evidence presented on both sides of the issue. The future direction of *T. pallidum* outer membrane protein identification lies in the development of reliable methods for accessing the constituents of the outer membrane, for it is only with this factual proof that the controversies can be resolved. The definitive identification of a resident outer membrane protein will ultimately depend on several independent means of experimental confirmation, including surface localization studies and functional and structural analyses.

The *T. pallidum* outer membrane can be likened to a politician: what it reveals is suggestive, but what it conceals is vital. In analyzing scientific research, one has to hunt for the hidden truth. This is especially relevant for the field of research on the constituents of the *T. pallidum* outer membrane. Despite intensive research performed on the *T. pallidum* outer membrane since the discovery of its existence over 30 years ago, we still understand little of its nature. However, due to the diligence, creativity and innovation of investigators in this field, we are gaining a constantly expanding knowledge of the treponemal surface and how it contributes to the remarkable pathogenic mechanisms exhibited by this bacterium. To quote the response of Sir Isaac Newton when questioned on how he made discoveries; "By always thinking unto them. I keep the subject constantly before me and wait till the first dawnings open little by little into the full light." Similarly, the mysteries of the *T. pallidum* outer membrane will be slowly revealed, and with this knowledge will come the understanding of how to combat this elusive pathogen that has plagued humankind for centuries.

Acknowledgment

Original research cited in this chapter from Cameron and coworkers was supported by grant AI-51334 from the National Institutes of Health.

References

Akins, D.R., Purcell, B.K., Mitra, M.M., Norgard, M.V., and Radolf, J.D. (1993). Lipid modification of the 17-kilodalton membrane immunogen of *Treponema pallidum* determines macrophage activation as well as amphiphilicity. Infect. Immun. *61*, 1202-1210.

Akins, D.R., Robinson, E., Shevchenko, D., Elkins, C., Cox, D.L., and Radolf, J.D. (1997). Tromp1, a putative rare outer membrane protein, is anchored by an uncleaved signal sequence to the *Treponema pallidum* cytoplasmic membrane. J. Bacteriol. *179*, 5076-5086.

Alder, J.D., Friess, L., Tengowski, M., and Schell, R.F. (1990). Phagocytosis of opsonized *Treponema pallidum* subsp. *pallidum* proceeds slowly. Infect. Immun. *58*, 1167-1173.

Alderete, J.F. and Baseman, J.B. (1979). Surface-associated host proteins on virulent *Treponema pallidum*. Infect. Immun. *26*, 1048-1056.

Alderete, J.F. and Baseman, J.B. (1980). Surface characterization of virulent *Treponema pallidum*. Infect Immun *30*, 814-823.

Bailey, M.J., Penn, C.W., and Cockayne, A. (1985), Evidence for the presence of lipopolysaccharide in *Treponema phagedenis* (biotype Reiterii) but not in *Treponema pallidum* (Nichols). FEMS Microbiol. Lett. 27, 117-121.

Baker-Zander, S.A. and Lukehart, S.A. (1992). Macrophage-mediated killing of opsonized *Treponema pallidum*. J. Infect. Dis. *165*, 69-74.

Baker-Zander, S.A., Shaffer, J.M., and Lukehart, S.A. (1993a). Characterization of the serum requirement for macrophage-mediated killing of *Treponema pallidum* ssp. *pallidum*: relationship to the development of opsonizing antibodies. FEMS Immunol. Med. Microbiol. *6*, 273-279.

Baker-Zander, S.A., Shaffer, J.M., and Lukehart, S.A. (1993b). VDRL antibodies enhance phagocytosis of *Treponema pallidum* by macrophages. J. Infect. Dis. *167*, 1100-1105.

Barbour, A.G. and Hayes, S.F. (1986). Biology of Borrelia species. Microbiol. Rev. *50*, 381-400.

Belisle, J.T., Brandt, M.E., Radolf, J.D., and Norgard, M.V. (1994). Fatty acids of *Treponema pallidum* and *Borrelia burgdorferi* lipoproteins. J. Bacteriol. *176*, 2151-2157.

Berg, H.C. (1976). How spirochetes may swim. J. Theor. Biol. *56*, 269-273.

Bishop, N.H. and Miller, J.N. (1976). Humoral immunity in experimental syphilis. II. The relationship of neutralizing factors in immune serum to acquired resistance. J. Immunol. *117*, 197-207.

Blanco, D.R., Champion, C.I., Dooley, A., Cox, D.L., Whitelegge, J.P., Faull, K., and Lovett, M.A. (2005). A monoclonal antibody that conveys *in vitro* killing and partial protection in experimental syphilis binds a phosphorylcholine surface epitope of *Treponema pallidum*. Infect. Immun. *73*, 3083-3095.

Blanco, D.R., Champion, C.I., Exner, M.M., Erdjument-Bromage, H., Hancock, R.E., Tempst, P., Miller, J.N., and Lovett, M.A. (1995). Porin activity and sequence analysis of a 31-kilodalton *Treponema pallidum* subsp. *pallidum* rare outer membrane protein (Tromp1). J. Bacteriol. *177*, 3556-3562.

Blanco, D.R., Champion, C.I., Exner, M.M., Shang, E.S., Skare, J.T., Hancock, R.E., Miller, J.N., and Lovett, M.A. (1996). Recombinant *Treponema pallidum* rare outer membrane protein 1 (Tromp1) expressed in *Escherichia coli* has porin activity and surface antigenic exposure. J. Bacteriol. *178*, 6685-6692.

Blanco, D.R., Champion, C.I., Lewinski, M.A., Shang, E.S., Simkins, S.G., Miller, J.N., and Lovett, M.A. (1999a). Immunization with *Treponema pallidum* outer membrane vesicles induces high-titer complement-dependent treponemicidal activity and aggregation of *T. pallidum* rare outer membrane proteins (TROMPs). J. Immunol. *163*, 2741-2746.

Blanco, D.R., Giladi, M., Champion, C.I., Haake, D.A., Chikami, G.K., Miller, J.N., and Lovett, M.A. (1991). Identification of *Treponema pallidum* subspecies *pallidum* genes encoding signal peptides and membrane-spanning sequences using a novel alkaline phosphatase expression vector. Mol. Microbiol. *5*, 2405-2415.

Blanco, D.R., Miller, J.N., and Lovett, M.A. (1997). Surface antigens of the syphilis spirochete and their potential as virulence determinants. Emerg. Infect. Dis. *3*, 11-20.

Blanco, D.R., Reimann, K., Skare, J., Champion, C.I., Foley, D., Exner, M.M., Hancock, R.E., Miller, J.N., and Lovett, M.A. (1994). Isolation of the outer membranes from *Treponema pallidum* and *Treponema vincentii*. J. Bacteriol. *176*, 6088-6099.

Blanco, D.R., Walker, E.M., Haake, D.A., Champion, C.I., Miller, J.N., and Lovett, M.A. (1990). Complement activation limits the rate of *in vitro* treponemicidal activity and correlates with antibody-mediated aggregation of *Treponema pallidum* rare outer membrane protein. J. Immunol. *144*, 1914-1921.

Blanco, D.R., Whitelegge, J.P., Miller, J.N., and Lovett, M.A. (1999b). Demonstration by mass spectrometry that purified native *Treponema pallidum* rare outer membrane protein 1 (Tromp1) has a cleaved signal peptide. J. Bacteriol. *181*, 5094-5098.

Blattner, F.R., Plunkett, G., III, Bloch, C.A., Perna, N.T., Burland, V., Riley, M., Collado-Vides, J., Glasner, J.D., Rode, C.K., Mayhew, G.F., Gregor, J., Davis, N.W., Kirkpatrick, H.A., Goeden, M.A., Rose, D.J., Mau, B., and Shao, Y. (1997). The complete genome sequence of *Escherichia coli* K-12. Science *277*, 1453-1474.

Bordier, C. (1981). Phase separation of integral membrane proteins in Triton X-114 solution. J. Biol. Chem. *256*, 1604-1607.

Bourell, K.W., Schulz, W., Norgard, M.V., and Radolf, J.D. (1994). *Treponema pallidum* rare outer membrane proteins: analysis of mobility by freeze-fracture electron microscopy. J. Bacteriol. *176*, 1598-1608.

Caimano, M.J., Bourell, K.W., Bannister, T.D., Cox, D.L., and Radolf, J.D. (1999). The *Treponema denticola* major sheath protein is predominantly periplasmic and has only limited surface exposure. Infect. Immun. *67*, 4072-4083.

Cameron, C.E. (2003). Identification of a *Treponema pallidum* laminin-binding protein. Infect. Immun. *71*, 2525-2533.

Cameron, C.E., Brouwer, N.L., Tisch, L.M., and Kuroiwa, J.M.Y. (2005). Defining the interaction of the *Treponema pallidum* adhesin Tp0751 with laminin. Infect. Immun. *73*, 7485-7494.

Cameron, C.E., Brown, E.L., Kuroiwa, J.M.Y., Schnapp, L.M., and Brouwer, N.L. (2004). *Treponema pallidum* fibronectin-binding proteins. J. Bacteriol. *186*, 7019-7022.

Cameron, C.E., Castro, C., Lukehart, S.A., and Van Voorhis, W.C. (1998). Function and protective capacity of *Treponema pallidum* subsp. *pallidum* glycerophosphodiester phosphodiesterase. Infect. Immun. *66*, 5763-5770.

Cameron, C.E., Castro, C., Lukehart, S.A., and Van Voorhis, W.C. (1999). Sequence conservation of glycerophosphodiester phosphodiesterase among *Treponema pallidum* strains. Infect. Immun. *67*, 3168-3170.

Cameron, C.E., Lukehart, S.A., Castro, C., Molini, B., Godornes, C., and Van Voorhis, W.C. (2000). Opsonic potential, protective capacity, and sequence conservation of the *Treponema pallidum* subspecies *pallidum* Tp92. J. Infect. Dis. *181*, 1401-1413.

Centurion-Lara, A., Castro, C., Barrett, L., Cameron, C., Mostowfi, M., Van Voorhis, W.C., and Lukehart, S.A. (1999). *Treponema pallidum* major sheath protein homologue Tpr K is a target of opsonic antibody and the protective immune response. J Exp. Med. *189*, 647-656.

Centurion-Lara, A., LaFond, R.E., Hevner, K., Godornes, C., Molini, B.J., Van Voorhis, W.C., and Lukehart, S.A. (2004). Gene conversion: a mechanism for generation of heterogeneity in the *tprK* gene of *Treponema pallidum* during infection. Mol. Microbiol. *52*, 1579-1596.

Centurion-Lara, A., Sun, E.S., Barrett, L.K., Castro, C., Lukehart, S.A., and Van Voorhis, W.C. (2000). Multiple alleles of *Treponema pallidum* repeat gene D in *Treponema pallidum* isolates. J. Bacteriol. *182*, 2332-2335.

Chamberlain, N.R., Brandt, M.E., Erwin, A.L., Radolf, J.D., and Norgard, M.V. (1989a). Major integral membrane protein immunogens of *Treponema pallidum* are proteolipids. Infect. Immun. *57*, 2872-2877.

Chamberlain, N.R., DeOgny, L., Slaughter, C., Radolf, J.D., and Norgard, M.V. (1989b). Acylation of the 47-kilodalton major membrane immunogen of *Treponema pallidum* determines its hydrophobicity. Infect. Immun. *57*, 2878-2885.

Champion, C.I., Blanco, D.R., Exner, M.M., Erdjument-Bromage, H., Hancock, R.E., Tempst, P., Miller, J.N., and Lovett, M.A. (1997). Sequence analysis and recombinant expression of a 28-kilodalton *Treponema pallidum* subsp. *pallidum* rare outer membrane protein (Tromp2). J. Bacteriol. *179*, 1230-1238.

Christiansen, S. (1963). Protective layer covering pathogenic treponemata. Lancet *1*, 423-425.

Claros, M.G. and von Heijne, G. (1994). TopPred II: an improved software for membrane protein structure predictions. Comput. Appl. Biosci. *10*, 685-686.

Corpet, F., Gouzy, J., and Kahn, D. (1998). The ProDom database of protein domain families. Nucleic Acids Res. *26*, 323-326.

Cox, D.L., Akins, D.R., Bourell, K.W., Lahdenne, P., Norgard, M.V., and Radolf, J.D. (1996). Limited surface exposure of *Borrelia burgdorferi* outer surface lipoproteins. Proc. Natl. Acad. Sci U. S. A. *93*, 7973-7978.

Cox, D.L., Akins, D.R., Porcella, S.F., Norgard, M.V., and Radolf, J.D. (1995). *Treponema pallidum* in gel microdroplets: a novel strategy for investigation of treponemal molecular architecture. Mol. Microbiol. *15*, 1151-1164.

Cox, D.L., Chang, P., McDowall, A.W., and Radolf, J.D. (1992). The outer membrane, not a coat of host proteins, limits antigenicity of virulent *Treponema pallidum*. Infect. Immun. *60*, 1076-1083.

Cox, D.L. and Radolf, J.D. (2001). Insertion of fluorescent fatty acid probes into the outer membranes of the pathogenic spirochaetes *Treponema pallidum* and *Borrelia burgdorferi*. Microbiology *147*, 1161-1169.

Cumberland, M.C. and Turner, T.B. (1949). Rate of multiplication of *Treponema pallidum* in normal and immune rabbits. Amer. J. Syph. *33*, 201-212.

Cunningham, T.M., Walker, E.M., Miller, J.N., and Lovett, M.A. (1988). Selective release of the *Treponema pallidum* outer membrane and associated polypeptides with Triton X-114. J. Bacteriol. *170*, 5789-5796.

Dallas, W.S., Ray, P.H., Leong, J., Benedict, C.D., Stamm, L.V., and Bassford, P.J., Jr. (1987). Identification and purification of a recombinant *Treponema pallidum* basic membrane protein antigen expressed in *Escherichia coli*. Infect. Immun. *55*, 1106-1115.

Deacon, W.E., Falcone, V.H., and Harris, A. (1957). A fluorescent test for treponemal antibodies. Proc. Soc. Exp. Biol. Med. *96*, 477-480.

Deka, R.K., Lee, Y.H., Hagman, K.E., Shevchenko, D., Lingwood, C.A., Hasemann, C.A., Norgard, M.V., and Radolf, J.D. (1999). Physicochemical evidence that *Treponema pallidum* TroA Is a zinc- containing metalloprotein that lacks porin-like structure. J Bacteriol *181*, 4420-4423.

Egli, C., Leung, W.K., Muller, K.H., Hancock, R.E., and McBride, B.C. (1993). Pore-forming properties of the major 53-kilodalton surface antigen from the outer sheath of *Treponema denticola*. Infect. Immun. *61*, 1694-1699.

Fehniger, T.E., Radolf, J.D., Walfield, A.M., Cunningham, T.M., Miller, J.N., and Lovett, M.A. (1986). Native surface association of a recombinant 38-kilodalton *Treponema pallidum* antigen isolated from the *Escherichia coli* outer membrane. Infect. Immun. *52*, 586-593.

Fenno, J.C., Muller, K.H., and McBride, B.C. (1996). Sequence analysis, expression, and binding activity of recombinant major outer sheath protein (Msp) of *Treponema denticola*. J. Bacteriol. *178*, 2489-2497.

Fitzgerald, T.J., Cleveland, P., Johnson, R.C., Miller, J.N., and Sykes, J.A. (1977). Scanning electron microscopy of *Treponema pallidum* (Nichols strain) attached to cultured mammalian cells. J. Bacteriol. *130*, 1333-1344.

Fitzgerald, T.J. and Johnson, R.C. (1979). Surface mucopolysaccharides of *Treponema pallidum*. Infect. Immun. *24*, 244-251.

Fitzgerald, T.J., Miller, J.N., Repesh, L.A., Rice, M., and Urquhart, A. (1985). Binding of glycosaminoglycans to the surface of *Treponema pallidum* and subsequent effects on complement interactions between antigen and antibody. Genitourin. Med. *61*, 13-20.

Fitzgerald, T.J., Miller, J.N., and Sykes, J.A. (1975). *Treponema pallidum* (Nichols strain) in tissue cultures: cellular attachment, entry, and survival. Infect. Immun. *11*, 1133-1140.

Flack, F.S., Loosmore, S., Chong, P., and Thomas, W.R. (1995). The sequencing of the 80-kDa D15 protective surface antigen of *Haemophilus influenzae*. Gene *156*, 97-99.

Fraser, C.M., Norris, S.J., Weinstock, G.M., White, O., Sutton, G.G., Dodson, R., Gwinn, M., Hickey, E.K., Clayton, R., Ketchum, K.A., Sodergren, E., Hardham, J.M., McLeod, M.P., et al. (1998). Complete genome sequence of *Treponema pallidum*, the syphilis spirochete. Science *281*, 375-388.

Genevrois, S., Steeghs, L., Roholl, P., Letesson, J.J., and van der, L.P. (2003). The Omp85 protein of *Neisseria meningitidis* is required for lipid export to the outer membrane. EMBO J. *22*, 1780-1789.

Haapasalo, M., Muller, K.H., Uitto, V.J., Leung, W.K., and McBride, B.C. (1992). Characterization, cloning, and binding properties of the major 53- kilodalton *Treponema denticola* surface antigen. Infect. Immun. *60*, 2058-2065.

Hansen, E.B., Pedersen, P.E., Schouls, L.M., Severin, E., and Van Embden, J.D. (1985). Genetic characterization and partial sequence determination of a *Treponema pallidum* operon expressing two immunogenic membrane proteins in *Escherichia coli*. J. Bacteriol. *162*, 1227-1237.

Hardham, J.M., Frye, J.G., and Stamm, L.V. (1995). Identification and sequences of the *Treponema pallidum fliM'*, *fliY*, *fliP*, *fliQ*, *fliR* and *flhB'* genes. Gene *166*, 57-64.

Hardham, J.M., Stamm, L.V., Porcella, S.F., Frye, J.G., Barnes, N.Y., Howell, J.K., Mueller, S.L., Radolf, J.D., Weinstock, G.M., and Norris, S.J. (1997). Identification and transcriptional analysis of a *Treponema pallidum* operon encoding a putative ABC transport system, an iron-activated repressor protein homolog, and a glycolytic pathway enzyme homolog. Gene *197*, 47-64.

Hardy, P.H., Jr. and Levin, J. (1983). Lack of endotoxin in *Borrelia hispanica* and *Treponema pallidum*. Proc. Soc. Exp. Biol. Med. *174*, 47-52.

Hardy, P.H., Jr. and Nell, E.E. (1957). Study of the antigenic structure of *Treponema pallidum* by specific agglutination. Am. J. Hyg. *66*, 160-172.

Hazlett, K.R., Cox, D.L., Decaffmeyer, M., Bennett, M.P., Desrosiers, D.C., La Vake, C.J., La Vake, M.E., Bourell, K.W., Robinson, E.J., Brasseur, R., and Radolf, J.D. (2005). Tp0453, a concealed outer membrane protein of *Treponema pallidum*, enhances membrane permeability. J. Bacteriol. *187*, 6499-6508.

Hazlett, K.R., Rusnak, F., Kehres, D.G., Bearden, S.W., La Vake, C.J., La Vake, M.E., Maguire, M.E., Perry, R.D., and Radolf, J.D. (2003). The *Treponema pallidum tro* operon encodes a multiple metal transporter, a zinc-dependent transcriptional repressor, and a semi-autonomously expressed phosphoglycerate mutase. J. Biol. Chem. *278*, 20687-20694.

Hazlett, K.R., Sellati, T.J., Nguyen, T.T., Cox, D.L., Clawson, M.L., Caimano, M.J., and Radolf, J.D. (2001). The TprK protein of *Treponema pallidum* is periplasmic and is not a target of opsonic antibody or protective immunity. J. Exp. Med. *193*, 1015-1026.

Henikoff, S., Pietrokovski, S., and Henikoff, J.G. (1998). Superior performance in protein homology detection with the Blocks Database servers. Nucleic Acids Res. *26*, 309-312.

Hoffman, C.S. and Wright, A. (1985). Fusions of secreted proteins to alkaline phosphatase: an approach for studying protein secretion. Proc. Natl. Acad. Sci. U.S.A *82*, 5107-5111.

Holt, S.C. (1978). Anatomy and chemistry of spirochetes. Microbiol Rev. *42*, 114-160.

Hovind-Hougen, K. (1983). Morphology. In Pathogenesis and immunology of treponemal infection, R.F.Schell and D.M.Musher, eds. (New York: Marcell Dekker Inc.), pp. 3-28.

Hovind-Hougen, K., Birch-Andersen, A., and Nielsen, H.A. (1979a). Electron microscopy of treponemes subjected to the *Treponema pallidum* immobilization (TPI) test. II. Immunoelectron microscopy. Acta. Pathol. Microbiol. Scand. *87C*, 263-268.

Hovind-Hougen, K., Nielsen, H.A., and Birch-Andersen, A. (1979b). Electron microscopy of treponemes subjected to the *Treponema pallidum* immobilization (TPI) test. I. Comparison of immunoimmobilized cells and control cells. Acta. Pathol. Microbiol. Scand. *87C*, 217-222.

Hubbard, C.L., Gherardini, F.C., Bassford, P.J., Jr., and Stamm, L.V. (1991). Molecular cloning and characterization of a 35.5-kilodalton lipoprotein of *Treponema pallidum*. Infect. Immun. *59*, 1521-1528.

Jackson, S. and Black, S.H. (1971). Ultrastructure of *Treponema pallidum* Nichols following lysis by physical and chemical methods. II. Axial filaments. Arch. Mikrobiol. *76*, 325-340.

Jepsen, O.B., Hougen, K.H., and Birch-Andersen, A. (1968). Electron microscopy of *Treponema pallidum* Nichols. Acta. Pathol. Microbiol. Scand. *74*, 241-258.

Johnson, R.C. (1983). Composition. In Pathogenesis and immunology of Treponemal infections, R.F.Schell and D.M.Musher, eds. (New York: Marcel Dekker, Inc.), pp. 29-37.

Johnson, R.C., Ritzi, D.M., and Livermore, B.P. (1973). Outer envelope of virulent *Treponema pallidum*. Infect. Immun. *8*, 291-295.

Jones, J.D., Bourell, K.W., Norgard, M.V., and Radolf, J.D. (1995). Membrane topology of *Borrelia burgdorferi* and *Treponema pallidum* lipoproteins. Infect Immun *63*, 2424-2434.

Kelley, L.A., MacCallum, R.M., and Sternberg, M.J. (2000). Enhanced genome annotation using structural profiles in the program 3D- PSSM. J. Mol. Biol. *299*, 499-520.

Koebnik, R., Locher, K.P., and Van Gelder, P. (2000). Structure and function of bacterial outer membrane proteins: barrels in a nutshell. Mol. Microbiol *37*, 239-253.

Lee, Y.H., Deka, R.K., Norgard, M.V., Radolf, J.D., and Hasemann, C.A. (1999). *Treponema pallidum* TroA is a periplasmic zinc-binding protein with a helical backbone. Nat. Struct. Biol. *6*, 628-633.

Lima, O.C., Figueiredo, C.C., Pereira, B.A., Coelho, M.G., Morandi, V., and Lopes-Bezerra, L.M. (1999). Adhesion of the human pathogen *Sporothrix schenckii* to several extracellular matrix proteins. Braz. J. Med. Biol. Res. *32*, 651-657.

Livermore, B.P. and Johnson, R.C. (1974). Lipids of the Spirochaetales: comparison of the lipids of several members of the genera *Spirochaeta, Treponema,* and *Leptospira*. J. Bacteriol. *120*, 1268-1273.

Loosmore, S.M., Yang, Y.P., Coleman, D.C., Shortreed, J.M., England, D.M., and Klein, M.H. (1997). Outer membrane protein D15 is conserved among *Haemophilus influenzae* species and may represent a universal protective antigen against invasive disease. Infect. Immun. *65*, 3701-3707.

Lugtenberg, B. and Van Alphen, L. (1983). Molecular architecture and functioning of the outer membrane of *Escherichia coli* and other Gram-negative bacteria. Biochim. Biophys. Acta. *737*, 51-115.

Lukehart, S.A. and Miller, J.N. (1978). Demonstration of the *in vitro* phagocytosis of *Treponema pallidum* by rabbit peritoneal macrophages. J. Immunol. *121*, 2014-2024.

Magnuson, H.J. and Eagle, H. (1948). The minimal infectious inoculum of *Spirochaeta pallida* (Nichols strain), and a consideration of its rate of multiplication *in vivo*. Amer. J. Syph. *32*, 1-18.

Manning, D.S., Reschke, D.K., and Judd, R.C. (1998). Omp85 proteins of Neisseria gonorrhoeae and Neisseria meningitidis are similar to *Haemophilus influenzae* D-15-Ag and *Pasteurella multocida* Oma87. Microb. Pathog. *25*, 11-21.

Matsuyama, S., Tajima, T., and Tokuda, H. (1995). A novel periplasmic carrier protein involved in the sorting and transport of *Escherichia coli* lipoproteins destined for the outer membrane. EMBO J. *14*, 3365-3372.

Matsuyama, S., Yokota, N., and Tokuda, H. (1997). A novel outer membrane lipoprotein, LolB (HemM), involved in the LolA (p20)-dependent localization of lipoproteins to the outer membrane of *Escherichia coli*. EMBO J. *16*, 6947-6955.

Matthews, H.M., Yang, T.K., and Jenkin, H.M. (1979). Unique lipid composition of *Treponema pallidum* (Nichols virulent strain). Infect. Immun. *24*, 713-719.

McMahon, J.P., Wheat, J., Sobel, M.E., Pasula, R., Downing, J.F., and Martin, W.J. (1995). Murine laminin binds to *Histoplasma capsulatum*. A possible mechanism of dissemination. J. Clin. Invest. *96*, 1010-1017.

Metzger, M. and Ruczkowska, J. (1964). Influence of lysozyme upon the reactivity of *Treponema pallidum* in the fluoresent antibody reaction. Arch. Immunol. Ther. Exp. (Warsz.) *12*, 702-708.

Miller, J.N., Fazzan, F.P., and Whang, S.J. (1963). Studies on immunity in experimental syphilis. II. *Treponema pallidum* immobilization (TPI) antibody and the immune response. Br. J. Vener. Dis. *39*, 199-203.

Morgan, C.A., Molini, B.J., Lukehart, S.A., and Van Voorhis, W.C. (2002). Segregation of B and T cell epitopes of *Treponema pallidum* repeat protein K to variable and conserved regions during experimental syphilis infection. J. Immunol. *169*, 952-957.

Nakai, K. and Horton, P. (1999). PSORT: a program for detecting sorting signals in proteins and predicting their subcellular localization. Trends Biochem. Sci. *24*, 34-36.

Nelson, R.A., Jr. and Mayer, M.M. (1949). Immobilization of *Treponema pallidum in vitro* by antibody produced by syphilitic infection. J. Exp. Med. 89, 369-393.

Nielsen, H., Engelbrecht, J., Brunak, S., and von Heijne, G. (1997). Identification of prokaryotic and eukaryotic signal peptides and prediction of their cleavage sites. Protein Eng *.10*, 1-6.

Norris, S.J. and Sell, S. (1984). Antigenic complexity of *Treponema pallidum:* antigenicity and surface localization of major polypeptides. J. Immunol. *133*, 2686-2692.

Osborn, M.J., Gander, J.E., Parisi, E., and Carson, J. (1972). Mechanism of assembly of the outer membrane of *Salmonella typhimurium*. Isolation and characterization of cytoplasmic and outer membrane. J. Biol. Chem. *247*, 3962-3972.

Penn, C.W. (1981). Avoidance of host defences by *Treponema pallidum* in situ and on extraction from infected rabbit testes. J. Gen. Microbiol. *126*, 69-75.

Penn, C.W., Cockayne, A., and Bailey, M.J. (1985). The outer membrane of *Treponema pallidum*: biological significance and biochemical properties. J. Gen. Microbiol *131*, 2349-2357.

Penn, C.W. and Lichfield, J. (1982). The outer membrane of *Treponema pallidum*: solubilization by detergents to release axial filaments. FEMS Microbiol. Lett. *14*, 61-64.

Penn, C.W. and Rhodes, J.G. (1982). Surface-associated antigens of *Treponema pallidum* concealed by an inert outer layer. Immunology *46*, 9-16.

Peterson, K., Baseman, J.B., and Alderete, J.F. (1987). Molecular cloning of *Treponema pallidum* outer envelope fibronectin binding proteins, P1 and P2. Genitourin. Med. *63*, 355-360.

Peterson, K.M., Baseman, J.B., and Alderete, J.F. (1983). *Treponema pallidum* receptor binding proteins interact with fibronectin. J. Exp. Med. *157*, 1958-1970.

Posey, J.E., Hardham, J.M., Norris, S.J., and Gherardini, F.C. (1999). Characterization of a manganese-dependent regulatory protein, TroR, from *Treponema pallidum*. Proc. Natl. Acad. Sci. U.S.A *96*, 10887-10892.

Purcell, B.K., Chamberlain, N.R., Goldberg, M.S., Andrews, L.P., Robinson, E.J., Norgard, M.V., and Radolf, J.D. (1989). Molecular cloning and characterization of the 15-kilodalton major immunogen of *Treponema pallidum*. Infect. Immun. *57*, 3708-3714.

Purcell, B.K., Swancutt, M.A., and Radolf, J.D. (1990). Lipid modification of the 15 kiloDalton major membrane immunogen of *Treponema pallidum*. Mol. Microbiol. *4*, 1371-1379.

Radolf, J.D. (1994). Role of outer membrane architecture in immune evasion by *Treponema pallidum* and *Borrelia burgdorferi*. Trends Microbiol. *2*, 307-311.

Radolf, J.D., Bourell, K.W., Akins, D.R., Brusca, J.S., and Norgard, M.V. (1994). Analysis of *Borrelia burgdorferi* membrane architecture by freeze- fracture electron microscopy. J. Bacteriol. *176*, 21-31.

Radolf, J.D., Chamberlain, N.R., Clausell, A., and Norgard, M.V. (1988). Identification and localization of integral membrane proteins of virulent *Treponema pallidum* subsp. *pallidum* by phase partitioning with the nonionic detergent triton X-114. Infect. Immun. *56*, 490-498.

Radolf, J.D., Fehniger, T.E., Silverblatt, F.J., Miller, J.N., and Lovett, M.A. (1986). The surface of virulent *Treponema pallidum:* resistance to antibody binding in the absence of complement and surface association of recombinant antigen 4D. Infect. Immun. *52*, 579-585.

Radolf, J.D., Moomaw, C., Slaughter, C.A., and Norgard, M.V. (1989a). Penicillin-binding proteins and peptidoglycan of *Treponema pallidum* subsp. *pallidum*. Infect Immun *57*, 1248-1254.

Radolf, J.D. and Norgard, M.V. (1988). Pathogen specificity of *Treponema pallidum* subsp. *pallidum* integral membrane proteins identified by phase partitioning with Triton X-114. Infect Immun *56*, 1825-1828.

Radolf, J.D., Norgard, M.V., and Schulz, W.W. (1989b). Outer membrane ultrastructure explains the limited antigenicity of virulent *Treponema pallidum*. Proc. Natl. Acad. Sci. U.S.A. *86*, 2051-2055.

Radolf, J.D., Robinson, E.J., Bourell, K.W., Akins, D.R., Porcella, S.F., Weigel, L.M., Jones, J.D., and Norgard, M.V. (1995). Characterization of outer membranes isolated from *Treponema pallidum*, the syphilis spirochete. Infect. Immun. *63*, 4244-4252.

Robb, C.W., Orihuela, C.J., Ekkelenkamp, M.B., and Niesel, D.W. (2001). Identification and characterization of an *in vivo* regulated D15/Oma87 homologue in *Shigella flexneri* using differential display polymerase chain reaction. Gene *262*, 169-177.

Robertson, S.M., Kettman, J.R., Miller, J.N., and Norgard, M.V. (1982). Murine monoclonal antibodies specific for virulent *Treponema pallidum* (Nichols). Infect. Immun. *36*, 1076-1085.

Ruan, M.R., Akkoyunlu, M., Grubb, A., and Forsgren, A. (1990). Protein D of *Haemophilus influenzae*. A novel bacterial surface protein with affinity for human IgD. J. Immunol. *145*, 3379-3384.

Ruffolo, C.G. and Adler, B. (1996). Cloning, sequencing, expression, and protective capacity of the *oma87* gene encoding the *Pasteurella multocida* 87-kilodalton outer membrane antigen. Infect. Immun. *64*, 3161-3167.

Schouls, L.M., Ijsselmuiden, O.E., Weel, J., and Van Embden, J.D. (1989a). Overproduction and purification of *Treponema pallidum* recombinant-DNA- derived proteins TmpA and TmpB and their potential use in serodiagnosis of syphilis. Infect. Immun. *57*, 2612-2623.

Schouls, L.M., Mout, R., Dekker, J., and Van Embden, J.D. (1989b). Characterization of lipid-modified immunogenic proteins of *Treponema pallidum* expressed in *Escherichia coli*. Microb. Pathog. *7*, 175-188.

Schouls, L.M., van der Heide, H.G., and Van Embden, J.D. (1991). Characterization of the 35-kilodalton *Treponema pallidum* subsp. *pallidum* recombinant lipoprotein TmpC and antibody response to lipidated and nonlipidated *T. pallidum* antigens. Infect. Immun. *59*, 3536-3546.

Seydel, A., Gounon, P., and Pugsley, A.P. (1999). Testing the '+2 rule' for lipoprotein sorting in the *Escherichia coli* cell envelope with a new genetic selection. Mol. Microbiol. *34*, 810-821.

Shang, E.S., Skare, J.T., Erdjument-Bromage, H., Blanco, D.R., Tempst, P., Miller, J.N., and Lovett, M.A. (1997). Sequence analysis and characterization of a 40-kilodalton *Borrelia hermsii* glycerophosphodiester phosphodiesterase homolog. J. Bacteriol. *179*, 2238-2246.

Shevchenko, D.V., Akins, D.R., Robinson, E., Li, M., Popova, T.G., Cox, D.L., and Radolf, J.D. (1997a). Molecular characterization and cellular localization of TpLRR, a processed leucine-rich repeat protein of *Treponema pallidum*, the syphilis spirochete. J. Bacteriol. *179*, 3188-3195.

Shevchenko, D.V., Akins, D.R., Robinson, E.J., Li, M., Shevchenko, O.V., and Radolf, J.D. (1997b). Identification of homologs for thioredoxin, peptidyl prolyl cis-trans isomerase, and glycerophosphodiester phosphodiesterase in outer membrane fractions from *Treponema pallidum*, the syphilis spirochete. Infect. Immun. *65*, 4179-4189.

Shevchenko, D.V., Sellati, T.J., Cox, D.L., Shevchenko, O.V., Robinson, E.J., and Radolf, J.D. (1999). Membrane topology and cellular location of the *Treponema pallidum* glycerophosphodiester phosphodiesterase (GlpQ) ortholog. Infect. Immun. *67*, 2266-2276.

Shotton, D.M. and Severs, N.J. (1995). An introduction to freeze fracture and deep etching. In Rapid freezing, freeze fracture and deep etching, D.M.Shotton and N.J.Severs, eds. (New York: Wiley-Liss, Inc.), pp. 2-30.

Stamm, L.V. and Bergen, H.L. (2000). The sequence-variable, single-copy *tprK* gene of *Treponema pallidum* Nichols strain UNC and Street strain 14 encodes heterogeneous TprK proteins. Infect. Immun. *68*, 6482-6486.

Stamm, L.V., Greene, S.R., Bergen, H.L., Hardham, J.M., and Barnes, N.Y. (1998). Identification and sequence analysis of *Treponema pallidum tprJ*, a member of a polymorphic multigene family. FEMS Microbiol. Lett. *169*, 155-163.

Stamm, L.V., Hodinka, R.L., Wyrick, P.B., and Bassford, P.J., Jr. (1987). Changes in the cell surface properties of *Treponema pallidum* that occur during *in vitro* incubation of freshly extracted organisms. Infect. Immun. *55*, 2255-2261.

Stebeck, C.E., Shaffer, J.M., Arroll, T.W., Lukehart, S.A., and Van Voorhis, W.C. (1997). Identification of the *Treponema pallidum* subsp. *pallidum* glycerophosphodiester phosphodiesterase homologue. FEMS Microbiol. Lett. *154*, 303-310.

Swancutt, M.A., Radolf, J.D., and Norgard, M.V. (1990). The 34-kilodalton membrane immunogen of *Treponema pallidum* is a lipoprotein. Infect. Immun. *58*, 384-392.

Swancutt, M.A., Riley, B.S., Radolf, J.D., and Norgard, M.V. (1989). Molecular characterization of the pathogen-specific, 34-kilodalton membrane immunogen of *Treponema pallidum*. Infect. Immun. *57*, 3314-3323.

Thomas, D.D., Baseman, J.B., and Alderete, J.F. (1985). Putative *Treponema pallidum* cytadhesins share a common functional domain. Infect. Immun. *49*, 833-835.

Thomas, K.L., Leduc, I., Olsen, B., Thomas, C.E., Cameron, D.W., and Elkins, C. (2001). Cloning, overexpression, purification, and immunobiology of an 85- kilodalton outer membrane protein from *Haemophilus ducreyi*. Infect. Immun. *69*, 4438-4446.

Timpl, R. and Dziadek, M. (1986). Structure, development, and molecular pathology of basement membranes. Int. Rev. Exp. Pathol. *29*, 1-112.

Vaczi, L., Kiraly, K., and Rethy, A. (1966). Lipid composition of treponemal strains. Acta. Microbiol. Acad. Sci. Hung. *13*, 79-84.

Van Embden, J.D., van der Donk, H.J., van Eijk, R.V., van der Heide, H.G., de Jong, J.A., van Olderen, M.F., Osterhaus, A.B., and Schouls, L.M. (1983). Molecular cloning and expression of *Treponema pallidum* DNA in *Escherichia coli* K-12. Infect. Immun. *42*, 187-196.

Vicentini, A.P., Gesztesi, J.L., Franco, M.F., de Souza, W., de Moraes, J.Z., Travassos, L.R., and Lopes, J.D. (1994). Binding of *Paracoccidioides brasiliensis* to laminin through surface glycoprotein gp43 leads to enhancement of fungal pathogenesis. Infect. Immun. *62*, 1465-1469.

Voulhoux, R., Bos, M.P., Geurtsen, J., Mols, M., and Tommassen, J. (2003). Role of a highly conserved bacterial protein in outer membrane protein assembly. Science *299*, 262-265.

Walker, E.M., Zampighi, G.A., Blanco, D.R., Miller, J.N., and Lovett, M.A. (1989). Demonstration of rare protein in the outer membrane of *Treponema pallidum* subsp. pallidum by freeze-fracture analysis. J Bacteriol *171*, 5005-5011.

Weinstock, G.M., Hardham, J.M., McLeod, M.P., Sodergren, E.J., and Norris, S.J. (1998). The genome of *Treponema pallidum*: new light on the agent of syphilis. FEMS Microbiol. Rev. *22*, 323-332.

Wiese, A., Brandenburg, K., Ulmer, A.J., Seydel, U., and Muller-Loennies, S. (1999). The dual role of lipopolysaccharide as effector and target molecule. Biol. Chem. *380*, 767-784.

Wile, U.J., Picard, R.G., and Kearney, E.B. (1942). The morphology of *Spirochaeta pallida* in the electon microscope. J. Am. Med. Assoc. *119*, 880-881.

Yan, S., Rodrigues, R.G., Cahn-Hidalgo, D., Walsh, T.J., and Roberts, D.D. (1998). Hemoglobin induces binding of several extracellular matrix proteins to *Candida albicans*. Identification of a common receptor for fibronectin, fibrinogen, and laminin. J. Biol. Chem. *273*, 5638-5644.

Yang, Y., Thomas, W.R., Chong, P., Loosmore, S.M., and Klein, M.H. (1998). A 20-kilodalton N-terminal fragment of the D15 protein contains a protective epitope(s) against *Haemophilus influenzae* type a and type b. Infect. Immun. *66*, 3349-3354.

Yen, M.R., Peabody, C.R., Partovi, S.M., Zhai, Y., Tseng, Y.H., and Saier, M.H. (2002). Protein-translocating outer membrane porins of Gram-negative bacteria. Biochim. Biophys. Acta. *1562*, 6-31.

Zeigler, J.A., Jones, A.M., Jones, R.H., and Kubica, K.M. (1976). Demonstration of extracellular material at the surface of pathogenic *T. pallidum* cells. Br. J. Vener. Dis. *52*, 1-8.

Zhang, H.H., Blanco, D.R., Exner, M.M., Shang, E.S., Champion, C.I., Phillips, M.L., Miller, J.N., and Lovett, M.A. (1999). Renaturation of recombinant *Treponema pallidum* rare outer membrane protein 1 into a trimeric, hydrophobic, and porin-active conformation. J. Bacteriol. *181*, 7168-7175.

12

Antigenic Variation in *Treponema pallidum*

*Arturo Centurion-Lara**

Abstract

The pathogenic treponemes include three subspecies of *Treponema pallidum* (*T. pallidum*), *T. carateum*, *T. paraluiscuniculi,* and the unclassified Fribourg-Blanc (simian) isolate. These treponemes are antigenically highly related, cause persistent infections in their respective hosts, and induce similar histopathological changes and immune responses. The host immune response is able clear the great majority of spirochetes from early active lesions, however, some organisms escape immune clearance and cause chronic infections. Several mechanisms, including antigenic variation, have been proposed to explain the ability of *T. pallidum* to disseminate and evade the immune response of the host. One member of the *tpr* gene family, *tprK*, undergoes extensive sequence diversification during infection, and its encoded protein induces an active immune response. The variable portions of the protein are antibody epitopes, and variation of these epitope sequences would potentially allow *T. pallidum* to avoid the immune response or, alternatively, to adapt to the changing micro-environments posed by dissemination to different tissue sites. We describe in this chapter a non-reciprocal gene conversion-like mechanism for the generation of *tprK* diversity in *Treponema pallidum* subsp. *pallidum*, in which new variants are created by unidirectional exchange between donor sites and the *tprK* expression site. This mechanism of antigenic variation is likely to be shared by all members of this group of pathogenic treponemes and is likely to contribute to the persistence of treponemal infections.

1. Introduction

The pathogenic non-cultivatable treponemes include *Treponema carateum*, *Treponema paraluiscuniculi* and three *Treponema pallidum* (*T. pallidum*) subspecies (*T. pallidum* subsp. *pallidum*, *T. pallidum* subsp. *pertenue* and *T. pallidum* subsp. *endemicum*). The three *T. pallidum* subspecies naturally infect humans but can also experimentally infect rabbits and other mammals. *Treponema carateum* naturally infects humans but is unable to multiply in rabbits or other non-primates

*For correspondence email acentur@u.washington.edu

and *T. paraluiscuniculi* naturally infects rabbits but is reportedly not infectious for humans (Graves and Downes, 1981; Turner and Hollander, 1957). One unclassified treponemal isolate is the Fribourg-Blanc or Simian treponeme, which is able to infect humans and cause an active infection (Smith et al., 1971). This strain was isolated from a baboon from a region in West Africa where *T. p. pertenue* infections are endemic (Fribourg-Blanc and Mollaret, 1969; Fribourg-Blanc et al., 1966). All these treponemes are morphologically indistinguishable (Hovind-Hougen, 1983), induce similar histopathological changes and antibody and cellular responses indicating a high level of relatedness. They are poorly studied except for *T. pallidum* subsp. *pallidum,* the etiologic agent of venereal syphilis. For this reason, we will mainly focus on this organism to describe an antigenic variation system that is likely to be common to all members of this group of pathogenic treponemes.

The syphilis treponeme is able to cause a life long infection in the infected individual. It contains a small genome of 1.21 Mb, and cannot be genetically manipulated or cultured continuously *in vitro* (Fraser et al., 1998). Consequently, most of the studies on pathogenesis, immunology and physiology of this organism are carried out in rabbit-propagated treponemes. However, the recent availability of the genome sequence of the Nichols strain has provided a wealth of information, and has helped in the identification of potential pathogenic determinants and in the formulation of working hypotheses about mechanisms by which this spirochete may cause such a persistent infection in its host. The infection is characterized by episodes of active clinical disease interrupted by periods of asymptomatic, latent infection. *T. pallidum* multiplies at the site of inoculation and immediately disseminates throughout the body. The primary ulcerative lesion contains very large numbers of treponemes and persists for several weeks. Then, it heals spontaneously and is followed by a secondary stage characterized by disseminated lesions, which disappear spontaneously over a period of a few weeks or months. The infected host enters the latent stage, which may be interrupted by recurrent secondary manifestations or later, by the tertiary stage. Few treponemes can be detected during the latent and tertiary stages of the infection. In early syphilis, *T. pallidum* induces a strong humoral and cellular immune response capable of eliminating millions of organisms, but unable to clear the few spirochetes that escape from primary and secondary lesions and survive during the latent or tertiary stages.

It is still unknown how this organism disseminates throughout the different organs, and how some treponemes avoid immune response of the host. Several mechanisms have been proposed to explain the ability of *T. pallidum* to survive in its host and cause persistent infection. These include: 1- Localization in immunologically privileged organs or tissues such as the central nervous system, the eye and the fetus. 2- Intracellular location of *T. pallidum* in non-phagocytic cells (Azar et al., 1970; Fitzgerald et al., 1975; Lauderdale and Goldman, 1972; Ovcinnikov and Delektorskij, 1972; Sykes et al., 1974). 3- Resistance of *T. pallidum* to high titers of specific treponemal antibodies due to the presence of a protective outer coat of host origin surrounding the bacterial cell (Alderete and Baseman, 1979; Christiansen, 1963; Turner and Hollander, 1957). 4- A poorly antigenic outer membrane due to the chemical composition of the bacterial outer surface and a low density of surface-exposed antigens (Cunningham et al., 1988; Medici, 1972; Penn,

1981; Penn and Rhodes, 1982; Radolf et al., 1988; Radolf and Norgard, 1988; Radolf et al., 1989; Walker et al., 1989). 5- Failure of the host immune response (Fitzgerald et al., 1991; Musher et al., 1975; 1974; Pavia et al., 1977; 1976; Tomai et al., 1989) and 6- Antigenic diversity or variation. It has been speculated for example, that the limited protection against heterologous *T. pallidum* isolates in rabbits and humans is due to significant antigenic diversity among isolates, but no candidate antigens had been identified until recently. The chronicity of this infection suggests that this spirochete can adapt very well and evade the host immune system using one or more of the above mechanisms.

2. Genetic diversity and antigenic variation

Despite the fact that the pathogenic treponemes have different host specificities and cause clinically distinct diseases, they are antigenically related. Immunological cross-reactivity is demonstrated by both cross-immunity and antigenic profiles studies. In cross-immunity experiments, animals are usually infected with one immunizing strain and then challenged with different treponemal strains. In the rabbit model the infection with a particular pathogenic treponeme will induce complete protection against re-infection with the homologous organism, while protection against other species or subspecies is either partial or non-existent. Failure to develop lesions after challenge indicates a substantial degree of cross immunity and a close antigenic composition. For example, rabbits infected with the Nichols strains develop variable degrees of cross-immunity when challenged with different strains of syphilis, *T. pertenue*, *T. endemicum* or *T. paraluiscuniculi* strains (Miller, 1973; Turner and Hollander, 1957; Lukehart et al. Unpublished). Overall, animals challenged with the homologous strain develop no lesions and challenge with *T. paraluiscuniculi* confers minimal resistance to challenge. Cross-reactivity in serological tests such as immobilization, cross-agglutination and immunofluorescence tests also indicate the presence the common antigens among treponemes. Antigenic profiles are usually studied by immunoblotting and immunoprecipitation techniques. There are only subtle differences in antibody specificities to *T. pallidum* antigens between syphilis, *T. pertenue* and *T. paraluiscuniculi* isolates (Baker-Zander and Lukehart, 1984; Martin et al., 1990; Noordhoek et al., 1990a; Thornburg and Baseman, 1983) indicating a similar antigenic composition. Their antigenic patterns are composed mainly by the 70, 47, 41, 37, 35, 30, 17 and the 15 kD antigens (Baker-Zander and Lukehart, 1983). Subtle differences (*e.g.* band intensity) may reflect slightly different antigenic determinants or differential expression patterns among isolates.

Comparative sequence analysis of *T. pallidum* genes reveals a high degree of sequence conservation among *T. pallidum* isolates (Cameron et al., 1999; Weinstock, 2000; Lukehart, et al. Unpublished). However, some genes vary among isolates and some of these genetic differences are used for syphilis strain typing or to differentiate syphilis from non-syphilis isolates. Pillay et al. (1998) reported different numbers of repeats in the *arp* gene and polymorphism in the Subfamily II of the *tpr* genes among syphilis strains . Other studies identified single nucleotide polymorphisms in the *tpF1*, *gpd, tp92* genes and the 5′ flanking region of the *tp15* gene (Cameron et al., 1999; 2000; Centurion-Lara et al., 1998; Noordhoek et al., 1989; 1990b). These single base pair changes differentiate syphilis from non-syphilis treponemes.

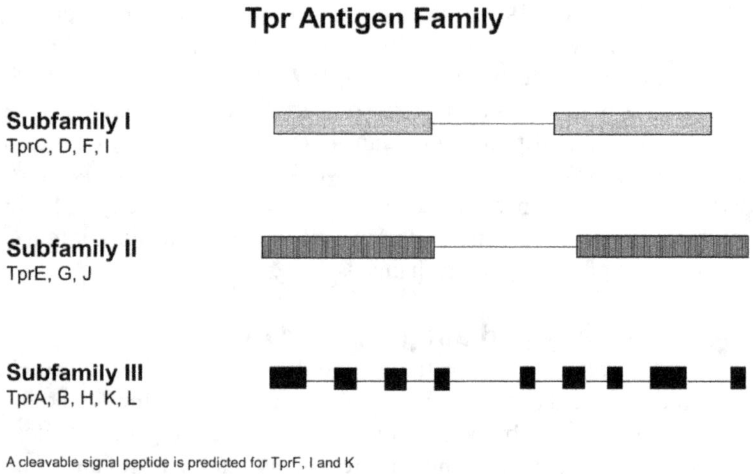

Figure 1. Schematic representation of the three subfamilies based upon alignments of the Tpr proteins from the published Nichols genome sequence. Boxes indicate the regions of homology between subfamily members. Lines indicate unique sequences. The amino and carboxyl regions are highly conserved in Subfamily I and II. The central regions are unique to each Tpr antigen. The Subfamily III Tprs have only small regions of homology.

A 12-membered gene family named *tpr* (*Treponema pallidum* repeat) has been identified and is being actively studied (Centurion-Lara et al., 1999; Fraser et al., 1998; Stamm et al., 1998). The Tpr antigens are homologous to the major sheath protein (Msp) of *Treponema denticola*, a protein reported to be surface-exposed and involved in cell attachment. Three Tpr subfamilies can be identified by their predicted amino acid homology: Subfamily I, TprC, D, F, and I; Subfamily II, TprE, G, and J; and Subfamily III, Tpr A, B, H, K, and L (Figure 1). Alignment of the amino acid sequences of the Tprs of Subfamily I and Subfamily II shows that the NH_2- and COOH-terminal regions of the predicted proteins are conserved, whereas the central domains are variable in terms of sequence and length. Within Subfamily I, Tpr C and D are identical, whereas TprF has a deletion and lacks a central variable domain and the conserved COOH-terminal domain. Subfamily III is composed of five members that are comparatively poorly homologous to each other or to the other Tpr proteins, but that still retain small areas of conserved sequences. Potentially cleavable signal peptides are predicted for all members of Subfamily I and TprK. TprK, F, and I are predicted to be located in the periplasmic space or in the outer membrane.

There is considerable diversity in the Tpr Subfamily I and II among treponemal isolates (Centurion-Lara et al., 2000; Sun et al., 2004). Table 1 provides a summary of some of the most distinctive allelic variations in the *tpr* genes among treponemal isolates. The *tprC* and *tprD* Subfamily I genes are identical alleles in the Nichols isolate; in other treponemal isolates two alternative *tprD* alleles (*tprD2*, and *tprD3*) are present in the *tprD* locus (Centurion-Lara et al., 2000). There is also diversity in the *tprC* gene among treponemal isolates. *T. pallidum* strains that have the *tprD* gene also have an identical *tprD* allele at the *tprC* locus. Those *T. pallidum* isolates

Table 1. Polymorphism in the loci of the Subfamily I and Subfamily II *tpr*s among treponemal species and subspecies

Species or subspecies		Subfamily I				Subfamily II		
	Subgroup Strain	***tprD* locus**	***tprC* locus**	***tprF* locus**	***tprI* locus**	***tprE* locus**	***tprG* locus**	***tprJ* locus**
T. pallidum subsp. *pallidum*	Subgroup 1 (Nichols) [5]	*tprD*	*tprD*	*tprF*	*tprI*	*tprE*	*tprG*	*tprJ*
	Subgroup 2 (Sea-81-3) [5]	*tprD2* [6]	*tprD* [1] like	*tprF*	*tprI*	*tprE* [2]	*tprG* [2]	*tprJ* [2]
	Subgroup 3 (Sea-81-4) [5]	*tprD2* [6]	*tprD* [1] like	*tprF*	*tprI*	*tprE* [2]	*tprG* truncated	*tprG* like
T. pallidum subsp. *endemicum*	Iraq B	*tprD2* [6]	*tprD* [1] like	*tprI* like	*tprI* like	*tprE*	*tprG* [2]	absent [2,7]
T. pallidum subsp. *pertenue*	Gauthier	*tprD3* [4]	*tprD3* [4]	*tprI* [8] like	*tprI* [8] like	*tprE*	*tprG*	*tprG* like
T. paraluiscuniculi	Cuniculi A	*tprD2* [6]	*tprD2* [6]	*tprG/I* [3] hybrid	absent	*tprG like*	*tprG/I* [3] hybrid	*tprG* like

[1] indicates 7 to 27 amino acid differences with the *tprD* allele. [2] as determined by PCR using gene specific primers. [3] a single *tprG/I* hybrid replaces the DNA fragment containing the *tprG* and *tprF* genes. [4] *tprD3* is 95% identical to *tprD*; with the major region of heterogeneity in the central region and scattered amino acid differences throughout the protein sequence. [5] indicates a representative strain for the subgroup. [6] *tprD2* differentiates from *tprD* by the presence of a 330-bp central unique region and three smaller heterogeneous regions at the 3' end. This allele is present in 50% of the syphilis isolates. [7] It is likely to carry the *tprG*-like ORF in this locus. [8] identical alleles.

that have the *tprD2* allele at the D locus have *tprD*-like alleles at the *tprC* locus. The Gauthier strain of *T. p. pertenue*, which contains the *tprD3* allele in the *tprD* locus, also has a *tprD3* allele in the *tprC* locus. In contrast to the *tprC* and *tprD* homologs, *tprF* and *tprI* are invariant among syphilis isolates. In the Gauthier strain of *T. p. pertenue*, the *tprF* gene is absent and a duplicated *tprI* ORF occupies this locus. No sequence diversity is present in the *tpr* members of the Subfamily II among syphilis isolates except for the Sea 81-4 strain. This isolate carries in the *tprJ* locus a *tprG*-like ORF, in which the central region of *tprJ* is replaced by the central portion of the *tprG* homolog. This *tprG*-like ORF is also present in *T. p. pertenue*, Gauthier strain (Stamm et al., 1998).

Of particular interest is the *tpr* gene family of *T. paraluiscuniculi* (Giacani et al., 2004). There are unique features in all members of the Subfamilies I and II (Table 1) and in some of the Subfamily III ORFs. The *tprC* and *tprD* loci contain two identical alleles homologous to the *T. pallidum tprD2* gene (Centurion-Lara et al., 2000). The *tprF*, *tprI* and *tprE* genes are absent. The *tprE* and *tprJ* loci contain two identical sequences (named *tprG1* and *tprG2*) homologous to the *tprG*-like genes present in the Sea 81-4 and Gauthier strains. A hybrid *tprG/I* ORF replaces the *tprG* and *tprF* loci. Remarkably, the *tprC*, *tprD*, *tprG1* and *tprG2* ORFs and the *tprG/I* hybrid have a frameshift, which introduces a premature stop in their predicted protein sequences. The changes observed in the Tpr Subfamilies I and II among treponemal isolates indicates genetic drift rather than true antigenic variation. Diversity is not observed within an isolate, and accumulation of diversity during infection is not seen. The significance of the sequence diversity is still unknown. These changes may be involved in functional diversification, immune evasion or isolate specific immunity. In *T. paraluiscuniculi*, all subfamily I, II and the *tprG/I* ORFs are predicted to encode truncated proteins strongly suggesting that these genes are not essential for survival in the rabbit. These sequence findings are further supported by studies on antibody and T cell responses in the rabbit model (Giacani et al., 2004)

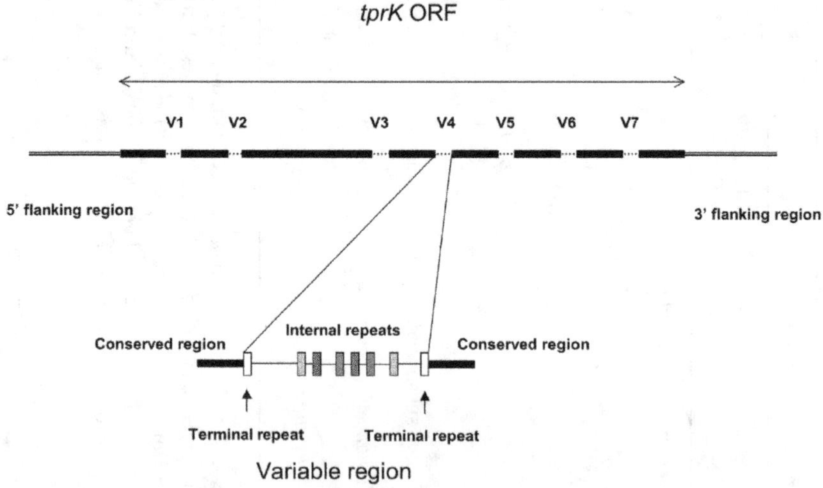

Figure 2. The *tprK* ORF and the architecture of its V regions. All seven V regions share the same sequence structure and are flanked by highly conserved sequences.

Among the Subfamily III *tprs*, no heterogeneity among human isolates is described except for *tprK* (Centurion-Lara et al., 1999). In *T. paraluiscuniculi*, *tprA* and *tprL* show differences when compared with the Nichols homologs (Giacani et al., 2004). The Nichols *tprA* has an authentic frameshift and a premature stop codon. This frameshift is absent in the *T. paraluiscuniculi tprA* generating a full-length predicted protein sequence. The rabbit *tprL* homolog has a unique region of 366 nucleotides with no sequence homology in the *T. pallidum* databases.

tprK in human isolates undergoes extensive antigenic diversification during infection providing, for the first time, evidence of an antigenic variation system and chromosomal plasticity in *T. pallidum*. Conceptually, antigenic variation in syphilis can be defined as an adaptive mechanism to create genetic flexibility of *T. pallidum* to survive environmental changes. This would allow *T. pallidum* to adapt efficiently to changing micro-environmental conditions in its host (*e.g.* different tissue locations) and/or to avoid the immune response, which ultimately, would facilitate its survival in an individual and secure transmission in human populations. The rest of this chapter will focus on the description of a recently discovered non-reciprocal gene conversion-like mechanism that accounts for the generation of TprK diversity during syphilis infection.

3. The TprK antigen

The published genome sequence revealed a single ORF in the *T. pallidum*, Nichols strain, chromosome coding for the TprK antigen (Fraser et al., 1998). Later studies demonstrated that the *tprK* gene is highly heterogeneous within and among human strains with heterogeneity focused in seven distinct variable (V) regions (V1 – V7) flanked by highly conserved sequences. Five of the V regions (V3-V7) are located in the second half of the *tprK* ORF, while only two (V1-V2) are located in the first half of the gene (Centurion-Lara et al., 2000) (Figure 2). All seven V regions share a unique sequence architecture of terminal repeats, internal repeats and non-repetitive sequences, and the V regions boundaries are defined by two identical repeats: one at the beginning and one at the end of each V region. Heterogeneity is due to apparent base changes, insertions, or deletions; however, none of these changes introduces frameshifts or stop codons in the ORFs (Centurion-Lara et al., 2004). In the rabbit treponeme, the *tprK* changes are similar to those observed in human isolates, but diversity is less impressive than in the human isolates. Distinct heterogeneity is present in only five V regions (V1, V2, V4, V6, and V7). V3 is not variable and V5 has very limited diversity (Giacani et al., 2004).

3.1. TprK as target of the humoral and cellular immune response

The nature of the immune response to *T. pallidum* is not fully characterized. Antibody and cellular immune responses are mounted against many antigens during the course of natural and experimental infection. Among these antigens are members of the Tpr antigen family such as TprF, TprI and TprK. TprK is the target of a strong and early antibody and cellular immune response. Morgan et al. (2002a) studied the immunogenic properties of three protein fragments spanning the entire length of TprK: fragment 1, the hydrophilic N-terminal portion downstream of the proposed signal sequence cleavage site (V1, V2 and V3); fragment 2, a hydrophobic middle region (V4, V5); and fragment 3, the hydrophilic C-terminal region (V6 and

V7). All three TprK fragments were able to elicit proliferative responses in spleen cells from rabbits infected for 10 to 280 days, but only fragments 1 and 3 are recognized by antibodies. Antibodies did not recognize fragment 2. These findings prompted more detailed epitope mapping studies. Using synthetic 20 amino acid peptides for proliferation and ELISA assays, Morgan et al. (2002b) determined that during experimental infection with the Nichols strain, T cell responses are directed to conserved regions, while the antibody responses are directed primarily to variable regions. In contrast to infection-induced immunity, antibodies from rabbits immunized with recombinant TprK recognized both conserved and variable regions, suggesting that the conserved regions are as immunogenic as the variable regions. Additional studies not only confirmed initial observations that TprK is a target of the immune response but also found that antibody responses to the V regions of one TprK molecule show little or no cross-reactivity with heterologous TprK V regions and that differences in the V regions of *tprK* in the challenge inocula correlate with reduced control of treponemal growth (Morgan et al., 2003). These findings strongly suggest an important role for the immune response to TprK V regions during infection.

3.2. A non-reciprocal recombination model for generation of *tprK* variants: Expression and donor sites.

The presence of multiple *tprK* variants within and among isolates strongly suggests that recombinatorial events may be involved in the generation of heterogeneity in the seven *tprK* V regions. The proposed model for generation of *tprK* diversity by non-reciprocal recombination in *T. pallidum* during infection includes a single expression site, the *tprK* ORF with its seven V regions, and donor sites in the flanking regions of *tprD*, another *tpr* homolog (Centurion-Lara et al., 2000). In this model, new *tprK* variants are generated by non-reciprocal exchange of stretches of DNA sequences from the donor sequences into the *tprK* V regions with sequence conservation of the donor regions (Figure 3). The first component in this system (the expression site) is located in the *T. pallidum* Nichols genome at positions 975833-974319 (Fraser et al., 1998) and codes for a protein containing a signal peptide (Hazlett et al., 2001). The strong immune response that TprK induces in the host indicates that this antigen is actively expressed during infection (Morgan et al., 2002a; 2003) . Although a large number of distinct *tprK* alleles are present in *T. pallidum* isolates, Southern blot and genomic DNA restriction digest analyses from the *T. pallidum* Nichols UNC and Street strain 14 strains suggest that *tprK* is also a single copy gene in these isolates (Stamm and Bergen, 2000). The second component of this model is the putative donor sites for the V regions. In the published genome sequence of the Nichols strain and in the Chicago isolate, the 5′- and 3′-flanking regions of the *tprD* locus, located at positions 152897-151104 in the *T. pallidum* genome, contain DNA stretches (donor sequences) homologous to *tprK* V regions (Centurion-Lara et al., 2004). The putative donor sites contain DNA sequences identical to partial or nearly complete *tprK* V regions. The *tprD* 5′-flanking donor region contains one donor site for V6, one donor site for V7 and 125 bases identical to a highly conserved *tprK* region. Downstream of the *tprD* gene of the NicholsSea and the Chicago isolates, there are 45 donor sites in a ~3.8 kB long

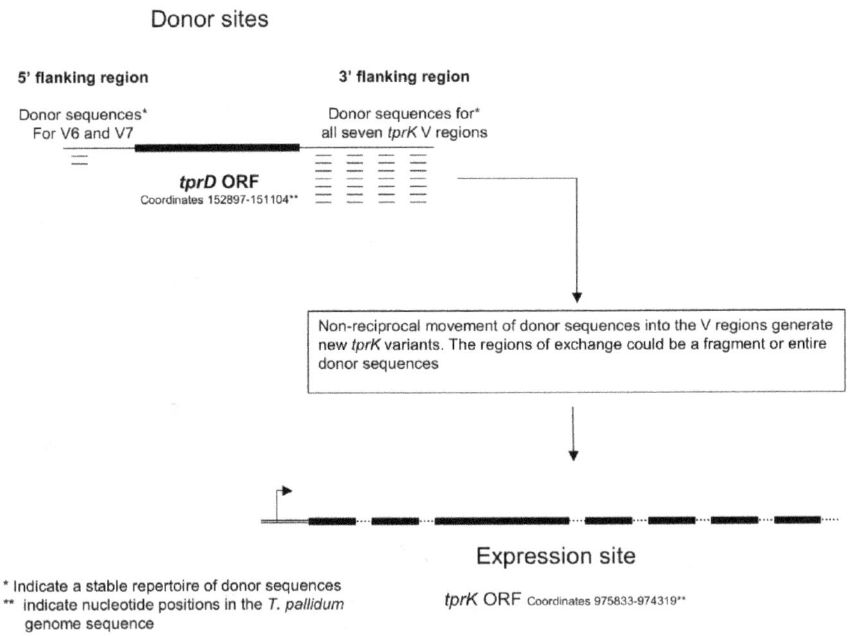

Figure 3. A non-reciprocal (gene conversion-like) recombination model for generation of heterogeneity in the *tprK* gene

region, which has five donor sites for V1 and V5, six for V2 and V7, nine for V6, ten for V3, four for V4 and one sequence with homology to a conserved region of *tprK* are spread throughout this region (Centurion-Lara et al., 2004).

3.3. tprK diversity is generated during infection

Amplification of *tprK* from rabbit-propagated isolates and directly from patient samples shows that the infecting organisms contain different populations of treponemes as defined by their heterogeneous *tprK* sequences. LaFond et al. (2003) has demonstrated that *tprK* diversity is present during syphilis infection in patients, that heterogeneity accumulates during rabbit passage and that individual V regions have different intrinsic rates of sequence change levels of diversification. Furthermore, clustering analysis showed that sequences derived directly from syphilis patients are diverse but cluster more tightly within infected individuals than rabbit-propagated isolates suggesting that diversity increases upon propagation. The ability of this bacterium to chronically infect its host is likely associated with the observed degree of heterogeneity. As in syphilis, it has been proposed that the ability of *Helicobacter pylori* to cause a chronic infection is due to its extreme genetic heterogeneity both within and among patient isolates generated by a combination of different mechanisms including antigenic variation (Israel et al., 2001; Suerbaum, 2000), which would facilitate adaptation to the complex and harsh gastric environment.

The use of a clonal isolate derived from the polyclonal Chicago strain (Chicago clonal C), as defined by the presence of identical *tprK* sequences, allowed the demonstration of accumulation of diversity during infection in the rabbit model by sequential passages with minimal adaptive immune pressure followed by passages

A

```
                       P  V  H  Y  K  V  L  K  A  R  A  R  A  G  A  A  V  P  A  A  A  D  D  I  Y  F  P
Progenitor *           CCTGTCCATTACAAAGTCCTAAAAGCCCGAGCCCGAGCCGGAGCCGCAGTGCCAGCCGCAGCCGACGACATCTACTTCCCG
Donor site @   CGTACATGCCCGTCCATTGGAACGCCTTCA---CCCAAGCCCGAGCCCTGCCCGGAGCCCCAGTCCCAGCC------ATCTACTTCCCGGT
New variant ^          CCCGTCCATTGGAACGCCTTCA---CCCAAGCCCGAGCCCTGCCCGGAGCCCCAGTCCCAGCC------ATCTACTTCCCG
                       P  V  H  W  N  A  F  T     Q  A  R  A  L  P  G  A  P  V  P  A        I  Y  F  P
```

B

```
                 Y  F  A  S  Q  A  S  N  V  F  Q  G  V  F  L  T  T  P  H  Q  K  D  D  C  A  A  Y  I
Progenitor *     TATTTCGCATCGCAGGCCAGCAATGTATTTCAGGGTGTCTTTCTCACCACACCCATGCAGAAGGACGACTGTGCTGCCTATATT
Donor site @                              TATTCGGGGGAGTATTTCTCACCAATAACATGCTGCAGCACGACTG
Donor site @          CGCATCGAAGGCCAGCAATGTATTCGAGGGAGTATTTCT
New variant ^    TATTTCGCATCGAAGGCCAGCAATGTATTCGAGGGAGTATTTCTCACCAATAACATGCTGCAGCACGACTGTGCTGCCTATATT
                 Y  F  A  S  K  A  S  N  V  F  E  G  V  F  L  T  N  N  H  L  Q  H  D  C  A  A  Y  I
```

Figure 4. Two examples of new V regions in the *tprK* expression generated by non-reciprocal replacement. * indicates the progenitor sequence in the *tprK* locus. @ indicates the donor site(s) in the flanking regions of *tprD*, which are copied and moved into the *tprK* V region in the expression site. ^ indicates the newly generated *tprK* V region in the expression site. Gray color highlights the regions of exchange and, nucleotide and amino acid changes. A: A single recombination event generates a new V6 region in the *tprK* locus. B: Two-step recombination is needed to generate a new V5 region in the *tprK* ORF. Underline indicates the terminal repeats of the progenitor and the new V regions.

in presence of adaptive immunity (Centurion-Lara et al., 2004). These studies confirmed that diversity accumulates over time and showed that variation in *tprK* cannot only arise spontaneously but is more impressive when immune pressure is present. The new variants show a wide range of sequence changes in the V regions including mosaic sequences suggesting single or multiple recombination steps templated by the donor sites. An example is provided in Figure 4. These findings have similarities with the antigenic variation system in *Anaplasma marginale* (Brayton et al., 2002; Meeus et al., 2003) supporting the hypothesis that *tprK* diversity is mediated by a segmental gene conversion-like mechanism, where new *tprK* V regions are generated by non-reciprocal exchange of DNA sequences from the donor sequences in the *tprD* flanking regions into the *tprK* V regions in the expression site. Analysis of the donor sites of the NicholsSea and the polyclonal Chicago strains as well as the clonal Chicago C and one Chicago isolate harvested after being passaged in presence of adaptive immune response shows no changes in the sequence composition of the donor sites supporting the theory of non-reciprocal exchange of information.

3.4. The recombination machinery of *T. pallidum*

The specific molecular mechanisms that generate *tprK* diversity are unknown. Major mechanisms involved in the generation of genetic diversity are lateral DNA transfer and antigenic variation. Lateral transfer, defined as the horizontal exchange and stable integration of DNA in the chromosome of the recipient, is a mechanism frequently used to generate diversity. To date, no circumstantial or hard evidence exists for horizontal exchange in *T. pallidum* and no systems for transduction, conjugation, natural transformation or transposition have been yet identified, though the machinery used by other organisms to stably integrate transferred DNA is also present in *T. pallidum* (Fraser et al., 1998). The Nichols strain DNA recombination system includes *recA*, the homologs of the *recF* pathway (*recFGJNR*), *recX*, and two ORFs with homology to the *codV* (*B. subtilis*) and *xprB* (*E. coli*) site-specific recombinases. Homologs of *recBCD* pathway are lacking. *T. pallidum* also has the major systems of *uvr* excision repair, *mutL*/*mutS* mismatch repair, *mutY*, and *dat* (Fraser et al., 1998).

In *N. gonorrhoeae,* antigenic variation in the *pil* system by gene conversion is mediated by RecA-and the RecF-like pathway (RecQ, RecO) (Koomey et al., 1987; Mehr and Seifert, 1998) with minimal sequence identity requirements upstream and downstream of the regions of exchange for crossover formation (Howell-Adams and Seifert, 2000). A similar unidirectional segmental gene conversion mechanism is postulated for the generation of antigenic variation in *Anaplasma marginale* (Brayton et al., 2002). The regions of homology flanking the fragments of exchange in some new *tprK* variants may function as crossover regions or anchors for enzymatic activity. In others, homology is limited to one of the flanking regions of the DNA fragment replaced by the incoming DNA cassette, suggesting that sequence homology both in the upstream and downstream regions might not be necessary for segmental replacement. There are no data available to support any specific mechanism, but it is likely that a combination of mechanisms (*recA*, the *recF* pathway, site-specific recombinases, or mismatch repair systems) mediates the non-reciprocal insertion from cassettes from the donor sites into the *tprK* V regions. Whether a single pathway or combined mechanisms are involved, this system is able to generate a very impressive number of heterogeneous *tprK* alleles. Analysis of the RecA and the RecF pathway genes in other *T. pallidum* strains shows no differences with the ORFs described in the Nichols genome sequence. Despite the fact that in other organisms the specific role of some recombinases in gene conversion has been established, Rec proteins may not be so specific regarding their participation in any particular pathway (Sawitzke and Stahl, 1994). In *Neisseria*, for example, it is proposed that the availability of particular recombinatorial functions, and the presence of specific DNA substrates, will determine the type of recombinases involved in a particular *pilE* recombination event (Meyer and Hill, 2003). Due to the lack of genetic systems in *T. pallidum* and the inability to transform these spirochetes, studies to decipher the role of each Rec molecule in TprK recombination and the mechanisms that initiate such processes are more cumbersome.

3.5. Other mechanisms of *tprK* variation

Gene conversion falls in the category of the so-called programmed antigenic variation. It seems to be the major mechanism for generation of diversity in TprK, although the presence of numerous sets of repeats in the V regions may suggest an additional mechanism similar to the one proposed for Neisseria, in which direct repeats flank deleted portions and a new *pilE* is created after repair of the deletions using another *pilS* sequence as a template (Hill et al., 1990). Alternative mechanisms of programmed antigenic variation such as simple inversion switch, reciprocal recombination, DNA modification, and transcriptional or translational control are, however, not excluded and remain unexplored.

Random generation of sequence heterogeneity also plays an important role in shaping the genetic make-up of microorganisms. Single base mutation or larger changes in DNA sequences are generated by errors in DNA replication or repair and recombination between genes, among others. Point mutations that do not introduce stops in the *tprK* sequence are frequently observed in the *tprK* V regions. The presence of such single base changes in donor sequences suggests that these mutations are templated by the donor sites. The lack of premature stops induced by these single nucleotide changes suggests an important role for TprK in the biology of *T. pallidum*.

4. Future directions

Although the work presented here focuses on TprK, ongoing studies in our laboratory and elsewhere are actively investigating mechanisms of antigenic variation in the other *tpr* members and *T. pallidum* genes. Specifically, these studies intend to identify phase variation mechanisms controlling expression of these genes at the transcriptional level. Homopolymeric and dinucleotide repeats exist in the 5′ flanking regions of the *tpr* Subfamilies I and II. These repetitive sequences vary in length within isolates and might be associated with ON-OFF mechanisms caused by variation in the length of the homopolymeric or dinucleotide tracts (insertions or deletions) or by slipped-strand mechanisms.

4.1. Implications of antigenic variation: Immune and biological selection

Genetic recombination plays an important role in maintaining genetic diversity and plasticity, and genome stability. Recombination between donor and expression sites has been described in several microorganisms including members of the spirochetes. The high degree of diversity seen in the *tprK* variants can be explained by the presence of seven V regions in the expression site as well as the 47 donor sites in the *tprD* flanking regions. The conservation of intact ORFs, despite the diversity observed, suggests that the TprK protein is essential for survival of *T. pallidum*.

Only few treponemes survive the interactions with macrophages, T cells and antibodies at the primary chancre site. However, the main issue is not merely survival of these few organisms, but rather their ability to disseminate, multiply and cause the subsequent stages of the infection. Because syphilis infection is a multifactorial process, it is reasonable to assume that many bacterial strategies are involved in the interplay, including antigenic variation. The TprK antigenic variation system represents for *T. pallidum*, an organism believed to have a rock stable genome, the first system detected in pathogenic treponemes that indicates the presence of genome plasticity in these organisms. Its biological role is not yet fully understood but it is likely to play a key role in immune evasion or adaptation to changing microenvironmental conditions during infection. Several functions could be proposed including binding to metabolic intermediates or to cell surface receptors, but whatever the function of TprK is, some alleles need to remain functional despite the strong humoral and cellular immune response of the host. Although selective immunological pressure may be a factor in driving accumulation of *tprK* diversity, spontaneous generation of variation of *tprK* occurs during infection in the rabbit model in the absence of adaptive immune response. This suggests biological roles for TprK, other than immune evasion, such as tissue tropism. The degree of diversity generated spontaneously might be strain and TprK-dependent as documented for the variable surface protein (VSP) switching in *Giardia lamblia* in cultures (Nash et al., 1990).

TprK proteins differ in size and composition of their unique V regions. These differences in protein sequences, solely due to the presence of distinct V regions, may confer different properties associated with a particular microenvironment where the treponemes are located; because of this, they might undergo positive or negative selection *in vivo*. To date, there are no data supporting a biological adaptive mechanism for TprK variation. However, the observation made by LaFond et al. (LaFond et al., 2003) that variability within isolates is more limited than among isolates, as determined by clustering analysis, suggests that different strains might express a specific set of *tprK* genes as result of a strong positive selection for expression of fully functional *tprK*s. This has been demonstrated for the Opa proteins of *N. gonorrhoeae* (Connell et al., 1988; Kupsch et al., 1993; Stern et al., 1986). An additional possible adaptive mechanism for persistence of the infection could be the generation of populations with different abilities for survival in either an intracellular or extracellular environment. The great majority of treponemes are located in the extracellular milieu (Lukehart et al., 1980), but *T. pallidum* can also localize intracellularly *in vivo* and *in vitro* in non-phagocytic cells (Azar et al., 1970; Fitzgerald et al., 1975; Lauderdale and Goldman, 1972; Ovcinnikov and

Delektorskij, 1972; Sykes et al., 1974). If treponemes move regularly between the intracellular and extracellular environments, this interaction might serve very well to protect the organism from the innate and adaptive host response. A good example is Neisseria, where invasiveness and serum resistance (i. e. better adaptation to survival in an intracellular or extracellular milieu) is associated with a low or higher LPS sialylation phenotype, respectively (Frosch et al., 1989; Meyer and Hill, 2003; Schneider et al., 1991). By analogy, some TprK proteins might facilitate entry of *T. pallidum* into non-phagocytic cells and, therefore, favor its persistence in the host.

A classic pattern of antigenic variation in response to immune selection is seen in relapsing fever (Barbour, 2002), where peaks of bacterial organisms expressing one particular antigenic variant are followed by immune elimination and a decrease in the numbers of bacteria. It is likely that similar patterns of expression for TprK may be seen during syphilis, if the infection proceeds long enough to generate an immune response that selects for variants. We are currently examining the development of antibodies to individual V regions of TprK during long-term infection, where new antibody specificities might be detectable. However the temporal patterns and the factors inducing expression of particular TprK variants remain uncharacterized.

Unlike most *T. pallidum* isolates that contain heterogeneous *tprK* alleles, the laboratory Nichols has only one *tprK* sequence and is considered a rabbit-adapted strain. The Nichols UNC strain is unique in that it is the only reported Nichols strain that contains multiple *tprK* alleles (Stamm and Bergen, 2000). Ongoing studies in our laboratories show that the Nichols isolate is indeed capable of undergoing diversification in the presence of immune pressure (LaFond et al., 2006). The Nichols strain may be so well adapted to the rabbit that spontaneous generation of TprK is no longer required as a biological adaptive response to short-term infection in the rabbit testicle, and it needs to vary only when different micro-environmental conditions (such as antibodies or T cell responses) develop. Alternative explanations, such as deficient recombination machinery in the NicholsGen and NicholsSea strains, may also explain the limited diversity in their *tprK* genes.

Although the cellular location of TprK is still controversial (Centurion-Lara et al., 1999; Hazlett et al., 2001), the focus of the humoral immune response on the V regions suggests an important role for TprK in pathogenesis (Morgan et al., 2003; Morgan et al., 2002b). Thus, generation of diversity in the V regions of TprK may be important in immune evasion and, consequently, persistence of the infection (Centurion-Lara et al., 2000). The presence of large numbers of variants generated during infection suggests that it could be difficult to use the variable regions of this antigen as targets for vaccines. The antibody response to individual V regions is, in general, not cross reactive to V region variants (Kupsch et al., 1993), and the antigenic variation system described here would be able to generate enough variants to permit the lifelong infection seen in untreated syphilis. Even though the immunization studies conducted by Morgan and colleagues (2002a) using a recombinant peptide representing the amino terminus of TprK did not confer complete protection against challenge, lesions appearing in immunized rabbits were less likely to ulcerate and healed faster. The effect of such vaccines in reducing syphilis transmission in human populations cannot be predicted at this stage, but it is likely that complete protection may not be achieved with TprK alone. Components of TprK may be useful in combination with other antigens in the development of more effective vaccines.

5. Conclusion

Antigenic variation of the TprK antigen by a gene conversion-like mechanism provides the first evidence of genome plasticity in *T. pallidum*. Major questions about other mechanisms, enzymatic requirements for recombination, and the role of TprK in the biology and pathogenesis of the infection still remains unclear. Future studies will examine spontaneous variation versus selection for new variants in the presence of immune pressure and will explore how this mechanism serves in immune evasion or biological adaptation to a changing environment in the host. However, due to the intractability of this microorganism, the development of new biological and genetic experimental approaches is imperative.

Acknowledgement

This work was supported by NIH grants AI34616, AI63940, AI42143, and the University of Washington Royalty Research Fund.

References

Alderete, J. F., and Baseman, J. B. (1979). Surface-associated host proteins on virulent *Treponema pallidum*. Infect. Immun. *26*, 1048-1056.

Azar, H. A., Pham, T. D., and Kurban, A. K. (1970). An electron microscopic study of a syphilitic chancre. Engulfment of *Treponema pallidum* by plasma cells. Arch. Pathol. *90*, 143-150.

Baker-Zander, S. A., and Lukehart, S. A. (1983). Molecular basis of immunological cross-reactivity between *Treponema pallidum* and *Treponema pertenue*. Infect. Immun. *42*, 634-638.

Baker-Zander, S. A., and Lukehart, S. A. (1984). Antigenic cross-reactivity between *Treponema pallidum* and other pathogenic members of the family Spirochaetaceae. Infect. Immun. *46*, 116-121.

Barbour, A. G. (2002). Antigenic variation by relapsing fever Borrelia species and other bacterial pathogens. In Mobile DNA II, R. C. N.A. Craig, M. Gellert and A.M. Lambowitz, ed. (Washington D.C., ASM Press), pp. 972-994.

Brayton, K. A., Palmer, G. H., Lundgren, A., Yi, J., and Barbet, A. F. (2002). Antigenic variation of *Anaplasma marginale msp2* occurs by combinatorial gene conversion. Mol. Microbiol. *43*, 1151-1159.

Cameron, C. E., Castro, C., Lukehart, S. A., and Van Voorhis, W. C. (1999). Sequence conservation of glycerophosphodiester phosphodiesterase among *Treponema pallidum* strains. Infect. Immun. *67*, 3168-3170.

Cameron, C. E., Lukehart, S. A., Castro, C., Molini, B., Godornes, C., and Van Voorhis, W. C. (2000). Opsonic potential, protective capacity, and sequence conservation of the *Treponema pallidum* subspecies *pallidum* Tp92. J. Infect. Dis. *181*, 1401-1413.

Centurion-Lara, A., Castro, C., Barrett, L., Cameron, C., Mostowfi, M., Van Voorhis, W. C., and Lukehart, S. A. (1999). *Treponema pallidum* major sheath protein homologue TprK is a target of opsonic antibody and the protective immune response. J. Exp. Med. *189*, 647-656.

Centurion-Lara, A., Castro, C., Castillo, R., Shaffer, J. M., Van Voorhis, W. C., and Lukehart, S. A. (1998). The flanking region sequences of the 15-kDa lipoprotein gene differentiate pathogenic treponemes. J. Infect. Dis. *177*, 1036-1040.

Centurion-Lara, A., Godornes, C., Castro, C., Van Voorhis, W. C., and Lukehart, S. A. (2000). The *tprK* gene is heterogeneous among *Treponema pallidum* strains and has multiple alleles. Infect. Immun. *68*, 824-831.

Centurion-Lara, A., LaFond, R. E., Hevner, K., Godornes, C., Molini, B. J., Van Voorhis, W. C., and Lukehart, S. A. (2004). Gene conversion: a mechanism for generation of heterogeneity in the *tprK* gene of *Treponema pallidum* during infection. Molecular Microbiology *52*, 1579-1596.

Christiansen, S. (1963). Protective layer covering pathogenic treponematosis. Lancet *1*, 423-425.

Connell, T. D., Black, W. J., Kawula, T. H., Barritt, D. S., Dempsey, J. A., Kverneland, K., Jr., Stephenson, A., Schepart, B. S., Murphy, G. L., and Cannon, J. G. (1988). Recombination among protein II genes of Neisseria gonorrhoeae generates new coding sequences and increases structural variability in the protein II family. Mol. Microbiol. *2*, 227-236.

Cunningham, T. M., Walker, E. M., Miller, J. N., and Lovett, M. A. (1988). Selective release of the *Treponema pallidum* outer membrane and associated polypeptides with Triton X-114. J. Bacteriol. *170*, 5789-5796.

Fitzgerald, T. J., Miller, J. N., and Sykes, J. A. (1975). *Treponema pallidum* (Nichols strain) in tissue cultures: cellular attachment, entry, and survival. Infect. Immun. *11*, 1133-1140.

Fitzgerald, T. J., Tomai, M. A., Trachte, G. J., and Rice, T. (1991). Prostaglandins in experimental syphilis: treponemes stimulate adherent spleen cells to secrete prostaglandin E2, and indomethacin upregulates immune functions. Infect. Immun. *59*, 143-149.

Fraser, C. M., Norris, S. J., Weinstock, G. M., White, O., Sutton, G. G., Dodson, R., Gwinn, M., Hickey, E. K., Clayton, R., Ketchum, K. A., et al. (1998). Complete genome sequence of *Treponema pallidum*, the syphilis spirochete. Science *281*, 375-388.

Fribourg-Blanc, A., and Mollaret, H. H. (1969). Natural treponematosis of the African primate. Primates Med. *3*, 113-121.

Fribourg-Blanc, A., Mollaret, H. H., and Niel, G. (1966). [Serologic and microscopic confirmation of treponemosis in Guinea baboons]. Bull Soc. Pathol. Exot Filiales *59*, 54-59.

Frosch, M., Weisgerber, C., and Meyer, T. F. (1989). Molecular characterization and expression in Escherichia coli of the gene complex encoding the polysaccharide capsule of Neisseria meningitidis group B. Proc. Natl. Acad. Sci. USA. *86*, 1669-1673.

Giacani, L., Sun, E. S., Hevner, K., Molini, B. J., Van Voorhis, W. C., Lukehart, S. A., and Centurion-Lara, A. (2004). Tpr homologs in *Treponema paraluiscuniculi* Cuniculi A strain. Infect. Immun. *72*, 6561-6576.

Graves, S., and Downes, J. (1981). Experimental infection of man with rabbit-virulent *Treponema paraluis- cuniculi*. Br. J. Vener Dis. *57*, 7-10.

Hazlett, K. R. O., Sellati, T. J., Nguyen, T. T., Cox, D. L., Clawson, M. L., Caimano, M. J., and Radolf, J. D. (2001). The TprK protein of *Treponema pallidum* is periplasmic and is not a target of opsonic antibody or protective immunity. J. Exp. Med. *193*, 1015-1026.

Hill, S. A., Morrison, S. G., and Swanson, J. (1990). The role of direct oligonucleotide repeats in gonococcal pilin gene variation. Mol. Microbiol. *4*, 1341-1352.

Hovind-Hougen, K. (1983). Morpholgy. In Pathogenesis and Immunology of Treponemal Infection, R. F. Schell, and D. M. Musher, eds. (New York, NY, Marcel Dekker, Inc.).

Howell-Adams, B., and Seifert, H. S. (2000). Molecular models accounting for the gene conversion reactions mediating gonococcal pilin antigenic variation. Mol. Microbiol. *37*, 1146-1158.

Israel, D. A., Salama, N., Krishna, U., Rieger, U. M., Atherton, J. C., Falkow, S., and Peek, R. M., Jr. (2001). Helicobacter pylori genetic diversity within the gastric niche of a single human host. Proc. Natl. Acad. Sci. USA. *98*, 14625-14630.

Koomey, M., Gotschlich, E. C., Robbins, K., Bergstrom, S., and Swanson, J. (1987). Effects of *recA* mutations on pilus antigenic variation and phase transitions in *Neisseria gonorrhoeae*. Genetics *117*, 391-398.

Kupsch, E. M., Knepper, B., Kuroki, T., Heuer, I., and Meyer, T. F. (1993). Variable opacity (Opa) outer membrane proteins account for the cell tropisms displayed by *Neisseria gonorrhoeae* for human leukocytes and epithelial cells. Embo J. *12*, 641-650.

LaFond, R. E., Centurion-Lara, A., Godornes, C., Rompalo, A. M., Van Voorhis, W. C., and Lukehart, S. A. (2003). Sequence diversity of *Treponema pallidum* subsp. *pallidum tprK* in human syphilis lesions and rabbit-propagated isolates. J. Bacteriol. *185*, 6262-6268.

LaFond, R.E., Centurion-Lara, A., Godornes, C., Van Voorhis, W.C., and Lukehart, S.A. (2006). TprK sequence diversity accumulates during infection of rabbits with *Treponema pallidum* subsp. *pallidum* Nichols strain. *Infect Immun. 74*, 1896-1906.

Lauderdale, V., and Goldman, J. N. (1972). Serial ultrathin sectioning demonstrating the intracellularity of *T. Pallidum*. An electron microscopic study. Br. J. Vener Dis. *48*, 87-96.

Lukehart, S. A., Baker-Zander, S. A., Lloyd, R. M., and Sell, S. (1980). Characterization of lymphocyte responsiveness in early experimental syphilis. II. Nature of cellular infiltration and *Treponema pallidum* distribution in testicular lesions. J. Immunol. *124*, 461-467.

Martin, P. M., Cockayne, A., Georges, A. J., and Penn, C. W. (1990). Immune response to *Treponema pertenue* and *Treponema pallidum* Nichols in patients with yaws. Res. Microbiol. *141*, 181-186.

Medici, M. A. (1972). The immunoprotective niche--a new pathogenic mechanism for syphilis, the systemic mycoses and other infectious diseases. J. Theor Biol. *36*, 617-625.

Meeus, P. F., Brayton, K. A., Palmer, G. H., and Barbet, A. F. (2003). Conservation of a gene conversion mechanism in two distantly related paralogues of *Anaplasma marginale*. Mol. Microbiol. *47*, 633-643.

Mehr, I. J., and Seifert, H. S. (1998). Differential roles of homologous recombination pathways in *Neisseria gonorrhoeae* pilin antigenic variation, DNA transformation and DNA repair. Mol. Microbiol. *30*, 697-710.

Meyer, T. F., and Hill, S. A. (2003). Genetic variation in the pathogenic *Neisseria* species. In Antigenic Variation, A. C. a. A.Scherf, ed. (London, Academic Press), pp. 142-164.

Miller, J. N. (1973). Immunity in experimental syphilis. VI. Successful vaccination of rabbits with *Treponema pallidum*, Nichols strain, attenuated by γ- irradiation. J. Immunol. *110*, 1206-1215.

Morgan, C. A., Lukehart, S. A., and Van Voorhis, W. C. (2002a). Immunization with the N-terminal portion of *Treponema pallidum* repeat protein K attenuates syphilitic lesion development in the rabbit model. Infect. Immun. *70*, 6811-6816.

Morgan, C. A., Lukehart, S. A., and Van Voorhis, W. C. (2003). Protection against syphilis correlates with specificity of antibodies to the variable regions of *Treponema pallidum* repeat protein K. Infect. Immun. *71*, 5605-5612.

Morgan, C. A., Molini, B. J., Lukehart, S. A., and Van Voorhis, W. C. (2002b). Segregation of B and T cell epitopes of *Treponema pallidum* repeat protein K to variable and conserved regions during experimental syphilis infection. J. Immunol. *169*, 952-957.

Musher, D. M., Schell, R. F., Jones, R. H., and Jones, A. M. (1975). Lymphocyte transformation in syphilis: an *in vitro* correlate of immune suppression *in vivo*? Infect. Immun. *11*, 1261-1264.

Musher, D. M., Schell, R. F., and Knox, J. M. (1974). In vitro lymphocyte response to *Treponema refringens* im human syphilis. Infect. Immun. *9*, 654-657.

Nash, T. E., Banks, S. M., Alling, D. W., Merritt, J. W., Jr., and Conrad, J. T. (1990). Frequency of variant antigens in Giardia lamblia. Exp. Parasitol *71*, 415-421.

Noordhoek, G. T., Cockayne, A., Schouls, L. M., Meleon, R. H., Stolz, E., and van Embden, J. D. (1990a). A new attempt to distinguish serologically the subspecies of *Treponema pallidum* causing syphilis and yaws. J. Clin. Microbiol. *28*, 1600-1607.

Noordhoek, G. T., Hermans, P. W., Paul, A. N., Schouls, L. M., van der Sluis, J. J., and van Embden, J. D. (1989). *Treponema pallidum* subspecies *pallidum* (Nichols) and *Treponema pallidum* subspecies *pertenue* (CDC 2575) differ in at least one nucleotide: comparison of two homologous antigens. Microb Pathog *6*, 29-42.

Noordhoek, G. T., Wieles, B., van der Sluis, J. J., and van Embden, J. D. (1990b). Polymerase chain reaction and synthetic DNA probes: a means of distinguishing the causative agents of syphilis and yaws? Infect. Immun. *58*, 2011-2013.

Ovcinnikov, N. M., and Delektorskij, V. V. (1972). Electron microscopy of phagocytosis in syphilis and yaws. Br. J. Vener Dis. *48*, 227-248.

Pavia, C. S., Baseman, J. B., and Folds, J. D. (1977). Selective response of lymphocytes from *Treponema pallidum*-infected rabbits to mitogens and *Treponema reiter*i. Infect. Immun. *15*, 417-422.

Pavia, C. S., Folds, J. D., and Baseman, J. B. (1976). Depression of lymphocyte response to concanavalin A in rabbits infected with *Treponema pallidum* (Nichols strain). Infect. Immun. *14*, 320-322.

Penn, C. W. (1981). Avoidance of host defences by *Treponema pallidum* in situ and on extraction from infected rabbit testes. J. Gen. Microbiol. *126*, 69-75.

Penn, C. W., and Rhodes, J. G. (1982). Surface-associated antigens of *Treponema pallidum* concealed by an inert outer layer. Immunology *46*, 9-16.

Pillay, A., Liu, H., Chen, C. Y., Holloway, B., Sturm, A. W., Steiner, B., and Morse, S. A. (1998). Molecular subtyping of *Treponema pallidum* subspecies *pallidum*. Sex Transm. Dis. *25*, 408-414.

Radolf, J. D., Chamberlain, N. R., Clausell, A., and Norgard, M. V. (1988). Identification and localization of integral membrane proteins of virulent *Treponema pallidum* subsp. *pallidum* by phase partitioning with the nonionic detergent triton X-114. Infect. Immun. *56*, 490-498.

Radolf, J. D., and Norgard, M. V. (1988). Pathogen specificity of *Treponema pallidum* subsp. *pallidum* integral membrane proteins identified by phase partitioning with Triton X-114. Infect. Immun. *56*, 1825-1828.

Radolf, J. D., Norgard, M. V., and Schulz, W. W. (1989). Outer membrane ultrastructure explains the limited antigenicity of virulent *Treponema pallidum*. Proc. Natl. Acad. Sci. USA. *86*, 2051-2055.

Sawitzke, J. A., and Stahl, F. W. (1994). The phage lambda orf gene encodes a trans-acting factor that suppresses *Escherichia coli recO*, *recR*, and *recF* mutations for recombination of lambda but not of *E. coli*. J. Bacteriol. *176*, 6730-6737.

Schneider, H., Griffiss, J. M., Boslego, J. W., Hitchcock, P. J., Zahos, K. M., and Apicella, M. A. (1991). Expression of paragloboside-like lipooligosaccharides may be a necessary component of gonococcal pathogenesis in men. J. Exp. Med. *174*, 1601-1605.

Smith, J. L., David, N. J., Indgin, S., Israel, C. W., Levine, B. M., Justice, J. J., McCrary, J. A., Medina, R., Paez, P., Santana, E., et al. (1971). Neuro-ophthalmological study of late yaws and pinta. II. The Caracas Project. Brit. J. Vener Dis. *47*, 226-251.

Stamm, L. V., and Bergen, H. L. (2000). The sequence-variable, single-copy *tprK* gene of *Treponema pallidum* Nichols strain UNC and street strain 14 encodes heterogeneous TprK proteins. Infect. Immun. *68*, 6482-6486.

Stamm, L. V., Greene, S. R., Bergen, H. L., Hardham, J. M., and Barnes, N. Y. (1998). Identification and sequence analysis of *Treponema pallidum tprJ*, a member of a polymorphic multigene family. FEMS Microbiol. Lett. *169*, 155-163.

Stern, A., Brown, M., Nickel, P., and Meyer, T. F. (1986). Opacity genes in *Neisseria gonorrhoeae*: control of phase and antigenic variation. Cell *47*, 61-71.

Suerbaum, S. (2000). Genetic variability within *Helicobacter pylori*. Int. J. Med. Microbiol. *290*, 175-181.

Sun, E. S., Molini, B. J., Barrett, L. K., Centurion-Lara, A., Lukehart, S. A., and Van Voorhis, W. C. (2004). Subfamily I *Treponema pallidum* repeat protein family: sequence variation and immunity. Microbes Infect. *6*, 725-737.

Sykes, J. A., Miller, J. N., and Kalan, A. J. (1974). *Treponema pallidum* within cells of a primary chancre from a human female. Br. J. Vener Dis. *50*, 40-44.

Thornburg, R. W., and Baseman, J. B. (1983). Comparison of major protein antigens and protein profiles of *Treponema pallidum* and *Treponema pertenue*. Infect. Immun. *42*, 623-627.

Tomai, M. A., Elmquist, B. J., Warmka, S. M., and Fitzgerald, T. J. (1989). Macrophage-mediated suppression of con A-induced IL-2 production in spleen cells from syphilitic rabbits. J. Immunol. *143*, 309-314.

Turner, T. B., and Hollander, D. H. (1957). Biology of the Treponematoses (Geneva, World Health Organization).

Walker, E. M., Zampighi, G. A., Blanco, D. R., Miller, J. N., and Lovett, M. A. (1989). Demonstration of rare protein in the outer membrane of *Treponema pallidum* subsp. *pallidum* by freeze-fracture analysis. J. Bacteriol. *171*, 5005-5011.

Weinstock, G. M. (2000). Genomics and Bacterial Pathogenesis. Emerg. Infect. Dis. *6*, 496-503.

13

Immunology of Syphilis

Justin D. Radolf and Sheila A. Lukehart*

Abstract

Though venereal syphilis is infectious in origin, there is now a clear consensus among syphilologists that tissue damage and the clinical manifestations of the disease result from the host immune response elicited by the spirochete and its constituents. Whereas acute syphilitic infection was once regarded as being immunosuppressive, there is now abundant evidence that *Treponema pallidum* evokes a robust humoral and cellular response soon after inoculation. It also has become evident that a cooperative interaction between the humoral and cellular arms of the immune response is required to clear treponemes from sites of infection as well as to establish long lasting protective immunity. Antibodies that promote uptake and killing of spirochetes by activated macrophages are believed to play an essential role in bacterial clearance, while their role in neutralization and immobilization in vivo appears to be limited. The mechanisms that enable a small number of spirochetes to establish persistent infection are poorly understood, although recent evidence indicates that the poor antigenicity of the spirochetal outer membrane, coupled with an ability to generate TprK antigenic variants, may be important for the organism's ability to evade the host's immune response. Efforts to develop a safe and effective syphilis vaccine have been hindered by uncertainty about the relative importance of humoral and cellular mechanisms to protective immunity and the fact that *T. pallidum* outer membrane proteins have not been unambiguously identified.

1. Introduction

The complex, sometimes anomalous, immunologic phenomena associated with syphilis have long fascinated and confused students of the disease. Beginning in the early years of the nineteenth century, decades before the discovery of "*Spirochaeta pallida*", researchers began inoculating individuals with lesional specimens in an effort to understand the disease, particularly the nature and timing of acquired

*For correspondence email JRadolf@up.uchc.edu

immunity (Chesney, 1926). These studies gave rise to the once prevailing notion that ongoing infection was a prerequisite for resistance to re-infection ("chancre immunity"). The discovery of the syphilis spirochete early in the twentieth century, coupled with the development of animal models shortly thereafter, enabled investigators to define the timeframe for the development of protective immunity and, with the availability of penicillin, to prove definitively that the protective state does not require persistent infection (Magnuson and Rosenau, 1948). Beginning with the pioneering studies of Wassermann and co-workers, investigators have worked assiduously to define the syphilis spirochete's antigenic components (Norris, 1988). The development of methods for detecting antibody responses elicited by *T. pallidum* also represented a milestone in modern clinical syphilology; serodiagnostic tests remain mainstays of syphilis diagnosis and are undergoing constant refinements to improve their accuracy and utility. In the late 1970s and early 1980s, technical advances in immunology enabled investigators to turn their attention to the problem of cellular immunity (Lukehart et al., 1980a; 1980b; Norris, 1988; Lukehart, 1992). These studies unequivocally established the pivotal roles of cellular mechanisms in spirochetal clearance and, by extension, in protective immunity. The recognition that syphilis remains a global public health problem (Hook, III and Peeling, 2004) and a major cofactor for the transmission of HIV (Greenblatt et al., 1988; Galvin and Cohen, 2004) has revitalized interest in understanding the disease's complex immunobiology.

The literature on the immunology of syphilis is vast, and one cannot hope to summarize it in a single chapter. In fact, given the many scholarly reviews of this topic that have appeared over the years (Chesney, 1926; Cannefax, 1965; Wigfield, 1965; Musher et al., 1976a; Perine, 1981; Sell and Norris, 1983; Norris, 1988; Lukehart, 1992), it is not necessary to do so. There is a pressing need, however, to familiarize investigators with the key observations and concepts that have historically motivated syphilis researchers and which continue to be relevant to the field. Equally important is the need to re-examine and, when possible, re-interpret key findings within the context of current immunologic paradigms.

2. Natural history of syphilitic infection

2.1. Syphilis in humans

Acquired syphilis begins when treponemes are inoculated onto skin or mucosal surfaces, usually in the genital area during sexual intercourse. Following an incubation period usually ranging from 7 to 21 days, the density of spirochetes becomes sufficient to induce the formation of a papule that goes on to ulcerate, creating the chancre. The primary lesion, classically painless, indurated, and teeming with spirochetes, is often accompanied by inguinal adenopathy, the first clinical sign of spirochetal dissemination and immune activation beyond the site of inoculation. Manifestations of secondary syphilis, the disseminated stage of disease, usually begin 4 to 10 weeks after the appearance of the chancre. Although cutaneous manifestations predominate, true to the highly invasive nature of *T. pallidum*, secondary syphilis can involve virtually any organ system, including the central nervous system. Diffuse lymphadenopathy, another common finding during secondary syphilis reflects the widespread dissemination of spirochetes to numerous tissues, including lymph nodes (Turner and Wright,

1973), and the intense, systemic inflammatory processes now underway. Although constitutional symptoms are common (Chapel, 1980; Lukehart, 2004), sepsis-like pathophysiology is rare during secondary syphilis, presumably reflecting, at least in part, the spirochete's lack of lipopolysaccharide (LPS) (Young et al., 1982; Hardy, Jr. and Levin, 1983; Fraser et al., 1998). During the early years of infection, patients experience recurrences of spirochetemia as well as secondary relapses; improved immunological control eventually causes a reduction in spirochetemia(s), a dramatic decline in total spirochetal burden, and an end to infectiousness--the stage called late latency. Nevertheless, viable and presumably replicating treponemes are scattered throughout the body, forming residual foci of inflammation (Rosahn, 1947). In approximately one-third of untreated patients, the poorly defined immunological mechanisms that contain the pathogen at these foci weaken, resulting in enhanced replication of organisms in parallel with an intensified local inflammatory response.

Cellular infiltrates composed of lymphocytes, macrophages, and plasma cells accompanied by vasculopathic changes of varying severity are the *sine qua non* of syphilitic lesions in all stages of disease and the histopathologic correlates of organ system dysfunction (Sell and Norris, 1983; Lukehart, 2004). A wide variety of histologic patterns can be superimposed upon these common abnormalities. Progression to endarteritis obliterans (Figure 1A) is believed to account for tissue necrosis and ulceration typical in primary syphilis and the rare fulminant forms of secondary lesions ("lues maligna"). The occasional presence of granulomata (Figure 1B) underscores the importance of cellular mechanisms in the local response to the bacterium and is consistent with the Th1 polarization of cytokine responses

A

B

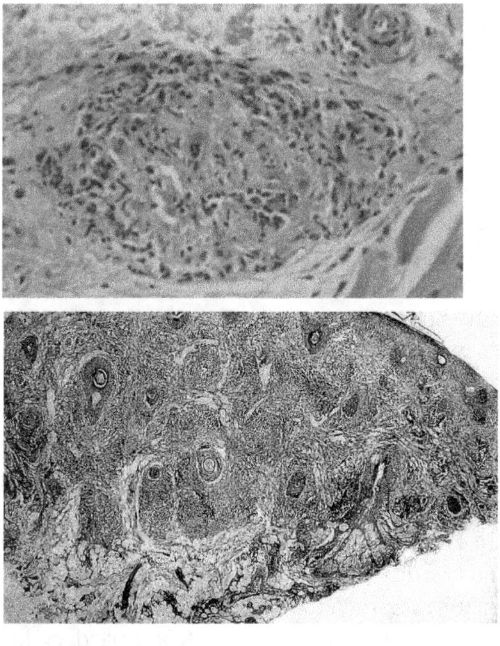

Figure 1. Histopathology of syphilis. A. Obliterative endertitis in an oral chancre (Barrett et al., 2004). B. Granulomatous infiltrate in a secondary skin lesion (Pandhi et al., 1995).

revealed by RT-PCR analysis of clinical specimens (Van Voorhis et al., 1996a). The necrotizing granulomata of gummatous disease are thought to represent an exaggerated delayed-type hypersensitivity response to treponemal antigens.

2.2. Animal models

Over the years, a number of animal species, including non-human primates, have been used as experimental models for syphilis (Turner and Hollander, 1957). Rabbits have become the model of choice because they (a) are highly susceptible to *T. pallidum* (Magnuson et al., 1948; 1956); (b) develop lesions that grossly and histologically resemble the chancres of primary human syphilis following intradermal inoculation (Turner and Hollander, 1957; Sell and Norris, 1983); (c) mount serological responses to treponemal infection highly similar to those of humans (Sell and Norris, 1983); and (d) develop immunity to re-infection that can be studied temporally, quantitatively, and microbiologically (Turner and Hollander, 1957). The major disadvantages of rabbits, besides cost, are the unavailability of inbred strains and the limited armamentarium of reagents for delineating leukocyte immunophenotypes in blood and tissues of infected animals. Experimental models employing inbred hamsters and guinea pigs have been developed to circumvent the limitations inherent in the use of outbred animals (Schell, 1983; Wicher and Wicher, 1989). Hamsters, however, are much less susceptible to the agent of venereal syphilis than they are to the *T. pallidum* subspecies that cause endemic syphilis and yaws (*endemicum* and *pertenue*, respectively) (Schell, 1983), while immunocompetent guinea pigs consistently develop ulcerative lesions only with inocula orders of magnitude greater than those required to elicit lesions in humans or rabbits (Wicher and Wicher, 1989). The absence of Th1 cytokines, the predominance of the anti-inflammatory cytokine IL-10, and the atypical histopathology (e.g., tissue eosinophilia) at intradermally infected sites (Wicher et al., 1998) also argue against the relevance of the guinea pig model. Mice are susceptible to infection with *T. pallidum* but have generally not been considered useful for syphilis research because they develop pathologically silent infection (Gueft and Rosahn, 1948).

Rabbits are inoculated most frequently via the intradermal or intratesticular routes. Cutaneous lesions begin as erythematous macules that become indurated and raised before undergoing central necrosis and ulceration. Intratesticular inoculation causes an orchitis in which the testes become enlarged, firm, nodular, and occasionally hemorrhagic (Turner and Hollander, 1957; Sell and Norris, 1983). In both instances, unchecked treponemal replication elicits perivascular infiltration of lymphocytes and macrophages that progresses to endarteritis obliterans (Turner and Hollander, 1957; Sell and Norris, 1983). Dissemination of treponemes shortly after infection (Brown and Pearce, 1920; Raiziss and Severac, 1935) results in marked hyperplasia of lymphoid organs which resolves within several weeks (Baker-Zander and Sell, 1980; Sell et al., 1980a). After lesions have resolved, rabbits remain latently infected for the remainder of their lifetimes (Sell et al., 1980b; Sell and Norris, 1983). Rabbits, unlike humans, do not consistently develop generalized manifestations following intradermal or intratesticular inoculation. Intravenous bolus inoculation can generate a secondary-like form of disease which, because of the raised and ulcerated appearance of the lesions, has been likened to disseminated primary infection (Turner and Hollander, 1957; Lukehart, 1992; Centurion-Lara et al., 2004).

3. An immunologic disorder with an infectious etiology

Acquired syphilis can be conceptualized as a contest between the extraordinary capacity of the spirochete to invade and disseminate and the ability of the host's "lumbering" defenses to "track down" the invader. The back and forth struggle between microbe and host is responsible for the waxing and waning nature of early infection, while the failure to eradicate the pathogen lays the groundwork for disease recrudescence (i.e., tertiary syphilis), sometimes decades after the initial infection. It was suggested at one time from *in vitro* studies that *T. pallidum* exerts cytotoxic activity against mammalian cells (Fitzgerald and Repesh, 1983). The biological relevance of this effect, however, was questioned in light of the extraordinarily high spirochete-to-cell ratios required to elicit it (Norris, 1988). Moreover, the *T. pallidum* genome does not encode orthologs for either well characterized bacterial toxins or the components for the secretory systems required to deliver them to the external milieu (Fraser et al., 1998). In their remarkably comprehensive textbook published more than six decades ago, Stokes et al. (1944) emphasized the importance of the inflammatory response to the development of disease manifestations, lesion resolution, and protective immunity in venereal syphilis. Turner and Hollander (1957) described the massive infiltration of lymphocytes, epithelioid cells (macrophages) and heterophiles (neutrophils), along with the subsequent necrotic reaction, elicited in rabbits following intradermal inoculation with *T. pallidum.* These perspectives clearly laid the groundwork for the current consensus among syphilologists that clinical manifestations of syphilitic infection emanate from inflammatory processes driven by the presence of treponemes and treponemal constituents within infected tissues (Norris, 1988; Lukehart, 1992; Salazar et al., 2002). These same inflammatory processes are presumed to generate B and T cells with the requisite specificities for the clearance of organisms during acute infection, promoting lesion resolution, and eventually conferring some degree of protection against subsequent encounters with the spirochete.

4. Humoral responses

Clinical and scientific considerations have prompted intensive examination of the humoral response to infection by *T. pallidum.* Clinical considerations have centered about the need for accurate serological tests for diagnosing a disease with a legendary proclivity for mimicking other disorders. Scientific considerations have centered about the search for serum factors mediating acquired immunity. As serological studies progressed, it became apparent that two distinct kinds of antibodies develop in response to treponemal infection. The first consists of so-called "nontreponemal" antibodies directed against a cardiolipin-lecithin-cholesterol complex. The other consists of "treponemal" antibodies directed against the spirochete's polypeptide constituents. Current clinical practice dictates that nontreponemal tests be used for screening and post-treatment followup and treponemal tests be used for confirmation of diagnosis (Miller, 1975; Lukehart, 2004). A detailed discussion of the use and interpretation of serodiagnostic tests for syphilis is beyond the scope of this chapter.

4.1. Anti-lipoidal (nontreponemal) antibodies

The complement fixation reaction reported by Wassermann and colleagues (Wasserman et al., 1906) was the first recognized serologic response to infection with *T. pallidum.* The term "reagin" was adopted by syphilologists of the day to denote Wassermann antibodies and should not be confused with allergy-associated IgE antibodies which were given the same name (Coca and Grove, 1925). Because Wassermann obtained his antigen from an extract of fetal syphilitic liver, an organ replete with treponemes, he reasoned that the patient sera were reacting with antigenic components of the recently discovered syphilis spirochete. It was not long before this idea was called into question with a report that extracts of normal tissues could replace those from syphilitic organs (Norris, 1988). The antigenic substance detected by the Wassermann reaction was clarified with reports that a new "phosphatide" extracted from beef heart, subsequently identified as cardiolipin, reacted with syphilitic sera when mixed with lecithin and cholesterol (Pangborn, 1941; Pangborn, 1945; De Bruijn, 1966). Wassermann's assay was based on the ability of a mixture of patient sera and the tissue extract to remove (or fix) complement from guinea pig serum, thereby diminishing the latter's ability to lyse sheep erythrocytes sensitized with rabbit anti-sheep erythrocyte antibodies. The complexity of the assay spurred numerous modifications, culminating in the VDRL and RPR tests in widespread usage today.

The mechanism(s) underlying the production of anti-phospholipid antibodies is one of the great enigmas of syphilis immunology and, at one time, was hotly debated (Turner and Hollander, 1957). Two theories have been proposed to explain this phenomenon. The prevailing viewpoint has held that cardiolipin complexes liberated from damaged cells at foci of infection provoke an autoimmune reaction stimulated by an adjuvant effect of the treponeme. Because, according to this thesis, the inciting antigen is of host origin, Wassermann antibodies were designated as "nontreponemal", a term still frequently used. Supporting this mechanism are the ubiquity of cardiolipin in mammalian tissues, the large number of acute and chronic conditions, autoimmune disorders most notably among the latter, that cause falsely reactive tests (formerly called "biologic false positive" tests) (Doniach et al., 1970; Schultz, 1997; Lukehart, 2004); and the concomitant production of antibodies to other self-antigens during syphilitic infection (Wicher and Wicher, 1990). On the other hand, several lines of evidence support the often discounted alternative theory that anti-phospholipid antibodies are actually elicited by treponemal lipids: (i) rabbits injected with heat-killed treponemes develop high titers of Wassermann antibodies (Eagle and Fleischman, 1948; McLeod and Magnuson, 1953); (ii) *T. pallidum* contains cardiolipin (Matthews et al., 1979; Belisle et al., 1994) and (iii) VDRL antibodies react with virulent treponemes, promoting agglutination (Hardy, Jr. and Nell, 1957) and phagocytosis by macrophages (Baker-Zander et al., 1993b) and conferring partial immunity against intradermal challenge (Baker-Zander et al., 1993b). Moreover, it is now well established that the anti-cardiolipin antibodies of syphilis and autoimmune diseases are antigenically and functionally distinct (Costello et al., 1990; Harris and Pierangeli, 1996; Schultz, 1997); thus it is no longer necessary to invoke an autoimmune etiology in order to explain the production of anti-phospholipid antibodies during syphilitic infection. Adding a final layer of complexity to this mystery is the fact that titers of nontreponemal antibodies generally correlate with disease activity (Singh and Romanowski, 1999), and, therefore, could reflect either the degree of tissue injury induced by the host response to the spirochete at a given point in the disease process or the body's total spirochetal burden. It is thought now that host-derived cardiolipin and other lipids are incorporated by the bacteria (see chapter on Metabolism of the *Treponema*). Unique epitopes may be formed during incorporation of cardiolipin and these may induce the "nontreponemal" antibody response seen during syphilis.

4.2. Treponemal antibodies

Improvements in methods for propagating *T. pallidum* paved the way for the development of assays to detect and measure antibodies that are unquestionably spirochete-specific. As noted, these studies were inextricably linked with the search for treponemicidal factors in serum that could be linked to protective immunity. Eberson (1921) was the first to describe an *in vitro in vivo* neutralization format in which the viability of spirochetes incubated with patient serum was assayed by intratesticular injection into rabbits. Two of his noteworthy observations were that treponemicidal activity correlated with the length of infection in humans and did not require Wassermann reactivity. Turner's group (Turner, 1939; Turner et al., 1948) refined Eberson's approach by substituting intradermal for intratesticular inoculation, a modification that made it possible to perform side-by-side comparisons of different sera and to obtain quantitative, statistically analyzable data by measuring parameters such as time to lesion appearance and resolution and lesion size. They found that incubation of treponemes with rabbit or human syphilitic serum attenuated or prevented lesion formation following intradermal inoculation; treponemicidal activity was diminished by heating to 56°C, suggesting a role for complement. Turner concluded that the activity detected in his assay was related to protective immunity, although, in retrospect, the description of the antibodies as "protective" seems an overstatement given their often modest effect on lesion development. Bishop and Miller (1976a) further expanded the study of neutralizing antibodies in the rabbit model.

Nelson and co-workers (Nelson, Jr. and Mayer, 1949; Nelson, Jr. and Diesendruck, 1951) refined these approaches into the *T. pallidum* immobilization or TPI test, the first entirely *in vitro* assay for detecting and quantitating complement-dependent immobilizing activity in human and rabbit syphilitic sera. An important distinction between the TPI test and its forerunners was the use of guinea pig complement in the incubation mixture as opposed to reliance on the variable levels of endogenous complement in the sample being assayed. Interestingly, not until the early 1980s was it actually demonstrated that IgG antibodies are responsible for complement-dependent treponemicidal activity (Blanco et al., 1984). Like their predecessors, Nelson and Mayer (1949) believed that the neutralizing antibodies detected in their assay mediated protective immunity. However, McLeod and Magnuson (1953) soon challenged this thesis by showing that rabbits immunized with killed *T. pallidum* developed TPI test reactivity in their sera but were not protected against low-dose intradermal challenge. Subsequently Miller (1973) demonstrated that rabbits immunized with γ-irradiated treponemes were completely protected against intradermal challenge when their sera were devoid of TPI test reactivity. Moreover, others have pointed to the obvious discrepancy between the high TPI test titers typically observed in secondary syphilitic sera and the fact that infectious spirochetes freely circulate in blood during this stage of disease (Turner et al., 1969; Musher et al., 1976b). By the 1970s, many authorities had come to believe that TPI test reactivity was caused by prolonged *in vitro* incubation ("aging") of the fastidious spirochete and was not reflective of pathogen-antibody interactions occurring *in vivo* (Perine et al., 1973).

Despite its importance as an advance for syphilis research, the TPI test was never practical for clinical laboratories because of its technical complexity, cost, and requirement for animals. Treponemal antibody testing on a large scale was made possible by the development of the FTA-ABS test in which serum antibodies reacting with acetone-fixed *T. pallidum* on glass slides are detected with an immunofluorescent conjugate (Knox et al., 1966; Deacon et al., 1966); simpler assays based on hemagglutination, particle agglutination, or ELISA formats, using treponemal lysates as antigen, followed thereafter. The specificity of treponemal assays was markedly improved with the discovery that *T. pallidum*-cross-reactive antibodies in normal sera can be removed by adsorption with a preparation of the Reiter treponeme (Hunter et al., 1964). With practical and accurate methodologies for treponemal and non-treponemal antibody testing in hand, it was possible by the 1970s to devise the algorithms and stage-associated serological profiles that remain central to syphilis serodiagnosis (Miller, 1975; Hart, 1986). The advent of one-and two dimensional immunoblotting (Lukehart et al., 1982; Hanff et al., 1982; Baker-Zander et al., 1985; Norris and Sell, 1984; Hensel et al., 1985), murine monoclonal antibodies (Mabs) (Lukehart et al., 1985; Marchitto et al., 1986), and recombinant DNA technology (Walfield et al., 1982; Stamm and Bassford, Jr., 1982; Norgard and Miller, 1983) in the 1980s spawned the intensive dissection of the molecular response to *T. pallidum* that continues today with the availability of the genomic sequence (Van Voorhis et al., 2003; McKevitt et al., 2005). Collectively, these studies have shown that IgM and IgG antibodies directed against a relatively small number of treponemal antigens are generated during primary syphilis and that both the intensity and spectrum of antigens recognized peak during secondary and early latent infection in accordance with the high spirochetal burdens characteristic of this stage of disease (Figure 2). In primary syphilis, *T. pallidum* antibodies are limited almost exclusively to IgG1 and IgM; in secondary syphilis, IgM levels are even further elevated, while IgG1 and IgG3 account for the preponderance of (approximately 90%) of the IgG antibodies in serum (Baughn et al., 1988). IgM reactivity tends to decline after therapy (Baker-Zander et al., 1985; Baughn et al., 1988), while IgG reactivity to a small number of polypeptides may persist years after infection (Miller, 1975; Hanff et al., 1982; Baker-Zander et al., 1985). The kinetics of the humoral response and the spectrum of antigens recognized are very similar in experimentally infected rabbits and humans (Lukehart et al., 1982; Hanff et al., 1983; Lukehart et al., 1986).

Based upon a comparison of two-dimensional electrophoretic silver stains and immunoblots, Norris (1988) concluded almost two decades ago that that nearly all of the major *T. pallidum* proteins are capable of eliciting an antibody response. The availability of the *T. pallidum* genomic sequence (Fraser et al., 1998) has made it possible to examine and refine this prediction. In a recent study, McKevitt et al. (2005) developed a protein array containing 882 of the 1039 polypeptides

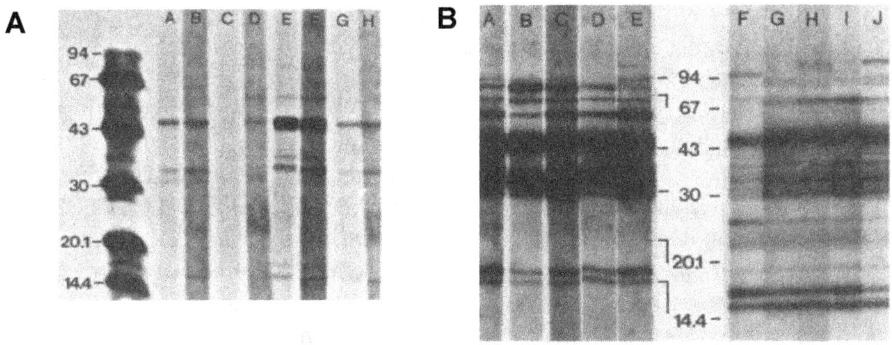

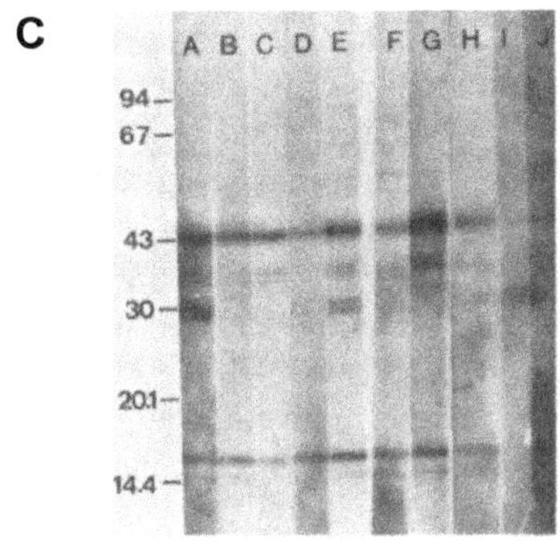

Figure 2. Humoral response to *T. pallidum* polypeptide antigens during acquired syphilis. A. Sera from four patients with primary syphilis were immunoblotted for *T. pallidum*-reactive IgM (strips A, C, E, and G) and IgG (strigs B, D, F, and H) antibodies. B. Secondary and early latent syphilis sera (left and right panels, respectively) were immunoblotted for *T. pallidum*-reactive IgG antibodies. C. Sera from late latent (strips A-E) and tertiary syphilis patients (strips F-J) were immunoblotted for *T. pallidum*-specific IgG antibodies. Panels A-C are from Hanff et al. (1982), copyright 1982, The American Association of Immunologists.

predicted to comprise the syphilis spirochete's proteome. Screening with rabbit syphilitic sera revealed that 106 of the 882 proteins induced detectable antibody responses. Of the 25 most immunogenic polypeptides, 10 were known or predicted lipoproteins and two, TpN17 (TP0435) and TpN47 (TP0574), have long been recognized as immunodominant *T. pallidum* antigens (Hanff et al., 1982; Lukehart et al., 1982; Jones et al., 1984; Baker-Zander et al., 1985; Hensel et al., 1985; Lukehart et al., 1985; Akins et al., 1993). These results are consistent with the longstanding observations that many of the spirochete's major immunogens are lipoproteins (Chamberlain et al., 1989; Schouls et al., 1989), as well as more recent reports demonstrating that ELISAs utilizing recombinant forms of TpN17, TpN47, and TpN15 (TP0171) are comparable in sensitivity to conventional treponemal tests (Young et al., 1998; Sambri et al., 2001). It is not well understood why lipid modification so dramatically enhances the antibody-inducing potential of these polypeptides. Studies with *Borrelia burgdorferi* in the murine model of Lyme disease suggest that lipid-modification enables protein antigens to bypass or short-circuit T cell-dependent interactions normally required for antibody production (Snapper et al., 1995; Fikrig et al., 1996; Mbow et al., 2001). These findings may, in turn, reflect the ability of lipoproteins to trigger activation of APCs and perhaps B cells via toll-like receptor 1 (TLR1) and TLR2 (Salazar et al., 2002; Alexopoulou et al., 2002).

4.3. Antibodies with functional activities

The term "functional antibodies" has been used by syphilis researchers to denote antibodies whose biological effects depend upon the recognition of antigenic determinants on the surface of live (i.e., motile) treponemes (Lukehart, 1992). Although conceptually straightforward, confirmation of the existence of such antibodies and identification of their antigenic targets has been highly problematic. Indeed, we now realize that the complexity of this issue reflects the difficulties in defining the antigenic components of the *T. pallidum* surface (see chapters on pathogenesis and outer membrane) and the precise role of antibodies in protective immunity (discussed below). As noted earlier, the notion that TPI test reactivity could be used as a surrogate marker for protective immunity was questioned soon after the assay was developed. The interpretation of TPI test reactivity became even more problematic in the molecular era as investigators attempted to use this assay, or variants thereof (i.e., *in vitro in vivo* neutralization), to confirm the surface location of *T. pallidum* antigens identified using recombinant DNA or Mab technologies. In the 1980s, complement-dependent immobilizing activity was reported for antibodies directed against proteins initially thought to be outer membrane proteins but now known to be intracellular (Fehniger et al., 1986; Radolf et al., 1986; Fraser et al., 1998; Jones et al., 1984; Radolf, 1995). In support of the surface specificity of this assay, Blanco and co-workers (1990) showed by freeze-fracture electron microscopy (EM) that immune rabbit sera could aggregate rare outer membrane proteins on treponemes incubated under TPI test conditions. They subsequently demonstrated that murine antisera generated against isolated *T. pallidum* outer membrane vesicles (OMVs) have high bactericidal titers and also aggregated rare OM proteins (Lewinski et al., 1999). Most recently, they described an anti-phosphorylcholine murine IgM Mab with *in vitro* bactericidal activity which surface labels intact treponemes and confers partial protection in rabbits following passive immunization (Blanco et al., 2005a). Unfortunately, these studies have not settled the issue. Using the same *in vitro* incubation conditions as Blanco et al., Bourell et al. (1994) were unable to induce aggregation of rare OM proteins with syphilitic serum. Furthermore, the putative surface-exposed proteins aggregated by OMV antibodies have yet to be identified. At the same time, it must be noted that complement-dependent killing is not the only functional activity detectable in syphilitic serum. Immune rabbit serum has been shown to enhance *in vitro* phagocytosis and killing of *T. pallidum* by macrophages (Lukehart and Miller, 1978; Alder et al., 1990; Baker-Zander and Lukehart, 1992) and to inhibit attachment of treponemes to eukaryotic cells (Fitzgerald et al., 1977; Hayes et al., 1977; Quist et al., 1983). To date, opsonic activity has been demonstrated for two candidate outer membrane proteins, Tp92 (TP0326) (Cameron et al., 2000) and TprK (TP0897) (Centurion-Lara et al., 1999), although not confirmed for the latter (Hazlett et al., 2001). Thus, the concept of functionally active antibodies directed against outer membrane proteins remains viable but requires further validation at the molecular level.

5. Cellular responses

5.1. The rise and fall of the immunosuppression hypothesis

Early in the twentieth century, studies of "tissue immunity" in syphilis were greatly influenced by the guinea pig model of experimental tuberculosis in which it was shown that the appearance of delayed type hypersensitivity ("allergy") correlated with immunity to re-infection. Noguchi (1911; 1912) described cutaneous responses to extracts of "pure cultures" of *T. pallidum* in hyper-immunized rabbits and patients with tertiary syphilis. Because his antigen preparation, designated "luetin", was derived from a cultivatable treponeme (probably *T. refringens*), rather than from *T. pallidum*, the results of these "luetin tests" have been difficult to interpret, although it is conceivable that the elicited responses were directed against cross-reactive *T. pallidum* antigens. Indeed, subsequent studies, which did employ antigen preparations derived from virulent *T. pallidum* (Nichols), corroborated Noguchi's finding that tuberculin-like reactions are not elicited until late latency or tertiary disease (though not necessarily in a majority of individuals) (Marshak and Rothman, 1951; Thivolet et al., 1953; Laird and Thorburn, 1966). These observations collectively gave rise to the belief that cellular immunity is impaired during early syphilis and that this state of "anergy" must be overcome for the host to make the transition from the infectious to the non-infectious stage of the disease. By the 1970s, this assertion had become firmly entrenched in the literature (Wigfield, 1965; Wright and Grimble, 1973; Pavia et al., 1978). Turner and Wright (1973) and Levene et al. (1971) reported the depletion of lymphocytes in the paracortical regions of lymph nodes from early syphilis patients, and they postulated that this histologic abnormality is associated with a selective impairment of cell-mediated immunity and the persistence of spirochetes within tissues. During this same period, investigators began to use *in vitro* assays to examine cell-mediated responses to *T. pallidum*. With a few exceptions (Badanoi et al., 1969; Janot et al., 1971; Friedmann and Turk, 1975), there appeared a spate of reports describing depressed blastogenic responses of peripheral blood lymphocytes from syphilitic humans and rabbits to mitogens or treponemal antigens and/or a suppressive effect of syphilitic serum on lymphocyte proliferation (Levene et al., 1969; 1971; Musher et al., 1974; 1975; From et al., 1976; Pavia et al., 1976; 1977; Wicher and Wicher, 1977a; 1977b; Friedmann and Turk, 1978; Ware et al., 1979; Bey et al., 1979; Maret et al., 1980). In a variation on this theme, Wicher et al. (1983) reported that mononuclear cells extracted from *T. pallidum*-infected rabbit testes inhibited mitogenic responses of normal peripheral blood mononuclear cells. In contrast, Schell, Musher and others (Schell et al., 1975b; Schell and Musher, 1975; Schell et al., 1975a) demonstrated in the mid-1970's that syphilitic rabbits had an increased resistance to unrelated intracellular bacteria (e.g., *Listeria monocytogenes*), suggesting up-regulation of cellular immune responses during syphilis infection. The applicability of this observation to resistance to syphilis infection was not recognized at the time, as their *in vitro* studies suggested that lymphocyte function was suppressed in early syphilis (Musher et al., 1975).

In the early 1980s, evidence began to emerge from the experimental rabbit model which systematically deconstructed the concept of a depression of cellular immunity during early syphilis. In contrast to prior *in vitro* studies which

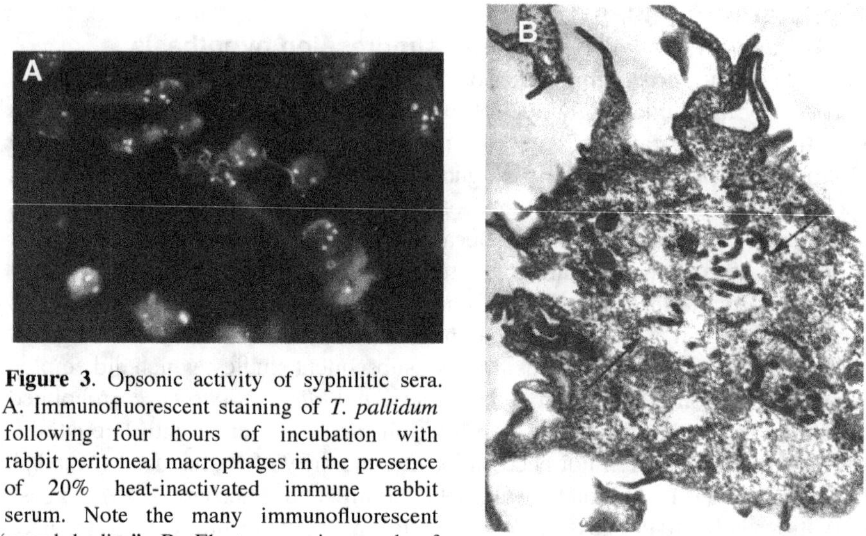

Figure 3. Opsonic activity of syphilitic sera. A. Immunofluorescent staining of *T. pallidum* following four hours of incubation with rabbit peritoneal macrophages in the presence of 20% heat-inactivated immune rabbit serum. Note the many immunofluorescent "round bodies". B. Electron micrograph of *T. pallidum* phagocytosed by a rabbit peritoneal macrophage; note that treponemes are within a phagolysosomal vacuole. Panels A and B are from Lukehart and Miller (1978), copyright 1978, The American Association of Immunologists.

examined proliferative responses of peripheral blood lymphocytes, Lukehart et al. (1980b) observed *T. pallidum*-induced proliferation of splenic and lymph node T cells harvested within days of infection; they also failed to obtain any evidence for a spirochete-induced depression of mitogen-induced T cell responses (Lukehart et al., 1980b). Additional evidence for a brisk cellular immune response during acute infection in rabbits came from *in situ* examination of treponeme-inoculated tissues. These studies clearly and convincingly demonstrated that replication of treponemes at the inoculation site elicits a progressively intense infiltrate of T cells and macrophages (Lukehart et al., 1980a; Baker-Zander and Sell, 1980; Sell et al., 1980b) and that the latter, activated by T cell-derived cytokines (Lukehart, 1982), phagocytose and destroy the invaders (Sell et al., 1982; Baker-Zander and Lukehart, 1992). With the bacteria cleared, lesions can then heal and the animal enters the latent stage of infection (Turner and Hollander, 1957; Collart et al., 1971; Sell et al., 1980b) characterized by long-lasting T cell memory to *T. pallidum* antigens (Baker-Zander and Sell, 1980; Baker-Zander et al., 1988; Arroll et al., 1999). *In vitro* studies demonstrated that the development of *T. pallidum*-reactive T cells in secondary lymphoid tissues parallels the appearance of T cells at the site of inoculation (Lukehart et al., 1980a), while the appearance of opsonic antibodies in serum immediately precedes bacterial clearance (Baker-Zander et al., 1993a). Other *ex vivo* studies supported the proposed critical role of macrophages and opsonic antibodies in bacterial clearance (Figure 3) (Lukehart and Miller, 1978; Alder et al., 1990; Baker-Zander and Lukehart, 1992). Analyses using SDS PAGE-purified native proteins and, subsequently, purified recombinant treponemal proteins moved these studies into the molecular realm by identifying *T. pallidum* polypeptides that evoke cell-mediated responses at the time of bacterial clearance (Baker-Zander et al., 1988; Arroll et al., 1999; Morgan et al., 2002b). Arroll et al.

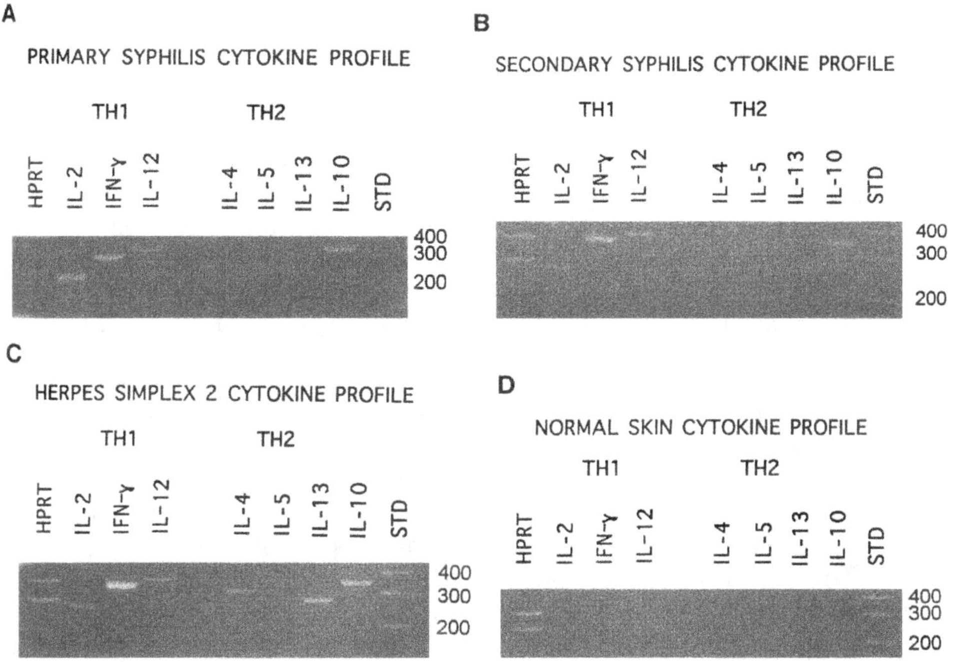

Figure 4. Detection of cytokine transcripts in primary and secondary syphilis lesions and controls (recurrent herpes genitalis and normal skin). Amplicons obtained by RT-PCR were separated by agarose gel electrophoresis and stained with ethidium bromide (Van Voorhis et al., 1996a).

(1999) clearly showed the Th1 bias in the rabbit cellular response by demonstrating high levels of IL-2 and IFNγ and low levels of IL-10 transcripts in splenocytes incubated with treponemal sonicates or recombinant proteins. The observation by this same group (Arroll et al., 1999) that Th1 responses could be elicited for as long as 6 months after infection (the latest time point examined) strongly refuted a proposal by Fitzgerald (1992) that a Th1 to Th2 switch is the pivotal event during syphilitic infection that enables the spirochete to become persistent.

While data on the cellular response of humans during syphilitic infection are comparatively sparse, they are generally in line with what has been gleaned from the rabbit model. Immunohistochemical and RT-PCR analyses of biopsy specimens have demonstrated that early syphilis lesions contain large numbers of macrophages and T cells expressing a Th1 cytokine profile (Figure 4) (Bjerke et al., 1981; Tosca et al., 1988; Van Voorhis et al., 1996a; McBroom et al., 1999). In humans, as in rabbits, therefore, it is likely that activated macrophages are primary effectors for spirochetal clearance. Surprisingly, opsonic antibodies have not been examined in human syphilitic sera, although one would anticipate their presence given the increasing predominance of *T. pallidum*-specific IgG1 and IgG3 antibodies as the disease progresses (Baughn et al., 1988). Truly provocative questions arise from the finding that CD8+ lymphocytes comprise a substantial proportion of T cells in early syphilis lesions and that they are activated *in situ* (Tosca et al., 1988; Engelkens et al., 1993; McBroom et al., 1999; Van Voorhis et al.,

1996b). The existence of *T. pallidum*-specific CD8+ T cells implies that spirochetal antigens taken up exogenously by APCs undergo some form of cross-presentation (Heath and Carbone, 2001; Rock, 2003). It is interesting to note in this regard that Bouis et al. (2001) showed that dendritic cells ingest virulent treponemes by coiling phagocytosis, an uptake mechanism that has been associated with cross-presentation (Rittig et al., 1994). The function of CD8+ cytolytic T cells at the local site also needs to be elucidated given that this subset responds to antigens presented via the class I MHC pathway and, therefore, typically is involved in the killing of viruses and intracellular bacterial pathogens (Stenger et al., 1997; Jiang et al., 2003). *T. pallidum* has been observed in nonphagocytic cells from primary lesions (Sykes and Miller, 1971; Sykes et al., 1974; Sykes and Kalan, 1975), and it has been speculated that cytolytic T cells are needed to eliminate what might serve as an intracellular reservoir for relapse (Van Voorhis et al., 1996b). Alternatively, the ability of CD8+ T cells to produce interferon-γ may be more important to the local immune response to *T. pallidum.*

Little is known about the systemic cellular response elicited by *T. pallidum* in humans. Three groups have used flow cytometry to examine circulating immunophenotypes during early syphilis. Both Pope et al. (1994) and Fan et al. (2004) reported significantly lower CD4+ T cells and significantly higher CD8+ T cell in syphilitic patents compared to non-infected controls. Fan et al. (2004) attributed the diminution in CD4+ T cells to Fas-mediated apoptosis and, in what amounts to a re-invocation of the immunosuppression hypothesis, proposed that the resulting immune dysfunction could be responsible for incomplete clearance of spirochetes during early infection. Salazar and co-workers (2005a), on the other hand, did not observe differences in circulating CD4/CD8 ratios between healthy controls and secondary syphilis patients. They did, however, observe significant increases in circulating memory and memory/effector T cells, changes consistent with an exuberant cellular response to treponemal antigens during the systemic phase of the disease.

5.2. The role of innate immunity

During the last several years, several lines of evidence have converged to establish a clearcut link between innate immune mechanisms and the host response to *T. pallidum.* The starting point for this work was the series of pioneering reports by Bessler, Jung, and collaborators (Johnson et al., 1983; Lex et al., 1986; Hoffmann et al., 1988) demonstrating that murein lipoprotein of *E. coli*, as well as synthetic lipopeptides based upon the molecule's N-terminus, activate monocytes/macrophages, B cells, and dendritic cells and also can serve as potent B and T cell adjuvants. The discovery that *T. pallidum* contains an abundance of lipoproteins (Chamberlain et al., 1989; Schouls et al., 1989), coupled with previous studies indicating that it lacks LPS (Young et al., 1982; Hardy, Jr. and Levin, 1983), suggested that treponemal lipoproteins might be major proinflammatory agonists during syphilitic infection. Using both purified treponemal lipoproteins and synthetic lipopeptides, this conjecture was confirmed in a series of *in vitro* experiments which also demonstrated that these activities were absolutely dependent upon the presence of covalently bound N-terminal lipids (Radolf et al., 1991; Akins et al., 1993; Radolf et al., 1995). Subsequent analyses revealed that, as with LPS, cellular activation by spirochetal lipoproteins/lipopeptides is facilitated by CD14 but involves a distinct, NF-κB-dependent signaling pathway (Norgard et al., 1996; Sellati et al., 1998; Wooten et al., 1998; Giambartolomei et al., 1999). This dichotomy was explained by the discovery that, after binding to CD14, bacterial lipoprotein/lipopeptides signal via toll-like receptor 2 (TLR2)/TLR1 heterodimers (Brightbill et al., 1999; Aliprantis et al., 1999; Hirschfeld et al., 1999; Lien et al., 1999; Alexopoulou et al., 2002), whereas LPS-mediated signaling typically proceeds via TLR4 (Poltorak et al., 1998; Akira and Takeda, 2004) (See Chapter 16, Schroder et al).

While cell culture systems are extremely useful for dissecting cell activation pathways, they provide only a glimpse into the inflammatory response that a pathogen or its constituents can trigger within a complex tissue environment. Thus, in order to understand the biological relevance of the innate responses triggered by lipoproteins, *in vivo* approaches must be utilized. To this end, rabbits initially were used to demonstrate that intradermal injection of lipopeptides elicits cellular infiltrates that histologically resemble those observed in experimental and human syphilis (Norgard et al., 1995). However, the paucity of reagents directed against rabbit immune cells precluded a detailed analysis of these lipopeptide-elicited infiltrates. To circumvent this limitation, Radolf and co-workers (Sellati et al., 2001; Salazar et al., 2005b) turned to an *in vivo* human skin model in which mild vacuum suction is used to elicit blister formation at the dermo-epidermal junction following intradermal injection with lipopeptides; fluid within the blisters contains extravasated immune cells which can be analyzed by multiparameter flow cytometry. Compared with peripheral blood, lipopeptide blister fluids were highly enriched for memory and memory/effector CD4+ and CD8+ T cells, as well as activated macrophages and dendritic cells. The enrichment for T lymphocytes expressing cutaneous lymphocyte antigen (CLA) and CCR5 was of interest given the association of these surface markers with a Th1 response (Teraki and Picker, 1997; Sallusto et al., 1998; Qin et al., 1998) and the well recognized role of CCR5 as a co-receptor for sexually transmitted macrophage tropic R5 strains of HIV-1 (Berger et al., 1999; Galvin and Cohen, 2004). Also noteworthy is that, compared with their peripheral blood counterparts, markedly greater percentages of blister fluid CD4+ and CD8+ T cells expressed activation markers (Salazar et al., 2005b). Taken as a whole, these studies are consistent with the notion that lipoproteins, possibly in concert with other spirochetal PAMPs, establish a proinflammatory milieu or "backdrop" that recruits the innate and adaptive cellular elements required for a coordinated and highly flexible response to the pathogen (Salazar et al., 2002).

5.3. A model for the integration of innate and adaptive responses to *T. pallidum*

Based upon the findings discussed above, one can propose a scenario for the ontogeny of cellular immunity during syphilitic infection that integrates innate and adaptive mechanisms and also takes into account the spirochete's notorious countermeasures against host defenses. According to this scheme, spirochetes initially replicate at the site of inoculation unchecked by the innate immune network surveillance system within skin (Pivarsci et al., 2005). It is believed that

T. pallidum's unique outer membrane architecture, most notably its lack of surface-exposed lipoproteins, helps the spirochete remain below the host's immunologic "radar" during this pre-clinical phase of the disease. At some point innate defenses must be triggered by the rapidly increasing local spirochetal burden, thereby sounding the "danger signal." How does this happen? One possible explanation is that resident phagocytes take up a small number of organisms even though this process is inefficient in the absence of opsonic antibodies (Lukehart and Miller, 1978). Once spirochetes are phagocytosed, degradation within phagolysomes would enable lipoproteins (or more likely lipopeptide degradation products) to gain access to pattern recognition receptors lining the vacuole (Underhill et al., 1999; Underhill and Gantner, 2004), thereby causing cellular activation and accelerating the pace of the inflammatory response. An alternative explanation is that TLRs on the surface of innate immune cells are triggered by lipoproteins released from dead organisms into the tissue milieu. With either mechanism, in a newly infected host, the memory/effector T cells recruited initially to the infected site by lipoprotein/lipopeptide-induced inflammation would not recognize *T. pallidum* antigens. A primary immune response would then ensue as antigen-presenting cells initially attracted to the infected site move to draining lymph nodes in order to present treponemal antigens to naïve T cells. Neo-sensitized T cells released into the circulation would then traffic from peripheral blood back into the chancre, where they would recognize their cognate antigens and release chemotactic and macrophage-activating cytokines (e.g. interferon-γ), thereby accelerating bacterial clearance and healing of the lesion. Clearance of organisms would be facilitated by the increasingly robust humoral response and the concomitant appearance of opsonic antibodies. As the infection progresses, there will be an increasing percentage of *T. pallidum*-specific memory/effector T cells in peripheral blood and a correspondingly increased percentage of pathogen-specific T cells recruited into secondary as opposed to primary lesions. This model is consistent with the well recognized observation that lesions of disseminated infection appear and progress when the chancre is resolving or healed. As with chancres, formation of secondary lesions would not occur until the organisms at a site of dissemination reach a threshold density capable of triggering the local inflammatory mechanisms that result in the recruitment of circulating innate and newly sensitized T cells (Lukehart, 1992). Rather than being due to immunosuppression, as was once believed, the "delay" in the evolution of secondary lesions would reflect the small numbers of spirochetes believed to be implanted at metastatic sites within skin, the bacterium's slow rate of replication (Magnuson et al., 1948; Cumberland and Turner, 1949), and its inherent ability to avoid triggering local danger signals on resident innate immune cells. With progressive T-cell sensitization and increasing titers of opsonic antibodies, the host immune response eventually gains the upper hand, secondary relapses become less frequent, and the patient progresses from early to late latency. In many cases, these mechanisms are not fully effective in eliminating organisms; persistence of spirochetes at a site elicits a smoldering local inflammatory response that, depending on a variety of factors, including anatomical location, may or may not become clinically evident at some point in the future. An age-related breakdown of a critical surveillance function in a persistently infected host would hasten the onset of recrudescent disease; HIV infection also has been postulated to serve this function. Exaggeration of cellular responsiveness in some individuals, perhaps in association with a booster effect caused by the increase in spirochetal replication, would cause the necrotizing granulomata typical of gummatous disease.

5.4. Immune evasion

One of the central paradoxes of syphilis is that the immune system has the capacity to eliminate millions of treponemes from the primary site of infection yet some organisms evade these defenses, giving rise to persistent, even lifelong infection. Over the years, several explanations for the spirochete's immunoevasiveness have been offered. One is that spirochetes take up residence in intracellular niches (Sykes and Miller, 1971; Sykes et al., 1974; Sykes and Kalan, 1975). A once popular notion was that the spirochete is shielded from circulating antibody by an outer coat comprised of serum proteins and mucopolysaccharides (Christiansen, 1963; Alderete and Baseman, 1979; Fitzgerald and Johnson, 1979). This idea has been supplanted by studies showing that the syphilis spirochete's outer membrane creates a barrier to antibody binding by virtue of a paucity of membrane-spanning proteins and lack of surface-exposed lipoproteins (Cox et al., 1992; Radolf, 1995; Blanco et al., 1997). Nevertheless, the functional activities of syphilitic serum, most notably its ability to promote uptake and killing of treponemes by macrophages, demonstrates that the organism's resistance to antibody binding cannot be absolute (Lukehart and Miller, 1978; Alder et al., 1990; Baker-Zander and Lukehart, 1992). The observation by Lukehart and co-workers (Lukehart et al., 1992) that a subpopulation of spirochetes resistant to the opsonic activity of syphilitic serum emerges during infection helps to reconcile the seemingly contradictory evidence in favor of both antibody binding and antibody resistance by virulent treponemes. Most recently, the same group has described antigenic variation of TprK among clinical isolates as another possible mechanism for evasion of host antibody responses (Centurion-Lara et al., 2004) (See Chapter 12, Centurion-Lara). A related question is whether the spirochete modulates its overall antigenic composition and surface molecular architecture during the course of infection as a means of evading clearance mechansisms. Recent successes in exploiting the *T. pallidum* genomic sequence and microarray technology to delineate the spirochetal transcriptome during acute infection in rabbits (Smajs et al., 2005) should help to set the stage for the gene profiling studies required to address this question. Our current inability to explain the conundrum of latency clearly underscores our fledgling understanding of the cellular and molecular biology of syphilitic infection.

6. Immunity in syphilis

6.1. Natural immunity

Because humans are the natural host for *T. pallidum*, it has generally been surmised that humans lack inherent immunity against syphilitic infection (Magnuson, 1948). The fact that disease-susceptible animals, such as a rabbits and monkeys, often fail to develop disseminated lesions comparable to those of secondary syphilis in humans also has been considered evidence that humans have little, if any, natural resistance to syphilis (Chesney, 1926). On the other hand, the existence of natural immunity has been proposed as an explanation (Bishop and Miller, 1983) for the epidemiologic observation that approximately 50% of individuals do not contract the disease following sexual contact with infectious syphilitic lesions. The experimental evidence for such resistance, however, rests entirely upon *in vitro* assays demonstrating the presence of treponemicidal activity in serum from healthy individuals (Bishop and Miller, 1983) along with immunoblot analyses showing that this cidal activity appears to be mediated by antibodies directed against subunits of the periplasmic flagellar apparatus (Blanco et al., 1986). The concept of natural immunity also must be reconciled with the human inoculation studies by Magnuson et al. (1956) which demonstrated that naïve individuals are highly susceptible to *T. pallidum* infection. The ID_{50} for the Nichols strain (57 organisms) in normal volunteers was essentially no different from that of similarly inoculated rabbits (23 organisms); in some volunteers, as few as ten organisms produced primary lesions. Thus, while one cannot entirely discount natural resistance as a background cause of diminished transmission efficiency, other variables, such as the degree and duration of contact, the site of inoculation (i.e., mucosal vs intact epithelial surfaces), and *T. pallidum*'s limited ability to withstand environmental stresses (e.g., dessication and atmospheric oxygen tensions) during the transmission process collectively provide a more plausible explanation.

6.2. Acquired immunity in syphilis

The concept of acquired immunity in syphilis originated during the early years of the nineteenth century when it was observed during local outbreaks that afflicted individuals rarely experienced second attacks of the disease. In his monumental review published in 1926, Chesney (1926) credits the great French syphilologist Ricord with playing a central role in helping to establish experimentally the notion of acquired immunity. In numerous auto-inoculation experiments, Ricord noted that material taken from syphilitic primary lesions ("hard chancres") did not cause new lesions when inoculated elsewhere on the same patient. By contrast, material taken from "soft chancres" (i.e., chancroidal ulcers) readily produced lesions when transferred to non-genital sites. In 1883, Mauriac summarized the consensus for the field when he stated that re-inoculation "is no longer possible when the chancre begins to develop." Although these investigators lacked the means for confirming their clinical diagnoses, Chesney gave considerable credence to their clinical and diagnostic acumen, the careful manner in which they recorded their findings, and the sheer volume of concordant evidence that they generated. He also maintained that their findings were corroborated by the comparatively small number of auto-inoculation experiments conducted in the years following the discovery of *T. pallidum* and the development of the Wassermann test. With the development of a primate model and the availability of anti-syphilis therapy, Neisser formulated an alternative theory of "chancre immunity" ("*Infektionsimmunität*") that dominated the syphilis field for many years. His experiments with primates led him to conclude that resistance to re-infection is dependent upon the presence of active foci of infection and that cure of the disease leads to the disappearance of resistance. According to Neisser's formulation, re-infection was, in and of itself, unequivocal evidence that the original infection was cured and that all spirochetes had been eliminated. The veracity of this theory with respect to humans ostensibly was supported by the auto-inoculation studies described above (not surprisingly since chancre-immune patients were nearly always infected at the time of challenge) and by a perceived increase in the number of re-infections that followed the introduction of salvarsan, the first effective anti-syphilis therapy (Driver, 1924).

With the publication of his monograph, Chesney mounted a challenge to Neisser's conception that eventually turned the tide of scientific opinion towards a viewpoint more in accord with contemporary immunologic thought. His "meta-analysis" of the available literature on challenge experiments in rabbits led him to propose that syphilitic infection does elicit a true and lasting immunity, *provided* the initial infection is of sufficient duration. More specifically, he concluded that if animals were infected for less than three months prior to treatment, subsequent challenge with the homologous strain usually resulted in clinical re-infection. Symptomatic re-infection, on the other hand, did not occur in animals infected for three months or longer before treatment. He also noted substantially less cross-immunity to challenge with heterologous strains of *T. pallidum*, an observation subsequently confirmed by Turner and Hollander (1957) using low-dose inocula and a much larger number of clinical isolates. One of the most impressive aspects of Chesney's argument was his re-analysis of Neisser's primate data. In addition to exposing major flaws in study design that undermined Neisser's interpretation of his own findings, he showed that Neisser's recorded observations for untreated and treated monkeys also revealed a correlation between the duration of infection and the likelihood of successful re-inoculation, with eighty days demarcating a discernible breakpoint between susceptibility or resistance.

A major criticism of the pre-penicillin-era studies supporting Chesney's position that animals retained their immune status following treatment was that the challenged animals may have retained residual amounts of the arsenical used to eradicate the immunizing infection. In a meticulously executed study involving 263 rabbits, Magnuson and Rosenau (1948) used graded challenge inocula, the recent availability of penicillin, and rabbit infectivity testing (RIT) to demonstrate unequivocally that immunity in syphilis was dose- and time-dependent and dissociable from concurrent infection. These investigators infected rabbits intratesticularly or intradermally for periods of time ranging from three to twenty-four weeks and then treated them with doses of mapharsen or penicillin exceeding the CD_{95} values for the respective drug. Six weeks after treatment, a period of time more than adequate to ensure that all drug had been excreted, the animals were re-challenged with inocula of varying sizes by the same route used for the initial infection. For each group of animals, they determined the challenge inocula that resulted in complete immunity (RIT-negative), asymptomatic (RIT-positive) or symptomatic infection (Figure 5). Identical results were obtained with both anti-syphilis drugs. A modestly increased threshold to challenge was observed at three weeks in the intratesticularly infected animals and at six weeks in animals infected intracutaneously. By twelve to 24 weeks of infection, animals were fully immune to inocula more than one thousand-fold greater than the infecting dose, and they failed to develop symptomatic infection when challenged with inocula one-million times greater. The investigators concluded that previous studies had simply failed to detect the early development of immunity because they had employed challenge inocula that were too large. They also concluded that asymptomatic infection is a state of partial immunity in which spirochetal multiplication at the site of

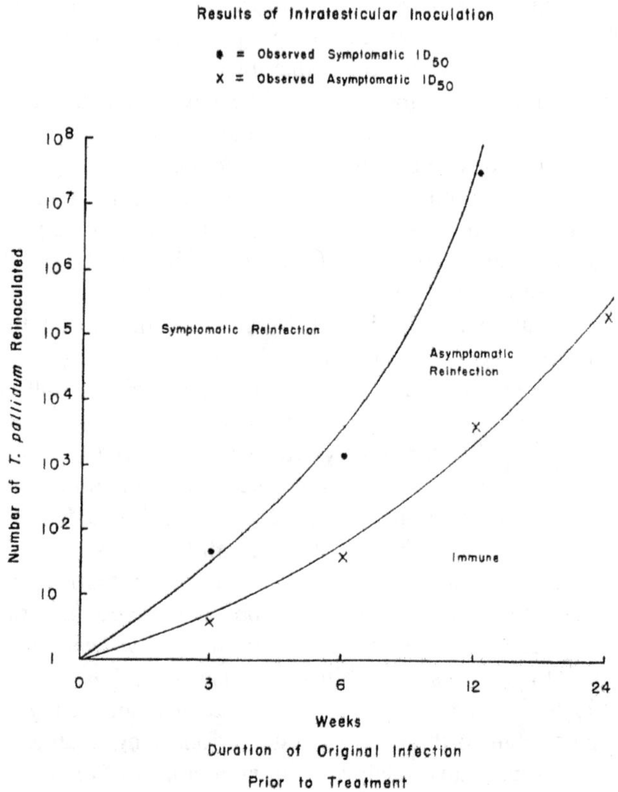

Figure 5. Relationship between the duration of original infection by intratesticular inoculation and the outcome of re-infection with varying doses of treponemes (administered six weeks after treatment of the immunizing infection) (Magnuson and Rosenau, 1948).

inoculation is sufficiently suppressed to prevent lesion formation even though some organisms manage to disseminate and persist. It follows from this conclusion that the term chancre immunity is inherently ambiguous since the phenomenon may reflect either asymptomatic infection or complete immunity. While their data did not permit them to conjecture on the mechanism for complete immunity, they pointed to implantation studies by Reynolds (Reynolds, 1941) which indicated that, in fully immune animals, spirochetes are contained locally and destroyed before they can disseminate. In a follow up study, the Magnuson group (Magnuson et al., 1949) further undermined the position of Neisser's proponents by showing that the immunity induced during a twelve week period of infection lasted for at least one year following the termination of the primary infection. Complementary results were described by Turner and Hollander (1957) in their monograph. They noted that measurable resistance to re-challenge was detectable as early as three weeks after infection and that animals infected for six months were "solidly" immune to very large inocula (5 x 10^6 organisms). They also observed that the partial immunity induced by short periods of infection (i.e., two months) waned over time, a finding with considerable clinical implications. In 1984, Strugnell and associates (1984) showed that immunity to low-dose homologous and heterologous challenge was detectable within seven weeks of intradermal inoculation of rabbits.

Once it was discovered that animals could be experimentally infected with *T. pallidum*, the deliberate inoculation of humans was largely abandoned. Recognition of the difficulty in extrapolating from animal data to the human situation was a primary motivation for Magnuson and colleagues (1956) to conduct their inoculation trial of human volunteers in Sing Sing prison, a study which undoubtedly will never be repeated. The trial consisted of two main arms. As noted already, in the first, non-syphilitic individuals were inoculated intradermally with graded doses of *T. pallidum* Nichols in order to confirm the virulence of the rabbit-passaged strain for humans and to determine its ID_{50}. In the second, volunteers with treated or untreated syphilis were inoculated with 10^5 spirochetes at a single intradermal site on the volar surface of the forearm. The results revealed a striking dichotomy in susceptibility between the early and late syphilis patients. Five patients with untreated late latent syphilis exhibited no skin reaction or serologic changes following infection; these results were interpreted as complete resistance. Of the 26 subjects previously treated for late latent syphilis, 13 showed neither clinical nor serologic change, while three developed darkfield negative lesions unassociated with increased serologic titers. In contrast, all 11 patients with a previous history of treated early syphilis developed lesions (nine of which were darkfield positive) and increased serologic titers. Unfortunately, despite its thoughtful design and meticulous execution, the Sing Sing Study cannot be regarded as definitive. Its principal flaw was that it employed a challenge inoculum 2000-fold greater than the Nichols strain ID_{50} and also likely to be much larger than that transmitted by sexual activity. The potential discrepancy between experimental and "real-world" inocula was compounded by the likelihood that heterologous challenge in many of the volunteers might have further "masked" a state of partial immunity. In the authors' own words: "It seems likely that the conditions of the present experiment represent unusually severe challenge to host resistance and, that, under conditions of normal exposure, many more patients might well have been able to resist that infection." Thus, while the study clearly showed that latent syphilis is associated with a strong and long lasting state of immunity and that active infection helps to maintain such resistance, it left unresolved the critical question of whether individuals with early syphilis are ever resistant. Magnuson's candid acknowledgement of this methodological flaw is rather ironic in light of his earlier rabbit work with Rosenau (Magnuson and Rosenau, 1948) which emphasized the importance of using small inocula to detect partial immunity. Equally surprising, this seminal work was not cited in the Sing Sing report despite its obvious conceptual and experimental linkage with the human trial.

In summary, immunity to re-infection develops in parallel, and undoubtedly as an integral part of, the host defenses that effect spirochetal clearance. Sometime between three and six months after infection (depending on the route and size of the inoculum), strong and long lasting immunity becomes established in rabbits; a substantial, though lesser, degree of resistance also is acquired against heterologous strains. Recent studies by Morgan and co-workers (2002b; 2003) indicate that antigenic diversity at the *tprK* locus may be important in determining these heterologous strain relationships. With respect to syphilis immunity in humans, we have followed Chesney's lead in accepting the findings of the auto-inoculation studies of the nineteenth century, fully recognizing that they fail to meet contemporary scientific or ethical standards for clinical research. We also concur with the authors of the Sing Sing study that the large challenge inocula employed almost certainly underestimated the level of immunity in their cohort. Working forward, we can propose that in humans, as in rabbits, immunity to syphilis exists on a continuum from partial to complete that begins with the systemic host response to *T. pallidum* during the primary stage of the infection. The Sing Sing data would suggest that, as in rabbits, the level of immunity continues to increase up to and into latency even as the clinical activity of the disease declines. In both rabbits and humans, it appears as if complete immunity requires a prolonged, yet, poorly understood immune interaction between virulent bacterium and mammalian host. It also seems likely that the partial immunity induced during early infection wanes following treatment, thereby rendering individuals susceptible to re-infection. The threshold for re-infection in partially immune, treated individuals would, in many instances, be even lower for heterologous isolates. Interestingly, the requirement for prolonged infection in order to achieve complete immunity points to an element of truth in the Neisserian concept of chancre immunity. The error of this school was its inability to conceive of immunity existing beyond the infectious state; we now can attribute long lasting immunity to immunologic memory even though, in the case of syphilis, the critical memory subpopulations and their cognate antigens have not been identified. The above scenario has interesting epidemiologic implications. Grassly et al. (2005) recently described a mathematical model for syphilis in the United States in which disease epidemics occur in oscillations with periods of eight to 11 years. They proposed that the oscillations are due to fluctuations in the level of herd immunity in at-risk populations. According to this model, widespread transmission of acute infection in an untreated sexual network would increase the level of partial immunity, curtailing further transmission, while the waning of immunity at the population level following treatment would gradually enhance the susceptibility of the population, setting the stage for the next upswing in the oscillatory pattern. In making immunologic, as opposed to behavioral and demographic factors central to the causality of syphilis epidemics, this innovative model has quickly become controversial. Unfortunately, there are no obvious experimental strategies to assess and compare the immune status of control and at-risk populations at various points in the putative oscillatory cycle.

6.3. Mechanisms of acquired immunity

The requirement for a collaboration between cellular and humoral mechanisms in order to clear spirochetes during acute infection would lead one to predict that acquired immunity also requires both arms of the immune response. Indeed, this prediction is borne out by experiments which demonstrate that neither passively administered antibodies nor T cells are capable of recapitulating the level of protection conferred by infection. A number of studies have shown that administration of immune serum or immunoglobulins purified from immune serum to naïve recipients delays lesion development following challenge with *T. pallidum* (Sepetjian et al., 1982; Perine et al., 1973; Turner et al., 1973; Bishop and Miller,

1976a; Titus and Weiser, 1979; Azadegan et al., 1983; Bishop and Miller, 1983; Pavia et al., 1985; Champion et al., 2005). Two recurrent observations underscore the limited degree of protection conferred by the administered antibodies and, as pointed out by Musher and colleagues (1976b) three decades ago, how dramatically syphilis immunity deviates from classical models of humoral immunity (e.g., *Streptococcus pneumoniae*): (i) large amounts of immune serum or purified immunoglobulins are needed to achieve a modest degree of lesion attenuation following challenge with relatively small challenge inocula (i.e., 500 to 1000 organisms), and (ii) spirochetal dissemination of spirochetes is not prevented by passive immunization. Because the time to lesion development inversely correlates with inoculum size (Turner and Hollander, 1957; Izzat et al., 1970), the partial protection observed in these studies has been interpreted to mean that immune serum causes a one- to two-log decrease in the effective size of the challenge inoculum (Turner et al., 1973; Blanco et al., 2005b). Whether it does so by killing a portion of the inoculum or suppressing replication of the population as a whole has not been determined. Arguing in favor of the latter is the finding that lesion formation can be suppressed by repeated administration of immune serum (Bishop and Miller, 1976a; Perine et al., 1973; Weiser et al., 1976). Recently, Champion et al. (2005) concluded, based upon quantitative PCR analysis of *T. pallidum* genome copy numbers in lesion biopsies, that delayed lesion formation is due to reduction in inoculum size. Their argument, however, overlooks the fact qPCR data does not provide information about the replication rate of the organisms being enumerated.

Proving a direct role for cellular immune processes in acquired immunity has been complicated by the fact that the commonly used outbred rabbit model is not amenable to cell transfer studies. Using endemic syphilis and yaws strains, Schell and co-workers (Chan et al., 1979; Schell et al., 1980) demonstrated that enriched populations of T cells from previously infected hamsters conferred partial resistance to naïve recipients, based upon attenuated lesion development and significant decreases in lymph node weights and nodal treponemal burdens. Subsequently, Pavia and Niederbuhl (1985) and Wicher et al. (1987) reported attenuated lesion development and asymptomatic infection, respectively, following challenge with *T. pallidum* subsp. *pallidum* in adoptive transfer studies conducted with inbred guinea pigs. Because the vast majority of treponemes are extracellular and virulent treponemes and treponemal lipoproteins do not activate T cells (Radolf et al., 1995; Sellati et al., 2000), it seems unlikely that the transferred lymphocytes exert their protective effect by acting directly on the spirochetes. Instead, one would predict that the transferred T cells are activated by *T. pallidum*- and lipoprotein-responsive APCs that phagocytose organisms in the challenge inoculum.

7. Development of a syphilis vaccine

7.1. Vaccines using whole treponemes

Pre-molecular attempts at vaccine development used antigen preparations derived from nonviable *T. pallidum* or non-pathogenic treponemes; these experiments were almost uniformly unsuccessful (*T. phagedenis* Reiter) (Magnuson et al., 1947; Eagle and Fleischman, 1948; Turner and Hollander, 1957; McLeod and Magnuson, 1953; Miller et al., 1963; Izzat et al., 1971). One notable exception, reported by Metzger and Smogor (1969) in 1969, involved repeated intramuscular injection of

T. pallidum inactivated by incubation *in vitro* at 4°C and/or 37°C in the presence or absence of penicillin. These investigators observed a dose-dependent effect in which five of the ten animals receiving the largest vaccine dose (total of 1.2×10^9 organisms) were fully protected against a sizeable (3×10^5 treponemes) intradermal challenge inoculum. The results of this study have not been independently confirmed, and it is fair to say that contemporary investigators are skeptical of its findings. In 1973, Miller (1973) reported complete protection by repeated intravenous injection of rabbits with γ-irradiated (i.e., avirulent but viable) *T. pallidum* suspensions. The immunized rabbits were completely resistant to intradermal challenge with 10^3 and 10^5 treponemes and remained immune one year after vaccination. Interestingly, the immunization regimen did not protect against infection with the Haiti B strain of *T. pallidum* subsp. *pertenue*, confirming earlier observations pointing to the existence of critical differences in protective antigenic determinants among these *T. pallidum* subspecies (Turner and Hollander, 1957). The lack of correlation between TPI antibody and protection was demonstrated by the observation that 8 of the 11 immune animals successfully protected from challenge one year post-immunization did not have detectable TPI titers at the time of challenge. While the Miller immunization regimen (60 injections given over a 37-week period) is clearly not practical, this meticulously performed study provides a benchmark for assessing the protective capacity of recombinant subunit vaccinogens. This is not a trivial issue given that protection studies in the recombinant era have relied almost exclusively upon alterations or delays in cutaneous lesion development as indicators of protection.

7.2. Recombinant vaccines

The advent of recombinant DNA technology in the early 1980s reinvigorated the search for a syphilis vaccine by promising essentially unlimited quantities of *T. pallidum* antigens for immunization trials. Initial efforts focused on antibody-based protection and were largely predicated on two assumptions: The first, supported by the significant, though partial, degree of protection conferred by passive immunization with IRS, is that antibodies directed against treponemal surface antigens would be protective. The second, subsequently obviated by the discovery of antigen processing and presentation (Janeway, Jr., 1989; Robinson and Delvig, 2002), is that immunodominant antigens would be surface-exposed. Lacking *a priori* information about *T. pallidum* outer membrane proteins, investigators in the early 1980s, in reality, had no choice but to use simple reactivity with syphilitic serum or *T. pallidum*-specific monoclonal antibodies as the basis for selection of vaccine candidates (Walfield et al., 1982; Stamm and Bassford, Jr., 1982; Norgard and Miller, 1983). As discussed in detail elsewhere in this book (see chapters on syphilis pathogenesis and outer membrane), none of the antigens identified by this type of antibody-based screening is surface-exposed (Radolf, 1995) and, consistent with the postulated importance of surface exposure, vaccinogens so identified usually have failed to confer significant protection (Wicher et al., 1991; Hsu et al., 1988; Centurion-Lara et al., 1997). Immunization with two recombinant immunogens, 4D/TpF1 (TP1038) and TmpB (TP0769), did, however, alter the course of experimental syphilis in rabbits and guinea pigs, respectively, resulting in delayed and/or attenuated lesion development (Borenstein et al., 1988; Wicher et al., 1991). Ironically, these two immunization trials inadvertently buttressed the cell transfer experiments described above, inasmuch as the elicitation of cell-mediated responses is the only possible explanation for the ability of periplasmic or, in the case of 4D/TpF1 (a bacteriferritin ortholog), cytoplasmic proteins to modulate the infectious process. The existence of a cellular component to vaccine-induced protective immunity is further supported by the finding that rabbits immunized with purified native flagella developed attenuated cutaneous lesions (Champion et al., 1990).

More recently, Cameron and co-workers (Stebeck et al., 1997) refined conventional antibody-based screening by devising a strategy to identify *E. coli* clones expressing likely targets for functional (i.e., opsonic) antibodies. Their differential screening approach, based on the fact that syphilitic infection induces opsonizing antibodies in rabbits whereas immunization with heat-killed treponemes does not, has yielded three vaccine candidates. The first is an ortholog for glycerophosphodiester phosphodiesterase (GlpQ/Gpd, TPO257), an enzyme that hydrolyzes deacylated phospholipids to alcohol plus glycerol-3-phosphate. Although Cameron et al. (1998) reported that rabbits immunized with GlpQ/Gpd developed atypical, nonulcerative lesions following intradermal challenge, Shevchenko et al. (1999) were unable to corroborate these findings. The second molecule identified by functional antibody screening, designated Tp92 (TPO326) (Cameron et al., 2000), is a homolog of the Omp85 family of Gram-negative bacterial outer membrane proteins involved in outer membrane biogenesis (Gentle et al., 2004; Voulhoux and Tommassen, 2004). It is worth noting that Tp92 is the only potential outer membrane protein in the *T. pallidum* genome with sequence homology to a *bona fide* Gram-negative bacterial outer membrane protein (Fraser et al., 1998). Although definitive localization data for Tp92 have not been published, an outer membrane location is supported by the observations that (i) Tp92 reactivity was not observed in treponemes washed extensively to remove their outer membranes; (ii) anti-Tp92 antiserum is opsonic, and (iii) rabbits immunized with Tp92 developed atypical, nonulcerative lesions (Cameron et al., 2000). The finding that Tp92 is highly conserved ($\geq$ 95%) among *T. pallidum* strains further enhances its attractiveness as a syphilis vaccine candidate. The third and most extensively studied antigen is TprK, a member of a 12-gene family in *T. pallidum* encoding proteins with sequence homology to the major sheath protein of *T. denticola* (Fraser et al., 1998; Centurion-Lara et al., 1999). In their initial publication, Centurion-Lara et al. (1999) reported that TprK is an opsonic target and that immunization with the protein elicits partial protection in rabbits; these findings subsequently were challenged by Hazlett et al. (2001) who contended that TprK is hydrophilic and resides in the periplasm. Lukehart and co-workers (Morgan et al., 2002a; 2002b; 2003) subsequently reinforced the status of TprK as a vaccine candidate by defining the respective roles of the antigen's conserved and variable domains in eliciting a protective response. Moreover, immunization with a recombinant peptide comprising the shared amino terminus of TprF and TprI, two other Tpr family members with predicted outer membrane locations, also altered lesion development following intradermal challenge (Sun et al., 2004).

The *T. pallidum* genomic sequence is now being mined intensively for candidate outer membrane proteins and vaccinogens. This bioinformatics approach is not easy given the spirochete's lack of orthologs for Gram-negative bacterial outer membrane proteins. As an alternative, investigators must employ computer algorithms to predict the likely cellular location and membrane topology of individual open reading frames. This rather tedious approach is also highly dependent upon the correct assignment of start codons, mistakes in which can determine whether or not the predicted protein does or does not appear to have a signal peptide, a prerequisite for cellular export (Centurion-Lara et al., 1999; Cameron et al., 2000; Hazlett et al., 2001). Despite these difficulties, the application of bioinformatics to the genomic sequence already has yielded a number of possible outer membrane proteins, including two that bind the extracellular matrix proteins fibronectin and laminin (Cameron, 2003; Cameron et al., 2004). Assessment of the protective capacity of these antigens is currently underway.

8. Future directions

The fundamental relationship between host responses and disease pathogenesis was suspected early in the history of syphilis research and has become a central tenet of the field in the post-genomic era. It also was well recognized, however, that, compared to other complex chronic infections, syphilis appears to play by its own rules. Stated another way, our understanding of syphilis immunology is such that we cannot easily discern the more subtle forms of interaction between *T. pallidum* and host immune system that underlie persistent infection. Presently, investigators have only two options to address these deficiencies in our knowledge base: continued investigations with the rabbit model and translational studies with humans. A strong argument can be made that the experimental rabbit model, though an outbred system, retains great potential to yield valuable new information, particularly with respect to latency and protective immunity. A re-examination of experiments conducted in the pre-molecular era, using post-genomics era technologies for evaluating spirochetal burdens in blood and tissues and gene expression profiling in both pathogen and host, in concert with traditional methodologies for visualizing spirochetes in tissues, would seem to be a reasonable point of departure along these lines. The recent study by Champion et al. (2005) represents a first step in this direction. Translational research with syphilis patients has undeniable possibilities for clarifying host-spirochete interactions although the unlikelihood that inoculation experiments will ever be performed in humans means that issues related to protective immunity will have to be addressed through surrogate *ex vivo* approaches and by extrapolation from animals. A secondary, but by no means insignificant, problem to be overcome in translational studies is the marked drop in syphilis rates in industrialized countries and the resultant need to develop international venues in order to pursue these avenues of investigation. Lastly, from the very outset, efforts to develop a recombinant or DNA-based syphilis subunit vaccine have been inseparable from the quest for *T. pallidum* outer membrane proteins (Lukehart, 1985; Radolf, 1995). Unequivocal identification of *T. pallidum* outer membrane proteins, now within the reach of syphilis investigators, will provide the answer to the critical unresolved question of whether antibodies directed against *T. pallidum* surface-exposed antigens are sufficient to confer protective immunity and whether cellular immunity also must be induced to achieve complete protection.

Acknowledgment

Supported by grants AI-26756 and AI-38894 (JDR), AI-10573 (KROH), and AI -34616, AI- 63940, AI-42143 (SAL) from the National Institutes of Health

References

Akins, D.R., Purcell, B.K., Mitra, M., Norgard, M.V., and Radolf, J.D. (1993). Lipid modification of the 17-kilodalton membrane immunogen of *Treponema pallidum* determines macrophage activation as well as amphiphilicity. Infect. Immun. *61*, 1202-1210.

Akira, S. and Takeda, K. (2004). Toll-like receptor signalling. Nat. Rev. Immunol. *4*, 499-511.

Alder, J.D., Friess, L., Tengowski, M., and Schell, R.F. (1990). Phagocytosis of opsonized *Treponema pallidum* subsp. *pallidum* proceeds slowly. Infect. Immun. *58*, 1167-1173.

Alderete, J.F. and Baseman, J.B. (1979). Surface-associated host proteins on virulent *Treponema pallidum*. Infect. Immun. *26*, 1048-1056.

Alexopoulou, L., Thomas, V., Schnare, M., Lobet, Y., Anguita, J., Schoen, R.T., Medzhitov, R., Fikrig, E., and Flavell, R.A. (2002). Hyporesponsiveness to vaccination with *Borrelia burgdorferi* OspA in humans and in TLR1- and TLR2-deficient mice. Nat. Med. *8*, 878-884.

Aliprantis, A.O., Yang, R.B., Mark, M.R., Suggett, S., Devaux, B., Radolf, J.D., Klimpel, G.R., Godowski, P., and Zychlinsky, A. (1999). Cell activation and apoptosis by bacterial lipoproteins through toll- like receptor-2. Science *285*, 736-739.

Arroll, T.W., Centurion-Lara, A., Lukehart, S.A., and Van Voorhis, W.C. (1999). T-Cell responses to *Treponema pallidum* subsp. *pallidum* antigens during the course of experimental syphilis infection. Infect. Immun. *67*, 4757-4763.

Azadegan, A.A., Schell, R.F., and LeFrock, J.L. (1983). Immune serum confers protection against syphilitic infection on hamsters. Infect. Immun. *42*, 42-47.

Badanoi, A., Gavrilesco, M., Nicolau, G., and Circiumaresco, T. (1969). Lymphoblastogenese immunoallergique et antigene treponemique dans la syphilis. Arch. Roum. Pathol. Exp. Microbiol. *28*, 419-427.

Baker-Zander, S. and Sell, S. (1980). A histopathologic and immunologic study of the course of syphilis in the experimentally infected rabbit. Demonstration of long-lasting cellular immunity. Am. J. Pathol. *101*, 387-413.

Baker-Zander, S.A., Fohn, M.J., and Lukehart, S.A. (1988). Development of cellular immunity to individual soluble antigens of *Treponema pallidum* during experimental syphilis. J. Immunol. *141*, 4363-4369.

Baker-Zander, S.A., Hook, E.W., III, Bonin, P., Handsfield, H.H., and Lukehart, S.A. (1985). Antigens of *Treponema pallidum* recognized by IgG and IgM antibodies during syphilis in humans. J. Infect. Dis. *151*, 264-272.

Baker-Zander, S.A. and Lukehart, S.A. (1992). Macrophage-mediated killing of opsonized *Treponema pallidum*. J. Infect. Dis. *165*, 69-74.

Baker-Zander, S.A., Shaffer, J.M., and Lukehart, S.A. (1993b). VDRL antibodies enhance phagocytosis of *Treponema pallidum* by macrophages. J. Infect. Dis. *167*, 1100-1105.

Baker-Zander, S.A., Shaffer, J.M., and Lukehart, S.A. (1993a). Characterization of the serum requirement for macrophage-mediated killing of *Treponema pallidum* ssp. *pallidum*: relationship to the development of opsonizing antibodies. FEMS Immunol. Med. Microbiol. *6*, 273-279.

Barrett, A.W., Villarroel, D.M., Hodgson, T.A., Porter, S.R., Hopper, C., Argiriadou, A.S., and Speight, P.M. (2004). The histopathology of syphilis of the oral mucosa. J. Oral Pathol. Med. *33*, 286-291.

Baughn, R.E., Jorizzo, J.L., Adams, C.B., and Musher, D.M. (1988). Ig class and IgG subclass responses to *Treponema pallidum* in patients with syphilis. J. Clin. Immunol. *8*, 128-139.

Belisle, J.T., Brandt, M.E., Radolf, J.D., and Norgard, M.V. (1994). Fatty acids of *Treponema pallidum* and *Borrelia burgdorferi* lipoproteins. J. Bacteriol. *176*, 2151-2157.

Berger, E.A., Murphy, P.M., and Farber, J.M. (1999). Chemokine receptors as HIV-1 coreceptors: roles in viral entry, tropism, and disease. Annu. Rev. Immunol. *17*, 657-700.

Bey, R.F., Johnson, R.C., and Fitzgerald, T.J. (1979). Suppression of lymphocyte response to concanavalin A by mucopolysaccharide material from *Treponema pallidum*-infected rabbits. Infect. Immun. *26*, 64-69.

Bishop, N.H. and Miller, J.N. (1976a). Humoral immunity in experimental syphilis. I. The demonstration of resistance conferred by passive immunization. J. Immunol. *117*, 191-196.

Bishop, N.H. and Miller, J.N. (1976b). Humoral immunity in experimental syphilis. II. The relationship of neutralizing factors in immune serum to acquired resistance. J. Immunol. *117*, 197-207.

Bishop, N.H. and Miller, J.N. (1983). Humoral immune mechanisms in experimental syphilis. In Pathogenesis and Immunology of Treponemal Infection, R.F.Schell and D.M.Musher, eds. (New York, N.Y.: Marcel Dekker, Inc.), pp. 241-270.

Bjerke, J.R., Krogh, H.K., and Matre, R. (1981). *In situ* identification of mononuclear cells in cutaneous infiltrates in discoid lupus erythematosus, sarcoidosis and secondary syphilis. Acta Dermatologica Venereologica *61*, 371-380.

Blanco, D.R., Champion, C.I., Dooley, A., Cox, D.L., Whitelegge, J.P., Faull, K., and Lovett, M.A. (2005a). A monoclonal antibody that conveys in vitro killing and partial protection in experimental syphilis binds a phosphorylcholine surface epitope of *Treponema pallidum*. Infect. Immun. *73*, 3083-3095.

Blanco, D.R., Champion, C.I., and Lovett, M.A. (2005b). Use of the skin protection assay in experimental syphilis to assess protective immunity against a specific *Treponema pallidum* surface epitope. FEMS Microbiol. Lett. *249*, 171-175.

Blanco, D.R., Miller, J.N., and Hanff, P.A. (1984). Humoral immunity in experimental syphilis: the demonstration of IgG as a treponemicidal factor in immune rabbit serum. J. Immunol. *133*, 2693-2697.

Blanco, D.R., Miller, J.N., and Lovett, M.A. (1997). Surface antigens of the syphilis spirochete and their potential as virulence determinants. Emer. Infect. Dis. *3*, 11-20.

Blanco, D.R., Radolf, J.D., Lovett, M.A., and Miller, J.N. (1986). Correlation of treponemicidal activity in normal human serum with the presence of IgG antibody directed against polypetides of *Treponema phagedenis* biotype Reiter and *Treponema pallidum*, Nichols strain. J. Immunol. *137*, 2031-2036.

Blanco, D.R., Walker, E.M., Haake, D.A., Champion, C.I., Miller, J.N., and Lovett, M.A. (1990). Complement activation limits the rate of *in vitro* treponemicidal activity and correlates with antibody-mediated aggregation of *Treponema pallidum* rare outer membrane protein. J. Immunol. *144*, 1914-1921.

Borenstein, L.A., Radolf, J.D., Fehniger, T.E., Blanco, D.R., Miller, J.N., and Lovett, M.A. (1988). Immunization with recombinant *Treponema pallidum* surface antigen 4D alters the course of experimental syphilis. J. Immunol. *140*, 2415-2421.

Bouis, D.A., Popova, T.G., Takashima, A., and Norgard, M.V. (2001). Dendritic cells phagocytose and are activated by *Treponema pallidum*. Infect. Immun. *69*, 518-528.

Bourell, K.W., Schulz, W., Norgard, M.V., and Radolf, J.D. (1994). *Treponema pallidum* rare outer membrane proteins: analysis of mobility by freeze-fracture electron microscopy. J. Bacteriol. *176*, 1598-1608.

Brightbill, H.D., Libraty, D.H., Krutzik, S.R., Yang, R.B., Belisle, J.T., Bleharski, J.R., Maitland, M., Norgard, M.V., Plevy, S.E., Smale, S.T., Brennan, P.J., Bloom, B.R., Godowski, P.J., and Modlin, R.L. (1999). Host defense mechanisms triggered by microbial lipoproteins through toll-like receptors. Science *285*, 732-736.

Brown, W.H. and Pearce, L. (1920). A note on the dissemination of *Spirochaeta pallida* from the primary focus of infection. Arch. Derm. Syph. *2*, 470-472.

Cameron, C.E. (2003). Identification of a *Treponema pallidum* laminin-binding protein. Infect. Immun. *71*, 2525-2533.

Cameron, C.E., Brown, E.L., Kuroiwa, J.M., Schnapp, L.M., and Brouwer, N.L. (2004). *Treponema pallidum* fibronectin-binding proteins. J. Bacteriol. *186*, 7019-7022.

Cameron, C.E., Castro, C., Lukehart, S.A., and Van Voorhis, W.C. (1998). Function and protective capacity of *Treponema pallidum* subsp. *pallidum* glycerophosphodiester phosphodiesterase. Infect. Immun. *66*, 5763-5770.

Cameron, C.E., Lukehart, S.A., Castro, C., Molini, B., Godornes, C., and Van Voorhis, W.C. (2000). Opsonic potential, protective capacity, and sequence conservation of the *Treponema pallidum* subspecies *pallidum* Tp92. J. Infect. Dis. *181*, 1401-1413.

Cannefax, G.R. (1965). Immunity in syphilis. Br. J. Vener. Dis. *41*, 260-274.

Centurion-Lara, A., Arroll, T., Castillo, R., Shaffer, J.M., Castro, C., Vanvoorhis, W.C., and Lukehart, S.A. (1997). Conservation of the 15-kilodalton lipoprotein gene among *Treponema pallidum* subspecies and strains and other pathogenic treponemes - genetic and antigenic analyses. Infect. Immun. *65*, 1440-1444.

Centurion-Lara, A., Castro, C., Barrett, L., Cameron, C., Mostowfi, M., Van Voorhis, W.C., and Lukehart, S.A. (1999). *Treponema pallidum* major sheath protein homologue TprK is a target of opsonic antibody and the protective immune response. J. Exp. Med. *189*, 647-656.

Centurion-Lara, A., LaFond, R.E., Hevner, K., Godornes, C., Molini, B.J., Van Voorhis, W.C., and Lukehart, S.A. (2004). Gene conversion: a mechanism for generation of heterogeneity in the *tprK* gene of *Treponema pallidum* during infection. Mol. Microbiol. *52*, 1579-1596.

Chamberlain, N.R., Brandt, M.E., Erwin, A.L., Radolf, J.D., and Norgard, M.V. (1989). Major integral membrane protein immunogens of *Treponema pallidum* are proteolipids. Infect. Immun. *57*, 2872-2877.

Champion, C.I., Blanco, D.R., and Lovett, M.A. (2005). Quantitative assessment of protection in experimental syphilis. Infect. Immun. *73*, 5923-5927.

Champion, C.I., Miller, J.N., Borenstein, L.A., Lovett, M.A., and Blanco, D.R. (1990). Immunization with *Treponema pallidum* endoflagella alters the course of experimental rabbit syphilis. Infect. Immun. *58*, 3158-3161.

Chan, J.K., Schell, R.F., and LeFrock, J.L. (1979). Ability of enriched immune T cells to confer resistance in hamsters to infection with Treponema pertenue. Infect. Immun. *26*, 448-452.

Chapel, T.A. (1980). The signs and symptoms of secondary syphilis. Sex. Transm. Dis. *7*, 161-164.

Chesney, A.M. (1926). Immunity in syphilis. Medicine *5*, 463-547.

Christiansen, S. (1963). Protective layer covering pathogenic treponemata. Lancet *1*, 423-425.

Coca, A.F. and Grove, E.F. (1925). Studies in hypersensitiveness. XIII. A study of the atopic reagins. J. Immunol. *10*, 445-464.

Collart, P., Franceschini, P., and Durel, P. (1971). Experimental rabbit syphilis. Br. J. Vener. Dis. *47*, 389-400.

Costello, P.B., Powell, G.L., and Green, F.A. (1990). The structural requirements for anti-cardiolipin antibody binding in sera from patients with syphilis and SLE. Clin. Immunol. Immunopathol. *56*, 393-400.

Cox, D.L., Chang, P., McDowall, A., and Radolf, J.D. (1992). The outer membrane, not a coat of host proteins, limits the antigenicity of virulent *Treponema pallidum*. Infect. Immun. *60*, 1076-1083.

Cumberland, M.C. and Turner, T.B. (1949). The rate of multiplication of *Treponema pallidum* in normal and immune rabbits. Am. J. Syph. *33*, 201-212.

De Bruijn, J.H. (1966). Chemical structure and serological activity of natural and synthetic cardiolipin and related compounds. Br. J. Vener. Dis. *42*, 125-128.

Deacon, W.E., Lucas, J.B., and Price, E.V. (1966). Fluorescent treponemal antibody-absorption (FTA-Abs) test for syphilis. JAMA *198*, 624-628.

Doniach, D., Delhanty, J., Lindqvist, H.J., and Catterall, R.D. (1970). Mitochondrial and other tissue autoantibodies in patients with biological false positive reactions for syphilis. Clin. Exp. Immunol. *6*, 871-884.

Driver, J.R. (1924). Reinfection in syphilis. JAMA *83*, 1728-1733.

Eagle, H. and Fleischman, R. (1948). The antibody response in rabbits to killed suspensions of pathogenic *T. pallidum*. J. Exp. Med. *87*, 369-384.

Eberson, F. (1921). Immunity studies in experimental syphilis. Arch. Derm. Syph. *4*, 490-511.

Engelkens, H.J., ten Kate, F.J., Judanarso, J., Vuzevski, V.D., van Lier, J.B.H., Godschalk, J.C.J., Van der Sluis, J.J., and Stolz, E. (1993). The localisation of treponemes and characterisation of the inflammatory infiltrate in skin biopsies from patients with primary or secondary syphilis, or early infectious yaws. Genitourin. Med. *69*, 102-107.

Fan, Y.M., Zeng, W.J., Wu, Z.H., and Li, S.F. (2004). Immunophenotypes, apoptosis, and expression of Fas and Bcl-2 from peripheral blood lymphocytes in patients with secondary early syphilis. Sex Transm. Dis. *31*, 221-224.

Fehniger, T.E., Radolf, J.D., Walfield, A.M., Cunningham, T.M., Miller, J.N., and Lovett, M.A. (1986). Native surface location of a recombinant 38-kDa *Treponema pallidum* antigen isolated from the *Escherichia coli* outer membrane. Infect. Immun. *52*, 586-592.

Fikrig, E., Barthold, S.W., Chen, M.C., Grewal, I.S., Craft, J., and Flavell, R.A. (1996). Protective antibodies in murine Lyme disease arise independently of CD40 ligand. J. Immunol. *157*, 1-3.

Fitzgerald, T.J. (1992). The Th1/Th2-like switch in syphilitic infection: is it detrimental? Infect. Immun. *60*, 3475-3479.

Fitzgerald, T.J. and Johnson, R.C. (1979). Surface mucopolysaccharides of *Treponema pallidum*. Infect. Immun. *24*, 244-251.

Fitzgerald, T.J., Johnson, R.C., Miller, J.N., and Sykes, J.A. (1977). Characterization of the attachment of *Treponema pallidum* (Nichols strain) to cultured mammalian cells and the potential relationship of attachment to pathogenicity. Infect. Immun. *18*, 467-478.

Fitzgerald, T.J. and Repesh, L.A. (1983). Toxic activities of *Treponema pallidum*. In Pathogenesis and Immunology of Treponemal Infection, R.F.Schell and D.M.Musher, eds. (New York: Marcell Dekker, Inc.), pp. 173-193.

Fraser, C.M., Norris, S.J., Weinstock, G.M., White, O., Sutton, G.G., Dodson, R., Gwinn, M., Hickey, E.K., Clayton, R., Ketchum, K.A., Sodergren, E., Hardham, J.M., McLeod, M.P., Salzberg, S., Peterson, J., Khalak, H., Richardson, D., Howell, J.K., Chidambaram, M., Utterback, T., McDonald, L., Artiach, P., Bowman, C., Cotton, M.D., and Venter, J.C. (1998). Complete genome sequence of *Treponema pallidum*, the syphilis spirochete. Science *281*, 375-388.

Friedmann, P.S. and Turk, J.L. (1975). A spectrum of lymphocyte responsiveness in human syphilis. Clin. Exp. Immunol. *21*, 59-64.

Friedmann, P.S. and Turk, J.L. (1978). The role of cell-mediated immune mechanisms in syphilis in Ethiopia. Clin. Exp. Immunol. *31*, 59-65.

From, E., Thestrup-Pedersen, K., and Thulin, H. (1976). Reactivity of lymphocytes from patients with syphilis towards *T. pallidum* antigen in the leucocyte migration and lymphocyte transformation tests. Br. J. Vener. Dis. *52*, 224-229.

Galvin, S.R. and Cohen, M.S. (2004). The role of sexually transmitted diseases in HIV transmission. Nat. Rev. Microbiol. *2*, 33-42.

Gentle, I., Gabriel, K., Beech, P., Waller, R., and Lithgow, T. (2004). The Omp85 family of proteins is essential for outer membrane biogenesis in mitochondria and bacteria. J. Cell Biol. *164*, 19-24.

Giambartolomei, G.H., Dennis, V.A., Lasater, B.L., and Philipp, M.T. (1999). Induction of pro- and anti-inflammatory cytokines by *Borrelia burgdorferi* lipoproteins in monocytes is mediated by CD14. Infect. Immun. *67*, 140-147.

Grassly, N.C., Fraser, C., and Garnett, G.P. (2005). Host immunity and synchronized epidemics of syphilis across the United States. Nature *433*, 417-421.

Greenblatt, R.M., Lukehart, S.A., Plummer, F.A., Quinn, T.C., Critchlow, C.W., Ashley, R.L., D'Costa, L.J., Ndinya-Achola, J.O., Corey, L., Ronald, A.R., and Holmes, K.K. (1988). Genital ulceration as a risk factor for human immunodeficiency virus infection. AIDS *2*, 47-50.

Gueft, B. and Rosahn, P.D. (1948). Experimental mouse syphilis, a critical review of the literature. Am. J. Syph. *32*, 59-88.

Hanff, P.A., Bishop, N.H., Miller, J.N., and Lovett, M.A. (1983). Humoral immune response in experimental syphilis to polypeptides of *Treponema pallidum*. J. Immunol. *131*, 1973-1977.

Hanff, P.A., Fehniger, T.E., Miller, J.N., and Lovett, M.A. (1982). Humoral immune response in human syphilis to polypeptides of *Treponema pallidum*. J. Immunol. *129*, 1287-1291.

Hardy, P.H., Jr. and Levin, J. (1983). Lack of endotoxin in *Borrelia hispanica* and *Treponema pallidum*. Proc. Soc. Exp. Biol. Med. *174*, 47-52.

Hardy, P.H., Jr. and Nell, E.E. (1957). Study of the antigenic structure of *Treponema pallidum* by specific agglutination. Am. J. Hyg. *66*, 160-172.

Harris, E.N. and Pierangeli, S.S. (1996). Functional effects of anticardiolipin antibodies. Lupus *5*, 372-377.

Hart, G. (1986). Syphilis tests in diagnosing and therapeutic decision making. Ann. Intern. Med. *104*, 368-376.

Hayes, N.S., Muse, K.E., Collier, A.M., and Baseman, J.B. (1977). Parasitism by virulent *Treponema pallidum* of host cell surfaces. Infect. Immun. *17*, 174-186.

Hazlett, K.R.O., Sellati, T.J., Nguyen, T.T., Cox, D.L., Clawson, M.L., Caimano, M.J., and Radolf, J.D. (2001). The TprK protein of *Treponema pallidum* is periplasmic and is not a target of opsonic antibody or protective immunity. J. Exp. Med. *193*, 1015-1026.

Heath, W.R. and Carbone, F.R. (2001). Cross-presentation, dendritic cells, tolerance and immunity. Annu. Rev. Immunol. *19*, 47-64.

Hensel, U., Wellensiek, H.J., and Bhakdi, S. (1985). Sodium dodecyl sulfate-polyacrylamide gel electrophoresis immunoblotting as a serological tool in the diagnosis of syphilitic infections. J. Clin. Microbiol. *21*, 82-87.

Hirschfeld, M., Kirschning, C.J., Schwandner, R., Wesche, H., Weis, J.H., Wooten, R.M., and Weis, J.J. (1999). Inflammatory signaling by *Borrelia burgdorferi* lipoproteins is mediated by toll-like receptor 2. J. Immunol. *163*, 2382-2386.

Hoffmann, P., Heinle, S., Schade, U.F., Loppnow, H., Ulmer, A.J., Flad, H.-D., Jung, G., and Bessler, W.G. (1988). Stimulation of human and murine adherent cells by bacterial lipoprotein and synthetic lipopeptide analogs. Immunobiol. *177*, 158-170.

Hook, E.W., III and Peeling, R.W. (2004). Syphilis control-a continuing challenge. N. Engl. J. Med. *351*, 121-124.

Hsu, P.L., Qin, M., Norris, S.J., and Sell, S. (1988). Isolation and characterization of recombinant Escherichia coli clones secreting a 24-kilodalton antigen of *Treponema pallidum*. Infect. Immun. *56*, 1135-1143.

Hunter, E.F., Deacon, W.E., and Meyer, P.E. (1964). An improved FTA test for syphilis, the absorption procedure (FTA-ABS). Public Health Rep. *79*, 410-412.

Izzat, N.N., Knox, J.M., Werth, J.A., and Dacres, W.G. (1970). Evolution of syphilitic chancres with virulent *Treponema pallidum* in the rabbit. Br. J. Vener. Dis. *47*, 67-72.

Izzat, N.N., Smith, E.B., Jackson, S.W., and Knox, J.M. (1971). Immunological studies on treponemal antigens. II. Serological changes and resistance to infection in rabbits immunized with culture supernatant of avirulent *Treponema pallidum*. Br. J. Vener. Dis. *47*, 335-338.

Janeway, C.A., Jr. (1989). Approaching the asymptote? Evolution and revolution in immunology. Cold Spring Harb. Symp. Quant. Biol. *54 Pt 1*, 1-13.

Janot, C., Grandidier, M., Pupil, P., Thomas, J.L., Beurey, J., and de Lavergne, E. (1971). Le test de transformation lymphoblatique au cours de la syphilis. Presse Med *79*, 1901-1904.

Jiang, J., Zenewicz, L.A., San Mateo, L.R., Lau, L.L., and Shen, H. (2003). Activation of antigen-specific CD8 T cells results in minimal killing of bystander bacteria. J. Immunol. *171*, 6032-6038.

Johnson, R.B., Köhl, S., Wiesmüller, K.-H., Jung, C., and Bessler, W.G. (1983). Synthetic analogues of the N-terminal lipid part of bacterial lipoprotein are B-lymphocyte mitogens in vitro and in vivo. Immunobiol. *165*, 27-35.

Jones, S.A., Marchitto, K.S., Miller, J.N., and Norgard, M.V. (1984). Monoclonal antibody with hemagglutination, immobilization, and neutralization activities defines an immunodominant, 47, 000 mol wt, surface-exposed immunogen of *Treponema pallidum* (Nichols). J. Exp. Med. *160*, 1404-1420.

Knox, J.M., Short, D.H., Wende, R.D., and Glicksman, J.M. (1966). The FTA-ABS test for syphilis. Performance in 1, 033 patients. Br. J. Vener. Dis. *42*, 16-20.

Laird, S.M. and Thorburn, A.L. (1966). Assessment of the "Luotest" in late syphilis. Br. J. Vener. Dis. *42*, 119-121.

Levene, G.M., Turk, J.L., Wright, D.J., and Grimble, A.G. (1969). Reduced lymphocyte transformation due to a plasma factor in patients with active syphilis. Lancet *2*, 246-247.

Levene, G.M., Wright, D.J., and Turk, J.L. (1971). Cell-mediated immunity and lymphocyte transformation in syphilis. Proc. R. Soc. Med. *64*, 426-428.

Lewinski, M.A., Miller, J.N., Lovett, M.A., and Blanco, D.R. (1999). Correlation of immunity in experimental syphilis with serum-mediated aggregation of *Treponema pallidum* rare outer membrane proteins. Infect. Immun. *67*, 3631-3636.

Lex, A., Wiesmuller, K.-H., Jung, G., and Bessler, W.G. (1986). A synthetic analogue of *Escherichia coli* lipoprotein, tripalmitoyl pentapeptide, constitutes a potent immune adjuvant. J. Immunol. *137*, 2676-2681.

Lien, E., Sellati, T.J., Yoshimura, A., Flo, T.H., Rawadi, G., Finberg, R.W., Carroll, J.D., Espevik, T., Ingalls, R.R., Radolf, J.D., and Golenbock, D.T. (1999). Toll-like receptor 2 functions as a pattern recognition receptor for diverse bacterial products. J. Biol. Chem. *274*, 33419-33425.

Lukehart, S.A. (1982). Activation of macrophages by products of lymphocytes from normal and syphilitic rabbits. Infect. Immun. *37*, 64-69.

Lukehart, S.A. (1985). Prospects for development of a treponemal vaccine. Rev. Infect. Dis. *7, Supplement 2*, S305-S313.

Lukehart, S.A. (1992). Immunology and pathogenesis of syphilis. In Advances in Host Defense Mechanisms. Sexually Transmitted Diseases, T.C.Quinn, ed. (New York: Raven Press), pp. 141-163.

Lukehart, S.A. (2004). Syphilis. In Harrison's Principles of Internal Medicine, E.Brauwald, A.S.Faucci, S.L.Hauser, D.L.Longo, and J.L.Jameson, eds. (New York: McGraw Hill), pp. 977-985.

Lukehart, S.A., Baker-Zander, S.A., Cheri Lloyd, R.M., and Sell, S. (1980a). Characterization of lymphocyte responsiveness in early experimental syphilis. II. Nature of cellular infiltration and *Treponema pallidum* distribution in testicular lesions. J. Immunol. *124*, 461-467.

Lukehart, S.A., Baker-Zander, S.A., and Gubish, E.R., Jr. (1982). Identification of *Treponema pallidum* antigens: Comparison with a nonpathogenic treponeme. J. Immunol. *129*, 833-838.

Lukehart, S.A., Baker-Zander, S.A., and Sell, S. (1980b). Characterization of lymphocyte responsiveness in early experimental syphilis. I. *In vitro* response to mitogens and *Treponema pallidum* antigens. J. Immunol. *124*, 454-460.

Lukehart, S.A., Baker-Zander, S.A., and Sell, S. (1986). Characterization of the humoral immune response of the rabbit to antigens of *Treponema pallidum* after experimental infection and therapy. Sex Transm. Dis. *13*, 9-15.

Lukehart, S.A. and Miller, J.N. (1978). Demonstration of the *in vitro* phagocytosis of *Treponema pallidum* by rabbit peritoneal macrophages. J. Immunol. *121*, 2014-2024.

Lukehart, S.A., Shaffer, J.M., and Baker-Zander, S.A. (1992). A subpopulation of *Treponema pallidum* is resistant to phagocytosis: possible mechanisms of persistence. J. Infect. Dis. *166*, 1449-1453.

Lukehart, S.A., Tam, M.R., Hom, J., Baker-Zander, S.A., Holmes, K.K., and Nowinski, R.C. (1985). Characterization of monoclonal antibodies to *Treponema pallidum*. J. Immunol. *134*, 585-592.

Magnuson, H.J. (1948). Current concepts of immunity in syphilis. Am. J. Med. *5*, 641-654.

Magnuson, H.J., Eagle, H., and Fleischman, R. (1948). The minimal infectious inoculum of *Spirochaeta pallida* (Nichols Strain), and a consideration of its rate of multiplication *in vivo*. Am. J. Syph. Gon. Vener. Dis. *32*, 1-19.

Magnuson, H.J., Halbert, S.P., and Rosenau, B.J. (1947). Attempted immunization of rabbits against syphilis with killed *Treponema pallidum* and adjuvants. J. Vener. Dis. Inf. *28*, 267-271.

Magnuson, H.J. and Rosenau, B.J. (1948). The rate of development and degree of acquired immunity in experimental syphilis. Am. J. Syph. Gon. Vener. Dis. *32*, 418-436.

Magnuson, H.J., Rosenau, B.J., and Clark, J.W., Jr. (1949). The duration of acquired immunity in experimental syphilis. Am. J. Syph. Gon. Vener. Dis. *33*, 297-302.

Magnuson, H.J., Thomas, E.W., Olansky, S., Kaplan, B.I., de Mello, L., and Cutler, J.C. (1956). Inoculation syphilis in human volunteers. Medicine *35*, 33-82.

Marchitto, K.S., Selland-Grossling, C.K., and Norgard, M.V. (1986). Molecular specificities of monoclonal antibodies directed against virulent *Treponema pallidum*. Infect. Immun. *51*, 168-176.

Maret, S.M., Baseman, J.B., and Folds, J.D. (1980). Cell-mediated immunity in *Treponema pallidum* infected rabbits: *in vitro* response of splenic and lymph node lymphocytes to mitogens and specific antigens. Clin. Exp. Immunol. *39*, 38-43.

Marshak, L.C. and Rothman, S. (1951). Skin testing with a purified suspension of *Treponema pallidum*. Am. J. Syph. Gon. Vener. Dis. *35*, 35-41.

Matthews, H.M., Yang, T.K., and Jenkin, H.M. (1979). Unique lipid composition of *Treponema pallidum* (Nichols virulent strain). Infect. Immun. *24*, 713-719.

Mbow, M.L., Zeidner, N., Gilmore, R.D., Jr., Dolan, M., Piesman, J., and Titus, R.G. (2001). Major histocompatibility complex class II-independent generation of neutralizing antibodies against T-cell-dependent Borrelia burgdorferi antigens presented by dendritic cells: regulation by NK and gammadelta T cells. Infect. Immun. *69*, 2407-2415.

McBroom, R.L., Styles, A.R., Chiu, M.J., Clegg, C., Cockerell, C.J., and Radolf, J.D. (1999). Secondary syphilis in persons infected with and not infected with HIV-1: a comparative immunohistologic study. Am. J. Dermatopathol. *21*, 432-441.

McKevitt, M., Brinkman, M.B., McLoughlin, M., Perez, C., Howell, J.K., Weinstock, G.M., Norris, S.J., and Palzkill, T. (2005). Genome scale identification of *Treponema pallidum* antigens. Infect. Immun. *73*, 4445-4450.

McLeod, C.P. and Magnuson, H.J. (1953). Production of immobilizing antibodies unaccompanied by active immunity to *Treponema pallidum* as shown by injecting rabbits and mice with the killed organisms. Am. J. Syph. Gon. Vener. Dis. *37*, 9-22.

Metzger, M. and Smogor, W. (1969). Artificial immunization of rabbits against syphilis. I. Effect of increasing doses of treponemes given by the intramuscular route. Br. J. Vener. Dis. *45*, 308-312.

Miller, J.N. (1975). Value and limitations of nontreponemal and treponemal tests in the laboratory diagnosis of syphilis. Clin. Obst. Gyn *18*, 191-203.

Miller, J.N. (1973). Immunity in experimental syphilis VI. Successful vaccination of rabbits with *Treponema pallidum*, Nichols strain, attenuated by gamma-irradiation. J. Immunol. *110*, 1206-1215.

Miller, J.N., WHANG, S.J., and FAZZAN, F.P. (1963). Studies on immunity in experimental syphilis. I. Immunologic response of rabbits immunized with Reiter protein antigen and challenged with *Treponema pallidum*. Br. J. Vener. Dis. *39*, 195-198.

Morgan, C.A., Lukehart, S.A., and Van Voorhis, W.C. (2003). Protection against syphilis correlates with specificity of antibodies to the variable regions of *Treponema pallidum* repeat protein K. Infect. Immun. *71*, 5605-5612.

Morgan, C.A., Lukehart, S.A., and Van Voorhis, W.C. (2002a). Immunization with the N-Terminal portion of *Treponema pallidum* repeat protein K attenuates syphilitic lesion development in the rabbit model. Infect. Immun. *70*, 6811-6816.

Morgan, C.A., Molini, B.J., Lukehart, S.A., and Van Voorhis, W.C. (2002b). Segregation of B and T cell epitopes of T*reponema pallidum* repeat protein K to variable and conserved regions during experimental syphilis infection. J. Immunol. *169*, 952-957.

Musher, D.M., Schell, R.F., Jones, R.H., and Jones, A.M. (1975). Lymphocyte transformation in syphilis: an in vitro correlate of immune suppression in vivo? Infect. Immun. *11*, 1261-1264.

Musher, D.M., Schell, R.F., and Knox, J.M. (1976a). Immunologic mechanisms of infectious syphilis. Cutis *17*, 739-742.

Musher, D.M., Schell, R.F., and Knox, J.M. (1974). In vitro lymphocyte response to *Treponema refringens* in human syphilis. Infect. Immun. *9*, 654-657.

Musher, D.M., Schell, R.F., and Knox, J.M. (1976b). The immunology of syphilis. Int. J. Dermatol. *15*, 566-576.

Nelson, R.A., Jr. and Mayer, M.M. (1949). Immobilization of *Treponema pallidum in vitro* by antibody produced in syphilitic infection. J. Exp. Med. *89*, 369-383.

Nelson, R.J., Jr. and Diesendruck, J.A. (1951). Studies on treponemal immobilizing antibodies in syphilis. I. Techniques of measurement and factors influencing immobilization. J. Immunol. *66*, 667-685.

Noguchi, H. (1911). A cutaneous reaction in syphilis. J. Exp. Med *14*, 557-568.

Noguchi, H. (1912). The luetin reaction. JAMA *59*, 1262-1263.

Norgard, M.V., Arndt, L.L., Akins, D.R., Curetty, L.L., Harrich, D.A., and Radolf, J.D. (1996). Activation of human monocytic cells by *Treponema pallidum* and *Borrelia burgdorferi* lipoproteins and synthetic lipopeptides proceeds via a pathway distinct from that of lipopolysaccharide but involves the transcriptional activator NF-κB. Infect. Immun. *64*, 3845-3852.

Norgard, M.V. and Miller, J.N. (1983). Cloning and expression of *Treponema pallidum*(Nichols) antigen genes in *Escherichia coli*. Infect. Immun. *42*, 435-445.

Norgard, M.V., Riley, B.S., Richardson, J.A., and Radolf, J.D. (1995). Dermal inflammation elicited by synthetic analogs of *Treponema pallidum* and *Borrelia burgdorferi* lipoproteins. Infect. Immun. *63*, 1507-1515.

Norris, S.J. (1988). Syphilis. In Immunology of Sexually Transmitted Diseases, D.J.M.Wright, ed. (Boston: Kluwer Academic Publishers), pp. 1-31.

Norris, S.J. and Sell, S. (1984). Antigenic complexity of *Treponema pallidum*: antigenicity and surface localization of major polypeptides. J. Immunol. *133*, 2686-2692.

Pandhi, R.K., Singh, N., and Ramam, M. (1995). Secondary syphilis: a clinicopathologic study. Int. J. Dermatol. *34*, 240-243.

Pangborn, M.C. (1941). A new serologically active phospholipid from beef heart. Proc. Soc. Exp. Biol. Med. *48*, 484-486.

Pangborn, M.C. (1945). A simplified preparation of cardiolipin with a note on purification of lecithin for serologic use. J. Biol. Chem. *161*, 71-82.

Pavia, C.S., Baseman, J.B., and Folds, J.D. (1977). Selective response of lymphocytes from *Treponema pallidum*- infected rabbits to mitogens and *Treponema reiteri*. Infect. Immun. *15*, 417-422.

Pavia, C.S., Folds, J.D., and Baseman, J.B. (1976). Depression of lymphocyte response to concanavalin A in rabbits infected with *Treponema pallidum* (Nichols strain). Infect. Immun. *14*, 320-322.

Pavia, C.S., Folds, J.D., and Baseman, J.B. (1978). Cell-mediated immunity during syphilis: a review. Br. J. Vener. Dis. *54*, 144-150.

Pavia, C.S. and Niederbuhl, C.J. (1985). Adoptive transfer of anti-syphilis immunity with lymphocytes from *Treponema pallidum*-infected guinea pigs. J. Immunol. *135*, 2829-2834.

Pavia, C.S., Niederbuhl, C.J., and Saunders, J. (1985). Antibody-mediated protection of guinea-pigs against infection with *Treponema pallidum*. Immunology *56*, 195-202.

Perine, P.L. (1981). Immunology of syphilis. In Immunology of Human Infection. Part I: Bacteria, Mycoplasmae, Chlamydiae, and Fungi, A.J.Nahmias and R.J.O'Reilly, eds. (New York: Plenum Medial Book Company), pp. 391-413.

Perine, P.L., Weiser, R.S., and Klebanoff, S.J. (1973). Immunity to syphilis. I. Passive transfer in rabbits with hyperimmune serum. Infect. Immun. *8*, 787-790.

Pivarsci, A., Nagy, I., and Kemeny, L. (2005). Innate immunity in the skin: how keratinocytes fight against pathogens. Curr. Immuno. Rev. *1*, 29-402.

Poltorak, A., He, X., Smirnova, I., Liu, M.-Y., Van Huffel, C., Du, X., Birdwell, D., Alejos, E., Silva, M., Galanos, C., Freudenberg, M., Ricciardi-Castagnoli, P., Layton, B., and Beutler, B. (1998). Defective LPS signaling in C3H/HeJ and C57Bl/10ScCr mice: mutations in *Tlr4* gene. Science *282*, 2085-2088.

Pope, V., Larsen, S.A., Rice, R.J., Goforth, S.N., Parham, C.E., and Fears, M.B. (1994). Flow cytometric analysis of peripheral blood lymphocyte immunophenotypes in persons infected with *Treponema pallidum*. Clin. Diagn. Lab. Immunol. *1*, 121-124.

Qin, S., Rottman, J.B., Myers, P., Kassam, N., Weinblatt, M., Loetscher, M., Koch, A.E., Moser, B., and Mackay, C.R. (1998). The chemokine receptors CXCR3 and CCR5 mark subsets of T cells associated with certain inflammatory reactions. J. Clin. Invest. *101*, 746-754.

Quist, E.E., Repesh, L.A., Zeleznikar, R., and Fitzgerald, T.J. (1983). Interaction of *Treponema pallidum* with isolated rabbit capillary tissues. Br. J. Vener. Dis. *59*, 11-20.

Radolf, J.D. (1995). *Treponema pallidum* and the quest for outer membrane proteins. Mol. Microbiol. *16*, 1067-1073.

Radolf, J.D., Arndt, L.L., Akins, D.R., Curetty, L.L., Levi, M.E., Shen, Y., Davis, L.S., and Norgard, M.V. (1995). *Treponema pallidum* and *Borrelia burgdorferi* lipoproteins and synthetic lipopeptides activate monocytes/macrophages. J. Immunol. *154*, 2866-2877.

Radolf, J.D., Fehniger, T.E., Silverblatt, F.J., Miller, J.N., and Lovett, M.A. (1986). The surface of virulent *Treponema pallidum*: Resistance to antibody binding in the absence of complement and surface association of recombinant antigen 4D. Infect. Immun. *52*, 579-585.

Radolf, J.D., Norgard, M.V., Brandt, M.E., Isaacs, R.D., Thompson, P.A., and Beutler, B. (1991). Lipoproteins of *Borrelia burgdorferi* and *Treponema pallidum* activate cachectin/tumor necrosis factor synthesis: analysis using a CAT reporter construct. J. Immunol. *147*, 1968-1974.

Raiziss, G.W. and Severac, M. (1935). Rapidity with which *Spirochaeta pallida* invades the blood stream. Arch. Derm. Syph. *35*, 1101-1109.

Reynolds, F.W. (1941). The fate of *Treponema pallidum* inoculated subcutaneously into immune rabbits. Bulletin John Hopkins University *69*, 53-60.

Rittig, M.G., Haupl, T., and Burmester, G.R. (1994). Coiling phagocytosis: a way for MHC class I presentation of bacterial antigens? Int. Arch. Allergy Immunol. *103*, 4-10.

Robinson, J.H. and Delvig, A.A. (2002). Diversity in MHC class II antigen presentation. Immunology *105*, 252-262.

Rock, K.L. (2003). The ins and outs of cross-presentation. Nat. Immunol. *4*, 941-943.

Rosahn, P.D. (1947). Autopsy studies in syphilis. J. Vener. Dis. Inf. *(suppl. 21)*, 1-67.

Salazar, J., Hazlett, K.R.O., and Radolf, J. (2002). The immune response to infection with *Treponema pallidum*, the stealth pathogen. Microbes. Infect. *4*, 1133.

Salazar, J. C., Cruz, A., Valderrama, L., Pope, C., Trujillo, R., Sombredero, A., Obando, A. M., Saravia, N., and Radolf, J. D. (2005a). Immunity in early syphilis: pathway to HIV infection. 16th meeting of the ISSTDR .

Salazar, J.C., Pope, C.D., Moore, M.W., Pope, J., Kiely, T.G., and Radolf, J.D. (2005b). Lipoprotein-dependent and -independent immune responses to spirochetal infection. Clin. Diagn. Lab Immunol. *12*, 949-958.

Sallusto, F., Lenig, D., Mackay, C.R., and Lanzavecchia, A. (1998). Flexible programs of chemokine receptor expression on human polarized T helper 1 and 2 lymphocytes. J. Exp. Med. *187*, 875-883.

Sambri, V., Marangoni, A., Simone, M.A., D'Antuono, A., Negosanti, M., and Cevenini, R. (2001). Evaluation of recomWell *Treponema*, a novel recombinant antigen-based enzyme-linked immunosorbent assay for the diagnosis of syphilis. Clin. Microbiol. Infect. *7*, 200-205.

Schell, R., Musher, D., Jacobson, K., and Schwethelm, P. (1975a). Induction of acquired cellular resistance following transfer of thymus-dependent lymphocytes from syphilitic rabbits. J. Immunol. *114*, 550-553.

Schell, R., Musher, D., Jacobson, K., Schwethelm, P., and Simmons, C. (1975b). Effect of macrophage activation on infection with *Treponema pallidum*. Infect. Immun. *12*, 505-511.

Schell, R.F. (1983). Rabbit and hamster model of treponemal infection. In Pathogenesis and Immunology of Treponemal Infection, R.F.Schell and D.M.Musher, eds. (New York: Marcel Dekker, Inc.), pp. 121-135.

Schell, R.F., Chan, J.K., LeFrock, J.L., and Bagasra, O. (1980). Endemic syphilis: transfer of resistance to *Treponema pallidum* strain Bosnia A in hamsters with a cell suspension enriched in thymus-derived cells. J. Infect. Dis. *141*, 752-758.

Schell, R.F. and Musher, D.M. (1975). Transfer of nonspecific resistance to *Listeria monocytogenes* using spleen cells from syphilitic rabbits (38574). Proc. Soc. Exp. Biol. Med. *148*, 516-518.

Schouls, L.M., Mout, R., Dekker, J., and Van Embden, J.D.A. (1989). Characterization of lipid-modified immunogenic proteins of *Treponema pallidum* expressed in *Escherichia coli*. Microb. Pathog. *7*, 175-188.

Schultz, D.R. (1997). Antiphospholipid antibodies: basic immunology and assays. Semin. Arthritis Rheum. *26*, 724-739.

Sell, S., Baker-Zander, S., and Powell, H.C. (1982). Experimental syphilitic orchitis in rabbits: ultrastructural appearance of *Treponema pallidum* during phagocytosis and dissolution by macrophages *in vivo*. Lab. Invest. *46*, 355-364.

Sell, S., Baker-Zander, S.A., and Lloyd, R.M. (1980a). T-cell hyperplasia of lymphoid tissues of rabbits infected with Treponema pallidum: evidence for a vigorous immune response. Sex Transm. Dis. *7*, 74-84.

Sell, S., Gamboa, D., Baker-Zander, S.A., Lukehart, S.A., and Miller, J.N. (1980b). Host response to *Treponema pallidum* in intradermally-infected rabbits: evidence for persistence of infection at local and distant sites. J. Invest Dermatol. *75*, 470-475.

Sell, S. and Norris, S.J. (1983). The biology, pathology, and immunology of syphilis. Int. Rev. Exp. Pathol. *24*, 203-276.

Sellati, T.J., Bouis, D.A., Kitchens, R.L., Darveau, R.P., Pugin, J., Ulevitch, R.J., Gangloff, S.C., Goyert, S.M., Norgard, M.V., and Radolf, J.D. (1998). *Treponema pallidum* and *Borrelia burgdorferi* lipoproteins and synthetic lipopeptides activate monocytic cells via a CD14-dependent pathway distinct from that used by lipopolysaccharide. J. Immunol. *160*, 5455-5464.

Sellati, T.J., Waldrop, S.L., Salazar, J.C., Bergstresser, P.R., Picker, L.J., and Radolf, J.D. (2001). The cutaneous response in humans to *Treponema pallidum* lipoprotein analogues involves cellular elements of both innate and adaptive immunity. J. Immunol. *166*, 4131-4140.

Sellati, T.J., Wilkinson, D.A., Sheffield, J.S., Koup, R.A., Radolf, J.D., and Norgard, M.V. (2000). Virulent *Treponema pallidum*, lipoprotein, and synthetic lipopeptides induce CCR5 on human monocytes and enhance their susceptibility to infection by human immunodeficiency virus type 1. J. Infect. Dis. *181*, 283-293.

Sepetjian, M., Salussola, D., and Thivolet, J. (1982). Attempt to protect rabbits against experimental syphilis by passive immunization. Br. J. Vener. Dis. *49*, 335-337.

Shevchenko, D.V., Sellati, T.J., Cox, D.L., Shevchenko, O.V., Robinson, E.J., and Radolf, J.D. (1999). Membrane topology and cellular location of the *Treponema pallidum* glycerophosphodiester phosphodiesterase (GlpQ) ortholog. Infect. Immun. *67*, 2266-2276.

Singh, A.E. and Romanowski, B. (1999). Syphilis: review with emphasis on clinical, epidemiologic, and some biologic features. Clin. Microbiol. Rev. *12*, 187-209.

Smajs, D., McKevitt, M., Howell, J.K., Norris, S.J., Cai, W.W., Palzkill, T., and Weinstock, G.M. (2005). Transcriptome of *Treponema pallidum*: gene expression profile during experimental rabbit infection. J. Bacteriol. *187*, 1866-1874.

Snapper, C.M., Rosas, F.R., Jin, L., Wortham, C., Kehry, M.R., and Mond, J.J. (1995). Bacterial lipoproteins may substitute for cytokines in the humoral immune response to T cell-independent type II antigens. J. Immunol. *155*, 5582-5589.

Stamm, L.V. and Bassford, P.J., Jr. (1982). Cloning and expression of *Treponema pallidum* protein antigens in Escherichia coli. DNA *1*, 329-333.

Stebeck, C.E., Shaffer, J.M., Arroll, T.W., Lukehart, S.A., and Van Voorhis, W.C. (1997). Identification of the *Treponema pallidum* susp. *pallidum* glycerophosphodiester phosphodiesterase homologue. FEMS Microbiol. Lett. *154*, 303-310.

Stenger, S., Mazzaccaro, R.J., Uyemura, K., Cho, S., Barnes, P.F., Rosat, J.P., Sette, A., Brenner, M.B., Porcelli, S.A., Bloom, B.R., and Modlin, R.L. (1997). Differential effects of cytolytic T cell subsets on intracellular infection. Science *276*, 1684-1687.

Stokes, J.H., Beerman, H., and Ingraham, N.R. (1944). Modern Clinical Syphilology. (Philadelphia: W.B. Saunders).

Strugnell, R.A., Faine, S., and Graves, S. (1984). Response of syphilitic rabbits to reinfection with homologous and heterologous *Treponema pallidum* strains. Infect. Immun. *45*, 561-565.

Sun, E.S., Molini, B.J., Barrett, L.K., Centurion-Lara, A., Lukehart, S.A., and Van Voorhis, W.C. (2004). Subfamily I *Treponema pallidum* repeat protein family: sequence variation and immunity. Microbes. Infect. *6*, 725-737.

Sykes, J.A. and Kalan, J. (1975). Intracellular *Treponema pallidum* in cells of a syphilitic lesion of the uterine cervix. Am. J. Obstet. Gynecol. *122*, 361-367.

Sykes, J.A. and Miller, J.N. (1971). Intracellular location of *Treponema pallidum* (Nichols strain) in the rabbit testis. Infect. Immun. *4*, 307-314.

Sykes, J.A., Miller, J.N., and Kalan, A.J. (1974). *Treponema pallidum* within cells of a primary chancre from a human female. Br. J. Vener. Dis. *50*, 40-44.

Teraki, Y. and Picker, L.J. (1997). Independent regulation of cutaneous lymphocyte-associated antigen expression and cytokine synthesis phenotype during human CD4+ memory T cell differentiation. J. Immunol. *159*, 6018-6029.

Thivolet, J., Simeray, A., Roland, M., and Challut, F. (1953). Étude de l'introderm-reaction aux suspension de treponemes fromoless (souch Nichols pathogene) chez les syphilitiques et les sujects normaux. Ann. Inst. Pasteur (Paris) *85*, 23-33.

Titus, R.G. and Weiser, R.S. (1979). Experimental syphilis in the rabbit: passive transfer of immunity with immunoglobulin G from immune serum. J. Infect. Dis. *140*, 904-913.

Tosca, A., Lehou, J., Hatjivasiliou, M., Varelzidis, A., and Stratigos, J.D. (1988). Infiltrate of syphilitic lesions before and after treatment. Genitourin. Med. *64*, 289-293.

Turner, D.R. and Wright, D.J.M. (1973). Lymphadenopathy in early syphilis. J. Pathol. *110*, 305-308.

Turner, T.B. (1939). Protective antibodies inthe serum of syphilitic rabbits. J. Exp. Med. *69*, 867-890.

Turner, T.B., Hardy, P.H., and Newman, B. (1969). Infectivity tests in syphilis. Br. J. Vener. Dis. *45*, 183-196.

Turner, T.B., Hardy, P.H., Newman, B., and Nell, E.E. (1973). Effects of passive immunization on experimental syphilis in the rabbit. Johns Hopkins Med. J. *133*, 241-251.

Turner, T.B. and Hollander, D.H. (1957). Biology of the Treponematoses. (Geneva: World Health Organziation).

Turner, T.B., Kluth, F.C., McLeod, C., and Winsor, C.P. (1948). Protective antibodies in the serum of syphilitic patients. Am. J. Hyg. *48*, 173-181.

Underhill, D.M. and Gantner, B. (2004). Integration of Toll-like receptor and phagocytic signaling for tailored immunity. Microbes. Infect. *6*, 1368-1373.

Underhill, D.M., Ozinsky, A., Hajjar, A.M., Stevens, A., Wilson, C.B., Bassetti, M., and Aderem, A. (1999). The Toll-like receptor 2 is recruited to macrophage phagosomes and discriminates between pathogens. Nature *401*, 811-815.

Van Voorhis, W.C., Barrett, L.K., Koelle, D.M., Nasio, J.M., Plummer, F.A., and Lukehart, S.A. (1996a). Primary and secondary syphilis lesions contain mRNA for Th1 cytokines. J. Infect. Dis. *173*, 491-495.

Van Voorhis, W.C., Barrett, L.K., Lukehart, S.A., Schmidt, B., Schriefer, M., and Cameron, C.E. (2003). Serodiagnosis of syphilis: antibodies to recombinant Tp0453, Tp92, and Gpd proteins are sensitive and specific indicators of infection by Treponema pallidum. J. Clin. Microbiol. *41*, 3668-3674.

Van Voorhis, W.C., Barrett, L.K., Nasio, J.M., Plummer, F.A., and Lukehart, S.A. (1996b). Lesions of primary and secondary syphilis contain activated cytolytic T cells. Infect. Immun. *64*, 1048-1050.

Voulhoux, R. and Tommassen, J. (2004). Omp85, an evolutionarily conserved bacterial protein involved in outer-membrane-protein assembly. Res. Microbiol. *155*, 129-135.

Walfield, A.M., Hanff, P.A., and Lovett, M.A. (1982). Expression of *Treponema pallidum* antigens in *Escherichia coli*. Science *216*, 522-523.

Ware, J.L., Folds, J.D., and Baseman, J.B. (1979). Serum of rabbits infected with *Treponema pallidum* (Nichols) inhibits in vitro transformation of normal rabbit lymphocytes. Cell Immunol. *42*, 363-372.

Wasserman, A., Neisser, F., and Bruck, C. (1906). Eine serodiagnostische reaktion bei syphilis. Dtsch. Med. Wochenschr. *32*, 745-746.

Weiser, R.S., Erickson, D., Perine, P.L., and Pearsall, N.N. (1976). Immunity to syphilis: passive transfer in rabbus using serial doses of immune serum. Infect. Immun. *13*, 1402-1407.

Wicher, K., Schouls, L.M., Wicher, V., Van Embden, J.D., and Nakeeb, S.S. (1991). Immunization of guinea pigs with recombinant TmpB antigen induces protection against challenge infection with *Treponema pallidum* Nichols. Infect. Immun. *59*, 4343-4348.

Wicher, K. and Wicher, V. (1989). Experimental syphilis in guinea pig. Crit Rev. Microbiol. *16*, 181-234.

Wicher, K. and Wicher, V. (1990). Autoimmunity in syphilis. Immunol. Ser. *52*, 101-124.

Wicher, K., Wicher, V., Nakeeb, S.M., and Dubiski, S. (1983). Studies of rabbit testes infected with *Treponema pallidum*. I Immunopathology. Br. J. Vener. Dis. *59*, 349-358.

Wicher, V., Scarozza, A.M., Ramsingh, A.I., and Wicher, K. (1998). Cytokine gene expression in skin of susceptible guinea-pig infected with *Treponema pallidum*. Immunology *95*, 242-247.

Wicher, V. and Wicher, K. (1977a). *In vitro* cell response of *Treponema pallidum*-infected rabbits. I. Lymphocyte transformation. Clin. Exp. Immunol. *29*, 480-486.

Wicher, V. and Wicher, K. (1977b). *In vitro* cell response of *Treponema pallidum*-infected rabbits. II. Inhibition of lymphocyte response to phytohaemagglutinin by serum of *T. pallidum*-infected rabbits. Clin. Exp. Immunol. *29*, 487-495.

Wicher, V., Wicher, K., Jakubowski, A., and Nakeeb, S.M. (1987). Adoptive transfer of immunity to *Treponema pallidum* Nichols infection in inbred strain 2 and C4D guinea pigs. Infect. Immun. *55*, 2502-2508.

Wigfield, A.S. (1965). Immunological phenomena in syphilis. Br. J. Vener. Dis. *41*, 275-285.

Wooten, R.M., Morrison, T.B., Weis, J.H., Wright, S.D., Thieringer, R., and Weis, J.J. (1998). The role of CD14 in signaling mediated by outer membrane lipoproteins of *Borrelia burgdorferi*. J. Immunol. *160*, 5485-5492.

Wright, D.J.M. and Grimble, A.S. (1973). Why is the infectious stage of syphilis prolonged. Br. J. Vener. Dis. *50*, 45-49.

Young, E.J., Weingarten, N.M., Baughn, R.E., and Duncan, W.C. (1982). Studies on the pathogenesis of the Jarisch-Herxheimer reaction: development of an animal model and evidence against a role for classical endotoxin. J. Infect. Dis. *146*, 606-615.

Young, H., Moyes, A., Seagar, L., and McMillan, A. (1998). Novel recombinant-antigen enzyme immunoassay for serological diagnosis of syphilis. J. Clin. Microbiol. *36*, 913-917.

14

The Oral Spirochetes: Their Ecology and Role in the Pathogenesis of Periodontal Disease

S. C. Holt and J.L. Ebersole*

Abstract

Nearly 350 years ago Antoni van Leeuwenhoek described "living animalcules aswimming… The biggest sort…bent their body into curves going forward..." Thus, the first written descriptions of what we now know to be spirochetes. These prokaryons are found both free-living in nature, and associated with a variety of animals, and humans. Both nonpathogenic and pathogenic species of spirochetes have been identified. Their participation as etiologic agents in human oral disease has been debated for many years. Numerous investigations have reported low numbers of spirochetes in the gingival sulcus at healthy sites, while disease progression with gingival inflammation, increased pocket depth, and a more anaerobic microenvironment correlated with an increase in the numbers and variety (small, medium, large) of spirochetes. Additionally, return to oral health resulted in a decrease in the number of spirochetes, *i.e.* to levels consistent with gingival health. Seminal contributions of Socransky and colleagues demonstrated that the microbial ecology in oral health and periodontitis was comprised of many individual genera and species; however, several specific microbial complexes or consortia appeared to reflect alterations in the clinical state of the site. In particular, one prominent consortium consists of *P. gingivalis, T. forsythia*, and *T. denticola.* This chapter provides an overview of the characteristics of oral spirochetes, and particularly *T. denticola,* as important members of oral biofilms and their relationship to periodontal disease. The content of the chapter also focuses on attempting to summarize the literature with regard to the numerous potential "virulence molecules" expressed by the spirochetes that might be essential for their *in vivo* survival and contribution to the progression of oral disease in susceptible hosts.

*For correspondence email sholt@forsyth.org

1. Introduction

Antonie van Leeuwenhoek recognized spirochetes in human plaque almost 350 years ago [Van Leeuwenhoek (1632-1723)]. He described the following occurring in human "spittle" (Dobell, 1932):

> "…an unbelievably great company of living animalcules aswimming more nimbly than any I had seen up to this time. The biggest sort (whereof there were a great plenty) bent their body into curves going forward … ."

This observation by van Leeuwenhoek was the first of what we recognize today as spirochetes (their anaerobic counterpart being *Treponema*) (Holt, 1978). While the genus *Treponema* represents, for the most part, anaerobic spirochetes, some species of the genus are capable of microaerophilic growth (*T. pallidum*), and some will grow and divide in tissue culture under aerobic conditions (*T. denticola*). A large body of evidence both in the historical literature, and in recent observations has considered these small, helical, rapidly moving bacteria to be indicators of periodontal inflammation and disease. For example, Keyes and Rams (1983) in their early work recognized the presence of oral spirochetes or oral treponemes with the onset of gingivitis and its progression to periodontitis. Their absence from the subgingival environment corresponded with periodontal health, and their presence in large numbers was an indication of gingivitis and active periodontal disease. Armitage et al. (1982) established a positive relationship between the percentage of spirochetes in periodontal pockets and clinical measures of inflammation, increased dental plaque, increasing gingival exudate, bleeding on probing, periodontal pocket depth, and loss of connective tissue attachment. Periodontal therapy (*i.e.* scaling and root planing, periodontal pocket reduction, or the use of selected antimicrobial agents) generally reduced gingival inflammation with a concomitant reduction in the number of spirochetes (Loesche and Laughon, 1982). Simonson and his colleagues (Simonson et al., 1988; 1990) extended these observations and demonstrated a positive correlation between periodontitis and the number of *T. denticola* found in a periodontal pocket. Spirochetes have also been associated with gingival tissues in acute necrotizing ulcerative gingivitis (Courtois et al., 1983; Listgarten, 1965; Riviere et al., 1991), chronic periodontitis (Frank, 1980; Saglie et al., 1982a; 1982b), and aggressive juvenile periodontitis (Carranza et al., 1983).

Employing DNA-DNA checkerboard hybridization techniques, Socransky et al. (1998) and Ximenez-Fyvie et al. (2000) have provided an understanding of the relationship between the distribution of the resident periodontal microbiota and the clinical progression of periodontitis. These investigators characterized a portion of the subgingival microbial ecology in health and periodontal disease. This technology permitted the examination of thousands of plaque samples, and established, with more detail than previously provided, that the subgingival plaque microbiota varied both in number and distribution from healthy sites to gingivitis to periodontitis. The distribution of the microbiota in these varied sites demonstrated some unique features, particularly in a representation of distinct microbial complexes. Socransky et al. (1998) designated these complexes as yellow, orange, green, purple, and red. The red complex, associated with the most destructive periodontal disease sites, consisted predominantly of three bacterial species: *Porphyromonas gingivalis*, *Tannerella forsythia* (formerly *Bacteroides forsythus*), and *T. denticola.*

2. Classification of the oral spirochetes

Paster and Dewhirst discuss the classification of the oral *Treponema* in their chapter on this subject. We provide here only a brief summary for contextual purposes. The order *Spirochaetales* has been divided into three major phylogenetic groups employing 16S rRNA or 16S rDNA sequences. The family Spirochaetaceae includes the genera *Borrelia, Brevinema, Cristispira, Spironema, Spirochaeta*, and *Treponema.* The Leptospiraceae contains *Leptonema* and *Leptospira.* The third proposed family Brachyspiraceae contains the genus *Brachyspira*, formerly known as *Serpulina* (Paster and Dewhirst, 2000). Many members of the genus *Treponema* have been speciated according to 16S rRNA sequence analyses (Paster and Dewhirst, 2000) and their phylogeny is addressed in Chapter 2.

Choi et al. (1994; 1996; 1997), Paster et al. (1991; 1998; 1999), and Dewhirst et al. (2000) have been leaders in the molecular classification of oral bacteria and have contributed to the recognition of a large number of "uncultivated spirochetes" identified by molecular techniques. Employing 16S rRNA-based analyses, these investigators and their colleagues have described and characterized *T. maltophilum* (Wyss et al., 1996); *T. medium* (Umemoto et al., 1997); *T. amylovorum* (Wyss et al., 1997); *T. lecithinolyticum* (Wyss et al., 1999), and *T. parvum* (Wyss et al., 2001). Recently, Wyss et al. (2004) have reported an additional treponeme, *T. putidum*, described as "a highly motile, medium-size, proteolytic but asaccharolytic spirochete cultured from human gingivitis, periodontitis, and acute necrotizing ulcerative gingivitis." Its content of trypsin-like enzymes and other APIZym™ detected enzymes was similar to that of *T. denticola.*

3. Association of oral treponemes with periodontal disease

The relationship of these oral treponemes to the inflammatory events in periodontal disease progression is extensive. *T. denticola* and other members of this group have been associated with gingivitis, chronic periodontitis (Armitage et al., 1982), acute necrotizing ulcerative gingivitis (Chung et al., 1983; Loesche et al., 1982), aggressive periodontitis, refractory periodontitis, and HIV-associated periodontal diseases (Chan and McLaughlin, 2000; Socransky and Haffajee, 1997; Moter et al., 1998; Willis et al., 1999; Dewhirst et al., 2000). Not only are the spirochetes important members of the periodontopathic microbiota, but they also have been associated with endodontic infections. In fact, Miller in 1894 was the first to describe these bacterial species in endodontic infections. Numerous investigators also have observed spirochetes in pulpal samples (Brown and Rudolph, 1957), and in root canals (Thilo et al. 1986; Nair, 1987; Dahle et al., 1993a; 1993b). Van Winkelhoff et al. (1985) reported the presence of spirochetes in the pus taken from periradicular abscesses. Recent reports employing molecular techniques have associated *T. denticola* and other *Treponema* species with infected root canals and endodontic infections (Smallwood et al., 1998; Siqueria et al., 2000a; 2000b, 2001a; 2001b, 2003; Rocas et al., 2001; Jung et al., 2001; Ashimoto et al., 1996; Sakamoto et al., 2001).

Treponema parvum is a small glucuronic or galacturonic acid dependent oral treponeme recently isolated and characterized from human periodontitis and acute necrotizing ulcerative gingivitis patients. As opposed to *T. pectinovorum*, this species has a different size, distinct protein and antigen profile, and cannot grow on pectin as the sole carbohydrate source. Multiple serotypes of both *T. parvum* and *T. pectinovorum* have been described (Wyss, 2001).

Umemoto et al. (1997) isolated a medium sized spirochete from the periodontal pocket of a periodontitis patient. They referred to this small spirochete as *T. medium*. Using nested PCR technology, Willis et al. (1999) were also able to identify seven *Treponema* species, including *T. medium*, in healthy and disease-associated dental plaques. These treponemes were found in all of the disease-associated plaques. Both *T. pectinovorum* and *T. vincentii* were routinely absent in disease-free plaques.

Wyss et al. (1996) developed a novel medium for the cultivation of several of the more fastidious oral treponemes. In addition to the already isolated and identified species of *T. denticola*, *T. pectinovorum*, *T. socranskii*, and *T. vincentii*, these investigators were able to isolate a small treponeme they named *T. maltophilum*. These anaerobic spirochetes have only 1 endoflagellum at each end of the cell. The species grew on a carbohydrate mixture and was distinguished from the other oral treponeme species by its 16S rRNA sequence, protein and antigenic patterns. Siqueira and Rocas (2003), using a molecular genetic approach, investigated the distribution of oral treponemes isolated from primary root canal infections. They employed 16S rDNA-based, nested PCR technology and examined samples taken from 31 infected root canals associated with either asymptomatic or symptomatic apical periodontitis. They found that *T. maltophilum* was the most prevalent found, being detected in 33% of the asymptomatic cases and 50% of the symptomatic ones. *T. lecithinolyticum* was the next most prevalent, followed by *T. amylovorum* and *T. medium*.

4. Spirochete structure

The reader is referred to Holt (1978) for a description of the anatomy of spirochetes. Three basic structures comprise all spirochetes: an outer sheath, a series of endoflagella, and a protoplasmic cylinder. As its name implies, the outer sheath encompasses the entire organism. It covers the cell, and functions both in the protection of the bacteria and in their interaction with host cells and other bacteria (adherence and co-aggregation, respectively). One of the earliest attempts at the isolation of the spirochete outer sheath was that of Masuda and Kawata (1982). The isolated and purified sheath of their strain, *Treponema* sp. E-21, was found to consist of a major 62 kDa hexagonal protein. Other outer sheath isolations and characterizations followed, and all of these isolated sheaths were found to contain at least one major outer sheath protein the molecular mass of which varied depending on the strain and the method of isolation. In *T. denticola* ATCC 33520, the major outer sheath protein (MSP) was found to have a molecular mass of 52 kDa (Umemoto et al., 1989), while Weinberg and Holt (1990), examining the outer sheath of *T. denticola* GM-1, characterized a 64 kDa MSP; other low passage clinical isolates had MSPs with molecular masses of ranging from 60 to 64 kDa. Interestingly, antibody against this protein did not cross-react with another outer sheath protein, described as a 58 kDa protein from *T. denticola* strains SR4 and ATCC 35404. The difference between the 58 and 64 kDa outer sheath proteins was investigated by Fenno et al. (1996), who determined the sequence of the genes coding for these major sheath proteins. It appears that the protein characterized by Weinberg and Holt (1990) from *T. denticola* GM-1 is a homologue of the major outer sheath proteins of the various *T. denticola* strains. Interestingly, the MSPs of several of the other treponeme species are different from those of *T. denticola* strains. For example, *T. pectinovorum* contains a 42 kDa protein (Walker et al., 1997), *T. maltophilum* contains a 62.3 kDa major outer sheath protein (MspA) coding gene (Heuner et al. 1999), and *T. lecithinolyticum* expresses a 65 kDa (MspTL) that is related to the MspA of *T. maltophilum* (Park, et al., 2002). This *mspA* sequence displayed little to no homology with the *msp* gene of *T. denticola* and was only weakly cross-reactive with that of *T. socranskii*. (See Adherence section for a discussion of the role of several of these major outer sheath proteins).

Each endoflagellum (also referred to as axial flagella, axial filaments, periplasmic flagella) is anchored at either end of the periplasmic sheath in a subterminal insertion pore (see chapters by Limberger and Izard, and by Charon et al.). These flagella overlap in the approximate center of the cell, and their number varies depending upon the species (Holt, 1987). Surrounding each endoflagellum is a protein sheath or endoflagellum sheath, which in *T. denticola* is a 38 kDa FlaA protein (Ruby et al., 1997). In addition to the 38 kDa protein, there are three other endoflagellar proteins, FlaB consists of 35 kDa proteins FlaB1, FlaB2, and a 34 kDa protein, FlaB3. Heinzerling et al. (1997) determined the sequence of these endoflagella genes, and Fraser et al. (1998) determined that at least seven of the *fla* genes were similar to those of *T. pallidum*. Functionally, the endoflagella are involved in at least two major activities: motility and the maintenance of cell shape. As proof, Li et al. (1996) were successful in developing a transformation system in *T. denticola* that demonstrated an endoflagella-less mutant was both nonmotile and irregular in morphology.

The cytoplasm is surrounded by the protoplasmic cylinder. It is a "unit" structure, consisting of the cell wall and an underlying cytoplasmic membrane. Pillot and Ryter (1965) have referred to this complex structure as "parietocytoplasmic" membrane, being similar to the classical Gram-negative cell wall-cell membrane (cell envelope). Structurally, this cylinder consists of two electron dense layers separated by an electron transparent layer. Chemical analysis of the protoplasmic cylinder revealed it to contain classical peptidoglycan molecules, including muramic acid and glucosamine.

5. Acquisition of the oral *Treponema*

The acquisition of the oral microbiota occurs at an early age, and is generally felt to result from direct person-to-person contact inasmuch as most of the members of this microbiota are obligately anaerobic and do not survive long periods in an aerobic environment. The interaction of the infant with the primary caregiver results in what appears to be an organized transmission of the oral microbiota to the child. While several of the members of the oral microbiota have been demonstrated to be transmitted at specific times in early childhood development (*i.e. Streptococcus mutans*), the process is less clear with microorganisms related to the subgingival ecology, *e.g.* oral *Treponema*. The oral spirochetes are distinct from those associated with genital and intestinal tracts and free-living spirochetes (Canale-Parola, 1977; Harwood and Canale-Parola, 1984). Loesche (1988) examined plaque from children (aged 3 to 12 years) and found that, in the younger cohort (3 to 5 yrs), spirochetes were observed in plaque approximately 40% of the time while, in older children (6 to 12 yrs), approximately 50% of the time. Importantly, although oral spirochetes were frequently detected, they appeared to represent $\leq$0.5% of the microbial population in healthy children. The species also appear to have a predilection for the subgingival environment, whether due to their anaerobic nature or to their morphological fragility.

6. Co-aggregation of oral *Trepomema*

Several investigators have observed that *T. denticola* not only interacts with host cells but also binds to other bacteria in close proximity (*i.e.* co-aggregation) (Grenier, 1992a; Onagawa et al., 1994; Kolenbrander et al., 1995; Yao et al., 1996). Their ability to co-aggregate with other bacteria (*e.g.*, *P. gingivalis* and *Fusobacterium nucleatum*) in the confines of the periodontal pocket provides an important nutritional and survival advantage to the pocket microbiota. In addition, the recent reports of the importance of plaque biofilms support co-aggregation as an important bacterial defense mechanism against physical/chemical removal by host cells and macromolecules (Socransky and Haffajee, 2002; Haffajee et al., 2003). In a very interesting study, Kolenbrander et al. (1995) examined 22 *Treponema* species (*e.g. T. denticola*, *T. socranskii*, *T. pectinovorum*, and *T. vincentii*) for co-aggregating ability with 7 strains of *Fusobacterium* spp., and 45 strains of other oral bacteria. They found that none of the treponemes was capable of co-aggregating with any of the 45 oral non-fusobacterial species or even with nonoral fusobacterial species; all of the treponemes (oral strains), however, co-aggregated with at least one of the oral *Fusobacterium* spp. Inactivation of the fusobacterial strains with heat (85°C, 30 min) abrogated the co-aggregating ability, while heating of the treponemes did not affect co-aggregation with unheated fusobacteria. Protease K treatment of the fusobacteria, but not the treponemes, also resulted in the loss of co-aggregating ability. *T. denticola* co-aggregation was inhibited by lactose and D-galactosamine, indicative of a lectin-type interaction.

The report of Socransky et al. (1998) suggesting that there are specific bacterial consortia linked to periodontitis (*e.g.* red, orange, yellow green complexes) and their emphasis on *T. denticola* as a major member of "red complex" support the likely importance of co-aggregation as a significant mechanism for the growth and survival of bacterial species within an organized biofilm (Grenier, 1992b; Onagawa et al., 1994).

7. *Treponema denticola* in a pathogenic consortium of periodontal disease

A large number of reports have unequivocally established the fact that the transition from health to gingivitis to periodontitis is marked by a shift in the resident microbiota from a "healthy" biofilm to a "diseased" microbial ecology. Keyes et al. (1978a; 1978b; 1983), Listgarten and Hellden (1978), Lindhe et al. (1980), Loesche and Laughon (1982), Riviere et al. (1995), and Loe et al. (1966) observed in their experimental gingivitis studies that as periodontal clinical measures increase (*i.e.* as tissue inflammation increases), the percentage of Gram-negative bacteria within the sulcus increases significantly. Reversal of this change (back to a predominant Gram-positive microbiota, and a reduction in the number of oral spirochetes) occurs very rapidly after the resumption of oral hygiene. Not only is *T. denticola* a member of this microbiota, but other oral treponemes, including *T. amylovorum*, *T. maltophilum*, *T. pectinovorum*, *T. vincentii*, *T. medium*, and *T. socranskii* are detected (Willis et al., 1999). The oral spirochetes make up a significant percentage (~40%) of the microbial population in dental plaques from untreated subjects with chronic periodontitis (Loesche et al., 1985) and pericoronitis (Weinberg et al., 1986). Severe periodontitis subjects also displayed significantly higher levels of particular *T. denticola* serovars in their deep pockets compared to "periodontally healthy" subjects or those diagnosed with moderate periodontitis (Simonson et al., 1990). Papapanou et al. (1997) employed a DNA-DNA checkerboard hybridization approach to examine the subgingival plaque of Chinese subjects who apparently never accessed dental care. Identical to observations from many other studies, these investigators documented a significant increase in *T. denticola*, *T. forsythia*, and *P. gingivalis* in both deep periodontal pockets, as well as in progressing periodontal sites. Finally, both Listgarten and Lai (1999) and Tanner et al. (1997) have observed an increase in spirochetes in subgingival plaque from failing implants.

Periodontitis is a polymicrobial infection with the three species of *Porphyromonas gingivalis*, *Tannerella forsythia* and *T. denticola* comprising the "red complex" (Socransky et al. 1998) that is highly associated with disease. *T. denticola* and *P. gingivalis* are often found in close proximity to each other in the periodontal pocket. While their individual roles in periodontopathogenesis remain to be determined, it is clear that these three bacterial species are generally found together in the subgingival plaque from active periodontal lesions. This may reflect their ability to enhance co-survival in the complex biofilm, as well as in their cooperation in host cell and tissue destruction. Simonson et al. (1992) observed that *T. denticola* only occurred in the subgingival environment when *P. gingivalis* was present, suggesting an important pathogenic synergism. Supporting the importance of the interaction of *T. denticola* with *P. gingivalis* is the report by Riviere et al. (1996) who employed monoclonal antibodies to selected periodontal bacterial species (*T. denticola*, *P. gingivalis*, *C. rectus*, *E. corrodens*, *T. socranskii*, and the pathogen related spirochete (PROS) to evaluate subgingival plaque from all teeth (except 3rd molars). They found co-infection of the subjects with *P. gingivalis*, which was associated with a distinct periodontal pathology. Kigure et al. (1995), employing immunoelectron microscopy, were able to localize a *P. gingivalis* – *T. denticola* complex in subgingival plaques from periodontitis subjects. These complexes occurred in the deeper aspects of the pockets, at depths >4 mm. Interestingly, *T. denticola* also was observed in what the investigators referred to as the surface layers of the subgingival plaque. In these regions *P. gingivalis* was also closely associated with *T. denticola* but juxtaposed directly below them. These observations may reflect the ability of *P. gingivalis* to supply growth factors for *T. denticola*. Supporting this synergism is the report of Nilius et al. (1993) who observed that cell-free culture filtrates of *P. gingivalis* stimulated the *in vitro* growth of *T. denticola*. The specific contribution of *P. gingivalis* that is important for enhanced *T. denticola* growth remains to be determined. However, it is clear that in deep periodontal pockets (*i.e.*> 4 mm), *T. denticola*, *T. forsythia*, and *P. gingivalis* are a routine consortium in the microbial ecology.

8. Potential *Treponema* virulence determinants

A large number of studies over many years have examined both the quantitative and qualitative aspects of the plaque microbiota. These observations gave rise to the nonspecific and specific plaque hypotheses. The former conveyed that the bacterial "load" was important in the establishment of oral disease, while the latter suggested that the presence of specific bacterial species is the underlying factor in the progression from health to disease. Be it microbial load or bacterial type (*e.g.* species), we now know that selected members of the resident periodontal pocket microbiota either act directly on the host or "trigger" the host to make cytokines and chemokines, which result in inflammation leading to hard and soft tissue destruction in the periodontium. Sampling of a periodontal pocket and microscopic examination of the plaque reveals an enormous mass of bacteria, reaching total numbers of bacteria of 1-2 x 10^8 cells/mg plaque. The number of bacterial taxa in the population have been estimated to include at least 500 different bacterial species, although this may be limited to 75-100 species in any individual oral cavity (personal communication, B. Paster). While many of these bacterial species have been isolated to purity, grown, and biochemically characterized, a significant number have been identified only by molecular techniques (see Choi et al., 1994; 1996; Paster et al., 1991; 1998; 1999; 2001).

In addition to vast numbers and diversity of bacterial species living in the confines of the periodontal pocket, *in vitro* studies have revealed that they have the ability to produce a variety of proteins and end products of metabolism which are detrimental or lethal to selected host cells (see Chapter 15) (Socransky, 1970; DeRenzis and Chen, 1983; Shenker et al., 1982, Fenno et al. 1998a; 1998b). Clearly, if these "virulence factors" are produced *in vivo* during the progression of periodontal infections, they would be detrimental to the host. Evaluation of the histopathologic data from human periodontitis tissues suggests that either these molecules are not produced in situ, or their effects on the host are minimized by the activity of other bacteria in the microbial biofilm, or they are produced in such small amounts that they can be processed and removed (detoxified) by the host. Thus, humans and animals, even during aggressive periodontitis, survive and thrive in the presence of these large bacterial populations. However, recent reports of the potential role of oral bacterial species in a variety of systemic diseases and conditions (heart disease, low birth weight, pulmonary disease, arthritis) have altered concepts of periodontitis and general health.

In vitro studies have demonstrated that *T. denticola* produces a variety of potential factors that could act as virulence determinants during its colonization and contribution to a pathogenic biofilm. These include components of motility and chemotaxis (Lux et al., 2001; Greene and Stamm, 1999; Stamm and Bergen, 1999), putative lipopolysaccharide/endotoxins (Rosen et al., 1999), proteases (Makinen et al., 1987; 1988; 1992; 1994; 1995a; 1995b; 1997; Mathers et al., 1996; Rosen, et al., 1995; Syed et al., 1993), adhesins (Rosen et al., 1999), molecules that trigger chronic inflammatory responses (Pollanen et al., 2000; Kesavalu et al., 1999), hemolysins, and cytotoxic factors (Mathers et al., 1996; Scott et al., 1993; Uitto et al., 1995), and polypeptidases/enzymes that can undermine the integrity of host tissues (Fiehn, 1986; Holt and Bramanti, 1991; Makinen et al., 1995b; Mathers et al., 1996; Uitto et al., 1995). Moreover, recent data linking chronic oral infections with various systemic sequelae suggest that the adverse effects of these factors may not be limited to the oral cavity or the local subgingival microenvironment. The data emerging from these studies remain preliminary. In this section of the review, we will, however, attempt to provide some information relevant to the production and activity of these macromolecules and metabolites on host cells, recognizing that data on their *in vivo* activity are either limited or nonexistent.

9. Motility and locomotion as potential virulence determinants

In spite of the poor resolution of the van Leeuwenhoek microscope, his writings indicate that he more than likely "saw or thought he saw the very rapid movement of small helical animalcules in the spittle [saliva?] of a human" (Dobell, 1932). These bacteria we describe as spirochetes, or oral treponemes. The ability of these bacteria to move very rapidly through the viscous environment of mucosal spaces and body secretions provides them with a distinct advantage relevant to their pathogenic and virulence potential (Canale-Parola, 1978; Prpic et al., 1981, Charon et al., 1992). In some of the earliest electron microscopic localization studies of bacteria in periodontally affected tissues, Saglie et al. (1982a; 1982b), Listgarten (1965), and Listgarten and Socransky (1965) showed that morphological entities resembling spirochetes were found in spaces in the junctional epithelium, in adjacent epithelium, as well as in connective tissue and on the surfaces of alveolar bone. Riviere and his colleagues (1991) demonstrated that spirochetes isolated from dental plaque were capable of translocating across a mouse abdominal wall and invading deeper tissues.

Among the most interesting and evocative observations are those of Greenberg and Canale-Parola (1977), Klitorinos et al. (1993); Pietrantonio et al. (1988), and Ruby and Charon (1998) regarding the movement of *T. denticola* in viscous environments. The rate of locomotion of *T. denticola* appears to be directly related to viscosity and temperature of the environment. In highly viscous fluids (*i.e.* blood, GCF) and at increased temperatures (*i.e.*, 35°C), *T. denticola* moved very rapidly as compared to its motility in water at lower temperatures. Interestingly, both the Lyme disease (*Borrelia burgdorferi*) and syphilis (*Treponema pallidum*) spirochetes have been reported to move more actively in host transudates as a function of their viscosity (Kimsey and Spielman, 1990; Prpic et al., 1981). The chapter by Charon et al. focuses on the motility of treponemes. Recent studies have generated molecular evidence for these early findings. Nonmotile and chemotaxis mutants of *T. denticola* were found to be defective in penetrating tissue layers. This activity was shown to be related to the *cheA* gene coding for a critical kinase in the chemotaxis pathway and the *dmcA/B* genes that are translated into the two chemoreceptors of this species (Lux et al., 2001). Moreover, human oral treponemes have been shown to be chemoattracted to rabbit serum. *T. denticola* showed a dose response to the serum that appeared related to 65 kDa and 70 kDa rabbit albumin-binding polypeptides located in the outer envelope of this species (Umemoto et al., 2001). By enhancing motility and contributing to a more efficient mechanism for penetrating epithelium and gingival connective tissue, these relationships might be relevant to the pathogenic potential of the treponemes in the subgingival sulcus bathed by gingival crevicular fluid (GCF).

10. Lipopolysaccharide/lipooligosaccharide (endotoxin) as a virulence determinant?

One of the major, and most studied, virulence determinants of Gram-negative prokaryotes is lipopolysaccharide (LPS). The various structures and chemistry of the LPSs can be found in any biochemistry and/or microbiology text. There has been much controversy over whether, in fact, spirochetes contain a LPS. Despite the studies of several investigators that described what might be considered a Gram-negative LPS equivalent (Westphal and Jahn, 1965; Yotis et al., 1995; Kurimoto et al., 1990; Dahle et al.,1996; Walker et al., 1999a), the consensus is that typical LPS is not present in oral treponemes. Schulz et al. (1998) described a new type of outer membrane lipid, separate from LPS, that was extracted from *T. denticola* using a protocol normally used for LPS. Importantly, the recently published *T. denticola* genome sequence revealed none of the enzymes necessary for synthesis of traditional Gram-negative LPS components, including KDO and Lipid A (Seshadri et al., 2004).

Nonetheless, biological activities have been associated with various extracts of oral spirochetes. Gopalsami et al. (1993) isolated outer sheaths from *T. denticola* strains 35405, 33521, and 35404. Incubating these various sheaths with the shafts, radii and ulnae of 19-day old fetal rats resulted in a significant increase in Ca-release as a function of sheath (outer membrane) concentration, suggesting that some "...LPS-like material is present in the outer membrane of *T. denticola* that may be responsible for bone resorption ..." Kesavalu et al. (2002) examined biological properties of the purified *T. pectinovorum* "LPS" of Walker et al. (1999a). As measured by the LAL assay, the *T. pectinovorum* "LPS" was approx. 4000 times less endotoxic than *E. coli* 055:B5 LPS. However, in an *in vivo* endotoxicity assay, infection of LPS-responder mice with live *T. pectinovorum* induced 100% lethality within 12 hrs in animals that had been sensitized with N-acetyl-galactosamine. The purified "LPS preparation" produced neither endotoxic symptoms nor lethality. Following stimulation with *T. pectinovorum,* HGFs secreted detectable levels of IL-1β, GMCSF, IFNγ, IL-6, IL-8, and MCP-1. In contrast, the purified "LPS" induced only IL-6, IL-8 and MCP-1 in amounts that were substantially lower than HGFs that were exposed to live *T. pectinovorum* (Nixon et al., 2000). A preparation from *T. denticola* increased tartrate-resistant acid phosphatase-positive (TRAP) cells, while increasing expression of osteoclast differentiation factor and decreasing osteoprotegerin mRNA in a mouse calvaria and bone marrow co-culture system. Additionally, this preparation increased expression of message for MMP-3, -8, -9, -10, -13, and -14. These results suggest that a component from this oral treponeme can contribute to both osteoclastogenesis and bone loss, as well as to altered soft tissue integrity noted in progressing periodontitis (Choi et al., 2003). A glycoconjugate from *T. medium* diminished the LPS-induced activation of human monocytes. Additionally, this material altered NFκB activation through TLR4 engagement, blocking binding of LPS to CD14 and LBP, and nitric oxide production related to TLR activation. The inhibitory effects were related to the lipophilic portion of the macromolecule and support the notion that the oral treponemes may have a general strategy of down-regulating host responses in their environment (Asai et al., 2003a; 2003b). These glycoconjugates also showed negligible endotoxic and immunobiologic properties (Hashimoto et al., 2003). Therefore, while the oral treponemes appear to produce lipids that have some physical and biological properties of LPS, the biological activities of these macromolecules are rather limited, and may, in fact, provide a strategy for the treponemes to "fly under the radar" of the host response.

11. Triggers of inflammation as virulence determinants

A major contributor to the tissue and bone destruction observed in periodontal disease progression more than likely comes about through the host itself. Selected members of the pathogenic microbiota (*i.e.* members of the red complex) "trigger" or activate the host to secrete proinflammatory cytokines and chemokines. When *T. denticola* and *T. pectinovorum* were incubated with HGFs, there was the rapid increase and secretion of a large number of these proinflammatory cytokines/chemokines, including IL-1, IL-6, IL-8, and IFNγ. The levels of IL-8, the chemokine responsible for the activation of PMNs, were significantly lower for *T. denticola* than *T. pectinovorum*. Deng et al. (2001) postulated that this low level of induction of IL-8 by *T. denticola* might be due to the production of proteases, such as CTLP. *T. denticola*-stimulated cultures were devoid of IL-8 unless a chymotrypsin inhibitor, chymostatin, was added to the cultures. However, the K1 dentilisin mutant of *T. denticola* also failed to induce IL-8. *Porphyromonas gingivalis*, another proteolytic periodontopathogen, has also been reported to attenuate the host inflammatory response through degradation of IL-8 (Huang et al., 1998; Steffen et al., 2000). *T. vincentii* and *T. medium* induced IL-8 production by human gingival epithelial cells (HGEC), apparently through engagement of the TLR2 receptor (Asai et al., 2003a,b). Similarly, treponemal membrane phospholipids inhibit pathogen-associated molecular pattern (PAMP) responses initiated through Toll-like receptors (TLR). The lipid, identified as a phosphatidylglycerol derivative, reduced immune responses and proinflammatory cytokine production in a murine model. Thus, these somewhat unique molecules of the treponemes may interfere with innate immune responses of the host (Hashimoto et al., 2003a).

The oral treponemes have also been shown to alter the functions of other cell types, critical for maintenance of homeostasis of the periodontium. Sonicates of *T. lecithinolyticum*, component(s) of which were heat-labile, stimulated osteoclast differentiation in a co-culture system via a PGE_2-dependent pathway (Choi, 2001a). The ability of *T. denticola* to reduce viability of periodontal ligament epithelial (PLE) cells has also been reported. The results demonstrated that while PLE cells were somewhat more resistant than other cell types, about 50% underwent an apoptotic event. The microorganism transiently activated ERK1/2 related to cell proliferation and both JNK and p38 kinases leading to apoptosis. Thus, it appears that *T. denticola* has the ability to control PLE proliferation and survival and may contribute to breaching the natural cellular barrier in the sulcus (Leung et al., 2002).

The majority of data currently available suggest that, rather than the pathogenic microbiota producing virulence factors, which have a direct effect on the host, components of the microbiota (*e.g.* LPS, secreted proteins, OMPs, etc) bind to host cells, up-regulating specific destructive inflammatory mediators in periodontal disease. The net result of induction of pro- and anti-inflammatory molecules, chronically, in the subgingival milieu tips the balance between health and tissue destruction. More recently, this concept has been increasingly linked to genetic control of phenotypic expression of inflammation as a determinant of the variation in periodontal disease severity.

12. Adherence of oral *Treponema*

The ability of a microorganism to maintain itself in an ecological niche requires a number of capabilities. It must be able to counter physical removal from the host, to "hide from" or evade the numerous chemical and biological defenses produced by other microorganisms or the host in the niche, and most importantly, to replicate using available nutrients in the environment. The ability of a microorganism to remain associated with a specific niche is generally associated with its ability to adhere to a structure in the environment. In animal or human hosts, this ability is related to attachment to mucosal cells and tissues, and in some cases, after this initial attachment, to the capacity to penetrate into or between selected host cells. Cytological and histochemical analyses have revealed that *T. denticola* favors an ecological niche in proximity to the junctional epithelium (Olsen 1984; Sela et al., 1987; Weinberg and Holt, 1990; Holt and Bramanti, 1991; Uitto et al., 1995; Leung et al., 1996; Thomas, 1996). *T. denticola* and the other oral treponemes contain significant amounts of outer sheath-associated proteases (see below); their proximity to epithelial cells must be important to the ability of these bacteria to penetrate into the deeper tissues of the host.

Most studies on the interaction of oral treponemes with host cells have involved *T. denticola* and human gingival fibroblasts (HGFs), human epithelial cells (*e.g.* KB, HEp-2) and, more recently, human umbilical vein endothelial cells (HUVEC). The interactions of *T. denticola*, *T. socranskii*, *T. vincentii*, *T. medium* and *T. pectinovorum* with respect to selected host cells also have been studied (Peters et al., 1999; Fitzgerald et al., 1977; Keulers et al., 1993a; 1993b; 1993c; Olsen, 1984; Reijntjens et al., 1986; Walker et al., 1999b; Asai et al., 2003a; 2003b). *T. denticola* was noted to be capable of adhering to a variety of HGFs and epithelial cells (Weinberg and Holt, 1990; Keulers et al., 1993a; 1993b; 1993c; Ellen et al., 1994a; Haapasalo et al., 1996), as well as to extracellular matrix proteins, and to the polysaccharide hyaluronan (hyaluronic acid) (Dawson and Ellen, 1990; Haapasalo et al. 1991). Weinberg and Holt (1990) also demonstrated that *T. denticola* adhered to HGFs *in vitro* under both aerobic and anaerobic environmental conditions. These interactions were lectin-mediated with an affinity for galactose and mannose. Proteinase K treatment resulted in an inhibition of adherence of the spirochetes to the HGFs; however, trypsin did not interfere with this interaction, suggesting the presence of a trypsin-resistant adhesin.

The majority of the *T. denticola* strains studied also appear to adhere tenaciously to buccal epithelial cells (ECs) and to purified fibronectin (Ellen et al., 1994b). Anti-fibronectin antibody interfered with, but did not abrogate, the adherence to HGFs (Ellen et al., 1994b), indicating the involvement of other matrix proteins in these interactions. *T. denticola* also adheres to laminin, type IV collagen, gelatin, and to fibrinogen. Treatment of *T. denticola* with the sulfhydryl reagent, *p*-chloromercurybenzoic acid resulted in an 80-90% reduction in attachment to HGFs. Heating of the spirochetes at 70°C for 10 minutes resulted in a 70% decrease in attachment to laminin and a 40-50% reduction in attachment to fibrinogen. In the presence of a mixed glycosidase, there was a decrease in attachment to HGFs by only 20%, indicating that the attachment to HGFs and ECs more than likely occurs via different proteins and different binding sites involving protein-sulfhydryl groups and/or carbohydrate residues (Haapasalo et al., 1991). Binding of *T. denticola* to both HGFs and ECs occurs through the recognition of sialic acid residues on the host cell surface (Keulers at al, 1993c). Walker et al. (1999b) also found that *T. denticola* adhered to, and caused pathology for HEp-2 epithelial cells. These investigators compared the interaction of *T. denticola* and *T. pectinovorum* with HEp-2 ECs. They found that the two treponemes attached to the ECs, but they did not compete for the same binding sites, suggesting the presence of different epithelial cell receptors. *T. denticola* (and their components, *i.e.* outer sheath) were also capable of adhering to collagen-coated beads (Battikhi et al., 1999). These beads, when incubated with HGFs were rapidly phagocytosed. Finally, *T. denticola* binds in higher numbers to fibronectin, laminin, collagen type I, gelatin, keratin, and lactoferrin than *T. vincentii*. Both species bound to the heparin I/fibrin I domain of fibronectin. *T. denticola* strains expressed similar outer membrane profiles, including CTLP, trypsin-like proteases and proline iminopeptidases, which correlated with co-aggregation with both *P. gingivalis* and *Streptococcus crista* (Edwards, 2003). Thus, *T. denticola* and selected other oral treponemes demonstrate the production of adhesins with some specificity for host cells and other bacteria, thus providing them the necessary characteristic to remain within the microbial biofilm.

In an interesting observation, Carranza et al. (1997) observed that the adherence of several oral treponemes to buccal ECs was a function of the ratio of bacteria to ECs. At low confluence (hence actively dividing), greater numbers of bacteria bound to the ECs than when challenged at high confluence. Olsen (1984) also demonstrated a relationship between the "age" of ECs and *T. denticola* binding. Employing electron microscopy, these authors observed that *T. denticola* bound specifically to ECs that had just completed mitosis. Treatment of these epithelial cells with hyaluronidase resulted in a significant decrease in *T. denticola*-EC interactions, suggesting the involvement of hyaluronan as a preferential receptor. Hyaluronan is high-molecular weight polysaccharide present in extracellular matrix, and is found in large amounts in stratified ECs of the oral mucosa and skin. The polysaccharide consists of repeating units of glucuronic acid and N-acetyl-D-glucosamine. It is also found in the stratified epithelium of the oral mucosa where it forms a hydrated gel maintaining the integrity of the intracellular spaces possibly functioning in the transport of nutrients into host cells.

Hyaluronan is also the major glycosaminoglycan produced by periodontal ligament ECs, accounting for almost 89% of all of the glycosaminoglycans produced. In an interesting series of observations regarding the role of hyaluronan as a potential specific EC receptor, Haapasalo et al. (1996) reported that *T. denticola* ATCC 35405 bound to hyaluronan but not to chondroitin-4-sulfate or dermatan sulfate proteoglycan. When *T. denticola* was treated with hyaluronan, it abrogated further interaction with the molecule. Heating *T. denticola*, or treating the bacteria with Na-periodate, phenylmethylsulfonyl fluoride, or *p*-chloromercurybenzoic acid, all blocked binding to hyaluronan. Importantly, all of these treatments also inhibited the chymotrypsin-like activity of *T. denticola*. Hyaluronan was also a strong competitive inhibitor for EC binding. As discussed above, *T. denticola* possesses a major 53-58 kDa outer sheath protein or outer membrane protein (OMP). This protein was capable of interacting with such mammalian proteins as laminin and fibrinogen. Uitto et al. (1995) have hypothesized that the 53 kDa OMP of *T. denticola* integrates into the host cell plasma membrane and functions in the transport of bacterial macromolecules into the host. *T. denticola* has also been shown to produce a complex of proteins, referred to as the outer membrane-associated chymotrypsin-like protease complex (CTLP), which is found in high concentration in the outer sheath and migrates on SDS-PAGE as a proteolytically-active doublet (unheated) with a molecular mass of 95 kDa. Heating and SDS-PAGE analysis results in the separation of the 95 kDa complexes into three fragments: a large 72 kDa fragment, and two smaller peptides migrating at molecular masses of 43 and 38 kDa (Correia et al., 2003; Ishihara et al., 2004). Haapasalo et al. (1996) reported that the binding of hyaluronan also occurred via the CTLP complex (purified protein).

Peters et al. (1999) observed the adherence and penetration of several oral treponemes (*T. denticola* strains GM-1, MS25, 33404, 33405, and 35531, *T. socranskii* subsp. *buccale* 35534, subsp. *paradis* 35535, subsp. *socranskii* 35536, and *T. vincentii* 35580 and the nonoral *T. phagedenis* biotype Reiter) when incubated with HUVEC (human umbilical vein endothelial cells) monolayers. All the bacteria except *T. phagedenis* attached to the monolayers in a dose-and time-dependent manner. Of those bacteria that did attach, all adhered to the endothelial cells at approximately the same rate, with maximum interaction with the cells at approx. 2 hours. Interestingly, electron microscopic observation revealed that treponemes attached to the endothelial cells both by their tips, and along their length. Some of the treponemes appeared to be closely associated with the endothelial microvilli. Penetration of the oral treponemes into the endothelial cells also occurred in a time dependent fashion, with the maximum number of bacteria found "inside" the cells after an overnight incubation. At 12 hr post-inoculation, approx. 5 to 6% of the input bacteria appeared to penetrate the HUVEC.

A large number of studies focusing on the effects of the oral treponemes on host cells have been performed with *T. denticola* and HGFs, ECs, HUVECs, PMNs, macrophages, lymphocytes, or red blood cells. Incubation of HGFs with *T. denticola* resulted in significant cytopathology: the fibroblasts became deformed, formed membrane blebs, rearranged their F-actin and detached from the growth substrate. The cells lost their proliferative capabilities and eventually died (Ellen et al., 1994a). Many other investigators have observed similar results. Recently, the

T. denticola component thought to be responsible for HGF cytoskeletal changes and pathology has been isolated and identified: the MSP described above. MSP is trypsin-resistant and appears to interfere with Ca^{++}-signaling by uncoupling store-operated channels from the host's endoplasmic reticulum and retarding both Ca^{++}-ion release and influx (Wang et al., 2001). The effect of oral treponemes on host cells is discussed extensively in the chapter by Ellen.

13. Proteases as virulence determinants

As an asaccharolytic microorganism, *T. denticola* produces a range of proteases and peptidases (trypsin-like protease, proline aminopeptidase, proline iminopeptidase) that have been reported to be active in *in vitro* host cell studies (Ohta et al., 1986; Makinen et al., 1992; 1994; 1995a; 1995b; 1996; 1997; Rosen et al., 1994; 1995; 1999; Makinen and Makinen, 1997; Mikx, 1991; 1997; Ishihara et al., 1998). Most of these secreted proteins were localized to the outer sheath or periplasmic space. As is the case with other treponemal virulence factors that have been characterized *in vitro*, the *in vivo* functions of these molecules remain to be determined.

The proline iminopeptidase may function in the termination of the peptidolytic cascade. The surface-associated proteases also displayed substrate specificity toward fibrinogen, and were susceptible to sulfhydryl group reagents. Rosen et al. (1995) using specific antibody in immunoblotting experiments, were able to identify 160, 190, and a 270 kDa outer sheath proteases in *T. denticola* ATCC 35404. These proteases were multiple forms of the 91 kDa phenylalanine protease (PAP), which was capable of degrading keratin but not native actin (Rosen et al., 1994). Biochemical analysis and inhibition studies resulted in the classification of PAP as a serine protease. Recently it has been shown that *T. denticola* can degrade Bio-Gide membranes used in guided-tissue regeneration of the periodontium. This activity appears to reside in the phenylalanine protease and could reflect the capacity of this species to degrade collagen membranes used clinically for guided-tissue regeneration (Sela, 2003). Recently, the *opdB* gene of *T. denticola* has been shown to produce a surface-associated oligopeptidase B family member. A strain lacking this enzyme has a slower growth rate, although it remains to be determined if this enzyme contributes to *T. denticola*-host cell or interbacterial interactions (Fenno, 2001).

Heuner et al. (2001) compared the proteolytic activity of 11 treponemal strains for casein, gelatin, fibronectin, and synthetic substrates. These results showed that all strains had some proteolytic capacity for one or more of these substrates, although variations in the distribution of trypsin-like and chymotrypsin-like activities were observed. The CTLP (chymotrypsin-like protease) complex is capable of lysing sheep RBCs (Uitto et al., 1995; Fenno et al., 1998a). The enzyme has broad proteolytic activity, functioning in the hydrolysis of fibrinogen, transferrin, α1-antitrypsin, IgA, IgG, gelatin, serum albumin, laminin, and substance P (Makinen et al., 1995b). The molecular mass of the enzyme, called dentilisin and encoded by the *prtP* gene, is 72 kDa. This protease appears to function in the structural organization of the outer sheath. The dentilisin is related to the subtilisin of *Bacillus subtilis*. Recent studies have identified paralogous proteases in a number of the oral treponemes, which contribute to the proteolytic phenotype of the species and interact in the host environment to undermine tissue integrity (Correia, 2003). *T. denticola* has the ability to disrupt transepithelial resistance reflecting penetration of the epithelial cell barriers. This was not due to cytotoxicity but appeared to reside in the ability of dentilisin to alter cellular tight junctions (Chi, 2003). The dentilisin of *T. denticola* also appears to be involved in fimbria-mediated coaggregation of this species with *P. gingivalis* (Hashimoto, 2003b).Together with the 53 kDa Msp (see above), the CTLP is involved in transport, (Fenno et al., 1998b), as well as in the adherence of *T. denticola* to the EC plasma membranes. Exposure of *T. denticola* to anti-CTLP IgG resulted in the inhibition of attachment of the treponemes to the ECs (Leung et al., 1996). The protease inhibitors phenyl methylsulfonyl fluoride (PMSF), disio propylfluorophosphate (DFP), and N-tosyl-L-phenyl alanine chloromethyl ketone (TLCK) all inhibit the activity of CTLP, suggesting that it functions as a true chymotrypsin-like protease. The protein also appears to have a significant toxic effect on ECs; incubation of ECs with either whole *T. denticola*, or purified CTLP complex resulted in plasma membrane blebbing and cell death. The CTLP was also able to degrade endogenous fibronectin in both ECs and in HGFs. The activity of CTLP on ECs might interfere with cell adhesion and locomotion. Uitto et al. (1995) found that interference with adhesion to fibronectin resulted in apoptosis of ECs. The association of the 53 kDa Msp with CTLP might also play a structural role in the organization of the *T. denticola* outer sheath (Ishihara et al., 1998; Bian et al., 2005). The ability of CTLP to hydrolyze such polymers as substance P, bradykinin, and angiotensin I might be important in the degradation of bioactive peptides and other peptidolytic catalyses in the inflamed periodontium.

In summary, the array of proteases produced by oral treponemes, and *T. denticola* in particular, likely reflects the basic physiological processes required for survival and multiplication of these species within the host environment. Degradation of barrier structures, cellular components and protective macromolecules of the host may represent collateral damage from the bacterium's efforts to satisfy its nutritional requirements. This provincial gathering of nutritional materials also may contribute to the overall features of the pathogenic biofilm by contributing to a food network and synergistic host evasion capacity of the biofilm ecology.

14. Hemolytic activity as a virulence determinant

The erythrocyte (RBC) is essential in providing iron for growth, multiplication and survival of the periodontal pocket microbiota at inflamed sites. Nearly all bacteria require iron for metabolism (excluding *Borrelia burgdorferi* and several lactobacilli which require a different e^- transfer molecule), and this iron can be obtained from the RBCs in periodontal pockets. The iron atom is specifically removed from the hemoglobin molecule and used by the periodontal microbiota. Iron (in the form of hemin) has been shown to be an important virulence up-regulator of selected periodontopathogens. The surface of the RBC (its plasma membrane) expresses a number of plasma membrane associated surface carbohydrates, which are recognized by *P. gingivalis* and *T. denticola*. *T. denticola*, for example, possesses both hemagglutinating and hemoxidizing activities (Grenier, 1991; Chu et al., 1994). Both of the RBC-interactive functions, hemoxidation and hemolysis, are upregulated under iron-limiting growth conditions. Initially, binding to the RBC occurs via a D-glucosamine–containing molecule followed by perturbation of the RBC plasma membrane. Chu et al. (1994) have characterized a 46 kDa protein from *T. denticola*, which displayed both hemoxidation and hemolytic activity. The destruction of the RBC occurred via a two step process: after attachment to the RBC plasma membrane there was a rapid hemoxidation of the hemoglobin molecule closely followed by hemolysis of the RBC. The enzyme appears capable of producing pores in the RBC membrane, an activity similar to that of MSP as described above. This protein, because of its cysteine desulfhydrase activity has been named cystalysin (Chu et al., 1994; 1997; 1999a; 1999b). Cystalysin attaches to the RBC plasma membrane, oxidizing hemoglobin to methaemoglobin and to sulphaemoglobin (Kurzban et al. 1999). Hemoxidation also results in the degradation of spectrin, a major RBC structural protein. Recently, these activities have been shown to be due to the production of H_2S from cysteine by *T. denticola* creating a toxic environment that rapidly lyses RBCs (Chu et al., 1999a). Chu and colleagues (1999b) have extended these initial observations by describing the enzymatic pathway for production of H_2S by *T. denticola*. They have determined that the ability to break down glutathione, the major human thiol molecule, involves a sequential three-step procedure requiring a gamma-glutamyltransferase (GGT; Chu, 2003), a cysteinylglycinase (CGase) and finally cytstalysin (Chu et al., 1997) originally identified as a hemolysin of this species. The resulting products, including H_2S at tissue and protein toxic levels, NH_3 (*i.e.* raise pocket pH), and pyruvate (growth nutrient), are examples of how physiologic by-products of *T. denticola* can enhance the pathogenic process, in effect, serving as virulence determinants.

Three additional surface-associated proteins from *T. denticola* also are active in RBC lysis. These include Msp, the CTLP complex, and a phospholipase C-like enzyme. Finally, recent studies have shown that *T. denticola* produces outer membrane hemin binding proteins that appear to contribute to acquisition of iron under low iron conditions. However, mutants for these proteins would effectively grow in the presence of lactoferrin (Xu et al. 2002). These findings, as a whole, support that *T. denticola* produces a repertoire of factors that enhance its ability to acquire iron, a critical nutrient, in the highly competitive environment of the microbial biofilm.

15. Enzymatic effects on host tissues as virulence determinants

T. denticola has also been found to produce several enzymes capable of degrading host macromolecules including extracellular matrix proteins (Uitto et al. 1988a,b). These proteases generally are found associated with extracellular outer sheath vesicles (Rosen et al., 1995). Treatment of gingival fibroblasts and periodontal ligament cells with sonicates, whole killed cells, or outer membrane fractions of *T. lecithinolyticum* induced MMP-2, which contributed to elevated collagenolytic activity and may help to degrade extracellular matrix in infected sites (Choi, 2001a; 2001b). Soluble extracts from *T. denticola*, *T. forsythia* and *P. gingivalis* adversely affected alkaline phosphatase activity of periodontal fibroblasts and altered the collagen content of the cultures. These results supported the ability of these oral pathogens to impact tissue regeneration by activating host inflammatory responses and by suppressing the growth and differentiation of fibroblasts (Marquez San Miguel, 2003). Recently, the pectate lyase of *T. pectinovorum* has been cloned, sequenced, and expressed in *E. coli*; this enzyme could target basement membrane integrity and contribute to alterations in the barrier functions of host tissues in the periodontium (Walker, 2003). Finally, Sela et al. (1997) and Ding et al. (1996), extracted a variety of proteins (*i.e.* 53 kDa MSP) from the outer sheath, interacted them with human PMNs, and ascertained the release of PMN-lysosomal enzymes. Lysozyme, the active forms of matrix metalloproteases 8 and 9, cathepsin G and elastase were activated and released, but β-glucuronidase and lactic acid dehydrogenase were not. This suggests that selected outer sheath proteins activate and propagate the release of selected enzymes from the PMNs.

16. Host antibody responses to *Treponema denticola* and oral treponema in periodontal disease

While IgG antibody responses to oral spirochetes (*e.g. T. denticola, T. vincentii*) have been reported, the data have generally shown normal to below normal levels in subjects with periodontal disease (Takahashi et al., 2001; Aukhil et al., 1988; Jacob et al., 1982; Lai et al., 1986; Murayama et al., 1988; Sela et al., 1987). A detailed study of serum antibody responses to various spirochetes showed a higher frequency of seropositive juvenile periodontitis patients, while those patients classified as having severe periodontitis (SP) had a uniformly low level or absence of antibody to the treponemes, although treponemes were cultivated from the SP patients (Smibert and Burmeister, 1983). The authors suggested that a significant exposure to periodontitis-associated treponemes may lead to a specific immune suppression. Therefore, while treponemal species are frequent in subgingival plaque and comprise a major proportion of some plaques, there does not appear to be an elevation in systemic antibody responses to the *Treponema* that can be isolated. Similarly, it was reported in subjects with rapidly progressive periodontitis that serum antibody to *T. denticola* was significantly lower than in matched controls (Smibert et al., 1984). Recently, serum IgG antibody to *T. denticola* was reported to be significantly decreased in periodontitis patients on maintenance who smoked compared to untreated patients. These results combined with other observations suggest that smoking may interrupt the immune maturation that has been shown to occur following treatment of periodontitis (Mooney, 2001). Finally, a study of patients with aggressive periodontitis indicated an elevation in total salivary IgA and IgA antibody to *T. denticola*, while the distribution of subclasses indicated that the IgA2 subclass was significantly decreased (Hagewald, 2002).

A study by Falkler et al. (1987) demonstrated that treponemes comprised a substantial proportion of the subgingival plaque in subjects with acute necrotizing ulcerative gingivitis (ANUG). An early study by Wilton and colleagues (1971) indicated that ANUG patients did not present with elevated antibody to various oral microorganisms suspected as pathogens in the disease. We have found similar results in patients with periodontitis and ANUG. Only a single report described an elevation in serum antibody directed against an "intermediate-sized spirochete" in ANUG patients (Linke et al., 2003). Therefore, while treponemal species are frequent in subgingival plaque and comprise a major proportion of some plaques in periodontitis and ANUG (Gmur *et al*, 2004; Papapanou et al., 2002), there does not appear to be an elevation in systemic antibody responses to the cultured treponemal species. A recent investigation into the potential relationship of pathogen-related spirochetes (PROS) to ulcerative gingivitis indicated that subjects with detectable PROS in subgingival plaque did not demonstrate a significantly increased frequency of antibody to PROS antigens as compared to PROS-negative subjects. In contrast, patients with ANUG demonstrated antibodies to *T. pallidum* antigens lacking in periodontitis patients (Riviere et al., 1991). We have also examined serum IgG antibody to a battery of oral pathogens including *T. denticola* in HIV^+ patients. While some patients with necrotizing ulcerative periodontitis (NUP) exhibited elevated antibody to suspected periodontopathogens, the antibody level to *T. denticola* isolates was low, and selected HIV^+ patients had minimal antibody to all of the oral microorganisms (unpublished data). More recent data suggested that total salivary IgA concentration is elevated in HIV^+ patients vs. controls, regardless of periodontal health. However, specific IgA antibody to selected oral bacteria was noted with highest levels in NUP and lowest levels in patients with healthy gingiva (Ebersole et al., 1991).

Apatzidou and Kinane (2004) examined human antibody responses to various oral bacteria, including *T. denticola*, after treatment by a variety of dental procedures. Their results showed that both scaling and root planing by quadrant or using a full-mouth same-day scaling lead to similar reductions in antibody levels. In some cases, this treatment also appeared to increase antibody avidity to selected bacteria. Similar types of studies of systemic humoral immune responses to oral bacteria have also been initiated in investigating oral-systemic disease linkages. Madianos et al. (2001) examined maternal IgG and fetal IgM antibody to a battery of oral bacteria. The results suggested that a lack of maternal IgG antibody to specific periodontal pathogens, including *T. denticola*, was associated with an increased risk of premature birth.

17. Oral treponema and animal models of infection and pathogenesis

The majority of animal models of infectious diseases have concentrated on an understanding of the types and function of specific microorganisms associated with human disease processes. These studies have, for the most part, attempted to satisfy the Koch-Henle postulates formulated to relate specific microorganisms to specific diseases. Relatively few studies (in comparison to the plethora of human-animal studies of other microbial diseases) have concerned themselves with the pathogenesis of human oral infections. However, these investigations provide important contributions to our understanding of the role of specific groups of microorganisms in oral diseases. Notable among these animal models are those involving the use of rats, mice, and hamsters in studies of dental caries (Linke et al., 2003; Koo et al., 2002); the nonhuman primate and dog models in describing parameters of gingivitis and periodontitis (Shou et al., 1993; Madden and Caton, 1994; Ebersole et al., 1991; Holt et al., 1988; Persson et al., 1994; Kornman et al., 1981); and most recently, the mouse model for determining the pathogenic mechanisms used by specific oral pathogens (Kesavalu et al., 1991; 1992; 1996; 1997a; 1997b, 1998; 1999a; 1999b; 2003; Genco et al., 1991; Gibson and Genco, 2002; Gonzalez et al., 2003).

The majority of animal models of oral infection have focused on induction of disease via the naturally occurring oral microbiota in the host species (*i.e.*

following ligation of the teeth) or after specific implantation of *P. gingivalis* or *A. actinomycetemcomitans* (Madden and Caton, 1994). At this point in time, only the nonhuman primate model of periodontitis has addressed the potential contribution of *T. denticola* to progressing periodontitis (Giardino et al., 1996; Moritz et al., 1998). Specifically, it is clear that *T. denticola* and other oral treponemes are detected in the subgingival microbiota of nonhuman primates. The distribution of these morphotypes/species appears to parallel the development of gingivitis and periodontitis in human populations. Interestingly, the immune response to the treponemes was rather low in the primates, as has been noted in humans (Moritz, et al., 1998; Mangan et al., 1982). According to Kesavalu et al. (1999b), this response is likely not due to a lack of antigenicity of the microorganisms but may be a result of the commensal nature of these bacteria in the oral microbial ecology.

The more recent animal model studies of periodontal pathogens and potential mechanisms that could lead to periodontal tissue destruction have employed a murine abscess model (Ebersole et al., 1995; Genco et al., 1991; 1994; Kesavalu et al., 1996; 1997a; 1997b; Schifferle, et al., 1993). The murine model has been a major contributor to investigating the virulence potential of both non-oral and oral bacteria (see, for example, Madden and Fujiwara, 1982). Various research groups have been responsible for the numerous studies employing the murine abscess model to study the virulence potential of oral pathogens, including *P. gingivalis* and *T. denticola* (Ebersole, et al., 1995; Evans et al., 1992; Genco et al., 1991; Loesche, 1988; Neiders et al., 1989; Taubman et al., 1984). Kesavalu and colleagues (1991; 1992a; 1992b; 1996; 1997; 1998; 1999a; 1999b) have made the major contribution to our understanding of the *in vivo* effects of the oral treponemes alone or as a polymicrobial infection with other periodontopathogens in murine models. Challenge of mice with viable *T. denticola*, *T. socranskii*, *T. pectinovorum*, and *T. vincentii* resulted in the induction of well-demarcated, dose-dependent, raised subcutaneous abscesses, which were similar in all respects to each other (Kesavalu et al., 1997). Interestingly, analysis of the trypsin-like protease (TLPase) activity revealed high levels in *T. denticola* but no activity in either *T. socranskii* or *T. vincentii*. The *T. pectinovorum* strains tested produced a variable amount of TLPase activity. The results suggested that abscess formation induced by these oral treponemes occurs by several mechanisms, one of which might relate to the trypsin-like protease. The size of the lesion formed was dose-dependent and only formed when live bacteria were injected. Protease inhibitors, such as Na-p-tosyl-L-lysine chloromethyl ketone (TLCK) or dithiothreitol (DTT), which blocked the activity of TLPase, did not affect the formation or size of the abscesses formed, indicating that the TLPase was not critical to abscess formation by the treponemes. Kesavalu et al. (1999) also used the murine model to investigate the role of the localized environment as a regulator of virulence of *T. denticola* and *T. pectinovorum* by varying both cysteine and iron during *in vitro* growth. Increases in cysteine resulted in significantly enhanced abscess formation and lesion size. Iron, a strong modulator of animal model virulence, was found to have no effect on abscess size. However, pretreatment of animals with iron prior to infection with *T. denticola*, resulted in significantly larger lesions. This group also evaluated the characteristics of the acquired immune responses to *T. denticola* following infection or immunization and the effect of this response on lesion formation. The results indicated that in the mouse, where *T. denticola* represents an exogenous pathogen, there is a robust systemic antibody response produced following primary infection and an even larger response noted to active immunization. Differences were noted in the IgG subclass distribution of antibody between these two types of challenges. Interestingly, the antibody level to *T. denticola* was significantly higher than the response to *P. gingivalis*, while the subclass distribution was similar; the immune response did not protect against lesion formation by the treponemes but was very effective against *P. gingivalis*. Finally, Kesavalu et al. (1998) also described the effects of polymicrobial infection with *T. denticola* and *P. gingivalis* in the murine lesion model. When *P. gingivalis* was injected into the backs of mice at low doses along with *T. denticola*, there was a significant increase in lesion size compared to infection with either species alone, including larger doses of *P. gingivalis*, supporting a potential synergistic virulence for this polymicrobial consortium.

More recent data by Kesavalu et al. (2003a; 2003b; 2004) have been obtained using microarray technology to analyze pathogen-induced responses in murine tissues following subcutaneous infection with *T. denticola* and *P. gingivalis*. The inflamed/infected tissue at the lesion site was excised, RNA isolated, and Murine GeneChip® arrays (U74v2; Affymetrix) were used to detect the level of message for the gene array. Of the 12,488 genes/ESTs assessed, approximately 7000 genes provided detectable signals utilizing a threshold for identifying meaningful differences of ≥2.0 fold change. Interestingly, 40% of the up-regulated genes were related to inflammatory or immune responses, while only 5% of the down-regulated genes were related to these types of host functions. Genes whose mRNAs were identified included inflammatory, proteolytic, prostaglandin/leukotriene synthetic, calcification/ossification, TH1 and TH2 adaptive immune responses, antigen presentation, immune receptor, immune inhibitors, apoptosis, growth factors and receptors, collagen regulation, tissue remodeling, and cell stress, as well as several uncharacterized genes regulated differentially in response to *T. denticola* infection *in vivo*. This is the first *in vivo* evidence that oral pathogens alter expression of a broad array of host cellular genes. These studies also provide a molecular description of the events that follow infection and offer opportunities for future investigation of differentially expressed host genes that may play a role in protective and/or pathologic responses to infection.

Thus, the results from the animal studies have provided a few insights into the virulence of oral treponemes and, in particular, of *T. denticola*. First, as compared to humans and nonhuman primates, the treponemes appear to be quite antigenic in rodents. This likely reflects the difference between immune responses to commensal versus exogenous bacterial challenge. The immune response that is produced by the intact bacteria appears rather ineffective in blocking the inflammation and tissue destructive properties of the treponemes. Similarly, the mice appear to demonstrate a robust host response to infection with *T. denticola* by up-regulating the expression of a broad array of genes linked to inflammation and possibly to systemic disease sequelae. Finally, the murine model has provided some ability to begin examining the polymicrobial nature of oral infections, although these are currently limited to parenteral challenge. A major limitation of the rodent model studies is the lack of reports documenting the ability to orally infect rodents with oral treponemes or consortia of bacteria that have been implicated in pathogenic biofilms. Further understanding of the role of the treponemes in the oral environment and their sufficiency or necessity for periodontal disease progression may require techniques and studies to enable reproducible oral infections of rodents better reflecting the complexity of the human oral microbial ecology.

18. Conclusion

Since the observations of van Leeuwenhoek in the 16th century, there has been a remarkable expansion in our knowledge of the relationship between the large number of bacterial species which occupy both the healthy sulcus and the inflamed periodontal pocket and their potential role in the progression of the healthy sulcus to gingivitis and ultimately to periodontal disease (*e.g.* loss of alveolar bone and associated tissue destruction). The early observations of Loesche and colleagues (1976; 1982; 1985; 1988) and the microscopic observations of Keyes and colleagues (1978a; 1978b; 1983) (and others too numerous to mention) postulated a positive relationship between the presence and number of spirochetes in the oral environment and oral health and diseases. While many reports have suggested that specific bacterial species are important to, or involved in, the progression from health to disease, it remained for the seminal studies of Socransky and Haffajee and their colleagues (*e.g.*, 1992; 1998) to provide the essential "roadmap" for understanding the importance of selected members of the oral microbiota to the maintenance of oral health and the progression to oral disease. It is clear from their work that the bacterial species involved in both the maintenance of health and the progression to gingivitis and periodontal disease do so in a complex and coordinated manner; specific groups (or complexes) function together in health and in disease. The "red complex", the group most associated with disease, possesses an impressive array of molecules (both cell-associated and secreted) which may result in the destruction of bone and tissue. Clearly these potential virulence factors are likely regulated by the localized host environment and at selected times under these environmental conditions. The mechanisms and characteristics of the environmental conditions that regulate these factors and the role played by the host in controlling this environment remains to be determined.

References

Apatzidou, D.A. and Kinane, D.F. (2004). Quadrant root planing versus same-day full-mouth root planing. I. Clinical Findings. J. Clin. Periodontol. *31*, 132-140.

Armitage, G.C., Dickenson, W.R., Jenderseck, R.J., Levine, S.M., and Chambers, D.W. (1982). Relationship between the percentage of subgingival spirochetes and the severity of periodontal disease. J. Periodontol. *53*, 550-556.

Asai, Y., Jinno, T. and Ogawa, T. (2003a). Oral treponemes and their outer membrane extracts activate human gingival epithelial cells through toll-like receptor 2. Infect. Immun. *71*, 717-25.

Asai, Y., Hashimoto, M. and Ogawa, T. (2003b) Treponemal glycoconjugate inhibits Toll-like receptor ligand-induced cell activation by blocking LPS-binding protein and CD14 functions. Eur. J. Immunol. *33*, 3196-204

Ashimoto, A., Chen, C., and Slots, J. (1996). Polymerase chain reaction detection of eight putative periodontal pathogens in subgingival plaque of gingivitis and advanced periodontitis lesions. Oral Microbiol. Immunol. *11*, 266-273.

Aukkhil I., Lopatin D.E., Syed S.A., Morrison E.C., and Kowalski C.J. (1988). The effects of periodontal therapy on serum antibody (IgG) levels to plaque microorganisms. J. Clin. Periodontol *15*, 544-550.

Battikhi, T., Lee, W., McCulloch, C.A., and Ellen, R.P. (1999). *Treponema denticola* outer membrane enhances the phagocytosis of collagen-coated beads by gingival fibroblasts. Infect. Immun. *67*, 1220-1226.

Bian, X., Wang, H., Ning, Y., Lee, S.Y., and Fenno J. C. (2005). Mutagenesis of a novel gene in the *prcA-prtP* protease locus affects expression of *Treponema denticola* membrane complexes. Infect. Immun. *73*, 1252-1255.

Brown, L.R.Jr, and Rudolph, C.E. (1957). Isolation and identification of microorganisms from unexposed canals of pulp-involved teeth. Oral Surg. Oral Med. Oral Pathol. *10*, 1094-1099.

Canale-Parola, E. (1977). Physiology and evolution of spirochetes. Bacteriol. Rev. *41*, 181-204.

Canale-Parola, E. (1978). Motility and chemotaxis of spirochetes. Annu. Rev. Microbiol. *32*, 69-99.

Carranza, F.A.J., Saglie, R., Newman, M.G., and Valentin, P.L. (1983). Scanning and transmission electron microscopic study of tissue-invading microorganisms in localized juvenile periodontitis. J. Periodontol. *54*, 598-617.

Carranza, N.Jr., Riviere, G.R., Smith, K.S., Adams, D.F., and Maier, T. (1997). Differential attachment of oral treponemes to monolayers of epithelial cells. J. Periodontol. *68*, 1010-1018.

Chan, E.C.S., and McLaughlin, R. (2000). Taxonomy and virulence of oral spirochetes. Oral Microbiol. Immunol. *15*, 1-9.

Charon, N.W., Greenberg, E.P., Koopman, M.B.H., and Limberger, R.J. (1992). Spirochete chemotaxis, motility, and the structure of the spirochetal periplasmic flagella. Res. Microbiol. *143*, 597-603.

Chi, B., Qi, M. and Kuramitsu, H. (2003). Role of dentilisin in *T. denticola* epithelial cell layer penetration. Res. Microbiol. *154*, 637-43

Choi, B., Jung, J., Suh, H., Yoo, Y., Cho, K., Chai, J. and Kim, C. (2001b). Activation of matrix metalloproteinase-2 by a novel spirochetal species *T. lecithinolyticum.* J. Periodontol. *72*, 1594-600.

Choi, B., Lee, H., Kang, J., Jeong, G., Min, C. and Yoo, Y. (2003). Induction of osteoclastogenesis and matrix metalloproteinase expression by the lipooligosaccharide of *T. denticola* Infect. Immun. *71(1)*, 226-33.

Choi, B.K., Nattermann, H., Grund, S., Haider, W., and Gröbel, U.B. (1997). Spirochetes from digital dermatitis lesions in cattle are closely related to treponema associated with human periodontitis. Int. J. Syst. Bacteriol. *47*, 175-181.

Choi, B., Ohk, S., Lee, H., Kang, J., Jeong, G. and Yoo, Y. (2001a). Effects of whole cell sonicates of *T. lecithinolyticum* on osteoclast differentiation. J. Periodontol. *72*, 1172-7.

Choi, B.K., Paster, B.J., Dewhirst, F.E., and Gobel, U. (1994). Diversity of cultivable and uncultivable oral spirochete from a patient with severe destructive periodontitis. Infect. Immun. *62*, 1889-1895.

Choi, B.K., Wyss, C., and Gobel, U.B. (1996). Phylogenetic analysis of pathogen related oral spirochetes. J. Clin. Microbiol. *34*, 1922-1925.

Chu, L., Ebersole, J.L., Kurzban, G.P., and Holt, S.C. (1997). Cystalysin, a 46-kilodalton cysteine desulfhydase from *T. denticola*, with hemolytic and hemoxidation activities. Infect. Immun. *65*, 3231-3238.

Chu, L., Ebersole, J.L., and Holt, S.C. (1999a). Hemoxidation and binding of the 46 kDa cystalysin of *T. denticola* leads to a cysteine-dependent hemolysis of human erythrocytes. Oral Microbiol. Immunol. *14*, 293-303.

Chu, L., Ebersole, J.L., Kurzban, G.P., and Holt, S.C. (1999b). Cystalysin, a 46-kilodalton cysteine desulfhydase from *T. denticola*, biochemical and biophysical characterization. Clin. Infect. Dis. *28*, 442-450.

Chu, L., Holt, S.C. (1994). Purification and characterization of a 45 kDa hemolysin from *T. denticola* ATCC 35404. Microb. Pathog. *16*, 183-195.

Chu, L., Xu, X., Dong, Z., Cappelli, D. and Ebersole J. (2003). Role for recombinant gamma-glutamyltransferase from *T. denticola* in glutathione metabolism. Infect. Immun. *71*, 335-342.

Chung P.C., Nisengard R.J., Slots J. and Genco R.J. (1983). Bacterial IgG and IgM antibody titers in acute necrotizing ulcerative gingivitis. J. Periodontol *54*, 557-562.

Correia, F., Plummer, A., Ellen, R., Wyss, C., Boches, S., Galvin, J., Paster, B. and Dewhirst, F. (2003). Two paralogous families of a two-gene subtilisin operon are widely distributed in oral treponemes. J. Bacteriol. *185*, 6860-6869.

Courtois, G.J.III., Cobb, C.M., and Killoy, W.J. (1983). Acute necrotizing ulcerative gingivitis, a transmission electron microscope study. J. Periodontol. *54*, 671-679.

Dahle, U., Tronstad, L., and Olsen, I. (1993a). Spirochaetes in oral infections. Endod. Dent. Traumatol. *9*, 87-94.

Dahle, U.R., Tronstad, L., and Olsen, I. (1993b). Observation of an unusually large spirochete in endodontic infection. Oral Microbiol. Immunol. *8*, 251-253.

Dahle, U.R., Tronstad, L., and Olsen, I. (1996). 3-Hydroxy fatty acids in a lipopolysaccharide-like material from *T. denticola* strain FM. Endod. Dent. Traumatol. *12*, 202-205.

Dawson, J.R. and Ellen, R.P. (1990). Tip-oriented adherence of *T. denticola* to fibronectin. Infect. Immun. *58*, 3924-3928.

Deng, Q.D., Han, Y., Xia, X., and Kuramitsu, H.K. (2001). Effects of the oral spirochete *T. denticola* on interleukin-8 expression from epithelial cells. Oral Microbiol. Immunol. *16*, 185-187.

DeRenzis, F.A., and Chen, S.Y. (1983) Ultrastructural study of cultured human gingival fibroblasts exposed to endotoxin. J. Periodontol. 54(2), 86-90.

Dewhirst, F.E., Tamer, M.A., Ericson, R.E., Lau, C.N., Levanos, V.A., Boches, S.K., Galvin, J.L., and Paster, B.J. (2000). The diversity of periodontal spirochetes by 16S rRNA analysis. Oral Microbiol. Immunol. *15*, 196-202.

Ding, Y., Uitto, V.J., Haapasalo, M., Lounatmaa, K., Konttinen, Y.T., Salo, T., Grenier, D., and Sorsa, T. (1996). Membrane components of *T. denticola* trigger proteinase release from human polymorphonuclear leukocytes. J. Dent. Res. *75*, 1986-1993.

Dobell, C. ed. (1932). Antoni van Leeuwenhoek and his "little animals, " (New York, Harcourt Brace and Co.).

Ebersole, J.L., Brunsvold, M., Steffensen, B., Wood, R., and Holt, S.C. (1991). Effects of immunization with Porphyromonas gingivalis and Prevotella intermedia on progression of ligature-induced periodontitis in the nonhuman primate Macaca fascicularis. Infect. Immun. *59*, 3351-3359.

Ebersole, J.L., Kesavalu, L., Schneider, S.L., Machen, R.L., and Holt, S.C. (1995). Comparative virulence of peridontopathogens in a mouse abscess model. Oral Dis. *1*, 115

Edwards, A., Dymock, D., Woodward, M. and Jenkinson, H. (2003). Genetic relatedness and phenotypic characteristics of treponemes associated with human periodontal tissues and ruminant foot disease. Microbiology. 149 *5*, 1083-1093.

Ellen, R.P., Dawson, J.R., and Yang, P.F. (1994a). *T. denticola* as a model for polar adhesion and cytopathogenicity of spirochetes. Trends Microbiol. *2*, 114-119.

Ellen, R.P., Song, M., and McCulloch, C.A. (1994b). Degradation of endogenous plasma membrane fibronectin concomitant with *T. denticola* 35405 adhesion to gingival fibroblasts. Infect. Immun. *62*, 3033-3037.

Evans, R.T., Clause, B. and Genco, R.J. (1992). Immunization with fimbrial protein and peptide protects against *Porphyromonas gingivalis* induced periodontal tissue destruction. Adv. Exp. Med. Biol. *527*, 255-262.

Falkler, W.A., Jr., Martin, S.A., Vincent, J.W., Tall, B.D., Nauman, R.K. and Suzuki, J.B. (1987). A clinical, demographic and microbiologic study of ANUG patients in an urban dental school. J. Clin. Periodontol *14*, 307-314.

Fenno, J.C., Hannam, P.M., Leung, W.K., Tamura, M., Uitto, V.J., and McBride, B.C. (1998b). Cytopathic effects of the major surface protein and the chymotrypsin-like protease of *T. denticola*. Infect. Immun. *66*, 1869-1877.

Fenno, J., Lee, S., Bayer, C. and Ning, Y. (2001). The opdB locus encodes the trypsin-like peptidase activity of *T. denticola*. Infect. Immun. *69*, 6193-6200.

Fenno, J., and McBride, B. (1998a). Virulence factors of oral treponemes. Anaerobe *4*, 1-17.

Fenno, J.C., Müller, K.-H, and McBride, B.C. (1996). Sequence analysis, expression and binding activity of recombinant major outer sheath protein (Msp) of *T. denticola* denticola. J. Bacteriol. *178*, 2489-2496.

Fiehn, N.E. (1986). Enzyme activities from eight small-sized oral spirochetes. Scand. J. Dent. Res. *94*, 132-140.

Fitzgerald, T.J., Johnson, R.C., Miller, J.N., and Sykes, A.J. (1977). Characterization of the attachment of *T. denticola* pallidum (Nichols strain) to cultured mammalian cells and the potential relationship of attachment to pathogenicity. Infect. Immun. *18*, 467-478.

Frank, R.M. (1980). Bacterial penetration in the apical pocket wall of advanced human periodontitis. J. Periodont. Res. *15*, 563-573.

Fraser, C.M., Norris, S.J., Weinstock, G.M., White, O., Sutton, G.G., Dodson, R., Gwinn, M., Hickey, E.K., Clayton, R., Ketchum, K.A., Sodergren, E., Hardham, J.M., McLeod, M.P., Salzberg, S., Peterson, J., Khalak, H., Richardson, D., Howell, J.K., Chidambaram, M., et al. (1998). Complete genome sequence of *T. pallidum*, the syphilis spirochete. Science *281*, 375-388.

Genco, C.A., Cutler, C.W., Kapczynski, D., Maloney, H., and Arnold, R.R. (1991). A novel mouse model to study the virulence of and host response to *Porphyromonas (Bacteroides) gingivalis*. Infect. Immun. *59*, 1255-1263.

Genco, C.A., Schifferle, R.E., Njoroge, T., Forng, R.Y., and Cutler, C.W. (1994). Resistance of a Tn4351 generated polysaccharide mutant of *Porphyromonas gingivalis* to polymorphonuclear leukocyte killing. Infect. Immun. *63*, 393-401.

Giardino, A., Ebersole, J.L., and Holt, S.C. (1996). Characteristics of systemic antibody responses of nonhuman primates following active immunization with *Porphyromonas gingivalis*, *Prevotella intermedia* and *Bacteroides fragilis*. Oral Microbiol. Immunol. *11*, 79-87.

Gibson, F C. 3rd, and Genco C.A. (2001) Prevention of Porphyromonas gingivalis-Induced Oral Bone Loss following Immunization with Gingipain R1. Infect. Immun*69*, 7959-7963.

Gmür, R, Wyss, C, Xue, Y, Thurnheer, T, and Guggenheim, B. (2004). Gingival crevice microbiota from Chinese patients with gingivitis or necrotizing ulcerative gingivitis. Eur. J. Oral Sci. *112*, 33-41.

Gonzalez, D., Tzianabos, A.O., Genco, C.A., and Gibson, F.C. 3rd. (2003) Immunization with *Porphyromonas gingivalis* capsular polysaccharide prevents *P. gingivalis*-elicited oral bone loss in a murine model. Infect. Immun.*71*, 2283-2287.

Gopalsami, C., Yotis, W., Corrigan, K., Schade, S., Keene, J., and Simonson, L. (1993). Effect of outer membrane of *T. denticola* on bone resorption. Oral Microbiol. Immunol. *8*, 121-124

Greenberg, E.P. and Canale-Parola, E. (1977). Relationship between cell coiling and motility of spirochetes in viscous environments. J.Bacteriol. *131*, 960-969.

Greene, S.R., and Stamm, L.V. (1999)Molecular characterization of a chemotaxis operon in the oral spirochete, *Treponema denticola*. Gene. *232*, 59-68.

Grenier, D. (1991). Characteristics of hemolytic and hemagglutinating activities of *T. denticola*. Oral Microbiol. Immunol. *6*, 246-249.

Grenier, D. (1992a). Demonstration of a bimodal coaggregation reaction between *Porphyromonas gingivalis* and *T. denticola*. Oral Microbiol. Immunol. *7*, 280-284.

Grenier, D. (1992b). Nutritional interactions between two suspected periodontopathogens, *T. denticola* and *Porphyromonas gingivalis*. Infect. Immun. *60*, 5298-5301.

Haapasalo, M., Hannam, P., McBride, B.C., and Uitto, V.J. (1996). Hyaluronan, a possible ligand mediating *T. denticola* binding to periodontal tissue. Oral Microbiol. Immunol. *11*, 156-160.

Haapasalo, M., Singh, U., McBride, B.C., and Uitto, V.J. (1991). Sulfhydryl-dependent attachment of *T. denticola* to laminin and other proteins. Infect. Immun. *59*, 4230-4237.

Haffajee, AD, Arguello, EI, Ximenez-Fyvie, LA, and Socransky, SS. (2003). Controlling the plaque biofilm. Int. Dent J. (53 Suppl 3), 191-199.

Hagewald, S., Bernimoulin, J., Kottgen, E. and Kage, A. (2002). Salivary IgA subclasses and bacteria-reactive IgA in patients with aggressive periodontitis. J. Periodontal Res. *37*, 333-339

Harwood, C.S., and Canale-Parola, E. (1984). Ecology of spirochetes. Ann. Rev. Microbiol. *38*, 161-192.

Hashimoto, M., Asai, Y., Jinno, T., Adachi, S., Kusumoto, S. and Ogawa, T. (2003c). Structural elucidation of polysaccharide part of glycoconjugate from *T. denticola* medium ATCC 700293. Eur. J. Biochem. *270*, 2671-9.

Hashimoto, M, Asai, Y. and Ogawa, T. (2003a). *T. denticola* phospholipids inhibit innate immune responses induced by pathogen-associated molecular patterns. J. Biol. Chem. *278*, 44205-13.

Hashimoto, M., Ogawa, S., Asai, Y., Takai, Y. and Ogawa, T. (2003b). Binding of *Porphyromonas gingivalis* fimbriae to *T. denticola* dentilisin. FEMS Microbiol. Lett. *226*, 267-271.

Heinzerling, H.F., Olivares, M., and Burne, R.A. (1997). Genetic and transcriptional analysis of flgB flagellar operon constituents in the oral spirochete *T. denticola* and their heterologous expression in enteric bacteria. Infect. Immun. *65*, 2041-2051.

Heuner, K., Bergmann, I., Heckenbach, K. and Gobel, U. (2001). Proteolytic activity among various oral *T. denticola* species and cloning of a prtP-like gene of *T. socranskii* subsp. *socranskii*. FEMS Microbiol. Lett. *201*, 169-176.

Heuner, K., Choi, B.K., Schade, R., Moter, A., Otto, A., and Göbel, U.B. (1999). Cloning and characterization of a gene (mspA) encoding the major sheath protein of *T. maltophilum* ATCC 51939(T). J. Bacteriol. *181*, 1025-1029.

Holt, S.C. (1978). Anatomy and chemistry of spirochetes. Microbiol. Rev. *42*, 114-160.

Holt, S.C., and Bramanti, T.E. (1991). Factors in virulence expression and their role in periodontal disease pathogenesis. Crit. Rev. Oral Biol. Med. *2*, 177-281.

Holt, S.C., Ebersole, J.L., Felton, J., Brunsvold, M., and Kornman, K.S. (1988). Implantation of *Bacteroides gingivalis* in nonhuman primates initiates progression of periodontitis. Science *239*, 55-57.

Huang, G.T.-J., Kinder Haake, S., Kim, J. -W, and Park, N.H. (1998). Differential expression of interleukin-8 and intercellular adhesion molecule-1 by human gingival epithelial cells in response to *Actinobacillus actinomycetemcomitans* or *Porphyromonas gingivalis* infection. Oral Microbiol. Immunol. *13*, 301-309.

Ishihara, K., Kuramitsu, H.K., Miura, T., and Okuda, K. (1998). Dentilisin activity affects the organization of the outer sheath of *T. denticola* . J. Bacteriol. *180*, 3837-3844.

Ishihara, K., Kuramitsu, H.K., and Okuda, K. (2004) A 43-dDa protein of *Treponema denticola* is essential for dentilisin activity. FEMS Microbiol. *232*, 181-188.

Jacob E, Meiller TF, and Nauman RK. (1982). Detection of elevated serum antibodies to *T. denticola* in humans with advanced periodontitis by an enzyme-linked immunosorbent assay. J. Perio Res. *17*, 145-153.

Jung, I.Y., Choi, B.K., Kum K.Y., Yoo, Y.F., Yoon, T.C., and Lee C.Y. (2001). Identification of oral spirochaetes at the species level and their association with other bacteria in endodontic infections. Oral Surg. Oral Med. Oral Pathol. *92*, 329-334.

Kesavalu, L., Ebersole, J. L., Machen, R.L., and Holt, S.C. (1992). *Porphyromonas gingivalis* virulence in mice, Induction of immunity to bacterial components. Infect. Immun. *60*, 1455-1464.

Kesavalu, L., Falk, C., Davis, K., Steffen, M., Xu, X., Holt, S. and Ebersole, J. (2002). Biological characterization of lipopolysaccharide from *T. pectinovorum*. Infect. Immun. *70*, 211-7.

Kesavalu, L., Holt, S.C., Crawley, R.R., Borinski, R., and Ebersole, J.L. (1991). Virulence of *Wolinella recta* in a murine abscess model. Infect. Immun. *59*, 2806-2817.

Kesavalu, L., Holt, S.C., and Ebersole, J.L. (1992). Proteases of *Porphyromonas gingivalis*, comparison in *in vitro* and *in vivo* functions. J. Dent. Res. *71*, 222.

Kesavalu, L., Holt. S.C., and Ebersole, J.L. (1996). Trypsin-like protease activity of *Porphyromonas gingivalis* as a potential virulence factor in a murine lesion model. Microb. Pathog. *20*, 1-10.

Kesavalu, L., Holt, S.C., and Ebersole, J.L. (1998). Virulence of a polymicrobial complex, *T. denticola* and *Porphyromonas gingivalis*, in a murine model. Oral Microbiol. Immunol. *13*, 373-377.

Kesavalu, L., Holt. S.C., and Ebersole, J.L. (1999a). Environmental modulation of oral treponemes virulence in a murine model. Infect. Immun. *67*, 2783-2789.

Kesavalu, L., Holt. S.C., and Ebersole, J.L. (1999b). Lack of humoral immune protection against *T. denticola* virulence in a murine model. Infect. Immun. *67*, 5736-5746.

Kesavalu, L, Holt SC, and Ebersole JL. (2003a) *In vitro* environmental regulation of *Porphyromonas gingivalis* virulence. Oral Microbiol. Immunol., *18*, 226-233.

Kesavalu, L, Kirakodu, S, Blalock, E, Ebersole, JL. (2004) Host responses to oral pathogens, Evaluation of atherosclerosis/vascular disease genes. J. Dent. Res. 83. Abstract #2267.

Keulers, R.A.C., Maltha, J.C., Mikx, F.H.M., and Wolters-Lutgerhorst, J.M.L. (1993a). Attachments of *T. denticola* strains ATCC *33520*, ATCC *35405*, B11 and Ny541 to a morphologically distinct population of rat palatal epithelial cells. J. Periodontol. *28*, 274-280.

Keulers, R.A.C., Maltha, J.C., Mikx, F.H.M., and Wolters-Lutgerhorst, J.M.L. (1993b). Attachment of *T. denticola* strains to monolayers of epithelial cells of different origin. Oral Microbiol. Immunol. *8*, 84-88.

Keulers, R.A.C., Maltha, J.C., Mikx, F.H.M., and Wolters-Lutgerhorst, J.M.L. (1993c). Involvement of *T. denticola* surface-located protein and carbohydrate moieties in the attachment of *T. denticola* ATCC 33520 to rat palatal epithelial cells. Oral Microbiol. Immunol. *8*, 236-241.

Kesavalu, L, Sreenatha, K, and Ebersole JL. (2003b) cDNA microarray analysis of differential gene expression during oral bacterial infection in mice. J. Dent. Res. *82*, Abstract #366.

Kesavalu, L., Walker, S., Holt, S.C., Crawley, R.R., and Ebersole, J.L. (1997a). Virulence characteristics of oral treponemes in a murine model. Infect. Immun. *65*, 5096-5102.

Keyes, P.H., and Rams, T.E. (1983). A rational for management of periodontal disease, rapid identification of microbial "therapeutic targets" with phase-contrast microscopy. J. Am. Dent. Assoc. *106*, 803-12.

Keyes, PH, Wright, W.E., and Howard, S.A. (1978a). The use of phase-contrast microscopy and chemotherapy in the diagnosis and treatment of periodontal lesions-a initial report (Ia). Quintessence Inter. *51*, 51-56.

Keyes, PH, Wright, W.E., and Howard, S.A. (1978b). The use of phase-contrast microscopy and chemotherapy in the diagnosis and treatment of periodontal lesions-a initial report (II). Quintessence Inter. *52*, 69-76.

Kigure, R., Saito, A., Seida, K., Aymada, S., Ishihara, K., and Okuda, K. (1995). Distribution of *Porphyromonas gingivalis* and *T. denticola* in human subgingival plaque at different periodontal pocket depths examined by immunohistochemical methods. J. Periodontal Res. *30*, 332-341.

Kimsey, R., and Spielman, A. (1990). Motility of Lyme disease spirochetes in fluids as viscous as the extracellular matrix. J. Infect. Dis. *162*, 1205-1208.

Klitorinos, A., Noble, P., Siboo, R., and Chan, E.C.S. (1993). Viscosity-dependent locomotion of oral spirochetes. Oral Microbiol. Immunol. *8*, 242-244.

Kolenbrander, P.E., Parrish, K.D., Andersen, R.N., and Greenberg, E.P. (1995). Intergeneric coaggregation of oral *T. denticola* spp. with *Fusobacterium* spp. and intrageneric coaggregation among *Fusobacterium* spp. Infect. Immun. *63*, 4584-4588.

Koo, H., Pearson, S.K., Scott-Anne, K., Abranches, J., Cury, J.A., Rosalen, P.L., Park, Y.K., Marquis, R.E., and Bowen, W.H. (2002) Effects of apigenin and tt-farnesol on glucosyltransferase activity, biofilm viability and caries development in rats. Oral Microbiol. Immunol. *17*, 337-433.

Kornman, K.S., Holt, S.C., and Robertson, P.B (1981) The microbiology of ligature-induced periodontitis in the cynomolgus monkey. J. Periodontal Res. *16*, 363-371.

Kurimoto, T., Suzuki, M., and Watanabe, T. (1990). Chemical composition and biological activities of lipopolysaccharides extracted from *T. denticola* and *T. vincentii*. Odontology (Tokyo) *78*, 208-232.

Kurzban, G.P., Chu, L., Ebersole, J.L., and Holt, S.C. (1999). Sulfhemoglobin formation in human erythrocytes by cystalysin, an L-cysteine desulfhydrase from *T. denticola*. Oral Microbiol. Immunol. *14*, 153-164.

Lai C-H, Listgarten MA, Evian CI, and Dougherty P. (1986) Serum IgA and IgG antibodies to *T. vincentii* and *T. denticola* in adult periodontitis, juvenile periodontitis and periodontally healthy subjects. J. Clin. Periodontol, *13*, 752-757.

Leung, W., Wu, Q., Hannam, P., McBride, B. and Uitto, V. (2002). *T. denticola* may stimulate both epithelial proliferation and apoptosis through MAP kinase signal pathways. J. Periodontal Res. *37*, 445-455.

Leung, W.K., Haapasalo, M., Uitto, V.J., Hannam, P.M., and McBride, B.C. (1996). The surface proteinase of *T. denticola* may mediate attachment of the bacteria to epithelial cells. Anaerobe *2*, 39-46.

Li, H., Ruby, J., Charon, N., and Kuramitsu, H.K. (1996). Gene inactivation in the oral spirochete *T. denticola* construction of an flgE mutant. J. Bacteriol. *178*, 3664-3667.

Lindhe, J., Liljenberg, B., and Listgarten, M. (1980). Some microbiological and histopathological features of periodontal disease in man. J. Periodontol. 51I, 264-269.

Linke, H.A., Kuyinu, E.O., Ogundare, B., Imam, M.M., Khan, S.H., Olawoye, O.O., and LeGeros, R.Z. (2003) Microbiological composition of whole saliva and caries experience in minority populations. Dent. Clin. North Am. *47*, 67-85

Listgarten, M.A. (1965). Electron microscopic observations on the bacterial flora of acute necrotizing ulcerative gingivitis. J. Periodontol. *36*, 328-339.

Listgarten, M.A., and Hellden, L. (1978). Relative distribution of bacteria at clinically healthy and periodontally diseased sites in humans. J. Clin. Periodontol. *5*, 115-132.

Listgarten, M., and Lai, C.H. (1999). Comparative microbiological characteristics of failing implants and periodontally diseased teeth. J. Periodontol. *70*, 431-437.

Listgarten, M.A. and Socransky, S.S. (1964). Electron microscopy of axial fibrils, outer envelope, and cell division of certain oral spirochetes. J. Bacteriol. *88*, 1087-1103.

Listgarten, M.A. and Socransky, S.S. (1965). Electron microscopy as an aid in the taxonomic differentiation of oral spirochetes. Arch. Oral. Biol. *10*, 127-138.

Loesche, W.J. (1976). Periodontal disease and the treponemes. In, The Biology of Parasitic Spirochetes, R.C. Johnson, ed. (New York, Academic Press) pp. 261-275.

Loesche, W.J. (1988). The role of spirochetes in periodontal disease. Adv. Dent. Res. *2*, 275-283.

Loesche, W.J., and Laughon, B.F. (1982). Role of spirochetes in periodontal disease. In, Host-parasite interactions in periodontal diseases, R.J. Genco and S.E. Mergenhagen, ed. (Washington, DC, American Society for Microbiology), pp. 62-75.

Loesche, W.J., Syed, S.A., Laughon, B.E., and Stoll, J. (1982). The bacteriology of acute necrotizing ulcerative gingivitis. J. Periodontol. *53*, 223-230.

Loesche, W.J., Syed, S.A., Schmidt, E., and Morrison, E.C. (1985). Bacterial profiles of subgingival plaques in periodontitis. J. Periodontol. *56*, 447-456.

Lux, R., Miller, J., Park, N. and Shi, W. (2001) Motility and chemotaxis in tissue penetration of oral epithelial cell layers by *T. denticola* . Infect. Immun. 69(10), 6276-83.

Madden, D.L., and Fujiwara, K. (1982). Selected bacterial diseases. In The mouse in biomedical research. Vol. IV. Experimental Biology, H.L. Faster, J.D. Small, J.G. Fox, ed. (New York, Academic Press), pp. 257-270.

Madden, T.E., and Caton, J.G. (1994). Animal models for periodontal disease. Methods Enzymol. *235*, 106-119.

Madianos, P.M., Lieff, S., Murtha, A.P., Boggess, K.A., Auten, R.L., Beck, J.D., and Offenbacher, S. (2001). Maternal periodontitis and prematurity. Pary II. Maternal infection and fetal exposure. Ann. Periodontol, *6*, 175-182.

Makinen, K., Chen, C.Y., and Makinen, P.L. (1996). Proline iminopeptidase from the outer cell envelope of the human oral spirochete *T. denticola* ATCC 35405. Infect. Immun. *64*, 702-708.

Makinen, K., Syed, S., Loesche, W., and Makinen P. (1988). Proteolytic profile *T. denticola* ATCC 35580 with special reference to collagenolytic and arginine aminopeptidase activity. Oral Microbiol. Immunol. *3*, 121-128.

Makinen, K., Syed, S., Makinen P., and Loesche, W. (1987). Dominance of aminopeptidase activity in the human oral bacterium *T. denticola* ATCC 35405. Curr. Microbiol. *4*, 341-346.

Mäkinen, K.K., Mäkinen, P.L., Loesche, W.J., and Syed, S.A. (1995a). Purification and general properties of an oligopeptidase from *T. denticola* ATCC 35405 – a human oral spirochete. Arch. Biochem. Biophys. *316*, 689-698.

Mäkinen, K.K., Mäkinen, P.L., and Syed, S.A. (1992). Purification and substrate specificity of an endopeptidase from human oral spirochete *T. denticola* ATCC *35405*, active on furylacryloyl-Leu-Gly-Pro-Ala and bradykinin. J. Biol. Chem. *267*, 14285-14293.

Mäkinen, P.L., and Mäkinen, K.K. (1997). Gamma-glutamyltransferase from the outer cell envelope of *T. denticola* ATCC 35405. Infect. Immun. *65*, 685-691.

Makinen P.L., Makinen, K., and Syed, S. (1994). An endo-acting proline-specific oligopeptidase from *T. denticola* ATCC *35405*, evidence of hydrolysis of human bioactive peptides. Infect. Immun. *62*, 4938-4947.

Mäkinen, P.L., Mäkinen, K.K., and Syed, S. (1995b). Role of the chymotrypsin-like membrane associated proteinase from *T. denticola* ATCC 35405 in inactivation of bioactive peptides. Infect. Immun. *63*, 3567-3575.

Malek, R., Fisher, J., Caleca, A., Stinson, M., van Oss, D., Lee, J., Cho, M., Genco, R., Evans, R., and Dyer, D. (1994). Inactivation of the *Porphyromonas gingivalis* fimA gene blocks periodontal damage in gnotobiotic rats. J. Bacteriol. *176*, 1052-1059.

Mangan, D.F., Laughon, B.E., Bower, B., and Lopatin, D.E.. (1982). *In vitro* lymphocyte blastogenic responses and titers of humoral antibodies from periodontitis patients to oral spirochete isolates. Infect. Immun. *37*, 445-451.

Marquez San Miguel, S., Rupf, S., Frenzel, J. and Eschrich, K. The effects of extracts from periodontopathic bacteria on human periodontal fibroblasts stimulated with mineralization supplements. J. Oral Sci. 2003 Sep; 45(3), 127-137

Masuda, K., and Kawata, T. (1982). Isolation, properties, and reassembly of outer sheath carrying a polygonal array from an oral treponemes. J. Bacteriol. *150*, 1405-1413.

Mathers, DA, Leung, WK, Fenno, JC, Hong, Y, and McBride, BC. (1996) The major surface protein complex of *Treponema denticola* depolarizes and induces ion channels in HeLa cell membranes. Infect. Immun. *64*, 2904-2910.

Mikx, F. (1991). Comparison of peptidase, glycosidase and esterase activities of oral and non-oral *T. denticola* species. J. Gen. Microbiol. *137*, 63-68.

Mikx, F.H. (1997). Environmental effects on the growth and proteolysis of *T. denticola* ATCC 33520. Oral Microbiol. Immunol. *12*, 249-253.

Mooney, J., Hodge, P. and Kinane, D. Humoral immune response in early-onset periodontitis, influence of smoking. J. Periodontal Res. 2001 Aug; 36(4), 227-232.

Moritz, A.J., Capelli D., Lantz, M.S., Holt S.C., and Ebersole, J.L. (1998) Immunization with Porphyromonas gingivalis cysteine protease, effects on experimental gingivitis and ligature-induced periodontitis in Macaca fascicularis. J. Periodontol. *69*, 686-97.

Moter, A., Hoenig, C., Choi, B.K., Riep, B., and Gobel, U.B. (1998) Molecular epidemiology of oral treponemes associated with periodontal disease. J. Clin. Microbiol. *36*, 1399-1403.

Murayama Y, Nagai A, and Okamura K. (1988). Serum immunoglobulin G antibody to periodontal bacteria. Adv. Dent Res. *2*, 339-345.

Nair, P.N.R. (1987). Light and electron microscopic studies of root canal flora and periapical lesions. J. Endo. *13*, 29-39.

Neiders, M.E., Chen, P.B., Suido, H. Reynolds, H.S., Zambon, J.J., and Genco, R.J. (1989). Heterogeneity of virulence among strains of *Bacteroides gingivalis*. J. Periodont. Res. *24*, 192-198.

Nilius, A.M., Spencer, S.C., and Simonson, L.G. (1993). Stimulation of *in vitro* growth of *T. denticola* by extracellular growth factors produced by *Porphyromonas gingivalis*. J. Dent. Res. *72*, 1027-1031.

Nixon, C.S., Steffen, M.J., and Ebersole, J.L. (2000). Cytokine Responses to *T. denticola* pectinovorum and *T. denticola* in human gingival fibroblasts. Infect. Immun. *68*, 5284-5292.

Ohta, K., Makinen, K., and Loesche, W. (1986). Purification and characterization of an enzyme produced by capable of hydrolyzing synthetic trypsin substrates. Infect. Immun. *53*, 213-220.

Olsen, I. (1984). Attachment of *T. denticola* to cultured human epithelial cells. Scand. J. Dent Res. *92*, 55-63.

Onagawa, M., Ishihara, K., and Okuda, K. (1994). Coaggregation between *Porphyromonas gingivalis* and *T. denticola* . Bull. Tokyo Dent. Coll. *35*, 171-181.

Papapanou, P.N., Baelum, V., Luan W.M., Madianos, P.N., Chen, X., Fejerskov, O., et al. (1997). Subgingival microbiota in adult Chinese, prevalence and relation to periodontal disease progression. J. Periodontol. *68*, 651-666.

Papapanou, PN, Neiderud, AM, Papadimitriou, A, Sandros, J, and Dahlen, G. (2000) "Checkerboard" assessments of periodontal microbiota and serum antibody responses, a case-control study. J. Periodontol. *71*, 885-897.

Park, K., Heuner, K., Gobel, U., Yoo, Y., Kim, C. and Choi, B. (2002) Cloning and characterization of a major surface protein (MspTL) of *T. lecithinolyticum* associated with rapidly progressive periodontitis. FEMS Microbiol. Lett. *207*, 185-92.

Paster, B.J., Boches, S.K., Galvin, J.L., Ericson, R.E., Lau, C.N., Levanos, V.A., et al. (2001). Bacterial diversity in human subgingival plaque. J. Bacteriol. *183*, 3770-3783.

Paster, B.J., Dewhirst, F.E., Weisburg, W.G., Tordoff, L.A., Fraser, G.J., Hespell, R.B., Stanton, T.B., Zablen, L, Mandelco, L., and Woese, C.R., (1991). Phylogenetic analysis of the spirochetes. J. Bacteriol. *173*, 6101-6109.

Paster, B.J., and Dewhirst, F.E. (2000) Phylogenetic foundation of spirochetes. J. Mol. Microbiol. Technol. *2*, 341-344.

Paster, B.J., Dewhirst, F.E., Coleman, B.C., Lau, C.N., and Ericson, R.L. (1998). Phylogenetic analysis of cultivable oral treponemes from the Smibert collection. Int. J. Syst. Bacteriol. *48*, 713-722.

Paster, B.J., Levanos, V.A., Tamer, M.A., Boches, S.K., Galvin, J.L., Lau, C.N., et al. (1999). Population distribution of cultivable and uncultivable treponemes in periodontal diseases. J. Dent Res. *78*, 504.

Persson, G.R., Engel, L.D., Whitney, C.W., Weinberg, A., Moncla, B.J., Darveau, R.P., Houston, L., Braham, P., and Page, R.C. (1994). *Macaca fascicularis* as a model in which to assess the safety and efficacy of a vaccine for periodontitis. Oral Microbiol. Immunol. *9*, 104-111.

Peters, S.R., Valdez, M., Riviere, G.R., and Thomas, D.D. (1999). Adherence to and penetration through endothelial cells by oral treponemes. Oral Microbiol. Immunol. *14*, 379-383.

Pietrantonio, F., Noble, P., Amsel, R., and Chan, E. (1988). Locomotory characteristics of *T. denticola*. Can. J. Microbiol. *34*, 748-752.

Pillot, J. and Ryter, A. (1964). Structure des spirochetes. I Etude des genres *T. denticola*, *Borrelia* et *Leptospira* au microscope electronique. Ann. Inst. Pasteur (Paris) *108*, 791-804.

Pollanen MT, Salonen JI, Grenier D, Uitto VJ (2000) Epithelial cell response to challenge of bacterial lipoteichoic acids and lipopolysaccharides *in vitro*. J. Med. Microbiol. *49*, 245-52.

Prpic, J.D., Trewartha, F., and Graves, S.R. (1981). Enhanced retention of motility and virulence of *T. pallidum* (Nichols strain) *in vitro* by the addition of gelatin to anaerobic medium. Sex. Transom. Dis. *8*, 1-4.

Reijntjens, F.M.J., Mikx, F.H.M., Wolters-Lutgerhorst, J.M.L., and Maltha, J.C. (1986). Adherence of oral treponemes and their effect on morphological damage and detachment of epithelial cells *in vitro*. Infect. Immun. *51*, 642-647.

Riviere, G.R., Smith, K.S., Carranza, N.Jr., Tzagaroulaki, E., Kay, S.L., and Dock, M. (1995). Subgingival distribution of *T. denticola* , *T. socranskii*, and pathogen-related oral spirochetes, prevalence and relationship to periodontal status of sampled sites. J. Periodontol. *66*, 829-837.

Riviere, G.R., Smith, K.S., Carranza Jr., N., Tzagaroulaki, E., Kay, S.L., Dock, M., Zhu, X., and DeRouen, T.A. (1996). Associations between *Porphyromonas gingivalis* and oral treponemes in subgingival plaque. Oral Microbiol. Immunol. *11*, 150-155.

Riviere, G.R., Wagoner, M.A., Baker-Zander, S.A., Weisz, K.S., Adams, D.F., Simonson, L., and Lukehart, S.A. (1991). Identification of spirochetes related to *T. pallidum* in necrotizing ulcerative gingivitis and chronic periodontitis. N. Engl. J. Med. *325*, 539-543.

Rôças, I.N., Siqueira, J.F. Jr., Santos, K.R.N., and Coelho, A.M.A. (2001). 'Red complex' (Bacteriodes forsythus, *Porphyromonas gingivalis*, and *T. denticola* in endodontic infections, a molecular approach. Oral Surg. Oral Med. Oral Pathol. *91*, 468-471.

Rosen, G., Naor, R., Kutner, S., and Sela, M.N. (1994) Characterization of fibrinolytic activities of Treponema denticola. Infect. Immun.*62*, 1749-1754.

Rosen, G., Naor, R., Rahamim, E., Yishai, R., and Sela, M.N. (1995). Proteases of *T. denticola* outer sheath and extracellular vesicles. Infect. Immun. *63*, 3973-3979.

Rosen, G., Sela, M.N., Naor, R., Halabi, A., Barak, V., and Shapira, L. (1999). Activation of murine macrophages by lipoprotein and lipooligosaccharide of *T. denticola*. Infect. Immun. *67*, 1180-1186.

Ruby, J., and Charon, N. (1998). Effect of temperature and viscosity on the motility of the spirochete *T. denticola*. FEMS Microbiol. Lett. *169*, 251-254.

Ruby, J.D., Li, H., Kuramitsu, H., Norris, S.J., Goldstein, S.F., Buttle, K.F., and Charon, N.W. (1997). Relationship of *T. denticola* periplasmic flagella to irregular cell morphology. J. Bacteriol. *179*, 1628-1635.

Saglie, F., Carranza F.J., Newman M., Cheng, L., and Lewin, K. (1982a). Identification of tissue-invading bacteria in human periodontal disease. J. Periodontal Res. *17*, 452-455.

Saglie, F., Newman, M., Carranza, F.J., and Pattison, G. (1982b). Bacterial invasion of gingival and advanced periodontitis in humans. J. Periodontol. *53*, 217-222.

Sakamoto, M., Takeuchi, Y., Umeda, M., Ishikawa, I., and Benno, Y. (2001). Rapid detection and quantification of five periodontopathic bacteria by real-time PCR, Microb. Immun. *45*, 39-44.

Schifferle, R.E., Chen, P.B., Davern, L.B., Aguirre, A., Genco, R.J., and Levine, M.J. (1993). Modification of experimental *Porphyromonas gingivalis* murine infection by immunization with a polysaccharide-protein conjugate. Oral Microbiol. Immunol. *8*, 266-271.

Schultz, C.P., Wolf, V, Lange, R., Mertens, E., Wecke, J., Naumann, D., and Zahringer, U. (1998). Evidence for a new type of outer membrane lipid in oral spirochete *Treponema denticola*. J. Biol. Chem. *273*, 15661-15666.

Scott, D., Siboo, I.R., Chan, E.C.S., Klitorinos, A., and Siboo, R. (1993). Binding of hemin and Congo red by oral hemolytic spirochetes. Oral Microbiol. Immunol. *8*, 245-250.

Sela, M.N., Bolotin, A., Naor, R., Weinberg, A., and Rosen, G. (1997). Lipoproteins of *T. denticola*, their effect on human polymorphonuclear neutrophils. J. Periodontal Res. *32*, 455-466.

Sela, M., Kohavi, D., Krausz, E., Steinberg, D. and Rosen, G. (2003). Enzymatic degradation of collagen-guided tissue regeneration membranes by periodontal bacteria. Clin. Oral Implants Res. *14*, 263-8

Sela, M.N., Kornman, K.S., Ebersole, J.L, , and Holt, S.C. (1987). Characterization of treponemes isolated from human and non-human primate periodontal pockets. Oral Micro Immunol. *2*, 144-149.

Seshadri, R., Myers, G.S., Tettelin, H., Eisen, J.A., Heidelberg, J.F., Dodson, R.J., Davidsen, T.M., DeBoy, R.T., Fouts, D.E., Haft, D.H., Selengut, J., et al. (2004). Comparison of the genome of the oral pathogen *Treponema denticola* with other spirochete genomes. Proc Nat. Acad. Sci. *101*, 5646-5651.

Shenker, B.J., McArthur, W.P., and Tsai, C.C. (1982) Immune suppression induced by Actinobacillus actinomycetemcomitans. I. Effects on human peripheral blood lymphocyte responses to mitogens and antigens. J. Immunol. *128(1)*, 148-54.

Schou, S., Holmstrup, P., and Kornman, K.S. (1993) Nonhuman primates used in studies of periodontal disease pathogenesis, a review of the literature. J. Periodontol. *64*, 497-508.

Siboo, R., Al-Joburi, W., Gornitsky, M., and Chan, E. (1989). Synthesis and secretion of phospholipase C by oral spirochetes. J. Clin. Microbiol. *27*, 568-570.

Simonson, L.G., Goodman, C.H., Bial, J.J., and Morton, H.E. (1988). Quantitative relationship of *T. denticola* to severity of periodontal disease. Infect. Immun. *56*, 726-728.

Simonson, L.G., Goodman, C.H., and Morton, H.E. (1990). Quantitative immunoassay of *T. denticola* serovar in adult periodontitis. J. Clin. Microbiol. *28*, 1493-1496.

Simonson, L.G., McMahon, K.T., Childers, D.W., and Morton, H.E. (1992). Bacterial synergy of *T. denticola* and *Porphyromonas gingivalis* in a multinational population. Oral Microbiol. Immunol. *7*, 111-112.

Siqueira, J.F. Jr., Rôças, I.N., Favieri, A., and Santos, K.R.N. (2000a). Detection of *T. denticola* in endodontic infections by 16S rRNA gene-directed polymerase chain reaction. Oral Microbiol. Immunol. *15*, 335-337.

Siqueira, J.F. Jr., Rôças, I.N., Favieri, A., Oliveira J.C., and Santos, K.R.N. (2001a). Polymerase chain reaction detection of *T. denticola* in endodontic infections within root canals. Int. Endod. J. *34*, 280284.

Siqueira, J.F., Jr., and Rocas, I.N. (2003). PCR-based identification of *T. denticola* maltophilum, *T. amylovorum*, *T. medium*, and *T. lecithinolyticum* in primary root canal infections. Arch. Oral Biol. *48*, 495-502.

Siqueira, J.F. Jr., Rôças, I.N., Souto, R., Uzeda, M., and Colombo, A.P. (2000b). Checkerboard DNA-DNA hybridization and analysis of endodontic infections. Oral Surg. Oral Med. Oral Pathol. *89*, 744-748.

Siqueira, J.F. Jr., Rôças, I.N., Souto, R., Uzeda, M., and Colombo, A.P. (2001b). Microbiological evaluation of acute periradicular abscesses by DNA-DNA hybridization. Oral Surg. Oral Med. Oral Pathol. *92*, 451-457.

Smallwood E., Gharbia, S.E., Gulabivala, K. and Shah, H.N. (1998). Isolation and direct nucleic acid detection of oral spirochaetes in root canal infections. Int. Endod. J. *31*, 189-220(ESE abstracts).

Smibert, R.M., and Burmeister. J/A. (1983). *T. pectinovorum* sp. nov. isolated from humans with periodontitis. Intl J. Syst Bacteriol. *33*, 852-856.

Smibert, R.M., Johnson, J.L., and Ranney, R.R.(1984). *T. socranskii* sp. nov.; *T. socranskii* subsp. *socranskii* subsp. nov; *T. socranskii* susp. *buccale* subsp. nov.; and *T. paredis* subsp. nov. isolated from humans with experimental gingivitis and periodontitis. Intl. J. Syst Bacteriol. *34*, 456-462.257.

Socransky, S.S. (1970) Relationship of bacteria to the etiology of periodontal disease. J. Dent Res. *49*, 203-222.

Socransky, S.S., and Haffajee A.D. (1992) The bacterial etiology of destructive periodontal disease, current concepts. J. Periodontol.63(4 Suppl), 322-331.

Socransky, S.S., Haffajee, A.D. (1997). Microbiology of periodontal disease. In Clinical Periodontology and Implant Dentistry, J. Lindhe, T. Karring, N.P. Lang, eds. (Copenhagen, Munksgaard) pp. 138-188.

Socransky, S.S. and Haffajee, A.D. (2002). Dental biofilms, difficult therapeutic targets. Periodontol 2000; *28*, 12-55.

Socransky, S.S., Haffajee, A.D., Cugini, M.A., Smith, C., and Kent, R.L., Jr. (1998). Microbial complexes in subgingival plaque. J. Clin. Periodontol. *25*, 134-144.

Stamm, L.V., and Bergen, H.L. (1999) Molecular characterization of a flagellar (fla) operon in the oral spirochete Treponema denticola ATCC 35405. FEMS Microbiol. Lett. *179*, 31-36.

Steffen, M.J., Holt, S.C., and Ebersole, J.L. (2000). *Porphyromonas gingivalis* induction of mediator and cytokine secretion by human gingival fibroblasts. Oral Microbiol. Immunol. *15*, 172-180.

Syed, S.A., Makinen, K.K., Makinen, P.L., Chen, C.Y., and Muhammed, Z. (1993). Proteolytic and oxidoreductase activity of *T. denticola* ATCC 35404 grown in an aerobic and anaerobic gaseous environment. Res. Microbiol. *144*, 317-326.

Takahashi, K., Nishimura, F., Kurihara, M., Iwamoto, Y., Takashiba, S., Miyata, T., and Murayama, Y. (2001) Subgingival microflora and antibody responses against periodontal bacteria of young Japanese patients with type 1 diabetes mellitus. J. Int. Acad. Periodontol *3*, 104-11.

Tanner, A., Maiden, M.F., Lee, K., Shulman, L.B., and Weber, H.P. (1997). Dental implant infections. Clin. Infect. Dis. *25*, Suppl-7.

Taubman, M.A., Yoshie, H., Ebersole, J.L. Smith, D.J., and Olson, C.L. (1984). Host response in experimental periodontal disease. J. Dent. Res. *63*, 455-460.

Tew, J.G., Marshall, D.R., Moore, W.E.C., Best, A.M., Palcanis, K.G., and Ranney, R.R. Serum antibody reactive with predominant organisms in the subgingival flora of young adults with generalized severe periodontitis. Infect. Immun. *1985*, *48*, 303-311.

Theilade, E., Wright, W.H., Jensen, S.B., and Loe, H. (1966) Experimental gingivitis in man. II. A longitudinal clinical and bacteriological investigation. J. Periodontal Res. *1*, 1-13

Thilo, B.E., Baehni, P., and Holz, J. (1986). Darkfield observation of bacterial distribution in root canals following pulp necrosis. J. Endod. *12*, 202-204.

Thomas, D.D. (1996). Aspects of adherence of oral spirochetes. Crit. Rev. Oral. Biol. Med. *7*, 4-11.

Uitto, V., Grenier, D., Chan, E., and McBride, B. (1988a). Isolation of a chymotrypsin-like enzyme from *T. denticola* . Infect. Immun. *56*, 2717-2722.

Uitto, V.J., Haapasalo, M. Meldal, M., and Breddam, K. (1988b). Degradation of basement membrane collagen by proteases from some anaerobic oral microorganisms. Oral Microbiol. Immunol. *3*, 97-102.

Uitto, V.J., Pan Y.M., Leung W., Larjava, H., Ellen, R., and Finlay B. et al. (1995). Cytopathic effects of *T. denticola* chymotrypsin-like proteinase on migrating and stratified epithelial cells. Infect. Immun. *56*, 3401-3410.

Umemoto, T., Jinno, T., Taiji, Y. and Ogawa, T. (2001). Chemotaxis of oral treponemes toward sera and albumin of rabbit. Microbiol. Immunol. *45*, 571-7.

Umemoto, T., Nakazawa, F., Hoshino, E., Okada, K., Fukunaga, M., and Namikawa, I. (1997). *T. medium* sp. nov., isolated from human subgingival dental plaque. Int. J. Syst. Bacteriol. *47*, 67-72.

Umemoto, T., Namikawa, I., Suido, H., and Asai, S. (1989). A major antigen on the outer envelope of a human oral spirochete, *T. denticola*. Infect. Immun. *57*, 2470-2474.

Van Winkelhoff, A.J., Carlee, A.W., and de Graaff, J. (1985). Bacteroides endodontalis and other black-pigmented *Bacteroides* species in odontogenic abscesses. Infect. Immun. *49*, 494-498.

Walker, S.G., Ebersole, J.L., and Holt, S.C. (1997). Identification, isolation, and characterization of the 42-kilodalton major outer membrane protein (MompA) from *T. pectinovorum* ATCC 33768. J. Bacteriol. *179*, 6441-6447.

Walker, S.G., Ebersole, J.L., and Holt, S.C. (1999b). Studies on the binding of *T. pectinovorum* to Hep-2 epithelial cells. Oral Microbiol. Immunol. *14*, 165-171.

Walker, S. and Ryan, M. (2003). Cloning and expression of a pectate lyase from the oral spirochete *T. pectinovorum* ATCC 33768. FEMS Microbiol. Lett. *226*, 385-90.

Walker, SG, Xu, X, Altman, E, Davis, KJ, Ebersole, JL, and Holt, SC. (1999a). Isolation and chemical analysis of a lipopolysaccharide from the membrane of the oral anaerobic *T. pectinovorum*. Oral Microbiol. Immunol. *14*, 304-8.

Wang, Q., Ko, K.S., Kapus, A., McCulloch, C.A., and Ellen, R.P. (2001). A spirochete surface protein uncouples store-operated calcium channels in fibroblasts, a novel cytotoxic mechanism. J. Biol. Chem. *276*, 23056-23064.

Weinberg, A., and Holt, S.C. (1990). Chemical and biological activities of a 64-kilodaltoon outer sheath protein from *T. denticola* strains. J. Bacteriol. *173*, 6935-6947.

Weinberg, A., Nitzan, D.W., Shteyer, A., and Sela, M.N. (1986). Inflammatory cells and bacteria in pericoronal exudates from acute periocoronitis. Int. J. Oral Maxillofacial Surg. *15*, 606-611.

Westphal, O., and Jahn, K. (1965). Bacterial lipopolysaccharides, extraction with phenol-water and further applications of the procedure. In Methods in carbohydrate chemistry, Whistler, R.L, ed. (New York, Academic Press), pp. 83-91.

Willis, S.G., Smith, K.S., Dunn, V.L., Gapter, L.A., Riviere, K.H., and Riviere, G.R. (1999). Identification of seven *T. denticola* species in health- and disease-associated dental plaque by nested PCR. J. Clin. Microb. *37*, 867-869.

Wilton, J.M., Ivanyi, L., and Lehner, T. (1971) Cell-mediated immunity and humoral antibodies in acute ulcerative gingivitis. J. Periodontal Res. *6*, 9-16.

Wyss, C., Choi, B., Schupbach, P., Guggenheim, B., and Gobel, U. (1996). *T. maltophilum* sp. nov., a small oral spirochete isolated from human periodontal lesions. Int. J. Syst. Bacteriol. *46*, 745-752.

Wyss, C., Choi, B.K., Schupbach, P., Guggenheim, B., and Gobel, U.B. (1997) Treponema amylovorum sp. nov., a saccharolytic spirochete of medium size isolated from an advanced human periodontal lesion. Int. J. Syst. Bacteriol. *47*, 842-5.

Wyss, C., Choi, B.K., Schupbach, P., Moter, A., Guggenheim, B., and Gobel, U.B. (1999) Treponema lecithinolyticum sp. nov., a small saccharolytic spirochaete with phospholipase A and C activities associated with periodontal diseases. Int. J. Syst. Bacteriol. *49*, 1329-39

Wyss, C., Dewhirst, F., Gmur, R., Thurnheer, T., Xue, Y., Schupbach, P., Guggenheim, B. and Paster, B. (2001). *T. parvum* sp. nov., a small, glucuronic or galacturonic acid-dependent oral spirochaete from lesions of human periodontitis and acute necrotizing ulcerative gingivitis. Int. J. Syst Evol. Microbiol.51(Pt 3), 955-62.

Wyss, C., Moter, A., Choi, B.K., Dewhirst, F.E., Xue, Yi, Schüpbach, P., Göbel, U.B., Paster, B.J., and Guggenheim, B. (2004) *T. putidum*. nov., a medium-size proteolytic spirochaete isolated from lesions of human periodontitis and acute necrotizing ulcerative gingivitis. Int. J. Syst. Evol. Microbiol. *54*, 1117-22.

Ximenez-Fyvie, L.A., Haffajee, A.D., and Socransky, S.S. (2000). Microbial composition of supra- and subgingival plaque in subjects with adult periodontitis. J. Clin. Periodontol *27*, 722-732.

Xu, X. and Kolodrubetz, D. (2002). Construction and analysis of hemin binding protein mutants in the oral pathogen *T. denticola* . Res. Microbiol. *153*, 569-77

Yao, E.S., Lamont, R.J., Leu, S.P., and Weinberg, A. (1996). Interbacterial binding among strains of pathogenic and commensal oral bacterial species. Oral Microbiol. Immunol. *11*, 35-41.

Yotis, W.W., Macaluso, F. and Gopalsami, C. (1995). Immunochemical features of macromolecule of *T. denticola*. J. Basic Microbiol. *35*, 255-268.

15

Virulence Determinants of Oral Treponemes

*Richard P. Ellen**

Abstract

Oral treponemes are indigenous bacteria that become opportunistic pathogens in the mixed microflora that colonizes the space between the teeth and inflamed gingival tissues (periodontal pocket). The extensively studied *Treponema denticola* binds to and degrades extracellular matrix and thereby probably contributes to the disruption of the gingival epithelial barrier and bacterial penetration of the underlying connective tissue. Well-conserved motility and chemotactic properties also promote local tissue invasion. The major outer sheath protein (Msp) of *T. denticola* is differentially toxic to a variety of host cells. It also perturbs cytoskeletal and calcium dynamics in host cells that retain viability (*e.g.* fibroblasts). The generation of volatile sulphur containing compounds from amino acid metabolism by the enzyme cystalysin is also cytotoxic. Some lipoproteins and outer sheath glycolipids trigger proinflammatory pathways of tissue damage. *T. denticola* may degrade chemokines of innate immunity pathways and activate host matrix metalloproteinases; yet, it is relatively resistant to β-defensins. Since 80% of the phylotypes of oral treponemes have yet to be cultivated *in vitro*, their potential pathogenicity has probably been underestimated.

1. Introduction

Virulence is defined as the capacity of a microorganism to cause damage to its host. Therefore, virulence determinants of a pathogen should include all the properties that foster its colonization, its emergence in the microflora, and its synthesis and delivery of noxious stimuli to a vulnerable host. Yet, these definitions are centered on the properties of the microorganism, and they seem to apply mostly to pathogens that cause single species, exogenous infections. Virulence should also encompass the issue of host susceptibility (Casadevall and Pirofski, 2001). When considering indigenous species that become opportunistic pathogens under

*For correspondence email richard.ellen@utoronto.ca

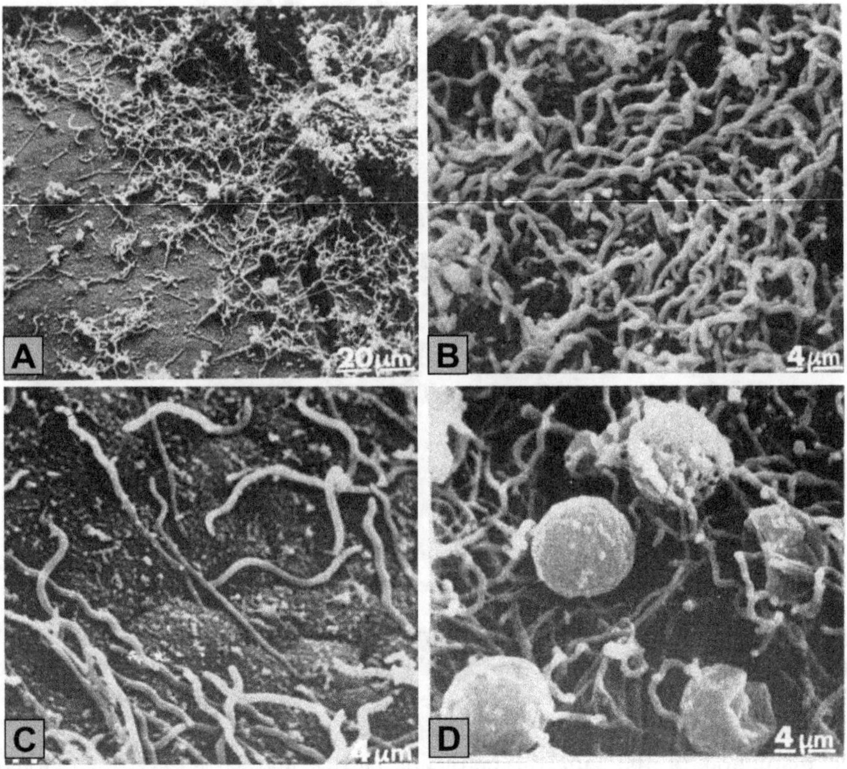

Figure 1. Scanning electron photomicrographs of the apical border of subgingival dental plaque biofilms on human teeth associated with chronic periodontitis. The mixed microflora has a variety of treponemal morphotypes (A, B), especially visible at higher magnification as cells of different size and wave periodicity (C). The spirochetes appear intimately associated with the root surface in the region near the epithelial attachment and with cells resembling erythrocytes (C, D). Photomicrographs reproduced from H.N. Newman (*British Dental Journal*, 1976), with permission from the Nature Publishing Group.

favorable environmental conditions, whether to include properties that determine colonization and survival *per se* as virulence determinants is debatable, since such microorganisms may develop a longstanding neutral or amphibiotic relationship with the host, without causing damage. Then, what determines the relative virulence of indigenous species? Must they express clearly defined cytotoxins, immunosuppressants, or hydrolytic enzymes that individually damage the host to be considered "virulent"? If so, there is very little known about virulence among the wide variety of treponemes that colonize the mouth. Hardly any have been investigated in sufficient depth beyond phylogenetic comparison of their 16S rRNA nucleotide sequence for the purpose of establishing taxonomic and evolutionary relationships (See Paster and Dewhirst, Chapter 1). Although there is clear evidence in the published literature that the severity of periodontal tissue inflammation is associated with an increase in the total number and proportional distribution of spirochetes in the adjacent periodontal pocket, whether all or most or some or very few of the species are pathogenic is unknown. It is assumed that further research will identify many pathogenic phylotypes and widely disseminated virulence factors among them, because 1) some extraoral species of treponemes are well-recognized pathogens; 2) the population density of some of the better known indigenous oral species fluctuates with the health/disease status of the periodontium (see Chapter 14, Holt and Ebersole); 3) several oral *Treponema* species can cause subcutaneous abscesses in a murine model (Kevasalu et al., 1997); and 4) studies of the best known oral treponeme, *Treponema denticola*, have already yielded some putative virulence determinants.

Oral treponemes have evolved to occupy a unique niche. Physically, they concentrate subgingivally, on the surface of dense bacterial biofilms, at the interface of the biofilms and the epithelium that lines the gingival sulcus. Therefore, they localize in a position where they may bind and parasitize other microorganisms, where they may bind and parasitize host tissues, and where they may forage for nutrients from the gingival crevicular fluid of the periodontal pocket and the extracellular matrix or lysed cells of the periodontium. Indeed, they have been detected at the deepest regions of the periodontal pocket, intimate with the junctional epithelium (Figure 1). Compared with most bacterial species that have been studied in research laboratories, treponemes are highly fastidious; for example, most have yet to be grown in even the most complex laboratory media or away from a reduced environment. Most of the cultivable oral spirochetes studied to date are asaccharolytic, and they derive energy by transporting and metabolizing peptides. They require iron and a complex variety of nutrient growth factors. Therefore, the capacity to inflict damage on the surrounding tissues is probably a closely linked consequence of their evolution to navigate and survive the conditions of the subgingival environment and to yield sufficient progeny to sustain the population of oral treponemes (Figure 2). The chronic nature of periodontal disease, which gives rise to deepened, stagnant, anaerobic pockets; prolonged

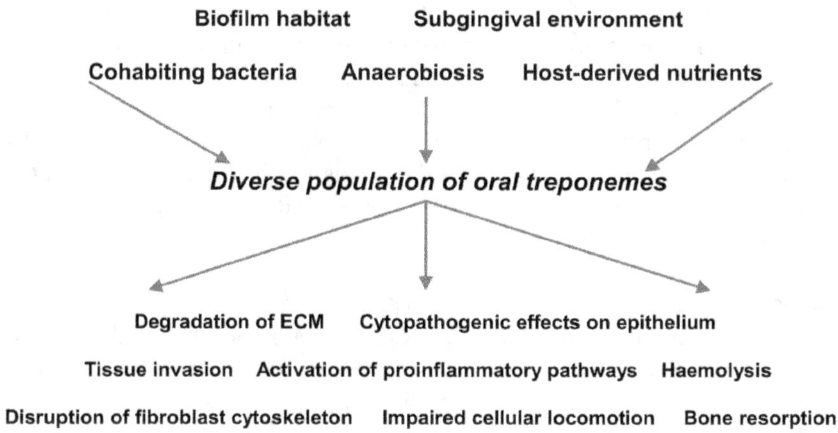

Figure 2. Flow diagram illustrating how the complex subgingival ecosystem supports the emergence of a diverse population of treponemes. Inflammatory responses and damage to the periodontium, as a consequence of bacterial metabolism and host cell/extracellular matrix contact with treponemal virulence determinants, perpetuates an environment conducive to both treponemal survival and active disease.

inflammatory degradation of host proteins and proteoglycans; and prolonged flow of a protein-rich serum transudate and blood into contact with subgingival biofilms, creates ideal conditions for the emergence and survival of diverse treponemes (Dewhirst et al., 2000). Therefore, for treponemes indigenous to the oral cavity, the capacity to survive is not only inseparable from their pathogenicity; virulence may be indispensable for survival of the treponemal colony. This chapter will address the putative virulence determinants of oral treponemes from an ecological perspective.

2. Opportune distribution at the gingival interface of subgingival biofilms

All treponemes are motile, and they express a highly conserved set of proteins that regulate and control chemotaxis and locomotion (See Charon et al., Chapter 6). Direct microscopy of freshly prepared samples of periodontal pocket debris most often reveals planktonic spirochetes of diverse size swimming among a rich variety of other highly motile bacterial species. The surface of the adjacent biofilm usually has a concentrated sessile population of these same morphotypes. There seems to be some structural organization to subgingival biofilms. Bacteria that are primary colonizers survive in a thick mat closer to the tooth surface. Successive groups of colonizers that depend on a combination of coadhesion, cross-feeding with other species, and uptake of host-derived nutrients for successful colonization populate the surface of the biofilm.

Treponemes often bind to the surface of biofilms while tethered by one pole; the cell may rotate on an axis about the fixed pole, perhaps creating a local vortex that may aid their contact with nutrients and their export of waste. Some who have observed this phenomenon in clinical samples have also noted community behavior among the multitudes of spirochetes that are crowded together and bound by their tips. Perhaps too often to be due to coincidence, the bacteria flex in concert, creating a wave, a treponemal version of the mass activity of fans who populate the stands at sporting events. One can imagine or hypothesize the evolutionary benefit. Adhesion to a surface *per se* is considered a major ecological determinant of oral bacterial colonization. Yet, the coincident movement of a population to create a wave motion could potentially serve a crucial ecological function by circulating solutes into more frequent contact with the porins and binding/transport proteins of such fastidious species. Perhaps it is a way of taking advantage of movement without expending the energy needed for locomotion.

3. Adhesion to extracellular matrix

Adhesion of bacteria to extracellular matrix (ECM) components is a characteristic shared by many pathogenic bacteria that penetrate peripheral host tissues. For example, the evolution of binding proteins that mediate specific adhesion of *Borrelia burgdorferi* to the proteoglycan decorin probably fosters the spirochete's localization upon tick transmission to the dermis and its trafficking to joints in Lyme arthritis (Guo et al., 1995; 1998). Analogously, *Treponema denticola* binds several components of the epithelial and connective tissue ECM, including laminin, fibronectin, fibrinogen, and heparin (Dawson and Ellen, 1990; Haapasalo et al.,

1991). Its polar orientation, binding perpendicular to ECM protein-conditioned surfaces, is similar to that of *Treponema pallidum*, and this characteristic function may provide an advantage that enhances growth (Ellen et al., 1998). *Treponema denticola* fibronectin-binding adhesins cluster, or cap, toward one pole when the bacterium migrates into contact with a fibronectin-coated surface (Dawson and Ellen, 1994). Therefore, adhesion probably helps *T. denticola* and other spirochetes localize close to a source of digestible proteins and peptides.

The repertoire of *T. denticola* adhesins that specifically target ECM components is not known, and the mechanisms that mediate adhesive interactions between oral treponemal adhesins and host ligands is poorly understood. Several *T. denticola* proteins have been found to bind soluble or surface-bound matrix components, but the impact of these interactions on invasion and virulence is presently obscure. Prominent among these is the major outer sheath protein (Msp), which probably has some domains that are oriented in an accessible location on the surface of the *T. denticola* cell and that have the ECM binding capacity that would be required for ecological significance. Both native and recombinant Msp have been shown to bind a variety of matrix proteins and to bind host cells (Haapasalo et al., 1992; Mathers et al., 1996; Fenno et al., 1996; 1998b). Although the deduced amino acid sequence of Msp has been known for almost a decade, its ECM-binding domains have not been identified. Indeed, there has even been some controversy about the extent of Msp surface exposure in the outer sheath (Caimano et al., 1999). Amino acid sequence analysis of Msp from the type strain ATCC 35405 (Chan et al., 1993) predicts a series of membrane-spanning regions, with a few domains that have characteristics of hydrophilicity and potential immunogenicity and that may be exposed externally. Moreover, antibodies raised against this Msp and the larger molecular mass Msp of strain GM-1 (Weinberg and Holt, 1991; Fenno et al., 1998a) bind to the surface of intact *T. denticola* cells and to isolated outer sheaths. The Msp of GM-1 and strain TD-4 (ATCC 35404) were also readily labeled by extrinsic ^{125}I labeling of whole cells, indicating their exposure on the surface of the bacterium (Weinberg and Holt, 1991). Whether Msp is the adhesin responsible for the observed polar orientation of *T. denticola* when it is bound to inanimate surfaces conditioned with fibronectin or to biofilms is as yet unknown. If Msp occurs in nature complexed with other outer sheath proteins in regularly spaced arrays (Fenno et al., 1997), it would presumably not have fluid properties in the outer sheath that would cause it to migrate toward and concentrate at the adherent pole upon ECM ligand engagement.

Since host cells communicate directly through integral membrane proteins with ECM components, treponemal adhesion to fibronectin and other ECM proteins may represent an initial step for their cytadhesion and localization near the host cell plasma membrane (Ellen, 2002). Yet, direct stereochemical adhesion to other cellular receptors that would initiate transmembrane signal transduction cascades is also feasible. For example, *T. denticola* Msp has been reported to bind a 65 kDa HeLa cell surface protein *in vitro* (Mathers et al., 1996), but the significance of the interaction in colonization, invasion, or perturbation of HeLa cell physiology is unknown. Msp's putative function in *T. denticola* survival is that of a large, nonrestrictive porin that presumably aids the entry of solutes like peptides and ions past the outer sheath and into the periplasm and the passage of waste metabolites in the opposite direction. Both enriched native Msp complex and recombinant Msp of the type strain ATCC 35405 have been found to form ion-permeable channels in black lipid bilayers *in vitro*, the most commonly used model to test bacterial porin activity (Egli et al., 1993; Fenno et al., 1998a). Hypothetically, polar adhesion to surfaces and the flexing motility of tip-bound treponemes may enhance the movement of solutes through such pores. Like porins of some other pathogenic bacteria, Msp has the capacity to insert in and transiently depolarize the HeLa cell membrane, presumably by establishing a short-lived ion-permeable channel (Mathers et al., 1996). Such activity would promote the acute influx of calcium ions into the cytoplasm of host cells (see below) upon their exposure to the shed *T. denticola* outer sheath (Ko et al., 1998a) or to Msp complex itself (Wang et al., 2001). While close approximation of whole *T. denticola* or shed outer sheath vesicles to host cells may conceivably foster the fusion of Msp complex with the plasma membrane, the actual transposition of Msp from the cytadherent parasite to the host cell membrane has not yet been demonstrated. Yet, the evolution of an outer sheath protein like Msp, which both binds ECM components and serves as a channel for the uptake of peptides, exemplifies how properties that foster survival and yet perturb host tissue homeostasis are often complementary.

There are a number of alternative ECM adhesins that are expressed by *T. denticola*. Several unidentified *T. denticola* polypeptides have been reported to absorb fibronectin from solution (Umemoto et al., 1993). Fenno and coworkers (2000) identified a 70 kDa *T. denticola* surface protein OppA, which binds soluble plasminogen and fibronectin but does not bind matrix proteins that are bound to surfaces. There is considerable interest in OppA both as a potential multifunctional binding protein associated with an ABC transporter for peptides and as a surface protein that may absorb host proteins and thereby decorate the bacterium with proteins that would help it evade immune recognition. The OppA homologue in *B. burgdorferi* is known to bind and activate plasminogen and thereby activate host proteases that may assist in tissue penetration (Hu et al., 1997). Though there is no direct evidence for an analogous OppA function in *T. denticola*, or in *Treponema vincentii,* which also has an OppA homolog (Fenno et al., 2000), the absorption and/or activation of bioactive host molecules could conceivably augment the virulence of oral spirochetes.

4. Adhesion combined with substrate degradation

Since enzymes recognize and bind their substrates selectively and stereochemically, it is likely that the subtilisin family serine protease of *T. denticola*, dentilisin (PrtP), may serve as an additional adhesin that targets ECM proteins. It is also implicated in the specific binding interaction with the surface fimbriae of *Porphyromonas gingivalis* (Hashimoto et al., 2003), one of the other proteolytic species with which *T. denticola* cohabits in biofilms associated with periodontitis. Dentilisin is a lipoprotein that complexes with two accessory proteins (gene products of *prcA*) in the outer sheath to form a "chymotrypsin-like" protease complex (CTLP, Uitto et al., 1988; Rosen et al., 1995; Lee et al., 2002). It is the major cell-associated protease of *T. denticola* that is known to degrade native ECM proteins and some other host proteins in the inflammatory cascade. For example, dentilisin hydrolyzes and inactivates substance P and angiotensin I (Mäkinen et al., 1995). It also has the capacity to activate matrix metalloproteinases (MMPs; Sorsa et al., 1992; Ding et al., 1996), which in turn may modulate chemokine and cytokine responses as well as degrade ECM proteins like collagen. The signature peptidase activity of dentilisin is to cleave prolyl-phenylalanine bonds, including a crucial peptide bond in the protein encoded by the *prcA* transcript, forming the two accessory proteins, and which is ultimately essential for expression of the active cell-associated dentilisin enzyme (Lee et al., 2002). During early and exponential growth of *T. denticola*, most of the dentilisin activity is bound to the bacterial surface, and it is released as part of vesicles or shedding outer sheath fragments into the growth medium in older or dying cultures. Its co-localization with Msp may provide the bacterium a readily accessible source of degraded peptides in proximity to a porin that fosters their uptake. Therefore, the adhesion of *T. denticola* to ECM proteins by either Msp or dentilisin would serve to promote the scavenging of peptides for the generation of energy as well as to promote colonization and subsequent gingival tissue invasion. Dentilisin's efficiency in generating small peptides for transport is probably augmented by *T. denticola*'s elaboration of additional endo-acting peptidases (Mikx et al., 1992), such as a prolyl oligopeptidase, a FALGPA peptidase (Mäkinen et al., 1995), and an arginyl-specific, "trypsin-like" specific peptidase.

Though a direct contribution to the virulence of *T. denticola* is unknown, it is notable that the three key bacterial species that cohabit and form the proteolytic cluster of species that is most associated with progressive, chronic periodontitis, *T. denticola*, *P. gingivalis*, and *Tannerella forsythia* (previously *Bacteroides forsythus*), share arginyl-specific peptidase activity. This property has given rise to a handy, inexpensive color-indicator test that can be used to monitor bulk population fluctuations of this triad of species in clinical plaque samples, the BANA Test, which is based on the rapid hydrolysis of N-α-benzoyl-DL-arginine-2-naphthylamide (Loesche et al., 1990). In the case of *P. gingivalis*, BANA-degrading activity is linked to cysteine proteases known as arg- and lys- gingipains (Rgp and Kgp), which are considered very important virulence determinants. In *T. denticola*, BANA degradation is mediated by the protein encoded by the gene *opdB* (Fenno et al., 2001). OpdB has a deduced molecular mass of 78 kDa and a peptide sequence that has considerable homology to prolyl oligopeptidases of several prokaryotic and eukaryotic parasites that infect humans. Insertional inactivation of *opdB* yielded a mutant totally defective in "trypsin-like" peptidase activity and no detectable pleiotropic effects other than a slightly slower maximum growth rate than its parent (Fenno et al., 2001).

5. Disruption of epithelial barriers and tissue penetration

Oral spirochetes can be found within the gingival connective tissue in cases of acute necrotizing ulcerative gingivitis (ANUG) and many cases of chronic periodontitis (Listgarten and Lewis, 1967; Riviere et al., 1991a; 1991c). Yet very little is known about the mechanisms that promote tissue invasion *in vivo*. Protease activity and directed motility would be two properties that are consistent with an invasive phenotype. Oral treponemes also elaborate hyaluronidase and chondroitinase activities that may contribute to degradation of the ECM (Scott et al., 1996b). Though *T. denticola* has been detected among mixed spirochete populations deep within gingival tissues in ANUG (Riviere et al., 1991c), it was found to be relatively less invasive than some other oral treponemes in a tissue transmigration system *in vitro* (Riviere et al., 1991a). Nevertheless, it has been used as a convenient model to study virulence properties that account for tissue penetration by oral treponemes. Dentilisin clearly has the capacity to promote the penetration of epithelial tissues by *T. denticola*. When *in vitro* monolayer or stratified cultures of oral epithelial cells were challenged with *T. denticola* or the native CTLP complex, the enzyme was subsequently found between cells and near the basement membrane. Such cultures became permeable to normally excluded molecules (Uitto et al., 1995). The CTLP has also been shown to foster the migration of *T. denticola* through an *in vitro* model of the ECM (Grenier et al., 1990). The effect of *T. denticola* and its major outer sheath protein complexes on the electrical resistance of epithelial monolayers was studied by cell-substrate impedance sensing, a biophysical system in which continuous real-time recordings of cell junctional resistance and cellular micromotion can be calculated mathematically (Ko et al., 1998b). *T. denticola* outer membrane (OM) preparations, which had substantial prolyl-phenylalanine degrading activity, destroyed junctional resistance among confluent HEp-2 cells. The protease inhibitor phenylmethylsulfonyl fluoride (PMSF) totally blocked the activity of the OM preparation. Enriched native CTLP had activity equivalent to the OM, and it caused a concentration-dependent reduction in junctional resistance and some effects on cellular micromotion. In contrast, protease-free enriched Msp complex had no significant effect on junctional resistance and, therefore, little direct effect on epithelial barrier function. Yet, Msp induced early changes in cellular micromotion, which may reflect its recognized perturbation of host cell cytoskeleton (see below). The importance of dentilisin in degrading cell-cell junctions was confirmed in subsequent experiments that found that an OM preparation of dentilisin-defective mutant K1 lost the wild type's ability to lower junctional resistance, and its inhibition of the spreading of epithelial cells into monolayers when epithelial cells were cultured in its presence (Ellen et al., 2000). Subsequent studies comparing Hep-2 monolayer resistance and penetration of whole wild type and dentilisin-deficient mutant K1 confirmed that dentilisin is crucial for disrupting cell-cell junctions and promoting epithelial penetration by *T. denticola* (Chi et al., 2003). Therefore, dentilisin would be expected to both damage epithelial integrity and to prevent or delay the migration of epithelial cells to heal infected gingival tissues.

Inhibition of virulence coincident with the inactivation of a specific gene is considered compelling evidence that the encoded bacterial product is a virulence determinant or is at least required for virulence. Since there is no appropriate and convenient small animal model of periodontitis in which treponemes survive to exacerbate lesions, investigators have resorted to a murine abscess model to determine oral treponemal virulence (Kevasalu et al., 1997; 1999). Though laborious, genetic transformation and gene inactivation of *T. denticola* is achievable, so far only by insertional mutagenesis (Li et al., 1996). Ishihara and coworkers (1998) were the first to inactivate the *prtP* gene, which yielded the dentilisin-defective mutant K1; the mutant's ability to cause subcutaneous abscesses was impaired relative to the wild type parent strain. Therefore, dentilisin (PrtP) is apparently crucial for the virulence of *T. denticola*, at least in terms of abscess formation in a confined wound. Yet, examination of the *prtP* mutant K1 yielded a rich variety of pleiotropic effects: 1) slow but dense growth of bacterial cells tightly coiled around each other; 2) reduced surface hydrophobicity; 3) reduced coaggregation with bacteria of the genus *Fusobacterium*; and 4) altered expression of other outer sheath proteins including Msp. Indeed, both Ishihara et al. (1998) and Fenno et al. (1998b) have reported that insertional inactivation of genes encoding either Msp or PrtP causes altered expression of the other surface proteins. Thus, dentilisin may affect virulence directly through its host tissue degradation [enhancing peptide acquisition], or indirectly by post-translational processing of other virulence factors, or by coupling within outer sheath complexes that have evolved an orientation or conformation of proteins that fosters the optimum expression of virulence.

6. Motility, chemotaxis, and their enhancement of gingival invasion

Since all sprirochetes are naturally motile, which fosters their colonization of protective environments, foraging for nutrients, and avoidance of noxious stimuli, motility *per se* does not discriminate between virulent and avirulent treponemes. Since spirochetal locomotion is influenced by the viscosity of the medium, it is likely that the rather advanced chemotactic and periplasmic flagellar system of treponemal motility evolved along with some of the species' invasiveness through the ECM. Experimental evidence indicates that the capacity to invade tissues differs among some groups of oral treponemes (Riviere et al., 1991b). Whether the difference depends on their motility or other factors like adhesion, proteolysis, or differential processing of chemotactic signals remains to be studied. The general capacity of oral spirochetes to penetrate tissues chemotactically through the [partially degraded] extracellular matrix probably accounts for the mixed nature of the treponemal morphotypes that have been observed at the forefront of acute periodontal lesions.

Recently, Lux and coworkers (2002) have begun to use molecular approaches to determine whether insertional disruption of genes encoding proteins that function in chemotaxis and motility diminishes tissue invasion by *T. denticola* 35405. They developed a tissue penetration assay using multilayers of tight-junctioned HOK-16B human gingival keratinocytes coincubated anaerobically in a Transwell® two-chamber system with *T. denticola* and mutant constructs that were either *flgE*- (Li et al., 1996), *dmcA*- (Kataoka et al., 1997), *dmcB*- (Li et al., 1999), or *cheA*- (Lux et al., 2001; 2002) defective. The *Treponema pallidum* subsp. *pallidum* Nichols strain and a *Treponema phagedenis* biotype Reiter strain were used as controls. *Treponema denticola* 35405 penetrated the tissue about half as effectively as *T. pallidum*, but *T. phagedenis* did not penetrate at all, even though its cells were motile. This finding alone suggests that treponemes differ in their innate capacity to penetrate tissues, as observed in the study by Riviere et al. (1991b), cited above, and infers that tissue penetration is not solely dependent on motility *per se*. Mutation of the flagellar hook gene of *T. denticola*, *flgE*, which rendered the cells nonmotile, completely inhibited their tissue penetration, as anticipated. Notably, mutation of the two genes that encode methyl accepting chemoreceptors greatly reduced tissue penetration without affecting motility. Similarly, the *cheA* mutant, which was deficient in the central phosphokinase of chemotactic signal transduction

(Lux et al., 2002), failed to penetrate the tissue model. Therefore, chemoattraction is apparently significant in the ability of *T. denticola*, and presumably other oral treponemes, to invade the gingival tissue. The proteins that comprise the structure, synthesis, export, orientation, motor, and energy systems for flagella, and their function must provide the means for the bacteria to propel themselves there.

Once within the tissues, flagellins from disrupted treponemes may feasibly interact with cognate [toll-like?] receptors on immunocompetent cells to stimulate innate immunity cascades that modulate inflammatory responses (Hayashi et al., 2001). Flagellin core proteins (FlaB) among *T. pallidum* subsp. *pallidum* Nichols strain and motile reference strains of most of the characterized oral treponemal species have been reported to be glycosylated (Wyss, 1998); these include *T. denticola*, *T. vincentii*, all three subspecies of *Treponema socranskii*, *Treponema amylovorum*, *Treponema maltophilum*, and *Treponema pectinovorum*. The periplasmic flagellar sheath proteins (FlaA) of *T. vincentii* and *Treponema medium* share an antigenic epitope that reacts with a monoclonal antibody, H9-2, originally raised against *T. pallidum* subsp. *pallidum* Nichols strain (Wyss, 1998). Such cross-reactivity gave rise last decade to the terminology "pathogen-related oral spirochetes" (PROS), first assigned to these treponemes by Riviere and coworkers (1991a). How flagellar glycosylation may affect innate and adpative immune responses that determine pathogenesis of lesions in which treponemes have invaded remains to be investigated.

It is also interesting to contemplate the theoretical evolution of treponemal flagellar complexes into virulence determinants of other pathogenic bacteria. Although treponemes are not known to use type III secretion systems to inject cytotoxins and other virulence factors into host cells upon contact, the proteins of the flagellar export apparatus are similar to and considered evolutionary antecedents of several proteins central to type III systems in enteric pathogens (Hardham et al., 1995). These are located in discrete operons in *T. pallidum* and in *T. denticola* (Stamm and Bergen, 1999, 2001). Futhermore, *T. denticola* expresses proteins MglA and MglB, encoded by an *mglBAC* operon (Lépine and Ellen, 2000), which closely resemble MglA and MglB in *T. pallidum* (Porcella et al., 1996; Stamm et al., 1996). Notably, though MglA and MglB comprise two of the component proteins of a putative ABC transporter, the deduced amino acid sequences of these proteins have rather high percent similarity to key proteins in the type III secretion systems of *Salmonella*, *Shigella*, and *Yersina* species (Lépine and Ellen, 2000). A combination of PCR and Southern hybridization detected *mglB* homologues in other oral treponemes: *T. vincentii* ATCC 35580, oral treponeme D66AR-1 (Smibert-2), and *T. socranskii* ATCC 35536, but not in *T. maltophilum* D4B-1.

7. Direct cytotoxicity

Motile treponemes that penetrate the gingival epithelium and gain access to the underlying connective tissues may encounter many host cell types. In the periodontal pocket, they may contact sulcular or junctional epithelial cells and neutrophils that have migrated chemotactically from the blood, through the epithelium, and into the pocket. Treponemes or their metabolic products may contact cells that maintain the homeostasis of connective tissues, like prominent gingival fibroblasts or their mesenchymal precursors. They may also interact with phagocytes such as macrophages, immunocompetent T and B cells, erythrocytes, and ultimately cells whose normally coupled functions maintain the integrity of the supportive alveolar bone. Toxic inhibition by lysing or affecting the growth of any of these cells may conceivably contribute to pathologic outcomes.

Fenno and coworkers (1998a) investigated cell lysis and metabolic suppression in a variety of cell types exposed to *T. denticola* native or recombinant Msp or CTLP complexes described above. These included the Chinese hamster ovary cell line (CHO-K1), rat osteogenic sarcoma cell line (ROS17/2.8), human gingival fibroblasts (HGFs), normal human epidermoid keratinocytes (NHEK), porcine periodontal ligament epithelial cells (PLE), the HeLa epithelial cell line, and human erythrocytes. Direct cytotoxicity varied with cell type; HGFs were most resistant. PLE cells were very sensitive to lysis with native and rMsp, as well as CTLP; erythrocytes were lysed only by Msp. Cytotoxic lysis, measured by release of lactate dehydrogenase, was concentration-dependent and seemed to follow the same pattern as adhesion of these molecules to the target cells.

How oral treponemes cause lysis of host target cells is not well understood. A number of species including *T. denticola*, *T. vincentii*, and *T. maltophilum* express varying degrees of phospholipase C (PLC) activity for substrate analogs *p*-nitrophenyl phosphorylcholine and 4-methylumbelliferyl phosphocholine, but *Treponema lecithinolyticum* is the only oral *Treponema* species known to express very strong PLC activity for the natural substrate egg or synthetic lecithin as well as strong phospholipase A activity (Wyss et al., 1999). Phospholipase activity combined with the wide substrate range of proteases expressed by some oral treponemes and cohabiting species in periodontal pockets should be sufficient to lyse the plasma membrane of target cells. *Treponema denticola* also has the capacity to lyse erythrocytes and to scavange hemin and iron (Grenier, 1991; Scott et al., 1996a). Hemolysis is mediated by the metabolic activity of a *T. denticola* cytoplasmic enzyme cystalysin, which evidently causes hemolysis by the formation of protrusions and ultimately irregular holes in erythrocyte membranes (Chu et al., 1994). Typical of putative virulence determinants of indigenous oral bacteria, the cystalysin of *T. denticola* ATCC 35404 has a key metabolic function; it acts as a L-cysteine desulfhydrase to catalyze the metabolism of sulphur-containing compounds, especially cysteine, with the release of H_2S, which can initiate lysis of erythrocytes and probably a variety of other cell types. (Chu et al., 1995; 1999; Krupka et al., 2000). H_2S is also known to increase the permeability of oral tissues, and it has been proposed to aid enzymatic digestion of collagen (Ng and Tonzetich, 1984). Since many of the oral spirochetes derive energy from amino acid metabolism, the generation of cytotoxic volatile sulphur-containing compounds from sulphated amino acids is likely to be widespread in the treponemal community. Notably, cultivation of *T. denticola* or *T. pectinovorum* at an elevated cysteine concentration increases their potential to induce abscesses in a murine model (Kesavalu et al., 1999).

While it is evident that oral spirochetes may cause lytic damage to some target cells, growth inhibition by interfering with critical metabolic processes is probably a more common outcome. In Fenno and coworkers' study of cytopathogenic effects of *T. denticola* Msp and CTLP (1998a), the epithelial cells (PLE, HeLa,

NHEK) were more sensitive than the HGFs to metabolic inhibition, as measured by a tetrazolium reduction kit. In another study, when the simple epithelial cells of the KB line from epidermoid carcinoma were challenged with high cell densities of whole *T. denticola* cells, two populations resulted, a nonviable pool, which detached from the substratum, and a pool that remained viable and recovered normal growth, although its cells had undergone significant shrinkage and aberrations in volume regulation (DeFilippo et al., 1995). Exposure of PLE cells to extremely high cell densities of *T. denticola* 35405 for 24 hours have also been reported to induce apoptosis in about 50% of PLE cells (Leung et al., 2002). In contrast, the inhibitory effects of *T. denticola* proteins, sonic extracts, and whole cells on the growth of HGFs seem to be transient and not related to cell lysis (Boehringer et al., 1984; Wang et al., 2001; Paes Batista da Silva et al., 2004). For example, HGFs exposed for hours to the same concentrations of *T. denticola* Msp that were used in the cytotoxicity studies of Fenno and coworkers (1998a) recover normal growth characteristics when they are washed free of Msp and incubated with fresh medium (Wang et al., 2001).

Treponema denticola, components of its outer sheath, some lipoproteins, and glycolipids have been found to affect the function of neutrophils, macrophages, and lymphocytes and to induce osteoclastic bone resorption *in vitro*. Neutrophils are able to phagocytose *T. denticola* cells (Lingaas et al., 1983; Olsen et al., 1984; Boehringer et al., 1986), but they do not necessarily respond as they would when phagocytosing other bacterial species that are common in dental plaque. Whole *T. denticola* cells have been reported to inhibit the oxidative responses of neutrophils (Sela et al., 1988) and to resist being killed by β-defensins in a cell-free system (Brissette and Lukehart, 2002). Yet, the components of the bacteria that mediate these effects have not been identified. Some lipoproteins that have been extracted from strain ATCC 35404 and Msp from 35405 have been reported to induce the oxidative response, and in the case of Msp, it did not alter the anticipated oxidative burst when the neutrophils were exposed to PMA, the positive control agonist. (Sela et al., 1997; J.Y. Lin et al., abstract #3672, Int. Assoc. Dent. Res. 2004). Both lipoproteins and glycolipids (lipooligosaccharides) extracted from *T. denticola* induce the expression of the pro-inflammatory cytokines tumor necrosis factor-α and interleukin-1 by macrophages; there is ample evidence that these cytokines contribute to progressive resorption of periodontal connective tissues, including bone.

Net bone loss in periodontitis has a complex pathogenesis that involves intersecting pathways of innate and adaptive immunity and the activation of inflammatory mediators that can resorb collagen, such as MMPs and cathepsins. Recent studies using NOD-SCID mice engrafted with human peripheral blood leukocytes and infected with the periodontal pathogen *Actinobacillus actinomycetemcomitans* have shown that specifically activated $CD4^+$ T cells are crucial for modulating the inflammatory response that activates osteoclastic bone resorption; RANKL (osteoprotegerin [OPG] ligand) was identified as a significant signaling molecule in this pathway (Teng, 2002; Teng et al., 2000). Outer membrane preparations from *T. denticola* ATCC 35405, 33521, and 35404 have been shown to induce resorption of fetal rat long bones *in vitro* (Gopalsami et al., 1993). The active components were heat stable, and their bone resorbing activity was enhanced by parathyroid hormone and prostaglandin E_2 (PGE_2), which are known to act synergistically with lipopolysaccharide (LPS). *Treponema denticola* is not known to produce classical LPS; yet, it contains 4-10 kDa glycosylated lipids (Schultz et al., 1998) that would presumably have the potential to perturb host cells through innate immunity receptors. Indeed, *T. denticola* outer membranes and glycolipids have been found to induce chemokines and cytokines in gingival epithelial cells and macrophages by activation of toll-like receptor 2 (Rosen et al., 1999; Opitz et al., 2001; Asai et al., 2003). Whole cells of *T. denticola* and, more potently, *T. pectinovorum* and its LPS-like glycolipid have been reported to induce elevated interleukin-6, interleukin-8 and [for *T. pectinovorum* and its LPS] macrophage-chemotactic protein from HGFs (Kesavalu et al., 2002). Recently, using a mouse calvaria and bone marrow cell co-culture system, Choi and coworkers (2003) found that glycolipids from *T. denticola* ATCC 33521 induced the differentiation of osteoclastic cells. Osteoblastic cells derived from the calvarial cultures and exposed to the glycolipid preparation yielded up-regulated mRNA expression for osteoblast differentiation factor (ODF) and down-regulated mRNA expression for OPG, the ODF receptor antagonist. The glycolipid preparation also induced the expression of PGE_2, an icosanoid that had been linked to the progression of connective tissue attachment loss in periodontitis and associated with the risk of systemic morbidity derived from longstanding periodontal inflammation (Offenbacher, 1998).

Alveolar bone matrix and the collagen rich matrix of the gingiva and periodontal ligament can be degraded by a variety of MMPs. MMPs are also thought to be important modulators of inflammatory responses by cleaving chemokines and their receptors (McQibban et al., 2002). In addition to activation of neutrophil and fibroblast MMPs by *T. denticola* dentilisin, mentioned above, the glycolipids, peptidoglycan, and a 53 kDa outer membrane protein induce the release of MMP-8 and MMP-9 from neutrophils (Ding et al., 1996). Glycolipids also caused increased expression of MMP-3, MMP-8, MMP-9, MMP-10, MMP-13, and MMP-14 from osteoblastic cells from the calvarial co-culture experiments cited above (Choi et al., 2003). It is evident, then, that several components associated with the cell wall and outer sheath of oral treponemes are possible candidates for virulence determinants that affect the integrity of the connective tissues that support the teeth, and they probably act by lysing host cells or by driving destructive inflammatory pathways that lead to resorption.

8. Perturbation of cytoskeletal dynamics and actin-dependent cellular functions

Bacteria and some of their products have the capacity to damage host cells without actually killing them. If prolonged infection interrupts pathways that normally remodel host tissues or that heal tissue injuries upon removal of an insult, then the microbial factors involved may be considered virulence determinants that contribute to disease chronicity. Since oral spirochetes penetrate the gingiva, live treponemes, shed outer sheath vesicles, and fragments of dying cells may engage connective tissue cells directly. Several laboratories have studied the responses of gingival fibroblasts to oral treponemes as a model to characterize the cytopathogenicity of the bacteria and to determine potential mechanisms by which they perturb cellular physiology. As indicated above, fibroblasts are relatively resistant to lytic killing by *T. denticola* dentilisin and Msp (Fenno et al., 1998a). Yet, exposure of HGFs to *T. denticola* cells causes the HGFs to undergo cytoskeletal reorganization, cellular rounding, formation of surface blebs, and eventual detachment from the substratum (Weinberg and Holt, 1990; Baehni et al., 1992). Whole treponemal cells, outer membrane extracts (OM), and Msp preparations all induce the disruption of phalloidin-stainable stress fibers in HGFs (Yang et al., 1998; Wang et al., 2001). Therefore, it is likely that a combination of *T. denticola* proteins contributes to the cytopathologic damage to the HGFs.

Whole *T. denticola* cells and the OM are highly proteolytic, and their dentilisin acts to degrade matrix proteins in the substratum as well as endogenous extracellular proteins, like fibronectin, that decorate the HGF plasma membrane (Ellen et al., 1994). This would tend to detach integrins of the HGFs from their cognate matrix components and contribute to cellular rounding, secondary cytoskeletal reorganization, and eventual detachment over hours of exposure. However, HGFs grown to confluence and challenged with bacteria on their dorsal surfaces undergo considerable cytoskeletal change starting within 30-45 minutes while remaining alive and attached to the substratum, even if dentilisin activity is inhibited with PMSF. Cytoskeletal rearrangement under these growth conditions is almost totally abrogated when HGFs are exposed to OM preparations that are heated at 60°C for 30 minutes, a temperature at which dentilisin remains active. Notably, protease-free native Msp preparations also induce actin filament reorganization and resultant perturbation of actin-dependent functions of fibroblasts (Wang et al., 2001; Paes Batista da Silva et al., 2004; Amin et al., 2004). Even in confluent epithelial cells, treatment with either OM or Msp induced transient micromotion of the cells, which is most likely a cytoskeletal response, when measured by cell substrate impedance sensing; the activity of the OM on micromotion was not prevented by inhibiting its proteolytic activity with PMSF (Ko et al., 1998b). These findings indicate that that dentilisin and nonproteolytic proteins of the *T. denticola* outer sheath, like Msp, can induce profound but not necessarily lethal cellular responses in host cells, and they most likely do it through different mechanisms.

Enriched native Msp complex of the type strain of *T. denticola*, ATCC 35405, has been found to have profound impact on actin dynamics in fibroblasts and some unprecedented downstream effects on actin-dependent functions. In addition to causing stress fiber disruption centrally and ventrally in the cell, exposure of fibroblasts to Msp complex induces coincident *de novo* assembly of a meshwork of subcortical actin filaments (Figure 3) (Wang et al., 2001; Amin et al., 2004). Concomitantly, the fibroblasts experience profound perturbation of store-operated calcium fluxes caused by the unresponsiveness of store-operated calcium channels (SOCs) to the normal signals that induce Ca^{2+} entry upon Ca^{2+} depletion of the endoplasmic reticulum (ER) stores (Wang et al., 2001). There is considerable evidence that subcortical actin assembly conformationally uncouples the SOCs and the ER. In this case, the uncoupling would be a downstream effect of Msp-induced subcortical actin assembly, which is a novel mechanism by which a bacterial protein would perturb the fine balance between calcium and actin dynamics in host cells.

Msp-induced subcortical actin filament assembly also affects the ability of fibroblasts to bind type 1 collagen, the major extracellular matrix protein in the

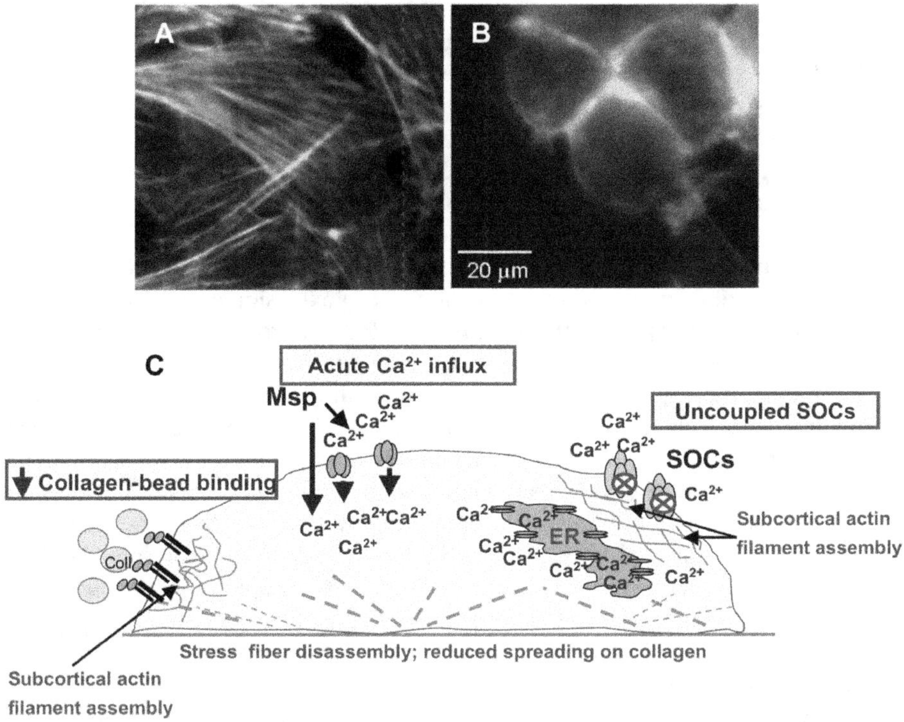

Figure 3. Fluorescence photomicrographs of Rat-2 fibroblasts exposed to A) αMEM (control vehicle) or B) Msp complex of *T. denticola* ATCC 35405 and stained with Alexa 488 phalloidin (Images provided by M. Amin and R.P. Ellen, unpublished but based on findings in Amin et al., 2004). Msp induces cell rounding and cytoskeletal reorganization that is characterized by disruption of stress fibers in the central and ventral regions of the cell and by *de novo* assembly of a network of subcortical actin filaments subjacent to the plasma membrane. C) Model depicting Msp-induced acute calcium influx and downstream effects on calcium and actin dynamics. Subcortical actin assembly is associated with the uncoupling of store-operated calcium channels (SOCs) from the release of endoplasmic reticulum (ER) Ca^{2+} stores, and restricted affinity modulation upon extracellular collagen engagement by β1 integrins, as seen experimentally in the i) reduced binding of collagen-coated fluorescent beads (pale grey) and ii) reduced cellular spreading on a collagen gel substratum.

gingiva. Paes Batista da Silva and coworkers (2004) have recently reported that pretreatment of HGFs and Rat-2 fibroblasts with native Msp inhibits the binding of collagen-coated beads by the fibroblasts and the spreading of the fibroblasts on a collagen substratum. Extensive experiments aimed to determine whether Msp altered the expression of β_1 integrins, which are the primary collagen receptors of these cells, the distribution of the integrins in relation to the collagen-coated beads, and the fluid migration of β_1 integrins in the plane of the plasma membrane found little if any change that was induced by Msp in comparison with an Msp-free vehicle. Yet, the Msp-pretreated fibroblasts repeatedly bound significantly fewer collagen-coated beads. A series of experiments designed to address the affinity of collagen-coated bead engagement by the cells determined that Msp-pretreatment of the fibroblasts prevented the beads from interacting in the minimal affinity range required to maintain stable adherence through a few rinse cycles.

Subsequent experiments that used β_1 integrin affinity-enhancing antibodies found that the antibodies reversed much of the inhibitory effect of Msp on the binding of collagen-coated beads. Similarly, prevention of Msp-induced subcortical actin filament assembly by an actin monomer-sequestering agent, latrunculin B, also reversed Msp's inhibition of collagen binding. Together, these findings suggest that Msp-induced subcortical actin assembly, in some way, prevents the inside-out modulation of β_1 integrin binding affinity that normally accompanies the engagement of extracellular collagen (Paes Batista da Silva et al., 2004). Again, this is a novel mechanism by which *T. denticola* Msp may perturb fibroblast function through its action on the actin cytoskeleton. As an extension of this finding, Amin and coworkers (2004) recently reported that Rat-2 fibroblasts that were treated with *T. denticola* Msp incorporated actin monomers at the fast-growing barbed end of subcortical actin filaments coincident with the depolymerization of actin filaments in the center of the cell. Msp-treated Rat-2 cells cultured as a confluent monolayer on a collagen matrix and then incised with a needle to create a model wound migrated more slowly than control cells to close the wound. Therefore, it is likely that Msp-induced actin reorganization inhibits fibroblast engagement of collagen by integrins at an affinity sufficient to support cellular spreading and migration on the matrix. These coordinated functions, which fibroblasts use in their normal remodeling of extracellular matrix, were greatly disturbed even though the cells maintained viability throughout the Msp exposure (Wang et al., 2001; Paes Batista da Silva et al., 2004).

The mechanisms by which Msp applied extracellularly induces actin reorganization are not well understood. Calcium is a ubiquitous messenger that regulates actin dynamics, and Msp-induced subcortical actin filament assembly was inhibited experimentally by BAPTA/AM, an intracellular Ca^{2+} chelator (Wang et al., 2001). Treatment of HGFs with either Msp or OM preparations is known to induce acute and prolonged cytoplasmic $[Ca^{2+}]$ transients due to immediate calcium entry from the extracellular pool (Wang et al., 2001). It is feasible that either the integration of the Msp into the host cell membrane (Mathers et al., 1996) or the interaction of Msp with an endogenous plasma membrane ion channel receptor could initiate acute calcium entry. Though compelling, it is not known whether Msp-induced Ca^{2+} influx is the initial trigger for the actin reorganization that determines the profound changes in cellular function described above, and the intermediate signaling pathways have not been identified and confirmed. Calcium is required to activate the binding of the actin capping and severing protein gelsolin. Our laboratory has preliminary evidence that exposure of HGFs to whole *T. denticola*, OM, and Msp preparations increases the association of gelsolin with a cytoskeletal fraction of the cells (Ellen et al., abstract #725, Int. Assoc. Dent. Res., 2002); yet the direct connection with acute Ca^{2+} influx and with the subsequent actin filament reorganization is so far elusive. There is also some evidence that whole *T. denticola* cells and OM preparations cause diminished inositol phosphate responses to extracellular agonists that normally induce ER Ca^{2+} store release (Yang et al., 1998). Thus it is likely that several signaling pathways are affected concurrently, and it will be difficult, indeed, to determine with precision which pathways are disabled or exploited by specific treponemal surface components.

Collagen phagocytosis, the major physiologic pathway by which collagen fibrils are remodeled in the gingiva, represents an example of this complexity. Though native Msp complex clearly inhibits β_1 integrin engagement of collagen (Paes Batista da Silva et al., 2004), the first and presumably rate-limiting step of collagen phagocytosis, treatment of HGFs with *T. denticola* OM preparations, which contain Msp, was actually found to enhance the phagocytosis of collagen-coated beads (Battikhi et al., 1999). The OM basically deregulated endocytosis in general, as it also enhanced phagocytosis of BSA-coated beads, which do not bind integrin receptors, and it even enhanced fluid phase pinocytosis. The mechanism of OM action is unknown. In addition to Msp, the OM preparations contained dentilisin and had considerable peptidase activity. It had numerous other proteins and probably associated glycolipids. There is little known about the architecture of the OM or the native outer sheath as it exists on the *T. denticola* surface. Therefore, inconsistencies in experimental findings using various native extracts from the bacteria are to be expected. For example, the inhibitory effect of Msp on collagen engagement and its effect on store-operated calcium fluxes were found to be transient and reversed by washing the cells free of Msp (Wang et al., 2001; Paes Batista da Silva et al., 2004). Yet, only short-term exposure to the OM was found to be sufficient for long-lasting enhancement of collagen phagocytosis (Battikhi et al., 1999). While *in vitro* cell challenge experiments are helpful to identify individual virulence determinants, their damaging effects on host cells, and some of the potential mechanisms by which they act, host cells are normally exposed *in vivo* to bacterial molecular complexes, and the bacteria may even be decorated as well with host macromolecules.

9. Proteins that are putative virulence determinants of oral treponemes

Though a logical model for the contribution of oral treponemes to the pathogenesis of mixed periodontal infections is presented above, almost all of the mechanisms raised are theoretical and based mostly on *in vitro* cytopathogenicity studies. A limited set of treponemal proteins has been characterized at the molecular level; even fewer have been mutated genetically and screened *in vivo* to determine whether they are essential for virulence of the parent species. Only one, cystalysin, has been crystallized and submitted to detailed structural analysis (Krupka et al., 2000). Virtually all of the limited genetic manipulation accomplished to date has centered on *T. denticola*, which is routinely cultivated and the first of the treponemes to be transformed in the laboratory (Li et al., 1996). All mutants generated to date have been accomplished by insertional mutagenesis using an *ermF-ermAM* gene cassette, a method that requires pre-conceived selection of the targeted gene. The field still lacks a transposon mutagenesis system, which would be useful for identifying the range of genes that encode or regulate a given virulence phenotype in treponemes. Nevertheless, substantial progress has been made to characterize some of the proteins mentioned above at the molecular level. The release of the annotated *T. denticola* genome (http://www.ncbi.nlm.nih.gov/genomes/framik.cgi?db=genomeandgi=381) and the partial genome sequence of *T. lecithinolyticum* in progress (F. E. Dewhirst, personal communication) should

Table 1. Putative virulence determinants of *Treponema denticola*

Putative virulence determinant	Function in physiology, bacterial survival	Putative contribution to pathogenicity
Chymotrypsin-like protease complex	Peptide acquisition	Degrade ECM proteins Activate host MMPs
Dentilisin, PrtP	Prolyl-phenylalanine peptidase	Promote migration through ECM Cytoskeletal reorganization secondary to ECM degradation
PrcA, PrcB	Accessory proteins Integration of PrtP in outer sheath, chaperone(s)?	Required for PrtP activity
Major outer sheath protein, Msp	Adhesin Porin	Cytoskeletal reorganization Stress fiber disruption Impairs volume regulation Subcortical actin assembly Inhibits collagen binding Impairs cell locomotion Acute calcium influx Uncouples store-operated channels Hemolysis
Cystalysin, Hly	Amino acid metabolism PLP– dependent amino-transferase	Generation of H_2S Hemolysis Hemoxidation
Chemotaxis protein A, CheA	Phosphokinase Energize chemotactic motility	Tissue invasion
Flagellar and chemotaxis proteins	Directed motility Scavenging nutrients Avoiding noxious agents	Tissue invasion

also highlight potential virulence determinants that have significant amino acid sequence similarity to virulence factors already identified and confirmed among other bacterial genera. The following paragraphs and Table 1 describe some of these putative treponemal virulence determinants in more detail.

9.1. Dentilisin and the PrcA – PrtP subtilisin families

Dentilisin (Ishihara et al., 1996) is the 72 kDa serine protease of the CTLP complex that clearly has properties of a key virulence determinant of *T. denticola*. Two accessory proteins, PrcA1 and PrcA2, were first identified biochemically and assigned a variety of masses (Uitto et al., 1988; Rosen et al., 1995). Clarity was brought to the relationship between these proteins in terms of biosynthesis and processing, and theoretically in terms of their putative functional relationship, through molecular biology and expression studies: 1) the detailed characterization of the nucleotide and deduced amino acid sequence of the gene encoding dentilisin (*prtP*) and the upstream open reading frame (*prcA*) (Ishihara et al., 1996); and 2) the likely series of steps for post-translational processing of the *prcA* gene product by dentilisin itself to form the accessory proteins of the CTLP (Lee et al., 2002). It is thought that the co-transcribed PrcA and PrtP have a reciprocal dependence, in that cleavage of the *prcA* gene product by PrtP is necessary to generate the accessory proteins that, in turn, are required for functional "maturation" of the CTLP in the outer sheath (Lee et al., 2002). Indeed, Ishihara and coworkers reported that insertional inactivation of the ORF now called *prcA* yields transcribed but functionally inactive dentilisin (K. Ishihara et al., abstract #941, Int. Assoc. Dent. Res, 2001). Whether the PrcA proteins serve as chaperones or in some other way sterically integrate dentilisin in a functional orientation in the outer sheath remains to be addressed specifically.

The cloning and characterization of dentilisin (Ishihara et al., 1996) and the combined molecular biology and physiology studies of PrtP and the PrcA proteins (Lee et al., 2002) provided key knowledge to seek homologous protease genes among other treponemes. Heuner and coworkers (2001) detected SAAPFNA-degrading activity among strains of *T. denticola*, *T. lecithinolyticum*, and *T. socranskii* subsp. *socranskii*. The *T. socranskii* subsp. *socranskii prtP* gene was cloned, and the nucleotide sequence yielded a deduced amino acid sequence that was 63% similar to that of *T. denticola* PrtP, with a gene of partial similarity to the immediate upstream ORF that had been reported by Ishihara and coworkers (1996). Partial sequences were also obtained for two partly similar *prtP* homologs in *T. lecithinolyticum*; one was more similar to *prtP* in *T. socranskii* subsp. *socranskii*. Recently, these significant but disparate findings came together with more clarity following the sequencing of *prcA* and *prtP* homologues for several additional species of oral treponemes by Correia and coworkers (2003). They found two paralogous families of subtilisin family serine proteases among the oral treponemes, the first lipoproteins reported among the subtilisins (Figure 4). Within each family cluster, the deduced amino acid sequence for both PrcA and PrtP were found to have the same high level of similarity. Notably, there were two *T. lecithinolyticum* sequences, and they segregated to each of the families. The representative strain of this species evidently has two copies of *prtP*, one from each cluster, which is compatible with the finding of Heuner et al. (2001). Only members of the family including *T. denticola* degraded SAAPFNA, and the *T. denticola* type strain was found to have a wide protein substrate-degrading range including casein, gelatin, fibronectin, fibrinogen, type VI collagen, and laminin. Significantly, the representative species of the other family, *T. vincentii* 861, had little if any obvious protease activity against these substrates. Yet, analysis of the deduced amino acid sequences for PrtP of both families found that the consensus sequences for catalytic domains typical of subtilisin family proteases were conserved (Figure 5). A possible explanation for undetectable proteolytic activity in one of the families may derive from the amino acid sequence of its *prcA* gene product, which is normally cleaved to form the two PrcA proteins (Lee et al., 2002). PrcA is evidently processed by dentilisin, itself, and therefore a favorable sequence for its cleavage would contain a conserved prolyl-phenylalanine peptide bond. Such a site is conserved among PrcAs in the family that contains *T. denticola*, but there are histidines substituted for the putative cleavage-site phenylalanine in the PrcAs of the strains examined

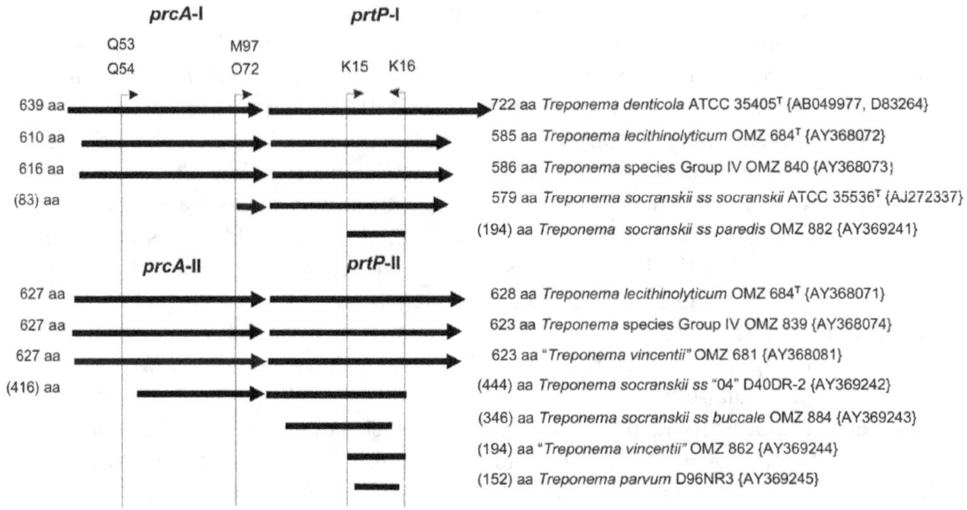

Figure 4. Schematic diagram of the two paralogous oral treponemal subtilisin family proteases encoded by the *prcA* - *prtP* operon. The relative length of genetic fragments that have been sequenced are shown as thick arrows. The primers that were used for amplifying the DNA fragments are shown in their approximate positions (thin bent arrows, dotted vertical lines, and letter codes). The number of deduced amino acids encoded by the genes is listed. The *prtP* gene encodes the active protease dentilisin, and the *prcA* gene encodes the two accessory proteins of the chymotrypsin-like protease complex of the treponemal outer sheath. Reprinted from Correia et al., J. Bacteriol. *185*, 6860-6869, 2003, with permission.

for the paralogous family that contains *T. vincentii*. If the *prtP* gene product of this latter family is proteolytic, its substrate range has yet to be identified.

9.2. Major outer sheath protein (Msp) and its homologs

The native major outer sheath protein of *T. denticola* is abundantly expressed as an immunogenic, detergent and protease-resistant, oligomeric complex on the surface of the bacterium (for details, see Weinberg and Holt, 1991; Fenno et al., 1996, 1997; 1998b). It shares some sequence similarity with the *T. pallidum* repeat (Tpr) proteins that are being investigated as vaccine candidates (See Chapters 2, 10, 11 and 12). Msps have evidently evolved as strain-specific molecular forms, with variable molecular mass, that are distributed among wild-type strains and that share between 50% and almost 100% amino acid identity (Weinberg and Holt, 1991; Fenno et al., 1997). Upon boiling, the complex dissociates into monomer polypeptides of 53 – 64 kDa, depending on the strain from which it was isolated (Weinberg and Holt, 1991; Fenno et al., 1997). The translated polypeptide has a typical prokaryotic signal sequence, which is its only strongly hydrophobic domain, and a signal peptidase cleavage site. The mature native protein has a molecular mass of ~56 kDa. Analysis of the amino acid sequence predicts several potential transmembrane regions in the form of a β barrel (Fenno et al., 1996). The overall structural characteristics resemble that of typical outer membrane porins. Indeed, Msp has been identified to function like a porin *in vitro* (Egli et al., 1993) and to

A. Alignment of conserved catalytic residues typical of subtilisins

PrtP-I

192YGDKTVVAGII**D**TGINMKHKDFK	247IPIGLNWDKGA**H**GTHCSGTIAAR	436GEDDYETMSGT**S**MATPFVTGTIA	*T. denticola*
175YGDKDVVAGII**D**TGINMLHEDFK	230VPLNSNWDADG**H**GTHCSGIMCAL	419DVNSYVSLSGT**S**MAAPFVTGFVS	*T. lecithinolyticum*
174YGTNEVVAGII**D**TGINMKHEDFK	226VPLDSNWDTDG**H**GTHCSGIMCAL	419STDSYVSMSGT**S**MAAPFVTGLIS	*T.* sp. OMZ 840
171YGDKDVVAGII**D**TGINMLHEDFN	224VPLGSNWDTDG**H**GTHCSGIICAL	412TASSYVSLSGT**S**MAAPFVTGLIS	*T. socranskii* ss *socranskii*
YG...VVAGII**D**TGINM.H.DF.	.P...NWD...**H**GTHCSG...A.	....Y...SGT**S**MA.PFVTG...	Consensus PrtP-I

PrtP-II

210DAKHPVLAAIV**D**SGVNGLHEDFY	260IPINDNWDDVG**H**GTHCAGTIAAV	458LRKSYTWMSGT**S**MATPFVTGTLT	*T. lecithinolyticum*
199AATNQVLVAIV**D**TGVNSLHEDLY	250VPINENWDNVG**H**GTHCSGTIAAV	448VKQSYLWMSGT**S**MATPFVTGTIA	*T.* sp. OMZ 839
200DAKYPVLAAII**D**TGVNSLHEDFY	251ISLDENWDNHG**H**GTHCSGTIAAV	449VKSSYRWMSGT**S**MATPFVTGVVT	*T. vincentii*
.A...VL.AI.**D**.GVN.LHED.Y	.....NWD..G**H**GTHC.GTIAAV	...SY.WMSGT**S**MATPFVTG...	Consensus PrtP-II

Dual consensus

.....V...I **D**.G.N..H.D..	.....NWD...**H**GTHC.G...A.	....Y...SGT**S**MA.PFVTG...	Dual consensus

B. Potential peptidase cleavage sites in PrcA deduced sequence

PrcA-I (↓ ... ↓)

SAAA**PL**SNAIFGRVFVIMERV**PF**TNNTFS	*Treponema denticola*
AAAS**DL**STCTFDKLSVNTERL**PF**SENLFS	*Treponema lecithinolyticum* gene I
NSTL**DL**SSCSFRNLSAFAERL**PY**SEGILS	*Treponema sp OMZ 840*

PrcA-II

PALTDISGSSVKDFRIEMHRF**PH**SLKLFS	*Treponema lecithinolyticum* gene II
PGLKSMTGARIKKFRVEMHRF**PH**SLKLFN	*Treponema sp OMZ 839*
PAIKDMNGASVKNFRVEMHRF**PH**SLKLFS	*Treponema socranskii ss "04"*

Figure 5. A) Aligned deduced amino acid sequences for the two treponemal subtilisin family proteases, PrtPI and PrtPII. Despite the conserved consensus sequence around the putative catalytic residues in both (bold, underlined Aspartatic acid, Histidine, Serine), neither protease nor peptidase activity has been identified for PrtPII. B) A possible explanation may lie in a nonconservative histidine substitution for phenylalanine in a prolyl-phenylalanine of accessory protein (chaperone?) PrcAII that is the signature target peptide in substrates (including PrcAI) digested by PrtPI subtilisins like dentilisin. Reprinted from Correia et al., J. Bacteriol. *185*, 6860-6869, 2003, with permission.

insert in and depolarize HeLa cell plasma membranes (Mathers et al., 1996; Fenno et al., 1998a).

Genes encoding Msp-like proteins are common among several oral treponemes and have been sequenced in a few strains of *T. denticola* and *T. maltophilum* ATCC 51939^{T} (Fenno et al., 1997, 1998b; Heuner et al., 1999). Many of these genes have well conserved nucleotide sequences (>90%) at the 5′ and 3′ flanking ends and at the signal peptide-coding region; yet some of the mature Msps do not cross-react serologically. The fully deduced amino acid sequence for ATCC 35405, 33520, and strain OTK have been compared by Fenno and coworkers (1997). Msps of the former two strains have close to the same predicted mass (~58.2 and ~58.7 kDa respectively) and identical amino acid sequences other than for 67 residues in the central region of the molecule, where they match with only 41% identity (61%, counting conserved substitutions). The deduced OTK Msp is larger at ~62.6 kDa, with only 43% identity (64% counting conserved substitutions) to the Msp of 35405; the signal peptide encoding region is the most conserved. The Msp of strain OTK was found to resemble that of the major outer sheath protein of *T. denticola* strain GM-1, reported earlier by Weinberg and Holt (1991). It does not react with anti-35405 Msp antibodies. While the various classes of Msps found among oral treponemal strains differ in antigenicity and amino acid sequence, especially in the central region of the molecule, they share a number of characteristics that suggest a common function. A secondary structure for the mature protein that is rich in potential transmembrane regions and virtually free of hydrophobic domains suggests a membrane topology similar to that of other bacterial porins and pore-forming toxins of Gram-negative bacteria (Fenno et al., 1997). It is likely that these fastidious species evolved such porins to optimize the uptake of nutrient solutes, and the potential toxic impact of Msps on host cells developed as a parasitic side effect.

9.3. Cystalysin

Cystalysin (Hly) is a 46 kDa cytoplasmic protein of *T. denticola* that has significant amino acid sequence similarity to the family of pyridoxal 5'-phosphate (PLP) – dependent aminotransferases (see Chu et al., 1994; 1995; 1997; Krupka et al., 2000 for details). It is the first protein of oral treponemes to be crystallized and its structure studied in three-dimensional detail. Cystalysin occurs as a homodimer in solution. Each monomer of cystalysin folds into two domains; the larger of the two binds the PLP cofactor covalently (Krupka et al. 2000). The small domain contains residues from both the N- (residues 1-47) and C- (residues 288-394) termini. The large domain has a central seven-stranded β-sheet, similar to that found in members of the aminotransferase family. The β-sheets are wedged between five and four helices at each end. The small domain has a three-stranded anti-parallel β-sheet bound by three and one helices. The PLP cofactor binds to a lysine residue at position 238, near the C-terminal end of the large domain's seven-stranded β-sheet, in a deep cleft formed by the two domains.

Cystalysin's structure is distinct from that of other known bacterial thiol-active hemolysins; it is less hydrophobic and has a higher content of charged amino acids. Furthermore, investigation of its amino acid sequence and synthesis have found no evidence of signal sequences or polypeptide structure that would indicate that it is secreted extracellarly and thereby in a position to contact erythrocytes or other host cells directly, as would pore-forming hemolysins and toxins. Based on structural evidence and similarity to other known PLP-dependent β-cystathionase enzymes, Krupka and coworkers (2000) proposed a model for cystalysin's hemolytic activity, which focuses on its catalysis of cysteine to H_2S, which would damage target cell membrane phospholipids and cause membrane protrusions and eventual lysis. A second putative mechanism that they proposed was based on cystalysin's generation of sulphane sulphur derivatives, like persulphides, that would have the potential to oxidize and transfer sulphur to essential enzymes of target cells and thereby cause irreversible damage and cell death. They also hypothesized that *T. denticola*'s hemagglutinating activity would position the bacteria in close proximity to erythrocytes, thereby increasing the local concentration of H_2S and also providing immediate access to the hemin and other cytoplasmic components of the lysed cells. Cystalysin has both hemolytic and hemoxidative activities (Chu et al., 1999). It is active over a wide pH range, with an optimum pH 7.8-8.0, and its enzymatic activity is enhanced by reducing agents. These are characteristics that would favor its activity in the periodontal pocket environment (Chu et al., 1997).

9.4. Motility and chemotaxis proteins

Knowledge of the *T. denticola* proteins that determine the bacterium's directional motility, and resultant tissue invasiveness, is derived mostly from studying the genetic organization of operons that have similarity to those of other motile bacteria, especially other spirochetes. As determined for *B. burgdorferi* and *T. pallidum*, whose genomes were sequenced earlier, motility and chemotaxis genes account for a significant segment of the genome, which suggests that these functions are central to the survival of the species. Genes that encode proteins previously assigned the motility functions of flagellum and hook biosynthesis, export, flagellar rotation motor and switch (Heinzerling et al., 1997; Stamm and Bergen, 1999; 2001; Limberger et al., 1999) are located in at least three regions in the *T. denticola* genome. A region of proteins designated flagellar biosynthesis consists of *fliQ, fliR, flhB*, and *flhA*; these genes have deduced amino acid sequences that suggest that they function in flagellar export (Stamm and Bergen, 2001). They are located among many genes for proteins of undesignated function. A second region includes *flgB* and *flgC*, basal-body rod proteins; *fliE*, hook-basal-body complex; *fliF*, flagellar M-ring protein; *fliG*, motor switch protein; TDE1217, putative *fliH* for flagellar assembly; *fliI*, flagella-specific ATP synthase; and TDE 1219, putative flagellar protein *FliJ*. The third region contains ten genes. The first three genes encode flagellar hook proteins involved in synthesis and assembly: *fliK* (*tap1*), *flgD*, and *flgE*. *FlgE* was the first treponemal gene to be mutated (Li et al., 1996). *MotA* and *motB* are genes that encode motility and motor rotation. They are followed by *fliL*, *fliM*, and *fliY*, the latter two assigned motor switch functions. Following *fliY* is a gene encoding a hypothetical protein TDE2761 and then *pliP*, which is assigned flagellar export. The gene that encodes the flagellar filament outer sheath protein, *flaA*, is located as an apparent orphan elsewhere in the *T. denticola* chromosome. Analogously, *Treponema vincentii flaA*-1 encodes a flagellar sheath protein. Its nucleotide sequence is close to that of *flaA*-1 in *T. pallidum* and a fragment of similar sequence could be amplified from the two *Treponema* species by PCR using primers based on the *T. vincentii* sequence (Sato and Kuramitsu, 2000). Notably, these same primers were able to amplify fragments from two samples among 6 patients with periodontitis, even though the samples were negative for amplification with a *T. vincentii* 16S rRNA primer set, suggesting that the samples contained other species of "pathogen-related oral spirochetes" mentioned above.

Proteins involved in chemotaxis comprise a sensing, signal transduction, and phosphotransferase system that guides and energizes directional motility. Genes that encode the major proteins in these pathways are located in two loci in the *T. denticola* genome. The genes that encode proteins involved in energetics, *cheA*, *che W-1*, *cheX*, and *cheY*, cluster in a single operon that is identical in organization to that of *T. pallidum*, and the *che* genes of these two *Treponema* species are phylogenetically close (Greene and Stamm, 1999). The genes that encode methyl-accepting chemotaxis receptor proteins, *dmcA* and *dmcB*, are located in an operon with an intervening protease gene, *prtB*. Li and coworkers (1999) characterized the *dmcB* locus, including insertional inactivation of the gene. Compared with their parent, the mutants failed to migrate in an agarose-layered chemotaxis assay. Moreover, the mutants yielded a protein methylation profile that lacked methylation of the prominent ~40 kDa bands and methylation of DmcA, which were typical of the parent strain. The way that the Che and Dmc protein pathways interact to mediate directional motility in spirochetes has not been addressed experimentally. Because the genes are well conserved in motile bacteria, it is likely that the methyl-accepting chemoreceptors modulate the activity of the central phosphokinase CheA. A recent review of bacterial locomotion presented a schematic model whereby CheW provides the interface between the chemoreceptors and the phosphotransfer pathway (Bourret et al., 2002). The proteins, genes, synthetic and regulatory pathways that control the physiology of treponemal directional motility are discussed in much more detail elsewhere in this volume (See Chapters 3, 6 and 8).

Are the proteins that regulate and function in treponemal chemotactic locomotion really “virulence determinants”? If so, then all motile bacteria would be expected to be pathogenic. All spirochetes, pathogens or not, are motile by virtue of this same set of proteins. Yet, there is currently no definitive system to categorize oral treponemes according to their innate pathogenicity. They are all parasites, meaning that they colonize, adapt, and live off the provisions scavenged in their human host’s oral environment. Oral spirochetes are fastidious survivors that have evolved directional motility, among other traits, to help individuals survive and their population to persist.

10. Summary and conclusion

Virtually all of the putative virulence determinants of oral treponemes studied to date provide the bacteria with survival strategies and perhaps selective advantages in their natural environment. Under conditions of periodontal health in a host who practices optimum oral hygiene, treponemes account for only a minor, almost undetectable, group within the bacterial population. They are fastidious anaerobes that would not find favorable conditions under these circumstances. Yet, in response to the growth in number and diversity of cohabiting species and the natural inflammation that would follow innate immune responses to an increase in biofilm mass that accompanies poor hygiene, the population of treponemes expands both in number and variety of phylotypes. Each of the putative virulence determinants discussed in this chapter would have a significant impact on the survival and expansion of the treponeme pool. Chemotactic receptors and energizers like Dmcs and CheA would orient the treponemes toward attractive solutes and away from noxious stimuli. Flagellar structural proteins, export proteins, and motors would provide the motility apparatus to move them in response to the chemotactic signals. Adhesins/porins like Msp would promote binding, acquisition, and uptake of nutrients. Proteinases like dentilisin would increase the local supply of required peptides in the vicinity of the bacterial surface. Through induction of cytoskeletal reorganization and proteolysis of cellular ligands and integrin receptors, these proteins may also deregulate host cell communication networks and migration, thereby prolonging the resolution of the infection. Surface glycoproteins and glycolipids, such as lipooligosaccharides, would stimulate inflammatory responses that provide a nourishing supply of host solutes in crevicular fluid and blood. Metabolic enzymes like cystalysin would help the bacteria derive energy from sulfated amino acids and generate end products that lyse erythrocytes, releasing hemin and a source of iron. Together, these factors contribute to the colonization, growth, adaptation, and procreation of the diverse phylotypes that comprise the oral treponemes. They are properties that promote the pathogenicity of some (most, all?) of the species that inhabit the oral cavity, and they can be modified by local conditions in the host. What scientists call "virulence determinants" indigenous treponemes call "lifestyle".

Acknowledgments

Original research from Ellen and coworkers, cited in this chapter, was supported by grant MOP-5619, the CIHR Group in Matrix Dynamics grant, and strategic training grant STP-53877 from the Canadian Institutes of Health Research.

References

Amin, M., A.C.S. Ho, A. Paes Batista da Silva, M. Glogauer, and R.P. Ellen. (2004). Induction of *de novo* subcortical actin filament assembly by *Treponema denticola* major outer sheath protein (Msp). Infect. Immun. *72*, 3650-3654.

Asai, Y., T. Jinno, and T. Ogawa. (2003). Oral treponemes and their outer membrane extracts activate human gingival epithelial cells through toll-like receptor 2. Infect. Immun. *71*, 717-725.

Baehni, P.C., M. Song, C.A.G. McCulloch, and R.P. Ellen. (1992). *Treponema denticola* induces actin rearrangement and detachment of human gingival fibroblasts. Infect. Immun. *60*, 3360-3368.

Battikhi, T., W. Lee, C.A.G. McCulloch, and R.P. Ellen. (1999). *Treponema denticola* outer membrane enhances the phagocytosis of collagen-coated beads by gingival fibroblasts. Infect. Immun. *67*, 1220-1226.

Boehringer, H., P.H. Berthold, and N.S. Taichman. (1986). Studies on the interaction of human neutrophils with plaque spirochetes. J. Periodontal Res. *21*, 195-209.

Boehringer, H., N.S. Taichman, and B.J. Shenker. (1984). Suppression of fibroblast proliferation by oral spirochetes. Infect. Immun. *45*, 155-159.

Bourret, R.B., N.W. Charon, A.M Stock, and A.H. West. (2002). Bright lights, abundant operons – Fluorescence and genomic technologies advance studies of bacterial locomotion and signal transduction, review of the BLAST meeting, Cuernavaca, Mexico, 14 to 19 January (2001). J. Bacteriol. *184*, 1-17.

Brissette, C.A. and S.A. Lukehart. (2002). *Treponema denticola* is resistant to human β-defensins. Infect. Immun. *70*, 3982-3984.

Caimano, M. J., K.W. Bourell, T.D. Bannister, D.L. Cox, and J.D. Radolf. (1999). The *Treponema denticola* major sheath protein is predominantly periplasmic and has only limited surface exposure. Infect. Immun. *67*, 4072-4083.

Casadevall, A. and L. Pirofski. (2001). Host-pathogen interactions, the attributes of virulence. J. Infect. Dis. *184*, 337-344.

Chan, E.C.S., R. Siboo, T. Keng, N. Psarra, R. Hurley, S.-L. Cheng, and I. Iugovaz. (1993). *Treponema denticola* (ex Brumpt 1925) sp. Nov., nom. Rev., and identification of new sprirochete isolates from periodontal pockets. Int. J. Syst. Bacteriol. *43*, 196-203.

Chi, B., M. Qi, and H.K. Kuramitsu. (2003). Role of dentilisin in *Treponema denticola* epithelial cell layer penetration. Res. Microbiol. *154*, 637-643.

Choi, B.-K., H.J. Lee, J.H. Kang, G.J. Jeong, C.K. Min, and Y.-J Yoo. (2003). Induction of osteoclastogenesis and matrix metalloproteinase expression by the lipooligosaccharide of *Treponema denticola*. Infect. Immun. *71*, 226-233.

Chu, L., A. Burgum, D. Kolodrubetz, and S.C. Holt. (1995). The 46- kilodalton-hemolysin gene from *Treponema denticola* encodes a novel hemolysin homologous to aminotransferases. Infect Immun. *63*, 4448-4455.

Chu, L., J.L. Ebersole, G.P. Kurzban, and S.C. Holt. (1997). Cystalysin, a 46-kilodalton cysteine desulfhydrase from *Treponema denticola*, with hemolytic and hemoxidative activities. Infect. Immun. *65*, 3231-3238.

Chu, L., J.L. Ebersole, G.P. Kurzban, and S.C. Holt. (1999). Cystalysin, a 46-kDa L-cysteine desulfhydrase from *Treponema denticola*, biochemical and biophysical characterization. Clin. Infect. Dis. *28*, 442-450.

Chu, L., W. Kennell, and S.C. Holt. (1994). Characterization of hemolysis and hemoxidation activities by *Treponema denticola*. Microb. Pathog. *16*, 183-195.

Correia, F.F., A.R. Plummer, R.P. Ellen, C. Wyss, S.K. Boches, J.L. Galvin, B.J. Paster, and F.E. Dewhirst. (2003). Two paralogous families of a two-gene subtilisin operon are widely distributed in oral treponemes. J. Bacteriol. *185*, 6860-6869.

Dawson, J.R. and R.P. Ellen. (1990). Tip-oriented adherence of *Treponema denticola* to fibronectin. Infect. Immun. *58*, 3924-3928.

Dawson, J.R. and R.P. Ellen. (1994). Clustering of fibronectin adhesins toward *Treponema denticola* tips upon contact with immobilized fibronectin. Infect. Immun. *62*, 2214-2221.

DeFilippo, A.B., R.P. Ellen, and C.A.G. McCulloch. (1995). Induction of cytoskeletal rearrangments and loss of volume regulation in epithelial cells by *Treponema denticola*. Arch. Oral Biol. *40*, 199-207.

Dewhirst, F.E., M.A. Tamer, R.E. Ericson, C.N. Lau, V.A. Levanos, S.K. Boches, S. K., J.L. Galvin, and B.J. Paster. (2000). The dirversity of periodontal spirochetes by 16S rRNA analysis. Oral Microbiol. Immunol. *15*, 196-202.

Ding, Y., V.-J. Uitto, M. Haapasalo, K. Lounatmaa, Y.T. Konttinen, T. Salo, D. Grenier, and T. Sorsa. (1996). Membrane components of *Treponema denticola* trigger proteinase release from human polymorphonuclear leukocytes. J. Dent. Res. *75*, 1986-(1993).

Egli, C., W.K. Leung, K.H. Müller, R.E.W. Hancock, and B.C. McBride. (1993). Pore-forming properties of the major 53 kilodalton surface antigen from the outer sheath of *Treponema denticola*. Infect. Immun. *61*, 1694-1699.

Ellen, R.P. (2002). Adhesion of oral spirochaetes to host cells and its cytopathogenic consequences. pp. 247-276 *in* Adhesion of Bacteria to Host Cells, Mechanisms and Consequences, M. Wilson (ed.). Cambridge, UK, Cambridge University Press.

Ellen, R.P., M. Song, and C.A.G. McCulloch. (1994). Degradation of endogenous plasma membrane fibronectin concomitant with *Treponema denticola* 35405 adhesion to gingival fibroblasts. Infect. Immun. *62*, 3033-3037.

Ellen, R.P., K.S.-C. Ko, C.-M. Lo, D.A. Grove, and K. Ishihara. (2000). Insertional inactivation of the *prtP* gene of *Treponema denticola* confirms dentilisin's disruption of epithelial junctions. J. Mol. Microbiol. Biotechnol. *2*, 581-586.

Ellen, R.P., M. Wikström, D.A. Grove, M. Song, and H. Elwing. (1998). Polar adhesion of *Treponema denticola* on wettability gradient surfaces. Colloids and Surfaces. B, Biointerfaces *11*, 177-186.

Fenno, J.C., P.M. Hannam, W.K. Leung, M. Tamura, M., V.-J. Uitto, and B.C. McBride. (1998a). Cytopathic effects of the major surface protein and the chymotrypsinlike protease of *Treponema denticola*. Infect. Immun. *66*, 1869-1877.

Fenno, J.C., S.Y. Lee, C.H. Bayer, and Y. Ning. (2001). The *opdB* locus encodes the trypsin-like peptidase activity of *Treponema denticola*. Infect. Immun. *69*, 6193-6200.

Fenno, J.C., K.-H. Müller, and B.C. McBride. (1996). Sequence analysis, expression, and binding activity of recombinant major outer sheath protein (Msp) of *Treponema denticola*. J. Bacteriol. *178*, 2489-2497.

Fenno, J.C., M. Tamura, P.M. Hannam, G.W. Wong, R.A. Chan, and B.C. McBride. (2000). Identification of a *Treponema denticola* OppA homologue that binds host proteins present in the subgingival environment. Infect. Immun. *68*, 1884-1892.

Fenno, J.C., G.W. Wong, P.M. Hannam, and B.C. McBride. (1998b). Mutagenesis of outer membrane virulence determinants of the oral spirochete *Treponema denticola*. FEMS Microbiol. Lett. *163*, 209-215.

Fenno, J.C., G.W.K. Wong, P.M. Hannam, K.-H. Müller, W.K. Leung, and B.C. McBride. (1997). Conservation of *msp*, the gene encoding the major outer sheath membrane protein of oral *Treponema* spp. J. Bacteriol. *179*, 1082-1089.

Gopalsami, C., W. Yotis, K. Corrigan, S. Schade, J. Keene, and L. Simonson. (1993). Effect of outer membrane of *Treponema denticola* on bone resorption. Oral Microbiol. Immunol. *8*, 121-124.

Greene, S.R. and L.V. Stamm. (1999). Molecular characterization of a chemotaxis operon in the oral spirochete, *Treponema denticola*. Gene *232*, 59-68.

Grenier, D. (1991). Characteristics of hemolytic and hemagglutinating activities of *Treponema denticola*. Oral Microbiol. Immunol. *6*, 246-249.

Grenier, D., V.-J. Uitto, and B.C. McBride. (1990). Cellular location of a *Treponema denticola* chymotrypsin-like protease and importance of the protease in migration through the basement membrane. Infect. Immun. *58*, 347-351.

Guo, B.P., E.L. Brown, D.W. Dorward, L.C. Rosenberg, and M. Höök. (1998). Decorin-binding adhesins from *Borrelia burgdorferi*. Mol. Microbiol. *30*, 711-723.

Guo, B.P., S.J. Norris, L.C. Rosenberg, and M. Höök. (1995). Adherence of *Borrelia burgdorferi* to the proteoglycan decorin. Infect. Immun. *63*, 3467-3472.

Haapasalo, M., K.-H. Müller, V.-J. Uitto, W.K. Leung, and B.C. McBride. (1992). Characterization, cloning, and binding properties of the major 53-kilodalton *Treponema denticola* surface antigen. Infect. Immun. *60*, 2058-2065.

Haapasalo, M., U. Singh, B.C. McBride, and V.-J. Uitto. (1991). Sulfhydryl-dependent attachment of *Treponema denticola* to laminin and other proteins. Infect. Immun. *59*, 4230-4237.

Hardham, J.M., J.G. Frye, and L.V. Stamm. (1995). Identification and sequences of the *Treponema pallidum fliM', fliY, fliQ, fliR,* and *flhB'* genes. Gene *166*, 57-64.

Hashimoto, M., S. Ogawa, Y. Asai, Y. Takai, and T. Ogawa. (2003). Binding of Porphyromonas gingivalis fimbriae to *Treponema denticola* dentilisin. FEMS Microbiol. Lett. *226*, 267-271.

Hayashi, F., K.D. Smith, A. Ozinsky, T.R. Hawn, E.C. Yi, D.R. Goodlett, J.K. Eng, S. Akira, D.M. Underhill, and A. Aderem. (2001). The innate immune response to bacterial flagellin is mediated by toll-like receptor 5. Nature *410*, 1099-1103.

Heinzerling, H.F., M. Olivares, and R.A. Burne. (1997). Genetic and transcriptional analysis of *flgB* flagellar operon constituents in the oral spriochete *Treponema denticola* and their heterologous expression in enteric bacteria. Infect. Immun. *65*, 2041-2051.

Heuner, K., I. Bergmann, K. Heckenbach, and U.B. Gobel. (2001). Proteolytic activity among various oral *Treponema* species and cloning of a *prtP*-like gene of *Treponema socranskii* subsp. *socranskii*. FEMS Microbiol. Lett. *201*, 169-176.

Heuner, K., B.-K. Choi, R. Schade, A. Moter, A. Otto, and U.B. Göbel. (1999). Cloning and characterization of a gene (*mspA*) encoding the major sheath protein of *Treponema maltophilum* ATCC 51939^T. J. Bacteriol. *181*, 1025-1029.

Hu, L.T., S.D. Pratt, G. Perides, L. katz, R.A. Rogers, and M.S. Klempner. (1997). Isolation, cloning, and expression of a 70-kildalton plasminogen binding protein of *Borrelia burgdorferi*. Infect. Immun. *65*, 4989-4995.

Ishihara, K., H.K. Kuramitsu, T. Miura, and K. Okuda. (1998). Dentilisin activity affects the organization of the outer sheath of *Treponema denticola*. J. Bacteriol. *180*, 3837-3844.

Ishihara, K., T. Miura, H.K. Kuramitsu, and K. Okuda. (1996). Characterization of the *Treponema denticola prtP* gene encoding a prolyl-phenylalanine-specific protease (dentilisin). Infect. Immun. *64*, 5178-5186.

Kataoka, M., H. Li, S. Arakawa, and H. Kuramitsu. (1997). Characterization of a methyl-accepting chemotaxis protein gene, *dmcA*, from the oral spriochete *Treponema denticola*. Infect. Immun. *65*, 4011-4016.

Kesavalu, L., C.W. Falk, K.J. Davis, M.J Steffen, X. Xu, S.C. Holt, and J.L. Ebersole. (2002). Biological characterization of lipopolysaccharide from *Treponema pectinovorum*. Infect. Immun. *70*, 211-217.

Kesavalu, L., S.C. Holt, and J.L. Ebersole. (1999). Environmental modulation of oral treponeme virulence in a murine model. Infect. Immun. *67*, 2783-2789.

Kevasalu, L., S.G. Walker, S.C. Holt, R.R. Crowley, and J.L. Ebersole. (1997). Virulence characteristics of oral treponemes in a murine model. Infect. Immun. *65*, 5096-5102.

Ko, K.S.-C., M. Glogauer, C.A.G. McCulloch, and R.P. Ellen. (1998a). *Treponema denticola* outer membrane inhibits calcium flux in gingival fibroblasts. Infect. Immun. *66*, 703-709.

Ko, K.S.-C., C.-L. Lo, J. Ferrier, P. Hannam, M. Tamura, B.C. McBride, and R.P. Ellen. (1998b). Cell-substrate impedance analysis of epithelial cell shape and micromotion upon challenge with bacterial proteins that perturb extracellular matrix and cytoskeleton. J. Microbiol. Methods *34*, 125-132.

Krupka, H.I., R. Huber, S.C. Holt, and T. Clausen. (2000). Crystal structure of cystalysin from *Treponema denticola*, a pyridoxal 5'-phosphate-dependent protein acting as a haemolytic enzyme. EMBO J. *19*, 3168-3178.

Lee, S.Y., X.-L. Bian, G.W.K. Wong, P.M. Hannam, B.C. McBride, and J.C. Fenno. (2002). Cleavage of *Treponema denticola* PrcA polypeptide to yield protease complex-associated proteins Prca1 and Prca2 is dependent on PrtP. J. Bacteriol. *184*, 3864-3870.

Lépine, G. and R.P. Ellen. (2000). *MglA* and *mglB* of *Treponema denticola*; similarity to ABC transport and *spa* genes. DNA Seq. *11*, 419-431.

Leung, W.K., Q. Wu, P.M. Hannam, B.C. McBride, and V.-J. Uitto. (2002). *Treponema denticola* may stimulate both epithelial proliferation and apoptosis through MAP kinase signal pathways. J. Periodontal Res. *37*, 445-455.

Li, H., S. Arakawa, Q.-D. Deng, and H.K. Kuramitsu. (1999). Characterization of a novel methyl-accepting chemotaxis gene, *dmcB*, from the oral spirochete *Treponema denticola*. Infect. Immun. *67*, 694-699.

Li, H., J. Ruby, N. Charon, and H.K. Kuramitsu. (1996). Gene inactivation in the oral spirochete *Treponema denticola*, contruction of a *flgE* mutant. J. Baceteriol. *178*, 3664-3667.

Limberger, R.J., L.L. Slivienski, J. Izard, and W.A. Samsonoff. (1999). Insertional inactivation of *Treponema denticola tap1* results in a nonmotile mutant with elongated flagellar hooks. J. Bacteriol. *181*, 3743-3750.

Lingaas, E., I. Olsen, T. Midtvedt, and B. Hurlen. (1983). Demonstration of the *in vitro* phagocytosis of *Treponema denticola* by human polymorphonuclear neutrophils. Acta Pathol. Microbiol. Immunol. Scand. [B] *91*, 333-337.

Listgarten M.A. and D.W. Lewis. (1967). The distribution of spirochetes in the lesion of acute necrotizing ulcerative gingivitis, an electron microscopic and statistical survey. J. Periodontol. *38*, 379-386.

Loesche W.J., W.A. Bretz, D. Kerschensteiner, J. Stoll, S.S. Socransky, P. Hujoel, and D.E. Lopatin. (1990). Development of a diagnostic test for anaerobic periodontal infections based on plaque hydrolysis of benzoyl-DL-arginine-naphthylamide. J. Clin. Microbiol. *28*, 1551-1559.

Lux, R. J.N. Miller, N.-N. Park, and W. Shi. (2001). Motility and chemotaxis in tissue penetration of oral epithelial cell layers by *Treponema denticola*. Infect. Immun. *69*, 6276-6283.

Lux, R. J.-H. Sim, J.P. Tsai, and W. Shi. (2002). Construction and characterization of a *cheA* mutant of *Treponema denticola*. J. Bacteriol. *184*, 3130-3134.

Mäkinen, P.-L., K.K. Mäkinen, and S.A. Syed. (1995). Role of the chymotrysin-like membrane-associated proteinase from *Treponema denticola* ATCC 35405 in inactivation of bioactive peptides. Infect. Immn. *63*, 3567-3575.

Mathers, D.A., W.K. Leung, J.C. Fenno, Y. Hong, and B.C. McBride. (1996). The major surface protein complex of *Treponema denticola* depolarizes and induces ion channels in HeLa cell membranes. Infect. Immun. *64*, 2904-2910.

McQuibban, G.A., J.H. Gong, J.P. Wong, J.L. Wallace, I. Clark-Lewis, and C.M. Overall. (2002). Matrix metalloproteinase processing of monocyte chemoattractant proteins generates CC chemokine receptor antagonists with anti-inflammatory properties *in vivo*. Blood *100*, 1160-1167.

Mikx, F.H.M., F. Jacobs, and C. Satumalay. (1992). Cell-bound peptidase activities of *Treponema denticola* ATCC 33520 in continuous culture. Infect. Immun. *60*, 1761-1766.

Newman, H.N. (1976). The apical border of plaque in chronic inflammatory periodontal disease. Brit. Dent. J. *141*, 105-113.

Ng, W. and J. Tonzetich. (1984). Effect of hydrogen sulfide and methyl mercaptan on the permeability of oral mucosa. J. Dent. Res. *63*, 994-997.

Offenbacher, S., Jared, H.L., O'Reilly, P.G., Wells, S.R., Salvi, G.E., Lawrence, H.P., Socransky, S.S., and Beck, J.D. (1998). Potential pathogenic mechanisms of periodontitis associated pregnancy complications. Ann. Periodontol. *3*, 233-250.

Olsen, I., E. Lingaas, B. Hurlen, and T. Midtvedt. (1984). Scanning and transmission electron microscopy of the phagocytosis of *Treponema denticola* and *Escherichia coli* by human neutrophils *in vitro*. Scand. J. Dent. Res. *92*, 282-293.

Opitz, B., N.W.J. Schröder, I. Spreitzer, K.S. Michelsen, C.J. Kirschning, W. Hallatschek, U. Zähringer, T. Hartung, and U.B. Göbel. (2001). Toll-like receptor-2 mediates *Treponema* glycolipid and lipoteichoic acid-induced NF-κB translocation. J. Biol. Chem. *276*, 22041-22047.

Paes Batista da Silva, A., W. Lee, E. Bajenova, C.A.G. McCulloch, and R.P. Ellen. (2004). The major outer sheath protein of *Treponema denticola* inhibits the binding step of collagen phagocytosis in fibroblasts. Cell Microbiol. *6*, 485-498.

Porcella, S.F., T.G. Popova, K.E. Hagman, C.W. Penn, J.D. Radolf, and M.V. Norgard. (1996). A *mgl*-like operon in *Treponema pallidum*, the syphilis spirochete. Gene *177*, 115-121.

Riviere, G.R., M.A. Wagoner, S.A. Baker-Zander, K.S. Weisz, D.F. Adams, L. Simonson, and S.A. Lukehart. (1991a). Indentification of spirochetes related to *Treponema pallidum* in necrotizing ulcerative gingivitis and chronic periodontitis. N. Engl. J. Med. *325*, 539-543.

Riviere, G.R., K.S. Weisz, D.F. Adams, and D.D. Thomas. (1991b). Pathogen-related oral spirochetes from dental plaque are invasive. Infect. Immun. *59*, 3377-3380.

Riviere, G.R., K.S. Weisz, L.G. Simonson, and S.A. Lukehart. (1991c). Pathogen-related spirochetes identified within gingival tissue from patients with acute necrotizing ulcerative gingivitis. Infect. Immun. *59*, 2653-2657.

Rosen, G., R. Naor, E. Rahamin, R. Yishai, and M.N. Sela. (1995). Proteases of *Treponema denticola* outer sheath and extracellular vesicles. Infect. Immun. *63*, 3973-3979.

Rosen, G., M.N. Sela, R. Naor, A. Halabi, V. Barak, and L. Shapira. (1999). Activation of murine macrophages by lipoprotein and lipooligosaccharide of *Treponema denticola*. Infect. Immun. *67*, 1180-1186.

Sato, T. and H.K. Kuramitsu. (2000). Polymerase chain reaction for the detection of *flaA*-1 genes of oral spirochaetes in human advanced periodontal pockets. Arch. Oral. Biol. *45*, 921-925.

Schultz, C.P., V. Wolf, R. Lange, E. Mertens, J. Wecke, D. Naumann, and U. Zahringer. (1998). Evidence for a new type of outer membrane lipid in oral spirochete *Treponema denticola*. Functioning permeation barrier without lipopolysaccharides. J. Biol. Chem. *273*, 15661-15666.

Scott, D., E.C.S. Chan, and R. Siboo. (1996a). Iron acquisition by oral hemolytic spirochetes, isolation of a hemin-binding protein and identification of iron reductase activity. Canad. J. Micriobiol. *42*, 1072-1079.

Scott, D., R. Siboo, and E.C.S. Chan. (1996b). An extracellular enzyme with hyaluronidase and chondroitinase activities from some oral anaerobic spirochetes. Microbiol. *142*, 2567-2576.

Sela, M.N., A. Bolotin, R. Naor, A. Weinberg, and G. Rosen. (1997). Lipoproteins of *Treponema denticola*, their effect on human polymorphonuclear neutrophils. J. Periodontal Res. *32*, 455-466.

Sela, M.N., A. Weinberg, R. Borinsky, S.C. Holt, and T. Dishon. (1988). Inhibition of superoxide production in human polymorphonuclear leukocytes by treponemal factors. Infect. Immun. *56*, 589-594.

Sorsa, T., T. Ingman, K. Suomalainen, M. Haapasalo, Y.T. Konttinen, O. Lindy, H. Saari, and V.-J. Uitto. (1992). Identification of proteases from periodontopathogenic bacteria as activators of latent human neutrophil and fibroblast-type interstitial collagenases. Infect. Immun. *60*, 4491-4495.

Stamm, L.V. and H.L. Bergen. (1999). Molecular characterization of the flagellar (*fla*) operon in the oral spirochete *Treponema denticola* ATCC 35405. FEMS Microbiol. Lett. *179*, 31-36.

Stamm, L.V. and H.L. Bergen. (2001). Molecular characterization of the *Treponema denticola fliQ* region. DNA Seq. *112*, 463-467.

Stamm, L.V., N.R. Young, J.G. Frye, and J.M. Hardham. (1996). Identification and sequences of the *Treponema pallidum mglA* and *mglC* genes. DNA Seq. *6*, 293-298.

Teng, Y.-T.A. (2002). Mixed periodontal Th1-Th2 cytokine profile in *Actinobacillus actinomycetemcomitans*-specific osteoprotegerin ligand (or RANK-L) – mediated alveolar bone destruction *in vivo*. Infect. Immun. *70*, 5269-5273.

Teng, Y.-T.A., H. Nguyen, X. Gao, Y.-Y. Kong, R.M. Gorczynski, B. Singh, R.P. Ellen, and J.M. Penninger. (2000). Functional human T-cell immunity and osteoprotegerin ligand control alveolar bone destruction in periodontal infection. J. Clin. Invest. *106*, R59-67.

Uitto, V.-J., D. Grenier, E.C.S. Chan, and B.C. McBride. (1988). Isolation of a chymotrypsinlike enzyme from *Treponema denticola*. Infect. Immun. *56*, 2717-2722.

Uitto, V.-J., Y.M. Pan, W.K. Leung, H. Larjava, R.P. Ellen, B.B. Finlay, and B.C. McBride. (1995). Cytopathic effects of *Treponema denticola* chymotrypsin-like proteinase on migrating and stratified epithelial cells. Infect. Immun. *63*, 3401-3410.

Umemoto, T., Y. Nakatani, Y. Nakamura, and I. Namikawa. (1993). Fibronectin-binding proteins of a human oral spirochete *Treponema denticola*. Microbiol. Immunol. *37*, 75-78.

Wang, Q., K.S. Ko, A. Kapus, C.A.G. McCulloch, and R.P. Ellen. (2001). A spirochete surface protein uncouples store-operated calcium channels in fibroblasts. J. Biol. Chem. *276*, 23056-23064.

Weinberg, A. and S.C. Holt. (1990). Interaction of *Treponema denticola* TD-4, GM-1, and MS25 with human gingival fibroblasts. Infect. Immun. *58*, 1720-1729.

Weinberg, A. and S.C. Holt. (1991). Chemical and biological activities of a 64-kilodalton outer sheath protein from *Treponema denticola* strains. J. Bacteriol. *173*, 6935-6947.

Wyss, C. (1998). Flagellins, but not endoflagellar sheath proteins, of *Treponema pallidum* and of pathogen-related oral spirochetes are glycosylated. Infect. Immun. *66*, 5751-5754.

Wyss, C., B.K. Choi, P. Schüpbach, A. Moter, B. Guggenheim, and U.B. Göbel. (1999). *Treponema lecithinolyticum* sp. nov., a small saccharolytic spirochaete with phospholipase A and C activities associated with periodontal diseases. Int. J. Syst. Bacteriol. *49*, 1329-1339.

Yang, P.F., M. Song, D.A. Grove, and R.P. Ellen. (1998). Filamentous actin disruption and diminished inositol phosphate response in gingival fibroblasts caused by *Treponema denticola*. Infect. Immun. *66*, 696-702.

16

Innate and Adaptive Immune Responses to Oral Treponemes

*Nicolas W.J. Schröder, Ralf R. Schumann, and Ulf B. Göbel**

Abstract

Innate immunity plays an important role in the pathogenesis of periodontitis. Toll-like receptors, a family of pattern recognition receptors recognizing a wide array of microbial ligands, are widely expressed on cells within the oral cavity. Oral treponemes exhibit at least two different types of molecules within their outer membranes interacting with TLRs: lipoproteins, exhibiting a common triacylated lipid motif, as well as glycolipids, sharing structural similarities with lipoteichoic acid (LTA) of Gram-positive bacteria. Both compounds have been found to interact with TLR-2, inducing cytokine release. Since pro-inflammatory cytokines have been shown to exhibit worsening effects within the course of periodontitis, the interaction of TLR-2 ligands with the host may be regarded as an event triggering disease. Next to innate immune responses, oral treponemes are also known to initiate adaptive immune responses via the production of antibodies. Since adaptive immune responses are initiated at least in part by dendritic cells also expressing TLR-2, the interaction of the host with TLR-2 ligands may be an important event for the induction of immunity.

1. Components of innate immunity within the oral cavity

1.1 General aspects

The innate immune system acts as a first line defence system against invading microorganisms (Medzhitov and Janeway, 1997). Next to neutrophils, the innate immune system on a cellular basis is mainly represented by monocytes/macrophages and dendritic cells, which are also referred to as antigen presenting cells (APCs). APCs continuously probe internal and external surfaces for molecular patterns indicating the presence of microorganisms. Such patterns are referred to as

*For correspondence email ulf.goebel@charite.de

pathogen associated molecular patterns (PAMPs) recognized by pattern recognition receptors (PRRs). Within the last years, the family of Toll-like receptors (TLRs) has been shown to play a major role in these processes. Up to now, eleven TLRs have been identified that are expressed on a variety of leukocytes as well as other cell types, such as epithelial and endothelial cells, initiating pro-inflammatory responses upon recognition of different ligands. Of particular interest for the oral cavity are TLR-4 as well as TLR-2.

TLR-4, a receptor expressed on monocytes and dendritic cells, has been shown to interact with lipopolysaccharide (LPS) of Gram-negative bacteria (Alexander and Rietschel, 2001) within a receptor complex including CD14 (Wright et al., 1990) as well as MD-2 (Shimazu et al., 1999). LPS is transferred to this complex via LPS binding protein (LBP), an acute phase protein synthesized in the liver (Schumann et al., 1990). On the other hand, TLR-2 recognizes different bacterial ligands, including lipoteichoic acid (LTA) and peptidoglycan (PGN) from Gram-positive bacteria as well as lipoproteins from spirochetes (Hirschfeld et al., 1999; Lien et al., 1999; Schröder et al., 2003b; Schwandner et al., 1999; Thoma-Uszynski et al., 2001). In order to act as a functional receptor, TLR-2 forms heterodimers with either TLR-1 or -6 (Ozinsky et al., 2000). The interaction of TLRs with their ligands initiates a signal transduction cascade involving the intracellular adapter molecule MyD88 leading to translocation of NF-κB into the nucleus and subsequent expression and release of cytokines, *i.e.* TNF-α, IL-1 and IL-6 (Barton and Medzhitov, 2003). In contrast to TLR-2, TLR-4 furthermore recruits a MyD88-independent pathway involving the TIR domain containing adapter inducing IFN-beta (referred to as TRIF), initiating the release of class I interferons (Hoebe et al., 2003; Yamamoto et al., 2003; 2002).

Next to initiating immediate responses against invading pathogens, *i.e.* opsonization and subsequent phagocytosis, the innate immune system also acts as a link to adaptive immunity (Medzhitov and Janeway, 1999). Dendritic cells phagocytose microorganisms and travel to regional lymph nodes, presenting antigen to $CD4^+$ T-cells. To mature, dendritic cells need an additional signal provided by the interaction of a TLR with its ligand (Schnare et al., 2001). It has been proposed that this additional stimulus is absolutely essential for the induction of adaptive immune responses (Pasare and Medzhitov, 2003). Furthermore, TLR-2 is present on T-helper cells and is upregulated during clonal expansion (Komai-Koma et al., 2004). It therefore acts like a co-stimulatory marker. $CD4^+/CD28^+$ regulatory T-cells, which in contrast to T-helper cells play a role in suppression of immune responses, have been shown to express TLR-4 (Caramalho et al., 2003).

1.2 Components of the innate immune system within the oral cavity

Macrophages are regarded to play an important role in the pathogenesis of periodontal disease (Page, 1991). They express TLR-2 in combination with TLR-1 and -6 (Shortman and Liu, 2002) as well as TLR-4 (Shortman and Liu, 2002), MD-2 (Shimazu et al., 1999) and CD14 (Goyert et al., 1988). Upon interaction with a ligand, macrophages secrete pro-inflammatory cytokines, *i.e.* TNF-α and IL-1, which have been shown to act in synergy causing inflammation and bone resorption (Stashenko et al., 1987). Furthermore, these cytokines induce the upregulation of adhesion molecule expression by endothelial cells, recruiting

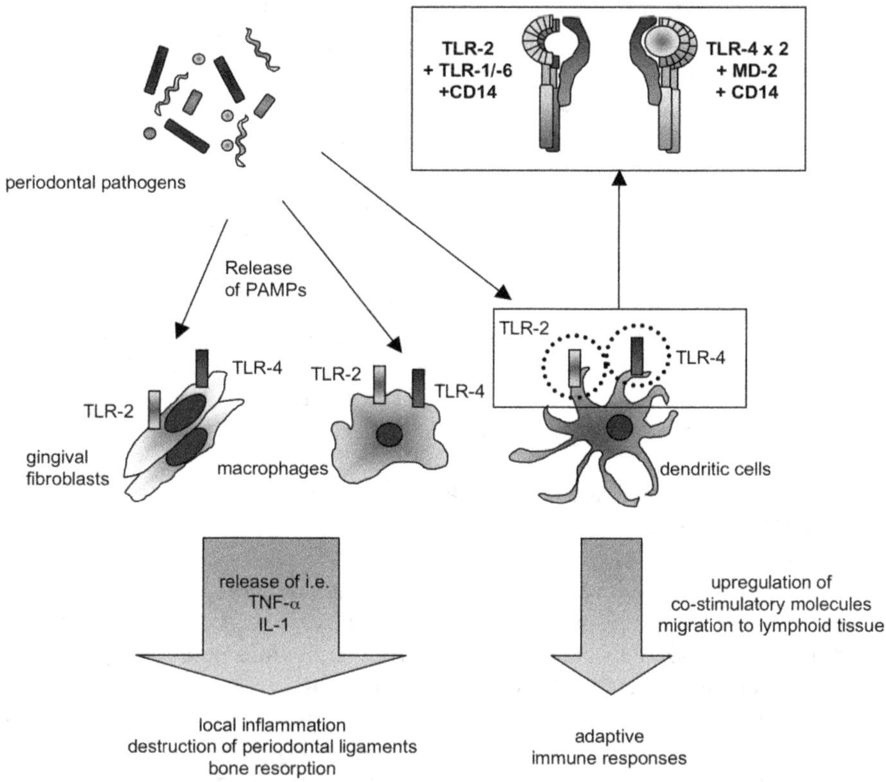

Figure 1. Cellular elements of innate immunity in the oral cavity. Within the oral cavity bacterial partial structures are released by oral pathogens during cell growth. These partial structures are referred to as pathogen associated molecular patterns (PAMPs). Several PAMPs are recognized by TLR-2 and -4 expressed by gingival fibroblasts, macrophages and dendritic cells. Upon the interaction with these ligands, macrophages and fibroblasts release pro-inflammatory cytokines, such as TNF-α or IL-1, inducing local inflammation, destruction of periodontal ligaments and bone resorption. Dendritic cells are involved in the initiation of adaptive immune responses by presenting antigen to T-cells after maturation. Note that the receptor complex recognizing lipoproteins and LTA is a heterodimer consisting of TLR-1 and -2, in combination with CD14.

leukocytes and helping to establish a local inflammatory reaction (Beutler, 1990). Gingival fibroblasts, a cell type abundantly present within the periodontal tissue, have also been shown to express TLR-2 and 4 (Morr et al., 2002; Wang et al., 2003), and are known to release cytokines, *i.e.* IL-6 and TNF-α (Wang and Ohura, 2002). Thus, interaction of bacterial ligands with TLRs is important to induce an immune response, but also may display harmful effects within the periodontal tissue by aggravating inflammation (Figure 1). This hypothesis was strengthened by a report describing that TLR-4-negative mice display less bone resorption in a mixed infection model as compared to their wild-type counterparts (Hou et al., 2000).

One recent study employing microarray and FACS analysis described that TLR-2, TLR-4 as well as CD14 are upregulated by human gingival fibroblasts derived from periodontal lesions, indicating an involvement in disease (Wang et al., 2003).

These data were supported in part by immunohistochemical studies showing that TLR-2 and TLR-4 are upregulated in macrophages present in inflamed periodontal tissue. However, these cells apparently down-regulated CD14 (Mori et al., 2003), indicating a potential negative feedback mechanism.

Dendritic cells have been reported to infiltrate the subgingival tissue, recruiting $CD4^+$ T-cells (Jotwani et al., 2001). Interestingly, dendritic cells stimulated with LPS derived from *Porphyromonas gingivalis*, which acts as a TLR-2 ligand, have been shown to release relatively higher levels of IL-10 in relation to IL-12 as compared to LPS from *Enterobacteriaceae* (Jotwani et al., 2001). IL-10 is known to act as an anti-inflammatory cytokine (de Waal Malefyt et al., 1991; Fiorentino et al., 1991), and IL-10 knock out mice have been shown to display more severe periodontitis as their wild-type counterparts (Al-Rasheed et al., 2003). Thus, the preferential induction of IL-10 by TLR-2 ligands may prevent harmful effects of inflammation within the periodontal tissue.

1.3. Genetic evidence for a linkage of innate immune responses and periodontal disease

1.3.1. General aspects

Smoking, poor oral hygiene as well as diabetes are commonly regarded as major risk factors for the development of periodontal disease, especially in its chronic form (Curtis, 2000; Papapanou, 1999). However, twin studies suggested that other, yet undefined genetic factors may also contribute to the pathogenesis of periodontitis (Michalowicz, 1994; Michalowicz et al., 1991). Regarding innate immune responses, data have been provided that single nucleotide polymorphisms (SNPs), genetic variations found within 1% of a certain population, sited within genes encoding cytokines, may act as risk factors for periodontitis. SNPs within the gene cluster of IL-1 causing elevated cytokine levels have been described to influence severity of aggressive and chronic disease (Kornman et al., 1997; McDevitt et al., 2000; Parkhill et al., 2000). However, data obtained so far are contradictory regarding the question whether these SNPs act as risk factors for non-smokers.

1.3.2. Genetic variants of TLRs

Within the gene encoding for TLR-4, two co-segregating SNPs, Asp299Gly and Thr399Ile, have been described (Arbour et al., 2000) leading to decreased cytokine release by epithelial cells upon LPS-stimulation *in vitro* (Arbour et al., 2000). The presence of these SNPs is furthermore associated with hyporesponsiveness to inhaled LPS (Arbour et al., 2000; Michel et al., 2003) as well as with a higher incidence of Gram-negative infections and septic shock (Agnese et al., 2002; Lorenz et al., 2002). On the other hand, Asp299Gly and Thr399Ile occur at a lower frequency among patients with carotid atherosclerosis, thus potentially indicating a protective effect of this genetic variant (Kiechl et al., 2002). Furthermore, within the gene encoding for human TLR-2 a SNP (Arg753Gln) was described within the signal transduction domain diminishing inflammatory responses to TLR-2 ligands *in vitro* (Lorenz et al., 2000; Schröder et al., 2003a). Both, the TLR-2 and -4 SNPs have been shown to be involved in disease susceptibility, as we have reviewed recently (Schröder and Schumann, 2005).

As mentioned above, TLR-4 negative C3H/HeJ mice have been reported to exhibit less bone resorption in a periodontitis model (Hou et al., 2000). Since this study involved infection models with Gram-negative as well as Gram-positive bacteria, the latter exhibiting predominantly TLR-2 ligands, it is tempting to speculate that SNPs of TLR-4 are associated with periodontitis, and that they are associated with the presence of treponemes. A recent study investigating TLR-4 and –2 polymorphisms and chronic periodontitis among a German population did not reveal any significant association, although a trend was observed towards a higher frequency of TLR-4 SNPs among the patients (Folwaczny et al., 2004). However, the study was performed among a relatively small group with 52 patients meeting the criteria for severe periodontitis. In contrast, we have shown recently that periodontal disease in fact is associated with TLR-4 SNPs (Schröder et al., 2005).

2. Treponemal compounds interacting with Toll-like receptors

2.1. Lipoproteins

Lipoproteins have been identified among a wide array of bacteria, including Gram-negative bacteria, mycobacteria and spirochetes. These compounds exhibit an N-terminal N-palmitoyl-S-[2,3-bis(palmitoyloxy)-(2RS)-propyl]-(R)-cysteinyl (Pam_3Cys) moiety, which was first described in the Braun lipoprotein of *E. coli* (Braun and Rehn, 1969). Thus, the protein is linked to the bacterial membrane *via* a lipid anchor consisting of three fatty acids (Figure 2). Cytokine-inducing properties of lipoproteins were first described for outer surface proteins (Osps) of *Borrelia burgdorferi* (Ma and Weis, 1993; Wooten et al., 1996), compounds which were previously known to be highly immunogenic (Brandt et al., 1990). Later on, immunostimulatory lipoproteins termed p17 and p47 were described in the outer membrane of *Treponema pallidum*. Although these compounds induced cytokines *via* a pathway depending on CD14 (Norgard et al., 1996; Selatti et al., 1998), they differed from lipopolysaccharide (LPS) of Gram-negative bacteria since they were able to activate immune cells derived from LPS-hyporesponsive C3H/HeJ mice (Norgard et al., 1996; Radolf et al., 1991).

Later studies showed that LPS-hyporesponsiveness in C3H/HeJ mice was caused by a mutation within Toll-like receptor (TLR)-4 (Hoshino et al., 1999; Poltorak et al., 1998a; 1998b), and lipoproteins from *Borrelia* spp. as well synthetic lipopeptides bearing the Pam_3Cys moiety were found to interact with TLR-2, explaining their activity towards cells isolated from C3H/HeJ mice (Hirschfeld et al., 1999; Lien et al., 1999; Schwandner et al., 1999). In order to interact with its ligands, TLR-2 forms heterodimers either with TLR-1 or TLR-6. TLR-2/-1 heterodimers recognize triacylated lipoproteins and lipopeptides (Alexopoulou et al., 2002; Takeuchi et al., 2002), while TLR-6 interacts with TLR-2 in order to detect diacylated lipopeptides, as present i.e. in *Mycoplasma* spp. (Takeuchi et al., 2001). Thus, spirochetal lipoproteins are recognized by a receptor complex including TLR-1 and -2 as well as CD14.

Next to *Treponema pallidum*, lipoproteins have been described in the oral spirochete *Treponema denticola* (Sela et al., 1997; Yotis et al., 1991). A following report on the biological characteristics of a stimulatory active lipoptotein derived

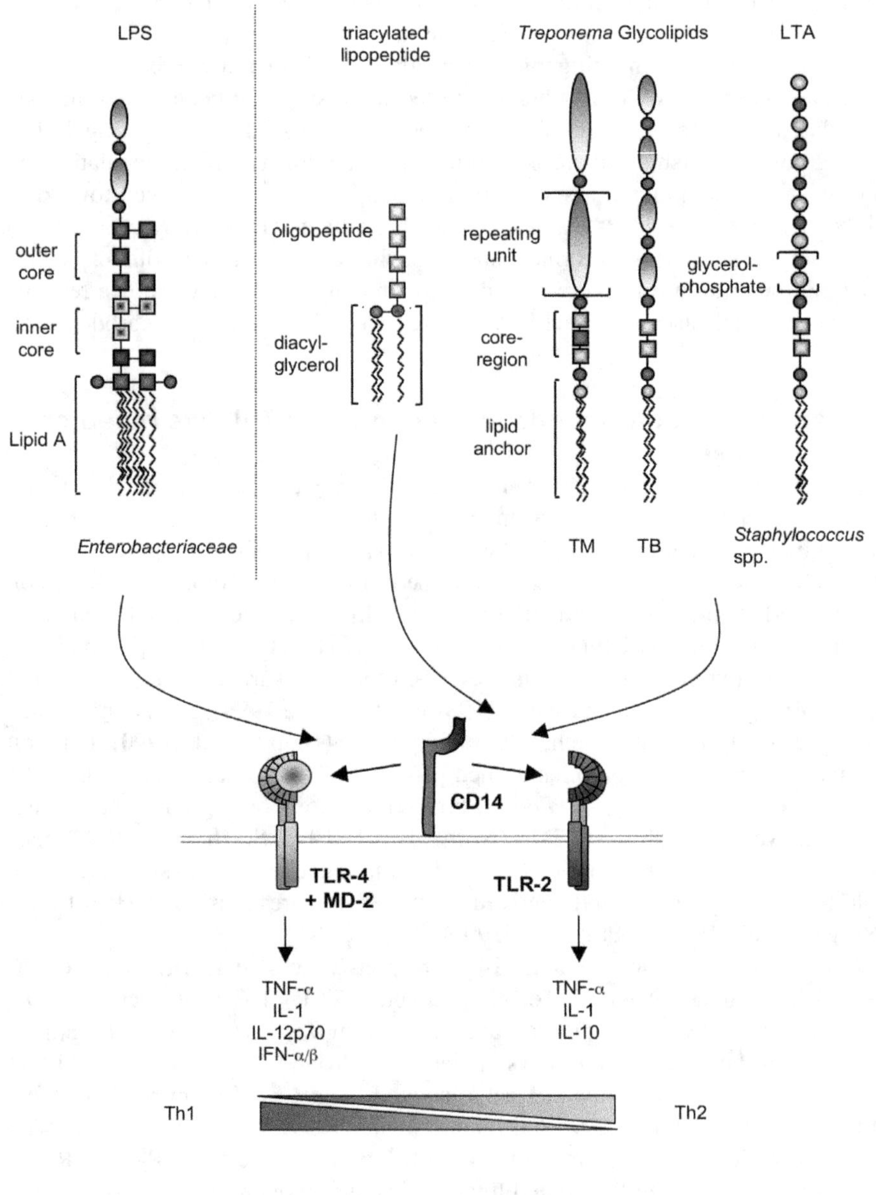

Figure 2. Structure function relationships of bacterial ligands. Lipopolysaccharide (LPS) of Gram-negative bacteria is a glycolipid consisting of a lipid portion termed LipidA, a core region and a hydrophilic chain. LPS interacts with TLR-4, together with MD-2, and induces cytokines. In contrast, lipopeptides corresponding to spirochetal lipoproteins as well as treponeme glycolipids interact with TLR-2, inducing a panel of cytokines yielding a Th2 bias, such as IL-10. Note the close structural relationship between treponeme glycolipids (TM : *T. maltophilum*, TB : *T. brennaborense*) and lipoteichoic acid (LTA) of Gram-positive bacteria.

from *T. denticola* ATCC 35404 found no differences regarding TNF-α and nitric oxide release by macrophages derived from C3H/HeJ mice and their wild-type counterpart (Rosen et al., 1999), implicating that these compounds also interact with TLR-2. However, this question has not been adressed in detail by now, and data on outer membrane lipoproteins from other oral treponemes is still lacking.

2.2. Glycolipids

2.2.1. LPS and Spirochetes

Starting with *Treponema pallidum* in 1973 (Jackson and Zey, 1973), LPS comparable to *Enterobacteriaceae* has been repeatedly described in treponemes as well as in other spirochetes. Following reports described this compound in the outer membranes of *Serpulina hyodyenteriae* (Nuessen et al., 1982; 1983), *B. burgdorferi* (Habicht et al., 1986), *T. phagedenis* (Petersen, 1986), *T. denticola* (Dahle et al., 1996) as well as *T. pectinovorum* (Walker et al., 1999). Evidence for LPS in these strains mainly consisted of hydrophilic compounds positive in silver staining, lacking protein and showing a positive limulus amoebocyte lysate (LAL) test. Furthermore, ß-hydroxy fatty acids as well as 2-keto 3 deoxy-octolusonic acid (KDO), both characteristic for LPS derived from *Enterobacteriaceae* have been described within spirochetes. However, the LAL test has been repeatedly reported to yield positive results with glycolipids other than LPS. Furthermore, ß-hydroxy fatty acids and KDO were found only in low amounts, raising the question whether these were contaminants only. However, the presence of LPS was ruled out for *T. pallidum* and *B. burgdorferi*, an observation supported by genome analyses (Fraser et al., 1997; 1998; Hardy and Levin, 1983; Norris and Weinstock, 2000; Seshadri et al., 2004; Takayama et al., 1987; Weinstock et al., 1998), indicating the absence of LPS in members of the genera *Treponema* and *Borrelia.*

In contrast, the presence of LPS has been demonstrated for *Leptospira interrogans*. First reports on LPS-like structures have been reported in the late eighties (Isogai et al., 1986) and in contrast to *Treponema*, genome analysis of *L. interrogans* yielded evidence for the presence of genes involved in LPS-biosynthesis (Bulach et al., 2000). LPS from *L. interrogans* is a ligand of TLR-2 (Werts et al., 2001), and recently the precise chemical structure of this compound was elucidated (Que-Gewirth et al., 2004), exhibiting an unusual lipid A with four fatty acids and a methylated phosphate residue. Thus, *L. interrogans*, in a comparable fashion to *P. gingivalis*, exhibits an LPS differing from that of *Enterobacteriacae* regarding structure and biological characteristics.

2.2.2. Evidence for diacylglycerol glycolipids in *T. denticola*

In 1998, Schultz et al. reported the presence of a glycolipid in the outer membrane of *T. denticola.* Employing more sophisticated chemical techniques such as combined gas/liquid chromatography-mass spectroscopy (GC-MS) and nuclear magnetic resonance (NMR) spectroscopy, they described a compound, termed outer membrane lipid (OML) being clearly different from LPS (Schultz et al., 1998). This glycolipid exhibits a diacylglycerol lipid anchor, a small linker-region of three carbohydrates and a hydrophilic chain of carbohydrates with repetitive sequences. KDO and ß-hydroxy fatty acids were only present as trace contaminations. Thus, *T. denticola* OML is clearly distinct from LPS. Schultz et al. focussed on the question how spirochetes manage to obtain a functional diffusion barrier without LPS or peptidoglycan in substantial amounts, and the potential interaction of OML with the host's innate immune system was not adressed. Later a glycolipid derived from *T. denticola* was described to induce cytokines and nitric oxide in immune cells (Rosen et al., 1999). Unfortunately, these reports did not involve chemical analysis.

2.2.3. Glycolipids in other treponeme strains

Aiming at a synthesis of chemical and immunological analysis of treponeme compounds, we investigated two treponeme strains, *T. maltophilum*, isolated from the periodontal pockets of a patient suffering from severe periodontitis (Wyss et al., 1996), and *T. brennaborense*, isolated from lesions of cattle suffering from digital dermatitis (Schrank et al., 1999). First evidence for cytokine-inducing compounds released by these bacteria was obtained by investigating the response of human monocytes to culture supernatants derived from these strains (Schröder et al., 2000). A marked secretion of TNF-α was observed, which could be inhibited by blocking CD14 antibodies (Schröder et al., 2000). Furthermore, LPS binding protein (LBP), a protein binding LPS and transferring it to CD14 (Schumann et al., 1990), enhanced the effects observed (Schröder et al., 2000). Based upon these results and previous reports on LPS in spirochetes, we employed a hot phenol/water extraction procedure suitable for the extraction of LPS from *Enterobacteriaceae*. However, chemical analysis revealed the presence of glycolipids exhibiting a diacylglycerol-lipid anchor and a small carbohydrate core region consisting of three (*T. maltophilum*) and two (*T. brennaborense*) hexose moieties, respectively (Figure 2). KDO and β-hydroxy fatty acids are lacking, and silver-staining provides evidence for the presence of carbohydrate repeating units in both glycolipids. Thus, these molecules are clearly distinct from LPS while resembling the glycolipid described in *T. denticola* (Schultz et al., 1998). A recent study investigating *T. medium* described a compound exhibiting a comparable structure, further indicating that this type of glycolipid is abundant among treponemes (Hashimoto et al., 2003a). It is obvious, however, that these structures also resembled lipoteichoic acid (LTA) of Gram-positive bacteria (Fischer et al., 1990, Figure 2), with the exception that glycerophosphate was not present in substantial amounts in the treponeme extracts.

2.2.4. Biological characteristics of treponeme glycolipids

Glycolipids from both treponeme strains act as strong cytokine-inducers in mononuclear cells of human and murine origin (Opitz et al., 2001; Schröder et al., 2000; Schröder et al., 2001). Activation is dependent on both CD14 and LBP (Schröder et al., 2000; Schröder et al., 2001). Furthermore, TLR-2 was identified to recognize both treponeme glycolipids by the use of several test systems including C3H/HeJ mice as well as transfection experiments in different cell lines (Opitz et al., 2001; Schröder et al., 2000). However, *T. brennaborense* derived glycolipids are also in part recognized by TLR-4 (Opitz et al., 2001; Schröder et al., 2000). Thus, although being structurally different from LPS and interacting with TLR-2, treponeme glycolipids exhibited biological characteristics considered to be specific for LPS, such as the interaction with LBP. As mentioned above, these molecules are chemically related to LTA of Gram-positive bacteria, raising the question whether the interaction with LBP is a common feature of amphiphilic molecules bearing a diacylglycerol lipid anchor. In subsequent studies, we were able to provide evidence that LBP also binds and transfers LTA of *Staphylococcus aureus* and *Streptococcus pneumoniae*, enhancing cytokine induction (Schröder et al., 2003b). Thus, treponeme glycolipids are not only chemically, but also biologically closely related to LTA of Gram-positive bacteria.

2.3. Phospholipids

The presence of phospholipids such as phosphatidylcholine has been repeatedly described in spirochetal outer membranes (Livermore et al., 1978; Livermore and Johnson, 1974; Radolf et al., 1995). A recent study reported that derivatives of phosphatidylcholine present in *T. medium* as well as *T. denticola* interfered with the host's innate immune system (Hashimoto et al., 2003b). These compounds, however, inhibit LPS-mediated signal transduction by direct interaction of LBP and CD14, thus preventing LPS from binding to its receptors (Asai et al., 2003; Hashimoto et al., 2003b). According to the authors, this inhibitory effect is an explanation for the chronicity of treponemal diseases, such as syphilis or periodontitis, by sustaining tolerance of the host against the pathogens.

3. Adaptive immune responses to oral treponemes

3.1. Adaptive immune responses in periodontitis

The importance of the adaptive immune system within the pathogenesis of periodontitis is illustrated by the occurrence of this disease in HIV patients (Alpagot et al., 2004; Winkler et al., 1989). It is well established that IgG as well as IgA antibodies against periodontal pathogens are found systemically in patients suffering from periodontitis, especially regarding the aggressive form (Chung et al., 2003; Hägewald et al., 2002; Lu et al., 1993). In the case of *P. gingivalis*, the fact that antibodies against RgpA are frequently found among patients (Inagaki et al., 2003) has lead to the development of a vaccine against this strain (Tsurumi et al., 2003). Evidence that antibodies against periodontal pathogens may be important for controlling these bacteria and preventing periodontitis can be drawn from data obtained on polymorphisms of Fc-gamma receptors yielding low affinity phenotypes, which have been shown to be associated with a higher risk of meningococcal disease (van der Pol et al., 2001). These variations have been associated with an increased risk for developing periodontitis (Kobayashi et al., 2000; Loos et al., 2003). Thus, antibodies are produced against periodontal pathogens and apparently play a role in the course of periodontal disease.

3.2. Antibody responses to oral treponemes

In sera of patients with chronic periodontitis, IgG and IgA immunoglobulins against *T. denticola* and *T. vincentii* have been described (Lai et al., 1986). The fact that periodontally diseased subjects raise a systemic adaptive immune response against oral spirochetes gained particular interest later on. Investigating sera of patients suffering from acute necrotizing ulcerative gingivitis, Riviere et al. found a frequent cross-reaction with antigens derived from *T. pallidum* thought to be unique for this pathogen (Riviere et al., 1991a). On the other hand, antibodies against these antigens, including p47 as well as a 37kDa endoflagellum sheath protein, detected treponemes in dental plaque (Riviere et al., 1991a). These antibodies were later on suggested as a tool for the detection of uncultivable oral treponemes, introducing the term pathogen related oral spirochetes (Barron et al., 1991; Riviere et al., 1991b). Later studies revealed that one of the antibodies tested was cross-reactive with a wide variety of spirochetes including several group I treponemes identified by comparative 16S RNA sequence analysis (Choi et al., 1994; 1996; Moter et al., 1998). However, these reports on systemic antibody responses against oral treponemes, particularly in severe forms of periodontitis, underlined both the role of spirochetes in periodontal disease as well as the involvement of adaptive immune responses. Meanwhile, antibodies against *T. denticola*, in combination with other strains including *i.e. P. gingivalis*, have been repeatedly studied in order to elucidate course of disease in different populations (Mooney et al., 2001; Morinushi et al., 1997). However, data on the contribution of antibodies against treponemes to eradication of these pathogens is rather scarce. One study described that IgG antibodies of different subclasses are formed against *T. denticola* during infection as well as immunization in mice (Kesavalu et al., 1999). However, data suggested that the antibodies were incapable of killing the bacteria.

Acknowledgements

The work by our group presented here was supported by grants given by the Deutsche Forschungsgemeinschaft (DFG) and the Bundesministerium für Bildung und Forschung (BMBF).

References

Agnese, D. M., Calvano, J. E., Hahm, S. J., Coyle, S. M., Corbett, S. A., Calvano, S. E., and Lowry, S. F. (2002). Human toll-like receptor 4 mutations but not CD14 polymorphisms are associated with an increased risk of Gram-negative infections. J. Infect. Dis. *186*, 1522-1525.

Alexander, C., and Rietschel, E. T. (2001). Bacterial lipopolysaccharides and innate immunity. J. Endotoxin Res. *7*, 167-202.

Alexopoulou, L., Thomas, V., Schnare, M., Lobet, Y., Anguita, J., Schoen, R. T., Medzhitov, R., Fikrig, E., and Flavell, R. A. (2002). Hyporesponsiveness to vaccination with *Borrelia burgdorferi* OspA in humans and in TLR1- and TLR2-deficient mice. Nat. Med. *8*, 878-884.

Alpagot, T., Duzgunes, N., Wolff, L. F., and Lee, A. (2004). Risk factors for periodontitis in HIV patients. J. Periodontal Res. *39*, 149-157.

Al-Rasheed, A., Scheerens, H., Rennick, D. M., Fletcher, H. M., and Tatakis, D. N. (2003). Accelerated alveolar bone loss in mice lacking interleukin-10. J. Dent. Res. *82*, 632-635.

Arbour, N. C., Lorenz, E., Schutte, B. C., Zabner, J., Kline, J. N., Jones, M., Frees, K., Watt, J. L., and Schwartz, D. A. (2000). TLR4 mutations are associated with endotoxin hyporesponsiveness in humans. Nat. Genet. *25*, 187-191.

Asai, Y., Hashimoto, M., and Ogawa, T. (2003). Treponemal glycoconjugate inhibits Toll-like receptor ligand-induced cell activation by blocking LPS-binding protein and CD14 functions. Eur. J. Immunol. *33*, 3196-3204.

Barron, S. L., Riviere, G. R., Simonson, L. G., Lukehart, S. A., Tira, D. E., and O'Neil, D. W. (1991). Use of monoclonal antibodies to enumerate spirochetes and identify *Treponema denticola* in dental plaque of children, adolescents and young adults. Oral Microbiol. Immunol. *6*, 97-101.

Barton, G. M., and Medzhitov, R. (2003). Toll-like receptor signaling pathways. Science *300*, 1524-1525.

Beutler, B. (1990). The complex regulation and biology of TNF (cachectin). Crit. Rev. Oncog. *2*, 9-18.

Brandt, M. E., Riley, B. S., Radolf, J. D., and Norgard, M. V. (1990). Immunogenic integral membrane proteins of *Borrelia burgdorferi* are lipoproteins. Infect. Immun. *58*, 983-991.

Braun, V., and Rehn, K. (1969). Chemical characterization, spatial distribution and function of a lipoprotein (murein-lipoprotein) of the *E. coli* cell wall. The specific effect of trypsin on the membrane structure. Eur. J. Biochem. *10*, 426-438.

Bulach, D. M., Kalambaheti, T., de la Pena-Moctezuma, A., and Adler, B. (2000). Lipopolysaccharide biosynthesis in *Leptospira*. J. Mol. Microbiol. Biotechnol. *2*, 375-380.

Caramalho, I., Lopes-Carvalho, T., Ostler, D., Zelenay, S., Haury, M., and Demengeot, J. (2003). Regulatory T cells selectively express toll-like receptors and are activated by lipopolysaccharide. J. Exp. Med. *197*, 403-411.

Choi, B. K., Paster, B. J., Dewhirst, F. E., and Göbel, U. B. (1994). Diversity of cultivable and uncultivable oral spirochetes from a patient with severe destructive periodontitis. Infect. Immun. *62*, 1889-1895.

Choi, B. K., Wyss, C., and Göbel, U. B. (1996). Phylogenetic analysis of pathogen-related oral spirochetes. J. Clin. Microbiol. *34*, 1922-1925.

Chung, H. Y., Lu, H. C., Chen, W. L., Lu, C. T., Yang, Y. H., and Tsai, C. C. (2003). Immunoglobulin G profiles in different forms of periodontitis. J. Periodontal Res. *38*, 471-476.

Curtis, J. W., Jr. (2000). Infections associated with diabetes mellitus. N. Engl. J. Med. *342*, 895-896.

Dahle, U. R., Tronstad, L., and Olsen, I. (1996). 3-hydroxy fatty acids in a lipopolysaccharide-like material from *Treponema denticola* strain FM. Endod. Dent. Traumatol. *12*, 202-205.

de Waal Malefyt, R., Abrams, J., Bennett, B., Figdor, C. G., and de Vries, J. E. (1991). Interleukin 10 (IL-10) inhibits cytokine synthesis by human monocytes: an autoregulatory role of IL-10 produced by monocytes. J. Exp. Med. *174*, 1209-1220.

Fiorentino, D. F., Zlotnik, A., Mosmann, T. R., Howard, M., and O'Garra, A. (1991). IL-10 inhibits cytokine production by activated macrophages. J. Immunol. *147*, 3815-3822.

Fischer, W., Mannsfeld, T., and Hagen, G. (1990). On the basic structure of poly(glycerophosphate) lipoteichoic acids. Biochem. Cell Biol. *68*, 33-43.

Folwaczny, M., Glas, J., Torok, H. P., Limbersky, O., and Folwaczny, C. (2004). Toll-like receptor (TLR) 2 and 4 mutations in periodontal disease. Clin. Exp. Immunol. *135*, 330-335.

Fraser, C. M., Casjens, S., Huang, W. M., Sutton, G. G., Clayton, R., Lathigra, R., White, O., Ketchum, K. A., Dodson, R., Hickey, E. K., et al. (1997). Genomic sequence of a Lyme disease spirochaete, *Borrelia burgdorferi*. Nature *390*, 580-586.

Fraser, C. M., Norris, S. J., Weinstock, G. M., White, O., Sutton, G. G., Dodson, R., Gwinn, M., Hickey, E. K., Clayton, R., Ketchum, K. A., et al. (1998). Complete genome sequence of *Treponema pallidum*, the syphilis spirochete. Science *281*, 375-388.

Goyert, S. M., Ferrero, E., Rettig, W. J., Yenamandra, A. K., Obata, F., and Le Beau, M. M. (1988). The CD14 monocyte differentiation antigen maps to a region encoding growth factors and receptors. Science *239*, 497-500.

Habicht, G. S., Beck, G., Benach, J. L., and Coleman, J. L. (1986). *Borrelia burgdorferi* lipopolysaccharide and its role in the pathogenesis of Lyme disease. Zentralbl. Bakteriol. Mikrobiol. Hyg. *263*, 137-141.

Hägewald, S., Bernimoulin, J. P., Köttgen, E., and Kage, A. (2002). Salivary IgA subclasses and bacteria-reactive IgA in patients with aggressive periodontitis. J. Periodontal Res. *37*, 333-339.

Hardy, P. H., Jr., and Levin, J. (1983). Lack of endotoxin in *Borrelia hispanica* and *Treponema pallidum*. Proc. Soc. Exp. Biol. Med. *174*, 47-52.

Hashimoto, M., Asai, Y., Jinno, T., Adachi, S., Kusumoto, S., and Ogawa, T. (2003a). Structural elucidation of polysaccharide part of glycoconjugate from *Treponema medium* ATCC 700293. Eur. J. Biochem. *270*, 2671-2679.

Hashimoto, M., Asai, Y., and Ogawa, T. (2003b). Treponemal phospholipids inhibit innate immune responses induced by pathogen-associated molecular patterns. J. Biol. Chem. *278*, 44205-44213.

Hirschfeld, M., Kirschning, C. J., Schwandner, R., Wesche, H., Weis, J. H., Wooten, R. M., and Weis, J. J. (1999). Cutting edge: inflammatory signaling by *Borrelia burgdorferi* lipoproteins is mediated by toll-like receptor 2. J. Immunol. *163*, 2382-2386.

Hoebe, K., Du, X., Georgel, P., Janssen, E., Tabeta, K., Kim, S. O., Goode, J., Lin, P., Mann, N., Mudd, S., et al. (2003). Identification of Lps2 as a key transducer of MyD88-independent TIR signalling. Nature *424*, 743-748.

Hoshino, K., Takeuchi, O., Kawai, T., Sanjo, H., Ogawa, T., Takeda, Y., Takeda, K., and Akira, S. (1999). Cutting edge: Toll-like receptor 4 (TLR4)-deficient mice are hyporesponsive to lipopolysaccharide: evidence for TLR4 as the Lps gene product. J. Immunol. *162*, 3749-3752.

Hou, L., Sasaki, H., and Stashenko, P. (2000). Toll-like receptor 4-deficient mice have reduced bone destruction following mixed anaerobic infection. Infect. Immun. *68*, 4681-4687.

Inagaki, S., Ishihara, K., Yasaki, Y., Yamada, S., and Okuda, K. (2003). Antibody responses of periodontitis patients to gingipains of *Porphyromonas gingivalis*. J. Periodontol. *74*, 1432-1439.

Isogai, E., Isogai, H., Kurebayashi, Y., and Ito, N. (1986). Biological activities of leptospiral lipopolysaccharide. Zentralbl. Bakteriol. Mikrobiol. Hyg. *261*, 53-64.

Jackson, S. W., and Zey, P. N. (1973). Ultrastructure of lipopolysaccharide isolated from *Treponema pallidum*. J. Bacteriol. *114*, 838-844.

Jotwani, R., Palucka, A. K., Al-Quotub, M., Nouri-Shirazi, M., Kim, J., Bell, D., Banchereau, J., and Cutler, C. W. (2001). Mature dendritic cells infiltrate the T cell-rich region of oral mucosa in chronic periodontitis: in situ, *in vivo*, and *in vitro* studies. J. Immunol. *167*, 4693-4700.

Kesavalu, L., Holt, S.C., and Ebersole, J. L. (1999). Lack of humoral immune protection against *Treponema denticola* virulence as a murine model. Infect. Immun. *67*, 5736-5741.

Kiechl, S., Lorenz, E., Reindl, M., Wiedermann, C. J., Oberhollenzer, F., Bonora, E., Willeit, J., and Schwartz, D. A. (2002). Toll-like receptor 4 polymorphisms and atherogenesis. N. Engl. J. Med. *347*, 185-192.

Kobayashi, T., Sugita, N., van der Pol, W. L., Nunokawa, Y., Westerdaal, N. A., Yamamoto, K., van de Winkel, J. G., and Yoshie, H. (2000). The Fc-gamma receptor genotype as a risk factor for generalized early-onset periodontitis in Japanese patients. J. Periodontol. *71*, 1425-1432.

Komai-Koma, M., Jones, L., Ogg, G. S., Xu, D., and Liew, F. Y. (2004). TLR2 is expressed on activated T cells as a costimulatory receptor. Proc. Natl. Acad. Sci. U S A *101*, 3029-3034.

Kornman, K. S., Crane, A., Wang, H. Y., di Giovine, F. S., Newman, M. G., Pirk, F. W., Wilson, T. G., Jr., Higginbottom, F. L., and Duff, G. W. (1997). The interleukin-1 genotype as a severity factor in adult periodontal disease. J. Clin. Periodontol. *24*, 72-77.

Lai, C. H., Listgarten, M. A., Evian, C. I., and Dougherty, P. (1986). Serum IgA and IgG antibodies to *Treponema vincentii* and *Treponema denticola* in adult periodontitis, juvenile periodontitis and periodontally healthy subjects. J. Clin. Periodontol. *13*, 752-757.

Lien, E., Sellati, T. J., Yoshimura, A., Flo, T. H., Rawadi, G., Finberg, R. W., Carroll, J. D., Espevik, T., Ingalls, R. R., Radolf, J. D., and Golenbock, D. T. (1999). Toll-like receptor 2 functions as a pattern recognition receptor for diverse bacterial products. J. Biol. Chem. *274*, 33419-33425.

Livermore, B. P., Bey, R. F., and Johnson, R. C. (1978). Lipid metabolism of *Borrelia hermsi*. Infect. Immun. *20*, 215-220.

Livermore, B. P., and Johnson, R. C. (1974). Lipids of the Spirochaetales: comparison of the lipids of several members of the genera *Spirochaeta*, *Treponema*, and *Leptospira*. J. Bacteriol. *120*, 1268-1273.

Loos, B. G., Leppers-Van de Straat, F. G., Van de Winkel, J. G., and Van der Velden, U. (2003). Fc-gamma receptor polymorphisms in relation to periodontitis. J. Clin. Periodontol. *30*, 595-602.

Lorenz, E., Mira, J. P., Cornish, K. L., Arbour, N. C., and Schwartz, D. A. (2000). A novel polymorphism in the Toll-like receptor 2 gene and its potential association with staphylococcal infection. Infect. Immun. *68*, 6398-6401.

Lorenz, E., Mira, J. P., Frees, K. L., and Schwartz, D. A. (2002). Relevance of mutations in the TLR4 receptor in patients with Gram-negative septic shock. Arch. Intern. Med. *162*, 1028-1032.

Lu, H., Califano, J. V., Schenkein, H. A., and Tew, J. G. (1993). Immunoglobulin class and subclass distribution of antibodies reactive with the immunodominant antigen of *Actinobacillus actinomycetemcomitans* serotype b. Infect. Immun. *61*, 2400-2407.

Ma, Y., and Weis, J. J. (1993). *Borrelia burgdorferi* outer surface lipoproteins OspA and OspB possess B-cell mitogenic and cytokine-stimulatory properties. Infect. Immun. *61*, 3843-3853.

McDevitt, M. J., Wang, H. Y., Knobelman, C., Newman, M. G., di Giovine, F. S., Timms, J., Duff, G. W., and Kornman, K. S. (2000). Interleukin-1 genetic association with periodontitis in clinical practice. J. Periodontol. *71*, 156-163.

Medzhitov, R., and Janeway, C. A., Jr. (1997). Innate immunity: the virtues of a nonclonal system of recognition. Cell *91*, 295-298.

Medzhitov, R., and Janeway, C. A., Jr. (1999). Innate immune induction of the adaptive immune response. Cold Spring Harb. Symp. Quant. Biol. *64*, 429-435.

Michalowicz, B. S. (1994). Genetic and heritable risk factors in periodontal disease. J. Periodontol. *65*, 479-488.

Michalowicz, B. S., Aeppli, D. P., Kuba, R. K., Bereuter, J. E., Conry, J. P., Segal, N. L., Bouchard, T. J., Jr., and Pihlstrom, B. L. (1991). A twin study of genetic variation in proportional radiographic alveolar bone height. J. Dent. Res. *70*, 1431-1435.

Michel, O., LeVan, T. D., Stern, D., Dentener, M., Thorn, J., Gnat, D., Beijer, M. L., Cochaux, P., Holt, P. G., Martinez, F. D., and Rylander, R. (2003). Systemic responsiveness to lipopolysaccharide and polymorphisms in the Toll-like receptor 4 gene in human beings. J Allergy Clin. Immunol. *112*, 923-929.

Mooney, J., Hodge, P. J., and Kinane, D. F. (2001). Humoral immune response in early-onset periodontitis: influence of smoking. J. Periodontal Res. *36*, 227-232.

Mori, Y., Yoshimura, A., Ukai, T., Lien, E., Espevik, T., and Hara, Y. (2003). Immunohistochemical localization of Toll-like receptors 2 and 4 in gingival tissue from patients with periodontitis. Oral Microbiol. Immunol. *18*, 54-58.

Morinushi, T., Lopatin, D. E., and Van Poperin, N. (1997). The relationship between gingivitis and the serum antibodies to the microbiota associated with periodontal disease in children with Down's syndrome. J. Periodontol. *68*, 626-631.

Morr, M., Takeuchi, O., Akira, S., Simon, M. M., and Mühlradt, P. F. (2002). Differential recognition of structural details of bacterial lipopeptides by toll-like receptors. Eur. J. Immunol. *32*, 3337-3347.

Moter, A., Hoenig, C., Choi, B. K., Riep, B., and Göbel, U. B. (1998). Molecular epidemiology of oral treponemes associated with periodontal disease. J. Clin. Microbiol. *36*, 1399-1403.

Norgard, M. V., Arndt, L. L., Akins, D. R., Curetty, L. L., Harrich, D. A., and Radolf, J. D. (1996). Activation of human monocytic cells by *Treponema pallidum* and *Borrelia burgdorferi* lipoproteins and synthetic lipopeptides proceeds via a pathway distinct from that of lipopolysaccharide but involves the transcriptional activator NF-kappa B. Infect. Immun. *64*, 3845-3852.

Norris, S. J., and Weinstock, G. M. (2000). The genome sequence of *Treponema pallidum*, the syphilis spirochete: will clinicians benefit? Curr. Opin. Infect. Dis. *13*, 29-36.

Nuessen, M. E., Birmingham, J. R., and Joens, L. A. (1982). Biological activity of a lipopolysaccharide extracted from *Treponema hyodysenteriae*. Infect. Immun. *37*, 138-142.

Nuessen, M. E., Joens, L. A., and Glock, R. D. (1983). Involvement of lipopolysaccharide in the pathogenicity of *Treponema hyodysenteriae*. J. Immunol. *131*, 997-999.

Opitz, B., Schröder, N. W., Spreitzer, I., Michelsen, K. S., Kirschning, C. J., Hallatschek, W., Zähringer, U., Hartung, T., Göbel, U. B., and Schumann, R. R. (2001). Toll-like receptor-2 mediates Treponema glycolipid and lipoteichoic acid-induced NF-kappaB translocation. J. Biol. Chem. *276*, 22041-22047.

Ozinsky, A., Underhill, D. M., Fontenot, J. D., Hajjar, A. M., Smith, K. D., Wilson, C. B., Schroeder, L., and Aderem, A. (2000). The repertoire for pattern recognition of pathogens by the innate immune system is defined by cooperation between Toll-like receptors. Proc. Natl. Acad. Sci. U S A *97*, 13766-13771.

Page, R. C. (1991). The role of inflammatory mediators in the pathogenesis of periodontal disease. J. Periodontal Res. *26*, 230-242.

Papapanou, P. N. (1999). Epidemiology of periodontal diseases: an update. J. Int. Acad. Periodontol. *1*, 110-116.

Parkhill, J. M., Hennig, B. J., Chapple, I. L., Heasman, P. A., and Taylor, J. J. (2000). Association of interleukin-1 gene polymorphisms with early-onset periodontitis. J. Clin. Periodontol. *27*, 682-689.

Pasare, C., and Medzhitov, R. (2003). Toll pathway-dependent blockade of CD4+CD25+ T cell-mediated suppression by dendritic cells. Science *299*, 1033-1036.

Petersen, C. S. (1986). Isolation and characterization of antigens from *Treponema phagedenis* biotype Reiter cross-reacting with *Treponema pallidum*. Dan. Med. Bull. *33*, 257-265.

Poltorak, A., He, X., Smirnova, I., Liu, M. Y., Van Huffel, C., Du, X., Birdwell, D., Alejos, E., Silva, M., Galanos, C., et al. (1998a). Defective LPS signaling in C3H/HeJ and C57BL/10ScCr mice: mutations in TLR4 gene. Science *282*, 2085-2088.

Poltorak, A., Smirnova, I., He, X., Liu, M. Y., Van Huffel, C., McNally, O., Birdwell, D., Alejos, E., Silva, M., Du, X., et al. (1998b). Genetic and physical mapping of the Lps locus: identification of the toll-4 receptor as a candidate gene in the critical region. Blood. Cells Mol. Dis. *24*, 340-355.

Que-Gewirth, N. L., Ribeiro, A. A., Kalb, S. R., Cotter, R. J., Bulach, D. M., Adler, B., Saint Girons, I., Werts, C., and Raetz, C. R. (2004). A methylated phosphate group and four amide-linked acyl chains in *Leptospira interrogans* lipid A: The membrane anchor of an unusual lipopolysaccharide that activates TLR2. J. Biol. Chem. *279*, 25420-25429.

Radolf, J. D., Norgard, M. V., Brandt, M. E., Isaacs, R. D., Thompson, P. A., and Beutler, B. (1991). Lipoproteins of *Borrelia burgdorferi* and *Treponema pallidum* activate cachectin/tumor necrosis factor synthesis. Analysis using a CAT reporter construct. J. Immunol. *147*, 1968-1974.

Radolf, J. D. (1995). *Treponema pallidum* and the quest for outer membrane proteins. Mol. Microbiol. *16*, 1067-1073.

Riviere, G. R., Wagoner, M. A., Baker-Zander, S. A., Weisz, K. S., Adams, D. F., Simonson, L., and Lukehart, S. A. (1991a). Identification of spirochetes related to *Treponema pallidum* in necrotizing ulcerative gingivitis and chronic periodontitis. N. Engl. J. Med. *325*, 539-543.

Riviere, G. R., Weisz, K. S., Simonson, L. G., and Lukehart, S. A. (1991b). Pathogen-related spirochetes identified within gingival tissue from patients with acute necrotizing ulcerative gingivitis. Infect. Immun. *59*, 2653-2657.

Rosen, G., Sela, M. N., Naor, R., Halabi, A., Barak, V., and Shapira, L. (1999). Activation of murine macrophages by lipoprotein and lipooligosaccharide of *Treponema denticola*. Infect. Immun. *67*, 1180-1186.

Schnare, M., Barton, G. M., Holt, A. C., Takeda, K., Akira, S., and Medzhitov, R. (2001). Toll-like receptors control activation of adaptive immune responses. Nat. Immunol. 2, 947-950.

Schrank, K., Choi, B. K., Grund, S., Moter, A., Heuner, K., Nattermann, H., and Göbel, U. B. (1999). *Treponema brennaborense* sp. nov., a novel spirochaete isolated from a dairy cow suffering from digital dermatitis. Int. J. Syst. Bacteriol. *49 Pt 1*, 43-50.

Schröder, N. W., Hermann, C., Hamann, L., Göbel, U. B., Hartung, T., and Schumann, R. R. (2003a). High frequency of polymorphism Arg753Gln of the Toll-like receptor-2 gene detected by a novel allele-specific PCR. J. Mol. Med. *81*, 368-372.

Schröder, N. W., Morath, S., Alexander, C., Hamann, L., Hartung, T., Zähringer, U., Göbel, U. B., Weber, J. R., and Schumann, R. R. (2003b). Lipoteichoic acid (LTA) of *Streptococcus pneumoniae* and *Staphylococcus aureus* activates immune cells via Toll-like receptor (TLR)-2, lipopolysaccharide-binding protein (LBP), and CD14, whereas TLR-4 and MD-2 are not involved. J. Biol. Chem. *278*, 15587-15594.

Schröder, N. W., Opitz, B., Lamping, N., Michelsen, K. S., Zähringer, U., Göbel, U. B., and Schumann, R. R. (2000). Involvement of lipopolysaccharide binding protein, CD14, and Toll-like receptors in the initiation of innate immune responses by *Treponema* glycolipids. J. Immunol. *165*, 2683-2693.

Schröder, N. W., Pfeil, D., Opitz, B., Michelsen, K. S., Amberger, J., Zähringer, U., Göbel, U. B., and Schumann, R. R. (2001). Activation of mitogen-activated protein kinases p42/44, p38, and stress-activated protein kinases in myelo-monocytic cells by *Treponema* lipoteichoic acid. J. Biol. Chem. *276*, 9713-9719.

Schröder, N. W. J., Meister, D., Wolff, V., Christan, C., Kaner, D., Haban, V., Purucker, P., Hermann, C., Moter, A., Göbel, U. B., and Schumann, R. R. (2005). Chronic periodontal disease is associated with single-nucleotide polymorphisms of the human TLR-4 gene. Genes Immun. *6*, 448-451.

Schröder, N. W. J., and Schumann, R. R. (2005). Single nucleotide polymorphisms of Toll-like receptors and susceptibility to infectious disease. Lancet Infect. Dis. *5*, 156-164.

Schultz, C. P., Wolf, V., Lange, R., Mertens, E., Wecke, J., Naumann, D., and Zähringer, U. (1998). Evidence for a new type of outer membrane lipid in oral spirochete *Treponema denticola*. Functioning permeation barrier without lipopolysaccharides. J. Biol. Chem. *273*, 15661-15666.

Schumann, R. R., Leong, S. R., Flaggs, G. W., Gray, P. W., Wright, S. D., Mathison, J. C., Tobias, P. S., and Ulevitch, R. J. (1990). Structure and function of lipopolysaccharide binding protein. Science *249*, 1429-1431.

Schwandner, R., Dziarski, R., Wesche, H., Rothe, M., and Kirschning, C. J. (1999). Peptidoglycan- and lipoteichoic acid-induced cell activation is mediated by Toll-like receptor 2. J. Biol. Chem. *274*, 17406-17409.

Sela, M. N., Bolotin, A., Naor, R., Weinberg, A., and Rosen, G. (1997). Lipoproteins of *Treponema denticola*: their effect on human polymorphonuclear neutrophils. J. Periodontal Res. *32*, 455-466.

Sellati, T. J., Bouis, D. A., Kitchens, R. L., Darveau, R. P., Pugin, J., Ulevitch, R. J., Gangloff, S. C., Goyert, S. M., Norgard, M. V., and Radolf, J. D. (1998). *Treponema pallidum* and *Borrelia burgdorferi* lipoproteins and synthetic lipopeptides activate monocytic cells via a CD14-dependent pathway distinct from that used by lipopolysaccharide. J. Immunol. *160*, 5455-5464.

Seshadri, R., Myers, G. S., Tettelin, H., Eisen, J. A., Heidelberg, J. F., Dodson, R. J., Davidsen, T. M., DeBoy, R. T., Fouts, D. E., Haft, D. H., et al. (2004). Comparison of the genome of the oral pathogen *Treponema denticola* with other spirochete genomes. Proc. Natl. Acad. Sci. U S A.

Shimazu, R., Akashi, S., Ogata, H., Nagai, Y., Fukudome, K., Miyake, K., and Kimoto, M. (1999). MD-2, a molecule that confers lipopolysaccharide responsiveness on Toll-like receptor 4. J. Exp. Med. *189*, 1777-1782.

Shortman, K., and Liu, Y. J. (2002). Mouse and human dendritic cell subtypes. Nat. Rev. Immunol. *2*, 151-161.

Stashenko, P., Dewhirst, F. E., Peros, W. J., Kent, R. L., and Ago, J. M. (1987). Synergistic interactions between interleukin 1, tumor necrosis factor, and lymphotoxin in bone resorption. J. Immunol. *138*, 1464-1468.

Takayama, K., Rothenberg, R. J., and Barbour, A. G. (1987). Absence of lipopolysaccharide in the Lyme disease spirochete, *Borrelia burgdorferi*. Infect. Immun. *55*, 2311-2313.

Takeuchi, O., Kawai, T., Mühlradt, P. F., Morr, M., Radolf, J. D., Zychlinsky, A., Takeda, K., and Akira, S. (2001). Discrimination of bacterial lipoproteins by Toll-like receptor 6. Int. Immunol. *13*, 933-940.

Takeuchi, O., Sato, S., Horiuchi, T., Hoshino, K., Takeda, K., Dong, Z., Modlin, R. L., and Akira, S. (2002). Cutting edge: role of Toll-like receptor 1 in mediating immune response to microbial lipoproteins. J. Immunol. *169*, 10-14.

Thoma-Uszynski, S., Stenger, S., Takeuchi, O., Ochoa, M. T., Engele, M., Sieling, P. A., Barnes, P. F., Röllinghoff, M., Bolcskei, P. L., Wagner, M., et al. (2001). Induction of direct antimicrobial activity through mammalian toll-like receptors. Science *291*, 1544-1547.

Tsurumi, Y., Hayakawa, M., Shibata, Y., and Abiko, Y. (2003). Production of antibody against a synthetic peptide of *Porphyromonas gingivalis* 40-kDa outer membrane protein. J. Oral Sci. *45*, 111-116.

van der Pol, W. L., Huizinga, T. W., Vidarsson, G., van der Linden, M. W., Jansen, M. D., Keijsers, V., de Straat, F. G., Westerdaal, N. A., de Winkel, J. G., and Westendorp, R. G. (2001). Relevance of Fc-gamma receptor and interleukin-10 polymorphisms for meningococcal disease. J. Infect. Dis. *184*, 1548-1555.

Walker, S. G., Xu, X., Altman, E., Davis, K. J., Ebersole, J. L., and Holt, S. C. (1999). Isolation and chemical analysis of a lipopolysaccharide from the outer membrane of the oral anaerobic spirochete *Treponema pectinovorum*. Oral Microbiol. Immunol. *14*, 304-308.

Wang, P. L., and Ohura, K. (2002). *Porphyromonas gingivalis* lipopolysaccharide signaling in gingival fibroblasts-CD14 and Toll-like receptors. Crit. Rev. Oral. Biol. Med. *13*, 132-142.

Wang, P. L., Ohura, K., Fujii, T., Oido-Mori, M., Kowashi, Y., Kikuchi, M., Suetsugu, Y., and Tanaka, J. (2003). DNA microarray analysis of human gingival fibroblasts from healthy and inflammatory gingival tissues. Biochem. Biophys. Res. Commun. *305*, 970-973.

Weinstock, G. M., Hardham, J. M., McLeod M. P., Sodergren, E. J., and Norris, S. J., (1998). The genome of *Treponema pallidum*: new light on the agent of syphilis. FEMS Microbiol. Rev. *22*, 323-332.

Werts, C., Tapping, R. I., Mathison, J. C., Chuang, T. H., Kravchenko, V., Saint Girons, I., Haake, D. A., Godowski, P. J., Hayashi, F., Ozinsky, A., et al. (2001). Leptospiral lipopolysaccharide activates cells through a TLR2-dependent mechanism. Nat. Immunol. *2*, 346-352.

Winkler, J. R., Murray, P. A., Grassi, M., and Hammerle, C. (1989). Diagnosis and management of HIV-associated periodontal lesions. J. Am. Dent. Assoc. Suppl., 25S-34S.

Wooten, R. M., Modur, V. R., McIntyre, T. M., and Weis, J. J. (1996). *Borrelia burgdorferi* outer membrane protein A induces nuclear translocation of nuclear factor-kappa B and inflammatory activation in human endothelial cells. J. Immunol. *157*, 4584-4590.

Wright, S. D., Ramos, R. A., Tobias, P. S., Ulevitch, R. J., and Mathison, J. C. (1990). CD14, a receptor for complexes of lipopolysaccharide (LPS) and LPS binding protein. Science *249*, 1431-1433.

Wyss, C., Choi, B. K., Schüpbach, P., Guggenheim, B., and Göbel, U. B. (1996). *Treponema maltophilum* sp. nov., a small oral spirochete isolated from human periodontal lesions. Int. J. Syst. Bacteriol. *46*, 745-752.

Yamamoto, M., Sato, S., Hemmi, H., Hoshino, K., Kaisho, T., Sanjo, H., Takeuchi, O., Sugiyama, M., Okabe, M., Takeda, K., and Akira, S. (2003). Role of adaptor TRIF in the MyD88-independent toll-like receptor signaling pathway. Science *301*, 640-643.

Yamamoto, M., Sato, S., Mori, K., Hoshino, K., Takeuchi, O., Takeda, K., and Akira, S. (2002). Cutting edge: a novel Toll/IL-1 receptor domain-containing adapter that preferentially activates the IFN-beta promoter in the Toll-like receptor signaling. J. Immunol. *169*, 6668-6672.

Yotis, W. W., Sharma, V. K., Gopalsami, C., Chegini, S., McNulty, J., Hoerman, K., Keene, J., Jr., and Simonson, L. G. (1991). Biochemical properties of the outer membrane of *Treponema denticola*. J. Clin. Microbiol. *29*, 1397-1406.

17

Treponema and Bovine Skin Disease: Papillomatous Digital Dermatitis and Ulcerative Mammary Dermatitis

Lola V. Stamm and Darren J. Trott*

Abstract

Papillomatous digital dermatitis (PDD) is an infectious, superficial dermatitis that primarily affects the feet of dairy cattle. PDD is associated with intensive housing of cattle and prolonged exposure of bovine feet to moist, unhygienic conditions. Experimental transmission of PDD has been achieved by inoculating the feet of cattle with lesion material. Treatment of the lesions with antibiotics results in rapid resolution, consistent with the bacterial etiology of the disease. PDD lesions contain a mixed population of predominantly anaerobic bacteria. A pathognomonic feature of PDD lesions is the presence of large numbers of spirochetes invading deep into the epidermis, whereas other bacteria are located more superficially. It is currently unclear whether the PDD-associated spirochetes are primary pathogens or secondary opportunists. A number of *Treponema* phylogenetic groups have been demonstrated in PDD lesions. Of the spirochetes that are cultivable, only *Treponema brennaborense* has been classified as a new species. The remaining cultivable *Treponema* are closely related to human-associated *T. denticola*, *T. medium/vincentii*, and *T. phagedenis*. Cattle affected by PDD develop an immune response to *Treponema* antigens. A bacterin incorporating cultivable PDD-associated *Treponema* has been shown to be effective at reducing lesion development. PDD-associated *Treponema* have been recently demonstrated in cases of bovine ulcerative mammary dermatitis and contagious ovine digital dermatitis. Little is known concerning the origin and pathogenic mechanisms of the PDD-associated *Treponema*. Further research is required to clarify the taxonomic position and basic biology of these spirochetes.

*For correspondence email lstamm@email.unc.edu

1. Introduction

Papillomatous digital dermatitis (PDD), also referred to as digital dermatitis (DD), interdigital papillomatosis, hairy heel warts, or hairy footwart, is a major cause of lameness in dairy cattle (Wells et al., 1999; Berry, 2001). PDD was first described in Italy by Cheli and Mortellaro in 1974. Since then, it has been observed in many countries throughout the world (Blowey et al., 1992; Van Amstel et al., 1995; Read and Walker, 1998c; Choi et al., 1997; Cruz et al., 2001). In 1980, Rebhun et al. (1980) reported the presence of PDD in dairy cattle in New York. During the next decade, sporadic PDD cases were observed in Atlantic and mid-western states (Whittier, 1988). In 1992, Read et al. (1992) reported an outbreak of PDD in southern California. By 1993, 89% of California dairy herds were affected, with intraherd morbidity estimates ranging from 0.5-12% per month (Read and Walker, 1998c). PDD is now present throughout most of the United States (U.S.) (Wells et al., 1999; Brown et al., 2000). Data from the National Animal Health Monitoring System Dairy '96 survey showed that 43.5% of U.S. dairy operations had observed PDD in their herds in the twelve months prior to the survey (Wells et al., 1999). In 78% of the affected herds, the first PDD cases occurred in 1993 or later. Wells et al. (1999) and others (Nutter and Moffit, 1990; Rodriguez-Lainz et al., 1996a; Read and Walker, 1998c) have proposed that the emergence and epidemic spread of PDD are related to changes in herd management practices.

2. Clinical presentation of PDD

The clinical presentation of PDD is well documented and distinct from other bovine foot diseases. In many countries, it has overtaken footrot as the most important infectious cause of lameness in dairy cattle (Murray et al., 1996). PDD typically appears in dairy herds as lameness outbreaks of variable severity (Read and Walker, 1998c; Wells et al., 1999). PDD lesions usually occur on the hind feet at or above the plantar interdigital ridge adjacent to the bulbs of the heel. Occasionally, lesions are found within the interdigital cleft. Lesions may occur on more than one foot and sometimes involve the front feet. Digital swelling is not usually a feature of the disease. Early lesions are initially 1-4 cm in diameter with distinct margins, have a red, granular ("strawberry-like") appearance, and are prone to hemorrhage (Figure 1). Mature lesions are more proliferative (2-6 cm), grey to whitish yellow in color, and are surrounded by long erect hairs. Wart-like brown/black papillary projections that vary in length (1 mm-3 cm) often protrude from the lesion surface, giving the disease its characteristic name (hairy heel warts). The surfaces of both early and mature lesions are moist with a serous exudate. A foul odor may be present. PDD is characterized by erosion of the superficial layers of the epidermis due to loss of keratin, epithelial hyperplasia and hypertrophy. A marked pain response is elicited when lesions are palpitated. Read and Walker (1998c) reported that cattle severely affected with PDD are reluctant to stand or walk. The most affected limb is often held shaking in partial flexion. The weight of less severely affected limbs is shifted on the toes causing clubbing of hooves and atrophy of the heel bulbs. Severe lameness is only observed when the lesions involve the horny structures of the digits.

Secondary effects of PDD-associated lameness are significant weight loss due to reduced feed intake, decreased milk production, and impaired reproductive

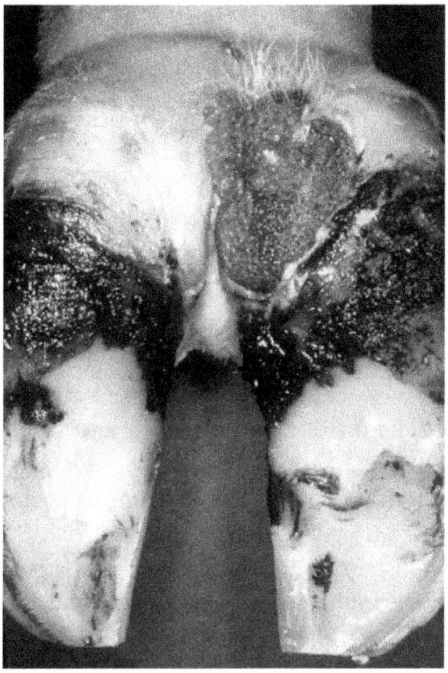

Figure 1. Early papillomatous digital dermatitis (PDD) lesion on the foot of a Holstein dairy cow. Note the circumscribed, granular appearance of the superficial ulcer. (Photograph courtesy of D.H. Read, University of California, Davis).

performance (Collick et al., 1989; Read and Walker, 1998c; Yeruham et al., 2000; Hernandez et al., 2001). If untreated, PDD can persist for months. Although the economic impact of PDD has not been adequately quantified, significant financial losses occur due to reductions in milk yield, premature culling of affected cattle, and costs associated with treatment and control efforts (Rebhun et al., 1980; Nutter and Moffit, 1990; Argaez-Rodriguez et al., 1997).

3. Epidemiology of PDD

PDD appears to be highly transmissible based upon the rapid intra- and inter-herd spread of the disease (Blowey et al., 1992; Read and Walker, 1998c; Wells et al., 1999). Male and female cattle of all breeds and ages are susceptible. However, first parity cows have the highest incidence of PDD and are more likely to show lameness than older animals (Rodriguez-Lainz et al., 1999). PDD is the major infectious cause of lameness in dairy cattle in the U.S. and Europe (Read, 1995; Murray et al., 1996: Berry, 2001), and the disease has been reported in Mexico, South America, Japan, Iran and Israel (Blowey et al., 1992; Argaez-Rodriguez et al., 1997; Rodriguez-Lainz et al., 1999; Brown et al., 2000; Cruz et al., 2001). In countries where intensive housing of dairy cattle is uncommon, such as Australia and New Zealand, PDD may only occur sporadically (McLennan et al., 1996; Milinovich et al., 2004; Vermunt and Hill, 2004). Wells et al. (1999) reported that the regions of the U.S. with the highest percent of herds affected included the Southwest, Northwest and Northeast. In the southern region of California, 75% of farms reported PDD in their herd, with up to 85% of cows affected on some individual farms (Rodriguez-Lainz et al., 1996a). Brown et al. (2000) reported that 29% of culled dairy cattle and 4% of culled beef cattle examined in one

Southeastern U.S. slaughterhouse had gross lesions on the left hind foot that were consistent with PDD. In regions, such as California, where throughout the year, dairy cows are kept in free stalls (where they never leave the barn or leave for only short periods) or drylots (where they are outside on a dirt lot that does not provide access to grazing), PDD is most severe during spring and summer months. In Europe, PDD is most severe during the winter housing months (Berry, 2001).

Prior to 1980, PDD occurred only sporadically in U.S. dairy herds (Rebhun et al., 1980). The epidemic spread of PDD in the early 1990s made it possible to identify herd-level or management practices associated with a high incidence of the disease (Wells et al., 1999). Risk factors for PDD include large herd size, reduced access to pasture, exposure of hooves to continual moisture, grooved concrete floors, introduction of replacement heifers from other operations, employment of hoof trimmers who also trimmed hooves on other operations, and failure to decontaminate hoof trimming equipment between cows. The latter factors confirm the contagious nature of PDD and indicate a potential role for fomites in disease transmission (Wells et al., 1999). In other regional studies, additional risk factors identified include free stall herds (as compared to tie stall herds or those confined to pasture) and breed disposition (Rodriguez-Lainz et al., 1996b; Rodriguez-Lainz et al., 1999; Berry, 2001).

4. Etiology of PDD

Read and Walker (1996;1998b) demonstrated that PDD is transmissible by inoculating the feet of healthy Holstein calves with fresh scrapings from active PDD lesions. The calves' feet were exposed to constant moisture and an inert wrap limited access to air. PDD transmission was 100% effective and did not require prior trauma. Although the histopathological character of PDD resembles that of a squamous papilloma, a variety of laboratory tests (electron microscopy, immunohistochemistry, cell culture, and PCR amplification) have yielded negative results for the presence of a viral agent (Rebhun et al., 1980; Read et al., 1992; Read, 1995; Borgmann et al., 1996). Additional tests have also been negative for parasitic protozoa and fungi (Read, 1995). The observed response of PDD lesions to parenteral and topical antibiotic treatment is strongly suggestive of a bacterial etiology. However, the nature and location of PDD lesions, coupled with the presence of contaminating fecal flora and environmental bacteria, have made identification of the specific etiologic agent(s) difficult. Culture and microscopic analyses of touch preparations of PDD lesions typically show a variety of different bacterial morphotypes including various Gram-negative rods (*Porphyromonas*, *Campylobacter*, *Prevotella*, *Fusobacterium*, etc.) and spirochetes (Read et al., 1992; Choi et al., 1997; Collighan and Woodward, 1997; Dopfer et al., 1997; Ohya et al., 1999). However, spirochetes are the only bacteria that are consistently present in large numbers in the superficial lesions (Figure 2) and also in deeper tissues (stratum spinosum and dermal papillae) where other bacteria are very rarely observed (Read et al., 1992; Borgmann et al., 1996; Choi et al., 1997; Moter et al., 1998; Read et al., 1998). Spirochetes are not present on the feet of healthy cattle (Blowey et al., 1992; Collighan and Woodward, 1997; Dopfer et al., 1997; Demirkan et al., 1998).

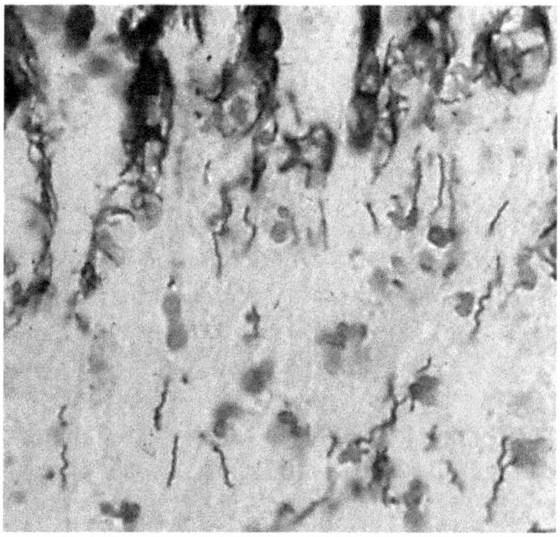

Figure 2. Silver stain (Parker modification of Steiner stain) of the epidermis from a PDD lesion showing invading *Treponema*. (Photograph courtesy of R.L. Walker, University of California, Davis).

5. Pathogenesis of PDD

The histological appearance of PDD lesions has been described as an acute, suppurative inflammation of the epidermis with superficial necrosis and hyperkeratosis (Rebhun et al., 1980; Read et al., 1992; Blowey et al., 1994a; Read and Walker, 1998c; El-Ghoul and Shaheed, 2001). Dopfer et al. (1997) graded PDD lesions into stage 1 (early, small, circumscribed granulomatous lesion), stage 2 (classic ulcerative lesion), stage 3 (ulcerative lesion in the process of healing), and stage 4 (classic proliferative lesion). Stage 1 lesions are characterized by the presence of long, thin spirochetes penetrating deep into the stratum spinosum, where they appear to evoke a significant perivascular inflammatory infiltrate dominated by neutrophils and monocytes. In stage 2 lesions, the inflammatory response is more pronounced and characterized by the presence of neutrophils, eosinophils, and plasma cells. At least two spirochete morphotypes are present in these lesions, with short thick spirochetes predominating over the long thin morphotype. Mumba et al. (1999) demonstrated that stage 2 lesions were consistently more positive in a *Treponema*-specific PCR test (Rijpkema et al., 1997) than were other disease stage lesions. Proliferation of the epithelium (stage 4) is a chronic response to inflammation with the purpose of eliminating invading organisms from the surface by desquamation. Proliferative lesions are characterized by ballooning degeneration of the epidermis, an increase in the number of infiltrating plasma cells, and the presence of at least two spirochete morphotypes in the stratum basale. Many of these features resemble the histopathological appearance of yaws and acute, necrotizing ulcerative gingivitis.

6. Isolation of PDD-associated *Treponema*

In 1995, Walker et al. (1995) reported the first successful *in vitro* cultivation of PDD-associated spirochetes from lesions of cattle from four different California dairies. They employed a complex isolation medium known as Oral Treponeme Isolation (OTI) broth that contains volatile fatty acids, various sugars, calf serum, and a reducing agent (Smibert et al., 1991). The medium was supplemented with rifampicin and enrofloxacin to promote selective enrichment of spirochetes (Stamm et al., 2001). Following incubation for 24-48 hours, aliquots of OTI enrichment broth were plated onto OTI agar to obtain single colonies of spirochetes free from contamination by other bacteria. Two groups of fastidious, anaerobic spirochetes were differentiated on the basis of morphologic, antigenic, and enzymatic properties. Both phenotypes of the PDD-associated spirochetes have characteristics consistent with the members of the genus *Treponema*, but are unique from bovine rumen treponemes (*T. bryantii*, *T. saccharophilum*, and a recently described novel *Treponema*) (Walker et al., 1995; Demirkan et al., 1998; Trkov et al., 2001). Sera from cattle with PDD showed a significant humoral response to cellular proteins of both PDD-associated *Treponema* phenotypes as assessed by reactivity in ELISA and Western blot (Walker et al., 1997; Demirkan et al., 1999b). Sera from control cattle in PDD-free herds were not significantly reactive in either immunoassay. These findings agree with the reported absence of PDD-associated *Treponema* in cattle with no history of PDD (Blowey et al., 1992; Dopfer et al., 1997; Demirkan et al., 1998). Read and Walker (1998a) showed by immunohistochemistry that spirochetes similar to the U.S. PDD-associated *Treponema* are present in the PDD lesions of cattle from 16 additional countries.

Due to difficulties associated with cultivation, only a limited number of PDD-associated *Treponema* isolates have been obtained from PDD lesions. The first cultivation of PDD-associated *Treponema* from European cattle was reported in 1999. Demirkan et al. (1999a) isolated a *Treponema* from PDD lesions of a cow in the United Kingdom (UK) using immunomagnetic separation that employed beads coated with polyclonal antibodies to *T. denticola* and *T. vincentii*. Subsequent phenotypic and genotypic analysis of this spirochete indicates that it is related to certain of the California PDD-associated *Treponema* (Edwards et al., 2003b; see Section 7). Schrank et al. (1999) used a chemically defined *Treponema* liquid culture medium (Wyss et al., 1996) supplemented with antibiotics and a limit dilution technique to isolate and characterize a novel *Treponema* (designated *T. brennaborense* sp. nov.) from PDD lesions of a dairy cow in Germany. Phenotypic and genotypic analysis indicated that this spirochete is different from all *Treponema* species described previously. Trott et al. (2003) reported cultivation of four *Treponema* isolates from PDD lesions of Iowa dairy cattle utilizing the technique of Walker et al. (1995). Further analysis of these genetically and antigenically diverse isolates indicated that they are closely related to certain of the California PDD-associated *Treponema* (see Section 7).

7. Genotypic and phenotypic characterization of PDD-associated *Treponema*

Comparative 16S ribosomal DNA (rDNA) sequence analysis of PDD-associated *Treponema* isolates from California dairy cattle (Walker et al., 1995) indicated that they represent three phylogenetic groups (phylotypes) that cluster most closely with the human-associated *T. denticola*, *T. medium/vincentii*, and *T. phagedenis* (Walker et al., 1998) (see *Treponema* Phylogeny, Chapter 1 for phylogenetic trees). *T. denticola* and *T. medium/vincentii* are strongly associated with human periodontal disease (Paster et al., 1998; Willis et al., 1999) and have been shown to be invasive and tissue-destructive in an *in vivo* animal model (Kesavalu et al., 1997). Trott et al. (2003) demonstrated that four PDD-associated *Treponema* isolates from Iowa cattle, together with two isolates from California cattle (Walker et al., 1995), have the same 16S rDNA sequence and cluster most closely with *T. phagedenis*. Stamm et al. (2002) applied 16S-23S rDNA intergenic spacer region (ISR) sequence analysis to both California and Iowa PDD-associated *Treponema*. Their results supported placement of the California PDD-associated *Treponema* into three phylotypes and confirmed the close genetic relationship between the Iowa PDD-associated *Treponema* and the *T. phagedenis*-like California PDD-associated *Treponema*.

Using comparative 16S rDNA sequence analysis, Choi et al. (1997) identified the presence of five phylotypes of PDD-associated *Treponema* in pooled punch biopsies obtained from the PDD lesions of four dairy cows in Germany. Three of these phylotypes cluster with *T. denticola*, *T. medium/vincentii*, or *T. phagedenis* and apparently correspond to the California PDD-associated *Treponema* phylotypes. The remaining two German phylotypes cluster with group IV human oral treponemes (ex. *T. maltophilum*) that are associated with periodontal disease. There is currently no evidence that the latter two German phylotypes are present in U.S. cattle. Moter et al. (1998) showed by fluorescence *in situ* hybridization (FISH) differential distribution of the as yet uncultivated German PDD-associated *Treponema* in lesion biopsies. Their results suggest that the PDD-associated *Treponema* phylotypes that are most closely related to *T. medium/vincentii, T. phagedenis,* and *T. maltophilum* are associated with invasion of deeper tissues, whereas those most closely related to *T. denticola* are located in the superficial epidermis.

Rijpkema et al. (1997), Demirkan et al. (1998), and Mumba et al. (1999) used PCR amplification of 16S rDNA to detect *Treponema* present in PDD lesions. However, these studies did not provide information concerning the diversity of *Treponema* phylotypes in the lesions. Rijpkema et al. (1997) and Collighan and Woodward (1997) sequenced PCR-amplified 16S rDNA from PDD lesions of UK dairy cattle and found *Treponema* sequences that were closely related to the *T. denticola*-like PDD-associated *Treponema*. Studies conducted by Edwards et al. (2003b) showed that a *Treponema* (UB1467) isolated from PDD lesions of a UK dairy cow (Demirkan et al., 1999a) is more closely related to *T. medium/vincentii*-like PDD-associated *Treponema* based on 16S rDNA sequence analysis and phenotypic properties.

With the exception of ultrastructural morphology, whole cell protein profiles and enzymatic activities, very few phenotypic characteristics have been determined for PDD-associated *Treponema*. Isolates that are closely related to *T. phagedenis* are 10-15 µm in length, 0.35-0.40 µm in diameter, and possess between 7-9 periplasmic flagella inserted subterminally at each end of the cell (14-18 flagella per cell) (Walker et al., 1995; Trott et al., 2003). In contrast, *T. denticola*-like and *T. medium/vincentii*-like isolates appear to be shorter and thinner and possess 2-5 periplasmic flagella per each end of the cell (4-10 flagella per cell) (Walker et al., 1995; Collighan et al., 2000). *T. brennaborense* possesses only two periplasmic flagella per each end of the cell (4 flagella per cell) and can also be distinguished from other PDD-associated *Treponema* on the basis of enzymatic activity in the API-ZYM system (Schrank et al., 1999). *T. phagedenis*-like PDD-associated *Treponema* produce acetate, proprionate, butyrate and formate as metabolic end products. By comparison, *T. phagedenis* produces acetate, prioprionate and butyrate, but not formate (D.J. Trott, unpublished data).

Due to the relatively small number of *Treponema* isolates that have been obtained from PDD lesions, it is difficult to determine genetic diversity within each of the five recognized phylogenetic groups. Walker et al. (1995) demonstrated that each of seven *T. phagedenis*-like PDD-associated *Treponema* from four different California farms had a unique restriction endonuclease analysis pattern, although two isolates from individual animals on the same farm were closely related. Based on pulsed field gel electrophoresis (PFGE) patterns obtained with three enzymes, each of four *T. phagedenis*-like PDD-associated *Treponema* isolated from the same farm in Iowa had a unique pattern. Interestingly, a California isolate possessed the same PFGE pattern as one of the Iowa isolates (Trott et al., 2003). These preliminary data suggest that *T. phagedenis*-like PDD-associated *Treponema* are genetically heterogeneous.

8. Pathogenic mechanisms of PDD-associated *Treponema*

It has not been definitively determined whether the PDD-associated *Treponema* are primary pathogens or opportunists. Attempts to reproduce PDD in dairy cattle by primary inoculation of *Treponema* isolates (either alone or in combination) have not been successful (R.L. Walker, personal communication). Although the PDD-associated *Treponema* are invasive and evoke strong immune responses in naturally infected cattle (Walker et al., 1997; Demirkan et al., 1999b; Trott et al., 2003), the pathogenic mechanisms of these spirochetes are poorly understood. Like human oral *Treponema* associated with periodontal disease, PDD-associated *Treponema* are motile and contain various cellular components (lipoproteins, glycolipids, lipopolysaccharide (LPS)-like material, peptidoglycan, etc.) that may elicit inflammatory responses (Opitz et al., 2001; Trott et al., 2003). For example, Opitz et al. (2001) reported that recognition of *T. brennaborense* glycolipid by murine macrophage Toll-like receptors (TLR2 and TLR4) activated a signaling pathway that resulted in production of TNFα, a proinflammatory cytokine. Additionally, some PDD-associated *Treponema* synthesize extracellular products such as hemolysin and proteases (Walker et al., 1995; Collighan and Woodward, 1997; Edwards et al., 2003b) that may promote lesion development and/or persistence. Kesavalu et al. (1998) demonstrated that a polymicrobial complex of *T. denticola* and *Porphyromonas gingivalis* has a synergistic effect on the formation and size of lesions in a murine model. Some investigators have speculated that a polymicrobial complex of PDD-associated *Treponema* and Gram-negative bacteria may be similarly important for the initiation of the superficial PDD lesion (Choi et al., 1997; Collighan and Woodward, 1997; Edwards et al., 2003a).

The relative contribution of each of the five PDD-associated *Treponema* phylotypes to lesion development is currently unknown. Interestingly, Moter et al. (1998) demonstrated by FISH that five PDD-associated *Treponema* phylotypes may be present in PDD lesions and that some of the phylotypes are more invasive than other phylotypes. PDD-associated *Treponema* closely related to *T. denticola* were only observed in the surface debris and superficial layers, whereas those related to *T. phagedenis, T. medium/vincentii* and group IV human oral treponemes were present deep within the epidermis next to the dermal papillae. A curved bacterial morphotype that reacted with the universal probe but not with any of the spirochete-specific probes was located even deeper within the tissues. Further studies are needed to define the contribution of the PDD-associated *Treponema* phylotypes (alone or in combination) to the initiation and progression of PDD. However, these studies are hampered by the lack of a small animal model that could be used to demonstrate differences in pathogenicity between the phylotypes and the inability to cultivate the full complement of *Treponema* that are present in PDD lesions.

9. Immune response of dairy cattle to PDD-associated *Treponema*

In the UK, serological studies initially established that cattle with PDD produced a seropositive response to *Borrelia burgdorferi* (Blowey et al., 1994b). Walker et al., (1997) first demonstrated by ELISA that cattle with PDD lesions developed specific serum antibodies to PDD-associated *Treponema*. The *Treponema* strains (1-9185MED and 2-1498) used as a source of antigens were from California dairy cattle and were closely related to *T. denticola* and *T. phagedenis*, respectively (Walker et al., 1995). Demirkan et al. (1999b) confirmed that the high antibody response to both *B. burgdorferi* and PDD-associated *Treponema* was due to the presence of strong cross-reacting epitopes, which are probably flagella proteins. Polyclonal antibodies to *T. denticola* and *T. vincentii* also identified a common 34 kDa protein in the PDD-associated *Treponema*, confirming their antigenic relationship to human oral treponemes. Sera from cattle with PDD lesions identified prominent antigens of 34, 41 and 55 kDa on PDD-associated *Treponema* strain 2-1498. Some cows also produced antibody to LPS-like material present in this strain.

Recently, Trott et al. (2003) confirmed that Iowa dairy cattle with ulcerative PDD lesions produced a strong humoral response to the LPS-like material of five of six *T. phagedenis*-like PDD-associated *Treponema*. Interestingly, California PDD-associated *Treponema* strain 2-1498 was poorly antigenic, being consistently unreactive with serum antibodies. Antigenic differences were detected between the reactive Iowa strains. Most notably, a strongly antigenic proteinase K-resistant band was identified in strain 4A. This strain may be useful in future immunoprophylaxis studies. Cattle with both ulcerative and papillomatous lesions also produced a significant cellular response to California PDD-associated *Treponema* strain 9-

3301, but not to strain 2-1498. Both humoral and cellular responses appeared to be of short duration and diminished rapidly as cattle became convalescent. These data suggest that if a PDD vaccine is shown to elicit a protective response, multiple doses would be required to sustain immunity.

10. Treatment and control of PDD

PDD has been treated in a variety of ways with a multitude of products. Some treatments have been tested and shown to be efficacious, while others have shown little or no efficacy or await objective evaluation (Britt et al., 1996; Shearer and Elliott, 1998; Shearer et al., 1998). Clinical observations suggest that anatomical location and stage of the PDD lesions may affect treatment efficacy, with lesions in the interdigital cleft less responsive to antibiotics than lesions on the bulb of the heel (Mumba et al., 1999; Hernandez et al., 2000). Early therapeutic approaches for PDD included surgical removal of lesions and topical treatment with various compounds, including antibiotics. Electrocautery and freezing were also used, but reoccurrence of lesions was common (Shearer et al., 1998). Treatment of PDD with antibiotics is considered to be extra-label use and therefore requires a prescription and often a period of milk withholding (Britt et al., 1999). Topical applications of oxytetracycline, lincomycin, or a combination of lincomycin and spectinomycin, followed by bandaging of the affected feet, have been shown to be highly effective (Shearer et al., 1998). However, this method is labor intensive and not likely to be practical in large herds in which the incidence of PDD is 5% or greater. Parenteral administration of antibiotics such as procaine penicillin G and ceftiofur has been used effectively in herds in California (Read and Walker, 1998c, Shearer and Elliott, 1998), but is economically unrewarding when many cows are affected. Additionally, some investigators claim that they have not observed therapeutic results following parenteral antibiotic treatment (Laven and Proven, 2000). Topical antibiotic sprays delivered directly on the lesions of cleaned, dry feet have been useful for treatment of PDD (Shearer and Elliott, 1998). This method requires daily treatment of all feet for one week and, in subsequent weeks, continued daily treatment of feet with visible lesions. Hernandez et al. (1999) demonstrated that a commercial formulation spray of soluble copper, peroxide compound and a cationic agent was as effective as oxytetracycline for treatment of PDD based on pain and lesion scores. Britt et al. (1996) showed that acidified ionized copper solution and acidified sodium chlorite applied topically at each milking were equally effective as tetracycline sprays for controlling PDD. However, non-antibiotic topical treatments such as formalin, chlorine bleach, or caustic substances are considered of limited usefulness and pose health risks for those applying the spray (Nutter and Moffit, 1990; Shearer et al., 1998).

Footbaths are commonly used in an attempt to treat and control PDD since they are less labor intensive than direct topical treatments. A number of antibiotic and non-antibiotic compounds have been employed with varied results (Shearer and Elliott, 1998; Shearer et al., 1998; Laven and Proven, 2000). Although footbaths may be useful for the temporary control of PDD, they do not eliminate the infection from the herd. Furthermore, there are unresolved issues concerning the changing frequency and disposal of footbath solutions, the exposure of farm personnel to hazardous materials, and the potential for emergence of drug resistant organisms as a consequence of suboptimal antibiotic treatment. Despite these problems, footbaths remain the method of choice for many farmers (Shearer and Elliott, 1998).

Reoccurrence of PDD is common regardless of treatment modality (Read and Walker, 1998c; Shearer and Elliott, 1998; Shearer et al., 1998). Van Amstel et al. (1995) reported a reinfection rate of 52% after seven months in a dairy herd that was treated with tetracycline. Similar results were observed in other studies. For example, Read and Walker (1998c) reported a combined rate of lesion reoccurrence and new lesion development of 48% in cows re-examined at 7-12 weeks after a therapeutic response. There are several possible reasons for reoccurrence of PDD. Deep PDD lesions may have limited exposure to topical treatments delivered by spray or footbath (Mumba et al., 1999). These unresolved lesions may harbor infectious *Treponema* that re-activate old lesions and initiate new lesions. The use of antibiotic treatments of limited efficacy or improper formulation and/or inconsistencies in application schedules could lead to selection of resistant organisms. Unfortunately, there are virtually no follow-up data on the cidal or static effect of antibiotic or non-antibiotic treatments on the PDD-associated *Treponema* or other bacterial species present in the lesions (Mumba et al., 1999). Finally, some "treatment failures" may actually represent new infections transmitted from cattle with untreated subclinical infections or from recently introduced replacement cattle.

The reoccurrence of PDD and the reduced humoral and cell-mediated immune responses demonstrated in convalescent animals suggests that natural immunity, if it develops, is short-lived (Trott et al., 2003). As observed with other spirochetal infections, it is possible that phase and/or antigenic variation of *Treponema* surface components facilitate immunoevasion (Stamm and Bergen, 2000). Novartis Animal Health US, Inc. markets the only fully USDA-licensed vaccine for PDD (http://www.livestock.novartis.com/diseases_heelwarts_dairy.html). This bacterin (Trep Shield HW) contains inactivated cultivable California PDD-associated *Treponema* and can be used in cattle six months of age or older. Novartis recommends a vaccination program employing three 4 ml doses of the bacterin spaced 21 days apart with a booster given every six months. Trials conducted by the company showed that vaccination of cattle with the *Treponema* bacterin results in a significant reduction of clinical disease (91% and 74%, respectively on two different farms). Some dairy producers who used Trep Shield HW reported a 50-80% reduction in the incidence of PDD. The success of the *Treponema* bacterin is likely to vary from farm to farm depending on herd management practices, the immunological status of the cattle and microbial factors. With regard to the latter, since the bacterin contains only cultivable PDD-associated *Treponema*, it may not evoke adequate cross-protection against as yet uncultivated *Treponema* that may also be present in PDD lesions (Choi et al., 1997; Moter et al., 1998; L.V. Stamm, unpublished data).

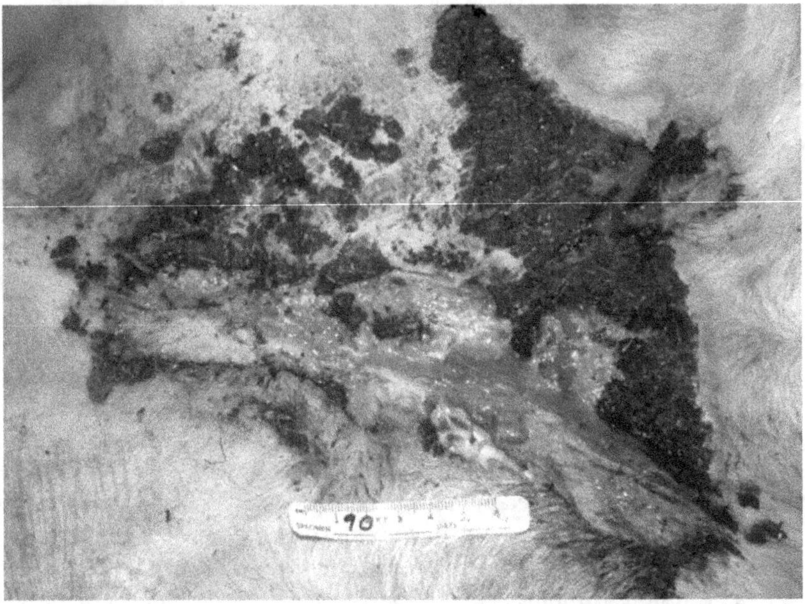

Figure 3. Ulcerative mammary dermatitis (UMD) lesion involving the anterior skin fold of the mammary gland (udder) of a Holstein dairy cow. (Photograph courtesy of D.H. Read, University of California, Davis).

11. PDD-associated *Treponema* and ulcerative mammary dermatitis

Outbreaks of ulcerative mammary dermatitis (UMD) have been observed in dairy cattle in England, Scotland, and the U.S. (Boyer and Singleton, 1998: Beattie and Taylor, 2000; Keil et al., 2002). Lesions are located anteriorly on the udder at the junction of the udder with the body wall and lay along the medium sulcus (Figure 3). UMD is characterized by rapid invasiveness and the presence of a foul necrotic odor. In untreated cases, 5-10 cm deep lesions have been observed (Boyer and Singleton, 1998). Treatment with oxytetracycline spray has been effective in early cases of UMD. Otherwise, parenteral tylosin has been required.

UMD is usually present in dairy herds where PDD is endemic. Beattie et al. (2000) reported the observation of several bacterial species, including spirochetes, in samples prepared from the surface of UMD lesions. Since these lesions resemble PDD lesions clinically and histopathologically, it is possible that PDD-associated *Treponema* are involved in the etiology of UMD. In support of this hypothesis, Keil et al. (2002) demonstrated that spirochetes present in paraffin-embedded UMD lesion biopsy tissue reacted with antisera directed against cultivable PDD-associated *Treponema*. Additionally, they isolated a *Treponema* from UMD lesions that is closely related, based on 16S rDNA sequence analysis, to the *T. phagedenis*-like PDD-associated *Treponema* (D. Keil, personal communication). Stamm et al. (2003) used 16S-tRNA(ile) ISR sequence analysis to investigate the genetic relatedness of three *Treponema* isolates from UMD lesions. Two of the isolates group with the *T.medium*-like PDD-associated *Treponema*, while the third isolate groups with the *T. phagedenis*-like PDD-associated *Treponema*. Analysis of cloned ISR amplicons from UMD lesion biopsy DNA identified sequences identical or virtually identical to the three cultivable PDD-associated *Treponema* phylotypes, as well as several unique *Treponema* ISR sequences. The results of Keil et al. (2002) and Stamm et al. (2003) support an etiologic link between PDD and UMD and provide further evidence of the pathogenic potential of these spirochetes.

12. PDD-associated *Treponema* and other livestock animals

Outbreaks of a severe virulent ovine footrot (SVOFR) emerged in the UK in the late 1990's (Harwood et al., 1997; Wassink et al., 2003). The new cases of footrot were clinically distinct from typical cases of footrot which is caused by a polymicrobial infection of *Dichelobacter nodosus*, *Flavobacterium necrophorum*, *Bacteroides* spp., and *Prevotella* spp. Davies et al. (1999) named the new condition contagious ovine digital dermatitis (CODD). Clinical signs of CODD include initial ulcerative and proliferative lesions at the coronet, followed by extensive underrunning of the toe. In some cases, complete detachment of the hoof occurs. The interdigital lesion that is usually present in footrot is not present in CODD. Both Naylor et al. (1998) and Demirkan et al. (2001) reported isolation of a fastidious, anaerobic spirochete from outbreaks of CODD. Comparative 16S rDNA sequence analysis showed that isolate UB1090 (Naylor et al., 1998) is more closely related to the *T. medium/ vincentii*-like PDD-associated *Treponema*; whereas isolate UB1466 (Demirkan et al., 2001) is more closely related to the *T. denticola*-like PDD-associated *Treponema* (Collighan et al., 2000; Edwards et al., 2003b). The genetic relatedness of PDD- and CODD-associated *Treponema* suggests that these spirochetes have broad host range. Results of serological studies conducted by Dhawi et al. (2005) in the UK suggest that several *Treponema* can infect the feet of both cattle and sheep resulting in severe lameness. Co-grazing of these livestock, which is a common practice in the UK, may have facilitated cross-species transmission of the PDD-associated and/or CODD-associated *Treponema* (Collighan et al., 2000).

Rashmir-Raven et al. (2000) reported a PDD-like condition with intraepidermal spirochetes in an U.S. horse with severe lameness. This disease (verrucous pododermatitis) has been observed in horses worldwide. Spirochetes have also been observed in pigs with New Zealand footrot (Osborne and Ensor, 1955). Interestingly, the latter disease can only be reproduced by injecting necrotic donor material containing actively motile spirochetes into the foot of a recipient animal. Further studies are necessary to determine if the invasive, as yet uncultivated, spirochetes present in PDD-like foot lesions of horses and pigs are genetically related to the PDD-associated *Treponema*.

13. Conclusion and future perspectives

PDD is an interesting example of how pathogenic *Treponema* can take advantage of an artificial ecological situation created by human activity. The intensive housing of dairy cattle in free stalls and drylots, where they consume more feed, can be milked more often and consequently give higher milk yields, promotes environmental conditions that are conducive to lesion development and rapid disease transmission. While much work has been conducted on defining the host-environment-agent complex involved with PDD, many questions remain concerning the PDD-associated *Treponema*. For example, the origin of these *Treponema* is currently unknown. However, their close genetic relationship to anaerobic human-associated treponemes suggests that the bovine oral cavity or intestinal tract may be a source of these spirochetes (Shibahara et al., 2002). Additional genotypic comparisons with *T. denticola, T. medium/vincentii* and *T. phagedenis*, particularly DNA-DNA reassociation studies, are needed to confirm the taxonomic position of the cultivable PDD-associated *Treponema* and to determine whether they warrant classification as new species. Comparative analysis of virulence factors that may be shared with human-associated *Treponema* and elucidation of the role of the PDD-associated *Treponema* phylotypes during lesion development are key to unraveling the complicated pathogenesis of PDD. The commercial availability of a bacterin that effectively reduces the incidence of PDD lesions in dairy herds, coupled with improved housing conditions, offers hope for non-antimicrobial control of this production-limiting disease. However, further analyses of the immune response to individual cultivable PDD-associated *Treponema* during various disease stages are needed to enhance our understanding of the host-parasite relationship and to improve vaccine efficacy. Finally, additional molecular epidemiological studies are essential to provide a more complete scope of *Treponema* diversity in PDD lesions, particularly from the standpoint of the as yet uncultivated *Treponema*.

Acknowledgment

Supported by the University of North Carolina Department of Epidemiology Infectious Diseases Trust Fund.

References

Argaez-Rodriguez, F.J., Hird, D.W., Hernandez de Anda, J., Read, D.H., and Rodriguez-Lainz, A. (1997). Papillomatous digital dermatitis on a commercial dairy farm in Mexicali, Mexico: incidence and effect on reproduction and milk production. Prev. Vet. Med. *32,* 275-286.

Beattie, K.G. and Taylor, D.J. (2000). An investigation into intertrigo (necrotic dermatitis or 'Foul Udder') in dairy cows. J. Brit. Cattle Vet. Assoc. *8*, 377-380.

Berry, S.L. (2001). Diseases of the digital soft tissues. Vet. Clinics North America: Food Animal Practice. *17*, 129-142.

Blowey, R.W., Sharp, M.W., and Done, S.H. (1992). Digital dermatitis. Vet. Rec. *131*, 9.

Blowey, R.W., Done, S.H., and Cooley, W. (1994a). Observations on the pathogenesis of digital dermatitis in cattle. Vet. Rec. *135,* 115-117.

Blowey, R.W., Carter, S.D., White, A.G., and Barnes, A. (1994b). *Borrelia burgdorferi* infections in UK cattle: a possible association with digital dermatitis. Vet. Rec. *135,* 577-578.

Boyer, P. and Singleton, G. (1998). Digital dermatitis, superfoul and severe necrotic dermatitis of the udder in dairy cows. Vet. Rec. *142*, 147-148.

Borgmann, I.E., Bailey, J., and Clark, E.G. (1996). Spirochete-associated bovine digital dermatitis. Can.Vet. J. *37*, 35-37.

Britt, J.S., Gska, J., Garrett, E.F., Konkle, D., and Mealy, M. (1996). Comparison of topical application of three products for the treatment of papillomatous digital dermatitis in dairy cattle. J. Am. Vet. Med. Assoc. *209*, 1134-1136.

Britt, J.S., Carson, M.C., von Bredow, J.D., and Condon, R.J. (1999). Antibiotic residues in milk samples obtained from cows after treatment for papillomatous digital dermatitis. J. Am. Vet. Med. Assoc. *215*, 833-836.

Brown, C.C., Kilgo, P.D., and Jacobsen, K.L. (2000). Prevalence of papillomatous digital dermatitis among culled adult cattle in the southeastern United States. 2000. Am. J. Vet. Res. *61*, 928-930.

Cheli, R. and Mortellaro, C. (1974). La dermatite digitale del bovino. In Proceedings of the 8th International Conference on Diseases of Cattle. p.208-213.

Choi, B.-K., Nattermann, H., Grund, S., Haider, W., and Gobel, U.B. (1997). Spirochetes from digital dermatitis lesions in cattle are closely related to treponemes associated with human periodontitis. Int. J. Syst. Bacteriol. *47*, 175-181.

Collick, D.W., Ward, W.R., and Dobson, H. (1989). Association between types of lameness and fertility. Vet. Rec. *125*:,103-106.

Collighan, R.J. and Woodward, M.J. (1997). Spirochetes and other bacterial species associated with bovine digital dermatitis. FEMS Microbiol. Lett. *156*, 37-41.

Collighan, R.J., Naylor, R.D., Martin, P.K., Cooley, B.A., Buller, N., and Woodward, M.J. (2000). A spirochete isolated from a case of severe virulent ovine foot disease is closely related to a *Treponema* isolated from human periodontitis and bovine digital dermatitis. Vet. Microbiol. *74*, 249-257.

Cruz, C., Driemeier, D., Cerva, C., and Corbellini, L.G. (2001). Bovine digital dermatitis in southern Brazil. Vet. Rec. *148*, 576-577.

Davies, I.H., Naylor, R.D., and Martin, P.K. (1999). Severe ovine foot disease. Vet. Rec. *145*, 646.

Demirkan, I., Carter, S.D., Murray, R.D., Blowey, R.W., and Woodward, M.J. (1998). The frequent detection of a treponeme in bovine digital dermatitis by immunocytochemistry and polymerase chain reaction. Vet. Microbiol. *60*, 285-292.

Demirkan, I., Carter, S.D., Hart, C.A., and Woodward, M.J. (1999a). Isolation and cultivation of a spirochete from bovine digital dermatitis. Vet. Rec. *145*, 497-498.

Demirkan, I., Walker, R.L., Murray, R.D., Blowey, R.W., and Carter, S.D. (1999b). Serological evidence of spirochaetal infections associated with digital dermatitis in dairy cattle. Vet. J. *157*, 69-77.

Demirkan, L., Carter, S.D., Winstanley, C., Bruce, K.D., McNair, N.V., Woodside, M. and Hart, C.A. (2001). Isolation and characterization of a novel spirochaete from severe virulent ovine foot rot. J. Med. Microbiol. *50*, 1061-1068.

Dhawi, A., Hart, C.A., Demirkan, I., Davies, I.H., and Carter, S.D. (2005). Bovine digital dermatitis and severe virulent ovine foot rot: a common spirochaetal pathogenesis. The Vet. J. *169*, 232-241.

Dopfer, D, Koopmans, A., Meijer, F.A., Szakall, I, Schukken, Y.H., Klee, W., Bosma, R.B., Cornelisse, J.L., van Asten, A.J.A.M., and ter Huurne, A.A.H.M. (1997). Histological and bacteriological evaluation of digital dermatitis in cattle, with special reference to spirochaetes and *Campylobacter faecalis*. Vet. Rec. *140*, 620-623.

Edwards, A.M., Dymock, D., and Jenkinson, H.F. (2003a). From tooth to hoof: treponemes in tissue-destructive diseases. J. Appl. Microbiol. *94*, 767-780.

Edwards, A.M., Dymock, D., Woodward, M.J., and Jenkinson, H.F. (2003b). Genetic relatedness and phenotypic characteristics of *Treponema* associated with human periodontal tissues and ruminant foot disease. Microbiol. *149*, 1083-1093.

El-Ghoul, W. and Shaheed, B.I. (2001). Ulcerative and papillomatous digital dermatitis of the pastern region in dairy cattle: clinical and histopathological studies. Dtsch. Tierärztl. Wschr. *108*, 216-222.

Harwood, D.G., Catell, J.H., Lewis, C.J., and Naylor, R. (1997). Virulent footrot in sheep. Vet. Rec. *140*, 687.

Hernandez, J., Shearer, J.K., and Elliott, J.B. (1999). Comparison of topical application of oxytetracycline and four nonantibiotic solutions for treatment of papillomatous digital dermatitis in dairy cows. J. Am. Vet. Med. Assoc. *214*, 688-690.

Hernandez, J. and Shearer, J.K. (2000). Efficacy of oxytetracycline for treatment of papillomatous digital dermatitis lesions on various anatomic locations in dairy cow. J. Am. Vet. Med. Assoc. *216*, 1288-1290.

Hernandez, J., Shearer, J.K., and Webb, D.W. (2001). Effect of lameness on the calving-to-conception interval in dairy cows. J. Am. Vet. Med. Assoc. *218*, 1611-1614.

Keil, D.J., Read, D., Sturgis, T.F., and Bockenstedt, C.R. (2002). Abstract 14. Isolation of a (papillomatous) digital dermatitis-associated *Treponema* spp. from bovine udder lesions.

83rd Annual Meeting of the Conference of Research Workers in Animal Diseases. St. Louis, MO.

Kesavalu, L., Walker, S.G, Holt, S.C., Crawley, R.R., and Ebersole, J.L. (1997). Virulence characteristics of oral treponemes in a murine model. Infect. Immun. *65*, 5096-5102.

Kesavalu, L., Holt, S.C., and Ebersole, J.L. (1998). Virulence of a polymicrobic complex, *Treponema denticola* and *Porphyromonas gingivalis*, in a murine model. Oral Microbiol. Immunol. *13*, 373-377.

Laven, R.A. and Proven, M.J. (2000). Use of an antibiotic footbath in the treatment of bovine digital dermatits. Vet. Rec. *147*, 503-506.

McLennan M.W. and McKenzie R.A., (1996). Digital dermatitis in a Friesian cow. Aust. Vet. J. *74*, 314-315.

Milinovich, G.J., Turner, S.A., McLennan, M.W., and Trott, D.J. (2004). Survey for papillomatous digital dermatitis in Australian dairy cattle. Aust. Vet. J. *82*: 20-24.

Moter, A., Leist, G., Rudolph, R., Schrank, K., Choi, B-K, Wagner, M., and Gobel, U.B. (1998). Fluorescence in situ hybridization shows spatial distribution of as yet uncultured treponemes in biopsies from digital dermatitis lesions. Microbiol. *144*, 2459-2467.

Mumba, T., Dopfer,D., Kruitwagen, C., Dreher, M., Gaastra, W., and van der Zeijst, B.A.M. (1999). Detection of spirochetes by polymerase chain reaction and its relation to the course of digital dermatitis after local antibiotic treatment in dairy cattle. J. Vet. Med. *46*, 117-126.

Murray, R.D., Downham, D.Y., Clarkson, M.J., Faull, W.B., Hughes, J.W., Manson, F.J., Merritt, J.B., Russell, W.B., Sutherst, J.E., and Ward, W.R. (1996). Epidemiology of lameness in dairy cattle: description and analysis of foot lesions. Vet. Rec. *138*, 586-591.

Naylor, R.D., Martin, P.K., Jones, J.R., and Burnell, M.C. (1998). Isolation of spirochaetes from an incident of severe virulent ovine footrot. Vet. Rec. *143*, 690-691.

Nutter, W.T.and Moffit, J.A. (1990). Digital dermatitis control. Vet. Rec. *126*, 200-201.

Ohya, T., Yamaguchi, H., Nii, Y., and Ito, H. (1999). Isolation of *Campylobacter sputorum* from lesions of papillomatous digital dermatitis in dairy cattle. Vet. Rec. *145*, 316-318.

Opitz, B., Schroder, N.W.J., Spreitzer, I., Michelsen, K.S., Kirschning, C.J., Hallatschek, W, Zahringer, U., Hartung, T., Gobel, U.B., and Schumann, R.R. (2001). Toll-like receptor-2 mediates *Treponema* glycolipid and lipotechoic acid-induced NF-κB translocation. J. Biol. Chem. *276*, 22041-22047.

Osborne, H.G. and Ensor, C.R. (1955). Some aspects of the pathology, aetiology, and therapeutics of foot-rot in pigs. New Zealand Vet. J. *3*, 91-95.

Paster, B.J., Dewhirst, F.E., Coleman, B.C., Lau, C.N., and Ericson, R.L. (1998). Phylogenetic analysis of cultivable oral treponemes from the Smibert collection. Int. J. Syst. Bacteriol. *48*, 713-722.

Rashmir-Raven, A.M., Black, S.S., Rickard, L.G., and Akin, M. (2000). Papillomatous pastern dermatitis with spirochetes and *Pelodera strongyloides* in a Tennessee walking horse. J. Vet. Diagn. Invest. *12*, 287-291.

Read, D.H. (1995). Papillomatous digital dermatitis. Proc.4-State Dairy Appl. Nat. Manage.Conf. 1995: 134-143.

Read, D.H., Walker, R.L., Castro, A.E., Sundberg, J.P., and Thurmond, M.C., (1992). An invasive spirochete associated with interdigital papillomatosis of dairy cattle. Vet. Rec. *130*, 59-60.

Read, D.H. and Walker, R. L. (1996). Experimental transmission of papillomatous digital dermatitis (footwarts) in cattle. Vet. Pathol. *33*, 607.

Read, D.H. and Walker, R.L. (1998a). Comparison of papillomatous digital dermatitis of cattle by histopathology and immunohistochemistry. Proc. 10th Int. Symp. Lameness in Ruminants, p. 268.

Read, D.H. and Walker, R.L. (1998b). Experimental transmission of papillomatous digital dermatitis (footwarts) in dairy cattle. Proc. 10th Int. Symp. Lameness in Ruminants. p. 270.

Read, D.H. and Walker, R.L. (1998c). Papillomatous digital dermatitis (footwarts) in California dairy cattle: clinical and gross pathologic findings. J. Vet. Diagn. Invest. *10*, 67-76.

Read, D.H., Nordhausen, R., and Walker, R.L. (1998). Pathogenesis of experimental papillomatous digital dermatitis (footwarts) in cattle: Bacterial morphotypes associated with early lesion development. Proc. 10th Int. Symp. Lameness in Ruminants. p. 271.

Rebhun, W.C., Payne, R.M., King, J.M., Wolfe, M., and Begg, S.N. (1980). Interdigital papillomatosis in dairy cattle. J. Am. Vet. Med. Assoc. *177*, 437-440.

Rijpkema, S.G.T., David, G.P., Hughes, S.L., and Woodward, M.J. (1997). Partial identification of spirochaetes from two dairy cows with digital dermatitis by polymerase chain reaction analysis of the 16S ribosomal RNA gene. Vet. Rec. *140*, 257-259.

Rodriguez-Lainz, A., Hird, D.W., Carpenter, T.E., and Read, D.H. (1996a). Case-control study of papillomatous digital dermatitis in southern Californian dairy farms. Prev. Vet. Med. *28*, 117-131.

Rodriguez-Lainz, A., Hird, D.W., Walker, R.L., and Read, D.H. (1996b). Papillomatous digital dermatitis in 458 dairies. J. Am. Vet. Med. Assoc. *209*, 1464-1467.

Rodriguez-Lainz, A., Melendez-Retamal, P., Hird, D.W., Read, D.H., and Walker, R.L. (1999) Farm- and host-level risk factors for papillomatous digital dermatitis in Chilean dairy cattle. Prev. Vet. Med. *42*, 87-97.

Schrank, K., Choi, B-K, Grund, S., Moter, A., Heuner, K., Nattermann, H., and Gobel, U.B. (1999). *Treponema brennaborense* sp. *nov*., a novel spirochete isolated from a dairy cow suffering from digital dermatitis. Int. J. Syst. Bacteriol. *49*, 43-50.

Shearer, J.K. and Elliott, J.B. (1998). Papillomatous digital dermatitis: Treatment and control strategies—Part I. The Compendium *20*, S158-S166.

Shearer, J.K., Hernandez, J., and Elliott, J.B. (1998). Papillomatous digital dermatitis: Treatment and control—Part II. The Compendium *20*, S213-S223.

Shibahara, T., Ohya, T., Ishii, R., Ogihara, Y., Maeda, T, Ishikawa, Y., and Kadota, K. (2002). Concurrent spirochaetal infections of the feet and colon of cattle in Japan. Aust. Vet. J. *80*, 497-502.

Smibert, R.M. (1991). Anaerobic spirochetes. In Manual of Clinical Microbiology 5th edition, A. Balows, W.J. Hausler Jnr., K.L. Herrmann, H.D. Isenberg and H.J. Shadomy, ed. (Washington D. C. U.S.A: American Society for Microbiology), pp 572-578.

Stamm, L.V., Bergen, H.L. and Shangraw, K.A. (2000). The sequence-variable, single-copy *tprK* gene of *Treponema pallidum* Nichols strain UNC and street strain 14 encodes heterogeneous TprK proteins. Infect. Immun. *68*, 6482-6486.

Stamm, L.V. and Bergen, H.L. (2001). Natural rifampin resistance in *Treponema* spp. correlates with presence of N531 in RpoB Rif cluster I. Antimicrob. Agents Chemother. *45*, 2973-2974.

Stamm, L.V., Bergen, H.L., and Walker, R.L. (2002). Molecular typing of papillomatous digital dermatitis-associated *Treponema* isolates based on analysis of 16S-23S ribosomal DNA intergenic spacer regions. J. Clin. Microbiol. *144*, 2459-2467.

Stamm, L.V., Walker, R.L., Read, D.H., Eigenheer, A.L., and Bergen, H.L. (2003). Phylogenetic analysis of bovine ulcerative mammary dermatitis-associated *Treponema*. Abstract D-166. In: Abstracts of the 103rd General Meeting of the American Society for Microbiology, p. 231. Washington, D.C. American Society for Microbiology.

Trkov, M., Accetto, T., Kostanjsek, M., and Avgustin, G. (2001). Preliminary characterization of a tentatively novel rumen bacterial species from the genus *Treponema*. Folia Microbiol. *46*, 91-95.

Trott, D.J., Moeller, M.R., Zuerner, R.L., Goff, J.P., Waters, W.R., Alt, D.P., Walker, R.L., and Wannemuehler, M.J. (2003). Characterization of *Treponema* phagedenis-like spirochetes isolated from papillomatous digital dermatitis lesions in dairy cattle. J. Clin. Microbiol. *41*, 2522-2529.

Van Amstel, S.R., Van Vuuren, S., and Tutt, C.L.C. (1995). Digital dermatitis: report of an outbreak. J.S. Afr. Vet. Assoc. *66*, 177-181.

Vermunt, J.J. and Hill, F.I. (2004). Papillomatous digital dermatitis in a Holstein-Friesian bull. N. Z. Vet. J. 52, 99-101

Walker, R.L., Read, D.H., Loretz, K.J., and Nordhausen, R.W. (1995). Spirochetes isolated from dairy cattle with papillomatous digital dermatitis and interdigital dermatitis. Vet. Microbiol. *47*, 343-355.

Walker, R.L., Read, D.H., Loretz, K.J., Hird, D.W., and Berry, S.L. (1997). Humoral response of dairy cattle to spirochetes isolated from papillomatous digital dermatitis lesions. Am. J. Vet. Rec. *58*, 744-748.

Walker, R.L., Read, D.H., Sawyer, S.J., and Loretz, K.L. (1998). Abstract #17. Phylogenetic analysis of spirochetes isolated from papillomatous digital dermatitis lesions in cattle. 79th Annual Meeting of the Conference of Research Workers in Animal Diseases. Chicago, IL.

Wassink, G.J., Moore, L.J., Grogono-Thomas, R., and Green, L.E. (2003). Exploratory findings on the prevalence of contagious ovine digital dermatitis in sheep in England and Wales during 1999 to 2000. Vet. Rec. *152*, 504-506.

Wells, S.J., Garber, L.P., and Wagner, B.A. (1999). Papillomatous digital dermatitis and associated risk factors in US dairy herds. Prev. Vet. Med. *38*, 11-24.

Whittier, W.D. (1988). Foot warts. Dairy 1988. June 12-13.

Willis, S.G., Smith, K.S., Dunn, V.L., Gapter, L.A., Riviere, K.H., and Riviere, G.R. (1999). Identification of seven *Treponema* species in health- and disease-associated dental plaque by nested PCR. J.Clin. Microbiol. *37*, 867-869.

Wyss, C., Choi, B.-K., Schüpbach, P., Guggenheim, B., and Gobel, U.B. (1996). *Treponema maltophilum* sp. nov., a small oral spirochete isolated from human periodontal lesions. Int. J. Syst. Bacteriol. *46,* 745-752.

Yeruham, I., Friedman, S., Elad, D., and Perl, S. (2000). Association between milk production, somatic cell count and bacterial dermatoses in three dairy cattle herds. Aust. Vet. J. *70*, 250-253.

18

Termite Gut Spirochetes

*John A. Breznak**

Abstract
Spirochetes comprise an unusually abundant component of the hindgut microbiota of termites, accounting for as many as one-half of all prokaryotic cells. However, for many years our understanding of termite gut spirochetes was meager owing, in part, to a lack of pure culture representatives. In recent years this situation has changed dramatically. Cultivation-independent (molecular biological) approaches have shown that termite gut spirochetes constitute a rich diversity of novel species of *Treponema*. Moreover, studies of the first pure cultures have revealed physiological capabilities (H_2/CO_2-acetogenesis and dinitrogen fixation) previously unknown in spirochetes which contribute to the nutrition and vitality of their host. This chapter will summarize our current understanding of termite gut treponemes and the extent to which they have already expanded our appreciation of spirochetal phylogenetic and physiological diversity.

1. Introduction
There are few natural habitats on earth in which spirochetes constitute such a prominent component of the microbial community as the hindgut of termites and of the wood-eating cockroach, *Cryptocercus punctulatus*, which is closely related to termites evolutionarily. The association of spirochetes with these insects is probably a long-standing one and may even go back to the first emergence of these insect lines about 140 million years ago (Nalepa and Bandi, 2000). In any case, the association is at least 15-20 million years old, as inferred from the presence of spirochete-like cells in hindguts of termites preserved in Miocene amber (Weir et al., 2002).

The occurrence of spirochetes in termite guts was first documented in the late 1800's by Joseph Leidy, a professor of anatomy at the University of Pennsylvania, during microscopic examination of hindgut contents of *Termes* (now *Reticulitermes*) *flavipes*, the common eastern subterranean termite (Leidy,

*For correspondence email breznak@msu.edu

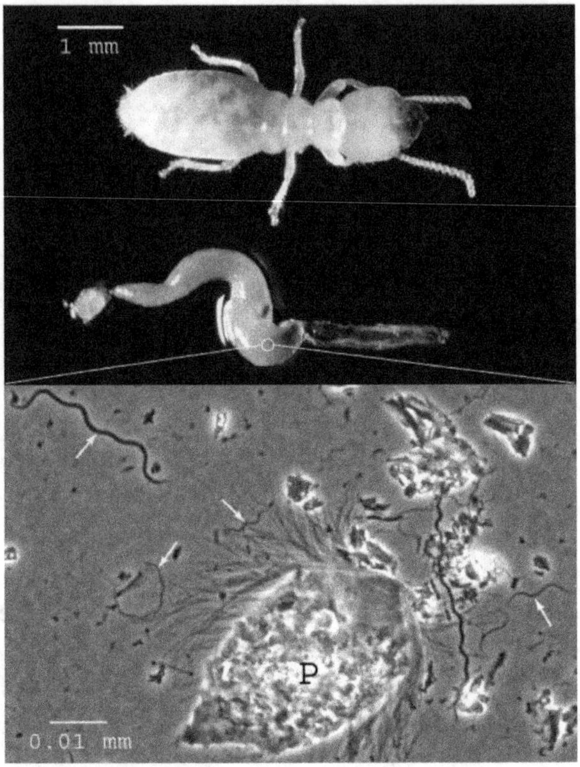

Figure 1. Top panel: A worker larva of *Reticulitermes flavipes* is positioned next to a gut extracted from a separate individual. The tubular midgut region (right) joins the larger hindgut region (left) at an enteric valve. Bottom panel: Phase contrast micrograph of diluted hindgut contents liberated by puncture of the bulbous "paunch" region of the hindgut (at the white circle, top panel). A hypermastigote protozoan (*Trichonympha agilis*; P) is surrounded by various morphological types of spirochetes (white arrows).

1874-1881; 1877) (Figure 1). Representation of spirochetes among the hindgut microbiota is impressive. Spirochetes can account for up to one-half of the direct (electron) microscopic count of all prokaryotic cells in hindguts (Paster et al., 1996), and their 16S rRNA genes can account for one-half of the clones in libraries generated by PCR amplification of termite gut DNA with general primers for 16S rRNA-encoding genes of *Bacteria* (Hongoh et al., 2003). Striking too is their morphological diversity. Cells range in size from 0.1 to 1 μm in diameter and from 3 to 100 μm in length and can bear anywhere from 2 periplasmic flagella to 100 or more in some of the larger forms. Some cells possess distinctive ultrastructural features, such as folds and grooves in the outer membrane, cytoplasmic tubules, and tube-like polar cell extensions. As a group, they exhibit a range of cell wavelengths and amplitudes (or cell body pitch) such that, even by using light microscopy alone, it is not difficult to distinguish a dozen or so morphotypes of spirochetes in termite hindgut contents merely on the bases of size and shape (Figure 1). Given the fairly wide distribution of termites in tropical and temperate regions, it is not surprising that termites have been recommended as an ideal source material to illustrate the morphological diversity and motility of spirochetes (and of other pro- and eukaryotic microbes) for educational purposes, and simple procedures for doing so in the classroom have been described (Beckwith and Light, 1927; Breznak and Leadbetter, 2002).

For more than a century following Leidy's first observations, our understanding of termite gut spirochetes remained limited, as none were ever isolated and studied in pure culture. Most published reports up through the mid-1990's dealt with their morphological diversity and ultrastructure, their distribution in various termite species, and the curious association of some of them as epibionts (surface-attached ectosymbionts) on hindgut protozoa. During this period, an assortment of new genus and species names was either created or revived for some of the spirochetes based primarily on morphological features. Over the past ten years, however, our understanding of termite gut spirochetes has taken a big leap forward. Cultivation-independent (molecular biological) approaches have provided a clearer picture of their phylogeny and phylogenetic diversity. Moreover, the isolation and study of the first pure cultures has revealed physiological activities (i.e. CO_2-reductive acetogenesis and N_2 fixation) hitherto unrecognized in the *Spirochaetes* phylum of the domain *Bacteria* which contribute directly to the carbon, nitrogen and energy requirements of their hosts. A number of previously published reviews provide a gateway to the early literature on termite gut spirochetes, including their ultrastructure and proposals for their taxonomy (Breznak, 1973; Breznak, 1984; Breznak and Leadbetter, 2002; Margulis and Hinkle, 1992; Wier et al., 2000). Therefore, following a general overview of termites and their symbiotic gut microbiota, this chapter will focus largely on research published within the past decade on termite gut spirochetes. A more comprehensive discussion of the biology and biochemistry of termites and their symbiotic interaction with various microbes can be found in *Termites: Evolution, Sociality, Symbioses, Ecology* (Abe et al., 2000) and in a recent review by Brune (2006).

2. Termites and their symbiotic gut microbiota

Termites belong to the order Dictyoptera, suborder Isoptera, the latter meaning "equal winged" and reflecting the fact that the fore and hind wings of sexually mature reproductives are similar in size and veination. They are among Earth's most abundant terrestrial insects, occurring in the tropics in biomass densities (up to ca. 100 g/m^2) frequently surpassing that of grazing mammalian herbivores, and they are important in the degradation of one of Earth's most abundant forms of biomass, i.e. lignocellulosic plant material and residues derived from it (humus). Termites are colonial insects, whose nests can be subterranean, epigeal or arboreal, and within which a division of labor exists among morphologically distinguishable castes. "Soldiers" and "reproductives" function in colony defense and production of young, respectively. Members of the "worker" caste (e.g. Figure 1) are typically the most abundant and dominate the nutrient dynamics of colonies. They are responsible for foraging and for nest construction and maintenance, as well as for distribution of food to nutrient-dependent castes, i.e. soldiers, whose mouthparts are differentiated for fighting, not feeding; reproductives; and recently hatched larvae, whose mouthparts may still be too soft to chew through plant material.

There are roughly 2600 recognized species of termites and their biologies, behaviors and nutritional ecologies are extremely diverse. However, all species examined possess a dense and diverse intestinal microbiota located mainly in the hindgut. The hindgut is a chitin-lined region that follows the foregut (which consists of the crop and muscular gizzard) and the tubular midgut (a microvilli-lined site of enzyme secretion and absorption of soluble nutrients). Malpighian tubules, which function like kidneys in invertebrates, discharge into the gut at or near the enteric valve, which separates the midgut from the hindgut. Hence, urinary metabolites (e.g. uric acid, a major excretory product of terrestrial insects), as well as food remaining after partial digestion in the midgut, are subject to metabolism by hindgut microbes before the products are either absorbed or voided to the exterior as feces. The enteric valve prevents refluxing of hindgut contents back to the midgut. Therefore, microbial metabolites produced in the hindgut and important for termite nutrition are absorbed directly through the hindgut wall. The volume of termite hindguts ranges from about one microliter to tens of microliters, depending on the species. In so-called "lower", i.e. evolutionarily basal, termites (families *Mastotermitidae*, *Kalotermitidae*, *Termopsidae*, *Hodotermitidae*, *Rhinotermitidae*, and *Serritermitidae*), which comprise about 25% of all species, the hindgut is fairly simple in structure. It consists of a dilated anterior portion termed the "paunch" (Figure 1), which then tapers down posteriorly through the colon and terminates at the rectum. In "higher", i.e. more derived, termites (family *Termitidae*), the hindgut trends toward further elongation and multi-compartmentalization.

Although relatively tiny in volume, termite guts possess marked physicochemical gradients of pH, E_h, O_2 and H_2. The pH of hindguts of wood-feeding termites is circumneutral, but the anterior portion in some higher termites can be somewhat alkaline owing to the influx of alkaline urine. In soil-feeding higher termites, the pH of the most anterior compartment of the hindgut is among the highest ever recorded in biological systems (pH = 11-12.5) (Brune and Kühl, 1996). The high pH is undoubtedly an adaptation that facilitates solubilization of humic components, from which these termites obtain their nutrition. Some of the oxygen supplied to the hindgut tissue via trachea also diffuses across the hindgut wall creating a zone of hypoxia in the peripheral region of the hindgut lumen. Studies with microelectrodes have shown that the hypoxic zone consists of an O_2 gradient steeply decreasing from about 50-100 μM (ca. 20-40% air saturation) at the gut wall to complete anoxia about 100 to 200 μm inward (Brune et al., 1995; Ebert and Brune, 1997). The depletion of O_2 owes to its consumption by the wall-associated microbiota, which is a major O_2 sink. Hence, only the central region of the lumen is anoxic and of low E_h. By contrast, the central region of hindguts of *Reticulitermes flavipes* (and probably other lower termites as well) has a relatively large pool of H_2 (ca. 40 μM dissolved H_2 in the center of the hindgut; equiv. 5% v/v or 50,000 ppmv H_2 in the gas phase) that is produced mainly by resident protozoa, which liberate much of the reduction equivalents generated during fermentation as H_2 (Brune et al., 1995; Ebert and Brune, 1997). The H_2 concentration decreases steeply with radial distance outward toward the hindgut wall, owing to its consumption by hydrogenotrophic microbes: very little of the produced H_2 escapes (discussed further in the section on *Treponema primitia*, below).

2.1. The gut microbiota

The gut microbiota of termites consists of *Bacteria*, *Archaea*, and, in some species, *Eukarya*. Among the more abundant *Bacteria* are members of the low G+C Gram-positive group (phylum *Firmicutes*, in particular clostridia and close relatives, as well as lacto- and enterococci), various subgroups of *Proteobacteria*, members of the *Cytophaga-Flavobacterium-Bacteroides* group (phylum *Bacteroidetes*), high G+C Gram-positives (phylum *Actinobacteria*), and of course spirochetes (phylum *Spirochaetes*). 16S rRNA gene clone libraries prepared from gut homogenates of the Japanese subterranean termite, *Reticulitermes speratus* (family *Rhinotermitidae*) contained roughly 268 different species-level "phylotypes" of bacteria per gut, i.e. 268 groups of 16S rRNA gene clones that bore ≤97% sequence similarity to each other (Hongoh et al., 2003). Roughly one-half of the clones in these libraries were of spirochetal origin. Also present in termite guts are members of the domain *Archaea*, most of which are methanogens of the families *Methanobacteriaceae* and *Methanosarcinaceae*. In lower termites, the gut microbiota includes an abundance of *Eukarya*, represented mainly by amitochondriate, anaerobic flagellate protozoa found virtually nowhere else in nature. Such protozoa belong to the orders *Trichomonadida*, *Hypermastigida* and *Oxymonadida* and are among the most deeply branching lineages on the eukaryotic line of descent (Inoue et al., 2000). In lower termites, protozoa actually comprise the bulk of the hindgut volume. Protozoa are not significant components in the gut of higher termites. Yeasts have also been isolated from guts of some termite species.

Non-spirochetal *Bacteria* and *Archaea* are found in the hindgut lumen and also heavily colonizing the gut wall, including chitinous spines that protrude into the lumen in some termite species. Some bacteria also colonize the ectoperitrophic space of the midgut. This is a space between the microvilliated midgut wall and the peritrophic membrane, which encloses the bolus of particulate food during passage down the midgut. Spirochetes, on the other hand, occur primarily in the lumen and rarely if ever appear to attach to the hindgut wall itself or to cells of the diverse, wall-associated, non-spirochetal prokaryotic microbiota. In lower termites, spirochetes occur as freely swimming cells in the lumen and also attached end-on, as epibionts (ectosymbionts), to the surface of protozoa. Epibiontic spirochetes may also be accompanied by nonspirochetal ecto- or endosymbiotic bacteria or methanoarchaea (Breznak, 2000; Stingl et al., 2004). Attachment of spirochetes usually involves morphologically-differentiated docking sites on the protozoan plasma membrane, as well as attachment structures at the poles of the spirochetal cells (Breznak, 1984). For *Mixotricha paradoxa*, a trichomonad found in the hindgut of the Australian termite *Mastotermes darwiniensis*, the consequence of spirochetal attachment is spectacular. It is the rhythmic beating of thousands of spirochete epibionts that serves to propel *M. paradoxa* through the gut fluid: the protozoan's own flagella are apparently only used for steering (Cleveland and Grimstone, 1964). Whether the rhythmic undulations of the attached spirochetes arise only passively, as a result of hydro-dynamic forces generated by the cells undulating in close proximity to each other (Machin, 1963), or whether they are somehow controlled by the protozoan, is not known. Apart from engagement in motility symbiosis, the metabolic/physiological interaction of epibiontic spirochetes with protozoa remains a matter of speculation (discussed further below). In higher termites, which do not possess symbiotic protozoa, spirochetes occur as free-swimming cells in one or more subsections of the hindgut proper as well as in the "mixed segment" (Bignell et al., 1983). The latter is a relatively short segment of the alimentary tract at the midgut-hindgut junction: it is bounded on one side by microvilliated midgut tissue and on the other by chitinous cuticle-lined hindgut tissue. A mixed segment is not found in lower termites.

Behavioral idiosyncrasies of termites ensure that all members of a colony will have a full complement of gut microbes characteristic of that particular termite species, as will new colonies founded by winged reproductives that leave the nest. Among the most important of these behaviors is trophallaxis, i.e. the process of food transfer from workers to nutrient-dependent castes (soldiers, reproductives, and young larvae). This occurs by regurgitation of already chewed food from the foregut (stomodeal trophyllaxis), as well as by expulsion from the rectum of food undergoing fermentation by hindgut microbes (proctodeal trophyllaxis). The latter serves as a rich inoculum of gut microbes. Proctodeal trophyllaxis is also an important mechanism for reinoculation of workers following periodic molts, during which much of the hindgut contents is lost. Secondary mechanisms of inoculation include coprophagy and cannibalism.

2.2. Symbiotic digestion in termites

Major structural and storage polysaccharides in plant material, i.e. cellulose, hemicelluloses (arabinoxylans, mannans, etc.) and starch, undergo digestion by the combined action of termites and their gut microbiota. Hydrolysis of polysaccharides (including cellulose) is initiated by various termite glycohydrolases present in salivary and midgut secretions. In lower termites, polysaccharides in food particles reaching the hindgut also undergo hydrolysis by enzymes present within food vacuoles of cellulolytic protozoa, after such particles are phagocytized. Hindgut bacteria also appear to participate in hydrolysis of noncellulosic polysaccharides, but there is no evidence that they are important in cellulose hydrolysis. Polysaccharides escaping hydrolysis in the fore- and midgut, as well as solubilized saccharides not taken up by termites during passage through the midgut, undergo anaerobic microbial fermentation in the hindgut leading to acetate as a major product, with smaller amounts of propionate and butyrate also formed. These volatile fatty acids are taken up from the hindgut and oxidized by termites for energy, as well as used as a carbon source for biosynthesis. Oxidation of the acetate produced by hindgut microbes can support up to 100% of the daily respiratory requirement of some termites. Reduction equivalents (as H_2 molecules) generated during the hindgut fermentation have two primary fates: they are used by methanogenic *Archaea* to reduce CO_2 to methane (CH_4), which is emitted by termites, and they are used by so-called acetogenic bacteria to reduce two molecules of CO_2 to acetate (CH_3COOH). In most anoxic habitats where CO_2 is a major terminal electron acceptor, methanogenesis almost always dominates CO_2-reductive acetogenesis as an H_2 "sink". However, the opposite is true in hindguts of many termites, especially wood- and grass-feeders. In fact, H_2-dependent reduction of CO_2 to acetate can account for 90% of all the H_2 consumed and as much as 5-30% of all the acetate produced during the hindgut fermentation of such termites. This anomalous fate of

H_2 has been the subject of numerous studies over the years, and its reconciliation now appears to be largely explained by the physical separation of the H_2-consuming populations, among which are spirochetes (below).

The other major structural polymer of terrestrial plant cell walls is lignin, which is especially abundant in woody plants and consists of ether-linked phenylpropanoid moieties. However, although some chemical modification of lignin may occur on passage through the termite gut (ring demethylation; side chain reduction), there is no evidence for its significant depolymerization, degradation or digestion.

Aside from being relatively difficult to digest, lignocellulosic plant material is low in combined nitrogen. Sound (undecayed) wood, for example, has a C/N ratio of about 1000/1 and contains only 0.03-0.1% N by weight, which is very low compared to that of most animal tissues (11-15% N). However, microbes can also contribute to termite N nutrition, and there are four general ways that this is accomplished. Some termites simply feed preferentially on plant material that is in some stage of decay by external fungi. Infiltration of the plant material by fungal mycelia and oxidation of some of the plant polysaccharide to CO_2 serves to lower the C/N ratio and effectively enrich the N content of the food subsequently consumed. Some termites even cultivate certain fungi (*Termitomyces* sp.) in elaborate gardens within the nest, on which they browse. Hindgut microbes facilitate the acquisition of new N by N_2 fixation, and this process can provide 30-60% of the N assimilated by some termite species (Tayasu et al., 1997; 1994). Likewise, hindgut microbes also help to conserve existing N by recycling excretory N (uric acid-N) back to termites for biosynthesis (Potrikus and Breznak, 1981). The general ability of microbes to assimilate poor sources of N (e.g. N_2, NH_3) and in the process upgrade them to more nutritious ones (e.g. amino acids and protein) is almost certainly another important contribution of gut microbes to termite N nutrition, but the extent to which gut microbes themselves are digested by termites has not yet been investigated in great detail. In any case, these processes are not mutually exclusive, and the last probably augments the other three.

Given the participation of the gut microbiota in digestion and nutrition, it is not surprising that its presence is important for termite survival. This has been inferred from numerous studies documenting an increased rate of mortality when some or all components of the gut microbiota are eliminated. However, owing to the diversity of gut microbes and the difficulty in selective elimination of specific microbes, as well as the difficulty in interpreting direct versus indirect effects of microbe elimination, it has not always been easy to attribute beneficial effects to a particular microbe or microbial group. Nevertheless, the classic studies of Cleveland (Cleveland, 1923; 1924; 1925) showed that cellulolytic protozoa were essential for survival of lower termites on a cellulose-based diet, even though it is now known that lower termites can synthesize and secrete their own cellulase enzymes (Nakashima et al., 2002; Watanabe and Tokuda, 2001). Presumably, the protozoa help increase the retention time of particulate food in the gut (by phagocytosis) and complement the cellulases secreted by the insect, which apparently are insufficient for complete digestion.

Elimination of gut bacteria, by feeding termites antibacterial drugs, results in guts becoming completely oxic (as inferred from the *in situ* color of redox dyes fed to termites after treatment), a condition deleterious for the anaerobic cellulolytic protozoa of lower termites. Such treatment also renders termite guts more susceptible to colonization by *Serratia marcescens*, an insect pathogen normally absent from, or present in low numbers in, normally colonized termites (Veivers et al., 1982). Of particular relevance to this chapter was the observation that elimination of spirochetes (without a major change in nonspirochetal prokaryotes, as judged microscopically) coincided with accelerated termite mortality implying that spirochetes were important to survival (Eutick et al., 1978). The remainder of this chapter will now review our recently enlightened understanding of termite gut spirochetes, which helps clarify their apparent role in termite vitality.

3. Phylogeny and phylogenetic diversity of termite gut spirochetes

In 1994, the first insight into the phylogeny of termite gut spirochetes was obtained by sequence analysis of a spirochetal 16S rRNA gene clone (clone MDS1) obtained from gut contents of *M. darwiniensis* (Berchtold et al., 1994). Clone MDS1 was affiliated with the genus *Treponema*, but it was not closely related to any known treponeme. This was followed soon after by the retrieval of spirochetal 16S rRNA gene clones from several other termite species, and similar conclusions were reached (Berchtold and König, 1996; Ohkuma and Kudo, 1996; 1998; Paster et al., 1996). As part of some of these investigations, fluorescent oligonucleotide probes were designed on the basis of the cloned sequences and were found to hybridize to spirochetes *in situ*, thereby confirming that the clones were indeed of spirochetal origin (Berchtold and König, 1996; Paster et al., 1996).

In 1999, Lilburn et al. (1999) analyzed 101 spirochetal 16S rRNA gene clones obtained by PCR amplification from gut homogenates of *R. flavipes* (family *Rhinotermitidae*), as well as 31 clones representing each of the major spirochetal ARDRA groups from *Coptotermes formosanus* (family *Rhinotermitidae*), and 3 clones from *Zootermopsis angusticollis* (family *Termopsidae*). These data were pooled with previously published sequences of spirochetal 16S rRNA gene clones from (termite family/no. of clones): *Reticulitermes speratus* (Rhinotermitidae/7), *M. darwiniensis* (Mastotermitidae/11), *Cryptotermes domesticus* (Kalotermitidae/3), and *Nasutitermes lujae* (Termitidae/1). Together, these termites represented five of the seven known families within the Isoptera (Lilburn et al., 1999). Results revealed that (i) all clones could be grouped within the genus *Treponema*, but none was closely related (i.e. all bore ≤ 91% sequence similarity) to any known treponeme or to any homologous clones from not-yet-cultured treponemes and (ii) nearly all of the clones fell into a distinct clade within the genus, termed by them the "Termite Cluster" of treponemes, that also contained *Spirochaeta stenostrepta* and *Spirochaeta caldaria* (Figure 2). A small number of clones fell within subgroup 1 of the traditional treponemes as defined by Paster et al. (1991) (Figure 2). Similar results were reported later that same year by Ohkuma and co-workers for 30 spirochetal phylotypes present in guts of four species of Japanese termites: *Pericapritermes niobei* and *Nasutitermes takasagoensis* (both *Termitidae*); *Glyptotermes fuscus* (*Kalotermitidae*); and *Hodotermopsis sjoestedii* (*Termopsidae*) (Ohkuma et al., 1999).

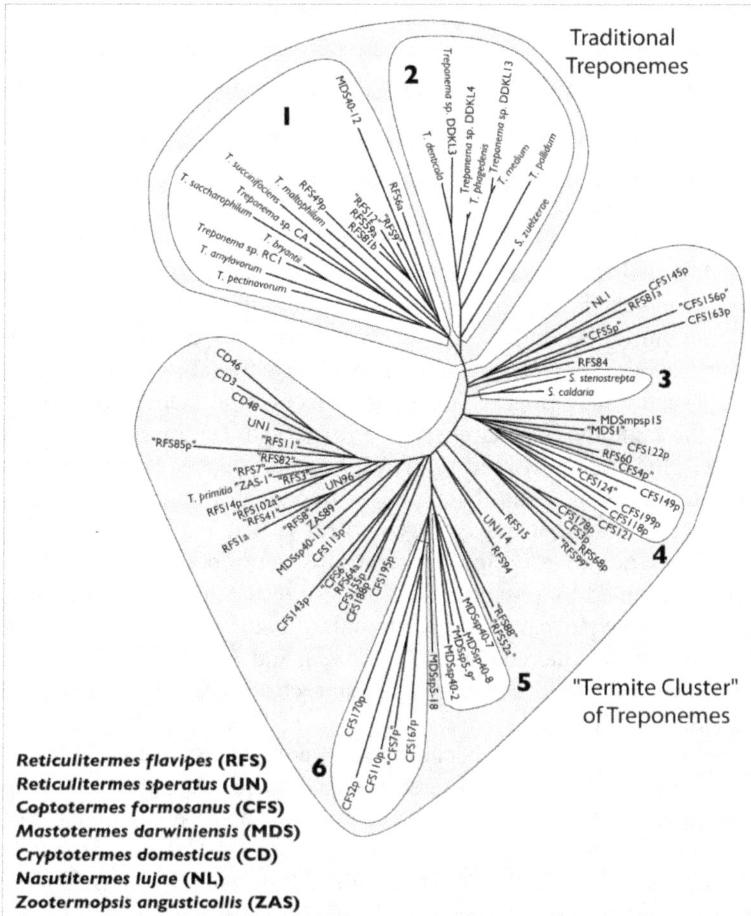

Figure 2. A radial phylogenetic tree based on 16S rRNA sequences depicting the relationship of the "Termite Cluster" of treponemes to the traditional treponemes. Pendant prefixes correspond to the host termite from which the cloned sequences were obtained, as encoded on the bottom left. Quotation marks around a pendant name indicate that it is a representative from a group of closely related clones. Subclusters 1 and 2 of the traditional treponemes (at top) were previously defined by Paster *et al.* (1991). Subcluster 3 is the *Spirochaeta* subcluster. Subclusters 4, 5 and 6 are groups of closely related clones obtained from *C. formosanus* and *M. darwiniensis. Leptospira noguchii* (not shown) was used as the outgroup. A scale bar is omitted, because addition of partial sequences (pendant suffixes a, b and p) to the framework tree constructed from near-full-length sequences distorts the branch lengths. Modified from Lilburn *et al.* (1999) and used by permission from Blackwell Publishing Co.

Four 16S rRNA signature nucleotides were identified by Lilburn et al. (1999) that almost perfectly distinguished members of the Termite Cluster from other treponemes. These occurred at (*E. coli* numbering) positions: 289-G•C-311; A at 812; and an inserted nucleotide at 1273. Results also revealed that, even by using conservative criteria for delineating *Treponema* species, individual termites such as *R. flavipes* contain as many as 21 different species of *Treponema* (Lilburn et al., 1999). Analysis of phylotype abundances revealed that their distribution was reasonably equitable (not excessively dominated by one or a few types). One inference from such a distribution is that there may not be much niche redundancy among the spirochete community in guts. Given the rather specific epibiontic associations of some of the spirochetes, as well as the physiological diversity seen among the few isolates now in hand (see below), this may be a valid inference. Some hints of co-evolution between termites and treponemes were also observed insofar as treponemal phylotypes from a particular termite species were, in general, more closely related to each other than to phylotypes from other termites (e.g. subclusters 4, 5 and 6; Figure 2). However, there were also clones from some termites that grouped among those obtained from termites of different species and even different families. A similar pattern was observed by Ohkuma et al. (1999) in their study of treponemal phylotypes in Japanese termites (above). It is not known to what extent such results reflect colonization of a common ancestor of two termite species with the same treponemal phylotype *versus* transfer, by some mechanism, of treponemal phylotypes between sympatric termite species. Hence, patterns of evolution between termites and their hindgut treponemes remain to be resolved. In any case, the picture that has emerged so far is that Earth's termites appear to constitute an extraordinarily rich reservoir of novel treponemal diversity, most of which is represented by phylotypes of the "Termite Cluster". To this writer's knowledge, no phylotypes have yet been found to group within subgroup 2 of the traditional treponemes, which includes *T. pallidum* and *T. denticola* (Figure 2).

It should be mentioned that the three spirochaetas shown in Figure 2 (*S. stenostrepta*, *S. caldaria* and *S. zuelzerae*) were named before their 16S rRNA gene sequences were determined and, because they were isolated as free-living forms and were not known to be host-associated, were assigned to the genus *Spirochaeta*. It has been suggested that these three "spirochaetas" may be descendents of the precursor to host-associated treponemes (Paster et al., 1991) or might actually be gut symbionts that were defecated by some animal host into the waters from which they were originally isolated (Leschine et al., 2003) (see also Chapter 1, Paster and Dewhirst). Evidence to support both of these possibilities has come from the recent discovery of a few *Spirochaeta* phylotypes in 16S rRNA gene libraries prepared from gut contents of the Japanese termite, *Reticulitermes speratus* (Hongoh et al., 2003).

3.1. Epibiontic spirochetes

Our understanding of spirochetal epibionts of termite hindgut protozoa is still meager. However, the 16S rRNA-based phylogeny of some of them has been determined after using micropipettes to harvest protozoan cells bearing them, then confirming the origin of sequences obtained by performing fluorescent *in situ* hybridization (FISH) with rRNA-targeted oligonucleotide probes. The few studies done so far indicate that (i) epibiontic spirochetes group within the genus *Treponema* and consist of representatives spread among the Termite Cluster as well as among subgroup 1 of the traditional treponemes, implying independent origins of the epibiontic lifestyle; (ii) as many as six treponemal phylotypes can be associated with particular species of protozoa; (iii) the distribution of specific treponemal phylotypes is often restricted to certain regions of the protozoan cell membrane; and (iv) epibiontic treponemes may be restricted to that lifestyle, as fluorescent oligonucleotide probes targeting such epibionts have not reacted with free-swimming spirochetes in the hindgut fluid (Iida et al., 2000; Noda et al., 2003; Wenzel et al., 2003).

4. The first pure cultures

In 1997, the first pure cultures of termite hindgut spirochetes were obtained in this writer's laboratory from successful enrichment cultures established with hindgut contents of the California dampwood termite, *Zootermopsis angusticollis* (Leadbetter et al., 1999). It is worth reviewing some of the factors that appear to have contributed to this success, as these might be useful to individuals seeking to isolate not-yet-cultured treponemes from other hosts. Among these were (i) the use of an anoxic, complex enrichment medium containing nutrient broth and pre-fermented, neutralized rumen fluid; (ii) incubation of the enrichment cultures under an atmosphere of 80% H_2 (balance CO_2); and (iii) inclusion in the medium of rifamycin and fosfomycin (antibacterial drugs to which many spirochetes are intrinsically resistant), as well as bromoethanesulfonate (to suppress the growth of H_2-consuming methanogens). Although the medium was complex, a decided effort was made to keep the concentration of nutrients relatively low, especially that of fermentable carbohydrate, to discourage overgrowth by undesired forms. The latter was a problem that plagued many previous enrichment or isolation attempts in our laboratory, even when antibacterial drugs were included in media, hence the use of pre-fermented rumen fluid (17% v/v) and only 0.2% nutrient broth. Enrichments were slow to develop (10-12 weeks) and, owing to the relatively low nutrient content, became only barely turbid to the naked eye. Nevertheless, periodic microscopic examinations revealed that spirochetes not only retained their motility in this medium, but also increased in length and exhibited division stages, occasionally forming short chains of independently moving cells with about 3-4 cells per chain. Eventually, spirochetes outnumbered non-spirochetal cells by about 50:1; from such enrichments pure cultures of the first two strains (strains ZAS-1 and ZAS-2) were isolated by repeated serial dilution in agar "shake tubes".

During the extended incubation of enrichment cultures, it was noticed that the growth of spirochetes coincided with the development of a vacuum in the headspace of culture vessels, and this was periodically returned to atmospheric pressure by addition of the H_2-CO_2 gas mixture. Companion analyses of culture fluids by HPLC revealed an accumulation of acetate in amounts that corresponded closely to the volume of H_2 consumed, according to the equation: $4\,H_2 + 2\,CO_2 \rightarrow CH_3COOH + 2\,H_2O$. These observations suggested that the spirochetes might be true "acetogens", a speculation that was ultimately borne out (below). Acetogens are not merely organisms that produce acetate as a fermentation product but are anaerobes that use CO_2 as a terminal electron acceptor via the acetyl-CoA pathway for production of acetate (and the acetyl moiety of acetyl-CoA) to obtain energy and cell carbon (Drake, 1994). Such organisms are sometimes referred to as "homoacetogens", because many produce acetate as the major, if not sole, product of anaerobic fermentation of a variety of substrates, including carbohydrates, organic acids and alcohols, the methyl group of methoxylated benzenoids and N-methyl compounds, one-carbon compounds (methanol, formate, CO), and, of course, $H_2 + CO_2$. Strains

Table 1. Major properties of *Treponema primitia* and *Treponema azotonutricium*[a]

Property	*Treponema primitia*		*Treponema azotonutricium*
	strain ZAS-1	**strain ZAS-2**	**strain ZAS-9**
Cell - size (μm)	0.2 x 3 to 7	0.2 x 3 to 7	0.2 to 0.3 x 10 to 12
- wavelength or body pitch (μm)	2.3	2.3	1.1
Energy-yielding substrates (doubling time at 30°C, h)[b]			
$H_2 + CO_2$	+ (26)	+ (48)	-
Glucose	+ (22)	+ (71)	+ (35)
Fructose	n.t.	n.t.	+ (37)
Maltose	+ (27)	+ (69)	+ (47)
Cellobiose	+ (27)	-	+ (45)
Ribose	-	-	+ (42)
Xylose	+ (26)	+ (101)	+ (35)
Arabinose	+ (26)	-	-
Mannitol	+ (23)	+ (70)	-
Trimethoxybenzoate	-.	+ (127)	n.t.
Syringate	-	+ (154)	n.t.
Ferulate	-	+ (315)	n.t.
Vanillate	-	+ (135)	n.t.
Fermentation products	Acetate (acetogen)	Acetate (acetogen)	Acetate, Ethanol, H_2, CO_2 (from maltose)
Genome properties			
- size (kb; mean ± SEM)	3,461 ± 14	3,835 ± 24	3,901 ± 47
- G+C content (mol%)	51.0	50.9	50.0
- *rrs* copy number	2	2	2

[a]All strains were catalase negative anaerobes that possessed two periplasmic flagella and grew optimally at 30°C.

[b]Abbreviations: +, utilized; -, tested, but not utilized; n.t., not tested. CO, formate, methanol, betaine, and choline are not used by *T. primitia.* Sucrose, trehalose and uric acid are not used by *T. azotonutricium.* Lactate, pyruvate and glycine are not used by either species.

ZAS-1 and ZAS-2 were recently assigned to the new species, *Treponema primitia* -- the specific epithet derived from the Latin noun meaning "first fruit" (of isolation after long work) (Graber et al., 2004). Subsequent attempts at direct isolation of spirochetes from hindguts of *Z. angusticollis* also were successful and led to the isolation of strain ZAS-9 (Lilburn et al., 2001). Strain ZAS-9 is the type strain of the new species, *Treponema azotonutricium* (Graber et al., 2004). Unlike *T. primitia*, *T. azotonutricium* is not an H_2-consuming acetogen; it is a saccharolytic treponeme that produces H_2 (as well as acetate, ethanol and CO_2) during growth. It is also capable of dinitrogen fixation and believed to contribute to the nitrogen nutrition of termites, hence the specific epithet *azotonutricium*, meaning "to nourish with nitrogen". The names *T. primitia* and *T. azotonutricium* have been validly published (Euzéby, 2004). By using similar procedures, six additional strains of *Treponema* have been isolated from *Z. angusticollis* and a few have been obtained from *R. flavipes*, but no experiments have been done with these owing to their very poor growth *in vitro* (Lilburn et al., 2001).

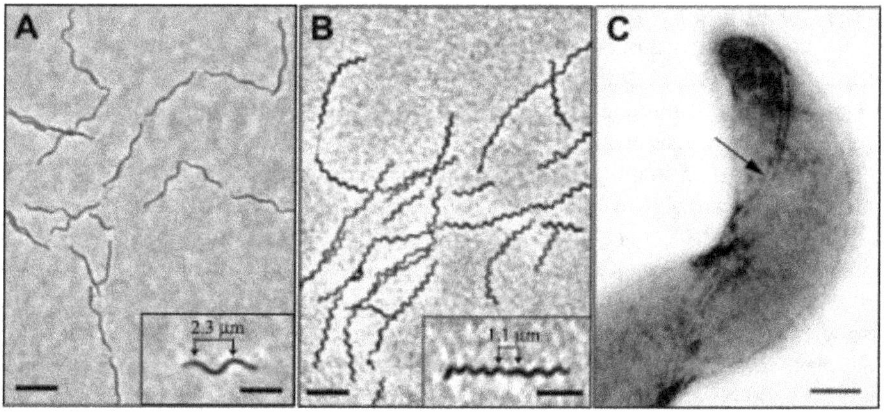

Figure 3. Morphology of *T. primitia* strain ZAS-2 (A) and *T. azotonutricium* strain ZAS-9 (B and C) by phase contrast (A and B) and transmission electron (C) microscopy. Insets in panels A and B show single cells of each strain, with wavelengths indicated by arrowed bars. One of the periplasmic flagella of strain ZAS-9 is seen in panel C (arrow). *T. primitia* strain ZAS-1 is indistinguishable from strain ZAS-2. Bars: 5 µm for panels A and B and 2.5 µm for insets; 0.1 µm for panel C. Modified from Graber *et al.* (2004) and used by permission of ASM Press.

5. Properties of *Treponema primitia* and *Treponema azotonutricium*

Major properties of *T. primitia* and *T. azotonutricium* are summarized in Table 1 and Figure 3 as a result of recent studies of these species (Graber and Breznak, 2004; Graber et al., 2004). It is interesting that the genomes of *T. primitia* and *T. azotonutricium* (3.5 to 3.9 Mb in size) are substantially larger than those of other host-associated spirochetes such as *T. pallidum* (1.1 Mb; (Fraser et al., 1998)), *T. denticola* (2.8 Mb; (Seshadri et al., 2004)) and *Borrelia burgdorferi* (1.5 Mb; (Casjens et al., 2000)). They are more similar in size to that of the free-living spirochete, *Spirochaeta aurantia* (3.7 Mb; (Graber et al., 2004)). This may imply that termite gut treponemes have been selected to retain genes not only necessary for growth *in situ*, but for production of metabolites important to the nutrition of the host (and perhaps also the nutrition of other members of the gut microbiota), for survival and re-colonization during vertical, non-germline-mediated transmission to offspring, and for lateral transmission to colony mates by proctodeal trophallaxis.

The relatively low copy number of the small subunit rRNA-encoding gene *rrs* (2 in each termite gut treponeme; Table 1) is similar to that of spirochetes in general (mean = 1.8; (Klappenbach et al., 2001)) and suggests that the fitness of spirochetes is tied to an ecological strategy characterized by efficient allocation of resources under relatively constant, slow-growth conditions (Klappenbach et al., 2000). For *T. primitia* and *T. azotonutricium* in particular, this would be consistent with their relatively slow growth rate *in vitro* (Table 1) and with their adaptation to life in the gut of an animal that feeds on a relatively refractory, N-poor resource. Following is a discussion of the salient physiological properties of *T. primitia* and *T. azotonutricium* and their implications for symbiosis with termites.

5.1. *Treponema primitia*

T. primitia is an acetogen, producing acetate as the major product from the fermentation of carbohydrates and of H_2 + CO_2. *T. primitia* is also capable of mixotrophy, i.e., simultaneous use of H_2 (+ CO_2) + organic compounds for energy. Strain ZAS-2 can also grow (slowly) by homoacetogenesis from the methyl group of methoxylated aromatic compounds, which are common in lignocellulosic plant material. Owing to a growth requirement for yeast extract/autolysate, initial studies were done with $^{14}CO_2$ (+H_2) to confirm that cell carbon and both C atoms of acetate could be derived from CO_2 (Leadbetter et al., 1999). Subsequent enzyme assays showed that CO_2-reductive acetogenesis in *T. primitia* proceeded by the well-known acetyl-CoA (Wood/Ljungdahl) pathway, and specific activities of pathway enzymes were consistent with the growth rates of cells on H_2 + CO_2 (Graber and Breznak, 2004; Leadbetter et al., 1999). Inasmuch as microbially-produced acetate is a major carbon and energy source for termites (capable of supporting up to 100% of the insects daily respiratory requirement) and 10-30% of that acetate arises from CO_2-reductive acetogenesis, an important role for *T. primitia* in termite nutrition was revealed.

The discovery of acetogenesis in *T. primitia* was additionally exciting because this was a previously unknown mode of metabolism in spirochetes. It also has helped reconcile the enigmatic dominance of CO_2-reductive acetogenesis over the more thermodynamically favorable process of CO_2-reductive methanogenesis in guts of wood-eating termites. It now seems most likely that the depletion of H_2 during its radial diffusion outward from the central region of the hindgut lumen is due to H_2/CO_2-utilizing acetogenic spirochetes like *T. primitia* that are abundant there. For reasons not yet clear, H_2/CO_2-utilizing methanogens (in *R. flavipes*, at least) are located primarily on or near the gut wall (Leadbetter and Breznak, 1996; Leadbetter et al., 1998) and hence are furthest downstream in the outwardly diffusing H_2 gradient, receiving what H_2 is left. H_2 concentrations in hindguts of *Z. angusticollis*, the termite from which *T. primitia* was isolated, have not been determined, but they are probably comparable to that in *R. flavipes* (above). Methanogens in *Z. angusticollis* are also associated with the hindgut wall (J. R. Leadbetter, personal communication), but some are also carried into the central region as endosymbionts within certain species of protozoa (Lee et al., 1987). As a result, rates of H_2/CO_2-methanogenesis in *Z. angusticollis* are much more comparable to those of H_2/CO_2-acetogenesis (Brauman et al., 1992). The emerging picture is that in hindguts of wood-feeding termites, there may not exist so much a head-to-head competition for H_2 between CO_2-reducing acetogens and methanogens as a sharing of H_2 between spatially separated populations of these H_2 consumers. As long as luminal concentrations of H_2 remain well above the H_2 threshold for acetogenesis (350-650 ppmv for *T. primitia* (Graber and Breznak, 2004) and ca. 300-5,000 ppmv for acetogens in general (Breznak, 1994)), *T. primitia* and perhaps other not-yet-isolated acetogenic spirochetes can be expected to process a significant fraction of that H_2.

The notion that spirochetes are among the major H_2/CO_2-utilizing acetogens in termite hindguts has been supported by the discovery that more than 75% of clones of formyltetrahydrofolate synthetase (FTHFS; a key enzyme of the acetyl-CoA pathway of H_2/CO_2-acetogenesis) retrieved from guts of *Z. angusticollis* by PCR form a novel, coherent cluster that contains FTHFS homologues from *T. primitia* (and *T. azotonutricium*) (Salmassi and Leadbetter, 2003). Results of that study also suggest that genes encoding FTHFS and other enzymes of acetogenesis in *T. primitia* (i.e., methenyltetrahydrofolate cyclohydrolase and methylenetetrahydr ofolate dehydrogenase) were acquired by horizontal transfer from clostridia. It will be extremely interesting to determine which of the various FTHFS genes in termite hindguts are actually being expressed, and to what extent.

5.1.1. Nutritional idiosyncrasies

Two nutritional idiosyncrasies of *T. primitia* deserve mention. One is the requirement for particular types of yeast autolysate or yeast extract in growth media. Initial studies indicated that the rumen fluid and nutrient broth components of the original enrichment and isolation medium could be replaced by "homemade" (i.e. noncommercial) yeast autolysate plus a defined mixture of vitamins and cofactors (Leadbetter et al., 1999). Both were required for growth. Though simple enough to prepare, it was surprising that homemade yeast autolysate could not be replaced by five other commercial extracts of yeast. Recently, it was found that certain yeast extract powders from Red Star Bioproducts of Juneau, Wisconsin, could replace homemade yeast autolysate. These included Tastone 900, Amberferm 5925, and Amberferm 5902 for *T. primitia* strain ZAS-1, and Tastone 154, Tastone 310, Amberex 1003, and Amberex 695 for strain ZAS-2. Curiously, no single product was effective for both strains (Graber and Breznak, 2004). The nutritional basis for this specificity is not yet known.

Another idiosyncrasy is the requirement for a folate compound. "All minus one" type nutritional experiments revealed that cells grew poorly or not at all if folinate (5-formytetrahydrofolate) was omitted from the vitamin/cofactor mixture. Subsequent experiments showed that the entire vitamin/cofactor mixture could be replaced by either folic acid, dihydrofolate, tetrahydrofolate (THF), or folinate for strain ZAS-2, whereas strain ZAS-1 grew only when provided with THF or folinate (Graber and Breznak, 2004). Although a folate requirement had previously been reported for *Treponema bryantii* (Stanton and Canale-Parola, 1980) and *Treponema phagedenis* (Steinman et al., 1954), a folate requirement by *T. primitia* was entirely unexpected, as THF is a key C1 carrier in the methyl group-forming arm of the acetyl-CoA pathway of acetogenesis. This suggested that the growth of *T. primitia in situ* was supported by secretion of a folate(s) by other members of the termite hindgut gut microbiota. Support for this hypothesis was obtained by the demonstration that strains of *Lactococcus lactis* and *Serratia grimesii*, which occur in hindguts of *Z. angusticollis* at a density of ca. 2×10^8 CFU/ml gut contents, secrete sufficient folinate to support *in vitro* growth of *T. primitia.* Moreover, folinate was the only folate compound detected in *Z. angusticollis* hindgut fluid, occurring at a concentration (1.3 μg/ml) 13-fold greater than that required to support half maximal growth of *T. primitia in vitro* (Graber and Breznak, 2005). Upon uptake by spirochetes, folinate is presumably converted to 5,10-methenyl-THF by the physiologically irreversible synthetase (EC 6.3.3.2), which is widely distributed in prokaryotes and eukaryotes and which serves to usher folinate into

THF-dependent C1 interconversion pathways (Huang and Schirch, 1995). In the original enrichment medium from which *T. primitia* was isolated, a suitable folate(s) was undoubtedly provided by inclusion of rumen fluid, which is known to contain such compounds (Herbeck and Bryant, 1974; Slyter and Weaver, 1977).

5.1.2. Yield coefficients

Substrate-specific biomass yields of *T. primitia* ZAS-1 and ZAS-2 *in vitro* (i.e. $Y_{substrate}$, expressed as g dry cell mass per mole substrate consumed) were surprisingly low: 0.1 and 0.2, respectively, for H_2 (plus CO_2) and 7.9 and 6.3, respectively, for glucose. $Y_{substrate}$ values for other (nonspirochetal) acetogens range from 0.5 to 4.2 with H_2 (plus CO_2) and from 32.7 to 53.0 with hexoses (Daniel et al., 1990; Eichler and Schink, 1984; Sharak-Genthner and Bryant, 1987; Tschech and Pfennig, 1984). However, this seemingly inefficient coupling of acetogenesis to biomass formation may actually contribute to the suitability of *T. primitia* as a symbiont. Inasmuch as acetate produced by hindgut microbes is a major carbon and energy source for termites, *T. primitia* can perform acetogenesis without a concomitantly large increase in biomass that might be detrimental to the host.

5.1.3. Oxygen tolerance

Although *T. primitia* is an anaerobe in the traditional sense, i.e. it is incapable of growth in air, it seemed likely that cells encounter the hypoxic periphery of hindguts periodically -- either by inadvertently swimming into that region or being bumped into it by the turbulent motility of hindgut protozoa. If so, then *T. primitia* might be expected to exhibit some level of tolerance to O_2 and the enzymatic capability to detoxify reactive oxygen species. This was shown to be the case: both strains could maintain growth upon addition of $\leq 0.5\%$ (v/v) O_2 to the gas phase (6 μM dissolved O_2); and both strains possessed NADH and NADPH peroxidase (strain ZAS-2 also possessed NADH oxidase) (Graber and Breznak, 2004).

5.2. *Treponema azotonutricium*

T. azotonutricium is easily distinguished from *T. primitia* morphologically (its wavelength or body pitch (1.1 μm) is half that of *T. primitia*; Figure 3) and physiologically (it is a saccharolytic spirochete that produces, not consumes, H_2 (as well as acetate, ethanol and CO_2)). Cells lack CO dehydrogenase/acetyl-CoA synthase (a key enzyme of the Wood-Ljungdahl pathway of acetogenesis); they cannot grow on H_2 (+CO_2) as energy source; and the presence of H_2 increases neither the growth rate nor the cell yield when included in media with a fermentable sugar (Graber et al., 2004). Moreover, the 16S rRNA gene sequence of *T. azotonutricium* shares only 92-93% similarity with those of *T. primitia* strains.

5.2.1. Dinitrogen fixation

Special interest attaches to *T. azotonutricium* owing to its ability to fix dinitrogen, another metabolic property not previously known in spirochetes (Lilburn et al., 2001). As discussed earlier in this chapter, N_2 fixation by gut microbes is an important mechanism of nitrogen acquisition in termites. Accordingly, the possible role of *T. azotonutricium* and other spirochetes in termite N nutrition, as well as acetate provision, was revealed. N_2 fixation was verified for *T. azotonutricium*

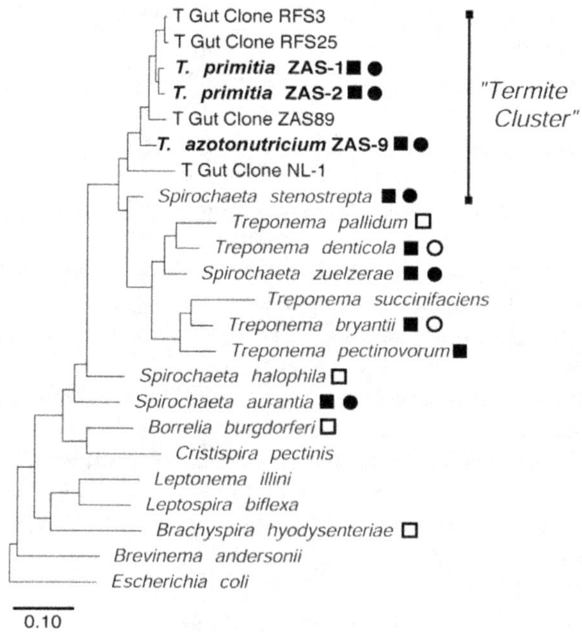

Figure 4. Phylogenetic tree inferred by maximum likelihood analysis of near-full-length 16S rRNA gene clones from *T. primitia* and *T. azotonutricium* strains (bold), representative non-termite-associated spirochetes, and spirochetal gene clones obtained directly from gut contents of the termites (clone prefix): *Zootermopsis angusticollis* (ZAS), *Reticulitermes flavipes* (RFS), and *Nasutitermes lujae* (NL). The homologous sequence from *E. coli* was used as the outgroup. Scale bar represents 0.1 changes per nucleotide. Symbols: *nifH* present (■) or not detected (❑); nitrogenase activity present (●) or not detected (○). Modified from Lilburn *et al.* (2001) and used by permission from Science.

by demonstration of: (i) NH_4^+-repressible acetylene reduction activity; (ii) N_2-dependent growth in media deficient in combined N; and (iii) fixation of $^{15}N_2$ into cellular N. Rates of N_2 fixation by *T. azotonutricium in vitro* were 10 to 100-fold greater than those exhibited by *T. primitia* and were entirely sufficient to account for the rates of N_2 fixation observed with live specimens of *Z. angusticollis*. Conservative estimates, based on the cell density of spirochetes in termite guts, the median rate of N_2 fixation exhibited by pure cultures of spirochetes *in vitro*, and the similarity of spirochete nitrogenase gene (*nifH*; encoding dinitrogenase reductase) sequences to those of transcripts known to be expressed *in situ*, suggested that spirochetes probably make a significant contribution to termite N_2 fixation.

The discovery of N_2 fixation in termite gut treponemes prompted a search for *nifH* and nitrogenase activity in other spirochetes. Results are summarized in Figure 4, which reveals that the gene or its activity was restricted to certain treponemes and spirochaetas (Lilburn et al., 2001). Curiously, *nifH* homologues were also identified in the ruminal treponeme, *T. bryantii*, and the oral treponeme, *T. denticola*, but N_2 fixation activity could not be demonstrated in either of these spirochetes. Neither *nifH* nor other nitrogenase structural genes were identified in the completely sequenced genomes of *T. pallidum* or *B. burgdorferi*.

The belief that *T. azotonutricium* and other treponemes may be important N_2 fixers in termite guts raises an interesting question. H_2 is a competitive inhibitor of nitrogenase: its K_i (i.e. the concentration of H_2 that doubles the K_m of nitrogenase and hence doubles the concentration of N_2 needed to achieve ½ V_{max}) ranges from 0.03 atm H_2 for nitrogenase activity of soybean nodule bacteroids to 0.5 atm H_2 for the conventional nitrogenase of *Clostridium pasteurianum* (Guth and Burris, 1983; Lockshin and Burris, 1965; Rasche and Arp, 1989). To what extent, then, are treponemes, like *T. azotonutricium*, adapted to catalyze nitrogen fixation in termite guts, whose H_2 concentrations can be as high as 0.05 atm, and when *T. azotonutricium* itself produces copious amounts of H_2 as a fermentation product (below)? In this regard, it is interesting that NifH1 from *T. azotonutricium* groups in "assemblage III" with the homologous protein of the conventional nitrogenase of *C. pasteurianum*, which has an unusually high K_i for H_2 (above) (Lilburn et al., 2001). It will be interesting to determine the K_i for H_2 with the *T. azotonutricium* enzyme directly. It is also possible that chemotactic mechanisms exist that enable N_2-fixing treponemes to position themselves in regions of the hindgut lumen that are least affected by the two opposing gradients of nitrogenase inhibitors (outwardly diffusing H_2 and inwardly diffusing O_2).

5.2.2. H_2 production and growth yield coefficient

T. azotonutricium produces a relatively large amount of H_2, i.e., 5.2 mol per mol maltose (equiv. 2.6 mol per mole hexose). In fact, assuming that maltose fermentation proceeds by way of traditional glycolysis to pyruvate, followed by pyruvate ferredoxin oxidoreductase-mediated cleavage of pyruvate en route to fermentation products (per mol maltose: 5.2 mol H_2, 3.6 mol CO_2, 2.8 mol acetate and 0.8 mol ethanol), only about 4.0 mol H_2 per mol maltose would be expected before further production of H_2 becomes thermodynamically unfavorable (Graber et al., 2004). This implies that production of an additional 1.2 mol H_2 per mol maltose requires expenditure of energy, an implication supported by the relatively low cell yield of *T. azotonutricium* on maltose ($Y_{maltose}$ = 50.3 g dry cell mass per mol). The yield is substantially lower than would be predicted (i.e. 71.4 g dry mass per mol) if 6.8 net ATP are derived from maltose (i.e. 4 via glycolysis and 2.8 accompanying the formation of each acetate via acetate kinase) and if Y_{ATP} is assumed to be 10.5 (a mean value derived from analyses of a variety of fermentative bacteria (Stouthamer, 1979)). The mechanism of such putative energy-dependent formation of H_2 remains to be elucidated. However, the limited production of biomass by *T. azotonutricium* during sugar fermentation parallels the low $Y_{substrate}$ values for *T. primitia* (above), and it is consistent with the notion that termite gut spirochetes have been selected as symbionts for catalysis (acetate production, nitrogen fixation), not for rapid growth or efficient coupling of substrate dissimilation to biomass production. By comparison, where quantitative analyses have been done, most free-living spirochetes that produce H_2 during carbohydrate fermentation (including *S. stenostrepta* and *S. zuelzerae*, which group within the treponemes by 16S rRNA gene sequences; above) produce far less H_2 than does *T. azotonutricium* (≤ 1.8 mol per mol hexose) (Canale-Parola, 1984; Harwood and Canale-Parola, 1981; Zhilina et al., 1996 and references therein). An exception is the thermophilic spirochete, *Spirochaeta thermophila* (2.9 mol per mole glucose; (Janssen and Morgan, 1992)).

6. Summary

Our understanding of termite gut spirochetes is still in its infancy. However, even the first few pure culture representatives have revealed physiological properties previously unknown in the *Spirochaetes* phylum of the domain *Bacteria* (i.e. H_2/ CO_2-acetogenesis and dinitrogen fixation), as well as traits seemingly important to their role as termite hindgut symbionts, i.e. relatively low growth rates and substrate-specific biomass yields. Nevertheless, important aspects of their symbiotic interaction with their host remain obscure. For example, we still have very little understanding of factors that favor such abundance of spirochetes in termite guts or of specific traits of treponemes that make them particularly well-suited for colonization of this habitat. Given the phylogenetic diversity of termite gut treponemes, one cannot help but feel certain that new and important principles of spirochete biology are awaiting discovery as more of these strains are coaxed into culture.

Acknowledgements

Much of the work described here was conducted by the author and his graduate students and postdoctoral research associates with support from the National Science Foundation (USA) and the Alexander von Humboldt Foundation (Germany).

References

Abe, T., Bignell, D. E., and Higashi, M., eds. (2000). Termites: evolution, sociality, symbioses, ecology (Dordrecht, NL, Kluwer Academic Publishers).

Beckwith, T. D., and Light, S. F. (1927). The spirals within the termite gut for class use. Science *66*, 656-657.

Berchtold, M., and König, H. (1996). Phylogenetic analysis and in situ identification of uncultivated spirochetes from the hindgut of the termite *Mastotermes darwiniensis*. System Appl Microbiol. *19*, 66-73.

Berchtold, M., Ludwig, W., and König, H. (1994). 16S rDNA sequence and phylogenetic position of an uncultivated spirochete from the hindgut of the termite *Mastotermes darwiniensis* Froggatt. FEMS Microbiol. Lett. *123*, 269-273.

Bignell, D. E., Oskarsson, H., Anderson, J. M., and Ineson, P. (1983). Structure, microbial associations and function of the so-called "mixed segment" of the gut in two soil-feeding termites, *Procubitermes aburiensis* and *Cubitermes severus* (Termitidae, Termitinae). J. Zool Lond *201*, 445-480.

Brauman, A., Kane, M. D., Labat, M., and Breznak, J. A. (1992). Genesis of acetate and methane by gut bacteria of nutritionally diverse termites. Science *257*, 1384-1387.

Breznak, J. A. (1973). Biology of nonpathogenic, host-associated spirochetes. CRC Critical Rev Microbiol. *2*, 457-489.

Breznak, J. A. (1984). Hindgut spirochetes of termites and *Cryptocercus punctulatus*. In Bergey's manual of systematic bacteriology, N. R. Krieg, and J. G. Holt, eds. (Baltimore, Williams and Wilkins), pp. 67-70.

Breznak, J. A. (1994). Acetogenesis from carbon dioxide in termite guts. In Acetogenesis, H. L. Drake, ed. (New York, Chapman and Hall), pp. 303-330.

Breznak, J. A. (2000). Ecology of prokaryotic microbes in the guts of wood- and litter-feeding termites. In Termites: evolution, sociality, symbiosis, ecology, T. Abe, D. E. Bignell, and M. Higashi, eds. (Dordrecht, NL, Kluwer Academic Publ.), pp. 209-231.

Breznak, J. A., and Leadbetter, J. R. (2002). Termite gut spirochetes. In The prokaryotes: an evolving electronic resource for the microbiological community, M. Dworkin, ed. (New York, Springer-Verlag).

Brune, A. (2006). Symbiotic associations between termites and prokaryotes. In The Prokaryotes (on-line version), M. Dworkin, ed. (New York, Springer-Verlag (***in press***).

Brune, A., Emerson, D., and Breznak, J. A. (1995). The termite gut microflora as an oxygen sink: microelectrode determination of oxygen and pH gradients in guts of lower and higher termites. Appl Environ Microbiol. *61*, 2681-2687.

Brune, A., and Kühl, M. (1996). pH profiles of the extremely alkaline hindguts of soil-feeding termites (Isoptera: Termitidae) determined with microelectrodes. J. Insect Physiol *42*, 1121-1127.

Canale-Parola, E. (1984). Family I. *Spirochaetaceae* Swellengrebel 1907, 581^{AL}. In Bergey's manual of systematic bacteriology, N. R. Krieg, and J. G. Holt, eds. (Baltimore, Williams and Wilkins), pp. 39-46.

Casjens, S., Palmer, N., van Vugt, R., Huang, W. M., Stevenson, B., Rosa, P., Lathigra, R., Sutton, G., Peterson, J., Dodson, R. J., *et al.* (2000). A bacterial genome in flux: the twelve linear and nine circular extrachromosomal DNAs in an infectious isolate of the Lyme disease spirochete *Borrelia burgdorferi*. Molecular Microbiology *35*, 490-516.

Cleveland, L. R. (1923). Symbiosis between termites and their intestinal protozoa. Proc. Nat. Acad Sci. USA. *9*, 424-428.

Cleveland, L. R. (1924). The physiological and symbiotic relationships between the intestinal protozoa of termites and their host, with special reference to *Reticulitermes flavipes* Kollar. Biol. Bull *46*, 178-227.

Cleveland, L. R. (1925). The effects of oxygenation and starvation on the symbiosis between the termite *Termopsis*, and its intestinal flagellates. Biol. Bull *48*, 309-326.

Cleveland, L. R., and Grimstone, A. V. (1964). The fine structure of the flagellate *Mixotricha paradoxa* and its associated micro-organisms. Proc. R Soc. London (serB) *159*, 668-686.

Daniel, S. L., Hsu, T., Dean, S. I., and Drake, H. L. (1990). Characterization of the H_2- and CO-dependent chemolithotrophic potentials of the acetogens *Clostridium thermoaceticum* and *Acetogenium kivui*. J. Bacteriol. *172*, 4464-4471.

Drake, H. L. (1994). Acetogenesis, acetogenic bacteria, and the acetyl-CoA "Wood/Ljungdahl" pathway: past and current perspectives. In Acetogenesis, H. L. Drake, ed. (New York, Chapman and Hall), pp. 3-60.

Ebert, A., and Brune, A. (1997). Hydrogen concentration profiles at the oxic-anoxic interface: a microsensor study of the hindgut of the wood-feeding lower termite *Reticulitermes flavipes* (Kollar). Appl Environ Microbiol. *63*, 4039-4046.

Eichler, B., and Schink, B. (1984). Oxidation of primary aliphatic alcohols by *Acetobacterium carbinolicum* sp. nov., a homoacetogenic anaerobe. Arch. Microbiol. *140*, 147-152.

Eutick, M. L., Veivers, P., O'Brien, R. W., and Slaytor, M. (1978). Dependence of the higher termite, *Nasutitermes exitiosus* and the lower termite, *Coptotermes lacteus* on their gut flora. J. Insect Physiol *24*, 363-368.

Euzéby, J. (2004). Validation list no. 97. Int. J. Syst Evol Microbiol. *54*, 631-632.

Fraser, C. M., Norris, S. J., Weinstock, C. M., White, O., Sutton, G. G., Dodson, R., Gwinn, M., Hickey, E. K., Clayton, R., Ketchum, K. A., *et al.* (1998). Complete genome sequence of *Treponema pallidum*, the syphilis spirochete. Science *281*, 375-388.

Graber, J. R., and Breznak, J. A. (2005). Folate cross-feeding supports symbiotic homoacetogenic spirochetes. Appl Environ Microbiol. *71*, 1883-1889.

Graber, J. R., and Breznak, J. A. (2004). Physiology and nutrition of *Treponema primitia*, an H_2/CO_2-acetogenic spirochete from termite hindguts. Appl Environ Microbiol. *70*, 1307-1314.

Graber, J. R., Leadbetter, J. R., and Breznak, J. A. (2004). Description of *Treponema azotonutricium* sp. nov. and *Treponema primitia* sp. nov., the first spirochetes isolated from termite hindguts. Appl Environ Microbiol. *70*, 1315-1320.

Guth, J. H., and Burris, R. H. (1983). Inhibition of nitrogenase-catalyzed NH_3 formation by H_2. Biochemistry *22*, 5111-5122.

Harwood, C. S., and Canale-Parola, E. (1981). Branched-chain amino acid fermentation by a marine spirochete: strategy for starvation survival. J. Bacteriol. *148*, 109-116.

Herbeck, J. L., and Bryant, M. P. (1974). Nutritional requirements of the intestinal anaerobe *Ruminococcus bromii*. Appl Environ Microbiol. *28*, 1018-1022.

Hongoh, Y., Ohkuma, M., and Kudo, T. (2003). Molecular analysis of bacterial microbiota in the gut of the termite *Reticulitermes speratus* (Isoptera; Rhinotermitidae). FEMS Microbiol. Ecol *44*, 231-242.

Huang, T., and Schirch, V. (1995). Mechanism for the coupling of ATP hydrolysis to the conversion of 5-formyltetrahydrofolate to 5,10-methenyltetrahydrofolate. J Biol. Chem. *270*, 22296-22300.

Iida, T., Ohkuma, M., Ohtoko, K., and Kudo, T. (2000). Symbiotic spirochetes in the termite hindgut: phylogenetic identification of ectosymbiotic spirochetes of oxymonad protists. FEMS Microbiology Ecology *34*, 17-26.

Inoue, T., Kitade, O., Yoshimura, T., and Yamaoka, I. (2000). Symbiotic associations with protists. In Termites: evolution, sociality, symbioses, ecology, T. Abe, D. E. Bignell, and M. Higashi, eds. (Dordrecht, Kluwer Academic Publishers), pp. 275-288.

Janssen, P. H., and Morgan, H. W. (1992). Glucose catabolism by *Spirochaeta thermophila* RI 19.B1. J. Bacteriol. *174*, 2449-2453.

Klappenbach, J. A., Dunbar, J. M., and Schmidt, T. M. (2000). rRNA operon copy number reflects ecological strategies of bacteria. Applied and Environmental Microbiology *66*, 1328-1333.

Klappenbach, J. A., Saxman, P. R., Cole, J. R., and Schmidt, T. M. (2001). rrndb: the Ribosomal RNA Operon Copy Number Database. Nucleic Acids Research *29*, 181-184.

Leadbetter, J. R., and Breznak, J. A. (1996). Physiological ecology of *Methanobrevibacter cuticularis* sp. nov. and *Methanobrevibacter curvatus* sp. nov., isolated from the hindgut of the termite *Reticulitermes flavipes*. Appl Environ Microbiol. *62*, 3620-3631.

Leadbetter, J. R., Crosby, L. D., and Breznak, J. A. (1998). *Methanobrevibacter filiformis* sp. nov., a filamentous methanogen from termite hindguts. Arch. Microbiol. *169*, 287-292.

Leadbetter, J. R., Schmidt, T. M., Graber, J. R., and Breznak, J. A. (1999). Acetogenesis from H_2 plus CO_2 by spirochetes from termite guts. Science *283*, 686-689.

Lee, M. J., Schreurs, P. J., Messer, A. C., and Zinder, S. H. (1987). Association of methanogenic bacteria with flagellated protozoa from a termite hindgut. Curr. Microbiol. *15*, 337-341.

Leidy, J. (1874-1881). The parasites of the termites. J. Acad. Nat Sci. (Phila) *8*, 425-447.

Leidy, J. (1877). On the intestinal parasites of *Termes flavipes*. Proc. Acad. Nat Sci. (Phila) *29*, 146-149.

Leschine, S., Paster, B. J., and Canale-Parola, E. (2003). Free-living saccharolytic spirochetes: the genus *Spirochaeta*. In The Prokaryotes (electronic version, release 3.5), M. Dworkin, ed. (New York, Springer-Verlag).

Lilburn, T. G., Kim, K. S., Ostrom, N. E., Byzek, K. R., Leadbetter, J. R., and Breznak, J. A. (2001). Nitrogen fixation by symbiotic and free-living spirochetes. Science *292*, 2495-2498.

Lilburn, T. G., Schmidt, T. M., and Breznak, J. A. (1999). Phylogenetic diversity of termite gut spirochaetes. Environmental Microbiology *1*, 331-345.

Lockshin, A., and Burris, R. H. (1965). Inhibitors of nitrogen fixation in extracts from *Clostridium pasteurianum*. Biochim Biophys Acta *111*, 1-10.

Machin, K. E. (1963). The control and synchronization of flagellar movement. Proc. Roy Soc. Lond B *158*, 88-104.

Margulis, L., and Hinkle, G. (1992). Large symbiotic spirochetes: *Clevelandina*, *Cristispira*, *Diplocalyx*, *Hollandina*, and *Pillotina*. In The prokaryotes, A. Balows, H. G. Trüper, M. Dworkin, W. Harder, and K.-H. Schleifer, eds. (New York, Springer-Verlag), pp. 3965-3978.

Nakashima, K., Watanabe, H., Saitoh, H., Tokuda, G., and Azuma, J.-I. (2002). Dual cellulose-digesting system of the wood-feeding termite, *Coptotermes formosanus* Shiraki. Insect Biochem. Mol Biol. *32*, 777-784.

Nalepa, C. A., and Bandi, C. (2000). Characterizing the ancestors: paedomorphosis and termite evolution. In Termites: evolution, sociality, symbioses, ecology, T. Abe, D. E. Bignell, and M. Higashi, eds. (Dordrecht, Kluwer Academic), pp. 53-75.

Noda, S., Ohkuma, M., Yamada, A., Hongoh, Y., and Kudo, T. (2003). Phylogenetic position and in situ identification of ectosymbiotic spirochetes on protists in the termite gut. Appl Environ Microbiol. *69*, 625-633.

Ohkuma, M., Iida, T., and Kudo, T. (1999). Phylogenetic relationships of symbiotic spirochetes in the gut of diverse termites. FEMS Microbiology Letters *181*, 123-129.

Ohkuma, M., and Kudo, T. (1996). Phylogenetic diversity of the intestinal bacterial community in the termite *Reticulitermes speratus*. Appl Environ Microbiol. *62*, 461-468.

Ohkuma, M., and Kudo, T. (1998). Phylogenetic analysis of the symbiotic intestinal microflora of the termite *Cryptotermes domesticus*. FEMS Microbiol. Lett. *164*, 389-395.

Paster, B. J., Dewhirst, F. E., Cooke, S. M., Fussing, V., Poulsen, L. K., and Breznak, J. A. (1996). Phylogeny of not-yet-cultured spirochetes from termite guts. Appl Environ Microbiol. *62*, 347-352.

Paster, B. J., Dewhirst, F. E., Weisburg, W. G., Tordoff, L. A., Fraser, G. J., Hespell, R. B., Stanton, T. B., Zablen, L., Mandelco, L., and Woese, C. R. (1991). Phylogenetic analysis of the spirochetes. J. Bacteriol. *173*, 6101-6109.

Potrikus, C. J., and Breznak, J. A. (1981). Gut bacteria recycle uric acid nitrogen in termites: a strategy for nutrient conservation. Proc. Nat. Acad Sci. USA. *78*, 4601-4605.

Rasche, M. E., and Arp, D. J. (1989). Hydrogen inhibition of nitrogen reduction by nitrogenase in isolated soybean nodule bacteroids. Plant Physiol *91*, 663-668.

Salmassi, T. M., and Leadbetter, J. R. (2003). Analysis of genes of tetrahydrofolate-dependent metabolism from cultivated spirochetes and the gut community of the termite *Zootermopsis angusticollis*. Microbiology *149*, 2529-2537.

Seshadri, G., Myers, G. S. A., Tettelin, H., Eisen, J. A., Heidelberg, J. F., Dodson, R. J., Davidsen, T. M., DeBoy, R. T., Fouts, D. E., Haft, D. H., *et al.* (2004). Comparison of the genome of the oral pathogen *Treponema denticola* with other spirochete genomes. Proceedings of the National Academy of Sciences of the United States of America *101*, 5646-5651.

Sharak-Genthner, B. R., and Bryant, M. P. (1987). Additional characteristics of one-carbon-compound utilization by *Eubacterium limosum* and *Acetobacterium woodii*. Appl Environ Microbiol. *53*, 471-476.

Slyter, L. L., and Weaver, J. M. (1977). Tetrahydrofolate and other growth requirements of certain strains of *Ruminococcus flavefaciens*. Appl Environ Microbiol. *33*, 363-369.

Stanton, T. B., and Canale-Parola, E. (1980). *Treponema bryantii* sp. nov., a rumen spirochete that interacts with cellulolytic bacteria. Arch. Microbiol. *127*, 145-156.

Steinman, H. G., Oyama, V. I., and Schulze, H. O. (1954). Carbon dioxide, cocarboxylase, citrovorum factor, and coenzyme A as essential growth factors for a saccharolytic treponeme. J Biol. Chem. *211*, 327-335.

Stingl, U., Maass, A., Radek, R., and Brune, A. (2004). Symbionts of the gut flagellate *Staurojoenina* sp. from *Neotermes cubanus* represent a novel termite-associated lineage of *Bacteroidales*: description of '*Candidatus* Vestibaculum illigatum'. Microbiology-Uk *150*, 2229-2235.

Stouthamer, A. H. (1979). The search for correlation between theoretical and experimental growth yields. In International Review of Biochemistry, J. R. Quayle, ed. (Baltimore, University Park Press), pp. 1-47.

Tayasu, I., Abe, T., Eggleton, P., and Bignell, D. E. (1997). Nitrogen and carbon isotope ratios in termites: an indicator of trophic habit along the gradient from wood-feeding to soil-feeding. Ecol Entomol *22*, 343-351.

Tayasu, I., Sugimoto, A., Wada, E., and Abe, T. (1994). Xylophagous termites depending on atmospheric nitrogen. Naturwissenschaften *81*, 229-231.

Tschech, A., and Pfennig, N. (1984). Growth yield increase linked to caffeate reduction in *Acetobacterium woodii*. Arch. Microbiol. *137*, 163-167.

Veivers, P. C., O'Brien, R. W., and Slaytor, M. (1982). Role of bacteria in maintaining the redox potential in the hindgut of termites and preventing the entry of foreign bacteria. J. Insect Physiol *28*, 947-951.

Watanabe, H., and Tokuda, G. (2001). Animal cellulases. Cellular and Molecular Life Sciences *58*, 1167-1178.

Weir, A., Dolan, M., Grimaldi, D., Guerrero, R., Wagensberg, J., and Margulis, L. (2002). Spirochete and protist symbionts of a termite (*Mastotermes electrodominicus*) in Miocene amber. Proc. Nat. Acad Sci. USA. *99*, 1410-1413.

Wenzel, M., Radek, R., Brugerolle, G., and König, H. (2003). Identification of the ectosymbiotic bacteria of *Mixotricha paradoxa* involved in movement symbiosis. Europ J. Protistol *39*, 11-23.

Wier, A., Ashen, J., and Margulis, L. (2000). *Canaleparolina darwiniensis*, gen. nov., sp. nov., and other pillotinaceous spirochetes from insects. Internatl Microbiol. *3*, 213-223.

Zhilina, T. N., Zavarzin, G. A., Rainey, F. A., Kevbrin, V. V., Kostrikina, N. A., and Lysenko, A. M. (1996). *Spirochaeta alkalica* sp. nov., *Spirochaeta africana* sp. nov., and *Spirochaeta asiatica* sp. nov., alkaliphilic anaerobes from the continental soda lakes in Central Asia and the East African Rift. Int. J. System Bacteriol. *46*, 305-312.

Afterword

Daniel M. Musher and Ronald F. Schell*

Afterword

If our book *Pathogenesis and Immunology of Treponemal Infections*, brought to mind the medieval metaphor of dwarfs standing on the shoulders of giants, the current work, *Pathogenic Treponema: Molecular and Cellular Biology*, makes us think that a new generation of giants has arisen. In our work of 1983, we tried to bring together what was known about treponemes – their biology and immunology, the pathogenesis of diseases that they cause, and the host response to them – all the time painfully aware that Hollander and Turner, in 1957, had stated, or at least hinted at, many of the same conclusions. It is exciting now to see what a new generation of scientists has done for the challenging, and now enlarged, field of treponemal research.

Perhaps the principal stumbling block to research on *T. pallidum* was the inability to cultivate the organism, despite a tremendous effort spanning most of the twentieth century. We looked through a glass darkly, our results clouded by insufficient understanding of treponemal metabolism, on the one hand, and by contamination of treponemal preparations by mammalian tissues from the intratesticular cultivation process, on the other. Thanks to genomic analysis, we now know that *T. pallidum* simply lacks genes that other free-living bacteria utilize for survival *ex vivo*.

The importance of insights gained from genomic analysis pervades the present volume. Imagine the sophistication of the following sentences: "The genomic sequence revealed that the syphilis spirochete relies upon glycolysis for energy production, that it lacks both the tricarboxylic acid cycle and oxidative phosphorylation pathways, and that it is unable to synthesize fatty acids, enzyme cofactors, and most amino acids. To compensate for its loss of biosynthetic capability, *T. pallidum* maintains a repertoire of transport systems predicted to have broad specificity for host-derived nutrients." Surely, the best confirmation of negative studies is the demonstration that the genetic material required to perform such syntheses is absent. How it came about that *T. pallidum* lost these genes

*For correspondence email dmusher@bcm.tmc.edu

and how the organism came to reside nearly exclusively in the human host is a fascinating question that awaits an answer.

In our earlier book, we wondered about the immune response to *T. pallidum* and the pathogenesis of syphilitic infection. These have been amply elucidated in the past two decades. Our findings that early syphilitic infection activates acquired cellular resistance appear to have been validated in numerous subsequent studies. Because *T. pallidum* expresses remarkably few proteins on its cell surface, the likelihood that the human host will recognize this organism as foreign is reduced, and there are fewer targets against which an immune response can be directed, although a number of immunodominant epitopes have been clearly identified. In any case, for all these reasons, the similarity between *T. pallidum* and gram negative bacilli appears to stop at the cytoplasmic membrane.

Identifying virulence factors will remain a major challenge. Such research, in the case of other microorganisms, has largely depended upon genetic engineering which eliminates one or another gene; without some dramatic technological breakthrough, these techniques can not be applied to a noncultivatable organism such as *T. pallidum*. Molecular strategies for studying virulence were only in their most rudimentary stages in 1983. A number of chapters in this book raise the exciting possibility that this impasse may be circumvented by using cultivatable treponemes, which can be genetically manipulated, as surrogates for the even more recalcitrant *T. pallidum*. Also entirely new since our textbook is the concept of innate immunity, which also has been studied, at least preliminarily, in the case of syphilis and periodontal disease.

We particularly admire the excellent section on the history of spirochetal disease which so nicely sets out the philosophic approach to medical history: "During the fifteenth century, diseases were identified according to a set of symptoms, not an underlying, specific causative agent. The way physicians and patients understand disease before and after the laboratory revolution is fundamentally different: it therefore makes little sense to trace the history of a category that did not exist ....Because historians rely on written records, language is central to our analysis. Paradoxically, as historians have grown to appreciate the impact of the laboratory revolution on the history of medicine, it has become more difficult for us to communicate with scientists interested in the history of specific, modern diseases, which are built on different definitions of what a disease is."

Finally, the present work gives a sense of how *T. pallidum* fits into the complex and rich family of treponemes, so many others of which have also not been successfully cultivated in vitro. The comparison of the structure and genetics of the newly described treponemes discussed in this volume will enhance our understanding of the better known species.

There will be many more goals, major and minor, to overcome before the progression of treponemal diseases is fully understood. We believe that the elegantly written chapters in *Pathogenic Treponema: Molecular and Cellular Biology* have defined many of the parameters within which relevant research will take place. Twenty years from now, a new edition, by new and younger investigators, will again summarize great advances.

Index

Written by Sara Lynn Eastler

D

E

www.ingramcontent.com/pod-product-compliance
Lightning Source LLC
LaVergne TN
LVHW011555200926
848484LV00001B/316
* 9 7 8 1 9 0 4 4 5 5 1 0 3 *